NOISE IN PHYSICAL SYSTEMS AND 1/f FLUCTUATIONS

AIP CONFERENCE PROCEEDINGS 285

NOISE IN PHYSICAL SYSTEMS AND 1/f FLUCTUATIONS

ST. LOUIS, MO 1993

EDITORS: PETER H. HANDEL
ALMA L. CHUNG

PHYSICS DEPARTMENT
UNIVERSITY OF MISSOURI

AIP PRESS

American Institute of Physics New York

L.C. Catalog Card No. 93-72575
ISBN 1-56396-270-5
DOE CONF-930866

Printed in the United States of America.

Contents

III. METALS AND MAGNETICS

IV. SUPERCONDUCTORS

V. LIQUID CONDUCTORS

VI. QUANTUM $1/f$ NOISE

VII. SEMICONDUCTOR DEVICES

VIII. MOSFET

IX. OPTICAL DEVICES AND OTHERS

X. LASERS

XI. QUANTUM SYSTEMS

XII. RANDOM PROCESSES AND STOCHASTIC SYSTEMS

XIII. CHAOS AND FRACTAL

XIV. MISCELLANEOUS 1/*f* FLUCTUATIONS

XV. MODELS AND SIMULATION OF 1/*f* FLUCTUATIONS

XVI. STATISTICAL PHYSICS OF NOISE IN SEMICONDUCTOR MATERIALS

XVII. MEMBRANES AND CELLS

XVIII. CARDIOVASCULAR SYSTEMS

XIX. BIOLOGICAL SYSTEMS

XX. NEURONAL NETWORKS

PREFACE

Fluctuations encountered in Physical Systems, mainly electric current and voltage fluctuations, are called noise in a generalized sense. The present Conference Proceedings also include biological systems and systems of any nature. Both stochastic and chaotic fluctuations are the subject of this Conference, which is a meeting place of physicists, engineers, and scientists of many other disciplines. It provides a bridge linking theorists and applied scientists involved in the design of new electronic devices. Most important, the Conference and these Proceedings provide a forum for new ideas and methods with great impact on the future development of science and technology on all continents.

One of the ideas in fashion today is expressed by the concept of stochastic resonance. Although it was known before that in nonlinear systems the transmitted signal-to-noise ratio can be an arbitrarily complicated function of the input signal and noise levels, the so-called stochastic resonance is an outright decrease of this ratio with decreasing noise input in some bistable systems at low levels of input noise. The concept is applied to the nervous system and to heterodyning in these Proceedings.

After 1925 the Nyquist theorem shaped the development of many-body physics and of the thermodynamics of irreversible processes. This basic theorem linking dissipation to equilibrium fluctuations exemplified the essential role which the phenomenon of noise in physical systems plays in all fields of science and engineering. At a more fundamental level, the quantum $1/f$ effect reshapes the notions of "cross section," "process rate," and "current" in general today, introducing new, physical, notions in their place, which describe and predict for the first time the technically important $1/f$ quantum fluctuations present in them. Van Vliet once called it a new aspect of quantum mechanics which stands out by its simplicity. After years of doubts and testing in the scientific community, its simple universal quantum $1/f$ formulas $2\alpha/\pi fN$ and $2\alpha A/fN$ validate, explain, and generalize not only Hooge's empirical formulation of the idea of fundamental $1/f$ noise, but also its initial (1966) expression in terms of the spectrum of homogeneous isotropic turbulence of the magnetic field.

The advent of the quantum $1/f$ theory, which is no model, but rather plain, straightforward quantum electrodynamics, marked a turning point in the field of noise in physical systems, dramatically increasing its relevance to high-technology applications and devices. Indeed, in most high-tech settings all classical sources of noise and instability have been eliminated or reduced to a minimum. Invariably, then, the fundamental quantum $1/f$ noise present in the elementary cross sections or process rates which control the performance of the device come to the forefront and become highly visible as the ubiquitous $1/f$ noise which determines the achievable stability limits. This is why the quantum $1/f$ effect limits the performance of most high-tech devices. If this is not the case in a device, we must improve it until the quantum limit is reached, as our great teacher Aldert van der Ziel used to say. Even after the quantum limit has been reached, we must use the quantum $1/f$ formulas to optimize the stability of the device on the basis of a reasonable figure of merit containing our quantum $1/f$ "fine structure" constant $\alpha = 1/137$, by modifying the blueprint within the restrictions set by all the other important device parameters. A Quantum $1/f$ Round Table evening is dedicated to both practical and theoretical aspects.

Quantum $1/f$ noise is as fundamental as time and space. In the long run, quantum $1/f$ research may thus clarify the fundamental dynamical nature of space, time, and quantum mechanics itself.

The groundbreaking work of Aldert van der Ziel who left us in 1991, stands out as a symbol for the often unrecognized or overlooked pioneering electronic noise research which made possible most of the experimental and technological discoveries of our century in physics and engineering. Due to the general scarcity of information, interest, and appreciation for the subject of electronic noise, the work of van der Ziel, Nyquist, Schottky, Johnson, and of other founders of this field has not received the recognition it deserved. We have to continue their tradition of noise research with increased confidence and stronger mutual cooperation.

The present Conference continues the series of International Conferences on Noise in Physical Systems started in Nottingham, England, 1968, and the series of International $1/f$ Noise Symposia initiated by Musha in Tokyo, 1977. The two series merged in Montpellier, France, in 1983 with the present title reflecting the steady increase of interest in $1/f$ noise since the mid-1970s. The 13th Conference will be held in Lithuania in 1995. An independent series of International Quantum $1/f$ Noise Symposia was initiated by van der Ziel in Minneapolis in 1985 and continued by him up to the Fourth Symposium* in Minneapolis, 1990. The Fifth Quantum $1/f$ Symposium, held for the first time in St. Louis, in May 1992, was named in the honor of van der Ziel.* The "Sixth International van der Ziel Symposium on Quantum $1/f$ Noise and Other Low-Frequency Fluctuations" will be held here in St. Louis, 1994.

We are grieving over the loss of Professor J. J. Brophy shortly after the 1991 Conference where he was represented by C. M. Van Vliet. He dedicated much of his scientific work to the study of $1/f$ noise.

We thank the members of the Scientific Committee for their help in the peer review of the submitted contributions. The accepted papers are grouped in 20 chapters. We are very indebted to the Organizing Committee for their support, in particular to Chancellor Blanche Touhill, Assistant Dean David Klostermann, and Professor Bernard Feldman. Last, but not least, acknowledgment goes to Michael Hennelly, Soraya Shalforoosh, and Maria Taylor from the American Institute of Physics Books Program, for preparing the Proceedings in time, and in the best technical conditions.

Peter H. Handel and Alma L. Chung
University of Missouri-St. Louis

*Proceedings copies are available at Registration.

Conference Committees

Permanent International Committee

Professor V. Bareikis	Lithuania
Professor L. J. DeFelice	USA
Professor P. H. Handel	USA
Professor F. N. Hooge	The Netherlands
Professor B. K. Jones	United Kingdom
Professor G. Lecoy	France
Professor V. I. Kozub	Russian Federation
Professor P. Mazzetti	Italy
Professor T. Musha	Japan
Professor M. Sikula	Czechoslovakia
Professor C. M. Van Vliet	USA
Professor D. Wolf	Germany

Scientific Program Committee

Professor G. Bosman
Professor L. J. DeFelice
Professor P. H. Handel, Chairman
Dr. R. P. Jindal
Dr. G. Kousik
Professor T. Musha
Professor C. M. Van Vliet
Professor D. Wolf

ICNF'93 Local Organizing Committee
University of Missouri-St. Louis

Campus Administration
- Dr. Blanche M. Touhill, Chancellor, Chairperson
- Ms. Kathy Osborn, Vice Chancellor for University Relations
- Mr. James R. Parke, Coordinator, Office of Research

College of Arts and Sciences
- Dr. E. Terrence Jones, Dean
- Dr. Sarapage McCorkle, Associate Dean
- Dr. Bernard Feldman, Department of Physics Chair and Director, Center for Molecular Electronics
- Dr. Peter H. Handel, Professor of Physics and Conference Chairman
- Mrs. Alma L. Chung, Research Assistant, Physics

Continuing Education—Extension Division
- Dr. Wendell Smith, Dean
- Mr. David Klostermann, Assistant Dean
- Mrs. Angeline Antonopoulos, Manager Marketing and Information

I. PLENARY LECTURES

NOISE AND ADMITTANCE OF HIGHLY TRANSMISSIVE CONDUCTORS

M. Büttiker
IBM T. J. Watson Research Center, Yorktown Heights, N. Y. 10598

ABSTRACT

A wide range of mesoscopic conductors exhibit open conduction channels which permit the transfer of carriers from one contact to another without scattering. At zero temperature open channels do not contribute to the shot noise. For non-interacting carriers we give the frequency and temperature dependent fluctuation spectra in terms of the scattering matrix of the conductor and discuss the frequency dependent admittance of multiprobe conductors. A self-consistent potential approach is invoked to treat interactions and to obtain current conservation.

INTRODUCTION

Conductance can be viewed as a scattering problem of carriers which impinge on the sample and are either reflected at the sample or are transmitted from one contact to another. Here I am interested in the fluctuation properties of currents and in the ac-admittance which such a conduction picture implies. A considerable body of literature exists which treats fluctuations in tunneling structures[1]. In these works a Bardeen tunneling Hamiltonian is taken as the starting point. The probability T for transmission of carriers is exponentially small and fluctuations are evaluated to linear order in T. In contrast in a Landauer discussion of conductance we are not limited to the case of small transmission probabilities but can investigate transport of highly transmissive conduction channels. Open conduction channels which permit transmission with a probability close to one occur in many situations: Quantum point contacts, resonant tunneling, ballistic conductors, and the quantum Hall effect. Even in metallic diffusive conductors in which at first sight one might assume that all conduction channels only exhibit a small transmission probability a fraction of the quantum channels can be viewed as being open. For a two-probe conductor described by transmission amplitudes $t_{\alpha\beta mn}$ and reflection amplitudes $r_{\alpha\alpha mn}$ where $\alpha = 1, 2$ and $\beta = 1, 2$ label the contacts and m and n label different transverse states (scattering channels) the shot noise at $kT = 0$ is given by[2]

$$< (\Delta I)^2 >_\nu = 2e\Delta\nu \mid V_1 - V_2 \mid \frac{e^2}{h} \mathrm{Tr}(\mathbf{r}^\dagger_{11}\mathbf{r}_{11}\mathbf{t}^\dagger_{21}\mathbf{t}_{21})$$

$$= 2e\Delta\nu \,|\, V_1 - V_2 \,|\, \frac{e^2}{h} \Sigma_n T_n(1 - T_n). \tag{1}$$

In Eq. (1) $\Delta\nu$ is a frequency interval, $|V_1 - V_2|$ is the voltage applied across the conductor and **r** and **t** are the reflection and transmission matrices. The last expression of Eq. (1) gives the shot noise in terms eigenvalues T_n of the product of the transmission matrices $\mathbf{t}^\dagger\mathbf{t}$. Eq. (1) applies for an arbitrary scattering matrix and is a generalization of the effective single channel result of Ref. 3. Eq. (1) expresses the shot noise in terms of a product of four scattering matrices. The current driven through the sample is $I = (e^2/h)\,\mathrm{Tr}(\mathbf{t}^\dagger\mathbf{t})\,|V_1 - V_2| = (e^2/h)\,\Sigma_n T_n \,|V_1 - V_2|$. Consequently the shot noise given by Eq. (1) is equal to the textbook result $< \Delta I^2 >_\nu = 2e\Delta\nu < I >$ only if all the transmission eigenvalues T_n are small compared to 1. Eq. (1) is a consequence of the fact that a carrier which strikes a scatterer can be either transmitted or reflected. There is more than one final state available. Eq. (1) correctly takes into account that this partition noise is maximal for a transmission probability $T = 1/2$. Both for $T = 1$ and $T = 0$ there is only one final state available and no partition noise is generated. For additional results on low-frequency current and voltage noise we refer the reader to Refs. 4-6.

CURRENT FLUCTUATION SPECTRA

Eq. (1) is valid in the zero-temperature and zero-frequency limit. Below I briefly indicate the derivation of this result and present a temperature and frequency dependent spectrum[7,4]. We deal with carriers which obey Fermi statistics. The statistics is in turn a consequence of the quantum mechanical indistinguishability of identical carriers. A discussion which starts from antisymmetrical many particle states or uses a second quantization language is thus in order. We introduce operators $a_{\alpha m}(E)$ and $b_{\alpha m}(E)$ which annihilate carriers in the incoming and outgoing channel m in probe α. For each probe we can group these operators into vectors $\mathbf{a}_\alpha(E)$ and $\mathbf{b}_\alpha(E)$ with $M_\alpha(E)$ components. $M_\alpha(E)$ is equal to the number of quantum channels with threshold below energy E. The $\mathbf{a}_\alpha(E)$ and $\mathbf{b}_\alpha(E)$ operators are related by the scattering matrices $\mathbf{b}_\alpha = \Sigma_\beta \mathbf{s}_{\alpha\beta}\mathbf{a}_\beta$. We are interested in the transfer of carriers from one contact to another but do not ask where exactly in the contact a carrier is created or annihilated. The current is then specified in terms of the vectors $\mathbf{a}_\alpha(E)$ and $\mathbf{b}_\alpha(E)$ and is found by evaluating[4]

$$I_\alpha(t) = \frac{e}{h} \int dE\,dE' [\mathbf{a}^\dagger_\alpha(E)\mathbf{a}_\alpha(E') - \mathbf{b}^\dagger_\alpha(E)\mathbf{b}_\alpha(E')] \exp(i(E - E')t/\hbar). \tag{2}$$

We can express the current operator in terms of the $\mathbf{a}_\alpha(E)$ operators alone with the help of the matrix[2]

$$\mathbf{A}_{\beta\gamma}(\alpha, E, E + \hbar\omega) = \mathbf{1}_\alpha \delta_{\alpha\beta}\delta_{\alpha\gamma} - \mathbf{s}^\dagger_{\alpha\beta}(E)\mathbf{s}_{\alpha\gamma}(E + \hbar\omega). \quad (3)$$

The frequency-dependent current fluctuations $< \Delta I_\alpha \Delta I_\beta >_\omega \equiv \Delta\nu S_{\alpha\beta}(\omega)$ are determined by the spectra[4,7]

$$S_{\alpha\beta}(\omega) = \frac{e^2}{h} \Sigma_{\gamma\delta} \int dE Tr[A_{\gamma\delta}(\alpha, E, E + \hbar\omega) A_{\delta\gamma}(\beta, E + \hbar\omega, E)] F_{\gamma\delta}(E, \omega), \quad (4)$$

$$F_{\gamma\delta}(E, \omega) = f_\gamma(E)(1 - f_\delta(E + \hbar\omega)) + f_\delta(E + \hbar\omega)(1 - f_\gamma(E)). \quad (5)$$

At equilibrium, when the Fermi functions in all reservoirs are taken at the same chemical potential, the function F is independent of the indices γ and δ and is given by $F(E, \hbar\omega) = 2[f(E) - f(E + \hbar\omega)]\ \varepsilon(\hbar\omega, kT)/\hbar\omega$. Here $\varepsilon(\hbar\omega, kT)$ is the energy of an harmonic (quantum) oscillator. Eq. (4) gives the Johnson-Nyquist noise at equilibrium and in the presence of transport describes the combined effects of thermal noise and partition noise. At $kT = 0$ for a two probe conductor, in the presence of a voltage difference, Eq. (4) leads to Eq. (1). In the zero-frequency limit, the spectra given by Eq. (4) can be shown to obey current conservation, $\Sigma_\alpha S_{\alpha\beta} = 0$, and $\Sigma_\beta S_{\alpha\beta} = 0$. For non-vanishing frequency current conservation is obeyed only in the limit where the energy dependence of the scattering matrix can be neglected. In that case the frequency dependence is entirely given by the Fermi functions. In general, for non-zero frequencies, charge can pile up inside the sample and the sum of all currents entering the sample need not be conserved. In such a case we should, however, consider not only the conductor itself but all nearby metallic bodies. It is the sum of the currents to all metallic bodies which is conserved.

AC-CONDUCTANCE

Eq. (4) is interesting from another point of view: At equilibrium Eq. (4) gives via the fluctuation dissipation theorem the real part of an ac-conductance. The imaginary part can be obtained via Kramers-Kronig relations. The combined real and imaginary part of the conductance is[8,9]

$$g_{\alpha\beta}(\omega) = \frac{e^2}{h} \int dE Tr[\mathbf{1}_\alpha \delta_{\alpha\beta} - \mathbf{s}^\dagger_{\alpha\beta}(E)\mathbf{s}_{\alpha\beta}(E + \hbar\omega)] \frac{f(E) - f(E + \hbar\omega)}{\hbar\omega}. \quad (6)$$

Eq. (6) gives the current response δI_α to an oscillating chemical potential $\delta\mu_\beta$. The current obtained from Eq. (6) is the response to a thermodynamic force (the chemical potential). This result, obtained in collaboration with Thomas[8] and with Prêtre and Thomas[9] should be contrasted with the usual linear response discussion which determines currents as a consequence of some unspecified external field (acting throughout the sample). Eq. (6) can be extended to the non-equilibrium case, if small time-dependent chemical potential oscillations occur on top of a dc potential difference. In this case we have to replace $f(E) - f(E + \hbar\omega)$ in Eq. (6) by $f_\beta(E) - f_\beta(E + \hbar\omega)$.

Like the current-current fluctuation spectra these conductances generally do not obey current conservation, $\Sigma_\alpha g_{\alpha\beta}(\omega) \neq 0$ and $\Sigma_\beta g_{\alpha\beta}(\omega) \neq 0$. A charge δQ_i can pile up on the conductor: $-i\omega\delta Q_i = \Sigma_\alpha \delta I_\alpha$ Current must be conserved, however, if we consider not only the conductor but all nearby metallic bodies. Below we consider for simplicity only one additional metallic conductor which we assume to be capacitively coupled to the conductor. The charge on this additional conductor is $Q_o = -Q_i$. The sum of all currents in the leads of the conductor, $\alpha = 1, 2, 3, ..$ and in the lead o connecting the gate is conserved, $\sum \delta I_\alpha + \delta I_o = 0$. Even if we treat the gate as a macroscopic body as we shall do below, this problem is difficult to solve since it requires finding the potential distribution $U_i(\mathbf{r}, t)$ throughout the mesoscopic conductor. Here we assume that it is sufficient to characterize the internal potential distribution by a single voltage $U_i(t)$. In a double barrier structure or in a quantum dot this internal potential could characterize the potential in the well or inside the dot. The current in probe α is $\delta I_\alpha = \Sigma_\gamma g_{\alpha\gamma}\delta U_\gamma - g_{\alpha i}\delta U_i$, where $g_{\alpha i}\delta U_i$ gives the current in probe α in response to the internal voltage. Since the current can depend only on voltage differences subtracting δU_i from all voltages gives, $\delta I_\alpha = \Sigma_\gamma g_{\alpha\gamma}\delta U_\gamma - (\Sigma_\gamma g_{\alpha\gamma})\delta U_i$. The net charge $-i\omega\delta Q_i = \Sigma_\alpha \delta I_\alpha \equiv -\delta I_o$ on the conductor gives rise to an induced potential $\delta U_i = \delta Q_i/C = -(i/\omega C)\delta I_o$. Solving these equations for the response of the conductor and the capacitor gives a for the interacting system (index I) admittances[9]

$$g^I_{\alpha\beta}(\omega) = g_{\alpha\beta}(\omega) - \left(\frac{(i/\omega C)\Sigma_\gamma g_{\alpha\gamma}(\omega)\Sigma_\delta g_{\delta\beta}(\omega)}{1 + (i/\omega C)\Sigma_{\gamma\delta} g_{\gamma\delta}(\omega)}\right). \tag{7}$$

$$g^I_{o\beta}(\omega) = -\frac{\Sigma_\delta g_{\delta\beta}(\omega)}{1 + (i/\omega C)\Sigma_{\gamma\delta} g_{\gamma\delta}(\omega)}. \tag{8}$$

$$g^I_{\beta o}(\omega) = -\frac{\Sigma_\gamma g_{\alpha\gamma}(\omega)}{1 + (i/\omega C)\Sigma_{\gamma\delta} g_{\gamma\delta}(\omega)}. \tag{9}$$

An oscillating potential δU_β gives a current at contact α of the conductor determined by Eq. (7) and gives a current δI_o flowing to the gate determined by Eq. (8). An oscillating gate voltage δU_o gives a current at contact α of the conductor determined by Eq. (9) and gives a current flowing to the gate given by $g^I_{oo}(\omega) = -\Sigma_\gamma g^I_{o\gamma}(\omega) = -\Sigma_\gamma g^I_{\gamma o}(\omega)$. The admittance matrix g^I is current conserving. The admittance matrix of the interacting system, Eq. (7,8), has the property that each element is a function of all the non-interacting conductances. An interacting system permits ac-currents for purely capacitively coupled conductors for which all off-diagonal admittances of the non-interacting system vanish. The degree to which Coulomb forces counteract deviations from a charge neutral state depends on the capacitance C. For a very large capacitance little energy is required to charge the sample and the admittances given by Eq. (7) exhibit only small departures from the admittance of the non-interacting system. In the limit of a small capacitance the current induced on the gate becomes small and the admittances Eq. (8) and (9) vanish. We conclude by emphasizing again that these results depend on the assumption that the internal potential can be described by a single voltage. However our discussion does demonstrate that the internal potential needs to be determined from the unscreened charges which are a consequence of the oscillations of the external potentials. That is in contrast to the typical linear response theory which determines a response to an external but unknown field.

FLUCTUATION SPECTRA WITH INTERACTIONS

For a conductor which can be described by a single induced voltage U_i the fluctuating current at probe α is given by $\Delta I_\alpha = \Sigma_\beta g_{\alpha\beta}\ (\delta U_\beta - \delta U_i) + \delta I_\alpha$ where δI_α are the fluctuations of the non-interacting system determined by Eq. (4). We are interested in the case where only spontaneous fluctuations occur and thus $\delta U_\beta \equiv 0$ at all contacts of the conductor. The induced voltage is thus determined by $\delta U_i = \delta Q_i / C = -(i/\omega C)\Delta I$ where $\Delta I = -\Sigma_\alpha \Delta I_\alpha$ is the sum of all fluctuating currents. We define

$$\chi_\alpha \equiv -\frac{i}{\omega C}\frac{\Sigma_\gamma g_{\alpha\gamma}(\omega)}{1 + (i/\omega C)\Sigma_{\gamma\delta} g_{\gamma\delta}(\omega)} \tag{10}$$

and define $\chi_o(\omega) \equiv 1 + \Sigma_\gamma \chi_\gamma(\omega)$ and find for the fluctuation spectra of the interacting system (conductor and capacitor)

$$S^I_{\alpha\beta}(\omega) = S_{\alpha\beta}(\omega) + \chi_\alpha(\omega)\Sigma_\delta S_{\delta\beta}(\omega) + \chi^*{}_\beta(\omega)\Sigma_\gamma S_{\alpha\gamma}(\omega)$$

$$+ \chi_\alpha(\omega)\chi^*{}_\beta(\omega)\Sigma_{\gamma\delta} S_{\gamma\delta}(\omega). \tag{11}$$

$$S^I_{o\alpha}(\omega) = \chi_o(\omega)\Sigma_\gamma S_{\gamma\alpha}(\omega) + \chi_o(\omega)\chi^*{}_\alpha(\omega)\Sigma_{\gamma\delta} S_{\gamma\delta}(\omega). \tag{12}$$

$$S^I_{\alpha o}(\omega) = \chi^*{}_o(\omega)\Sigma_\gamma S_{\alpha\gamma}(\omega) + \chi_o(\omega)\chi^*{}_\alpha(\omega)\Sigma_{\gamma\delta} S_{\gamma\delta}(\omega), \tag{13}$$

The current across the capacitance C is $\Delta I_0(\omega) = -i\omega\delta Q_o(\omega)$. It is correlated to the current fluctuations at the contacts of the conductor with a spectrum given by Eqs. (12,13). The spectrum $S^I_{oo}(\omega)$ can be obtained by summation over the correlations, $S^I_{oo}(\omega) = -\Sigma_\gamma S_{o\gamma}(\omega)$. For a very large capacitance $\chi_\alpha \approx 0$ and the fluctuation spectra are given by those of the non-interacting system, Eq. (4). In the zero-capacitance limit χ_α is determined by a frequency-dependent ratio of conductances leading to spectra which differ from Eq. (4). The interacting spectra obey current conservation $S^I_{\alpha o} + \Sigma_\gamma S^I_{\alpha\gamma} = 0$ and $S^I_{o\beta} + \Sigma_\delta S^I_{\delta\beta} = 0$.

CONCLUSIONS

As a function of capacitance Eqs. (7-9) and Eqs. (10-13) describe a transition in the admittances and the fluctuation spectra from a sample that can be charged at little energy expense (large C limit) to a sample that remains in a charge neutral state (zero capacitance limit). We have emphasized the distinction between external potentials applied to the contacts and internal potentials which are a consequence of deviations from the charge neutral state. The discussion presented here can be extended, first by treating the gate on equal footing with the conductor (mesoscopically) and can be extended by calculating with the help of Poisson's equation the space and time dependent actual voltage distribution[10].

REFERENCES

1. D. Rogovin and D. J. Scalapino, Ann. Phys. (N. Y.) **86**, 1, (1974); J. R. Tucker, IEEE J. Quantum Electron. **QE-15** , 1234 (1979).
2. M. Büttiker, Phys. Rev. Lett. **65**, 2901 (1990).
3. V. A. Khlus, Zh. Esp. Teor. Fiz. **93**, 843 (1987); (Sov. Phys. JETP **66** , 1243 (1987)); G. B. Lesovik, Pis'ma Zh. Eksp. Teor. Fiz. **49**, 513 (1989); (JETP Lett. **49**, 592 (1989)); B. Yurke and G. P. Kochanski, Phys. Rev. **B41**, 8184 (1990).
4. M. Büttiker, Phys. Rev. **B46**, 12485 (1992).
5. Th. Martin and R. Landauer, Phys. Rev. **B45** , 1742 (1992).
6. M. J. M. de Jong and C. W. J. Beenakker, Phys. Rev. **B46**, 13400, (1992): K. E. Nagaev, Phys. Lett. **A169** , 103 (1992); A. Levy Yeyati et al., Phys. Rev. **B47**, 10543 (1993).
7. M. Büttiker, Phys. Rev. **B45**, 3807 (1992).
8. M. Büttiker and H. Thomas, in: Quantum-Effect Physics, Electronics and Applications, edited by K. Ismail et al., (Institute of Physics Conference Series Number 127, Bristol, 1992). p. 19.
9. M. Büttiker, A. Prêtre and H. Thomas, "Dynamic conductance and the scattering matrix of small conductors", (unpublished).
10. M. Büttiker, H. Thomas and A. Prêtre, "Current partition in multiprobe conductors in the presence of slowly oscillating external potentials", (unpublished).

Hooge Parameter Determined by Impurity Scattering

Munecazu TACANO
Kyocera Corporation, Higashino Kitainoue 5-22,
Yamashina, Kyoto, 607, JAPAN

ABSTRACT

The Hooge parameter α_H of a doped semiconductor microstructure is shown to give α_{imp}, the noise parameter determined by the impurity scattering. The electric field dependence of the noise parameter at 77K is interpreted in terms of the impurity scattering. The temperature dependence of the noise parameter experimantally obtained below 100K is favorably compared with α_{imp} derived from the quantum 1/f noise theory. The cross-correlation model of the quantum 1/f noise theory is numerically analyzed for n-GaAs.

1. INTRODUCTION

In the last conference at Kyoto Hooge discussed the relation between α_H and α_{latt}.[1] This is successfully applied to a very pure sample at room temperature,[2] where the optical phonon scattering of the electrons is dominant in compound semiconductors. The cross correlation formula of the quantum 1/f mobility fluctuations was also discussed by Handel.[3] The dependence of α_H on temperature was favorably compared with that expected from the quantum model.[4] As an extension of these studies the contribution of the impurity scattering on the Hooge parameter α_H is discussed here.

First the dependence of the Hooge parameter on the electric field is analyzed theoretically, and is compared with that obtained from the experiments. This apparently indicates that the impurity scattering determines the Hooge parameter at 77K. The temperature dependence of α_H is also interpreted in terms of the impurity scattering at low temperature. This is quantitatively compared with the experiments. Finally some numerical analyses on the cross correlational formula of the quantum 1/f model are presented, and compared with those obtained by the KVBH model.[5]

2. Electric Field Dependence of α_H

The electric field dependence of α_H has been studied by Kleinpenning[6-8], Bosman[9] and others.[10-11] α_H of Si remarkably decreased above the field of 50 kV/m at 78K, and was interpreted in terms of the mobility fluctuations. α_H of GaAs did not change much below 3kV/m at 300K. Above that field α_H increased rapidly due to the carrier number fluctuations. These earlier studies assume α_H as a single function of the lattice scattering. In the following we intend to consider the effect of the impurity scattering as well as that of the lattice scattering. The hot electron transport in n-GaAs(12) indicates that, at 77K, the lattice mobility μ_{latt} decreases rapidly with increasing the electric field while the impurity mobility μ_{imp} remains virtually constant below 4kV/cm. The mobility is determined by the lattice scattering in a very pure sample, and by the impurity scattering in a doped n-GaAs. α_H will be expressed as:

$$\alpha_H(E) = (\mu(E)\mu_{sum}/\mu(E_o)/\mu_{lat})^2\,\alpha_{lat} + (\mu(E)\mu_{sum}/\mu(E_o)/\mu_{imp})^2\,\alpha_{imp} \qquad (1)$$

where μ_{sum}, $\mu(Eo)$, and $\mu(E)$ are the total mobility, the low field and the high field mobility, respectively. The 1st term is dominant in a very pure sample at 300K, and the 2nd term becomes dominant in a doped sample at 77K. The mobility μ_{sum}, μ_{lat}, μ_{imp} and the Hooge parameter α_{lat}, α_{imp} are obtained from the cross correlation model(3), and $\mu(E)$, $\mu(E_o)$ from the Monte Carlo simulation(13). We obtained from these results the field dependence of α_H as shown in Fig. 1. In a conventionally doped sample α_H is given rather independent of the electric field.

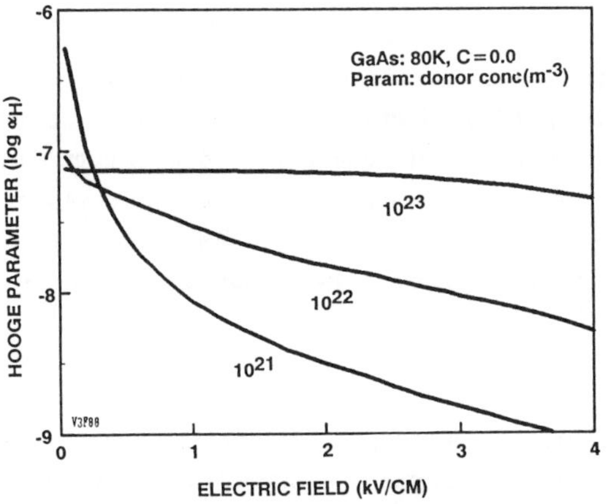

Fig. 1 Electricic field dependence of α_H, n_d as a parameter.

Figure 2 shows the voltage noise spectrum densities of a FIB filament(4) of n-GaAs as a function of the electric field at 77K. Small generation-recombination (GR) noise bulges, which correspond to the activation energy of 0.2 eV, were superposed on fairly pure $f^{-1.0}$ characteristics.

The measured parameter α_H was calculated from $\alpha_H = S_v fN/V^2$ on each electric field and plotted in Fig. 3. α_H changed little with the electric field up to 2 kV/cm and decreased slightly between 2 and 5 kV/cm. The electron mobility determined by the lattice scattering is known to decrease rapidly with increasing the electric field, while that determined by the impurity scattering remains virtually constant. To know which scattering is more dominant in the present case, the low field value of α_H was normalized by the impurity scattering $(\mu_{imp}(E)/\mu_{imp}(Eo))^2$ or the lattice scattering $(\mu_{lat}(E)/\mu_{lat}(Eo))^2$, and drawn by the solid lines in Fig. 3. It is the consequence of the impurity scattering that α_H is independent of the electric field.

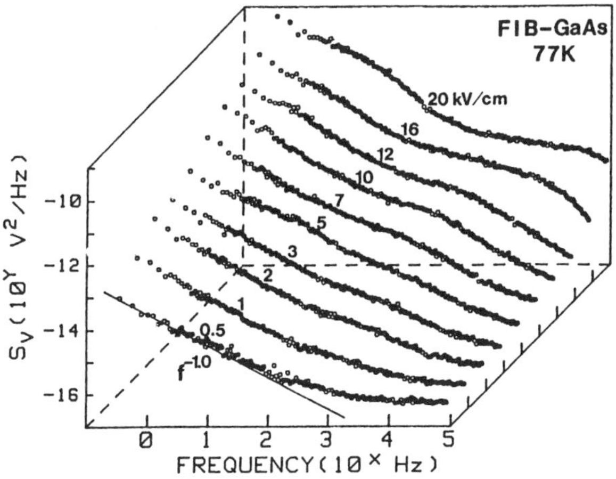

Fig. 2 Voltage Noise Spectra of FIB-GaAs, electric field as a parameter.

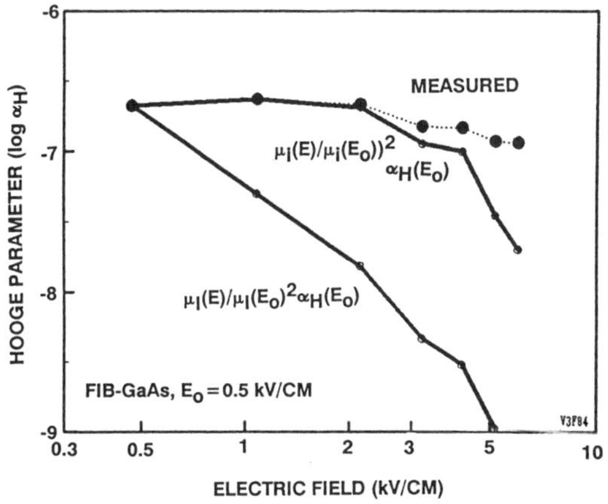

Fig. 3 Measured α_H at 77K, indicating impurity scattering deterministic.

3. Temperature dependence of α_H below 100K

The Hooge noise parameter α_H is also written in the form:

$$\alpha_H = (\mu_{sum}/\mu_{lat})^2\alpha_{lat} + (\mu_{sum}/\mu_{imp})^2\alpha_{imp}, \qquad (2)$$

The numerical analysis of the quantum fluctuation model makes clear of the relation between α_H and scattering mechanisms. Figure 4 shows the 1st, 2nd terms and their sum in eq. (2) as a function of the temperature. With the electron concentration of $1x10^{21}$ m^{-3}in n-GaAs, α_{latt} was $1.0x10^{-8}$ and α_{imp} was $7.1x10^{-8}$ at 300K, and the 1st term is $7x10^4$ times larger than the 2nd term at 300K. The 2nd term increases with decreasing the temperature, and becomes comparable to the 1st term at 100K. The 2nd term, the Hooge parameter due to the impurity scattering, becomes larger than the 1st term at the higher temperature when the electron concentration becomes larger.
The mobility is determined by the impurity scattering below 100K, and $\mu_{sum}/\mu_{imp} \simeq 1$ and $\mu_{sum}/\mu_{latt} << 1$. In this temperature range we are able to assume the relation

$$S_{\mu imp}/\mu^2_{imp} = \alpha_{imp}/fN. \quad (3)$$

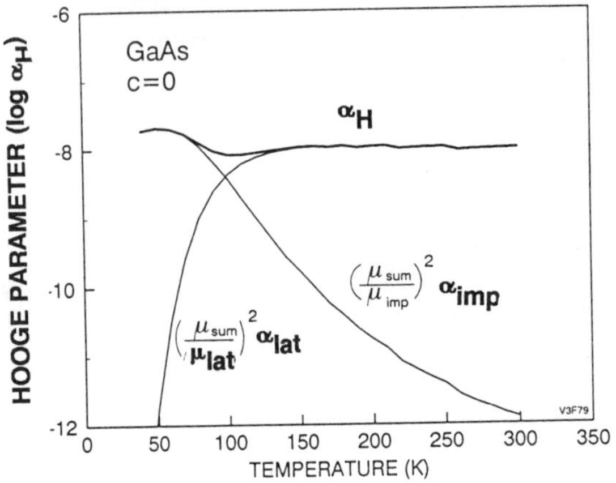

Fig. 4 Temperature dependence of α_H. $\alpha_H \simeq \alpha_{imp}$ at Low temperature.

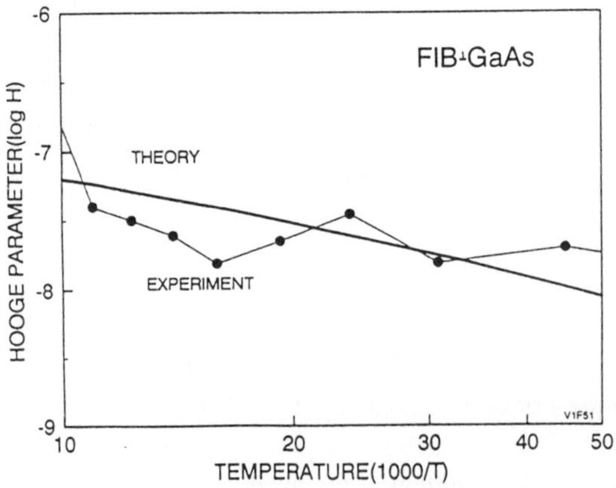

Fig. 5 α_H of FIB-GaAs. Solid line: quantum 1/f model, plots: experiments

The spectrum densities of a FIB filament of n-GaAs also show fairly pure f^{-1} characteristics. The noise parameter α_H was calculated from $\alpha_H = S_v fN/V^2$ on each temperature, and plotted in Fig. 5. The solid line in Fig. 5 was obtained from the quantum 1/f noise model. The parameters in the numerical analysis were determined so that the mobility and the carrier concentration were equal to those obtained from the measurements.

The experimental values and their temperature dependence of α_H in Fig. 5 agree quantitatively with the theoretical estimation. The absolute values of α_H scatter within one order from sample to sample. The impurity scattering was dominant with our quarter-micron samples below 100K, and we are able to obtain α_{imp} in this temperature range.

4. α_H of GaAs determined by Cross Correlational Quantum 1/f Model

The cross correlation model expects always larger α_H than the KVBH model. From experimental point of view we need the absolute values and their functional dependence. The α_H for the bulk n-GaAs was analysed numerically based on the CHX[3] model and compared with that obtained from the KVBH[5] model. The finite compensation ratio, which must be considered in the actual III-V compound semiconductors, was assumed in the analysis.[13] Figure 6 shows the cross correlational α_H at the donor concentration of $n = 2.4 \times 10^{23}$ m^{-3} at 80K and the compensation ratio of $c = 0.04$. The corresponding electron mobility μ_{sum} is also plotted in Fig. 6. The carrier concentration and the electron mobility correspond to those used in the experiments. α_H then must be compared with those obtained in experiments. The Hooge parameter α_H expected by the KVBH model($c=0$) is shown for comparison.

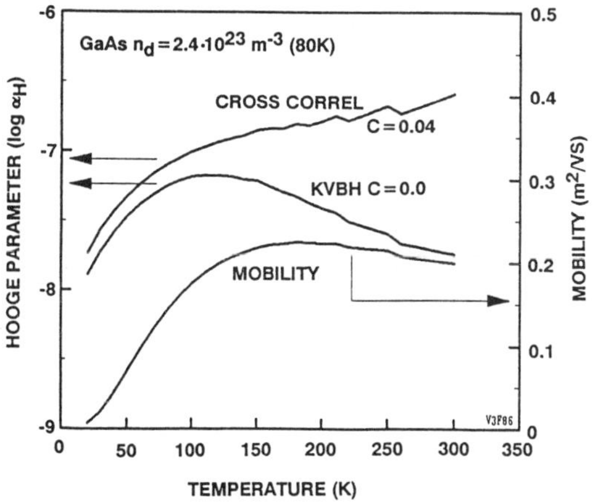

Fig. 6 α_H determined by Cross correlation and KVBH model, together with μ_{sum}.

It must be noticed that the cross correlational α_H has the similar dependence on temperature as that expected by the KVBH model(c≠0). The cross correlational α_H of n-GaAs has the following characteristics.

1) α_H of the doped sample is almost independent of the compensation ratio throughout the temperature between 300 and 30K, while it increased considerably at 300K by KVBH model.
2) The α_H is about twice of that expected from the KVBH model throughout the temperature range.
3) The α_H decreases with the increasing ionized impurity concentration at 80K, while it does not change at 300K.

Conclusion

The Hooge parameter α_H will be determined by α_{imp} when the impurity scattering is the dominant scattering mechanism in a semiconductor. The electric field dependence of α_H in n-GaAs at 77K might prove this. The temperature dependence of α_H below 100K was further interpreted in terms of the impurity scattering. Some numerical analyses on the cross correlational α_H were added.
The Hooge parameter α_H will be termed now as a new fundamental transport parameter in semiconductors.

References

1) F.N. Hooge: Proc. ICNF'91 7(1991).
2) L. Ren: Proc. ICNF '91 55(1991).
3) T.H. Chung, P.H. Handel, J. Xu: Proc. ICNF'91 163(1991).
4) M. Tacano, H. Tanoue and Y. Sugiyama: Proc. ICNF'91, 167(1991), Japan. J. Appl. Phys. **31** L316(1992).
5) G.S. Kousik, C.M. van Vliet, G. Bosman and P.H.Handel: Adv. Phys. 34(1985) 663
6) T.G.M. Kleinpenning: Physica, **103B**, 345(1981).
7) T.G.M. Kleinpenning: Physica, **113B**, 189(1982).
8) T.G.M. Kleinpenning: Physica, **142B**, 229(1986).
9) G. Bosman, R.J.J. Zijlstra, A. van Rheenen: Physica, **112B**, 188(1982).
10) M.E. Levinshtein, S.L. Rumyantsev: Sov. Phys.Semicond. **19**, 1015(1985).
11) M. Tacano, Y. Sugiyama: IEEE trans. ED, **ED-38**, 2548(1991).
12) J.G. Ruch and W. Fawcett: J. Appl. Phys. **41**, 3843(1970).
13) W. Walufiewicz, L. Lagowski, L. Jastrzebski, M. Lichtensteiger,

II. FILM CONDUCTORS AND SEMICONDUCTORS

EXCESS NOISE AND RELIABILITY OF AL-BASED THIN FILMS

T. M. Chen
Electrical Engineering Department
University of South Florida, Tampa, FL 33620

ABSTRACT

Noise measurement have been considered as a promising tool for reliability testing in the electronic industry. This paper discusses the characteristics of noise sources in Al-based thin films and their relationships to the reliability of VLSI interconnections. A technique of applying noise measurements in detecting existing defects/damage in the films, determining electromigration parameters and predicting the time of failure of VLSI interconnects is presented. This new technique is fast enough for wafer-level reliability testing and is nondestructive in nature. The problems associated with implementing wafer-level noise measurements for reliability testing were also discussed.

I. INTRODUCTION

Thin Al-based films have been widely used for interconnection in semiconductor VLSI circuits. Advancement of VLSI technology has resulted in continuous down scaling of device and interconnection dimensions in the chips. The minimum feature size used in the manufacturing of ICs has been reduced from 25 microns in 1960 to a fraction of micron today. This dramatic rate of size reduction creates problems associated with circuit reliability. Two of the most important reliability problems associated with high density small size circuits are electromigration failure and stress induced void failure of metallic interconnections in VLSI circuits. Characterization and prediction of electromigration and stress induced voids in VLSI circuits therefore become increasingly important for a newly developed circuit. The semiconductor industry urgently needs a testing method which can be used for wafer-level testing for quick adjustment of the VLSI interconnection processing parameters.

Our research group at University of South Florida, has been studying the relationship between electrical noise and reliability of Al-based thin films for several years. The studies have shown that noise measurements can be used as a sensitive method for testing reliability as well as determining electromigration parameters of thin film interconnections. Since noise measurements require much less time than conventional MTF (Median Time to Failure) measurement, and is non-destructive in nature, it can be used for wafer-level reliability testing. This paper will present some important relationships between electrical noise and reliability for $A\ell$-based thin films and techniques for implementing wafer-level noise measurements for reliability testing.

II. ELECTRICAL NOISE COMPONENTS AND RELIABILITY OF AL-BASED THIN FILMS

Our measurements have shown that the electrical noise in Al-based thin films consists of a $1/f$ noise component, a component with a $1/f^2$ frequency dependance and $\alpha \geq 2$ (to be called $1/f^2$ noise component hereafter) and the thermal noise component. Each of these three noise components has quantifiable relationships to defects/the electromigration process of the films. Our experimental results have shown that measurements of the thermal noise can be used for detection of hot spots/void formation in Al-films [1] and the magnitude of $1/f$ noise is closely related to the d. c. current distribution in the film. Stress induced voids, cracking, mechanical or electromigration damage will cause current crowding in the films and thereby increase the $1/f$ noise. We have previously reported [2] that the magnitude of $1/f$ noise in a thin film interconnection is extremely sensitive to electromigration damage. A few percentage change in the resistance of the film due to electromigration damage is, in some case, accompanied by as much as 1000% increase in the films observed $1/f$ noise. A film with stress induced voids or physical damage caused by scratches or improper wire bonding often generates a higher $1/f$ noise than a normal film. These defective films usually fail early in the accelerated lifetime test. Measurement of $1/f$ noise can therefore be used to detect early failure or existing defects/damage in the film.

For all the $A\ell$-based films measured, we have observed $1/f$ noise spectra at low frequencies when the film temperature is low and/ or the biasing current is small. As the film temperature and/or the biasing current is increased to a level commonly used for accelerated electromigration testing, the low frequency noise spectra change from $1/f$ to $1/f^2$ frequency dependance. The magnitude of $1/f^2$ noise has been found to be closely related to the electromigration rate of the films. We have found experimentally that the voltage spectral intensity of the $1/f^2$ noise follows the relation.

$$S_v(f) = \frac{Aj^3}{Tf^2}\exp\left(-\frac{Ea}{kT}\right) \tag{1}$$

where j is the biasing current density. T is the absolute film temperature, A is a constant and Ea is an activation energy which has been shown to have values close to those determined from MTF testing [3,5]. The median time to failure (MTF) of a film interconnection subject to electromigration is often described empirically by the following formula developed by Black [6]

$$MTF = t_{50} = K_1 Tj^{-n}\exp\left[\frac{Ea}{kT}\right] \tag{2}$$

where t_{50} is the time to 50% failure of a group of samples made under identical conditions. K_1 is a constant of proportionality and n is a constant having a value between 1 and 3. Since the MTF is a statistical result obtained from a large number of samples, one can reasonably assume that the time to failure, TTF, of a sample is proportional to MTF so that

$$[TTF]^{-1} = \frac{Kj^n}{T} \exp\left(-\frac{Ea}{kT}\right) \tag{3}$$

where K is a constant. $[TTF]^{-1}$ can be considered to be proportional to the average rate of electromigration damage. Equation (1) and (3) show that the magnitude of $1/f^2$ noise spectrum, $S_v(f)$, and the rate of electromigration damage, $(TTF)^{-1}$, have a similar dependence on T and j. This suggests that there is a close relationship between $1/f^2$ noise and TTF. By combining equations (1) and (3), one can relate the TTF to the mean square of $1/f^2$ noise voltage, $\overline{v^2}$ by

$$\log(TTF) = a + b \quad \log \overline{v^2} \tag{4}$$

where a and b are constants depending on the measurement conditions (T and j values) of TTF and $S_v(f)$. To experimentally confirm this relationship between the TTF and $\overline{v^2}$, two sets of films were fabricated by sputtering pure aluminum (99.99%) onto SiO_2 substrates of 10000Å thickness at two different substrate temperatures, 125° C and 250° C. These two sets of films will be called L-type and H-type corresponding to the deposition temperatures of 125° C and 250° C respectively. Our measurements and analysis showed that the two group of films had different TCR and different grain sizes and distribution, thereby providing an intentional variation in their expected lifetime.

$1/f^2$ Noise measurements were performed at a current density $j = 1.9 \times 10^6$ A/cm^2 and a film temperature $T = 505$k. The magnitude of the spectrum for the films varied several orders of magnitude at 1 Hz. Following the noise measurements, the films were stressed to failure using a constant current accelerated electromigration test. The test current density and film temperature were $j = 1.25 \times 10^6$ A/cm^2 and $T = 426$k respectively. The results of this experiment were plotted in Figure 1 which shows a linear relationship between log (TTF) and log $\overline{v^2}$ in agreement with equation (4). Therefore, one can predict TTF from the $1/f^2$ noise measurements.

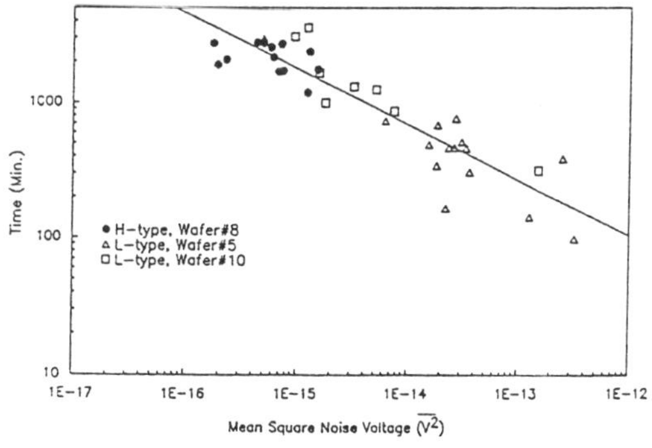

Fig. 1 Film time to failure versus mean square $1/f^2$ noise voltage

III. DETERMINATION OF ACTIVITIES ENERGY FROM NOISE MEASUREMENTS

In the electronic industry, the activation energy, Ea, in equation (2) (Blacks's equation) is often determined from the MTF measurement. In this measurement, a large number of identically prepared thin film interconnects are tested at elevated temperature and current density until failure. Due to the statistical nature of electromigration, a range of failure times occur at a given temperature and current density. Typically t_{50} is measured at a constant current density at several different temperatures from which Ea is determined. This is a tedious, expensive measurement and destroys all of the samples.

The magnitudes of both $1/f$ noise and $1/f^2$ noise components in Al-based films are temperature dependent and this temperature dependency can be used to determine the activation energies associated with each noise generation process. Our experimented results [3] as well as other published results [4,5] have shown that the activation of Al-based films determined from noise measurements have values close to electromigration energies obtained from the conventional MTF measurements. However, unlike MTF measurement, noise measurements can be used to determine Ea from individual sample. The noise measurement is much faster than the MTF measurement and is non-destructive in nature.

The $1/f$ noise in $A\ell$-based films is associated with thermally activated atomic motion at the grain boundary. Therefore,, the activation energy, Ea, associated with the $1/f$ noise can be determined by measuring the normalized noise spectral density (noise spectral density divided by the square of the bias voltage) as a function of the film temperature. Dutta and Horn has shown that [7]

$$-Ea = kT_p \ln (2\pi f \tau_o) \tag{5}$$

where T_p is the temperature at which the normalized $1/f$ noise spectral intensity, at angular frequency $2\pi f$, shows a peak and τ_o is the attempt time for the thermal activated process of atoms. Our measurement of Ea using this method [3] showed a close agreement with the values obtained from conventional MTF measurements.

To determine the activation energy associated with the $1/f^2$ noise, one has to recall that the generation of $1/f^2$ noise spectrum can be attributed to the following two different sources; (1) the rate fluctuation of vacancy diffusion around the grain boundaries and (2) a linear drift of the film resistance during noise measurement. However, our measurement system used a large condenser and a low frequency transformer at the input of the preamplifier to couple the film noise. These large inductance and capacitance components attenuate the $1/f^2$ spectrum due to a linear drift by more than 70 dB in our measurement circuit. Details of this effect will be explained in the other paper to be presented in this conference [8]. This large attention of the linear resistance drift effect has eliminated the linear drift as the possible same of $1/f^2$ noise spectra observed in our experiment. From equation (1), one can see that Ea can be determined from the slope of Arrhenius plots of T S_v (f) versus $(T)^{-1}$. The experimental results will be presented in the conference.

IV. WAFER LEVEL RELIABILITY TESTING

Our research findings have clearly shown that noise measurements can be used as a new technique for wafer-level reliability testing. However, since electrical noise levels to be detected in the reliability tests of metallic interconnections are low, accurate noise measurements in the production environment require special considerations. The problems to be considered include the design of low noise contact probes to the test structure and technique for reducing or avoiding vibration and interference of other wafer-level tests on the noise measurements. In addition, since each noise component (thermal noise, $1/f$ noise or $1/f^2$ noise) plays a different role in the reliability analysis, determination or separation of the magnitude of each noise component in the resultant sample noise measurement data is often necessary in this study. The solutions of these problems are discussed below.

Reduction of Probe Contact Noise: In wafer-level testing, contact probes are used to make electrical connections between the test sample and the measurement system. Since these connections are mechanical contacts, contact noise is generated at each current-carrying contact. If the magnitude of the contact noise is not negligible compared to the sample current noise to be measured, the sample noise cannot be accurately determined. Therefore, in order to successfully apply the noise measurement techniques to wafer-level testing, it is necessary to design low-noise contact probes and to develop techniques for reducing the effect of probe contact noise on the film noise measurements [9].

Separation of $1/f$ Noise and $1/f^2$ Noise component: Current noise spectra can often be measured by using the conventional noise measurement system. When the film is at room temperature and the biasing current density is less than $10^5\ {}^{A}/_{cm}{}^2$, the low frequency noise is dominated by the $1/f$ noise component. However, in many cases, this $1/f$ noise is very low and could be masked by the preamplifier or system noise. To increase the $1/f$ noise level, one can increase the biasing current since the $1/f$ noise level is proportional to the square of the biasing current. But the increase of the biasing current could be complicated by generation of $1/f^2$ noise in the film. The coexistence of $1/f$ noise and $1/f^2$ noise will make the determination of the magnitude of the $1/f$ noise component difficult. This difficulty can be alleviated by using the a.c. bridge noise measurement system [10,11] shown in Figure 2 for $1/f$ noise measurement. This bridge system is much more sensitive than the conventional system and can be used to accurately measure the $1/f$ noise component at a low biasing current and temperature under which $1/f^2$ noise generation is negligible. To measure the $1/f^2$

noise the wafer needs to be heated to an elevated temperature and the biasing current also has to be increased to the level at which the $1/f^2$ noise component dominates the $1/f$ noise component. When the film temperature and/or biasing current is increased to a level commonly used for accelerated electromigration testing, $1/f^2$ levels are usually high enough that the conventional system can be successfully used for detection. However, the noise or harmonic pick up associated with heating source and the air circulation around the probes have to be carefully reduced/controlled for accurate measurements.

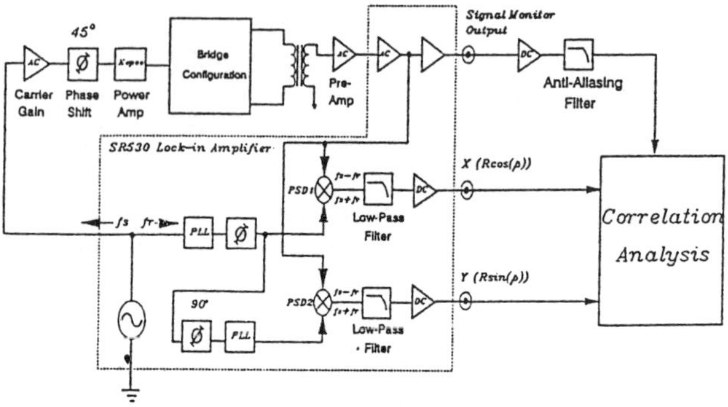

Fig. 2 Block diagram of a.c. bridge noise measurement system

REFERENCES

1. G.H. Massiha, T.M. Chen and G.J. Scott, IEEE Elect. Dev. Lett. Vol 10, 58 (1989)
2. T.M. Chen, T.P. Djeu and R.R. Moore, Proc of 23rd Int. Rel. Phys. Symp. 87 (1985)
3. J.G. Cottle, T.M. Chen and K.P. Rodbell, Proc of 26th Int. Rel Phys Symp 203 (1988)
4. R.H. Koch, J.R. Lloyd and J. Cronin, Phys Rev. Lett vol. 55, 2487 (1985)
5. B. Neri, A. Diligenti and P.E. Bagnoli, IEEE Trans Elect-Dev vol ED-34, 2317 (1987)
6. J.R. Black, Proc IEEE, Vol 57, 1587 (1969)
7. P. Dutta and P.M.Horn, Rev Mod Phys 53, 3 (1981)
8. J. Hall and T.M. Chen, Proc of 12th Int. Cong on Noise in Phys Syst. and $1/f$ Fluct(1993)
9. A.M. Yassine and T.M. Chen, IEEE Elect Dev Lett Vol 12, 200, (1991)
10. J.H. Scofield, Rev Sci Instrum. 58 (6), 985 (1987)
11. A.H. Verbruggen, H. Stoll, K. Heeck and R.H. Koch, Appl Phys A, 48, 223 (1989)

JOHNSON–NYQUIST NOISE SPECTRUM FOR 2D ELECTRON GAS IN A NARROW CHANNEL

O. M. Bulashenko
Institute of Semiconductor Physics, Academy of Sciences of Ukraine
Pr Nauki 45, Kiev 252650, Ukraine

ABSTRACT

The equilibrium current fluctuations (thermal noise) for 2D degenerate electron gas bounded in a narrow channel have been calculated within the semiclassical Boltzmann–equation approach. The associated noise spectrum has a non-Lorentzian shape with geometrical resonances at high frequences caused by the restriction on electron motion in the transverse direction (classical size–effect). Measurements of the size-dependent noise spectrum would give an additional information about the edge scattering of electrons.

INTRODUCTION

The noise properties of electron gas in small–size conductors have attracted considerable interest during last years.[1–4] The transition from diffusive to ballistic transport was found to be accompanied with new interesting phenomena: (i) the suppression of shot noise,[1,2] (ii) the noise redistribution toward higher frequencies ("blue–shift"), depending on the geometrical size of the sample,[3,4] (iii) the geometrical resonances in the spectrum,[4] and so on.

In this communication the Johnson–Nyquist noise characteristics for a two-dimensional (2D) electron gas bounded in a narrow channel are presented. Within the semiclassical approach the autocorrelation function of the current fluctuations is calculated both analytically and by use of the Monte Carlo technique. The spatial correlation of the fluctuations is taken into account, which is essential in the small–size conductors.[5]

CORRELATION FUNCTION FOR 2D ELECTRON GAS

Consider 2D electron gas in the xy–plane laterally restricted by the diffusely reflected boundaries at $y = 0$ and $y = d$. The channel width d is assumed to be much wider than the Fermi wavelength. The electrons are scattered both in the bulk and at the boundaries. The length L in the x–direction is much greater than the electron mean free path λ and terminated by contacts for measurement of the equilibrium current fluctuations. Thus, the electron transport is characterized by the parameter $\gamma = \lambda/d$, which is the "degree of ballistic transport".[3,4] Varying γ from $\gamma \ll 1$ to $\gamma \gg 1$ we are going from entirely bulk scattering (diffusive regime) to entirely boundary scattering of electrons (ballistic or Knudsen regime of electron transport).[3]

The instantaneous short–circuit current $I(t)$ through the sample of length L can be expressed as a sum of the instant velocities of all carriers presented at the time t in the sample [6]

$$I(t) = \frac{e}{L}\sum_{i=1}^{N} v_{xi} = \frac{e}{L}\int d\vec{r}\int \frac{2}{(2\pi)^2}d\vec{k}\; v_x\, f(\vec{r},\vec{k},t) \tag{1}$$

Here e is the electron charge, $\vec{r} \equiv (x, y)$ is the radius–vector, $\vec{k} \equiv (k_x, k_y)$ is the wave–vector, $v_x = \hbar k_x/m$ is the electron velocity component and m is the effective mass. The integration over $\vec{r}$ is taken over the channel area $L \times d$. $f(\vec{r},\vec{k},t)$ represents the electron distribution function (occupation numbers).

Hence, the current autocorrelation function can be expressed through the fluctuations in the occupation numbers by

$$C_I(t) = \frac{e^2}{L^2}\frac{1}{(2\pi^2)^2}\int d\vec{r}\int d\vec{r}'\int d\vec{k}\int d\vec{k}'\; v_x v_x'\; \langle \delta f(\vec{r},\vec{k},t)\; \delta f(\vec{r}',\vec{k}',0)\rangle \tag{2}$$

where $\delta f(\vec{r},\vec{k},t) = f(\vec{r},\vec{k},t) - f_0(\vec{k})$, $f_0(\vec{k})$ is the Fermi–Dirac distribution function. In equation (2) the angle brackets indicate averaging over the initial time moment $t = 0$ (for fixed value of t).

The correlation function $\langle \delta f(\vec{r},\vec{k},t)\; \delta f(\vec{r}',\vec{k}',0)\rangle$ satisfies for $t > 0$ the Boltzmann kinetic equation in the first set of variables.[2,7] By neglecting interaction between electrons the solution for the Boltzmann equation in the relaxation-time approximation takes the form

$$\langle \delta f(\vec{r},\vec{k},t)\delta f(\vec{r}',\vec{k}',0)\rangle = 2\pi^2 e^{-t/\tau}\delta(\vec{r}-\vec{r}'-\vec{v}t)\;\delta(\vec{k}-\vec{k}')\; f_0(\vec{k})\;[1-f_0(\vec{k})], \tag{3}$$

where $\tau = \lambda/v_F$ and the factor $(1-f_0)$ is included due to the Fermi statistics under view. We consider the case of fully diffuse electron scattering at the boundaries destroying any correlation between incident and reflected electrons. The function $\delta(\vec{r}-\vec{r}'-\vec{v}t)$ represents the spatial correlation of the fluctuations. Thus, $C_I(t)$ becomes

$$C_I(t) = \frac{2e^2}{\pi^2}\frac{d}{L}e^{-t/\tau}\int_0^{md/\hbar t} dk_y \int_0^{\infty} dk_x\; v_x^2(1-\frac{v_y t}{d})\; f_0(\vec{k})[1-f_0(\vec{k})] \tag{4}$$

For the complete degenerate case, where $f_0(1-f_0) = k_B T\cdot\delta(\epsilon-\epsilon_F)$, the current autocorrelation function (4) can be evaluated analytically

$$C_I(t) = \frac{e^2 n_2}{m}\frac{d}{L}\; k_B T\; e^{-t/\tau}\; f(u), \quad \text{where } u = \frac{\gamma t}{\tau} = \frac{v_F t}{d} \tag{5}$$

$$f(u) = \begin{cases} 1 - \frac{4}{3\pi}u, & 0 < u < 1 \\ \frac{2}{\pi}[\arcsin u^{-1} + \frac{1}{3}(1-u^{-2})^{1/2}(2u+u^{-1}) - \frac{2}{3}u], & u > 1 \end{cases}$$

$n_2 = k_F^2/2\pi$ is 2D electron concentration, k_B is the Boltzmann constant, T is the temperature, ϵ_F and k_F is the Fermi energy and momentum. The parameter u is the time in units of the transit time of electrons between the boundaries.

The random motion of a single electron with a Fermi velocity was simulated in the 2D channel on the base of the Monte Carlo algorithm.[3] The obtained velocity autocorrelation function $C_v(t)$, normalized to $\frac{1}{2}v_F^2$, coincides nicely with $C_I(t)$ calculated from (5) and normalized to the bulk value. Hence, the Monte Carlo technique can be fruitfully applied to obtain the thermal noise characteristics for degenerate case. But in order to get $C_I(t)$ from $C_v(t)$, estimated by the MC method for a single electron with the Fermi velocity, one must multiply the result by the factor k_BT/ϵ_F (in addition to the dimension term ne^2d/L), which is the fraction of electrons near the Fermi surface, taking part in the fluctuations.

SIZE DEPENDENT NOISE SPECTRUM

The noise spectral density $S_I(\omega)$ was calculated as a Fourier transform of the autocorrelation function $C_I(t)$ (see Figure). When $\gamma \to 0$, $u \to 0$, and $f(u) \to 1$, we have the ordinary Johnson–Nyquist noise with the Lorentzian spectrum. With increasing γ the low–frequency noise is suppressed with a redistribution toward higher frequencies and a remarkable deviation from the Lorentzian shape. One can also observe the oscillations at frequencies correspondent to the time of flight between the boundaries d/v_F (see Inset). The origin of the oscillations is caused by the restriction on the electron motion in the y–direction (the classical size–effect), and their nature is similar to those in Ref.4, where size–effect occurs between the contacts.

Due to the Nyquist theorem the formula for the conductivity or the diffusion constant for the bounded electron gas [8] is easily obtained by integrating $C_I(t)$ given by (5) over the time t. It should be noted that the frequency-dependent $\sigma(\omega)$ can be obtained from (5) by the same manner.

CONCLUSIONS

The autocorrelation function for the thermal (Johnson–Nyquist) noise and the correspondent noise spectrum have been calculated for 2D degenerate electron gas in a narrow channel with diffusely reflected boundaries. By diminishing the channel width the noise is redistributed toward higher frequencies with a suppression of its low-frequency magnitude. The noise spectrum has a non-Lorentzian shape with geometrical resonances at high frequences. The Monte Carlo approach discussed in the present and the previous papers [3] may be usefully applied for calculating both the conductivity and the thermal noise for the electron gas in the region with more complicated and partially reflected boundaries.

The work is supported by the Grant from the Ukrainian Council on Progress in Science and Engineering.

REFERENCES

1. G.B.Lesovik, JETP Lett. 49, 592 (1989).
   M.Büttiker, Phys.Rev.Lett. 65, 2901 (1990).
   Th.Martin and R.Landauer, Phys.Rev. B45, 1742 (1992).
   M.Büttiker, Phys.Rev. B45, 3807 (1992).
   C.W.J.Beenakker and M.Büttiker, Phys.Rev. B46, 1889 (1992).
2. C.W.J.Beenakker and H.van Houten, Phys.Rev. B43, 12066 (1991).
3. O.M.Bulashenko, O.V.Kochelap and V.A.Kochelap, Phys.Rev. B45, 14308 (1992).
4. T.Kuhn, L.Reggiani and L.Varani, Nuovo Cimento D14 509 (1992).
5. J.P.Nougier, J.C.Vaissière and C.Gontrand, Phys.Rev.Lett 51 513 (1983).
6. B.Pellegrini, Phys.Rev. B34 5921 (1986).
7. S.V.Gantsevich, V.L.Gurevich and R.Katilius, Riv.Nuovo Cimento 2 1 (1979).
8. C.W.J.Beenakker and H.van Houten, Phys.Rev. B38, 3232 (1988).

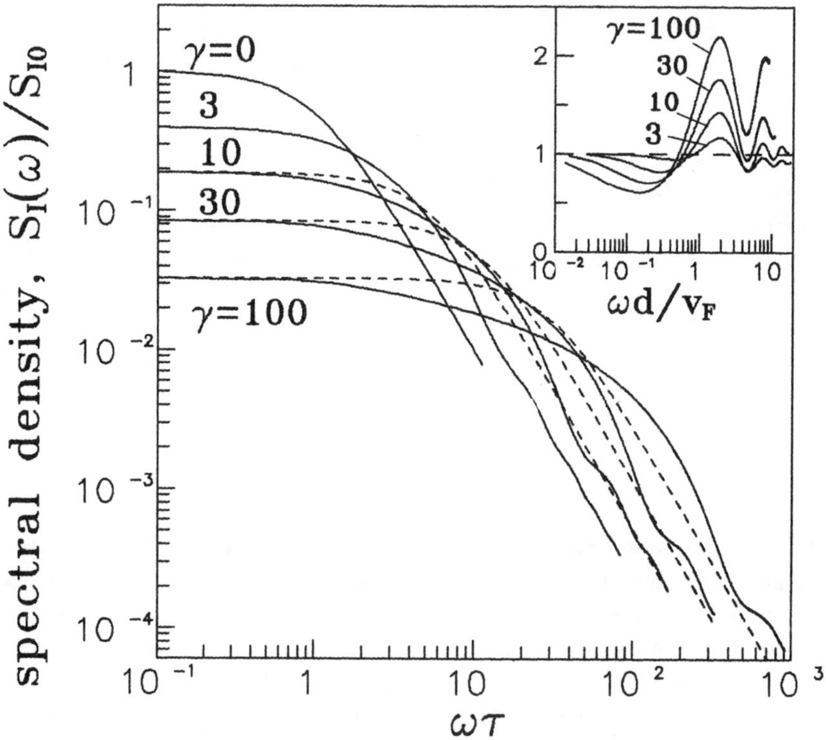

Figure. Spectral density of current fluctuations in the degenerate limit for different γ (solid lines) with the correspondent Lorentzian curves (dashed lines). The normalization constant is $S_{I0} = 4k_BT(n_2e^2\tau/m)(d/L)$. Inset: $S_I(\omega)$ normalized to the correspondent Lorentzians.

DETERMINATION OF $Al_{0.25}Ga_{0.75}As$ LONGITUDINAL DIFFUSION COEFFICIENT D(E) FROM H.F. NOISE MEASUREMENTS. COMPARISON WITH GaAs RESULTS.

M. de Murcia, D. Gasquet, E. Richard, P. Wolff,
Centre d'Electronique de Montpellier (Laboratoire associé au C.N.R.S. U.A. n°391),
Université de MontpellierII, 34095 Montpellier Cedex 5,France.

J.Zimmermann
L.P.C.S.-ENSERG. 23,Rue des Martyrs.BP 257. 38016 Grenoble Cedex.

J.Vanbremeersch
I.E.M.N.,Dept.Hyperfréquences et Semiconducteurs, Université des Sciences et Technologies, 59655 Villeneuve d'Ascq Cedex.

ABSTRACT

High frequency noise measurements in $Al_{0.25}Ga_{0.75}As$ material for hot electron transport, are reported. The longitudinal diffusion coefficient D(E) is derived from white noise measurements The results are compared with the values obtained in GaAs.

I-INTRODUCTION

The purpose of the present paper is to investigate the bulk Si doped $Al_{0.25}Ga_{0.75}As$ H.F. noise.The interest of such a study is two-fold. From the point of view of physics transport, it was noted[1] that in GaAs the dependence of the noise temperature $T_n(E)$ with the electric field could be a sensitive indicator for electron transition from the Γ band to the upper bands. This mean of investigation is usable in order to analyze the influence of the Al content x on the scattering mechanism controlling hot electron transport in $Al_xGa_{1-x}As$. From the point of view of Modfet's devices, the evaluation of diffusion coefficient D(E) is necessary to simulate the HF noise when parasitic conduction takes place in the AlGaAs layer.

II- DESCRIPTION OF THE DEVICE STRUCTURES

1- The planar devices used in the experiments had the following structure :
a GaAs buffer layer was grown on an GaAs -SI substrate followed by an undoped AlGaAs graded in composition to prevent any modulation-doping effects. Then a Si-doped $Al_xGa_{1-x}As$ (x=0.25) 1µm thick layer was grown. Growth is finished with a n^+ GaAs cap layer on which AuGe-Ni ohmic contacts were processed. The epitaxial layers were grown by MBE. The device structure is sketched on Fig.1.

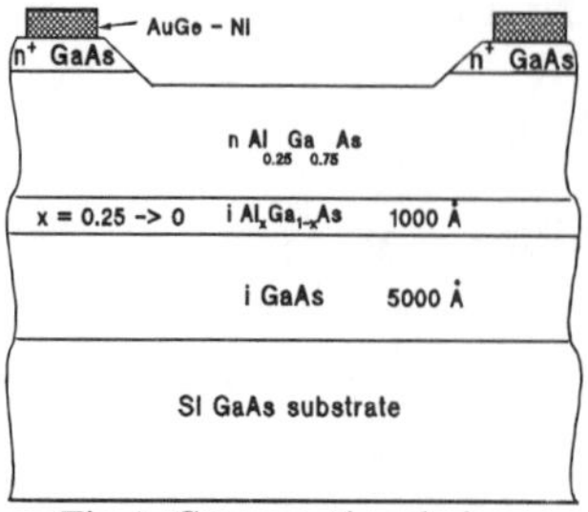

Fig.1. Cross-sectional view of AlGaAs structure.

2- The contact resistance R_C associated with the ohmic contact to the active region was determined by the conventional transmission line model (TLM). Four different sample lenghts were available: L=5, 10, 20, 50μm. The extracted $2R_C$ value was 10Ω.

III- EXPERIMENTAL RESULTS

Noise measurements were performed as a function of the electric field strength at the following frequencies : 50, 220, 320, 460, 650, 850 MHz and 4GHz.The bias voltage was applied in short pulses of width 1μs with a repetition rate of 70Hz in order to avoid lattice heating effect.

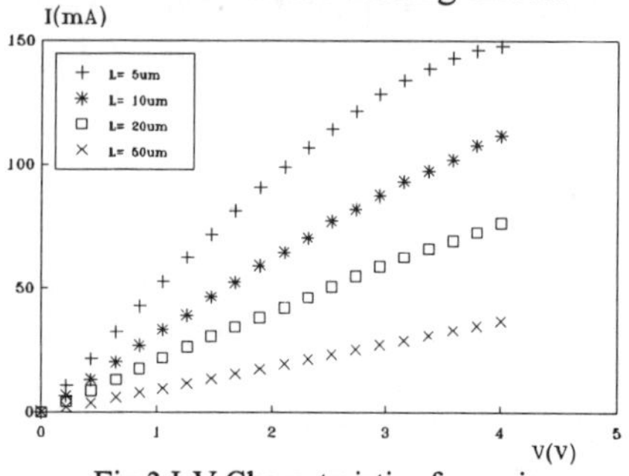

Fig.2.I-V Characteristics for various sample lenghts.

Nonohmic regime of the I-V characteristics can be observed at high fields in Fig.2 The excess noise temperature measurements are displayed in Fig.3 as a function of frequency. In the white noise region f>850 MHz, the longitudinal diffusion coefficient D(E) is evaluated from the generalized Einstein relation:

$$D(E) = \left[kT_n(E)/q\right]\mathrm{Re}\{\mu'\} \tag{1}$$

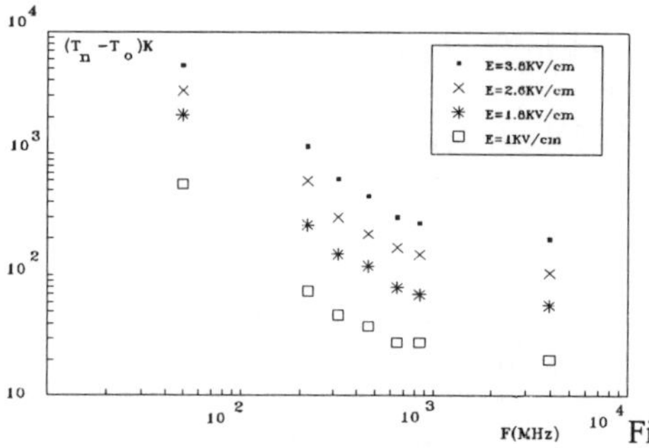

Fig. 3. Excess noise temperature vs. frequency.

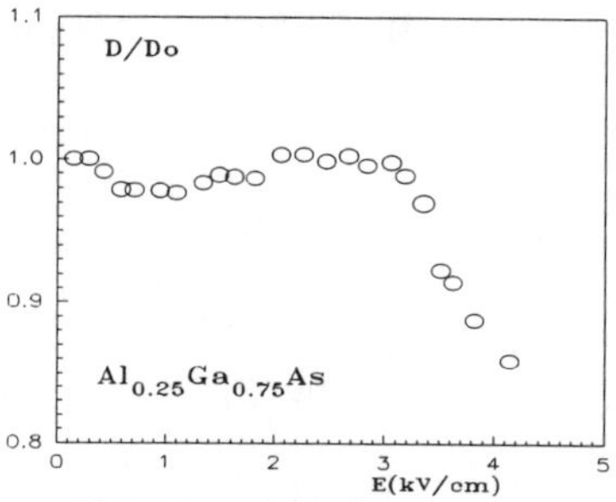

Fig.4. Dependence of D(E) on electric field.

where $\mu'(E)=dv/dE$ is the differential mobility. Assuming the differential impedance to be real up to these frequencies, from (1) we get:

$$D(E)/D_o=R_o(dI/dV)(T_n(E)/T_o) \quad (2)$$

where R_o is the ohmic resistance, T_o the lattice temperature and dI/dV the differential conductance corresponding to the average field E=V/L. The results are plotted in Fig.4.

IV-DISCUSSIONS

1/ AlGaAs results

The analysis of D(E) with E together with the observed behaviour of $T_n(E)$ and I(V) shows a linear-response region for E< 400V/cm, followed by a region where the noise temperature $T_n(E)$ increases and the differential mobility $\mu'(E)$ decreases with E. The coefficient D(E) proportional to the product $T_n(E)\mu'(E)$ remains almost constant. Above 3000V/cm, the drift velocity deviates strongly from linearity and the diffusion coefficient decreases rapidly.

2/ Comparison of GaAs and $Al_{0.25}Ga_{0.75}As$ diffusion coefficients

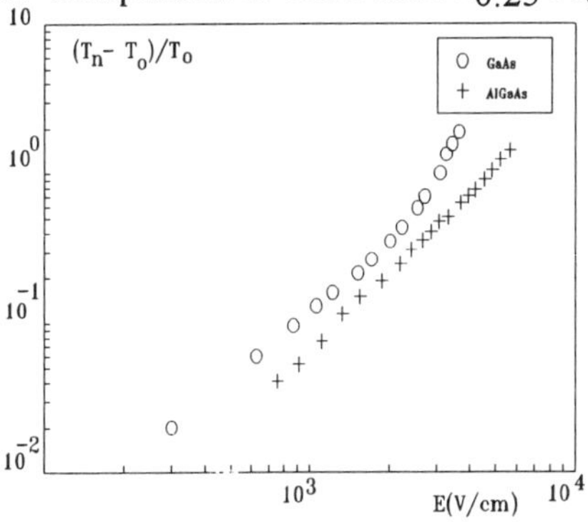

Fig.5 Dependence of the normalized temperature vs. Electric field at 4 GHz.

The excess noise temperature measurements as a function of E for GaAs and AlGaAs are reported and compared in Fig.5, at a frequency of 4GHz. The noise temperatures for both materials may be approximated by a same empirical formulas as:

$$T_n = T_o(1+ a(E/E_o)^n) \quad (3)$$

At the same field, noise temperatures in GaAs are higher than in AlGaAs. Two piece-wise linear segments are needed to fit the GaAs data in the whole electric field range. The behaviour of the noise temperature with electric field strength can be divided into three ranges. At fields lower than E<400V/cm $T_n(E)$ is equal to the lattice temperature T_o. Between 400< E <2500V/cm, both temperatures increase nearly at the same rate with E. In this field range intravalley velocity fluctuations[2] dominate the main noise source in GaAs. In AlGaAs electron heating by the field is substantially weaker[3] probably due to a lower mobility. The mean electron energy $<\varepsilon>$ or equivalent electron temperature T_e that we define as $3/2kT_e= <\varepsilon> - 1/2m^*v^2$, should be lower than in the GaAs[4]. This is also observed with the measured noise temperatures $T_n(E)$.

Above the threshold field E>2500V/cm, GaAs noise temperatures increase very fast with E probably due to intervalley transfers as is conjectured in ref.5. This feature is not observed in AlGaAs.

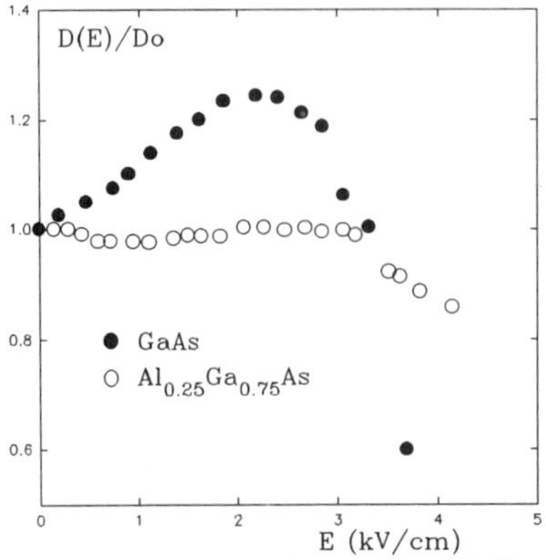

Fig.6. Dependence of normalized D(E) on Electric field for GaAs and AlGaAs.

In Fig.6 the diffusion coefficient measurements[1] of GaAs are reported together with those made in AlGaAs. It is worth noting that the initial increase of D(E) followed by a maximum value at the threshold field observed in GaAs material completly disappeared in the alloy AlGaAs.

V- CONCLUSIONS

The conclusions can be summarized as follows:

1- Noise is frequency dependent for f<850Mhz in AlGaAs[6].

2- In the white noise region, the noise temperatures in AlGaAs are lower than in GaAs grown with the same doping and at the same electric field.

3-In the electric field range 0<E<3000V/cm where intravalley collisions dominate the diffusion coefficient remains nearly constant in AlGaAs.

4- The steep increase of $T_n(E)$ due to intervalley transfers above the threshold electric field E_{th}=2500 V/cm present in GaAs is not observed in AlGaAs where the barrier height $\Delta E_{\Gamma L}$ controlling intervalley transfers is substantially reduced.

REFERENCES

1-M.de Murcia, D.Gasquet, A.Elamri, J.P.Nougier, J.Vanbremeersh. IEEE Trans.Electron Devices ,vol.38,n°11,p.2531,1991

2-C.Canali, F.Nava, L.Reggiani."Hot Electron Transport in Semiconductors". Edited by L.Reggiani Springer Verlag Berlin 1985.Chap.3.

3-S.Adachi.J.Appl.Phys.58,(3),R1,1985.

4-M.V.Fischetti.IEEE Trans.Electron Devices,vol.38 n°3, p.634,1991.

5-J.Pozela, A.Reklaitis.Sol.State Elec.vol.23,927,1980.

6-M.de Murcia, F.Pascal, G.Lecoy, L.K.J.Vandamme. "Exces noise in AlGaAs epitaxial layers". This conference.

EXCESS NOISE IN $Al_{0.25}Ga_{0.75}As$ EPITAXIAL LAYERS

M. de Murcia, F.Pascal, G.Lecoy
Centre d'Electronique de Montpellier (Laboratoire associé au C.N.R.S. U.A. n°391),
Université de MontpellierII, 34095 Montpellier Cedex 5,France.

L.K.J. Vandamme
Eindhoven University of Technology,Dep. El. Eng 5600 MB Eindhoven,
The Netherlands

ABSTRACT

Low frequency noise measurements have been performed in $Al_{0.25}Ga_{0.75}As$ test structures at 300K. The excess noise spectra show two Lorentzian shaped g-r contribution and 1/f noise. The results are analysed as a function of the sample length in order to distinguish between bulk and interface noise contributions.

I-INTRODUCTION

The low frequency noise properties have been studied by several authors in AlGaAs bulk material[1], AlGaAs/GaAs Modfet's [2,3], and AlGaAs/GaAs heterojunction bipolar transistors [4,5]. The noise spectra show generally several generation-recombinaison (g-r) noise components and 1/f noise. Most of these studies assign the physical origin of the g-r noise source to trapping and detrapping carriers at the DX centers present in AlGaAs althougth the magnitude of the deduced activation energies E_t show large scattering.
The aim of this contribution is to investigate conductance fluctuations in n type $Al_{0.25}Ga_{0.75}As$, and to distinguish between contact and bulk contributions by using different geometries in which the bulk and interface contribution scale differently.

II- EXPERIMENTAL PROCEDURE

The samples used in our experiments are planar devices. The epitaxial layers are grown by molecular beam epitaxy. The growing sequences are as follows: an undoped GaAs buffer layer of thickness 500nm is grown over a semi insulating GaAs substrate. The buffer is followed by a graded in composition (x=0 to x=0.25) undoped AlGaAs layer, 100nm thick. This to avoid any modulation -doping effects and electron accumulation in GaAs buffer. Then a 1 µm thick $Al_{0.25}Ga_{0.75}As$ layer is grown, Si doped at 4 $10^{17}cm^{-3}$. The top layer consists of an $n^{+}= 2\ 10^{18}cm^{-3}$ GaAs cap layer on which AuGe-Ni contacts were processed and AlGaAs surface passivated. The devices are planar resistors (Transmission Line Model) of 5, 10, 20, 50µm length and 80µm width. The device structures are sketched in Fig.1 The contact resistance R_c associated with the ohmic contact to the active AlGaAs layer is determined from the following equation:

$$R_t = \frac{\rho l}{A} + 2R_c \qquad (1)$$

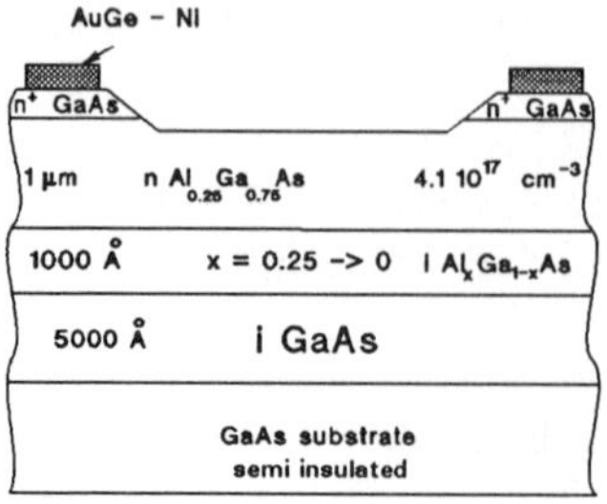

Fig.1- Layer configuration

where R_t is the resistance between the two electrodes, ρ the resistivity of the active layer AlGaAs of length l and cross section A. The linear extrapolation to l=0 in eqn. 1 gives $2R_c=10\Omega$. Only $2R_c=1.5\Omega$ was found from the same contact plugs to GaAs layers[6]. This points to a serious contribution of the heterojunction interface (n^+GaAs/ AlGaAs) in the devices.

III- EXPERIMENTAL RESULTS AND DISCUSSION

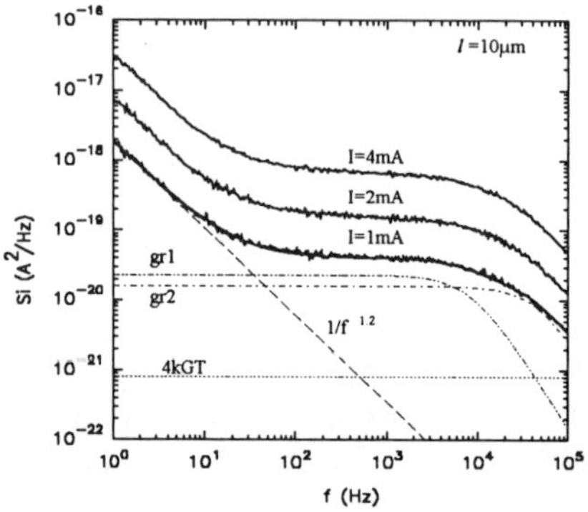

Fig.2- Current spectral density for a l=10µm AlGaAs device at 3 different bias currents.

Noise measurements were performed at room temperature in the frequency range 1Hz- 100KHz for different dc bias current I. The current fluctuations Si(f) are shown in Fig.2 with I as a parameter for a sample length of 10µm. The excess noise spectra are proportional to I^2 and involve 1/f noise and a broad g-r noise which can be fitted with two lorentzian shaped generation recombinaison contributions. The current noise is shown in Fig.3 for different sample lengths at constant bias current I=1ma. The experimental results can be summarized therefore by the expression:

$$S_I = \left[\frac{C_{1/f}}{f^{\gamma}} + \frac{C_{gr1}}{1+(f/f_{c1})^2} + \frac{C_{gr2}}{1+(f/f_{c2})^2} \right] I^2 \tag{2}$$

where $1<\gamma<1.2$, $f_{c1}= 8 10^3$Hz and $f_{c2}= 4.3 10^4$Hz are the corner frequencies, $C_{gri}.I^2$ is the low frequency plateau value of the g-r noise.

For conductance fluctuations in the ohmic region the relation below, always holds:

$$\frac{S_I}{I^2} = \frac{S_G}{G^2} = \frac{S_R}{R^2} \tag{3}$$

Considering eq.(1), the resistance noise S_R can be described[7] by an addition of a bulk and contact contribution given by:

$$S_R = \frac{\alpha R_{bulk}^2}{Nf} + \sum_{i=1}^{2} \frac{C_{gri} R_{bulk}^2}{1 + (f / f_{ci})^2} + 2S_{RC} \quad (4)$$

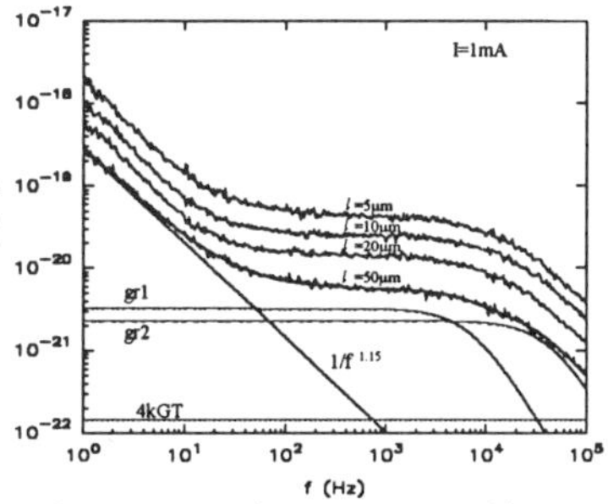

Fig.3- Current noise spectra Si versus frequency f from 4 samples of different length at an average current of I=1mA

where α is the 1/f noise parameter and N the total free charge carriers number.
When traps are distributed homogeneously over the bulk or only over the surface or interface of the epi-layer with the spacer, the term $C_{gri} = <\Delta N^2>/N^2.4\tau_i$ is expected to vary as $1/Nf_{ci}$ and hence $C_{gri}R^2_{bulk}$ is proportional to l. For the contact noise term S_{Rc},we expect no dependence on l. The dependence of $C_{1/f}$, C_{gr1}, C_{gr2} with N can be deduced from the plot of the current noise as a function of sample length at the frequencies f=3Hz, 1KHz, 10KHz and 100KHz (see Fig.4).

The results does not show an appreciable deviation from the $1/l$ dependency.

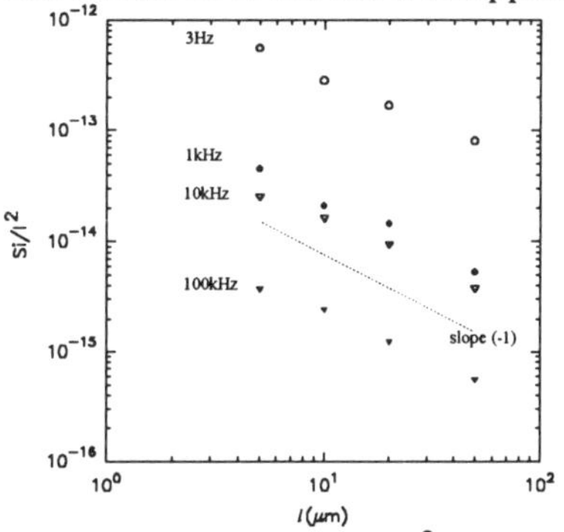

Fig.4- Relative noise Si/I^2 versus length l at f=3Hz, 1kHz, 10kHz, 100kHz.

These results point to the fact that both 1/f and generation-recombinaison noise stem from an AlGaAs layer. The contact plug with its heterojunction has a negligible low noise contribution. From the 1/f noise, the Hooge coefficient α_H has been determined. The value found is about $4 10^{-4}$ which is an indication for the quality of the layer[8]. Examination of the g-r noise expression with the device geometry, show that fluctuation stemming from the AlGaAs bulk or from the AlGaAs surface or the AlGaAs/spacer interface, scale all with l and cannot be distinguisted.

In this way, current noise comparisons have been performed in devices with and without AlGaAs passivated surface.

Experimental results plotted Fig.5, show no appreciable difference which means that the g-r noise and the 1/f noise do not stem from the AlGaAs surface. Moreover we measured the L-F noise of devices with Al mole fraction of x=0.2. Both devices (x=0.2 and x=0.25) were compared. While the x=0.25 and x=0.2 exhibited comparable 1/f noise, the g-r noise was eliminated for the x=0.2 devices as shown in Fig.6. These results give a strong support for the bulk origin of the g-r noise. The g-r noise source can be associated with traps in the AlGaAs bulk.

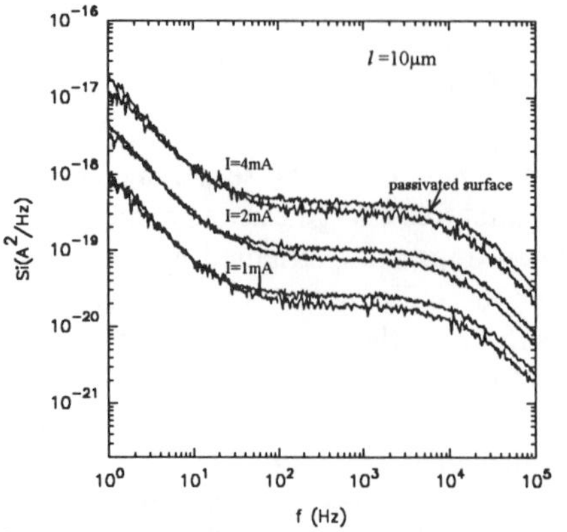

Fig.5- Comparison of current noise spectra Si of a l=10μm device, with and without AlGaAs surface passivation.

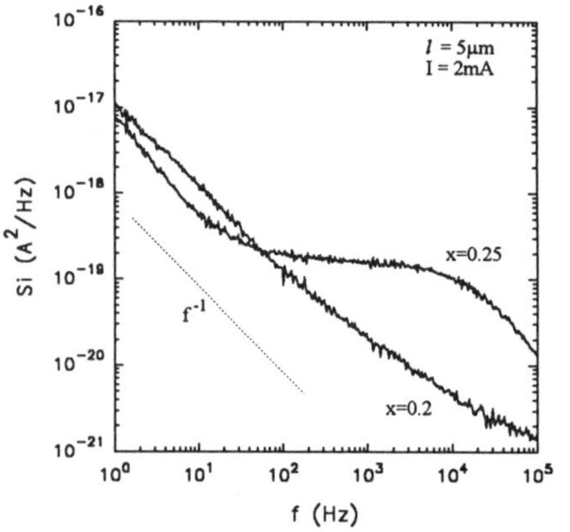

Fig.6- Comparison of current noise spectral density Si of l=5μm devices for two different Al mole fractions.

IV-CONCLUSIONS

The dominant source of 1/f in our AlGaAs devices was shown to vary as N^{-1} and is associated with AlGaAs bulk. The measurements also suggest that the origin of the g-r noise is due to fluctuations of traps distributed in AlGaAs bulk.

ACKNOWLEDGMENT

The authors wish to thank J.Vanbremeersch for growing the AlGaAs epitaxial layers.

REFERENCES

1-F.Hofman, R.J.J.Zijltra, J.MBettencourt De Freitas, J.C.M. Henning. Solid State Electronics vol34,n°1,p.23,1991

2-S.Kugler.IEEE Trans. Elec.Dev. vol 35,n°5,p.623,1988

3-M.Tacano, Y.Sugiyama. Sol.Sta. Elec.vol 34,n°10,p.1049,1991

4-D.Costa, J.S.Harris. IEEE Trans.Elec.Dev. vol 39,n°10,p2383,1992

5-F.Pascal, S.Jarrix, C.Delseny, G.Lecoy, C.Dubon-Chevallier, J.Dangla, H.Wang. This Conference

6-M.de Murcia, F.Pascal, D.Gasquet, G.Lecoy, J.Vanbremeersch. Proc.Int.Conf.on Noise in Physical Systems and 1/f fluctuations 1991, Kyoto, Japan; Eds. T.Musha, S.Sato, and M. Yamamoto,p199

7-F.N.Hooge, T.G.M.Kleinpenning, L.K.J.Vandamme. Rep.Prog.Phys.,vol 44, p.479, 1981.

8-L.K.J.Vandamme. Solid State Phenomena, vol 1, 2, p.153,1988.

FLUCTUATING DEEP-LEVEL TRAP OCCUPANCY MODEL FOR HOOGE'S 1/f NOISE PARAMETER FOR SEMICONDUCTOR RESISTORS

Patrick A. Folkes
Army Research Laboratory
Fort Monmouth, NJ 07703-5601

ABSTRACT

A theoretical expression for Hooge's 1/f noise parameter α, for a Schottky barrier field-effect transistor (MESFET), which has been biased at a small drain-source voltage (a gate-controlled semiconductor resistor), has been derived. The theory is based on the fluctuating occupancy of deep level traps in the depletion region. The theory shows that α varies approximately as $n^{-7/2}$, where n is the electron density, and that α is sensitive to the trap concentration, the gate (or semiconductor surface) potential, the thickness of the semiconductor conducting layer and the low-field electron mobility-depletion depth profile. We obtain excellent agreement between the theoretical and experimental dependence of α on the gate voltage.

INTRODUCTION

Hooge's empirical formula is widely used in discussing 1/f noise in semiconductor devices. S_f, the spectral density of the 1/f fluctuations in the current I, is given by the expression[1]

$$\frac{S_f}{I^2} = \frac{\alpha}{Nf} \tag{1}$$

where N is the number of charge carriers in the semiconductor resistor, f is the frequency and α which was initially chosen to be 2×10^{-3} to fit experimental data on semiconductor resistors, is Hooge's parameter. Numerous experimental results confirm the validity of Hooge's formula if α, is allowed to vary from sample to sample[2]. For room temperature semiconductor resistors, α decreases sharply with increasing electron density n, with observed values ranging from 10^{-2} to 10^{-7}. The dependence of α on electron concentration was attributed to the reduction in carrier mobility due to impurity scattering. However recent measurements on GaAs MESFETs do not support this conclusion[3,4]. To the best of my knowledge, the observed large variations in α remain unexplained, heretofore. In this paper we derive a theoretical expression for Hooge's parameter for a MESFET which is biased at a small voltage based on the recently developed fluctuating deep-level trap occupancy model for bulk semiconductor 1/f noise[5,6] and present experimental results which show excellent agreement with theory.

THEORY

Using the gradual channel approximation, I is given by the equation[7]:

$$I=-en\mu\frac{Z}{L}\int_{b_s(V_g,Q)}^{b_D(V_D,V_g,Q)}2\frac{W_0}{a}b\left(1-\frac{b}{a}\right)db \tag{2}$$

where Z and L are the width and length of the gate respectively, Q is the charge on the gate, a is the depth of the active layer, b,b_S and b_D are the steady-state channel heights at some point in the channel, the source and drain respectively, n and μ are the average electron density and mobility in the channel respectively, e is the electron charge, ε is the dielectric constant, V_D and V_g are the applied drain and gate voltages and $W_0 = nea^2/2\varepsilon$ is the channel potential needed to completely deplete the channel. $W_S= |V_b|-V_g$ and $W_D= W_S+V_D$ are the channel potentials relative to the gate at the source and drain respectively; V_b is the Schottky built-in potential. At any point in the channel b is related to the the channel potential W, by the equation:

$$W=W_0\left(1-\frac{b}{a}\right)^2 \tag{3}$$

A detailed one-dimensional analysis[5,6] shows that the MESFET's gate charge fluctuation results in 1/f noise over a spectral range which depends on the trap energy distribution. For small V_D the 1/f noise spectral density S_f can be approximated by[8]

$$S_f=\frac{1}{f}\frac{\mu(d)^2e^2Z\lambda_DN_t}{4L^3}\frac{F^2F_1^4}{(F_1F_2-F_3F_4)^2} \tag{4}$$

$$F(V_D,V_g)=(W_SW_0)^{1/2}-W_S-\frac{1-(W_D/W_0)^{1/2}}{1-(W_S/W_0)^{1/2}}[(W_DW_0)^{1/2}-W_D] \tag{5}$$

$$F_1=\frac{ab_S^2}{2}-\frac{b_S^3}{3}-\frac{ab_D^2}{2}+\frac{b_D^3}{3} \tag{6}$$

$$F_2=ab_S^2-b_S^3-ab_D^2(W_S/W_D)^{1/2}+b_D^3(W_S/W_D)^{1/2} \tag{7}$$

$$F_3=\frac{ab_S^3}{3}-\frac{b_S^4}{4}-\frac{ab_D^3}{3}+\frac{b_D^4}{4} \tag{8}$$

$$F_4=ab_S-b_S^2-ab_D(W_S/W_D)^{1/2}+b_D^2(W_S/W_D)^{1/2} \tag{9}$$

where f is the frequency, N_t is the average trap density, $\lambda_D=(\varepsilon kT/e^2n)^{1/2}$ is the Debye length, k is Boltzmann constant, T is the lattice temperature and $\mu(d)$ is the electron mobility at the channel/depletion region boundary. Integrating (2) we obtain

$$I = \frac{2e\mu nZW_0F_1}{La^2} \tag{10}$$

The number of electrons in the channel under the gate is given by[9]

$$N = nZL\frac{F_3}{F_1} \tag{11}$$

Combining (1), (4), (10) and (11) we obtain the following expression for α

$$\alpha = H(V_D,V_g)\frac{\varepsilon^{5/2}(kT)^{1/2}}{4e^3}\left(\frac{\mu(d)}{\mu}\right)^2\frac{N_t}{n^{7/2}} \tag{12}$$

$$H(V_D,V_g) = \frac{F_1F_3F^2}{(F_1F_2-F_3F_4)^2} \tag{13}$$

COMPARISON WITH EXPERIMENTAL RESULTS

1/f noise, current-voltage, electron concentration and low-field mobility profile measurements, were carried out on a MESFET which exhibits 1/f noise over the frequency range .02 - 100 MHz at 300K. The device parameters are: L=1 μm, Z=500 μm, a= 0.16μm, $n=10^{17}$ cm^{-3} and W_0=1.76V. Details of the experimental technique used in measuring the electron density and mobility $\mu(d)$ have been published[10]. $\mu(d)$ and $S_f(f = 0.15\text{MHz})$ were measured with V_D = 0.1V. Using (11) to calculate N, α can be accurately determined from measurements of S_f and I. Fig.1 shows the measured values of α as a function of V_g with V_D=0.1V. The experimental data shows that for a constant V_D, α increases as the conducting channel is depleted by varying V_g. The measured value of S_f at V_g=0 is used to determine that N_t =6.9 x 10^{13}cm^{-3} for this device. The parameter $(\mu(d)/\mu)^2$, which is plotted in Fig.1, is obtained by numerical integration of the $\mu(d)$ data. Using these and the above device parameters in (12), α is calculated and plotted in Fig.1 as a function of V_g. It should be emphasized that the theoretical values for α are obtained with the use of the single fitting parameter N_t =6.9 x 10^{13}cm^{-3}. The theory predicts that α= 1.52 x 10^{-5} at V_g= 0 which agrees with the measured value of 1.35 x 10^{-5}. Fig.1 also shows that the observed variation of α over the range -0.3V < V_g < 0.2V is in excellent agreement with the theoretical dependence.

Equation (12) shows that α is proportional to the trap density and does not depend on the resistor width or length. Fig.2a shows that $H(V_D,V_g)$ and hence α is insensitive to V_D for voltages small compared to W_0. If we vary n over the range 5 x 10^{16}cm^{-3} < n < 2.5 x 10^{17}cm^{-3} while keeping the other device parameters constant Fig.2b shows that H changes only slightly. This implies that over the same range of electron density, α is approximately $\propto n^{-7/2}$. This result could explain the observed reduction in samples with increased carrier density[2]. The theory also shows that α decreases by a factor of eight as the thickness of the semiconductor is changed from .15μm to .25μm as shown in Fig.2c. Note that α has a $(kT)^{1/2}$ temperature dependence. In concluding, we point out that the fluctuating deep level trap occupancy 1/f noise model readily explains the large variations in the observed values of α.

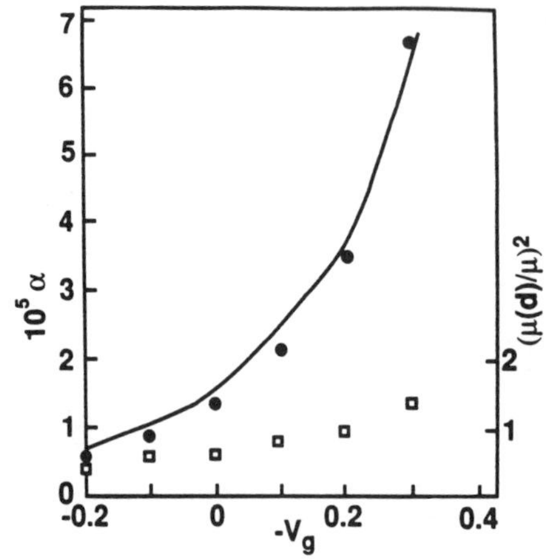

Fig.1 Experimental (circles), theoretical (solid line) α and $(\mu'(d)/\mu)^2$ (square dots) as a function of V_g with V_D=0.1V.

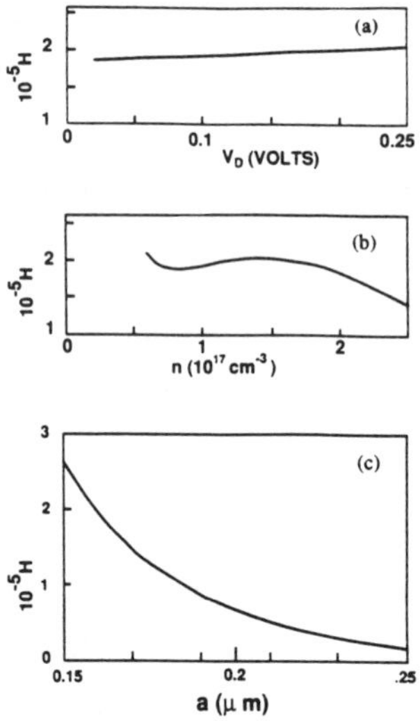

Fig.2 Theoretical plots of $H(V_D,V_g)$ vs (a) V_D (b) electron density and (c) active layer thickness.

REFERENCES

1. F.N. Hooge, Phys. Lett. 29A, 139 (1969).
2. L.J.K. Vandamme in Noise in Physical Systems and 1/f Noise, edited by M. Savelli, G. Lecoy and J-P. Nougier (Elsevier Science Publishers B.V., 1983), p. 183.
3. A. van der Ziel, P.H. Handel and K.H. Duh, IEEE Trans. on Electron Dev. ED-32, 667 (1985).
4. K.H. Duh and A. van der Ziel, IEEE Trans. on Electron Dev. ED-32, 662 (1985).
5. P.A. Folkes, Appl. Phys. Lett. 55, 2217 (1989).
6. P.A. Folkes, J. Appl. Phys. 68, 6279 (1990).
7. W. Shockley, Proc. I.R.E. 40, 1365 (1952).
8. P.A. Folkes, Solid State Electr. 36, 483 (1993).
9. P.A. Folkes, M. Taysing-Lara, W. Buchwald, P. Newman and L. Poli, IEEE Electron Dev. Lett. 12, 215 (1991).
10. P.A. Folkes, Appl. Phys. Lett. 48, 431 (1986).

DISORDER-INDUCED FLICKER NOISE IN SMALL STRUCTURES. NORMAL AND SUPERCONDUCTING STATE.

Yu.M.Galperin and V.I.Kozub
A.F.Ioffe Physico-Technical Institute, St.-Petersburg, Russia

ABSTRACT

A short review of theoretical and experimental results concerning flicker noise in small metallic structures with a special emphasis on telegraph noise related to separate "fluctuators". It is demonstrated that the behavior observed including temperature, bias and magnetic field behavior can be explained within a framework of the model of "soft" double-well interatomic potentials. The applications to superconducting state (including high-T_c materials are discussed.

No doubt that the most important achievements in the problem of flicker noise in recent years have been connected with the noise studies in small size systems and low-dimensional systems. These studies (see, e.g [1-5] have lead to a decomposition of the flicker noise to a sum of telegraph processes and proved the concept of "elementary fluctuator" following the ideas by Bernamont [6] and McWhorter [7].

In the case of MOSFET-like structures there are strong evidences (see review article [2]) that the fluctuators can be identified with some deep traps situated in the insulating layer while the relaxation time scatter necessary for the flicker noise formation is due to difference in spatial separation between the traps and the semiconductor. (Note however that as it has been shown by D'yakonova and Levinshtein [8] the deep traps in the bulk of semiconductors under some additional suggestions can also contribute to the flicker noise). However this mechanism of noise (although very important from the technical point of view) seems to be not general enough because 1/f noise exists in purely metallic systems as well.

From this point of view it appears that the fluctuators of most general nature in solids are those provided by some structural disorder and related to some slow atomic motion in double-well disorder-induced potentials. On the other hand, the necessary scatter of relaxation times is an inherent property of any disorder. Note that the connection of flicker noise with structural disorder has been evidenced by experiments by Pelz and Clarke [9].

At low temperatures such fluctuators can be identified with well-known two-level systems (TLS). Kogan and Nagaev [10] and Ludviksson, Kree and Schmid [11] were the first to suggest TLS as the source of the flicker noise while the detailed theory of this mechanism applicable to small-size systems was developed by one of the authors [12]. At higher temperatures, however, TLS model is not valid because the higher levels can be activated as well. However the concept of "soft" double-well interatomic potentials still holds. Having in mind that the low frequency noise is due interwell transitions while large relaxation times imply high and wide barriers one sees that the presence of many levels in each of the well lead to no dramatic effect. Indeed, on a large time scale (corresponding to interwell transitions) each well can be described by some averaged characteristics.

The general theory of the noise produced by these objects valid for the wide temperature range has been developed by Galperin, Karpov and Kozub [13] within a framework of soft-potential approach (SPA) [14]. It has been shown in particular that there should be relatively sharp crossover from tunneling type of fluctuator relaxation to activation one at some temperature $T_{cr} \sim w$ (where $w \sim 10 - 30K$ is some characteristic energy of SPA of the order of the interlevel spacing).

For the bulk samples the Hooge constant α has been estimated. The main conclusions have been the following: 1) α may depend on ω and T while the exact forms of such behavior are sensitive to the distribution of the barriers parameters, 2) typically α depends on ω only logarithmically, however power corrections to $1/\omega$ behavior are in principle possible, 3) for $T < T_{cr}$ $\alpha \propto T$ while for $T > T_{cr}$ α exhibits in general a non-monotonous behavior, 4) the quantitative estimates of α using the fluctuators densities typical for amorphous materials (greatest degree of disorder) are in agreement with greatest experimental values ($10^{-3} - 10^{-1}$). Thus the results seems to be consistent with existing experimental data.

The experiments by Ralls and Buhrman [3,15] (and later experiments by Holweg et al. [16]) on the metallic point contacts have given a possibility to study the parameters of separate fluctuator (activation energy, barriers heights and strengths, scattering cross-sections etc). It appears that at small and moderate temperatures (¡ 100 - 150 K) the experimental results are in agreement with SPA model of separate fluctuator [13]. However at higher temperatures pronounced effects of interfluctuator interactions leading to much more complex picture of noise have been observed. In a view that in metals such interaction can be mediated only by deformational coupling strongly decreasing with increase of the spacing ($\propto r^{-3}$ one may suggest that this fact implies that the fluctuators are formed within some close disordered clusters of atoms e.g. near dislocation lines. However until now the absence of microscopic model of the fluctuators prevents the detailed analysis of this problem.

In experiments [15,16] the effect of bias on the telegraph noise has been studied as well. In addition to increase of fluctuators relaxation rate τ^{-1} with the bias increase (that has been prescribed to electronic mechanism of fluctuators activation) even more surprising behavior like dependence of τ on the polarity of the bias has been reported. To explain the latter behavior the model of electromigration (involving some additional parameters has been suggested [15,16].

However recently it has been shown [17] that one can semiquantitatively explain the whole picture observed within a framework of SPA taking into account that at the biases larger than $\sim 1mV$ fluctuators within the nanostructure are coupled mainly with nonequilibrium phonons emitted by nonequilibrium electrons rather than with the electrons themselves. A direction of the phonon emission being sensitive to a current direction, one explains the effect of the bias polarity as a result of spatial asymmetry of nonequilibrium phonon "cloud" depending on the polarity in question.

We would like to mention as well instructive results of telegraph noise studies in bismuth microbridges [18] revealing in particular a surprising magnetic field effect on the noise that as has been proved in [18] can not be related to some localized magnetic moments. Authors of [18] suggested the possible role of spatial mesoscopic redistribution of electronic density influencing fluctuators parameters which is sensitive to magnetic field. However the picture observed (oscillations of the TLS interlevel spacing E reproduced for different fluctuators and increase of E at higher fields seems to be a systematic one rather than random one typical for mesoscopic effects. We believe that one can explain such a behavior basing on 2 facts: 1) the TLS interlevel spacing is adiabatically renormalized by the electrons in a way considered in [12], 2) in Bi magnetic fields of the order of $\sim 5T$ used in [18] appear to correspond to the extreme quantum limit implying a pronounced Landau quantization, that, in its turn, affects the interlevel spacing. We are going to analyze this behavior in more detail elsewhere. In addition, an anomalous temperature behavior of τ (increase of τ with temperature increase at $T < 1K$ has been observed for the same systems in [19]. Such an anomaly was explained with a help of "dissipative tunneling" model pioneered by Kondo [20].

These facts to our point of view allows one to identify the fluctuators at least in metals with soft interatomic potentials produced by some structural disorder (e.g. due to structural defects and their aggregates).

Note that the picture of the resistivity fluctuations can be the classical one (when the scattering events due to different scatterers are independent) or the quantum one - when the quantum interference effects are important (so-called universal conductance fluctuations - see review article [21]). According to our estimates ([22]) for metallic point contacts the classical effects dominate while for 2D electron systems in semiconductor structures quantum effects can be of importance.

We believe that the theory [13] can also explain the experimental data obtained for small tunnel junctions [4,5] when the fluctuators are situated in an insulating barrier. In particular it explains a crossover from tunneling to activation at $T \sim 15K$ observed in [4]. Sometimes the telegraph noise in tunnel junctions is prescribed to electronic traps within the barrier region (see e.g.[4]). However we would like to note that to explain large relaxation times observed the probability of electron tunneling to the trap from the *closest* of the metal electrodes; on the other hand such a suggestion seems to be incompatible with the fact that the tunneling through *thewholebarrier* (responsible for the junction conductance) should be effective enough. The model of "structural" fluctuators discussed above does not meet such a problem. As for anomalously large noise amplitude observed in [4,5] we believe that one can explain it taking into account the fluctuations of the barrier thickness; in a view of corresponding exponential dependence of the barrier transparency an effective cross-section of the contact appears to be much smaller than geometrical one that emphasizes the noise [22].

As it is well-known the flicker noise is of special importance for superconducting devices because it is these devices (like superconducting quantum magnetometer) that provide the record levels of sensitivity. Here one should first discuss the noise in Josephson structures (see e.g. the review article by Clarke [23]). As it has been shown [12] the disorder-induced fluctuators in Josephson tunnel junctions and other structures (point contacts, microbridges, etc) lead to the flicker noise in critical Josephson current. On the other hand, in a closed superconducting loop containing the junction the fluctuations of a magnetic flux through the loop should take place [12].

One can expect the disorder-induced flicker noise to be pronounced in high-T_c superconductors because the disorder (e.g. in oxygen vacancies arrangement) seems to be an inherent property of these materials. The theory of the noise has been generalized for high-T_c Josephson structures in actual temperature region in papers [22,24] The estimates of the noise obtained with the use of the independent data gives the upper limit of the sensitivity of high=T_c magnetometer to be $\sim 10^{-5}\Phi_o$ (Φ_o being the flux quantum). A detailed theory of disorder-induced flicker noise in high-T_c SQUIDs has been developed in paper [25] where different regimes important from the practical point of view has been considered. Theoretical estimates seems to be in agreement with existing experimental situation concerning the noise in high-T_c weak links (see e.g. [26,27]).

The other possible source of noise in superconducting devices can be connected with vortices jumps; the experimentalists succeeded in observation of telegraph noise connected with jumps of single vortex [28]. Such a magnetic noise can also be considered to be related to disorder providing the scatter of pinning energies.

We would like to note as well that one should expect the fluctuators coupling with superconductivity not only in Josephson structures but in "bulk" samples as well e.g. due to dependence of superconducting parameters on local strains. As a result the parameters in question (like T_c, "superconducting electrons density" N_s, magnetic field penetration length etc.) should exhibit local non-stationary fluctuations leading in particular to resistivity fluctuations within the superconducting transition region. An interesting scenario seems to be related to local fluctuations of critical magnetic field H_{c1} that may affect the vortices structure behavior. Then, one may expect that due to the coupling discussed the very fluctuators states are dependent on the state of the superconducting condensate.

That may impose the features of structural relaxation (like memory effects and slow relaxation phenomena) on the condensate responce. These phenomena are discussed in more detail in Ref.29.

Thus one can conclude that according to recent experimental and theoretical achievements the structural disorder may provide an effective source of the flicker noise especially pronounced in small structures in normal as well as in superconducting state.

REFERENCES

1. K. S. Ralls et al. , Phys. Rev. Lett. , **52**, 2434 (1984)
2. Kirton M. J. , Uren M. J. Adv. in Phys. , **38**,367 (1989)
3. K. S. Ralls and R. A. Buhrman Phys. Rev. Lett. , **60**,2434 (1988)
4. C. T. Rogers, R. A. Buhrman. Phys. Rev. Lett. , **53**,1272 (1984)
5. Huignard Jiang,M. A. Dubson,J. C. Garland Phys. Rev. B, **42**, 5427 (1990)
6. J. Bernamont. Ann. Phys. **7**, 71 (1937)
7. A. L. McWhorter Semiconductor surface physics. University of Pennsylvania Press, Philadelphia (1957)
8. N. V. D'yakonova, M. E. Levinshtein Fiz. Tekhn. Poluprov. **23** 283 (1989)
9. J. Pelz, J. Clarke Phys. Rev. Lett. , **55**, 738 (1985)
10. Sh. M. Kogan,K. E. Nagaev Sol. State Commun. ,**49**, 387 (1984)
11. A. Ludviksson,R. Kree,A. Schmid Phys. Rev. Lett. , **52**, 950 (1984)
12. V. I. Kozub Sov. Phys. JETP, **59**,1303 (1984); **60**, 810 (1984); Sov. Phys. -Solid State, **26**, 1186; 1851 (1984)
13. Yu. M. Galperin,V. G. Karpov,V. I. Kozub Sov. Phys. -JETP, **68**, 648 (1989)
14. V. G. Karpov,M. I. Klinger,F. N. Ignat'ev Sol. State Commun. , **43**, 274 (1982)
15. K. S. Ralls,D. C. Ralph,R. A. Buhrman Phys. Rev. B, **40**,11561 (1989)
16. P. A. M. Holweg et al. , Phys. Rev. B, **45**, 9311 (1992)
17. V. I. Kozub,A. M. Rudin Phys. Rev. B (to be published)

18. N. M. Zimmerman, B. Golding, W. H. Haemmerle Phys. Rev. Lett. **67**,1322 (1991)

19. B. Golding, N. Zimmerman, S. N. Coppersmith. Phys. Rev. Lett. , **68** 998 (1992)

20. J. Kondo Physica **123 B**, 175 (1984)

21. M. Giordano: in "Mesoscopics Physics", North Holland

22. Yu. M. Galperin,V. L. Gurevich,V. I. Kozub Europhys. Lett. **10**, (1989)

23. J. Clarke Adv. Supercond. Press NATO Adv. Study Inst. , Erice, July, 3-15,1982. New York, 1983, p. 13.

24. Yu. M. Galperin, V. L. Gurevich,V. I. Kozub Sov. Phys. -Solid State, **31**, 807 (1989)

25. Yu. M. Galperin. Journ. of Appl. Phys. , to be published

26. B. Golding et al. Phys. Rev. B, **36**, 5606 (1987)

27. V. M. Zakosarenko, E. V. Il'ichev Fiz. Tverd. Tela, **34**, N 5 (1992)

28. M. J. Ferrari et al. , Phys. Rev. Lett. **64**, 72 (1990)

29. V. I. Kozub, the same volume.

HYDRODYNAMICAL FLUCTUATIONS IN HOT ELECTRON GAS OF SEMICONDUCTORS: CROSS-CORRELATION EFFECTS

V.A.Kochelap and N.A.Zakhleniuk
Institute for Semiconductor Physics, Academy of Sciences of Ukraine
Pr. Nauka 45, Kiev-28 252650, Ukraine

ABSTRACT

We develop the theory of the hydrodynamic fluctuations in nonequilibrium electron gas which can be described by electron temperature. For this case the characteristic spatial-time parameters are the electric charge decay time τ_M, the electron temperature relaxation time τ_T and corresponding with them two diffusional lengths L_M and L_T. The spectral densities of the fluctuations are found and investigated at arbitrary relationships between the fluctuation frequency ω and times τ_M, τ_T for wave vectors $\vec{q}$ and the lengths L_M, L_T. In the first it is established the effect of the cross-over correlation of the electron density $\delta n(\vec{q},\omega)$ and temperature $\delta T(\vec{q},\omega)$ fluctuations. The cross-correlation depends on ω and changes its sign that indicates the existence of the frequence regions of correlation and anti-correlation of $\delta n(\vec{q},\omega)$ and $\delta T(\vec{q},\omega)$. In general case, spectral densities of the fluctuations have non-Lorentz form and that holds with thermal equilibrium too. At the equilibrium the cross-correlation effect leads to only redistribution of the fluctuation intensity over the frequency region. Under nonequilibrium conditions the cross-correlation changes also integral intensities of the fluctuations and directly related with additional kinetic correlation of the hot electrons caused by electron-electron interaction. The results are applied to the calculation of the light scattering by the electron plasma fluctuations.

INTRODUCTION

Hydrodynamical fluctuations are long-range and low-frequency stochastic excitations of the physical system with respect to stationary state of the system. It is assumed that for typical wave vector $\vec{q}$ and frequency ω of fluctuations the following conditions are satisfied

$$\omega\tau \ll 1, \quad qL \ll 1, \tag{1}$$

where L and τ are *microscopic* characteristics of space and time parameters of the system. In this case the physical object behaves as a continuous medium and can be described by a system of macroscopic equations. Nevertheless, that does not means that description of fluctuations by such a system of macroscopic equations is complete. Sources of all fluctuations are determined by microscopic random events, i.e. by collisions of the particles in the system. The evolution of these fluctuative excitations has two stages.

At the first stage the fast relaxation of initial excitation to local stationary distribution takes place. This stage of relaxation occurs as process in momentum space and is governed by Boltzmann-Langevin type equation. Local stationarity is reached after microscopic time τ, moreover, this process occurs independently in each element of volume of the physical system. At times greater than τ the detailed features of initial distribution of particles in momentum space disappear and the distribution takes its usual form in momentum space. In general case this distribution depends not only on momenta of particles but also on some *macroscopic* parameters of the physical system. These macroscopic parameters are functions of the space coordinates that leads to local flows in real space. That is, random processes in momentum space give rise to the dependence of the macroscopic parameters of the system on coordinates.The following relaxation of these macroscopic parameters occurs during the second slow stage of the evolution of the fluctuations to the final uniform steady state. This stage specifies the hydrodynamical evolution of the fluctuations including space transfer of electron density, energy, etc, and it is described by hydrodynamic type equations for the macroscopic parameters.

The hydrodynamical fluctuations of such plasma were subject of the numerical studies. However, the most advanced results have been formulated in two following cases:

$$\tau_T \ll \tau_M, \quad \omega\tau_T \ll 1, \quad q^2 L_T^2 \ll 1, \tag{2}$$

$$\tau_M \ll \tau_T, \quad \omega\tau_M \ll 1, \quad q^2 L_M^2 \ll 1. \tag{3}$$

In this work we develop the theory of the hydrodynamical fluctuations of the hot electron plasma for more general case when it is possible to drop the limitations (2) and (3). Such a generalisation is important because of different dependences of the parameters τ_M and τ_T on external electric field, that means both inequalities (2) and (3) can be realised for the same physical system.

BASIC EQUATIONS FOR HYDRODYNAMICAL FLUCTUATIONS

Our consideration is based on the Boltzmann-Langevin kinetic equation for the fluctuation of the electron distribution function $\delta F_{\vec p}(\vec q, \omega)$:

$$\hat{B}_{\vec p}(\vec q, \omega)\delta F_{\vec p}(\vec q, \omega) \equiv [-i\omega + i\vec q\vec v + \hat{L}_{\vec p}]\delta F_{\vec p}(\vec q, \omega) - i\vec q U(\vec q)\frac{\partial F_{\vec p}}{\partial \vec p} \times$$

$$\times \sum_{\vec p} \delta F_{\vec p}(\vec q, \omega) = y_{\vec p}(\vec q, \omega)\,, \tag{4}$$

where $y_{\vec p}(\vec q, \omega)$ is the stochastic microscopic force with the known correlation properties, $\vec v$ is the electron velocity, e is charge of electron, $\hat{L}_{\vec p}$ is the linearized operator of the Boltzmann equation:

$$\hat{L}_{\vec{p}} = e\vec{E}_0 \frac{\partial}{\partial \vec{p}} + \hat{I}^{th}_{\vec{p}} + \hat{I}^{ee}_{\vec{p}}\{F_{\vec{p}}\} . \tag{5}$$

Here $\vec{E}_0$ is external d.c. electric field, $\bar{F}_{\vec{p}}$ is hot electron distribution function, $\hat{I}^{th}_{\vec{p}}$ is the integral operator describing interaction with thermal bath, $\hat{I}^{ee}_{\vec{p}}\{F_{\vec{p}}\}$ is the linearized electron-electron collision operator and $\delta\vec{E}(\vec{q},\omega)$ being the self-consistent electric field caused by the spatial redistribution of electrons. This electric field is determined by Poisson equation

$$\delta\vec{E}(\vec{q},\omega) = -\frac{i}{e}\vec{q}U(\vec{q})\sum_{\vec{p}} \delta F_{\vec{p}}(\vec{q},\omega) \, , \quad U(\vec{q}) = \frac{4\pi e^2}{\varepsilon_0 V_0 q^2} \, , \tag{6}$$

where ε_0 is the lattice dielectric constant, V_0 is the volume of crystal.

We will seek the solution of the equation (4) in the low-frequency and long-range limit (1) using the Chapman-Enskog method under typical criteria for the hot electron plasma:

$$\tau_{\vec{p}} \ll \tau_{ee} \ll \tau_{\epsilon}, \tag{7}$$

where $\tau_{\vec{p}}$ and τ_{ϵ} are respectively the momentum $\vec{p}$ and energy $\varepsilon_{\vec{p}}$ relaxation times for the electrons interacting with the thermostat, τ_{ee} being electron-electron scattering time.

It is shown that the fluctuation of $\delta F_{\vec{p}}(\vec{q},\omega)$ can be expressed via two fluctuating parameters:$\delta n(\vec{q},\omega)$ and $\delta T(\vec{q},\omega)$. The parameters $\delta n(\vec{q},\omega)$ and $\delta T(\vec{q},\omega)$ which appear into this solution can be obtained from conditions of solvable of equation (4). These conditions are the continuity equation and the energy transfer equation for the fluctuations:

$$\sum_{\vec{p}} \hat{B}_{\vec{p}}(\vec{q},\omega)\delta F_{\vec{p}}(\vec{q},\omega) = 0 \, , \tag{8}$$

$$\sum_{\vec{p}} \varepsilon_{\vec{p}}\hat{B}_{\vec{p}}(\vec{q},\omega)\delta F_{\vec{p}}(\vec{q},\omega) = \sum_{\vec{p}} \varepsilon_{\vec{p}} y_{\vec{p}}(\vec{q},\omega) \, . \tag{9}$$

SPECTRAL DENSITY OF HYDRODYNAMICAL FLUCTUATIONS

Solving Eqs.(8)-(9) we have received spectral densities of fluctuations for general case. Here we show some of them which have the simplest form.

a) Equilibrium fluctuations: $\vec{E}_0 = 0$:

$$(\delta n^2)_{\vec{q}\omega} = \frac{2n^2}{N}\mathcal{D}_0 q^2 \frac{\omega^2 + \nu_T \nu_T^0}{(\omega^2 - \nu_M \nu_T^0)^2 + \omega^2(\nu_M + \nu_T)^2} \, , \tag{10}$$

$$(\delta T^2)_{\vec{q}\omega} = \frac{4T^2}{3N}\frac{\omega^2 \nu_T + \nu_M^2 \nu_T^0}{(\omega^2 - \nu_M \nu_T^0)^2 + \omega^2(\nu_M + \nu_T)^2} \, , \tag{11}$$

$$Re(\delta n\delta T)_{\vec{q}\omega} = \frac{4nT}{3N}\gamma_0\mathcal{D}_0 q^2 \frac{\omega^2 - \nu_M\nu_T^0}{(\omega^2 - \nu_M\nu_T^0)^2 + \omega^2(\nu_M + \nu_T)^2}, \tag{12}$$

$$Im(\delta n\delta T)_{\vec{q}\omega} = 0 \, . \tag{13}$$

Here n is electron density, $N = nV_0$, $\mathcal{D}_0(T)$ is diffusion coefficient, ν_M, ν_T, ν_T^0 are some parameters depending on τ_M, τ_T, $\mathcal{D}_0(T)$ and q, $\gamma_0 = \partial \ell n \mathcal{D}_0(T)/\partial \ell n T$.

Consideration of the two fluctuative degrees of freedom leads to the new results even under thermal equilibrium conditions: the existence of the cross-correlation of fluctuations $\delta n(\vec{q},\omega)$ and $\delta T(\vec{q},\omega)$, non-Lorentz form of spectra, $Re(\delta n\delta T)_{\vec{q}\omega}$ is nonmonotonous function of ω and changes sighn into some frequency region. His behaviour indicates the existence of regions of "correlation and anticorrelation" of the fluctuations of δn and δT.

b) Transverse fluctuations: $\vec{q} \perp \vec{E}_0$:

$$(\delta n^2)^{\perp}_{\vec{q}\omega} = \frac{2n^2}{N}\mathcal{D}_0 q^2 \frac{\omega^2 + \nu_T\nu_T^0 - \frac{2}{9}\gamma_0^2\mathcal{D}_0 q^2 \frac{R(T)}{NT^2}}{(\omega^2 - \nu_M\nu_T^0)^2 + \omega^2(\nu_M + \nu_T)^2}, \tag{14}$$

$$(\delta T^2)^{\perp}_{\vec{q}\omega} = \frac{4T^2}{3N}\frac{\nu_T\omega^2 + \nu_M^2\nu_T^0 - \frac{R(T)}{3NT^2}(\omega^2 + \nu_M^2)}{(\omega^2 - \nu_M\nu_T^0)^2 + \omega^2(\nu_M + \nu_T)^2}, \tag{15}$$

$$Re(\delta n\delta T)^{\perp}_{\vec{q}\omega} = \frac{4nT}{3N}\gamma_0\mathcal{D}_0 q^2 \frac{\omega^2 - \nu_M\nu_T^0 - \frac{R(T)}{3NT^2}\nu_M}{(\omega^2 - \nu_M\nu_T^0)^2 + \omega^2(\nu_M + \nu_T)^2}, \tag{16}$$

$$Im(\delta n\delta T)^{\perp}_{\vec{q}\omega} = \frac{4nT}{3N}\gamma_0\mathcal{D}_0 q^2 \frac{\frac{R(T)}{3NT^2}\omega}{(\omega^2 - \nu_M\nu_T^0)^2 + \omega^2(\nu_M + \nu_T)^2}. \tag{17}$$

It is known that for fluctuations with $\vec{q} \perp \vec{E}_0$ the similarity of both nonequilibrium spectra and equilibrium spectra (with arbitrary orientation of $\vec{q}$) takes place. In our case, as it follows from (14)-(17), the nonequilibrium state of system shows radical violation of such similarity. The reason of this change is the extra-correlation of the carriers that is caused by the electron-electron interaction. The measure of the extra correlation is the quantity $R(T)$ and which can be presented as $R(T) = \sum_{\vec{p}_1} \epsilon_{\vec{p}}\epsilon_{\vec{p}_1}\hat{I}^{ee}_{\vec{p}\vec{p}_1}\{\bar{F},\bar{F}\}$. In equilibrium steady state $R(T) = 0$ and extra correlation disappears. Depending on sign of quantity $R(T)$ the extra correlation can increase (if $R(T) < 0$) or decrease (if $R(T) > 0$) the intensity of fluctuations.

The above results were applied to the calculation of the light scattering by the hot electron plasma fluctuations. It was shown that the cross-correlation effects give important contributions into the cross-section of the light scattering. The cross-correlation contribution has the same order of magnitude with that from the density and temperature fluctuations. Moreover, the contributions in the cross-section given by the real and imaginary parts of the cross-correlation term can be separated because of their different dependences on the frequency of the incident light. The cross-effects bring about as the additional anisotropy of the light scattering by the plasma, as the qualitatively new phenomenon: a correlation-induced shift of the peak of the spectral line of the scattered light.

ELECTRON NOISE DUE TO PHONON SCATTERING IN QUANTUM WIRES

R. Mickevičius and V. Mitin
Wayne State University, Detroit, MI 48202

ABSTRACT

We have employed a novel Monte Carlo technique for the self-consistent simulation of electron transport and noise in GaAs quantum wires at low temperatures. The electron noise at low electric fields is completely controlled by acoustic phonon scattering. It is demonstrated that the efficiency of acoustic phonon scattering decreases as the electric field increases. As the result electron mobility and noise increase as the field initially increases. With further increase in electric field the role of optical phonon scattering increases and the electron noise starts decreasing and finally collapses at the streaming frequency.

INTRODUCTION

Low dimensional semiconductor structures and particularly quasi-one-dimensional (1D) quantum wires (QWI) presently attract much attention due to possibilities to achieve very high electron mobilities at low temperatures[1] and possible high-technology applications. A crucial device characteristic is its electric noise because it sets lower limits to the accuracy of any measurement. So far there exist almost no studies of electron noise in QWIs. The reason is that understanding of noise comes with the understanding of electron scattering. (In turn, noise and fluctuations carry all the information about scattering processes in solids). In QWIs, where electron-electron and impurity scattering is virtually eliminated[1], the only inherent scattering processes are scattering by all kinds of phonons. There exist rather elaborated models for electron interaction with 1D confined longitudinal optical (LO) phonons and by localized surface (interface) optical (SO) phonons in QWIs[2,3]. However, so far there was a considerable gap in understanding of some essential aspects of 1D electron scattering by acoustic phonons. It is common practice to treat acoustic phonon scattering within elastic or quasi-elastic approximations (see, e.g. Ref.4) as in bulk materials. However, it has been shown[5] that acoustic phonon scattering in low dimensional structures and particularly in QWIs is essentially inelastic and is far more important than in bulk materials, where acoustic phonon scattering is quasi-elastic. Due to peculiarities of phonon scattering in QWIs, electron kinetic and noise parameters may be considerably different from those in bulk materials.

The aim of the present paper is to get an insight into the electron transport and noise in QWIs at low temperatures and in a wide range of electric fields through the study of electron noise spectral density. We employ recently developed Monte Carlo code which efficiently includes electron scattering by acoustic as well as by optical phonons and permits self-consistent simulation of electron transport and noise in 1D structures.

RESULTS AND DISCUSSION

We have considered rectangular GaAs QWI embedded into AlAs with several different cross sections. We consider non-degenerate electron gas which corresponds to electron concentration of the order of $10^5 cm^{-1}$ or less. Electron scattering by confined longitudinal optical (LO) phonons and localized surface (interface) optical (SO) phonons[2] as well as by bulk-like acoustic phonons[5] has been taken into account. Ionized impurities are assumed to be located sufficiently remote from the QWI so that they do not affect the electron motion inside the wire. We have considered 7 subbands but only the first 2 or 3 influence to electron transport at low temperatures.

Figure 1 shows electron drift velocity as a function of applied electric field. For thick QWIs (curves 1 and 2 in Fig.1) the superlinear region appears on velocity-field dependence at electric field of the order of 10 V/cm. As the thickness of a QWI decreases this superlinear region goes up in electric fields and finally disappears for 40x40 $\AA^2$ QWI. The occurrence of the superlinear region is related to the beginning of electron heating by electric fields. As we already mentioned in Introduction the superlinear dependence is caused by reduction of the efficiency of acoustic phonon scattering in QWIs as the electron gas gets heated. This is purely 1D effect.

To examine the superlinear region we have calculated electron diffusion coefficient. It is plotted in Fig. 2 as a function of electric field. The superlinear region appears on diffusivity-field dependence as a broad maximum. Note that diffusion coefficient is equivalent to a noise spectral density at zero frequency (see discussion below). The maximum of diffusivity is well pronounced for thick QWIs and disappears for $40 \times 40\AA^2$ QWI. The reason for this is that electron scattering by acoustic phonons in this thin QWI is so strong[5] that it prevents electron heating by an electric field up to very high electric fields where optical phonon emission starts dominating and electrons enter the streaming regime. The transition from superlinear region to electron streaming manifests itself as the decrease in diffusivity.

The transition from diffusive electron transport to streaming can be precisely revealed from the analysis of velocity autocorrelation function:

$$C(T) = \langle \delta v(t) \delta v(t+T) \rangle , \tag{1}$$

where angular brackets stand for an average over time t, $\delta v(t) = v(t) - v_d$ is the deviation from the drift velocity v_d at time t.

Figure 3 shows autocorrelation functions plotted versus delay time and calculated for different electric fields corresponding to the ohmic electron transport (zero electric field), the superlinear electron transport regime (20 V/cm) and the near-streaming regime (200 V/cm). The characteristic decay time increases as the electric field increases from zero to 20 V/cm reflecting the decrease in acoustic phonon scattering efficiency at that fields. The negative autocorrelator which appears at 20 V/cm turns to damping oscillations when electrons reach the streaming regime (200 V/cm). The oscillation period coincides with the period of electron motion in **k**-space: $t_s = \sqrt{2\hbar\omega m^*}/eE$.

The calculated autocorrelation functions have been used to calculate the frequency dependences of electron diffusion coefficient (velocity noise spectral density) related with autocorrelation function through Wiener-Khintchine theorem:

$$D(\omega) = \int_0^\infty dT e^{-i\omega T} C(T). \quad (2)$$

The results are presented in Figure 4. On the ordinates we plot the normalized diffusion coefficient which in fact is merely the normalized noise power spectral density of conservative electron system. The relationship between the spectral density and diffusion coefficient is given by[6]: $S(\omega) = 4D(\omega)$.

The frequency dependence of diffusion coefficient at zero field has the Lorenzian shape: constant value up to some critical frequency, and then the rapid step-like decrease. (The same Lorenzian dependence with virtually the same critical frequency is obtained for 20 V/cm, i.e. at the maximum of diffusivity). The critical frequency is related to electron scattering rate which increases effectively with the onset of optical phonon scattering. The effective time of electron scattering by optical phonons is determined primarily by the electric field, i.e. is equal to the streaming time, because electron penetration into the active region is still negligible. At higher electric fields when electron streaming takes over electron diffusive motion, the critical frequency is shifted up and the peak related to the streaming frequency separates from the step-like diffusivity-frequency dependence (see Fig. 4). With the further increase in electric field, the peak related to the streaming increases while the plateau of constant diffusivity is going down. In that case almost all diffusivity (noise) collapses to the streaming frequency $f_s = 1/t_s$ and frequencies which are multiples of the streaming frequency.

Acknowledgement: This work was supported by the NSF and ARO.

REFERENCES

1. H. Sakaki, Japan J. Appl. Phys. **19**, L735 (1980).
2. K.W.Kim, M.A.Stroscio, A.Bhatt, R.Mickevičius, and V.V.Mitin, J. Appl. Phys. **70**, 319 (1991).
3. H. Rücker, E. Molinari, and P. Lugli, Phys. Rev. B **44**, 3463 (1991).
4. F. Comas, C. Trallero Giner, and J. Tutor, phys. stat. solidi (b) **139**, 433 (1987).
5. R. Mickevičius and V. Mitin, submitted to Phys. Rev. B.
6. J.Pozela, *Hot Electron Diffusion, Ed. by J.Pozela* (Mokslas, Vilnius 1981), p.210.

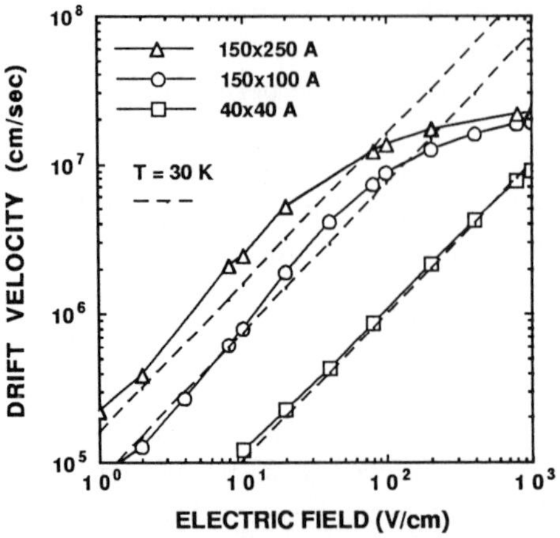

Fig. 1. Drift velocity as a function of electric field for three cross sections of a QWI.

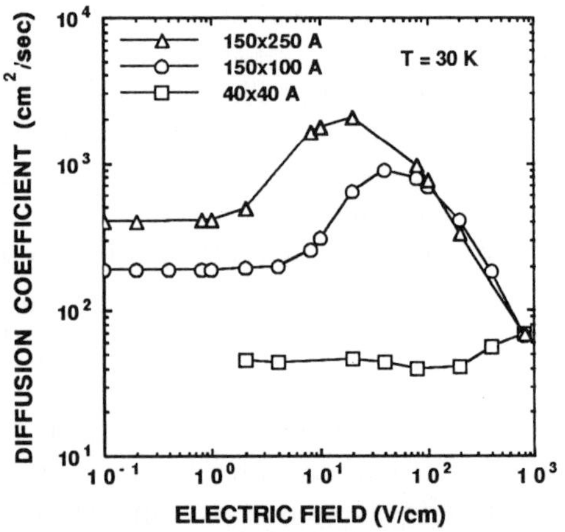

Fig. 2. Diffusion coefficient as a function of electric field for three cross sections of a QWI.

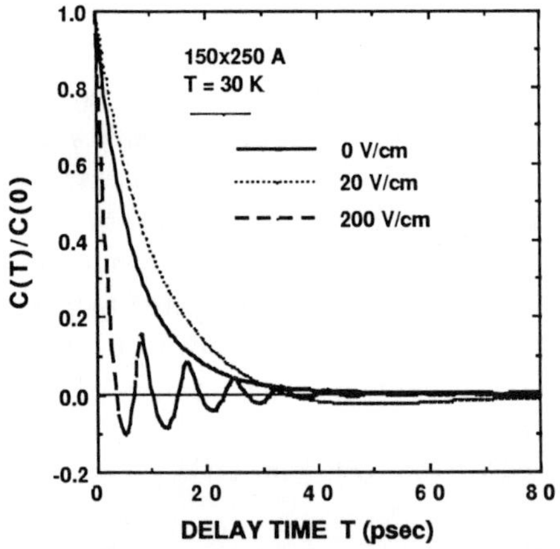

Fig. 3. Normalized autocorrelation function versus delay time at three electric fields.

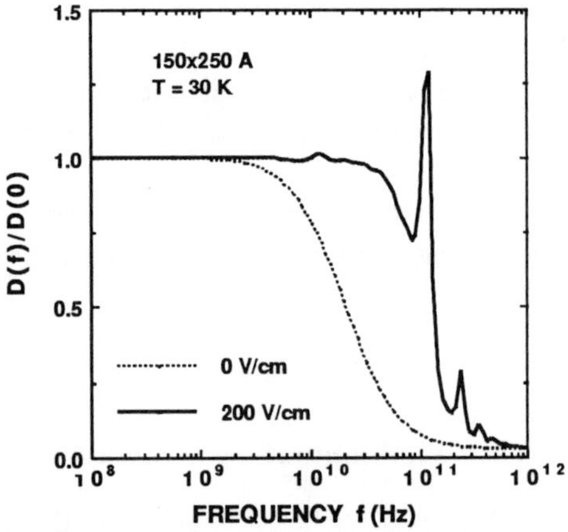

Fig. 4. Normalized diffusion coefficient as a function of frequency at two electric fields.

ADSORBATE-INDUCED INFRARED SHIFT IN THE 1/f NOISE OF DISCONTINUOUS PLATINUM FILMS

M. Mihaila
ICCE-Bucharest, str. E.I. Nicolae 32B, 72996 Bucharest, Romania

A. Masoero
Department of Physics, University of Modena, Via Campi 213/A, 41100 Modena

A. Stepanescu
Department of Physics, Politechnical Institute of Turin, 10129 Turin, Italy

ABSTRACT

In discontinuous platinum films, the 1/f noise temperature dependence shows a peak located at about 133K(11.4meV) which is attributed to a platinum 11.1meV surface phonon. As a consequence of the surface phonon hardening effect brought about by residual adsorbates, the peak shifts to 144K(12.3meV). This infrared shift is confirmed by leaking hydrogen in the high vacuum system and stands for a possible new proof for the surface phonons participation in the 1/f noise generation.

INTRODUCTION

Tunneling between metal islands is considered to be the main mechanism producing 1/f noise in discontinuous platinum films[1,2](DPF). Recently, we have reported that the phonon density of states fits well the dependence of the 1/f noise intensity on temperature in DPF[3]. The question why the phonon density of states should be mirrored in the noise could be answered if one considers, as usually in an overwhelming number of physical systems[4] including scanning tunneling microscope[5], that tunneling process is an inelastic one, hence phonon assisted. In DPT the tunneling between metal islands is in essence a surface-effect, thus surface phonon assisted. The presence of an adsorbate(e.g.,hydrogen) between metal islands bears on the surface lattice dynamics and usually a "stiffening"[6] or lowering of the force constants within the atoms of the surface layer is observed. Consequently, surface phonon energy hardening or softening is brought about by adsorbates. We report in this contribution the fact that under the influence of the residual adsorbates a peak existing in the 1/f noise of a DPF vs. temperature exhibits a displacement corresponding to a phonon hardening effect(infrared shift). This effect could stand for a possible new proof of surface phonons participation in the 1/f noise generation.

EXPERIMENTS AND DISCUSSION

Noise measurements in the temperature range (90-200)K were performed on DPF evaporated on sapphire substrate, in an ultrahigh-vacuum(10^{-9}mbar), electron beam evaporation system. After evaporation and some annealing cycles, the resistance of the samples was determined as a function of temperature. A negative thermal observed, 10K(≈0.86meV), is in agreement with the reported values for other H_2/metal film systems[10]. The effect has been observed in all samples(in one of them the peak was downward shifted, however).

To verify the infrared shift hypothesis, a small amount($\Delta p \approx 6.10^{-9}$mbar) of hydrogen has been introduced in the vacuum system at room temperature, the sample was cooled then till about 90K, and the noise measurement repeated. In accordance with other existing data in niobium[11] and palladium[12], excess 1/f noise has been found by hydrogen adsorption(Fig.1, curve 3,dashed). We were surprised to observe that hydrogen produced an enhanced fine structure in the noise(curve 3, Fig.1). It is hardly to attribute such a fine structure to desorption processes because, except for small steps, no desorption peak was evident in p(T)(curve 3, Fig.2). The hydrogen-metal system is a system with strong proton-lattice coupling[13] and also with a strong proton-electron interaction[14] so as, in producing the noise the proton acts as an intermediary between the electron and the lattice and, in this way, some of the lattice dynamics properties should be mirrored into the noise. That is why many of the peaks of the curve 3 can be easily associated with platinum phonon energies. Few of them cannot; for instance, the sharp peak located at 173K corresponds to a clear diffusion noise spectrum($f^{-3/2}$), while for 174K we observed an almost pure 1/f noise spectrum.Adopting a self-trapped state model for the hydrogen atom into the host lattice[13], a diffusion spectrum can appear by thermal activation of the proton over the barrier, folowed by surface diffusion, while an 1/f noise spectrum could appear by a phonon-assisted quantum tunneling mechanism[15]. Between (115-145)K the noise structure of the curve 3 is quasi-similar to the structure of the curve 2, except for the fact that in the curve 3 all peaks are slightly(5K) shifted

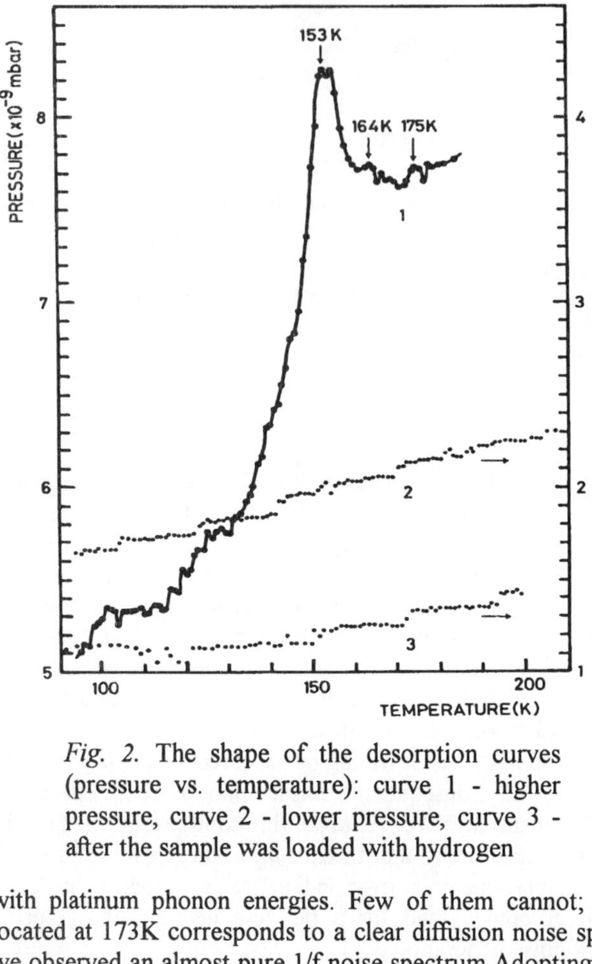

Fig. 2. The shape of the desorption curves (pressure vs. temperature): curve 1 - higher pressure, curve 2 - lower pressure, curve 3 - after the sample was loaded with hydrogen

coefficient of resistance has been found for all the investigated samples(five) which is an indicative of the discontinuous structure of the films. In all samples, $1/f^{\alpha}$ noise($\alpha \approx 1$) has been found and the dependence of the relative voltage noise spectral density(S_V/V^2) on temperature has been investigated in detail. During each noise measurement run, the temperature dependence of the residual gas pressure in the evaporation system(p(T)) was monitored.

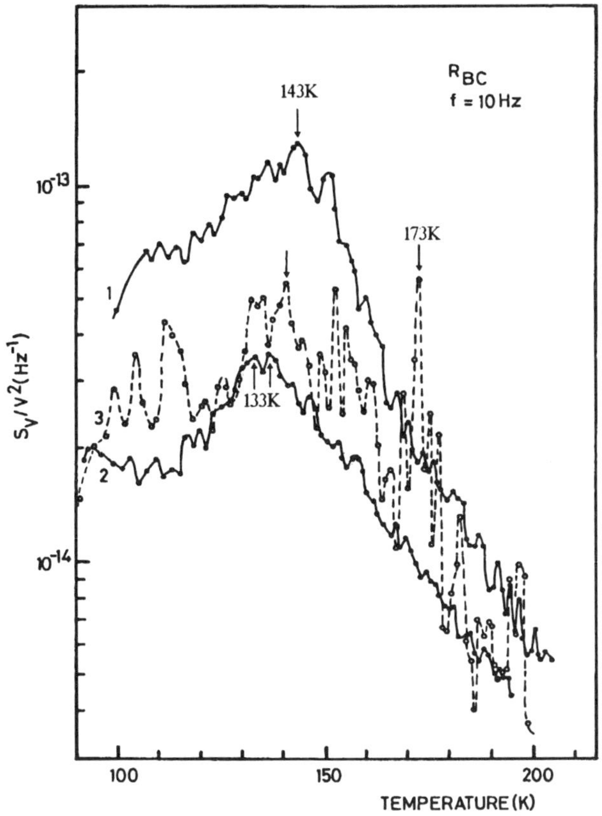

Fig. 1. S_V/V^2 vs. T in R_{BC} film: curve 1 - higher pressure, curve 2 - lower pressure, curve 3 - sample loaded with hydrogen

Curve 1 of figure 1 shows that the dependence of the S_V/V^2 on temperature for a given film(R_{BC}), at a frequency of 10Hz, presents a fine structure and a peak located at about 143K. Each point in Fig.1 has been obtained by averaging 15 snap-shot spectra. The corresponding p(T) dependence(curve 1 of Fig.2) features some desorption peaks at 153K, 164K and 175K which were attributed either to H_2 desorption[7](153K) or to water desorption[8](164K and 175K) at platinum surface. The vacuum system was pumped down further and the noise measurement repeated; the dependence p(T)(curve 2, Fig. 2) shows no desorption peak this time, while S_V/V^2 vs. T(curve 2 of Fig. 1) goes again through a maximum. One notes that noise intensity depends on coverage: at lower coverage the noise intensity decreases. Since no desorption peak appears in this situation, it results that the noise peak does is not induced by adsorbate but is an intrinsic property of the material itself. According to our earlier interpretation[3], for a cleaner surface(curve 2, Fig. 1), the peak located at about 133K(11.4meV) corresponds to a platinum surface phonon of 11.1meV, in the ΓK direction[9]. Figure 1 shows that under adsorbate influence, the peak located at 133K shifts to 143K(12.3meV). We believe that this infrared shift is an expression of the surface phonon hardening in the ΓK direction by the adsorbed hydrogen. The displacement upward. This observation confirms the existence of the infrared shift as a real effect; similar results were obtained for other samples.

CONCLUSIONS

In conclusion, we have measured 1/f noise in discontinuous Pt films in the temperature range (90-200)K. We have found that a noise peak appearing at about 133K and assigned to a platinum surface phonon energy shifts upward when residual gases are adsorbed on the platinum surface. The noise showed a similar behaviour when hydrogen was leaked into the high vacuum system. This infrared shift was atributted to a phonon hardening effect brought about by adsorbate which bears on the surface lattice dynamics. This effect could be a possible new proof of the surface phonons participation into the 1/f noise generation.

REFERENCES

1. J.L. Williams, I.L. Stone, J. Phys.C:Solid State Phys. 5, 2105(1972).
2. M.Celasco,A Masoero,P Mazzetti, A.Stepanescu, Phys. Rev. B17,2553(1978).
3. M. Mihaila, A. Stepanescu, A. Masoero, Proc. Int. Conf. on Noise in Phys. Syst., T. Musha,S. Sato,M. Yamamoto(editors),(Ohmsha Ltd.,1991), p.17.
4. E. L. Wolf, Principles of Electron Tunneling Spectroscopy(Oxford Univ. 1985).
5. Douglas P.E. Smith, Gerd Binnig, Calvin Quate,Appl. Phys. Lett.49,1641(1986)
6. J. M. Rowe, N. Vagelatos, J.J. Rush, H.E. Flotow, Phys. Rev. B12, 2959(1975).
7. B. Poelsema, L.S. Brown, K. Lenz, L.K. Verheij, G. Comsa, Surf. Sci. 171, L395(1986).
8. G. B. Fisher, I.L. Gland, Surf. Sci. 94, 446(1980).
9. U. Harten, J.P. Toennies, C. Woll, G. Zhang, Phys. Rev. Lett. 55,2308(1985).
10. E. Hulpke, J. Lüdecke, Phys. Rev. Lett. 68, 2846(1992).
11. J. H. Scofield, W. W. Webb, Phys. Rev. Lett. 54, 353(1985).
12. N. H. Zimmerman, W.W. Webb, Phys. Rev. Lett. 61, 889(1988).
13. K. W. Kerr, in Hydrogen in Metals I, G. Alefeld and J. Völkl(editors), (Springer Verlag, 1978), p.197.
14. T. Springer, in Hydrogen in Metals I, G. Alefeld and J. Völkl(editors), (Springer Verlag, 1978), p.75.
15. X. D. Zhu, A. Lee, A. Wong, V. Linke, Phys. Rev. Lett. 68, 1862(1992).

HIGH-FREQUENCY MOBILITY - NOISE TEMPERATURE

J.P. NOUGIER, L. HLOU, J.C. VAISSIERE
Centre d'Electronique de Montpellier. Université Montpellier II, 34095 Montpellier Cedex 5, France.
(Laboratoire associé au CNRS, UA 391)

ABSTRACT

The purpose of that paper is to compute the noise temperature through the Boltzmann and Einstein equations. Using the scattered packet method, described in [1], we can solve the Boltzmann equation and compute the diffusion coefficient when an electric field is applied. In order to obtain the noise temperature it is then necessary to know the differential (a.c.) mobility of the carriers. At low frequency we can calculate the slope of the velocity field characteristic but this method is not enough accurate at low frequency and even not valid at all at high frequency. We have developed a program enabling us to solve the small signal Boltzmann equation and to obtain the a.c. mobility.

INTRODUCTION

The Boltzmann equation can be solved in transient [2] and steady [3] state with a good accuracy but for small signal parameters it is better to compute directly the differential distribution function of the carriers.

METHOD

When we apply a small sinusoidal perturbation, the total electric field is given by:

$$E(t) = E_s + \delta E \exp(j\omega t)$$

If $f_s(\mathbf{k})$ is the stationary distribution function of the carriers when the stationary electric field E_s is applied, $f(\mathbf{k},t)$ solution of the Boltzmann equation for E(t) writes:

$$f(\mathbf{k},t) = f_s(\mathbf{k}) + \delta f \exp(j\omega t)$$

δf is complex and depends on **k** and not on t.
The first order series expansion of the Boltzmann equation gives:

$$j\omega\, \delta f + (e\, E_s / \hbar)\, \nabla_k\, \delta f + (e\, \delta E / \hbar)\, \nabla_k\, f_s = C\, \delta f$$

C is the collision operator and f_s is the stationary distribution function which can be computed using the matrix method [3].
This set of two equations (one for the real part and one for the imaginary part) is discretized in the **k**-space and the linear system of 2*600 equations is easily solved. The solution gives δf which is the perturbation (at frequency $\nu=\omega/(2\pi)$) of the hole distribution function around its stationary value $f_s(\mathbf{k})$.

The a.c. mobility is easily obtained as:

$$\mu'(E_S,\omega) = (1/\delta E) \int v(\mathbf{k})\ \delta f(\mathbf{k}, E_S, \omega)\ d\mathbf{k}$$

RESULTS

Results are obtained on p type silicon at 300 K using a single non-parabolic band model. At 10 kV/cm the perturbation of the distribution function δf is shown in Fig. 1 (real part) and Fig. 2 (imaginary part) at a frequency $\nu=5.10^{11}$Hz. Each radial curve gives the variation of δf versus k at a given value of the angle θ between the wave vector **k** and the electric field **E**. The value used for δE is 1 V/cm and the results are plotted using an arbitrary scale.

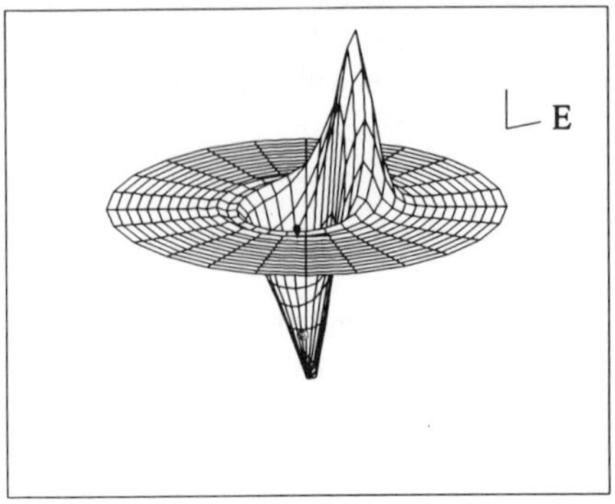

Fig. 1
Real part of the perturbation of the distribution function. p-Si, Na=0, T=300K, E=10 kV/cm, $\nu=5.10^{11}$Hz.

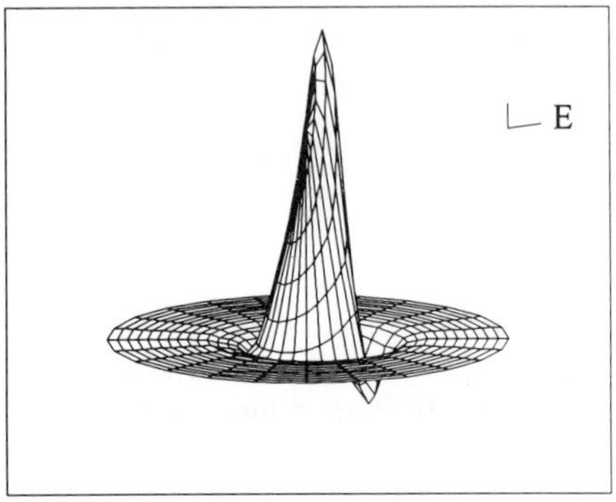

Fig. 2
Imaginary part of the perturbation of the distribution function. p-Si, Na=0, T=300K, E=10 kV/cm, $\nu=5.10^{11}$Hz.

The real part of the mobility obtained for two doping levels is plotted Fig. 3 and Fig. 4 versus frequency for five electric fields between 0 and 50 kV/cm. At low field the longitudinal differential mobility decreases at high frequencies. When the field is high, above 10 kV/cm there is a hump, just before the cut-off frequency. This is similar to what is obtained for the frequency dependent diffusion coefficient.

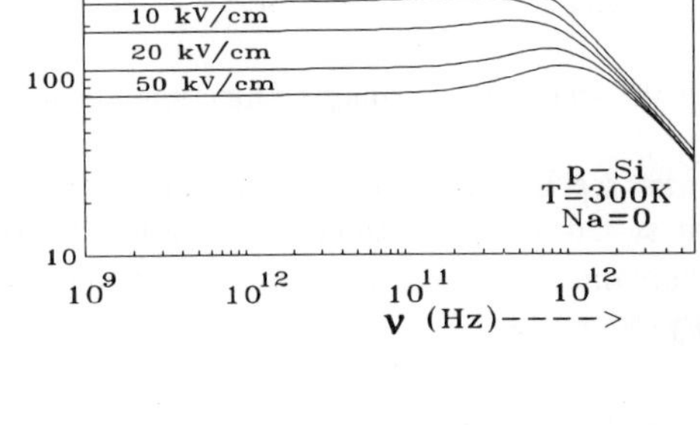

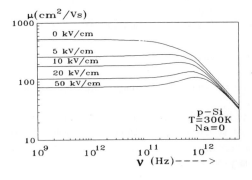

Fig. 3
Dynamic mobility versus frequency Na=0.

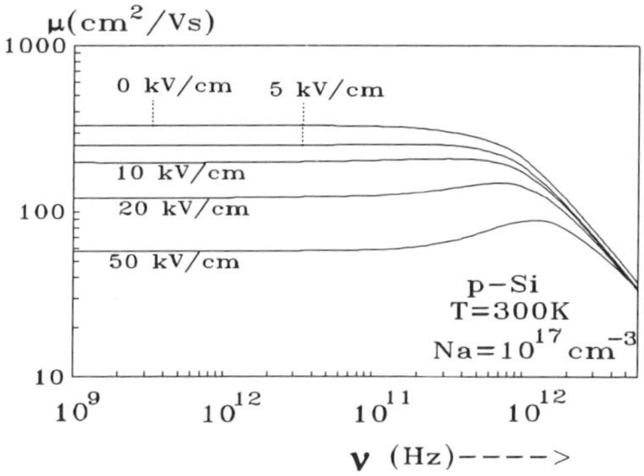

Fig. 4
Dynamic mobility versus frequency
$Na=10^{17}cm^{-3}$.

Using the frequency and electric field dependent diffusion coefficient given by the scattered packet method, the noise temperature is computed by:

$$k_B\, T_n(E_s,\omega)/q = D(E_s,\omega)/\mu'(E_s,\omega)$$

Using the frequency dependent ohmic temperature we can compute the excess noise temperature. Results are plotted Fig. 5 and Fig. 6 for two doping levels.

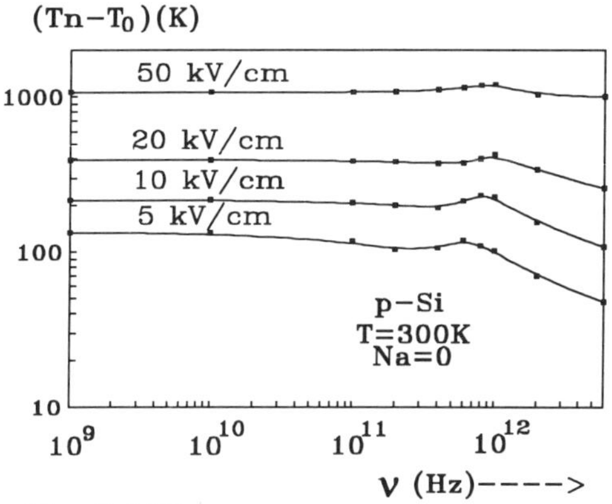

Fig. 5
Excess noise versus frequency.
Na=0

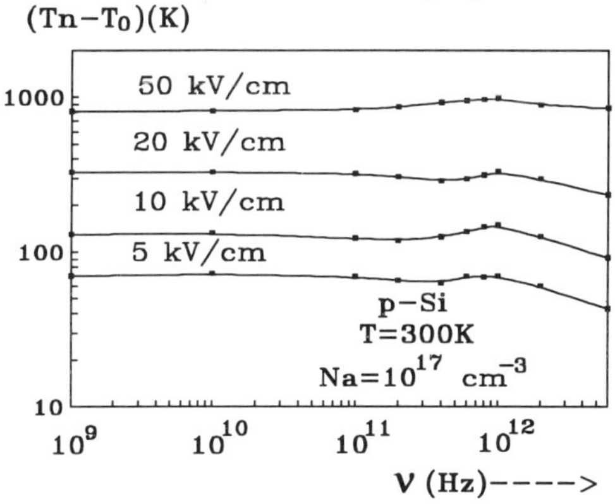

Fig. 6
Excess noise versus frequency.
$Na=10^{17}cm^{-3}$

Figure 7 shows that there is a good agreement between excess noise experiments using a pulse technique [4] and theoretical values given by this method for p-type silicon at room temperature and for a doping level $N_d=2.5*10^{14}$ cm^{-3}.

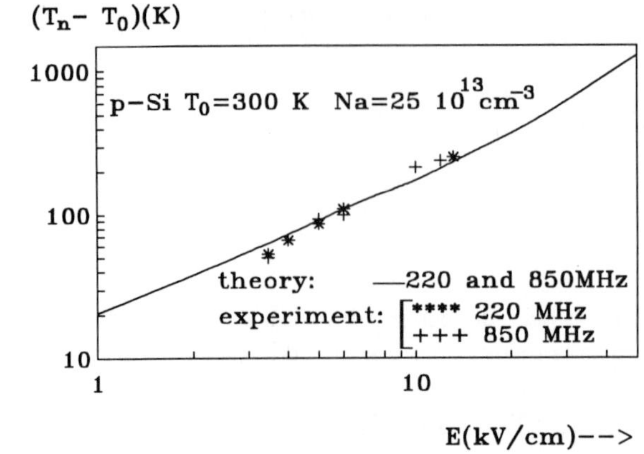

Fig. 7
Excess noise versus electric field for holes in p-type silicon.

Full curve: present method. Symbols refers to experiments.

CONCLUSIONS

The frequency dependent dynamic mobility is computed using the perturbed distribution function $\delta f(\delta E, \mathbf{k}, E_S, \omega)$ around its stationary value $f_S(\mathbf{k}, E_S)$. δf is obtained by solving the small signal Boltzmann equation. The excess noise temperature is then calculated through the Einstein relation. Results are in good agreement with "low" frequency ($\cong$ 1 GHz) noise experiments.

ACKNOWLEDGMENT

This work was partially supported by the French Groupement Circuit Intégrés Silicium, and ELEN CEE contract.

REFERENCES

1. J.C. VAISSIERE, L. HLOU, J.P. NOUGIER, A. ACHACHI, Proceedings 12th Int. Conf. on Noise ICNF'93, St Louis (1993).

2. J.C. VAISSIERE, M. ELKSSIMI, J.P. NOUGIER Semicond. Sci. Technol., **7** (1992), pp B308-B311.

3. J-P. AUBERT, J-C. VAISSIERE, J-P. NOUGIER J. A. P., Vol. 56 (4), pp 1128-1132 (1984).

4. D. GASQUET, J.C. VAISSIERE, J.P. NOUGIER Proceedings 6th Int. Conf. on Noise, Washington (1981), pp 305-308.

$1/f$ NOISE IN THIN METAL FILMS: THE ROLE OF STEADY AND MOBILE DEFECTS

V.V. Potemkin, A.V. Stepanov
Physics Department, Moscow State University, Moscow, GSP 119899, Russia

G.P. Zhigal'skii
Electronics Technology Institute, Moscow, 103498, Russia

ABSTRACT

The results of experimental research aimed at determining the role of mobile and steady defects of the lattice in $1/f$ noise generation in thin metal films are presented. There were studied technological factors and external effects (thermal and current annealing, natural aging and γ irradiation) which influence concentration of microdefects and the noise. Both equilibrium and nonlinear resistance fluctuations at high current density were investigated. The obtained results demonstrate that $1/f$ noise in metal films is caused by mobile defects.

The purpose of our experimental research was to clear up the role of mobile and steady defects in $1/f$ noise generation in thin metal films. There was studied the dependence of noise characteristics on different technological factors and on external effects (thermal and current annealing, natural aging and γ - quantum irradiation) which affect concentration of microdefects. Al, Mo, Nb and Cr films 10 nm - 1 μm thick were investigated. The main experimental results are given below.

As a rule low noise level corresponding to the Hooge parameter value $\alpha \approx 10^{-3}$ was observed in films with low concentration of defects. The resistivity of such films was close to that of bulk pure metal. On the other hand films with high concentration of steady defects (high values of resistivity ρ in comparison with resistivity of bulk metal ρ_0) also revealed low noise level ($\alpha \approx 10^{-3} \ldots 10^{-5}$). It is illustrated in Fig.1 where noise level vs current density experimental curve for Cr film with high concentration of defects ($\rho/\rho_0 \approx 10$) is presented and compared to the curve calculated from the Hooge relation[1] ($\alpha \approx 10^{-3}$).

Noise reduction was also observed in alloys. For example the noise level of Al+Si films with 1 atomic percent of Si ($\alpha \approx 10^{-3}$) was significantly lower than that of pure Al films ($\alpha \approx 10^{-2}$) when they were evaporated under the same technological conditions and the resistivity values of both the samples were approximately equal (within 10%). The noise reduction in Al+Si films is explained by low vacancy concentration at the grain boundaries achieved when Si atoms occupy the vacant lattice sites of Al.

Another feature of imperfect films is that at sufficiently large current density j the noise level increase ($\sim j^4$) deviates from the j^2 law obtained when resistivity fluctuations are independent of the current (equilibrium resistance fluctuations). As it is shown in Fig.1 some samples did not reveal ordinary j^2 dependence at all. Noise measurements of films with various thickness imply that these nonlinear resistivity fluctuations prevail over equilibrium resistivity fluctuations when dc current probing the sample heats the film up to some definite temperature. Above this temperature additional mobile defects are generated in the lattice and hence the 1/*f* noise increases. The nonlinear resistivity fluctuations were exhibited more explicitly in thick films because of heat removal worsening.

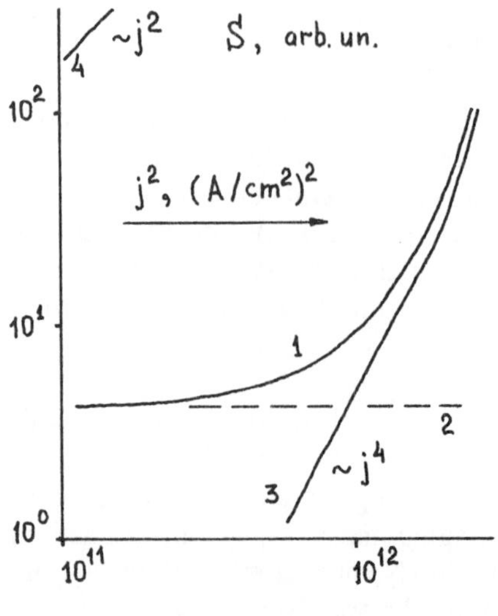

Fig.1. Noise power (S) vs current density (j) for Cr film with high defects concentration. 1-total noise, 2-thermal noise level, 3-thermal noise subtracted, 4-Hooge's relation.

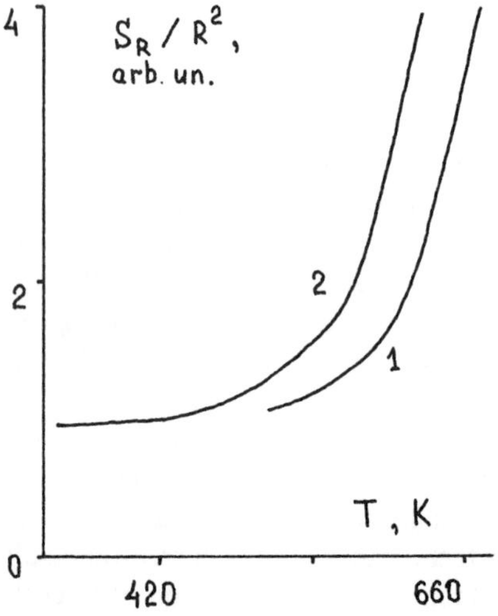

Fig.2. Relative noise power (S_R/R^2) vs temperature (T) for Cr film 1-thermal heating in an oven, 2-current heating.

In Fig.2 the effects of thermal heating in an oven and heating by the current on the noise level are presented (the temperature of current heating was estimated from the measured film resistance function of temperature). The curves are the same but for the earlier noise increase in case of current heating. This is explained by the appearance of local overheated areas of the film due to nonuniform distribution of steady defects causing nonuniform resistivity and current distributions in the film. That is why in case of current heating the local temperature at which additional mobile defects are generated is achieved at lower average temperature of the film in comparison with the case of thermal heating.

The activation type dependence of noise on temperature with the activation energy $E \approx 0.1 \ldots 0.4$ eV was observed in films with high concentration of mobile defects. These E values correspond to the energy which is required for a vacancy creation when 1-2 lattice bonds are split. The noise function of temperature registered in films with high concentration of steady defects was significantly weaker.

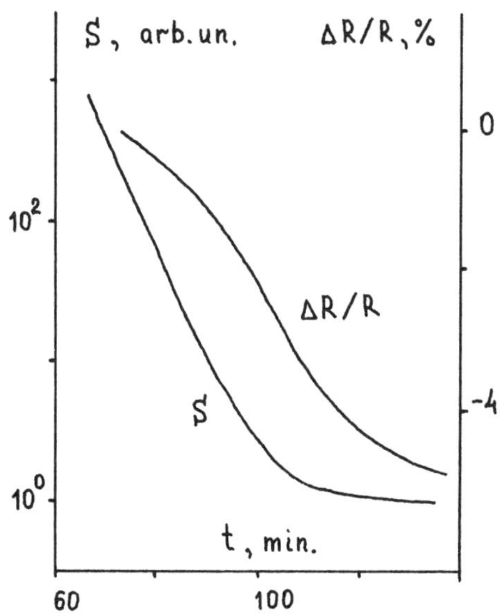

Fig.3. Time evolution of noise power (S) and relative resistance deviation ($\Delta R/R$) for Al film.

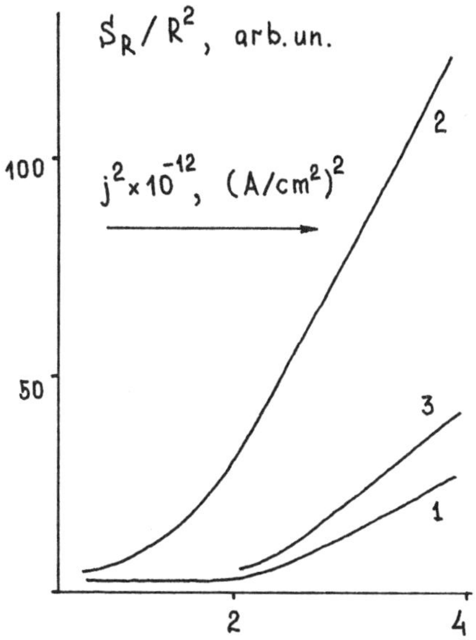

Fig.4. Relative noise power (S_R/R^2) vs current density (j) for Nb film 1-initial, 2-after irradiation, 3-after current annealing.

As a result of natural aging, thermal and current annealing noise reduction caused by annealing of defects was observed. The time evolution of noise and resistance of Al film after the end of condensation is shown in Fig.3. These samples had strongly non-equilibrium structure and high $1/f$ noise level. There was exponential decline of the noise level that is characteristic of microdefects annealing by diffusion process. It must be noted that the noise level relaxation to the stationary value is shorter than the relaxation of the resistance. It demonstrates that relaxation time of microdefects contributing to the $1/f$ noise is smaller than that of defects contributing to the resistance. With the increase of film thickness and average grain size the increase of the relaxation time was observed.

Thermal annealing of samples was performed at temperatures exceeding the temperature of film condensation. During a few thermocycles consisting of film heating and slowly cooling to the room temperature the unreversable decrease of noise and resistance to some stationary levels was observed and further annealing did not lead to noise reduction. But the increase of annealing temperature resumed the decrease of noise and resistance. This result can be explained by the assumption that there are microdefects with various activation energies and hence the defects with high activation energy are annealed only at sufficiently high temperature.

Analogous unreversable exponential decrease of noise and resistance with time was observed when films with high defects concentration were heated by dc current applied to a sample and its value exceeded some threshold value (current annealing). During the current annealing some Cr films revealed complex relaxation composed of two processes having different relaxation time constants. It is interesting to note that the noise function of temperature for such films also consisted of two regions with different activation energies. Finally it must be emphasized that in all cases of thermal and current annealing of defects the decrease of noise (orders of magnitude) was sufficiently larger than that of resistance (several percent).

Fig.4 displays the effect of γ irradiation on $1/f$ noise of Nb films. The noise level increased because additional lattice defects were introduced as a result of irradiation. When the samples were annealed by dc current the noise decreased down to approximately initial level. Analogous effect was observed when Cu films were irradiated and annealed[2].

It should be noted that the noise variation after irradiation and annealing was much more explicit in the region of nonlinear resistance fluctuations compared to the variation of equilibrium resistance fluctuations. This is explained by the fact that at high current density local overheating occur at steady defects and additional mobile defects are generated in overheated areas. Although steady defects themselves can not produce resistivity fluctuations they influence the noise level in the region of nonlinear fluctuations. Hence the variation of steady defects concentration of the sample is revealed at high current density. Because of the same reason films with approximately equal noise level at small current had different level of nonlinear fluctuations.

CONCLUSION

Our results demonstrate that $1/f$ noise in metal films is caused by mobile defects (mainly by the vacancies) giving rise to resistivity fluctuations and that the noise level is determined by quasi-equilibrium concentration of mobile defects. This concentration is controlled by defects sources and sinks in the film and depends on the film microstructure, in particular on the average grain size. The external effects increasing (γ irradiation) or decreasing (annealing) mobile defects concentration lead accordingly to increase or decrease of the $1/f$ noise.

REFERENCES

1. F. N. Hooge, Physica B162, 344 (1990).

2. J. Pelz, J. Clarke, Phys.Rev.Lett., 55, 783 (1985).

INTRINSIC AND EXTRINSIC 1/f NOISE SOURCES IN IRRADIATED n-GaAs

L. Ren and F.N. Hooge
Department of Electrical Engineering
Eindhoven University of Technology, P.O. Box 513
5600 MB Eindhoven, The Netherlands

ABSTRACT

In this contribution we report on our recent experimental results of 1/f noise in electron and proton irradiated n-GaAs. Two branches were found in plots of lnα versus 1/T, which corresponded to two different noise mechanisms.

INTRODUCTION

It has been demonstrated that an empirical relation[1] is successful in describing the 1/f noise in homogeneous semiconductors and metals. This relation relates the relative 1/f noise power density S_R/R^2 of the fluctuations in the resistance R to the total number of free charge carriers N in the sample:

$$\frac{S_R}{R^2} = \frac{\alpha}{fN} \qquad (1)$$

α is often called the 1/f noise parameter. For semiconductors, α was found to be scattering in a wide rage of 10^{-7}-10^{-2}, depending on the quality of crystal lattice[2]. In order to obtain a better understanding of the quality dependence of α, we have studied high-quality n-GaAs grown by molecular beam epitaxy (MBE). Two different temperature dependences of α were found[3,4]. These dependences cannot be explained by the Dutta-Dimon-Horn model[5]. In this contribution we will show that the two different temperature dependences of α in epitaxial n-GaAs are caused by two different noise mechanisms.

EXPERIMENTAL TECHNIQUE

Epitaxial n-GaAs doped with Si (~2×10^{16}cm^{-3}) was grown by a VARIAN MOD 3" molecular beam epitaxy (MBE) system. The thickness of the epitaxial layers was 3.2μm. Hall bar structures with six side contacts were prepared using conventional photolithography and wet-etching procedures. The samples were irradiated with 3 MeV electrons (e^-) and with 3 MeV protons (H^+). In this way, point lattice defects and clusters of lattice defects[6] were introduced by e^- and H^+ irradiation, respectively. The penetration depth for 3 MeV electrons is quite large, about several mm. The projected range R_p and the straggling ΔR_p for 3 MeV protons[6] are estimated to be about 50μm and 1μm, respectively. Consequently, the protons were stopped deep in the substrate. In view of the thickness of our epitaxial layers, the defect production can thus be regarded to be homogeneous in the epitaxial layers.

RESULTS AND DISCUSSION

Excess resistance noise was measured both before and after irradiations. The noise was measured as a function of the irradiation dose and temperature between 77 and 300 K in a frequency range of 1 Hz to 20 kHz. Several bias voltages were applied in order to check whether the measured noise stemmed from resistivity fluctuations. The results always showed a quadratic dependence of the noise on the bias voltage. Before the H^+-irradiation, all noise spectra had a good 1/f shape at all temperatures. After the H^+-irradiation, in addition to the 1/f noise and thermal noise, some generation-recombination noise was observed in both the e^-- and H^+-irradiated samples. Here we shall consider the 1/f noise only.

For the e^--irradiated samples, it was found that α and its temperature dependence were little affected by the irradiation despite of a significant change in the sample resistivity. However, the results for H^+-irradiation were quite different. Fig. 1 shows α, the 1/f noise parameter, as a function of inverse temperature. The α-values were evaluated at 1 Hz from NS_V/V^2, where N is the number of charge carriers as determined from the Hall effect, S_V is the voltage spectral power density and V is the bias voltage. After irradiation, the striking result is that at high temperatures where α is strongly temperature dependent, the 1/f noise is not affected by the proton irradiation. While at low temperatures, where α is weakly dependent on temperature, α increases with the irradiation doses. In Fig. 2, we plot the α-values at 78K (representing the temperature-independent branch of α) and α-values at 300 K (representing the temperature dependent branch of α) versus the irradiation dose ϕ. Fig. 2 shows that α-values at 78 K are almost linearly proportional to ϕ while α-values at 300 K are almost independent of ϕ.

The noise data in Fig. 1 clearly reveal that the two branches of the temperature dependence of α correspond to two different noise mechanisms: at the high temperatures, the 1/f noise seems to be dominated by an unknown source of an intrinsic origin, while at low temperatures the noise is obviously dominated by an extrinsic noise source induced by the H^+-irradiation. To test this conclusion, we have examined the relation between the temperature dependence of the noise spectral density $S_V(f,T)$ and the slope of the spectra $\gamma(T)$. According to Dutta-Dimon-Horn model[5],

$$\gamma(T) = 1 - \frac{1}{\ln(2\pi f\tau_0)}\left[\frac{\partial \ln S_V(f,T)}{\partial \ln T} - 1\right]. \tag{2}$$

The directly measured values of γ for the samples of a dose $1.5\text{x}10^{13} H^+/cm^2$, using the least square fit to the spectra S_V at the low frequencies where no obvious "knees" or "bendings" appear, are plotted in Fig. 3 as a function of inverse temperature. The values of γ calculated from eq. 2 at different temperatures, using the spline fit to $\alpha(T)$ and $\tau_0 = 10^{-12}$s, are also shown in Fig. 3. It can be seen that the general pattern of calculated $\gamma(T)$ follows that of the experimental data, except near room temperature. The disagreement between the trend of calculated and experimental $\gamma(T)$ near room temperature is a good indication of the domination of the intrinsic noise source at high temperatures.

Considering difference in defect types induced by the e^-- and H^+-irradiations, the results for the extrinsic 1/f noise source are clearly consistent with the "local-interference" model[7,8] due to defect motion. The result $\alpha \propto \phi^{1.0}$, shown in Fig. 2, indicates that the moving defects are associated with the clusters, since the number of clusters is expected to be directly proportional to ϕ. Here, we have implied that also the extrinsic source in n-GaAs is due to mobility fluctuations. In order to check this, we have studied the extrinsic 1/f noise in samples with different original doping under a fixed proton dose of about $1.9\text{x}10^{12}\ H^+/cm^2$. The argument for mobility fluctuations is simple. If there are two different scattering mechanisms, Matthiessen's rule gives

$$\frac{1}{\mu} = \frac{1}{\mu_1} + \frac{1}{\mu_2}. \qquad (3)$$

Then, if the 1/f noise stems from one scattering mechanism only, one will find[4] that $\alpha \propto \mu^2$. The experimental results presented in Fig. 4 prove this trend indeed.

CONCLUSIONS

In conclusion, we have shown that the two different temperature dependences of α in epitaxial n-GaAs correspond to two different noise mechanisms: intrinsic and extrinsic noise sources. The "local-interference" model is very likely the mechanism of the extrinsic noise source.

This work has been supported by the research programme of the "Stichting voor Fundamenteel Onderzoek der Materie" (FOM).

REFERENCES

1 F.N. Hooge, T.G.M. Kleinpenning and L.K.J. Vandamme, Rep. Prog. Phys. **44**, 479(1981).
2 F.N. Hooge, Physica B **162**, 344(1990).
3 L. Ren, in *Proc. of the Int. on Noise in Physical Systems and 1/f Noise*, edited by T. Musha, S. Sato and M. Yamamoto, Kyoto, Ohmsa, Tokyo, 55(1991).
4 L. Ren and F.N. Hooge, Physica B **176**, 209(1992).
5 P. Dutta and P.M. Horn, Rev. Mod. Phys. **53**, 497(1981).
6 H. Ryssel and I. Ruge: *Ion Implantation*, John Wiley & Sons, Chichester-New York-Brisbane-Toronto-Singapore(1986).
7 J. Pelz and J. Clarke, Phys. Rev. B **36**, 4479(1987).
8 S. Hershfield, Phys. Rev. B **37**, 8557(1988).

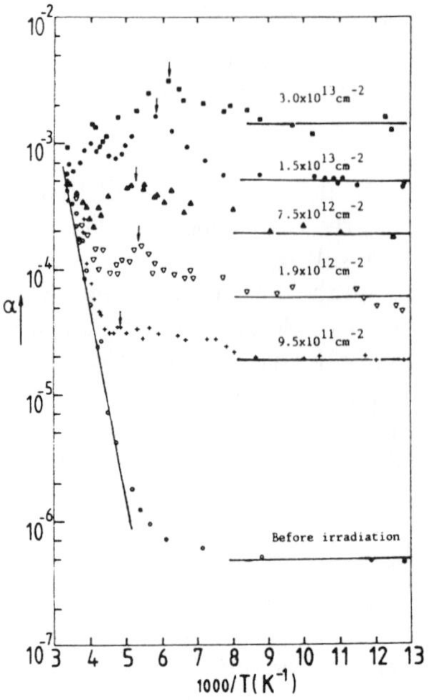

Fig. 1 α as a function of inverse temperature. The doses are given at the horizontal parts.

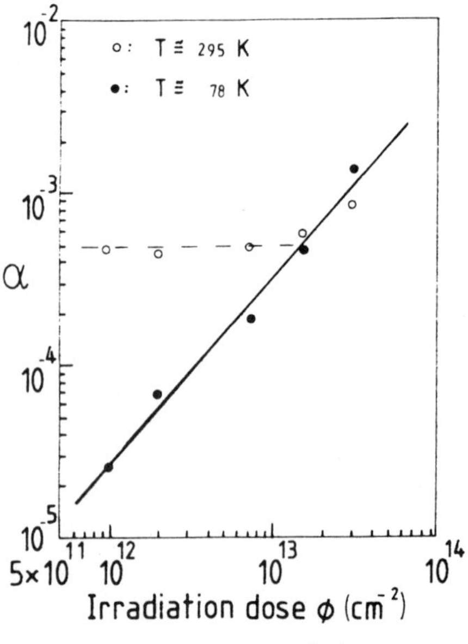

Fig. 2 α-values at 300 K and 78 K as a function of the doses φ. The solid line is the best fit to the data at 78 K. The broken line represents dose-independent α-values at 300 K. Obviously, the relatively higher α-values at 300 K of the samples with the two highest doses are caused by the contribution from the extrinsic 1/f noise source.

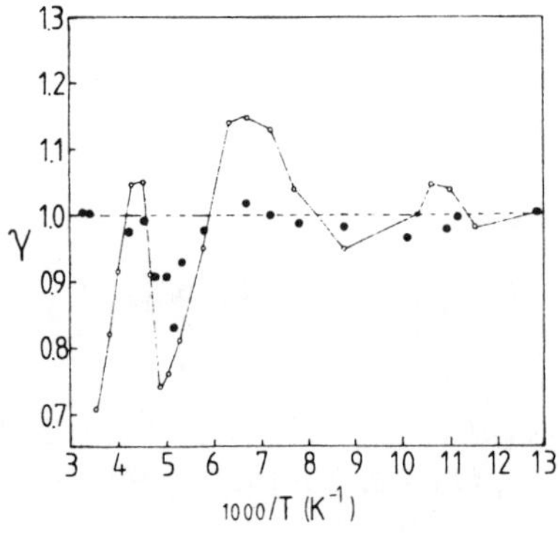

Fig. 3 γ as a function of inverse temperature for the samples with a dose $1.5x10^{13}cm^{-2}$. The solid circles are the experimental data and the open circles connected by the solid lines are the results calculated from eq. (2).

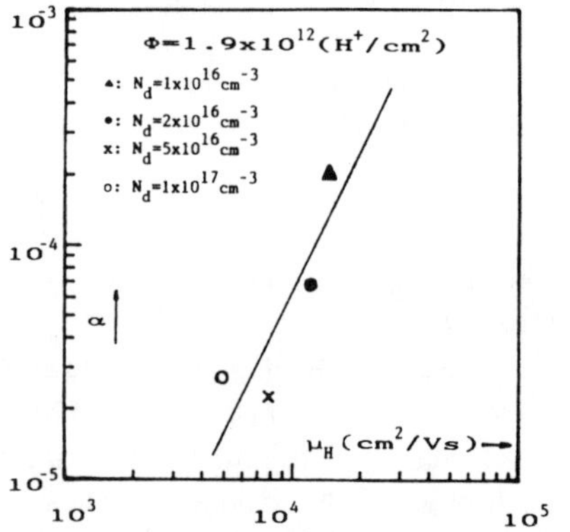

Fig. 4 α of the extrinsic 1/f noise versus μ_H at 78 K. The solid line indicates a quadratic dependence.

ELECTRICAL CONDUCTIVITY NOISE IN INTRINSIC a-Si:H AT LOW FREQUENCIES

L. Tóth, M. Koós and I. Pócsik

Research Institute for Solid State Physics
H-1525 Budapest, P.O.Box 49, Hungary

ABSTRACT

Current noise properties of intrinsic a-Si:H samples, prepared at various substrate temperatures (T_s), were investigated and their noise power spectra were found to exhibit $1/f^{\beta}$ frequency dependence ($0.8<\beta<1.2$) in low frequency region (10^{-2}-5 Hz) at room temperature. The absolute noise power was found to increase with increasing T_s, and its variation on DC current was found to show power law dependence with exponents over 1.6.

INTRODUCTION

However the lack of general consensus in the origin of 1/f fluctuations[1,2] increases the need for systematic experimental investigations, the experimental data are insufficient even in the most promising amorphous material: viz. amorphous silicon (a-Si:H)[3,4].

In this paper we should like to report the preliminary results of conductivity noise measurements carried out on a-Si:H films.

EXPERIMENTS

Current fluctuations were studied in intrinsic a-Si:H films deposited from pure silane by the glow discharge method at different substrate temperatures. Sandwich-type specimens with NiCr electrodes were used. All measurements were performed at 300 K in vacuum, the computer controlled sample temperature regulator made access also to rather low frequencies.

RESULTS AND DISCUSSION

Measurements were carried out on sandwich type (NiCr-i-NiCr) sample. Metal–semiconductor interface showed better ohmic behaviour for samples deposited below $T_s \leq 200\ ^{o}C$. Attempts to obtain better ohmic behaviour by heat treatment or by using n^{+}-i-n^{+} sample structure, specially for samples deposited at higher T_s, have not been successful. The current noise measurements were performed in the voltage range, in which the deviation from Ohmic behaviour seemed to be negligible.

In Fig. 1 we show a power spectrum, the slope of the fitted line is very close to -1. The power spectra measured in samples of different deposition temperatures show $1/f^{\beta}$ dependence with $0.8 < \beta < 1.2$, and no systematic variation of the exponent with deposition temperature has been found.

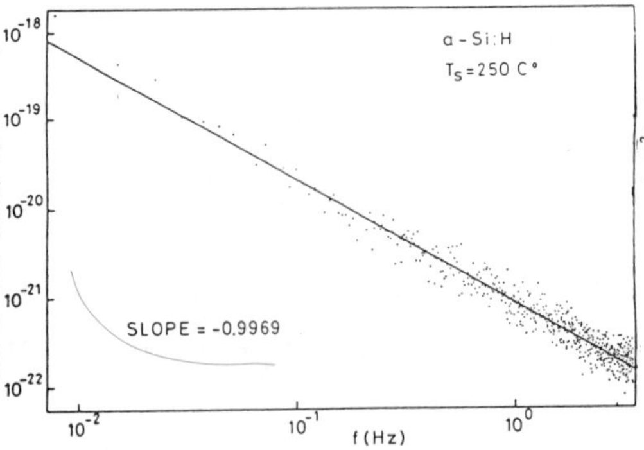

Figure 1. Current noise power spectrum for a-Si:H sample of T_s = 250 °C.

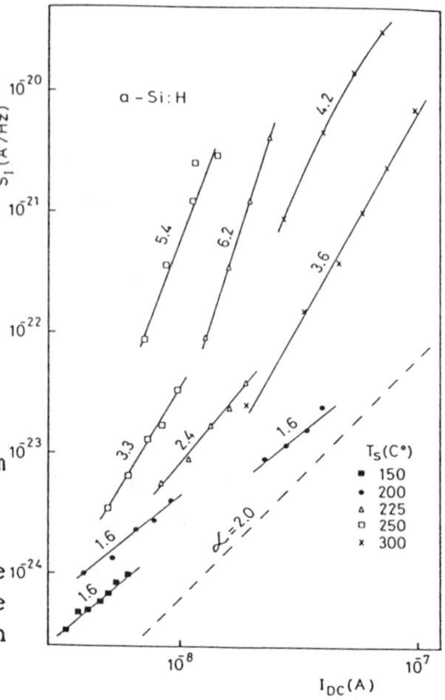

Figure 2. DC current dependence of noise intensity calculated at 1 Hz from noise power spectra measured at 300 K in samples of different T_s.

The variation of noise intensity with DC current was also investigated for samples with different T_s (Fig. 2). To determine noise intensity we used the results of the same line fitting, which was used also to determine the β exponent. The value of this straight line function at 1 Hz frequency was used as a measure of the intensity of the given spectrum. The DC current dependence of these noise intensity values follows the power-law $S_I(1\ Hz) \approx I^{\alpha}_{DC}$, which gives a straight line on a log-log plot, as can be seen in Fig. 2. The samples have different slopes ($1.6 < \alpha < 6.2$) depending on the deposition temperature. It can also be seen from Fig. 2 that there is tendency not only for exponent α to increase with increasing substrate temperature, but also the noise intensity. The dashed line in Fig. 2 represents $\alpha = 2$, which would correspond to the empirical Hooge law[5]. The main results of our experiments are the following:

i. in the range of 10^{-2} - 10 Hz a $1/f^{\beta}$-type current noise ($0.8 < \beta < 1.2$) can be detected in intrinsic a-Si:H films at room temperature.

ii. the noise intensity increases with increasing deposition temperature, but the 1/f character remains unchanged.

iii. in each sample the intensity of the 1/f noise increased with increasing DC current, I_{DC}, going through the sample as $\approx I^{\alpha}_{DC}$, where $1.6 < \alpha < 6.2$

iv. higher α values and higher variances were found for samples prepared at higher deposition temperature.

To interpret these results we proposed a general model for current noise[6], which would be applicable both for a-Si:H and for the other amorphous materials. The structure of amorphous materials can be built up from clusters of lognormal size distribution[7] and the electrons are localized on these. The electronic conductivity of such cluster systems shows a percolation feature: one dimensional conducting lines are formed by a continuous series of conducting junctions, which may be branching. Inter-cluster conductivity rises or falls as a consequence of the atomic rearrangement of the clusters and this process creates the fluctuation. A hierarchy can be supposed for these junctions, caused by the branching structure of the conducting lines. The 1/f frequency dependence is a consequence of the structure of this hierarchy[6].

The existence of clusters in a-Si:H films is based on experimental evidence from SEM micrographs[8] and EBIC measurements[9]. It is reasonable to suppose that the cluster sizes change when substrate temperature is varied: at lower T_s smaller clusters are expected; higher T_s is expected to prefer larger clusters.

In the case of smaller cluster-size we expect a larger number of such independent hierarchical tree-like structures, which are coupled parallel between the electrodes in the case of the sandwich structure. The contribution of these trees to the conductivity is additional, so their statistics follows normal (Gaussian) distribution (the non-Gaussian behaviour within the tree[10] remains unchanged). The amplitude of the 1/f fluctuation is expected to decrease by the inverse square root of the number of trees. Thus, the intensity of the 1/f current noise becomes smaller in the low T_s samples, where the clusters should be smaller; this finding is in qualitative agreement with our experimental observation.

The noise intensity dependence on DC current would be quadratic if noise were to be caused by resistance fluctuations (Ohm's law is valid). Deviation from this behaviour reflects slight or strong non-linearity in the resistance. Although our measurements were performed in the Ohmic range of applied biases, the exponents observed differ from $\alpha = 2$. The values of $\alpha < 2$, which were measured in a-Si:H samples of $T_s < 200\ ^{o}C$, were observed by Main and Owen[11] in vitreous semiconductors. This behaviour of amorphous materials is, in our opinion, very likely the consequence of their structure, which is built up from clusters of different conductance which results in current inhomogeneity throughout the sample.

The steeper increasing noise intensity with DC current in samples of $T_s > 200\ ^oC$ may be the consequence of extra noise generated by space-charge breakdown at the metal-semiconductor interface. The behaviour of this interface strongly depends on deep defect states localized in the gap[12], hence the properties of the same metallic contact vary with the structure of a-Si:H films. The I-V characteristics show the weakly Ohmic behaviour of the interface, though it should be noted that the difference between I-V characteristics of samples deposited at various temperatures is not so large as to give the reason for the observed deviations of noise intensity dependence on DC current.

Acknowledgements. The work was supported by the Hungarian National Science Foundation under contract No.:OTKA-1975.

REFERENCES

1. P. Dutta and P. M.Horn, Rev. Mod. Phys. **53**, 497 (1981).
2. M. B. Weismann, Rev. Mod. Phys. **60**, 537,(1988).
3. J. C. Anderson, Phil. Mag, B **48**, 31, (1983)
4. F Z. Bathei and J C. Anderson, Phil. Mag. B **55**, 87 (1987)
5. F. N. Hooge, Phys. Lett. **29**, 139, (1969)
6. I. Pócsik, M. Koós and L. Tóth, in: *Int. Conf. on Physical Systems and 1/f Fluctuations* Ed. by T. Musha, S. Sato and M. Yamamoto, Ohmsa Ltd, Tokyo, 1991.
7. I. Pócsik and M. Koós, Solid State Comm. **74**, 1253 (1990).
8. S. T. Kshirsagar, N. R. Khaladkar, J. B. Mamdapurkar and A. P. B. Sinha, Jap. J. of Appl. Phys. **25**, 1788, (1986)
9. M Füstöss-Wegner, L. Pogány, M. Koós, L. Tóth, and Gy. Zentai, Mater. Sci. Forum **38-41**, 1487 (1989).
10. I. Pócsik, M. Koós and L. Tóth, Solid State Comm. **74**, 1253 (1990).
11. C. Main and A. E. Owen, Phys. Stat. Sol. (a), **1**, 297 (1970).
12. K. Winer and L. Ley, in: *Advances in Disordered Semiconductors* Vol. 1 ed. by H. Fritzsche, World Sci. Singapore, 1990. p. 366.

NOISE OBTAINED FROM THE SCATTERED PACKET METHOD

J.C. VAISSIERE, L. HLOU, J.P. NOUGIER
Centre d'Electronique de Montpellier. Université Montpellier II, 34095 Montpellier Cedex 5, France

A. ACHACHI
L.G.E.O. Faculté des Sciences de Kénitra, Maroc

ABSTRACT

A numerical method is developed for solving the transient Boltzmann equation. The *scattered packet method* (SPM) gives the carrier distribution function when an electric field is applied. Using that method we can also follow the evolution and the spreading in the **k**-space of a packet of carriers, due to the field and the collisions with the lattice and ionized impurities. The current correlation function and the diffusion coefficient can then be computed in a simple way.

SCATTERED PACKET METHOD

We discretize the **k**-space. The number n(**k**) of carriers inside each mesh is then computed at thermal equilibrium and allocated to the central point **k** of the mesh. We can compute numerically the probability for a carrier initially in a mesh **k** at time t to be in an other mesh **k'** at time t+dt. For this, we assume that the packet of carriers initially in the state **k** undergoes a free flight during dt. At the end of the flight the packet is scattered and distributed in all the meshes of the k-space. The probabilities are calculated for each mesh and placed in a matrix **B** allowing us to compute n(**k**,t+dt) in each mesh if n(**k**,t) is known.

$$n(\mathbf{k}, t + dt) = \mathbf{B} * n(\mathbf{k}, t)$$

The resulting algorithm (similar to the lattice-gas cellular-automaton method [1]) is physically equivalent to the ensemble Monte Carlo method.

CORRELATION AND DIFFUSION

A single packet initially in one mesh centered in **k** at time t is distributed over all the meshes (over all the **k'** states) at time t+dt by the matrix product of **B** by the initial distribution, and then at time t+2*dt by an other matrix multiplication and so on. Using that method we can follow, for each initial packet, its evolution and spreading in the **k**-space due to the field and the collisions with the lattice and ionized impurities. The current correlation function and the diffusion coefficient are then computed in a simple way.

At time t=0 the number of carriers in the state k_L is n_L chosen equal to the steady state value of the number of carriers in this state. At time t≠0 the number of carriers in the state k_M (coming from the state k_L at time t=0) is n_{LM}. The longitudinal correlation function is then:

$$C(t) = q^2 \Sigma_L \Sigma_M [n_{LM}(t) (v_L(0)-v_d) (v_M(t)-v_d)]$$

where q is the electron charge, v_d, v_L and v_M are the projections along the electric field of the drift velocity and the carrier velocities in states L and M.

RESULTS

The material chosen for that study is p-type silicon at room temperature. As concerns the scattering mechanisms, our study included acoustic deformation-potential, optical and impurity scattering. The valence band taken for p-Si is a non parabolic, spherical single band model. Using the number n(**k**, t) of carriers in each mesh **k** at time t we can obtain the transient distribution function of the carriers. Figure 1 shows a comparison between the steady state distribution function given by this method and the "*exact solution*" of the Boltzmann equation [2].

Figure 2 shows the centred correlation function of the current densities obtained at different electric fields.

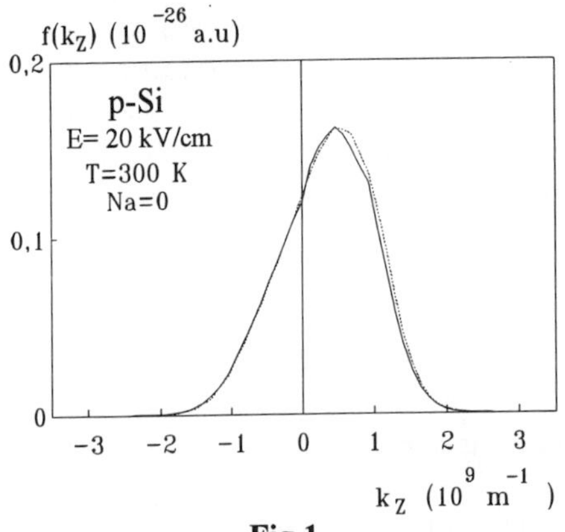

Fig 1

Stationary distribution function along the field. The full and dashed curves represent SPM and matrix results.

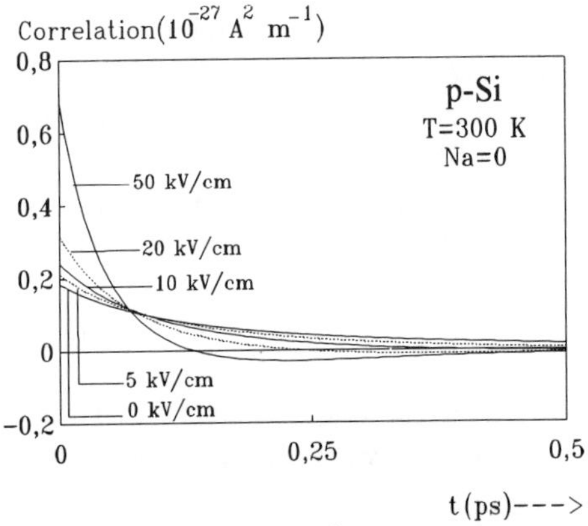

Fig 2

Longitudinal correlation versus time. Na=0

The diffusion coefficient (or the current spectral density) is given by the Fourier transform of the correlation function:

$$D(E,\nu) = \frac{q^2}{n} \int_0^{\infty} C(t) \cos(2\pi\nu t)\, dt$$

The diffusion coefficients computed versus frequency are plotted at different electric fields on figure 3. The longitudinal diffusion coefficient exhibits, just below the cut-off frequency, a bump, the amplitude of which increases with increasing field, when the velocity correlation function is negative [3,4,5].

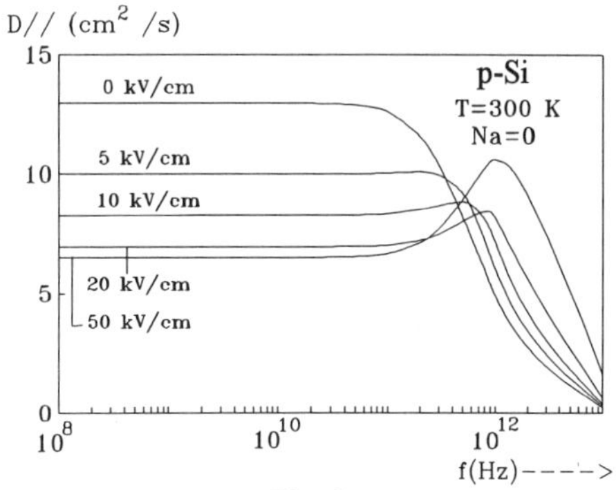

Fig 3

Longitudinal diffusion coefficient versus frequency. Na=0

Using the same band structure model we compare Monte Carlo and SPM results on figure 4 and 5. We can see that the agreement is good between the two methods at 300K and 77K.

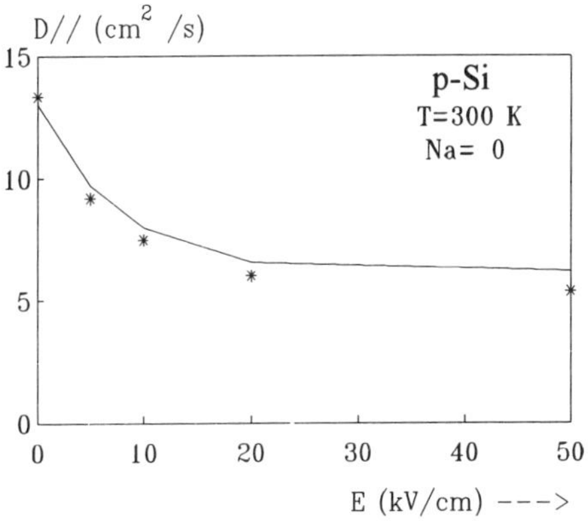

Fig 5

—— ***Scattered Packet Method***

**** ***Monte Carlo***

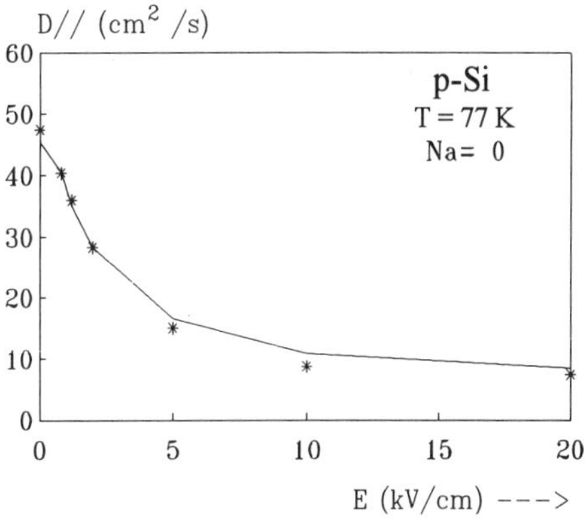

Fig 6

—— ***Scattered Packet Method***

**** ***Monte Carlo***

Using the frequency dependent diffusion coefficient and the mobility given in [7] we can compute the noise temperature through the Einstein relation. Comparisons with experimental data given by pulsed noise measurements on p-Si are given figure 7. We can see that the agreement is good.

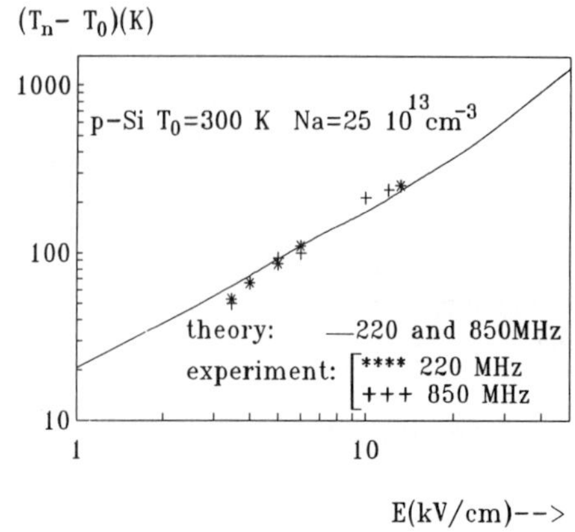

Fig. 7
Excess noise temperature versus electric field for holes in p-type silicon.
Full curve: present method.
Symbols refers to pulsed noise experiments.

CONCLUSIONS

Using the scattered packet method, we can solve the Boltzmann equation. We have also the possibility to follow the spreading in the k-space of a packet of carriers initially in a state **k**. The SPM method is equivalent to use an ensemble Monte Carlo simulation with an infinite number of carriers in the same initial state **k**, the difference being that the scattered carriers are located at the nodes of the mesh instead of at any point in **k** space. Changing the number of carriers and the initial state, with respect to the stationary distribution function, we can get the correlation function and the diffusion coefficient.

ACKNOWLEDGMENT

This work was partially supported by the French Groupement Circuit Intégrés Silicium, and ELEN CEE contract.

REFERENCES

1. K. KOMETER, G. ZANDLER and P. VOGL, Phys. Rev. **B46**, 1382 (1992).
2. J-P. AUBERT, J-C. VAISSIERE, J-P. NOUGIER J. A. P., Vol. 56 (4), pp 1128-1132 (1984).
3. J. ZIMMERMANN, E. CONSTANT, Sol. STAT. Electronics, **23**, 915 (1980).
4. E. CONSTANT, B. BOITTIAUX, Journal de Physique, C7, N10, **42**, C7-73 (1981).
5. R. BRUNETTI, C. JACOBONI, Phys. Rev. B29, 5739 (1984).
6. J.C. VAISSIERE, A. MOATADID, J.P. NOUGIER, Proc. 9th Int. Conf. on Noise in Physical Systems, Montreal, 97 (1987)
7. J.P. NOUGIER, L. HLOU, J.C. VAISSIERE, Proceedings 12th Int. Conf. on Noise ICNF'93, St Louis (1993).

CARRIER SPREADING AND DIFFUSION COEFFICIENT IN n-Si.

J.C. VAISSIERE, O. CHAPELON, L. HLOU, J.P. NOUGIER, D. GASQUET
Centre d'Electronique de Montpellier (Laboratoire associé au CNRS, UA 391)
Université Montpellier II, 34095 Montpellier Cedex 5, France.

ABSTRACT

We have solved the transient {**k**} and space dependent Boltzmann equation in n-type silicon. The transient distribution function f(**k**,z,t) is obtained in the k space and along the direction of the applied electric field (along the z axis). The material is assumed to be homogeneous so that the Poisson equation is not taken into account. Hence we can follow the spreading of a packet of carriers initially located around z=0. By a simple integration over the whole k space we obtain f(z,t), v(z,t) and ε(z,t). The longitudinal diffusion coefficient is computed using the slope of the variance of the carriers' positions.

INTRODUCTION

The knowledge of the diffusion coefficient is important for device simulation. However, only few experimental data are available in the literature. Some values [1] have been obtained using either time of flight measurements (related to the spreading of a packet of carriers) on lightly doped materials, or thermoelectric microwave technique [1,2], or noise measurement's [1,3,4] (using a pulse technique [5]).

In the present simulation we select some carriers initially located around z=0. It is equivalent to paint them in *red* color (or any other color) and to follow their spreading in the k-space and along the z axis. The material is assumed to be homogeneous so that the electric field is constant and the Poisson equation not needed. The total number of carriers in each mesh is constant but the number of the selected (*red*) carriers in each mesh is never constant. When the global system is stationary the *red* system is always in transient regime due to the spreading of the carriers in the phase space.

At time t≤0 the electric field is null, the *red* distribution is at thermal equilibrium in the k-space. At time t=0 the electric field is applied and all the quantities defined hereafter are related only to the *red* carriers, which are chosen at t=0 according to a gaussian distribution along z centred at z=0.

The material used is n-type silicon at T=300K. The microscopic model consists of six ellipsoïdal non parabolic bands. Parameters are taken from [6].

DISTRIBUTION FUNCTION

Using the transient distribution function $f_i(\mathbf{k}, z, t)$ the distribution $f(z, t)$ along the electric field is defined as:

$$f(z, t) = \sum_{i=1}^{2} \iiint f_i(\mathbf{k}, z, t)\, d^3k$$

where the subscript i is related to cold (i=1) and hot (i=2) groups of valleys.

On figure 1 we can see that the packet is drifted by the field and spreads due both to the field and to the scattering mechanisms. We can see that some electrons are going in the direction of the field (E=5kV/cm) while the majority of them are going in the opposite direction.

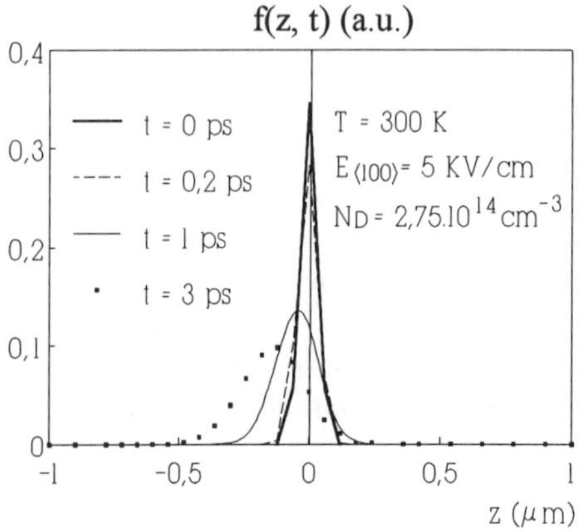

Fig. 1
Spreading of a packet of carriers along z at fixed times in n-Si.
Na=2.75 $10^{14}cm^{-3}$, T=300K, E=5 kV/cm.

This is better shown on figure 2 on which we have plotted the mean velocity of the carriers versus z at fixed times. Velocities are assumed to be positive for electron going in the direction of -E (for z<0). At t=0 the carriers are at thermal equilibrium, mean velocities are equal to zero in all the meshes. At the beginning t=0.2 ps, the mean velocity in the meshes near z=0 is strongly modified (meshes for great values of |z| are empty). The shape is practically linear versus z. Carriers with high k values move very fast in the real space and are located at the front of the packet where the mean velocity is great. These carriers have not been yet scattered, their high velocity is due to the free flight and to the acceleration by the field. For positive values of z the velocity is lower and some values at the beginning are negative. Then, for a fixed z, the mean velocity decreases versus time, due to collisions.

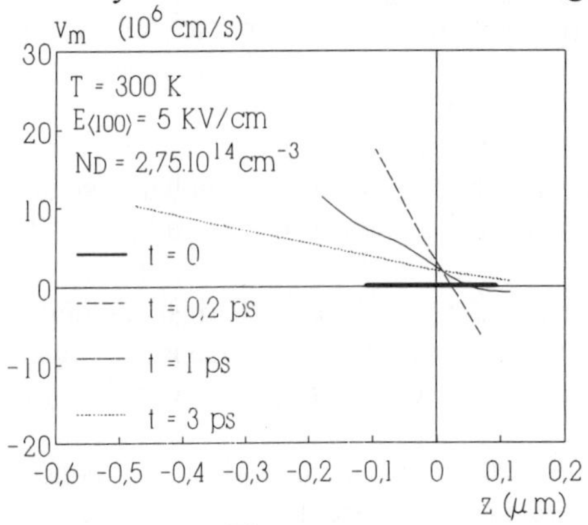

Fig. 2
Mean velocity of the carriers versus z at fixed times in n-Si.
Na=2.75 $10^{14}cm^{-3}$, T=300K, E=5 kV/cm.

The mean velocity of all the carriers, integrated over **k** and z (drift velocity) is plotted figure 3 versus time. Values are compared with those given by a direct solution of the homogeneous Boltzmann equation in the k-space. We can notice that the results are practically the same. At t>1.5 ps the steady state is obtained for the global system. The steady state velocity is reached (figure 3) while the local velocities are still in transient regime (figure 2).

DIFFUSION COEFFICIENT

The longitudinal diffusion coefficient is computed using the slope of the variance of the carriers' positions:

$$D = \frac{1}{2}\frac{d}{dt} < (z(t) - < z >)^2 >$$

$$< z > = \frac{1}{N} \int_{-\infty}^{+\infty} z \; f(z,t) \; dz$$

Figure 4 gives the transient diffusion coefficient computed for an applied electric field of 5kV/cm along the <100> direction at room temperature. After a transient regime of about 3 ps the steady state value is obtained. The doping level used in that simulation is $2.75 * 10^{14}$ cm^{-3}.

Figure 5 shows a comparison between theoretical and experimental values of the diffusion coefficient obtained for the same impurity concentration. Experimental data comes from high frequency noise measurements performed in the laboratory using a pulse technique in order to avoid thermal heating of the material. For electric fields between 100 and 5000 kV/cm the agreement is quite good.

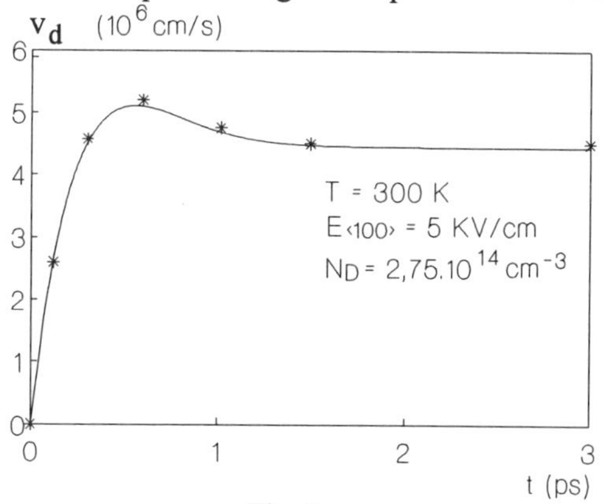

Fig. 3
Transient mean velocity.
Na=2.75 10^{14}cm^{-3}, T=300K, E=5 kV/cm.
—— transient method
******* present work***

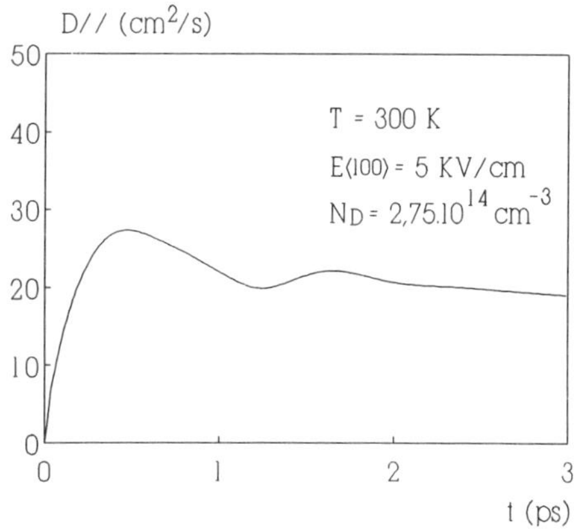

Fig. 4
Transient diffusion coefficient computed using the slope of the variance of the carriers.

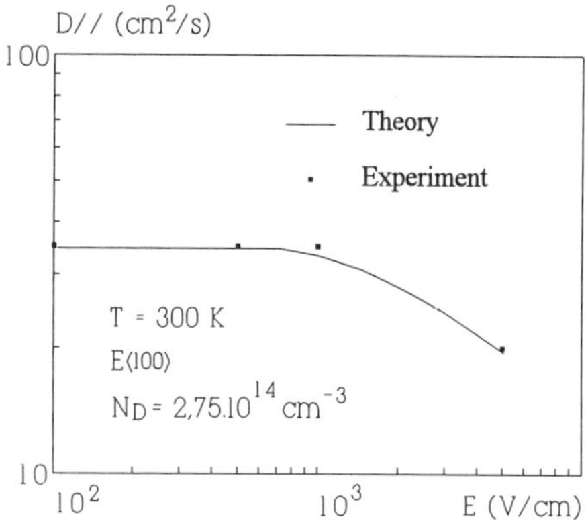

Fig. 5
Steady state diffusion coefficient versus electric field.
Full line: theory and dots: noise experiments

CONCLUSION

We can solve the one dimensional space-dependent transient Boltzmann equation in n-type silicon. The material is assumed to be homogeneous and we have computed the transient distribution function of a packet of carriers initially centred in z=0. Local distribution functions of these carriers are always in transient regime even when the integral over all the space is in steady state. Local densities and transport parameters are obtained with a good accuracy. The diffusion coefficient computed using the spatial distribution function of the carriers is then easily obtained. The comparison with the diffusion coefficient derived from noise temperature measurements is good.

ACKNOWLEDGMENT

This work was partially supported by the French Groupement Circuit Intégrés Silicium, and ELEN CEE contract.

REFERENCES

1. C. JACOBONI, L. REGGIANI, Advances in Physics, **28**, 493 (1979)
2. S.P. ASHMONTAS, Y.K. POZHELA, L.E. SUBACHYUS, Soviet Phys. Semicond., **11**, 206 (1977)
3. F. NAVA, C. CANALI, L. REGGIANI, D. GASQUET, J.C. VAISSIERE, J.P. NOUGIER, J. Apll. Phys. **50**, 922 (1979).
4. D. GASQUET, H. TIJANI, J.P. NOUGIER, A. VAN DER ZIEL, Proc. 7th Int. Nat. Conf. on Noise, Montpellier, 165 (1983)
5. D. GASQUET, J.C. VAISSIERE, J.P. NOUGIER, Proc. 6th Int. Nat. Conf. on Noise, Washington, 165 (1981)
6. C. CANALI , G. OTTAVIANI, J. Phys. Chem. Solids, **32**, 1707 (1971)

AN EFFECT OF SRUCTURE FACTORS AND MECHANICAL STRESS ON 1/f NOISE IN METAL FILMS

G.P.Zhigal'skii

Electronic Technology Institute
Moscow. 103498. Russia

ABSTRACT

A study is made of the effect on the $1/f^{\gamma}$ noise of microstructure (grain diameter, degree of cristallinity), mechanical stress (internal and created by the external force) and temperature in metal films obtained under various technological conditions. It was established that films with higher tensile stress and smaller average size of grains show greater $1/f^{\gamma}$ noise levels. The exponent γ is increased with the increment of the stress. For the Al films having mechanical stress $\sigma \le 10^8$ Pa the noise level is derived from the Hooge correlation with $\alpha_H \approx 2 \cdot 10^{-3}$.

1. SPECIMENS AND METHOD OF EXPERIMENTS

The test samples were Al, Cr and Mo films deposited on glass and oxidized silicon substrates. The film thickness was varied from 50 nm to 1 μm. The spectral power density of the 1/f noise was measured over the frequency range from f = 2 Hz to 10 kHz, at a direct current density $j \le 10^5$ A/cm^2. The stress was determined by measuring the inflection of a cantilever-mounted substrate [1]. Films with different properties were produced by varying the conditions. The film samples for the stress and electrical measurements were deposited on the substrate in the same batch.

In order to perform direct experiments to determine the effect of enternal controlable mechanical stresses on the character and level of 1/f noise, a special device was developed to generate both tensile and compressive stresses within a Cr film deposited on glass substrate by deflecting a cantilever - mounted substrate holding the film. Tensile stress was created by applying an external bending force to the free end of the substrate perpendicular to the plane of the latter from the side carrying the film, while compression was produced by applying the force in the other direction. The Al films were deposited on elastic substrates and controlable mechanical stress was created by applying external tensile force to the unfastened end of a substrate.

2. EXPERIMENTAL RESULTS

The spectral density of noise power in the studied frequency range obeys a law $S(f) \sim f^{-\gamma}$.

For Cr films internal stress,resistivity and 1/f noise level were measured in vacuum chamber directly after film deposition at a substrate temperature equal to the condensation temperature, as well as during the process of film aging in vacuum and air. Directly upon condensation, Cr films have a high noise level which upon 30-60 min of maintenance in vacuum at the condensation temperature, or with cooling, decreases by several orders of magnitude. At the same time there was a reduction in intrinsic mechanical stress and resistivity. Reduction of the 1/f noise with time during the process of film aging occurs on the exponential law (relaxation time $\tau \approx 5$ min), characteristic for the annealing of nonequilibrium vacancies in the film due to migration to various sinks. Estimates of the exponent γ in films directly after condensation give a value of γ = 2.5-3. After maintenance in a vacuum for 30-60 min γ decreases to values of $\gamma \approx 1$. The noise level relaxes to the stationary margin quicker than the magnitude resistivity. This result confirms that arising of the 1/f noise is due to mobile defects [2].

It was found that films with higher internal stress have higher levels of 1/f noise. The spectral power density of the 1/f noise is proportional to the $\exp(-\sigma V/kT)$, where σ is the stress, and V is the activation volume. For Cr and Mo films we found that values V are close to the atomic volumes Ω of bulk metals [3]. This results point to a vacancy mechanism for the 1/f noise in these films. Figure 1 shows the dependencies of the Hooge's parameter α [4] on the internal mechanical stress for Al films with thickness

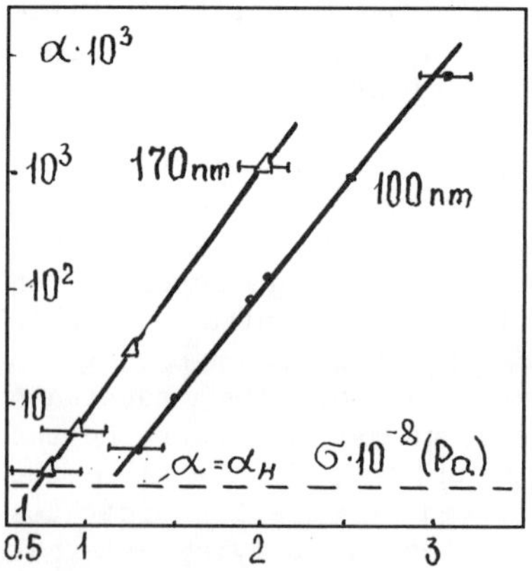

Fig.1. Dependences of the parameter α on the internal mechanical stress for Al films.

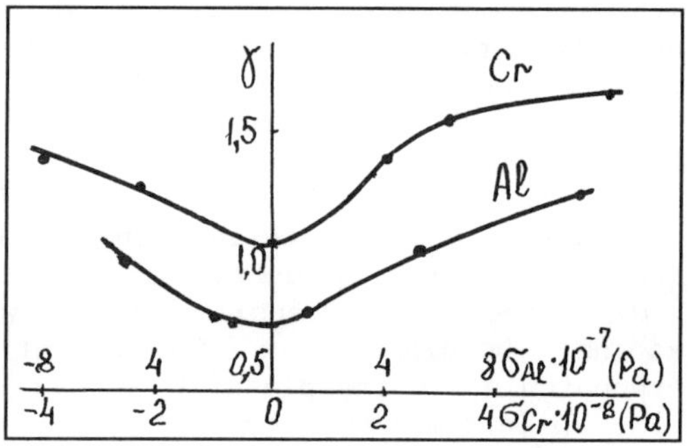

Fig.2. Dependences of the exponent γ on the internal mechanical stress for the Cr and Al films.

h=100±10 nm and 170±10 nm (carrier density is taken $n=1.8\cdot10^{23}$ cm^{-3}). From these curves we found $V\approx 0.2\ nm^3$.

This value is close to the magnitude of $12\Omega_{Al}$ ($\Omega_{Al}=1.66\cdot10^{-2} nm^3$). The latter corresponds to the volume of atoms displaced by a single vacancy in the face-centered cubic lattice or to the volume of a vacancy aggrigate. This vacansy aggregate can contain as much as 12 vacansies. Direct observation of the defects in quenched aluminium has shown existance of different vacancy aggrigate [5].

For Al films with the internal mechanical stress $\sigma\leq10^8$Pa the noise level is low and is given by the Hooge formula [4] with a $\alpha_H \approx 2\cdot10^{-3}$. Theaverage size of grains in this films is l≥200 nm [6].

Figure 2 shows the dependencies of the exponent γ on the internal mechanical stress for the Cr and Al films. This result allows to explain a difference between values of the exponent γ for the different films.

The experiments on determining the effect of external stresses on the character and level of 1/f noise have shown that the noise spectral density increases with the increase in tensile stress both for Al and Cr films. When external compression is applied a reduction in noise (for Cr films) is observed. Upon transition to the range of small compressions, the noise level reaches its minimum. In this case the noise is of the same order of magnitude for the films produced with differing techniques and having different internal mechanical stress and initial noise level. As tensile stress is increased the quantity γ increases from $\gamma \approx 1$ to $\gamma \approx 2.5$.

For significant deformations ($\varepsilon \geq 0,5\%$) irreversible changes occur in the film. Then the reverse path of the noise curves as functions of deformation do not coincide with the forward paths. In this case the noise increases when deformaion is applied and then decreases with relaxation time of $\tau\approx10-50$ min, that is explaned by a capture of the vacancies by the moving dislocations under a plastic deformation.

Investigation of structure influence factors on 1/f noise has shown that as the grain diameter and grade of cristallinity increase, the magnitude of 1/f noise decreases [7]. For the films with the high noise level the temperature noise dependence is exponential [3] with activation energy E_a depending on grain diameter and cristallinity grade at the same frequensy [7].

We also discovered that the activation energy E_a for noise power spectral density increases with the frequensy decrease. We observed values of the activation energy $E_a \approx (0.2-0.7)$ eV at the frequencies ranging from 10 kHz to 2 Hz for the Cr and Mo films. The broad and quasicontinuous distribution of the E_a can be explained by microstresses at the grain boundaries which are randomly distributed and can surpass significatly macrostress σ (sometimes for more than an order of magnitude)[8]. Changing of activation energy at local regions due to mechanical stress σ_m for Cr film when $\sigma_m = 2\cdot10^9$Pa and $V = \Omega_{Cr} = 1.2\cdot10^{-2}$ nm^3 is $\Delta E_a = \sigma_m V \approx$ 0.24eV (we consider, that Guck's law is hold true here). This value is comparable with the energy per bond in the crystall lattice.

The results obtained confirm that a vacancy mechanism is responsible for the 1/f noise in continuous metal films with an elevated concentration of mobile defects [3,6]. This mechanism involves the creation and annihilation of vacancies at various types of sinks.

REFERENCES

1. R.W.Hoffman, Physics of Thin Films, Vol.3 (ed. G.Hass and R.E.Thun), Academic, New York, (1966).
2. J.Pelz and J. Clarke, Phys. Rev. Lett. 55, 783 (1985).
3. G.P.Zhigal'skii, Proc. of Int. Conf. on Noise Phys. Syst. Kyoto (1991), p.39.
4. F.N. Hooge, Physica B 162, 344 (1990).
5. Lattice defects in quenched metals. Int. Conf. held at Argonne Nat. Lab. (ed. R.M.J. Cotterill), N.Y.-Lond. (1965).
6. V.V.Potemkin, I.S.Bakshi, and G.P.Zhigal'skii, Radiotekh, Elektron. 28, 2211 (1983).
7. G.P.Zhigal'skii, A.V.Karev, I.Sh. Siranashvili. et al. Izv. Vyssh. Uchebn. Zaved., Radiofiz., 33, N10, 1181 (1990).
8. L.S.Palatnik, M.Ya.Fuks, V.M.Kosevich, Formation Mechanism and Substructure of condensed Films [in Russian], Nauka, Moscow (1972).

III. METALS AND MAGNETICS

RECENT ADVANCES IN THE INTERPRETATION OF THE BARKHAUSEN NOISE IN FERROMAGNETIC SYSTEMS

G. Bertotti
Istituto Elettrotecnico Nazionale Galileo Ferraris and GNSM-INFM
Corso M. d'Azeglio, 42, I-10125 Torino, Italy

ABSTRACT

Barkhausen noise statistical properties are derived from a set of stochastic differential equations describing the motion of a magnetic domain wall in a randomly perturbed medium. The noise amplitude distribution obeys the law $P_0(v) \propto v^{c-1}\exp(-v)$, with $c > 0$. When $c < 1$, the noise is characterized by the presence of Barkhausen jumps distributed in duration Δu and size Δx according to the scaling laws $(\Delta u)^{-\alpha}$ and $(\Delta x)^{-\beta}$, with $\alpha \approx 2$ and $\beta \approx 1.5$. The theory gives a natural interpretation of the non-linear dependence of the noise power spectrum on magnetization rate and permeability. Extensions of the theory aimed at describing the influence on Barkhausen noise of magnetic viscosity effects and of diffusive-like propagation of local magnetization changes are discussed.

INTRODUCTION

The Barkhausen noise (BN) can provide rich and valuable information on the mechanisms responsible for ferromagnetic hysteresis, once a suitable interpretation of its origin is worked out. There have been widespread attempts in the past to describe the most evident feature of BN, the existence of burst events often termed Barkhausen jumps, in terms of clustering of elementary domain wall jumps triggered by some local instability.[1-3] More recently, there has been renewed interest in Barkhausen jumps as an example of self-organized-criticality,[4,5] although the interpretation of measured quantities by self-organized-criticality concepts is still at a preliminary stage.

An aspect which makes the interpretation of BN particularly difficult is the fact that BN is strongly non-stationary along each magnetization half-cycle, as a result of the different nature of the various magnetization processes responsible for the noise, domain wall (DW) motion in the central part of the hysteresis loop and magnetic domain creation and annihilation near saturation.[6] This is the main reason preventing a clear physical interpretation of those experiments where BN properties are averaged out over the whole magnetization loop. Definite progress can in fact be achievd by restricting BN measurements to a limited part of the hysteresis loop around the coercive field point.[7] Here BN is approximately stationary and is associated with a well-defined magnetization mechanism, DW motion. The BN signal is a measure of the velocity fluctuations of active DWs and the interpretation of BN is reduced to the problem of developing a convenient theory of DW dynamics in a randomly perturbed medium.

STOCHASTIC DIFFERENTIAL EQUATIONS FOR DOMAIN WALL DYNAMICS

DW dynamics in metallic ferromagnets is governed by the simple linear relation $v_{DW} \propto H_a - H_m - H_p$, where v_{DW} is the DW velocity, H_a is the applied field, H_m describes internal magnetostatic interactions, and H_p characterizes the pinning effects due to microstructural perturbations. The proportionality between field and velocity originates from the viscous-like action of the electric currents induced by DW motion.[8] H_a is usually a known function of time and H_m is to a good approximation proportional to the DW position. H_p is instead some random function of the DW position, as a consequence of microstructural disorder. These H_p space variations induce DW velocity fluctuations that give rise to the BN signal. In terms of convenient dimensionless variables (u for time, x and v = dx/du for DW position and velocity, $h_a(u)$ and $h_p(x)$ for applied and pinning fields), we have the equation

$$v = h_a(u) - x - h_p(x) \, , \quad v \equiv \frac{dx}{du} \tag{1}$$

We see that the DW behaves like a zero-mass single degree of freedom, subject to a set of forces (fields) balancing to zero (the term v is just the eddy-current field). Quantitative BN predictions are obtained by making specific assumptions on the random process $h_p(x)$.[9] The behavior of $h_p(x)$ has been experimentally investigated for some special systems, where a single DW is responsible for magnetization changes.[10-12] These studies have shown that $h_p(x)$ statistical properties are approximately those of the Wiener-Lévy (WL) process w(x). This result must necessarily be approximate, because the WL process has unphysical features in many respects: i) it is non-differentiable at any point, which is in contrast with the physical fact that a well-defined reversible susceptibility is associated with small-amplitude DW displacements around any given DW stable position; ii) it is non-stationary, which means that h_p can reach arbitrarily large values. These drawbacks are overcome by refining the WL description with the introduction of two correlation lengths α_1 and α_2 ($\alpha_1 < \alpha_2$), according to the equations:

$$\frac{dh_p}{dx} + \frac{h_p}{\alpha_2} = s(x) \, , \qquad \frac{ds}{dx} + \frac{s}{\alpha_1} = \frac{1}{\alpha_1}\frac{dw}{dx} \, , \qquad \langle |dw|^2 \rangle = 2dx \tag{2}$$

Equations (1)-(2) represent our theoretical frame for the interpretation of BN properties.

SCALING PROPERTIES

The basic physical consequences of Eqs.(1)-(2) are well illustrated by the limiting case where $h_p(x)$ is a pure WL process, i.e., $\alpha_1 \to 0$ and $\alpha_2 \to \infty$. By taking the time derivative of Eq.(1) and by assuming, as is commonly the case in BN experiments, that $h_a(u)$ increases at a constant rate $dh_a/du = c$, we obtain

$$\frac{dv}{du} + (v - c) = \frac{dw}{du} \, , \qquad \langle |dw|^2 \rangle = 2dx \equiv 2v du \tag{3}$$

v(u) is a Markov process, so that its statistical properties are fully determined by the transition density $P(v,u|v_0)$. P obeys the Fokker-Planck equation associated with Eq.(3):

$$\frac{\partial P}{\partial u} - \frac{\partial[(v-c)P]}{\partial v} - \frac{\partial^2[vP]}{\partial v^2} = 0 \tag{4}$$

The solution of Eq.(4) can be expressed as a series of Laguerre polynomials:[9]

$$P(v,u|v_0) = v^{c-1}\exp(-v)\sum_{n=0}^{\infty}\exp(-nu)\frac{n!L_n^{c-1}(v)L_n^{c-1}(v_0)}{\Gamma(c+n)} \tag{5}$$

Equation (5) shows, in particular, that the stationary amplitude probability density $P_0(v)$ and the autocorrelation function $R(\Delta u)$ of the v(u) process are given by

$$P_0(v) = \frac{1}{\Gamma(c)}v^{c-1}\exp(-v) , \qquad R(\Delta u) = c\exp(-|\Delta u|) \tag{6}$$

According to Eq.(6), the behavior of $P_0(v)$ for $v \to 0$ changes drastically, from $P_0 \to \infty$ to $P_0 \to 0$, when $c = \langle v\rangle$ crosses the value $c = 1$. This value represents the boundary between two quite different regimes. For $c < 1$, there is a significant probability to find values $v \sim 0$. The DW motion has a jerky character and proceeds by Barkhausen jumps separated by random waiting times during which the DW is nearly motionless ($v \sim 0$). On the other hand, when $c > 1$ the Barkhausen jumps merge into each other and the motion becomes continuous, with a negligible probability to find values $v \sim 0$. The law $P_0(v) \propto v^{c-1}\exp(-v)$ and the transition from burst-like to continuous regime have been fully confirmed by numerous BN experiments.[7]

The power-law divergence in $P_0(v)$ when $c < 1$ suggests the presence of self-similar properties in the noise behavior. This is confirmed by the structure of Eq.(3). In fact, this equation reduces to $dv/du - c \approx dw/du$ when $v \ll 1$. This approximate relation is invariant when we change both v and u by the same scale factor k ($v \to kv$, $u \to ku$, which implies $x \to k^2x$ and $\langle|dw|^2\rangle \equiv 2dx \to k^2\langle|dw|^2\rangle$, i.e., $w \to kw$). This result can be exploited to derive information on the scaling properties of the distribution of Barkhausen jump durations Δu and amplitudes Δx. First of all, let us notice that the term Barkhausen jump needs to be carefully defined, since, owing to the self-similar nature of the process v(u), there are jumps on any, even arbitrarily small, scale. Deciding whether the DW is jumping ($v > 0$) or not ($v \sim 0$) implies the introduction of some coefficient r, measuring our ability to resolve v and u details not smaller than r, which means, in particular, that $v \sim 0$ whenever $v < r$. The mean duration $\langle\Delta u\rangle$ of the Barkhausen jumps associated with the resolution coefficient r will be proportional to the probability that $v > r$, which, according to Eq.(6), for sufficiently small r and c, will be

$$\langle\Delta u\rangle \propto Prob[v>r] \equiv 1 - \int_0^r dv\, P_0(v) \simeq 1 - r^c \tag{7}$$

Let us now consider the distribution $P(\Delta u;r)$ of Barkhausen jump durations. The self-similar nature of v(u) implies that $P(\Delta u;r)$ must be a function of $\Delta u/r$ only, with a scaling structure of the form $P(\Delta u/r) \sim (\Delta u/r)^{-\alpha}$. The resolution r permits us to detect jumps of minimum duration of the order of r. On the other hand, the characteristic relaxation time of Eq.(3), equal to unity, forbids jump durations $\Delta u \gg 1$. This means that

$$<\Delta u> \propto \int_r^1 d\Delta u \ \Delta u^{-\alpha+1} \propto 1 - r^{2-\alpha} \tag{8}$$

By comparing Eq.(7) with Eq.(8) we conclude that the scaling exponent for jump durations is $\alpha = 2 - c$. Similar considerations lead to the conclusion that also the distribution of jump amplitudes Δx obeys a scaling law of the form $P(\Delta x/r^2) \sim (\Delta x/r^2)^{-\beta}$, with $\beta = 1.5 - c/2$. The scaling properties of BN have been recently investigated in connection with the search for self-organized-criticality effects in BN.[4] Values of (α, β) equal to (1.82, 1.74) for amorphous ribbons, (1.64, 1.88) for iron, (2.1, 1.78) for alumel, have been found, which, given the difficulties of such type of measurements, seem to be in reasonable agreement with the present predictions.

POWER SPECTRUM

It has been known for a long time that the BN power spectrum $F(\omega)$ depends in a complicated, non-linear way on the average magnetization rate dI/dt and on the presence of demagnetizing effects.[13] The spectrum usually exhibits a maximum at some frequency ω_0 which progressively shifts towards higher values with increasing dI/dt or decreasing permeability. At high ω, the spectrum approaches the law $F(\omega) \approx k(dI/dt)/\omega^2$, where the proportionality constant k is controlled by the microstructural properties of the material considered.

These features, which have been the subject of detailed analysis in the frame of the models based on clustering of elementary DW jumps,[3] can be quite naturally interpreted by Eqs.(1)-(2). Let us first notice that, according to Eq.(6), the approximation where $h_p(x)$ is a pure WL process gives an exponential autocorrelation and thus a lorentzian spectrum, $F(\omega) = 4c/(\omega^2 + 1)$. Since $c \equiv <v> \propto dI/dt$, no non-linearities of any sort are found. This result drastically changes when we release the approximation $\alpha_2 \rightarrow \infty$ and we include α_2 in the treatment. This becomes particularly evident in the limit of high mean DW velocities, where v fluctuations are so small with respect to $<v> \equiv c$ that we can assume $x \approx cu$ in Eq.(2) (we will keep on assuming $\alpha_1 \rightarrow 0$, i.e., no reversible magnetization contributions). Under these approximations, the h_p spectrum, calculated from the Fourier transform of Eq.(2), is equal to $4c/(\omega^2 + \tau_c^{-2})$, with $\tau_c = \alpha_2/c$. By making use of this result in the calculation of the Fourier transform of the time derivative of Eq.(1), we find

$$F(\omega) = 4c \frac{\omega^2}{\left[(\omega^2 + 1)(\omega^2 + \tau_c^{-2})\right]} \tag{9}$$

According to Eq.(9), $F(\omega) \approx 4c/\omega^2$ at high ω. $F(\omega)$ has a maximum at the frequency $\omega_0 = \tau_c^{-1/2} = (c/\alpha_2)^{1/2}$. Here $F(\omega_0) = 4c/(c/\alpha_2 + 1)^2$. Since $c \propto dI/dt$, we find that the spectrum maximum is shifted to higher frequencies with increasing dI/dt, and that its absolute value in this region increases less than linearly with dI/dt. These non-linearities originate from the fact that α_2 is a correlation length in space and not in time. A distance of the order of α_2 is swept by the DW in a time inversely proportional to its mean velocity c, and this gives rise to a magnetization-rate-dependent cut-off $\tau_c^{-1} = c/\alpha_2$ in h_p time fluctuations, which is responsible for the non-linear spectrum behavior. All these predictions (as well as others, concerning the spectrum dependence on permeability, not discussed here) are in excellent quantitative agreement with BN experiments performed under controlled magnetization rate and permeability values.[7]

EXTENSIONS OF THE THEORY

We have previously mentioned that Eq.(1) has the structure of a balance equation for the forces acting on a zero-mass single degree of freedom. This naturally suggests that additional features of BN could be taken into account by simply adding proper additional field terms to Eq.(1). This viewpoint proves particular effective is the study of BN in the presence of magnetic viscosity effects, i.e., of DW energy variations originating from thermally-activated atomic diffusion processes taking place inside the DW. It was shown by Néel,[14] that these processes just give rise to an additional viscosity field h_v acting on the DW. h_v obeys the evolution equation

$$\frac{dh_v}{du} + \frac{h_v}{\tau_v} = v(u)\frac{h_M}{\Delta x_v}\int_{-\infty}^{u}\frac{du_1}{\tau_v}\exp\left[-\frac{u-u_1}{\tau_v}\right]f'\left[\frac{x(u)-x(u_1)}{\Delta x_v}\right] \tag{10}$$

where h_M, Δx_v, τ_v are suitable parameters, f(x) describes the h_v behavior in the case of an instantaneous DW displacement[14] and f'(x) means df/dx. The basic equations for the problem are Eq.(10), Eq.(2) and Eq.(1) in the modified form $v = h_a(u) - x - h_p(x) - h_v(u;\{x(u)\})$, where the notation $h_v(u;\{x(u)\})$ reminds that h_v depends on the whole DW past history $\{x(u)\}$.

This theoretical description has been applied to the interpretation of the BN behavior in 12% Al-Fe alloys, where strong magnetic viscosity phenomena occur around 500 C in correspondence of atomic pair ordering processes.[15] A good estimate of the activation energy of the process can be obtained from the analysis of the temperature dependence of the maximum of the BN spectrum.[16]

Another interesting generalization of Eq.(1) is suggested by the fact that Eq.(1) does not take into account the possible role of DW internal degrees of freedom, in particular of the fact that the DW behaves like a flexible, rather than a rigid surface. Distortions of the DW profile in a given cross-section of the system can be induced by the pressure of eddy-current counterfields, which is inhomogeneous as a consequence of the boundary conditions imposed by the system geometry. Yet, these distortions, usually termed DW bowing, become relevant only at large DW velocities, when the eddy-current fields reach a strength comparable with the DW surface tension, and are not expected to have significant influence on BN. On the other hand, an aspect that can have much more important consequences is the fact that any magnetization change originating in a given cross-section of the system must eventually propagate along the applied field direction z. Perfect propagation is obtained if the DW behaves like a rigid surface along z, as is implicitly assumed in Eq.(1), but this is too an idealized situation. The microstructural perturbations giving rise to the pinning field h_p will have a finite correlation length along z, so that the net pinning force experienced by the DW will fluctuate along z. This will induce DW longitudinal distortions and will partially inhibit the longitudinal propagation of the magnetic flux. In order to deal with such mechanisms, we need to introduce in Eq.(1) some additional term describing how the local field balance is altered by DW z distortions. It has been known for a long time and it has been stressed by several authors[1] that longitudinal magnetic flux propagation is expected to obey some diffusion-like equation, a fact supported by experimental findings and by the very structure of Maxwell equations. A natural generalization of Eq.(1) in this direction is

$$\frac{\partial x}{\partial u} = D\frac{\partial^2 x}{\partial z^2} + h_a(u) - x - h_p(x,z) \tag{11}$$

where $h_p(x,z)$ is a proper bidimensional stochastic process, characterized by suitable correlation lengths along x (see Eq.(2)) and z.

Equation (11) has been successfully applied to the interpretation of the time and space propagation of individual Barkhausen jumps,[17] and to the prediction of the BN spectrum in NiFe alloys.[18] The diffusive term in Eq.(11) modifies the spectrum shape described by Eq.(9) by shifting the position of the maximum towards lower frequencies and by introducing an intermediate frequency region beyond the maximum where the spectrum attains a slope in between -1.5 and -2. We believe that a deeper investigation of the properties of Eq.(11), supplemented with a proper description of the process $h_p(x,z)$, could lead to important progress in the comprehension of magnetization dynamics in ferromagnetic systems.

REFERENCES

1. H. Bittel, IEEE Trans. Magn. **5**, 359 (1969).
2. G. Montalenti, Z. Angew. Phys. **28**, 295 (1970).
3. M. Celasco, F. Fiorillo, and P. Mazzetti, Nuovo Cimento **B23**, 376 (1974).
4. P.J. Cote and L.V. Meisel, Phys. Rev. Lett. **67**, 1334 (1991); Phys. Rev. **B46**, 10882 (1992).
5. P. Bak, C. Tang, and K. Wiesenfeld, Phys. Rev. Lett. **59**, 381 (1987); Phys. Rev. **A38**, 364 (1988).
6. G. Bertotti, F. Fiorillo, and M.P. Sassi, J. Magn. Magn. Mater. **23**, 136 (1981).
7. B. Alessandro, C. Beatrice, G. Bertotti, and A. Montorsi, J. Appl. Phys. **64**, 5355 (1988); J. Appl. Phys. **68**, 2901 (1990); J. Appl. Phys. **68**, 2908 (1990).
8. H.J. Williams, W. Shockley, and C. Kittel, Phys. Rev. **80**, 1090 (1950).
9. G. Bertotti, Magnetic Excitations and Fluctuations II, U. Balucani et al., eds. (Springer, Berlin, 1987), p. 135; Phys. Rev. **B39**, 6737 (1989); Models of Hysteresis, Pitman Lectures in Mathematics, A. Visintin, ed. (Longman, London), in press.
10. J.A. Baldwin, Jr. and G.M. Pickles, J. Appl. Phys. **43**, 4746 (1972).
11. W. Grosse-Nobis, J. Magn. Magn. Mater. **4,** 247 (1977).
12. R. Vergne, J.C. Cotillard, and J.L. Porteseil, Rev. Phys. Appl. **16**, 449 (1981).
13. P. Mazzetti and G. Montalenti, Proceedings of the ICM Conference, (The Institute of Physics, London, 1964), p. 701.
14. L. Néel, J. Phys. Rad. **13**, 249 (1952).
15. A. Ferro, P. Mazzetti, and G. Montalenti, Nuovo Cimento **56**, 111 (1968).
16. C. Beatrice and G. Bertotti, J. Magn. Magn. Mater. **104-107**, 324 (1992); Noise in Physical Systems and 1/f Fluctuations, T. Musha et al., ed. (Ohmsha, Tokyo, 1991), p. 95.
17. G. Bertotti and F. Fiorillo, IEEE Trans. Magn. **25**, 3970 (1989).
18. C. Couderchon, J.L. Porteseil, G. Bertotti, F. Fiorillo, and G.P. Soardo, IEEE Trans. Magn. **25**, 3973 (1989).

METHODS OF MAGNETIC NOISE DEPRESSION

V. V. Potemkin

Physical Department, Moscow State University,
Moscow, 119899 GSP, Russia

ABSTRACT

Three experimental methods of magnetic noise depression in thin magnetic permalloy films are presented. The films were remagnetized by sinusoidal magnetic field with frequencies ranging from 100 kHz to 16 MHz. Fluctuation spectral density $S_V(f)$ of EMS was measured in indicator winding at low frequencies. Experiments were carried out on the thin magnetic films, containing 80% Ni and 20% Fe, with thickness 500-1500 A, prepared by the method of thermic evaporation on the glass substrate. Each method of noise depression (25 dB - 80 dB) is explained on the basis of magnetics.

INTRODUCTION

The studies of magnetic noise depression have been made not only for the importance in device engineering, but also for purely scientific interest in physics of magnetics. The problem of noise depression in any physical system is one of the most important. By now there were no experiments which proved the cooling influence on the magnetic noise by cycle remagnetization of ferromagnetic material. The aim of the present paper is to obtain reliable method of noise depression in thin magnetic films and to discuss the physical origin explaining the noise decrease.

RESULTS AND DISCUSSION

The first method of magnetic noise depression consists of superposition of light constant magnetic field (part of Oersted) $H_\perp$ which is orthogonal to remagnetization field.[1] Alternating magnetization was carried out along the light or hard magnetize axis of the film. The constant magnetic field $H_\perp$ created an advantageous direction of the turning the magnetization vector under the influence of the pumping fields towards the constant fields. Depression is achieved on account that the most part of the film becomes remagnetized due to the one-side rotation, but not by displacement of domains boundaries. At the certain value $H_\perp$, wiping off the angular dispersion, all the film is remagnetized due to the one-side rotation, which results in a large depression (Fig. 1). The noise depression by remagnetization of one-side rotation is the result of the time remagnetization decreasing. And secondly the critical field dispersion decreases, because in this case the transition over the state with the maximum energy required the less field energy, i.e. lesser critical field. This method made possible to reduce the noise by 80dB, Magnetic permeability abating under 10%.

The second method consists of deep cooling of the oxidized films down to the liquid helium temperature.[2] Experimental results are shown on the Fig. 2. Presented curves correspond to different times of films oxidation with the air heated up to 150°C. By cooling to the temperature close to liquid nitrogen temperature there was no observed noise change, independently on direction of remagnetization. But starting from certain, for any given specimen, characteristic temperature a strong decreasing of noise intensity with further cooling comes. Noise depression by this method achieves 30 dB. We call temperature depression T_d the temperature at which the temperature dependence changing of magnetic noise takes place.

This effect of low temperature depression is connected to the irregular domain boundaries behavior, which was first put forward by Neel and Gorelic.

To explain magnetic noise depression effect it is necessary to take into account the presence of three magnetic phases, which have a different value of interchange interaction constant. These phases are as

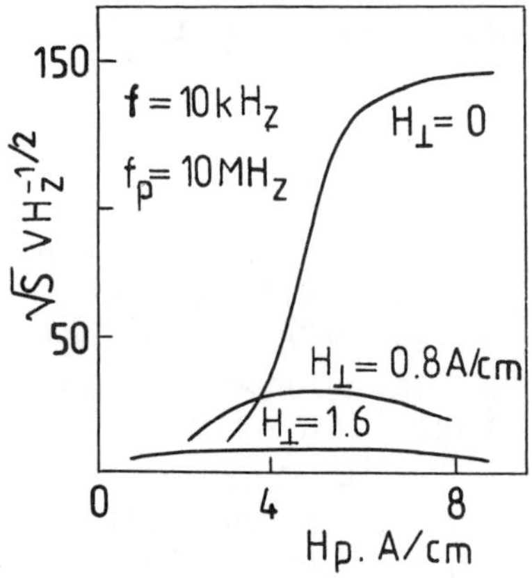

Fig. 1.

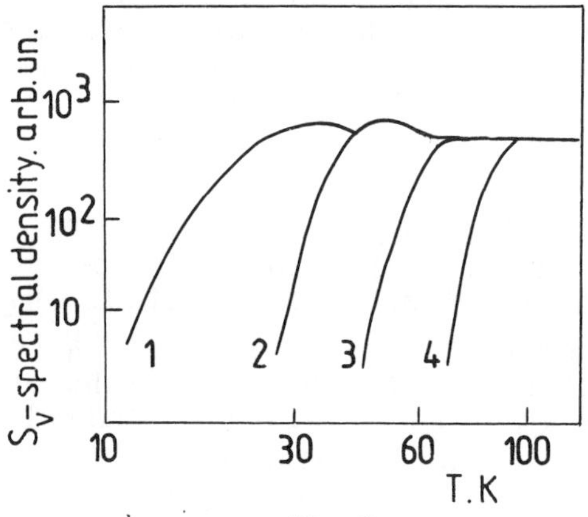

Fig. 2.

Fig. 1. Noise depression caused by constant magnetic field $H_\perp$ orthogonal to remagnetization field.

Fig. 2. Temperature dependence of magnetic noise for thin magnetic film. Curve 1 - without oxidation, curve 2 - the oxidation time is 4 hours, curve 3 - the oxidation time is 16 hours, curve 4 - the oxidation time is 32 hours.

follows. The basic ferromagnetic layer of film, small dispersion oxide phase inclusions, and transition domains between basic layer and oxide inclusions. The transition domains have a small interchange constant in comparison with basic ferromagnetic layer, because of their nonordering crystal lattice.

For the analysis of experimental results it is necessary to attract the notion of time constant τ for the relaxation process of including spin axis after the basic layer magnetization. Owing to the including parameters dispersion (energy anisotropy, coupling energy with basic layer) there is a temperature distribution of time constant τ depending on temperature. This distribution has an exponential character

$$\tau = \tau_0 \exp(KV/k_B T)$$

where τ_0 is the response time, K - anisotropy constant, V - volume of oxide phase.

At the temperature $T<T_d$ the domains which have a large τ, can be considered as frozen local domains. They become the centers of domain structure formation and make the potential relief essentially heterogeneous. With high frequency remagnetization of the film magnetic noise is caused by the not repeatable character of such process as domain structure formation and domain remagnetization, occurring in different cycles. Homogeneity of potential relief preconditions high levels of magnetic noise because of the nonreproducible domain structure. When cooling the film below the depression temperature, the potential relief becomes heterogeneous and the noise reduces abruptly. In certain samples this reduction amounted as high as 180 dB per one temperature octave.

Monotonous raise of T_d by increasing film oxidizing is caused by increasing of oxide phase volume, because the effect of magnetization fastening on the boundary line with oxide grows.

The third method to obtain magnetic noise depression involves superposition of the additional high frequency field on the remagnetized sample. The additional high frequency field causes narrowing of fluctuation zones of all the barriers and leads to reduction of the Barkhausen high jumps which are responsible for low frequency component of the noise spectrum, which, in its turn, makes the noise go lower by 25 dB.

REFERENCES

1. V.V. Potemkin, Proc. III Soviet Conf. on Fluctuation in Phys. systems, 1983, 94-97.
2. U.V. Gorohov, V.V. Potemkin, Fizika Met. Metalloved., 1979, v. 48, N 2, 426-430.

MAGNETIC NOISE IN SAMPLES WITH RECTANGULAR HYSTERESIS LOOP

Vera V. Kolatchevsky
Nickolay N. Kolatchevsky
Moscow Institute of Physics and Technology
Dolgoprudny, 141700, Moscow Region, Russia

ABSTRACT

Experimentally determined the group of ferromagnetic materials possesing the oscillating magnetic noise spectrum. The spectral distribution shape is connected with appearance of Barkhausen discontinuity correlated bundles in different remagnetization cycles. The phenomenological theory of stochastical effects in such samples is created.

INTRODUCTION

The spectral analysis of e.m.f. power arising in the sensor coil wounded around the ferromagnetic sample in the external periodically changing magnetic field shows the idea of Barkhausen discontinuity independent fluctuations (Poisson model) is not accurate. Building the adequate theory one has to take into consideration the mutual fluctuation correlation. Experimental investigations of temporal oscillogrames of Barkhausen discontinuities sequences confirm this suggestion.

Correlated "bundles" and "tandems" often appear in oscillogrames. The correlation is also observed in adjacent remagnetization half-periods. The estimative number of remagnetization half-periods with the correlated amplitude fluctuation of Barkhausen discontinuities is valid for the magnetic noise statistic determination including "1/f" noise.

THEORY

Imagine, that in every half-period the single Barkhausen discontinuity appears at the fixed moment with the stochastic amplitude. Using such a suggestion, the e.m.f. in the sensor coil can be presented as :

$$E(t) = \sum_{k=-\infty}^{\infty} (-1)^k a_k v(t - kT_0/2). \tag{1}$$

Where a_k is the stochastic amplitude, $v(t)$ is the function determining the impulse shape, T_0 is the period of the external sinusoidal field. The general expression for spectral density $g_\varepsilon(\omega)$ is given by exp. (2) in case amplitudes are correlated :

$$g_\varepsilon(\omega) = 2\omega_0\sigma^2|S(\omega)|^2 F(\omega/\omega_0). \tag{2}$$

where $F(\omega/\omega_0) = 1 + 2\sum_{k=1}^{\infty}(-1)^k R(k)\,cos((\pi\omega/\omega_0)k)$, $\omega = 2\pi/T_0$, $\sigma^2 = \overline{(a-\overline{a})^2}$.

$R(k)$ is the coefficient of the amplitude correlation, $S(\omega)$ is the result of $v(t)$ Fourier transformation. The condition $g(0) = 0$ imposes on $R(k)$ the following restriction:

$$\sum_{k=1}^{\infty}(-1)^k R(k) = -0.5. \tag{3}$$

Take $R(k)$ for $k > 0$: $R(k) = r\,exp(-k/n)$, $(n > 0)$ and obtain

$$F(x) = 1 + (r/2)\sum_{k=1}^{\infty}(-1)^k\Big(exp(-k\alpha_1) + exp(-k\alpha_2)\Big), \tag{4}$$

where $x = \pi\omega/\omega_0$, $\alpha_1 = ix - 1/n$, $\alpha_2 = -ix + 1/n$. After the calculation of the sum (4) with condition (3), the coefficient r is determined as: $r = 1 + exp(1/n)$. And finally:

$$F(\omega/\omega_0, n) = 1 - \frac{\Big(1 + exp(1/n)\Big)\Big(1 + (exp(1/n))\,cos(\pi\omega/\omega_0)\Big)}{1 + 2\,(exp(1/n))cos(\pi\omega/\omega_0) + exp(2/n)}. \tag{5}$$

The curves $F(\omega/\omega_0, n)$ are presented on Fig. 1. One can see, that $F(\omega/\omega_0, n)$ becomes sharper when n increases and concentrates in the vicinity of odd harmonics (near the third in the calculated case). The $F(\omega/\omega_0, n)$ function also modulates the spectral distribution $g_\varepsilon(\omega)$ and determines the magnetic noise spectrum in case $S(\omega)$ is a "white noise" (short discontinuities).

The specific correlation kind of magnetic moment changing in adjacent remagnetization half-periods takes place in samples with the near-to-rectangular hysteresis loop. In this samples the remagnetization occurs practically instantly in comparison with T_0 and all discontinuities are fused. The total magnetic moment of the sample fluctuates itself. The oscillogram of the magnetic induction flow fluctuating component is represented as near-to-rectangular pulses of different signs with stochastic amplitudes appear at $kT_0/2$.

Write down the expression for fluctuating component:

$$\varphi(t) = \sum_{k=-\infty}^{\infty}(-1)^k \xi_k F(t - kT_0/2).$$

The spectral density of the e.m.f. power fluctuating component is given by the equation:

$$g_\varepsilon(\omega) = 2(\sigma_\xi^2/\pi^2)\omega_0\,sin^2(\pi\omega/2\omega_0). \tag{6}$$

According to (6) the noise spectral distribution has the oscillating nature and vanishes on all even harmonics including the zero one.

EXPERIMENT

The state of the sample with the near-to-rectangular hysteresis loop is reached by producing in this sample an induced magnetic anysotropy. Such kind of anysotropy arises when the permalloy film is deposited in vacuum on the clean hot layer in the presence of magnetic field. The latter is applied in the film plain and has the intensity about 100 Oe. Under such conditions the easy-magnetic axis is formed in the film plain.

There were carried out the investigations using round plain Mo-alloyed Fe-Ni film with 10 mm diameter and 950 Å thickness. The spectral curve of e.m.f. power fluctuations under 1kHz cyclic remagnetization is presented on Fig. 2 (the experimental data are marked with the crosses).

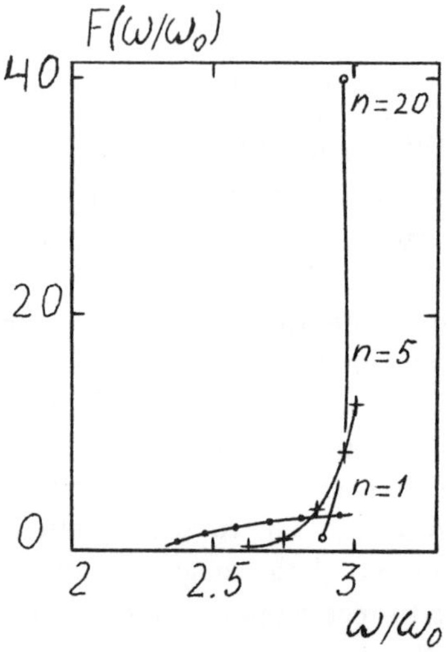

Fig 1. Function $F(\omega/\omega_0, n)$.

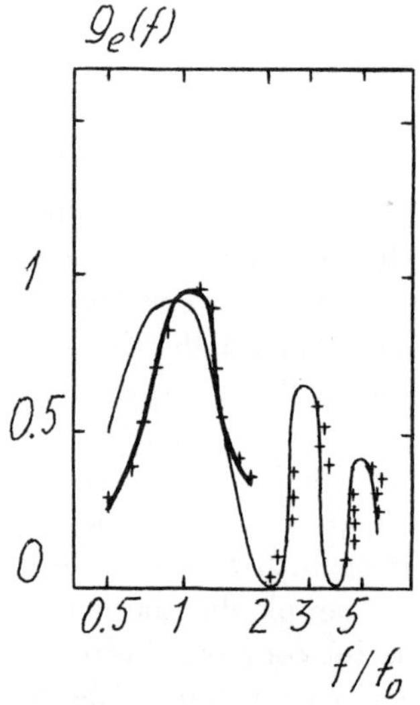

Fig 2. Power spectrum density of pair nonindependent pulses.

Analyzing Fig.2 one can notice, that $g_\varepsilon(f)$ considerably decreases on a characteristic high frequency. Taking into consideration that the steep part of $\varphi(t)$ pulse has the finite average duration ϑ, the characteristic frequency of the high-frequency cut is equal to $f_b = \omega_b/2\pi = (2\pi\vartheta)^{-1}$. The g_ε slope location coincides with the steep up-slope location of "$1/f$" noise on harmonics. Due to this fact, the integral $g_\varepsilon(f)$ slope occures only on tuning out from even harmonics and $g_\varepsilon(f)$ in the vicinity of even harmonics doesn't turn into zero. The $g_\varepsilon(f)$ slopes on even harmonics correspond to the definite Barkhausen discontinuity correlation type on increasing and decreasing hysteresis loop parts. If $\varphi(t)$ function consists of adjacent rectangular pulses, then e.m.f. pulses are short equal in pairs pulses of different signs. They arise around the points of the external field changing its sign. Every such pair can be treated as a single pulse. When these pairs do not correlate to each other, the e.m.f. power spectrum is expressed:

$$g_\varepsilon(\omega) = (4\omega_0\sigma_0^2\vartheta^2/\pi^2)(sin(\omega\vartheta/2)/(\omega\vartheta/2))sin^2(\pi\omega/2\omega_0). \quad (7)$$

When the remagnetization frequency changes, the correlation character changes too. There were investigated experimental curves of magnetic noise spectral distribution in the region from the second to the third harmonic on $f = 0.5$, 1, 5, 20 and 50 kHz.

The correlation parameter for these frequencies is proved to be equal to 1, 1, 1, 2, 12 respectively. Therefore, for small f_0 values only the two-impulse correlation takes place. In this experiment the e.m.f. pulse looks like very short equal in pairs impulses of different signs arising around points when the external field changes its sign. The e.m.f. power spectrum is well-approximated by the curve presented on Fig.2 (thin curve).

Analyzing the obtained results one can make a suggestion that observed correlation effects due to relaxation processes of domain boundaries displacements are connected with the magnetic viscosity. The microcurl effect influence is negligible when the samples with reported thickness are used. It is essential to remark, that the Fe-Ni alloys with the high permeance posses the abnormal magnetic viscosity and the maximal viscosity is observed in structures provides high magnetic properties.

References

N.N. Kolathevsky, *Magnetic noise*, Moscow "Nauka" 1971

IV. SUPERCONDUCTORS

DISORDER-INDUCED FLICKER NOISE IN HIGH-T_c SUPERCONDUCTORS

V.I.Kozub
A.F.Ioffe Physico-Technical Institute, St.-Petersburg, Russia

ABSTRACT

Effect of structural defects with internal degree of freedom and exponential scatter of relaxation times ("fluctuators") on superconducting parameters like critical temperature T_c and superconducting electrons density N_s is considered within a phenomenological approach. The parameters in question are shown to exhibit local non-stationary fluctuations providing in particular a resistance flicker noise peak within a superconducting transition region while in a presence of external magnetic field magnetic noise is expected even in a pure Meissner state. In addition, local fluctuations of H_{c1} value are predicted. It is shown as well that fluctuators coupling with the superconducting condensate provides in its turn a dependence of fluctuators occupation numbers on the condensate e.g. on the presence of a supercurrent. Slow relaxation and memory effects imposed on superconductivity by structural relaxation are expected as well.

There is a lot of experimental data evidencing that a structure of high-T_c superconducting materials even for monocrystalline samples is by no means a perfect one. Among different aspects of this fact one can note a presence of specific defects with internal degree of freedom and exponentially-broad distribution of relaxation times. It has been revealed by internal friction experiments [1] as well as by heat-release experiments [2]. The most probable candidates for the role of such defects seem to be oxygen vacancies, their movement being affected by some structural imperfections. Thus one can expect a formation of effective double-well potentials of the type known to be responsible for two-level systems (TLS) in amorphous solids (a correlation between TLS density and oxygen vacancies concentration has been reported in [1].

It has been shown earlier that such defects ("fluctuators") can be responsible for 1/f resistance noise in normal state actual for a wide temperature range [3]. In Josephson structures fluctuators lead to a critical current noise [4]. Experimental data for weak link noise in high-T_c SQUIDs [5] seem to be in agreement with these predictions.

Here we would like to point at the fact that the structural relaxation produced by the fluctuators may affect most of superconducting parameters like critical temperature T_c, magnetic field penetration length λ, "superconducting carriers density" N_s etc. because all the parameters in question are sensitive to structure (and some of them like λ and N_s are as well sensitive to normal resistivity).

We have studied the effect of the fluctuators on superconducting parameters in question within a framework of phenomenological approach taking into account a contribution to Ginzburg - Landau free energy due to coupling with fluctuators:

$$\delta F = \sum_i \gamma_i \mid \Psi \mid^2 a^3 n_i \tag{1}$$

(here γ_i is a dimensionless parameter describing the relative influence of i_{th} fluctuator on the superconducting condensate, a is a characteristic size of fluctuator (supposed to be of the order of a lattice constant), Ψ is an order parameter, n_i is an occupation number of upper of the fluctuator states. It has been shown that the parameters in question exhibit local non-stationary fluctuations with a flicker frequency spectrum.

It was shown in particular that the very value of the critical temperature T_c exhibit local fluctuations, its spatial correlation length ζ_c being given by one of the static disorder. The fluctuations spectral density is estimated as

$$\frac{< (\delta T_c)^2 >_\omega}{T_c^2} \sim (\frac{1}{\omega})\gamma^2(\frac{a}{\zeta_c})^6 PT\zeta_c^3 \tag{2}$$

P being the fluctuators density of states. It is natural that these fluctuations are most important at temperatures close to T_c. Due to the fact that a resistivity near the superconducting transition is a function of the quantity $(T-T_c)$ the fluctuations of T_c lead to fluctuations of resistance R having a peak at the transition region (where a derivative (dR/dT) has a peak). Note that the flicker noise peak at $T \sim T_c$ seem to be typical for superconducting bolometers (see, e.g. [6]) and sometimes is prescribed to temperature fluctuations. It can be shown however that the later can not explain the effect in wide region of relatively low frequencies. On the other hand the estimates within a framework of our model (which readily accounts for the 1/f spectrum) a noise level predicted is in agreement with existing experimental data.

The fluctuations of T_c lead in its turn to spatially inhomogeneous fluctuations of N_s (and respectively of λ. Besides one should note that N_s and λ are sensitive in general to fluctuations $\delta\rho$ of normal resistivity. Taking these considerations into account one obtains for the mean fluctuations δN_s, $\delta\lambda$ for the volume V:

$$\frac{2\delta\lambda}{\lambda} \mid_V \sim \frac{\delta N_s}{N_s} \mid_V \sim (\frac{\delta T_c}{T_c} + \frac{\xi_0}{l}\frac{\delta\rho}{\rho}) \mid_V \tag{3}$$

where ξ_0 is a coherence length, l is an electron mean free path. The fluctuations of N_s and λ manifest themselves in the presence of external magnetic field as some redistribution of magnetic flux and can provide magnetic noise actual even in purely Meissner state that is in the absence of vortices jumps.

On the other hand, fluctuations of N_s providing local changes of supercurrent energy density $mv_s^2N_s/2$ (v_s being superconducting condensate velocity) modify the value of energy necessary for the vortex formation which can be considered as local fluctuation of critical magnetic field H_{c1}. For the films that are not extremely thin that is for which the vortex line is influenced by many fluctuators the corresponding mean quadratic fluctuation can be estimated as

$$\frac{<(\delta H_{c1})^2>_\omega^2}{H_{c1}^2} \sim (\frac{1}{\omega})\frac{(PT)^{5/3}}{d}\frac{1}{\log\kappa}(\gamma a^3)^2 \tag{4}$$

Here κ is Ginzburg - Landau parameter while d is the film thickness. This factor providing non-stationary fluctuations of pinning forces imply an additional source of noise in vortices structure. Non-exponential fluctuator-induced magnetic relaxation has been considered as well.

Until now we have considered fluctuators to be completely independent from the superconducting condensate thus being some "external" factor with respect to the superconductivity. However one sees that the coupling given by Eq.1 provides in general a dependence of fluctuator occupation number on a local value of the order parameter. As a fact one deals with a coupled system fluctuators - condensate. It can be shown in particular that in the presence of a supercurrent with a superconducting velocity v_s the energy spacing between fluctuator states is renormalized by an addition

$$\Delta E_i \sim \frac{mv_s^2}{2}N_s a^3 \gamma_i \tag{5}$$

In a view that v_s is directly related to magnetic field this allows one to prescribe to fluctuators some "magnetic susceptibilities". As a result, magnetic response of the material can manifest slowly relaxing component related to structural relaxation connected with the fluctuators. E.g. one can expect the presence of "polaron-like" dressing of the vortices by fluctuators rearranging their occupation numbers due to renormalization of E in the v_s field related to vortices.

The factor discussed above may provide memory and slow relaxation effects of different sorts. So we believe that at least some of "glass-like" behavior observed for high-T_c materials may be due to structural disorder rather than due to "glassy" vortices structure.

In addition, it is shown that the fluctuators coupling with the superconducting condensate provide a dependence of fluctuators states on the state of condensate e.g. on a supercurrent. As a result specific "memory effects" can exist due to structural relaxation in question.

References

[1] P.Esquinazi et al. Phys.Rev.B, **37**, 545 (1988); B.Golding et al. Phys. Rev.B, **36**, 5606 (1987)

[2] A.Sahling, S.Sahling J.Low Temp.Phys., **77**,399 (1989)

[3] Yu.M.Galperin,V.G.Karpov,V.I.Kozub. Adv. in Phys., **38**, 669 (1989)

[4] Yu.M.Galperin,V.L.Gurevich,V.I.Kozub Europhys.Lett., **10**, 753 (1989)

[5] V.M.Zakosarenko, E.V.Il'ichev Fiz.Tverd.Tela, **34**, N 5, (1992)

[6] A.Hirai et al., Cryogenics, **30**, 910 (1990); R.Leoni, Ph.Lerch, J.Clarke IEEE Trans. on Magn., **25**, 973 (1989)

A STATISTICAL PERCOLATIVE MODEL OF THE H T_c SUPERCONDUCTOR TRANSPORT PROPERTIES

M. Celasco[1], G. Cone[4], D. Grobnic[4], A. Masoero[2], P. Mazzetti[3], I. Pop[4], A. Stepanescu[3], I. Stirbart[4]

1 Dip. Fisica Università di Genova, Via Dodecaneso 33, 16100 Genova - Italy
2 Dip. Fisica Università di Modena, Via Campi 213/A, 41100 Modena - Italy
3 Dip. Fisica Politecnico di Torino, C.so Duca Abruzzi 24, 10129 Torino - Italy
4 Dept. of Physics, Polyt. Inst. of Bucharest, Splaiul Independentei 313, 77206 Bucharest - Romania

ABSTRACT

A percolative model which simulate the behaviour of transport properties of a ceramic superconductor HT_c (HTS) when subjected to an electrical current and a magnetic field is presented. The model is a percolative one and takes into account the statistical distribution of the normal Josephson junction resistance and critical temperature distribution of the grains. The model consider the superconductive sample as a 3-dimensional random resistor network with the value of the resistance computed in terms of temperature, magnetic field and current. Comparison between experimental and theoretical results concerning current noise and resistance hysteresis under magnetic field is given and discussed.

1. INTRODUCTION

As it is well known, as a consequence of their internal structure, the new HTS display strong granular features such as degradation of the critical current density, the broadening of the resistive transition, and a large current noise near the percolation threshold. It seems that the granularity influences to a large extent the behaviour of the ceramic superconductor, and great effort has been devoted to its mechanism of action[1-3]. This paper is dedicated to the same task. We have tried to gain information about data and dependencies that characterize the weak-links which couple the islands of strongly superconducting material, which we assume as being Josephson junctions. This was performed by fitting the dependency of the electrical resistance and noise on magnetic field, given by the pericolative model, with the corresponding experimental data. We consider the HTS sample as consisting of a large collection of superconducting grains linked through weak-links of the Josephson junction type. For each junction the resistance shunted junction model[4] was considered, with the value of the parallel resistor determined by the temperature, current and magnetic field. Considering a distribution of the coupling strengths, inferred from the distribution of the normal state resistance of the Josephson junction and the critical temperature distribution inside the grains, the model allows us to simulate the dependency of the resistance and noise on a looped magnetic field, and to make quantitative and qualitative assumption on the microscopic characteristic of the weak-links and on the sources of the noise.

2. THE MODEL

We consider the superconductor as a system of superconductive islands arranged in a three dimensional network, connected by the Josephson junction links. The key feature of the model is the randomness of the coupling between the superconductive islands. From the many stochastic parameters that define the coupling of the grains, we have chosen the normal resistance of the Josephson junction and the critical temperature of the grains, that we think as being directly determined by the fundamental independent variable of the Josephson junction, size and oxygen content. We assume that the normal state resistance obeys a lognormal distribution and that the critical grain temperature a gaussian one. It was farther assumed that the variances of these distributions are proportional to their mean values.

Considering that the key feature of the transition region from normal conductance to superconductive state is the granularity, we assume that each grain becomes superconductive when the temperature drops below their own critical temperature and the intergrain junction allows the current to flow. This happens when the Josephson coupling energy[5] exceeds the thermal energy $k_B T$ and the current through the junction does not exceeds a critical current density. This condition defines the critical temperature of the Josephson junction. Taking into account the distribution of normal state resistance of the junction and of the critical temperature of the grains, the coupling of the grains follows a certain distribution function and the superconductor transition becomes a bond percolation problem[6,7]. To solve this problem we need to know the fraction of the superconducting bonds out of the total number of the junctions and therefore the distribution function of the coupling events between two grains.

Three simultaneous conditions have to be met for the occurrence of such an event. The two adjacent grains must be in a superconducting state, the coupling Josephson energy to exceed the thermal energy[8] and the current through the junction to be less than a critical value. For a given temperature T_c grater than T they will be in a superconducting state. Considering a gaussian random distribution of the critical temperature, with mean value T_{cmed} and variance σ, the probability for two adjacent grains to be in a superconducting state is:

$$P = \left[1 - \frac{1}{\sqrt{2\pi}\sigma}\int \exp\left(-\frac{(T_c - T_{cmed})^2}{2\sigma^2}\right) dT_c\right]^2 , \qquad (1)$$

for $H = 0$. When $H \neq 0$, we have to evaluate the critical current I_c, which enters the Josephson energy

$$E_c = \frac{\phi_0 I_c}{2\pi} \qquad \text{with}^{5,9-12} \quad I_c(T,H) \propto \frac{T_c^2}{T \cdot R_n}\left(1 - \frac{T}{T_c}\right)^2 \cdot \frac{1}{1 + 2\frac{H}{H_0}} \qquad (2)$$

where the characteristic field H_0 is given by $H_0 = \phi_0/\mu d L$ with $d = 2\lambda + t$, the effective junction thickness, λ the London penetration depth, ϕ_0 flux quanta, t the barrier thickness and L the junction length[13] and R_n is the normal state resistance of the junction. From eq. (2) it is now possible to evaluate the condition for a Josephson junction to be open ($E_c > k_B T$) and thus also the probability that a current would flow through the junction. The value of the individual intergrain critical current depends strongly on the actual value of the magnetic field in the junction region. But, due to the high magnetic field used, the results would be complicated by the penetration of the flux line in the grains themselves. So that the local field in the junction depends not only on the value of the external field but on the actual distribution of superconducting islands and the intergrain flux too. Since the critical intergrain current is very low, several order of magnitude lower then the critical current density inside the grains, it is affected to a negligible extent by its own magnetic field and we may neglect the diamagnetic induced shielding currents in closed

loops that include the junction.
The intergrain field, which depends on the external field and on the probability that a junction is closed, can be calculated by an iterative procedure using a simplified model[14]. Once the percolation probability was computed, the objective functions of resistance and current noise are computed in a straightforward way. For each bond i,j of the random resistor network the Kirchhoff law gives: $(V_i - V_j)\sigma_{ij} = I_{ij}$. By addition of all the equations for the node i, it follows that:

$$\Sigma_j(V_i - V_j)\sigma_{ij} = I_i \tag{3}$$

where V_i, I_{ij}, σ_{ij} are the voltage of the node i, the current through the bond i-j and the conductivity of the bond i-j respectively, while, I_i is the total current through the node i. The expression (3) defines a set of n-1 equations that give as a solution the voltage of each nodes relative to one node considered as grounded. The equivalent resistance is computed as $R_{eq} = (\Sigma r_\alpha I_\alpha^2)/I^2$, where the summing is performed over all the bonds in the network. Noise is the general term for random fluctuation of power in the electrical systems. Considering that in our percolating network each resistance fluctuates in time around an average value I_{med} independent from the other resistances, which may be true due to the short coherence length in the ceramic superconductors, then $r_\alpha = r_{med} + \delta r_\alpha$. The noise power spectrum can then be calculated from the expression[15]:

$$S_R(\omega) = \Sigma_\alpha S_r(\omega)\frac{I_\alpha^4}{I^2} \tag{4}$$

where $S_R(\omega)$ is the power spectrum of the total resistance fluctuation ΔR, while $S_r(\omega)$ is the power spectrum of the individual junction resistance fluctuation δr_α.

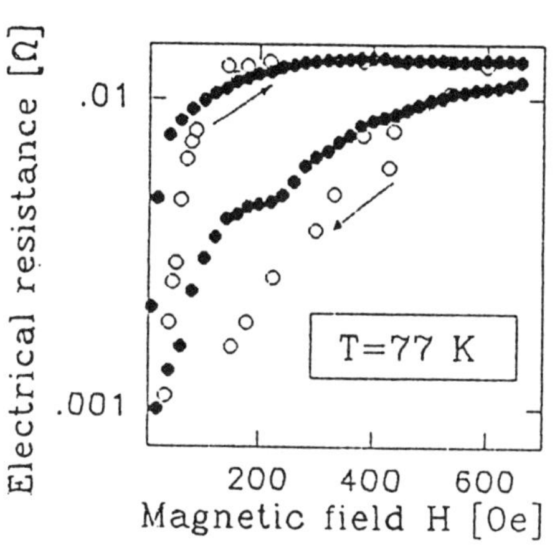

Fig.1. Experimental (•) and simulated (°) values of the resistence versus magnetic field.

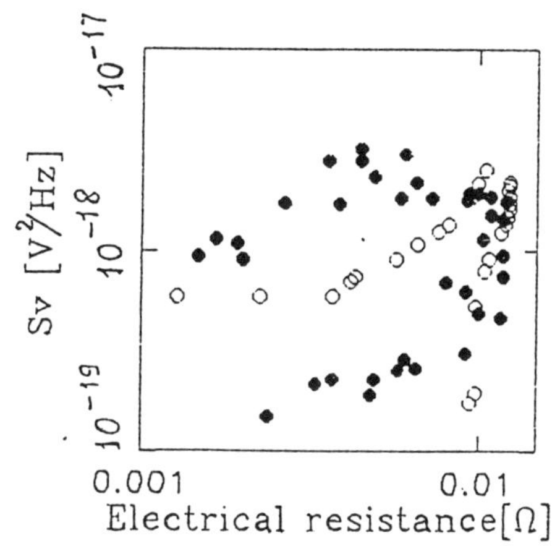

Fig.2. Experimental (•) and simulated (°) values for the relative resistence fluctuation noise versus resistance.

3. RESULTS

The experimental results[16] for a superconductive specimen(YBCO), that display histeresis behaviour for resistivity and noise in magnetic field, were interpreted using the above presented model. For simulation purpose a critical temperature of the grains of 90 K was used and current density of $10^4 A/m^2$ at 77 K[16]. The magnetic field was looped between 0 and 700 gauss. The estimates of the junction parameters as grain size and junction thickness was taken from the literature[13]. The relative variance of the junction resistance, the variance of the critical temperature distribution and the phenomenological parameter of grain inductance were taken as free parameters and fitted by trial and error method.

Multiple trials using the gaussian normal distribution of the normal junction resistance were unsuccessfully for a large range of the variance values. Only when a lognormal distribution was chosen an acceptable fit of the resistence versus field data was obtained for value in the range (3.0-3.2) for relative variance $\varepsilon_{lnR} = \sigma_{lnR}/ < lnR >$ of the lognormal distribution of the normal junction resistance and around 3.5K for the σ_{Tc}, the variance of the critical temperature resistance.

The intergrain field, which depends on the external field and on the probability that a junction is closed, can be calculated by an iterative procedure using a simplified model[14]. Some preliminary results concerning the resistance hysteresis and the relative resistance fluctuation noise are reported in fig.1 and fig.2 respectively. The simulated results are averaged values over the 100 independent runs of a 300 grains sample.

REFERENCES

1. K.H. Lee and D. Stroud, Phys. Rev. B45, 2417 (1992).
2. C. Mee, A.I.M. Rae, W.F. Vinen and C.E. Gough, Phys. Rev. B43, 2946 (1991).
3. Z.Q. Wang and D. Stroud, Phys. Rev. B44, 9643 (1991).
4. K.H. Lee and D. Stroud, Phys. Rev. B43, 5280 (1991).
5. C.J. Lobb, D.W. Abraham and M. Tinkham, Phys. Rev. B27, 150 (1983).
6. M. Celasco, M. Lo Bue, A. Masoero, P. Mazzetti and A. Stepanescu, in :Noise in Physical Systems and 1/f Fluctuation, eds T. Musha, S. Sato and M. Yamamoto, (Ohmsha, 1991) p.111
7. L.B. Kiss, T. Larsson, P. Svedlindh, L. Lundgren, H. Ohlsén, M. Ottosson, J. Hudner and L. Stolt: Physica C 207, 318 (1993).
8. L. Solymer, Superconducting Tunneling and Application (Chaoman and Hall, London), 1972.
9. V. Ambegaokar and A. Baratoff, Phys. Rev. Lett. 10, 486 (1963).
10. V. Ambegaokar and A. Baratoff, Phys. Rev. Lett. 11, 104 (1963).
11. G. Deutscher and K.A. Müller, Phys. Rev. Lett. 59, 1745 (1987).
12. K.H. Müller, M. Nikolo, R. Driver, Phys. Rev. B43, 7976 (1991).
13. R.L. Peterson and J.W. Ekin, Phys. Rev. B37, 9848 (1988).
14. C.E. Goug in "High Temperature Superconductivity" Proceeding of the Thirty-Ninth Scottish Universities Summer School in Physics, St. Andrews, June 1991, p. 37.
15. R.R. Tremblay, G. Albinet, A.M.S. Tremblay, Phys. Rev. B45, 755 (1992).
16. M. Celasco, A. Masoero, P. Mazzetti and A. Stepanescu, Phys. Rev. B44, 5366 (1991).

RESISTANCE DRIFT AND $1/f^2$ VOLTAGE SPECTRA IN THE HIGH-T_C SUPERCONDUCTING TRANSITION REGION

Hall, J. and Hall, N.J.
Microelectronic Systems Research Center (MSRC), West Virginia University, ECE Dept., Morgantown, WV 26505-6101
Chen, T.M.
University of South Florida, Electrical Engineering Department, Tampa, FL 33620

ABSTRACT

In the past many debates have arisen over the appearance of $1/f^2$ voltage spectra in the process of measuring electrical noise. In this paper, we will look at the theory of resistance drift causality to the $1/f^2$ spectra and how the measurement system affects these measurements. A superconductor operating in its transition region is a perfect test case for examining this area since vacancy fluctuations do not seem to occur. We can generalize that an acquired $1/f^2$ signal is a result of the drifting resistance and proceed to analyze the data. Our measurements indicate that the traditional theoretical case of a $1/f^2$ spectra resulting from a simple voltage ramp due to resistance drift is not entirely valid in a real-world measurement system because the input circuitry to the amplifier causes attenuation. An mathematical solution for the effect of drifting resistance when coupled with the input circuitry was developed and fit quite well to the measured signals.

INTRODUCTION

The cause of $1/f^2$ signals in electrical noise measurements has been a topic of hot debate in the reliability field [1-3] with one school of thought advancing the notion that the drifting resistance present during the sample measurement will cause a $1/f^2$ spectra and the other side theorizing that vacancy fluctuations are the cause of the $1/f^2$ spectra. When we were measuring noise in aluminum interconnects, the resulting $1/f^2$ spectra was compared to both theories, and if the magnitude of the noise was near the resistance drift theoretical model then we would assume that the noise of the vacancy fluctuations was masked out by the resistance drift. On the other hand, if the $1/f^2$ signal had a much larger magnitude than the drifting resistance theory then we would assume that the vacancy fluctuations were the cause of the signal. A problem arose when the measured $1/f^2$ signal was two orders of magnitude below the resistance drift theoretical case. This theory is very simple to prove (and is developed later in the paper) and the results were puzzling. In order to verify if the resistance drift theory was true for our measurements, a method of causing resistance change in the transition region in the superconductor was developed, where we could control a fairly large resistance change without worrying about side effects such as vacancy fluctuations [4–5]. These results were also an enigma until the theory of the resistance change causing a $1/f^2$ signal was rethought and applied to our measurement system. This research also has a second added benefit because a practical superconducting device, the bolometer, will operate in the transition region on the principal of changing resistance and traditional noise measurements for bolometer characterization have been in the static case where the temperature is at equilibrium. Our results show that the noise measurements in the dynamic case of the bolometer under operation should be performed since a different type of signal arises in the operational phase.

EXPERIMENTAL METHOD

In order to measure the superconductor in its transition region under a linear resistance change, we had to find the most linear part of the transition curve and operate our measurements in this area. Our sample had a fairly linear dR/dT in the center of the transition region which we called Region 1 and $\Delta T = 1\text{K}$. In order to assure that the measurement system itself was not the cause of the signal, a Region 2 in the normal state of the superconductor was selected where the dR/dT is very small and in this the ΔT was also 1K. Figure 1 shows the resistance–temperature (R–T) curve for our thin–film sample with the Regions 1 & 2 defined.

To perform the measurements, we used a c–axis YBaCuO thin–film sample with a thickness of 3000Å and deposited on a $SrTiO_3$ substrate. The sample was placed in an evacuated chamber in a Janis 6NDT Cryostat. A Lake Shore 805 Temperature Controller was used for the crucial part of regulating the temperature to ensure a linear resistance change and a computer was connected to delicately control the temperature. The sample was biased using ultra–low–noise gel cells along with a wire–wound resistor for current control. The signal was fed into a PAR 1900 transformer coupled to a PAR 113 low–noise amplifier for impedance matching, and the output of the amplifier went to an HP3561 Spectrum Analyzer for data acquisition. Figure 2 shows the noise traces for Regions 1 & 2 which were measured at 10mA. The Region 1 traces shows a $1/f^2$ signal and Region 2 has a $1/f$ signal. The signal in Region 1 is about 3–4 orders of magnitude lower than what was expected from the resistance change theory leading us to pursue a different solution to the $1/f^2$ signal and its relation to resistance change.

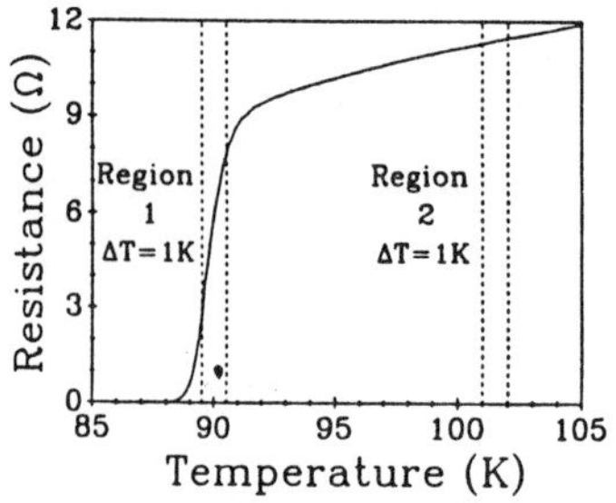

Figure 1. R vs. T curve for the sample used in this experiment. Region 1 indicates a large $\partial R/\partial T$ in the superconducting transition region. Region 2, in the normal conducting region, has a $\partial R/\partial T$ which is relatively small.

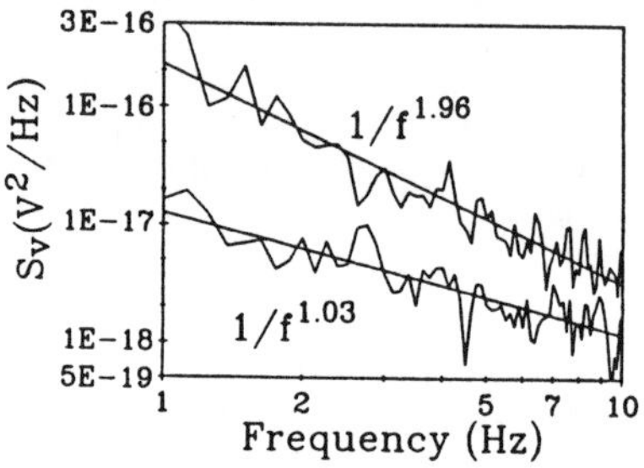

Figure 2. Noise voltage spectral density (S_V) vs. Frequency (f) plot for the two regions shown in Figure 1. Region 1 where $\partial R/\partial T$ is large has $S_V \propto 1/f^{1.96}$, whereas Region 2 where $\partial R/\partial T$ is small has $S_V \propto 1/f^{1.03}$.

THEORETICAL DEVELOPMENT OF $1/f^2$ SIGNAL

1. Fourier Analysis

The development for the resulting signal from a resistance change is developed from simple Fourier analysis [1]. If we consider a sample that has a base resistance of R_0 and the resistance change linearly with time (dR/dT=constant) is given as ΔR. Then we can say that the total resistance, $R(t)$, during the measurement at any given time, t, is represented as $R(t) = R_0 + \Delta Rt$. When a constant current, I, is passed through the sample, the resulting voltage is given by $v(t) = IR_0 + I\Delta Rt)$. Applying the complex coefficient part of Fourier analysis, we get for c_n,

$$c_n = \frac{1}{T_0}\int_0^{T_0} I\Delta Rt \exp[-jn\omega_0 t]\, dt = \frac{I\Delta R}{\omega_n}$$

The portion of the signal resulting from the IR_0 constant will be a spike at 0Hz and is essentially ignored. Using general noise theory, the voltage spectrum of a signal can be written in the terms of the the Fourier coefficients,

$$S_V(f) = \left(\frac{1}{2\pi}\right)^2 2T_0\overline{c_n^* c_n} = 2T_0\frac{I^2(\Delta R)^2}{(2\pi f)^2}$$

This equation shows that the spectrum is dependent upon three variables T_0 (measurement time, usually constant), I (current, usually a control variable), and ΔR (the change in resistance over time, the active variable). This development is valid, but it is very hard to find where it has been proved by experiment successfully. In the case of the superconductor in the transition region we had I=10mA, T_0=80 sec, and we measured $\Delta R = 0.05\ \Omega/s$. Applying these numbers to the theory we would get $S_V(1\ \text{Hz}) = 1.01 \times 10^{-7}\ \text{V}^2/\text{Hz}{=}{-69.9}$ dB. Yet, looking at Figure 2, one can see that the measured signal is not anywhere near that number, the experimental magnitude was $S_V(1\ \text{Hz}) = 1.92 \times 10^{-15}\ \text{V}^2/\text{Hz}{=}{-147.2}$ dB. This is a very clear discrepancy, where $S_{V_{diff}} = -69.9 - 147.2 = -77.3$ dB which is between 3 and 4 orders of magnitude lower.

2. Circuit Analysis

Since we saw discrepancies in both the aluminum interconnect and the superconducting thin-film $1/f^2$ measurements, a look at the defining theory was in order. From observing the Fourier analysis, it is assumed that the analysis is done on the original ramp signal, but a ramp signal has DC components that will be affected by the input circuitry to the amplifier. For our noise measurements, our input circuit to the amplifier consists of an equivalent resistance made up of the reistance of the sample plus the changing resistance, a large blocking capacitor is used to keep DC current in the sample from "magnetizing" the transformer, and the transformer which is an equivalent inductance to the amplifier. Figure 3 shows the equivalent circuit for the input to the amplifier, with I = 10mA, $\Delta R = 0.001\Omega/s$, C_b=25000μF and R_b=bias resistor=2500 Ω.

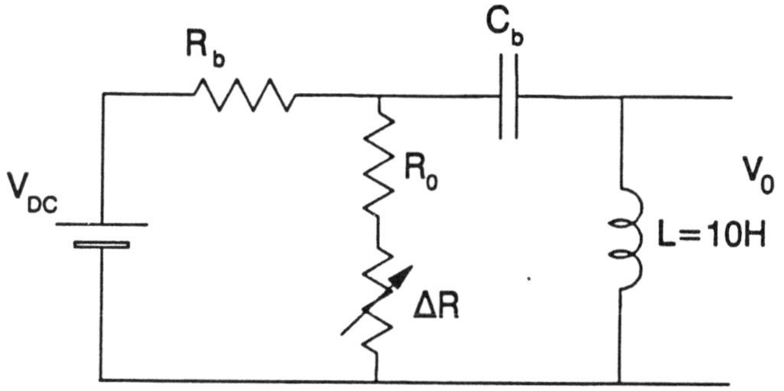

Figure 3: Typical circuit for resistance change experiment.

From Figure 3, if we take the voltages across the circuit elements and lump them together, we get

$$V_T(t) = V_{DC}(r(t)/R_b) = r(t)i + \frac{1}{C}\int i\ dt + L\frac{di}{dt}$$

where $r(t) = R_0 + \Delta Rt$. If we take the time derivative of both sides and rearrange the variables, we get a second order nonlinear equation of the form,

$$\frac{d^2i}{dt^2} + \frac{r(t)}{L}\frac{di}{dt} + \frac{1}{LC}i = V_{DC}\frac{\Delta R}{LR_b}$$

This equation is not of a "special" form, so the equation has to be developed for a solution to this PDE. A method which adapts itself well to this equation is the WKBJ approximation [6] where a change of variable is used to get a linear second order equation of the form $i'' + 2P(t)i' + R^2(t)i = Q(t)$ where

$$P(t) = \frac{r(t)}{2L} \quad R(t) = \sqrt{\frac{1}{LC}} \quad Q(t) = \frac{\Delta RV_{DC}}{LR_b}$$

and the change of variable gives $i = y\exp\left[-\int P(t)\ dt\right]$. This equation is then applied to the variation of parameters method which will give a solution for y of

$$y'' + \left[\frac{1}{LC} - \frac{R_0^2 + \Delta R^2t^2}{4L^2} - \frac{R_0A}{L}\right]y = 0$$

This equation is then solved by finding a complementary, y_c, and a particular, y_p, solution. In this case, y_c for our measurements times is negligible when transferred back to i for the solution. The particular solution, i_p, is the saving grace to this analysis where

$$y_p = \left(-\int \frac{Ny_2}{y_1y_2' - y_1'y_2}dt\right) y_1 + \left(-\int \frac{Ny_1}{y_1y_2' - y_1'y_2}dt\right) y_2$$

where $N = (\Delta RV_{DC}/LR_b)\exp[(R_0/2L)t + (\Delta R/4L)t^2 = 10^{-6}\exp[0.5t + 2.5 \times 10^{-5}t^2$. After working through the equations and converting y back to i, our solution becomes

$$i = 9.45 \times 10^{-7} - 3.37 \times 10^{-11}t + 9.78 \times 10^{-11}t^2$$

and

$$v_L = v_o = L\frac{di}{dt} = -3.37 \times 10^{-10} + 1.96 \times 10^{-9}t$$

The DC portion of v_L is filter out by the analyzer, so our modified signal is given as $v_L = 1.76 \times 10^{-9}t$. So we still have a ramp signal to perform an FFT on, it's just that the signal is much different than the ramp signal in the DUT. Applying our Fourier solution for the ramp signal, we get

$$S_V(f) = 2T\left(\frac{1.76 \times 10^{-9}}{2\pi f}\right)^2 = \frac{1.26 \times 10^{-17}}{f^2} \ (\mathrm{V^2/Hz})$$

$$S_V(1\ \mathrm{Hz}) = 1.26 \times 10^{-17}\ (\mathrm{V^2/Hz}) = -169\ \mathrm{dB}$$

Comparing this to our original ramp where $\Delta R = 0.001\Omega/\mathrm{s}$, $T = 80$ sec, and $I = 10\mathrm{mA}$, we get

$$S_{V_{orig}} = 2T\frac{\Delta R^2 I^2}{(2\pi f)^2} = 4.05 \times 10^{-10}\ (\ \mathrm{V^2/Hz}) = -93.9\ \mathrm{dB}$$

This gives a difference in expected over measured noise voltage to be

$$S_{V_{diff}} = -93.9 - (-169.0) = 75.1\ \mathrm{dB}$$

These results were clarified using measurements on aluminum thin–films at cryogenic temperatures to keep out vacancy fluctuations. In all five samples measured, all differences were with 5% of 75.1dB. The superconductor also proved to be a good example, as reported in the earlier section, there was a difference of about 77dB between the traditional resistance drift theory and the experimental data, and this seems to fit quite well with the analysis presented here.

This analysis can be found in full development in Appendix A of Ref [2]. It should also be noted that the 75dB benchmark is for our particular measurement system. Since the analysis was long, numerical values were plugged in early in the development to assist in approximations and bookkeeping.

REFERENCES

1. A. van der Ziel, T.M. Chen, P. Fang, *Sol St Elec*, **12**:11, 1989.
2. D.M. Liou, J. Gong, C.C. Chen, *Jpn Jrnl App Phys*, **29**:1283, 1990.
3. J.G. Cottle, N.S. Klonaris, M. Borderlon, IEEE Elec Dev Lett, **EDL–11**:523, 1990.
4. J. Hall, H. Hickman, T.M. Chen, *Sol St Comm*, **76**:921, 1990.
5. J. Hall, "Noise Measurements of YBaCuO Thin–Film High–Temperature Superconductors," PhD Thesis, *Univ of South Florida*, Tampa, FL, May 1992.
6. W.J. Cunninghan, *Intro to Nonlinear Analysis*, McGraw–Hill, 1958.

APPLICATIONS OF NOISE MEASUREMENTS IN $YBa_2Cu_3O_7$ SUPERCONDUCTING THIN–FILMS WITH DIFFERING PREFERRED AXES OF ORIENTATION

Hall, J.
Microelectronic Systems Research Center (MSRC), West Virginia University, ECE Dept., Morgantown, WV 26505–6101
Chen, T.M.
University of South Florida, Electrical Engineering Department, Tampa, FL 33620

ABSTRACT

In this paper we will present a comparison of DC and electrical noise characteristics of high–temperature superconductor thin–films with differing preferred axes of orientation. Four thin–film samples were thoroughly studied, two of *c*–axis and two of *a*–axis orientation. All films were YBaCuO superconducting material on strontium titanate ($SrTiO_3$) substrates. The *R–T*, *I–V*, and electrical noise properties were measured for each of the films and comparisons were made between the films. The experimental data was compared with current hypotheses on conduction and crystal structure. Our preliminary results show that from the DC and electrical noise measurements that the *c*–axis films are much better than *a*–axis films in applications where sensitivity and sharp transition is an important factor.

INTRODUCTION

The issue of noise measurements in high–temperature superconductors becomes more critical as practical devices such as bolometers and magenetometers are put into operation. These types of devices are basically detection devices where the minimum sensitivity is defined by the noise present in their operational region. The YBaCuO superconducting material is a great candidate for such devices and with its anisotropic properties, the issue of differing properties in the differing axes of orientation becomes important. Characterizing those samples based upon their preferred axis may help determine which type of film will be better for applications where sensitivity is a defining criterium for device selection. Several researchers [1–7] have measured electrical noise in YBaCuO superconductors and we will attempt to quantify their findings, along with our own, in defining device quality and performance.

SAMPLE SELECTION

The samples used for this study consisted of 2 *c*–axis oriented YBaCuO films (one with 1400Å thickness, the other with 3000Å thickness) deposited on $SrTiO_3$ substrates with (100) orientation. The other 2 samples were *a*–axis oriented YBaCuO films (both with 3000Å thickness) deposited on (110) $SrTiO_3$ substrates. The films were inspected using X–ray diffraction and found to be >90% of the preferred axis in all films used. The films all had four contacts (arranged linearly) made of silver paste and added during the final anneal process. The contacts were arranged in Kelvin connection fashion for the resistance, voltage, and electrical noise measurements. Figure 1(a) shows a typical sample layout. Table 1 shows a list of the samples used and their associated properties. Samples #2 & #3 were both fabricated from the same piece of bulk material with the same thickness and planar dimensions, and made consecutively in the same chamber in order to keep all parameters the same except the preferred axis of orientation.

EXPERIMENTAL METHOD

The range of experiments made on these samples ranged from resistance–temperature (*R–T*) over a wide current range, current–voltage (*I–V*) throughout the temperature range of the

Table 1: YBaCuO thin–film samples used

Film No.	Film Label	Axis Orient.	Thickness (Å)	Substrate Material	Substrate Orient.
1	AU15901	*c*	1400	$SrTiO_3$	(100)
1	DE109	*c*	3000	$SrTiO_3$	(100)
1	AP109D	*a*	3000	$SrTiO_3$	(110)
1	JU27A	*a*	3000	$SrTiO_3$	(110)

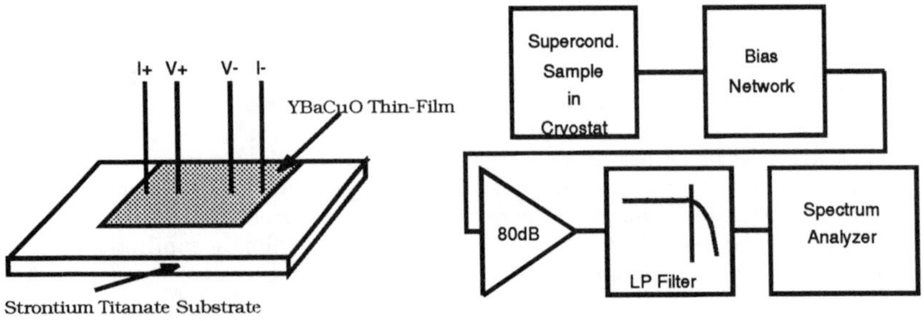

Figure 1: Sample configuration (a) and experimental setup (b) for electrical noise measurements.

transition region, and electrical noise measurements at several current levels also throughout the temperature range of the transition region and beyond.

The *R–T* measurements were made at current levels ranging from 1 mA up to the critical current of each sample. The results from Film #1 show a rather low transition temperature compared to Film #2 but it still had a relatively sharp transition.The discrepancies in these results are probably due partially to the rather small thickness of Film #1 and the fact that Film #1 was made a couple of months before the other films when the film growth process was in its start–up phase. The results for Films #3 & #4 were nearly identical as expected since they were of the same thickness and axis orientation, but had much more broader transitions than the *c*–axis films. Figure 2 shows a comparison of *R–T* curves at 10mA for Films #2 & #3 (the two films with all similar parameters except preferred axis.) The graph clearly shows that the transition temperature of the *c*–axis film is greater than that of the *a*–axis film, the *c*–axis film has a much sharper transition, and it also has a much lower normal state resistance. These three phenomena related to the films were predicted by Hickman, et al [7] using the Kosterlitz–Thouless theory showing the formation of elliptical vortices in anisotropic thin–films. Also, Penney, et al [8] did some preliminary measurements on single crystal YBaCuO and he found that the *a–b* plane (which is the *c*–axis film conduction plane) had a much lower resistivity than that of the *b–c* plane (*a*–axis conduction plane.)

The electrical noise measurements of these samples is probably the most important from a device development standpoint. The electrical noise measurements were taken as power spectra over a range from 1–51 Hz at various current levels and temperatures for each film. This process is the most time consuming and the most delicate so many precautions were taken to ensure that all the data used for analysis was "good". The current was biased in the circuit using a series combination of Gel cells at 24V with a large bias resistor (wire–wound) to block out the effects (if any) of contact noise. The current was applied and the measurement temperature was allowed to come to equilibrium before the power spectra was taken.

After all the data was collected for each sample, the noise voltage spectral density (in V^2/Hz) was extracted at 10Hz from each trace and plotted as a function of temperature to get a look at how the electrical noise levels change in the transition region. Figure 3 shows a typical graph of noise voltage spectral density (S_V) vs. temperature (T) for Film #1 at various current levels.

The graph shows that the S_V is not very "stable" in the transition region and this phenomena may be due to vortex "modes" in this part of the transition process. These peaks were evident in the other samples but not nearly as pronounced as in the thinnest film. Plotting the normalized noise voltage spectral density (S_V/V^2, where V^2 is the square of the voltage applied across the device under test) vs. temperature allows a better picture for comparison between samples since it helps to nullify the effects of geometry and resistance that affect the magnitude of the noise. Figure 2 shows a comparison between S_V/V^2 vs. T (along with the R–T curves for guide) of Films #2 & #3. From the graph, it is evident that the a-axis film has a much higher normalized noise than the c-axis film (about 4 orders of magnitude higher) which means the sensitivity of an a-axis film would be much less than that of a similar c-axis film. This result was predicted by Testa, et al [4], who hypothesized that the conduction in a c-axis film takes place in the metallic-like a–b planes whereas in an a-axis film the conduction takes place in the a/b–c planes where the mechanism of conduction is a semiconductor-like hopping behavior and semiconductors generally exhibit much higher noise than metallic conductors.

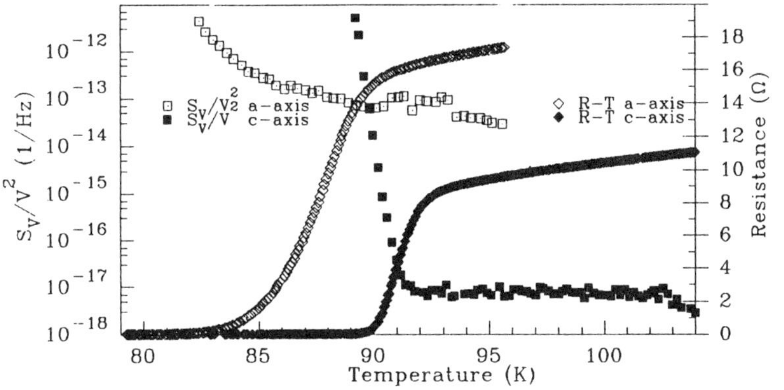

Figure 2: Comparison of R–T (at 10mA) and S_V/V^2 vs. T (at 10mA) graphs for YBaCuO thin-film with differing axis orientation.

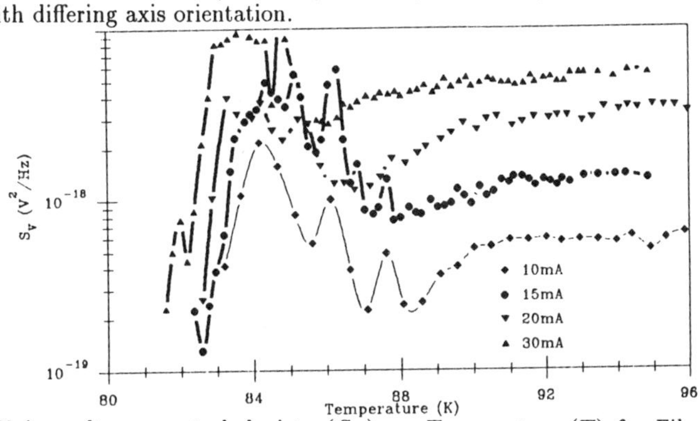

Figure 3: Noise voltage spectral denisty (S_V) vs. Temperature (T) for Film #1 at various current levels.

Much debate has also arisen from researchers (Black, et al [5]) who claim that the electrical noise observed in the films is merely a consequence of resistance fluctuations. In order to see if this is true, the normalized noise is plotted against β where $\beta = (1/R)\partial R/\partial T$ on a log–log scale. If the slope of this line is around two, then one can make the assumption that the measured noise is a result of resistance fluctuations. Figure 4 shows a plot of the typical S_V/V^2 vs. β for Film #1 at 10mA. The slope is much greater than 2 (about 5.68) and this was true for all samples measured. Table 2 shows the β values at 10mA for each of the samples. Rosenthal, et

Table 2: Table of β powers.

Film No.	Beta Power
1	5.68
2	4.40
3	7.68
4	8.46

Figure 4: S_V/V^2vs. β for Film #1 at 10mA.

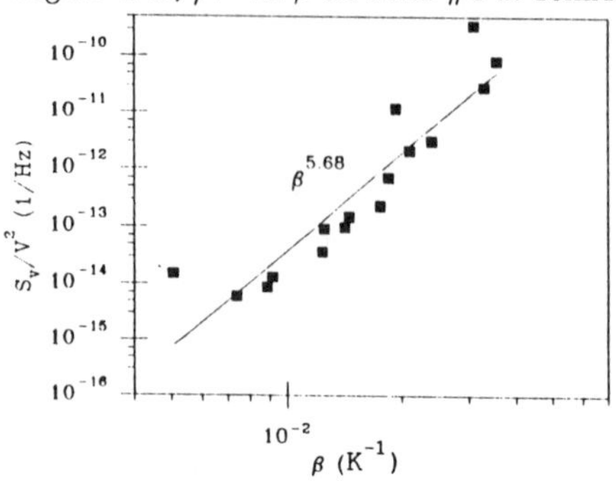

al [6] found slopes in his a-axis sample to be in the neighborhood of 30. So, it is reasonable to say that the noise in our samples was not a result of resistance fluctuations.

The data collected for the I-V measurements is very interesting and seems to fit the Kosterlitz-Thouless theory, but currently no relation with electrical noise measurements has been found, though we are pursuing theoretical work in this area.

CONCLUSIONS

Measurements were made on four YBaCuO superconducting thin-films in the R-T, I-V, and electrical noise areas. As predicted our a-axis films had a much lower transition temperature, a broader transition region, and much, much higher noise than comparable c-axis films. For this reason, our conclusion is that c-axis films are much better for applications such as bolometers and magnetometers where minimal noise is a defining property for device quality and a sharp transition is necessary. Also, the mechanism for the electrical noise is not clearly understood at this time, though current work with the I-V data and the Kosterlitz-Thouless theory may shed some light in that area. We are confident that the electrical noise is not simply a result of resistance fluctuations.

This work was partially supported by the DARPA High-Temperature Superconductor Initiative. The authors would like to thank Drs. Lou Testardi and C.L. Chang who provided the samples from the Florida State University. We would also like to thank Dr. H. Hickman and Dr. A.J. Dekker who helped provide the theoretical foundation for our experiments.

REFERENCES

1. J. Hall. *Noise Measurements of $YBa_2Cu_3O_7$ Thin-Film High-Temperature Superconductors.* PhD thesis, University of South Florida, Tampa, FL, May 1992.
2. H. Hickman, T.M. Chen, S.T. Liu, W.J. Wallace, "Low Frequency Electrical Noise in Bulk $YBa_2Cu_3O_7$". In *Proceedings of the 1989 IEEE SOUTHEASTCON*, p. 1057, Columbia, SC, April 1989.
3. J. Hall, H. Hickman, T.M. Chen. "Resistance Drift Noise in High T_C Superconductor Bolometers". *Solid State Communications*, **76**(7):921, 1990.
4. J.A. Testa, et al. "Noise Power Spectrum of Copper Oxide Superconductors in the Normal State". *Materials Research Society*, **99**:357, 1988.
5. R.D. Black, et al. "Thermal Fluctuations and $1/f$ Noise in Oriented and Unoriented YBaCuO Films". *Applied Physics Letters*, **55**(21):2233, 1989.
6. P. Rosenthal, et al. "Low Frequency Resistance Fluctuations in Films of High Temperature Superconductors". *IEEE Transactions on Magnetics*, **MAG-25**(2):973, 1989.
7. H. Hickman, T.M. Chen, A.J. Dekker. "Elliptical Flux Vortices in Polycrystalline YBaCuO". In *International Conference on the Advances in Materials Science and Applications of High-Temperature Superconductors*, NASA-GSFC, Greenbelt, MD, April 1990.
8. T. Penney, et al. "Strongly Anisotropic Electrical Properties of Single Crystal YBaCuO". *Physical Review B*, **38**:2918, 1988.

LOW-FREQUENCY EXCESS NOISE IN YBCO THIN FILMS NEAR THE TRANSITION TEMPERATURE

Sisi Jiang, Peter Hallemeier, and Charles Surya
Department of Electrical and Computer Engineering
Northeastern University
409 Dana Research Building
Boston, MA 02115

Julia M. Phillips
AT&T Bell Laboratories
600 Mountain Ave.
Murray Hill, NJ 07974

ABSTRACT

We conducted detailed systematic studies on the properties of low frequency noise in YBCO thin films near the transition region. Detailed studies on the dependences of the low frequency noise on the biasing current, $\partial R/\partial T$, and spatial correlation were conducted. It was shown that the measured voltage noise power spectra were proportional to I^2 and $(\partial R/\partial T)^2$, and were correlated over a spatial separation of 300 μm. Also, the voltage noise power spectral densities exhibit a lower cutoff frequency of 5 Hz, in excellent quantitative agreement with the cutoff frequency predicted by the Thermal Fluctuation model. The experimental results provide strong evidence that the low frequency excess noise in the device originated from equilibrium temperature fluctuations.

INTRODUCTION

In this paper we examine the underlying mechanism of low frequency excess noise in YBCO thin films on $LaAlO_3$ substrates near the transition temperature. Prompted by the success of the Temperature Fluctuation model [1, 2] in conventional superconductors at the transition region, we conducted detailed studies to examine the validity of the model in our samples.

The Thermal Fluctuation model states that low frequency excess noise arises from local temperature fluctuations in the device which modulate the device conductance according to the parameter $\beta = \partial R/\partial T$. First principle calculation of the temperature noise power spectral density was shown to be

$$S_T(f) \propto \begin{cases} Constant & f \ll f_1 \\ \ln(1/f) & f_1 \ll f \ll f_2 \\ 1/f^{0.5} & f_2 \ll f \ll f_3 \\ 1/f^{1.5} & f_3 \ll f, \end{cases} \quad (1)$$

in which

$$f_i = D/4\pi L_i^2, \quad i = 1,2,3 \quad (2)$$

where L_1, L_2 and L_3 are the length, width and thickness of the device respectively, and D is the thermal diffusivity of the material. The voltage noise power spectral density for a device under constant current bias was shown to be

$$S_V(f) \propto \frac{I^2\beta^2}{\Omega} S_T(f) \tag{3}$$

where I is the current bias, and Ω is the volume of the sample. In addition, it was shown that temperature fluctuations follow a diffusion mechanism and the noise is thus correlated over a distance given by $L_c = \sqrt{D/\pi f}$, where L_c is the correlation length.

EXPERIMENT AND RESULTS

The samples were provided by AT&T Bell Laboratories, fabricated by e-beam co-evaporation on $LaAlO_3$ substrates. Systematic measurements of voltage noise from two segments of a YBCO micro-strip line were conducted with the device under constant current bias. The fluctuating voltages were coupled to two PAR1900 transformers before being amplified by the PAR113 low noise pre-amplifiers. The amplified noise was then fed to the HP35665A dual channel dynamic signal analyzer for spectral and correlation analyses. The entire experimental set up was placed on a vibration isolation table located in a shielded room. Temperature stability of the samples was maintained to within 50 mK.

The T_C's of the samples we studied were about 91.8 K. Typical R–T and $\partial R/\partial T$–T characteristics are shown in Fig. 1. In which $\partial R/\partial T$ was found to peak at $\sim$ 91.35 K. Typical voltage noise power spectral densities exhibit a lower cutoff frequency at $\sim$ 5 Hz, below which the noise power spectra become constant as shown in Fig. 2.

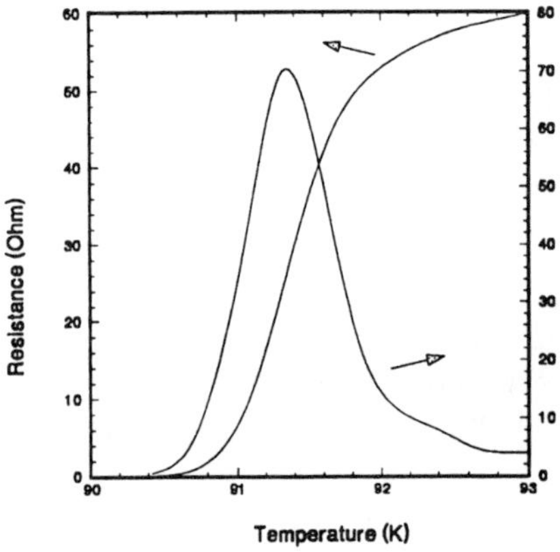

Fig. 1 The temperature dependence of the resistance, R, and $\partial R/\partial T$ in YBCO thin film at $I = 50\mu$A.

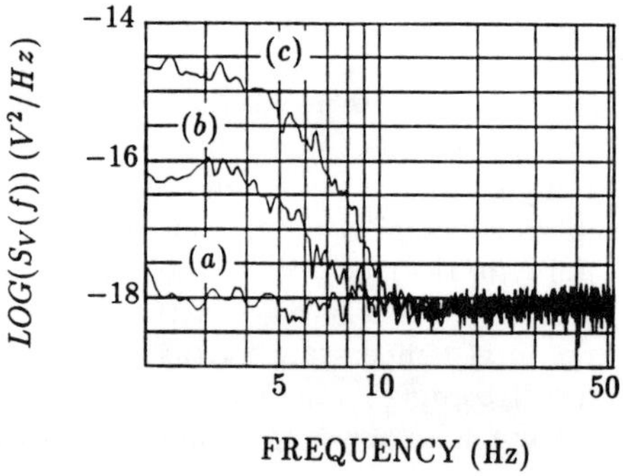

Fig. 2 Voltage noise power spectral densities $S_V(f)$ at T = 91.65 K for (a) the background noise; (b) $I = 10\mu$A; and (c) $I = 50\mu$A.

Little change was observed in the cutoff frequency over the entire temperature and biasing range of the experiment. The device temperature for this case was 91.65 K and curves (a), (b) and (c) correspond to voltage noise power spectral densities at I = 0, 10, and 50μA respectively. Using published values of the thermal diffusivity,

$D \approx 2\times 10^{-2}\mathrm{cm^2s^{-1}}$ at 90 K [3], we computed the lower cutoff frequency following Eq. 2 for a segment length $L_1 = 150\mu\mathrm{m}$. We found that $f_1 \approx 7$ Hz, which has an excellent agreement with our experimentally observed cutoff frequency. The magnitudes of the voltage noise power spectra were found to vary as I^2 as shown in Fig. 3 in which we have presented the results for three different temperatures — 91.04, 91.18, and 92.16 K. The I^2 dependence of the voltage noise power spectra shows that the noise originated from an equilibrium process as in the case of local temperature fluctuations. To further investigate the validity of the model we studied the dependence of $S_V(f)$ on β at different current biases. Typical results are shown in Fig. 4 in which β^2 dependence of the voltage noise power spectra for a wide range of current biases was observed.

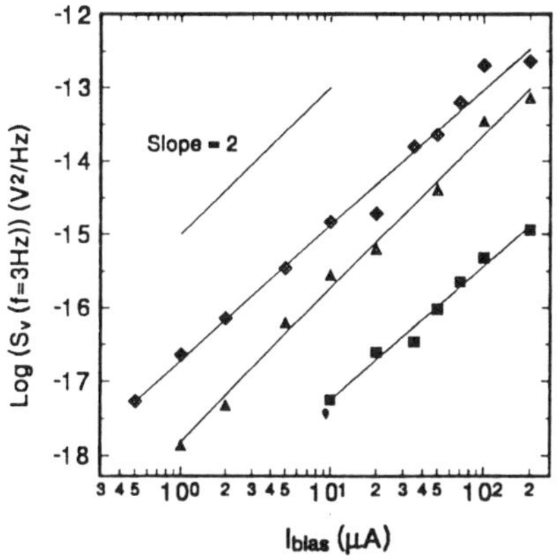

Fig. 3 Voltage noise power spectral densities $S_V(3Hz)$ versus I^2 at $T = 91.04$ K (solid triangle), 91.18 K (solid diamond) and 91.65 K (solid square).

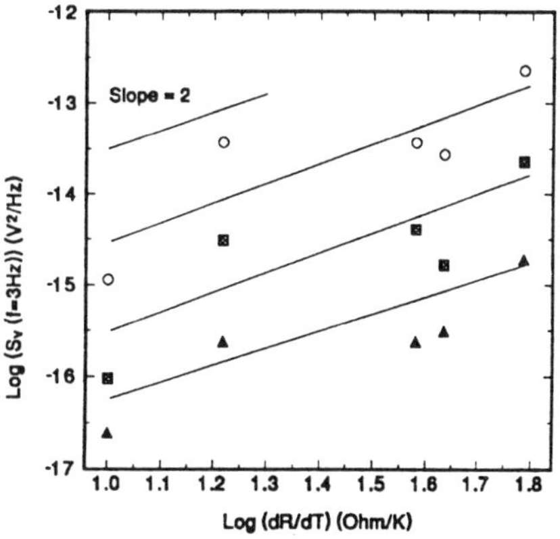

Fig. 4 Log $(S_V(3Hz))$ versus Log $(\partial R/\partial T)$ for $I = 20\mu$A (solid triangle), 50 μA (solid square), and 200 μA (open circle).

One of the most important tests for Temperature Fluctuation model is the measurement of spatial correlation in the noise. This is because equilibrium temperature fluctuation was shown to be governed by a diffusion mechanism [2], resulting in the spatial correlation of the fluctuations over a distance, L_c. To investigate this phenomenon we conducted detailed experiments to study the coherence factor of the low frequency noise over two different segments in a sample separated by a distance of 300 μm. The frequency dependent coherence factor is given by

$$C(f) = \frac{S_{VC}^2(f)}{S_{V1}(f)S_{V2}(f)}, \tag{4}$$

where $S_{VC}(f)$ is the voltage cross-spectral density, and $S_{V1}(f)$ and $S_{V2}(f)$ are the voltage noise power spectral densities of the two segments. The coherence factor of the low frequency noise presented in Fig. 5, in which curves (a) and (b) correspond to coherence factors of the device at $T = 91.18$ and 91.65 K respectively for $I = 2$ μA. The drop in the coherence factor for $f > 7$ Hz was due to the domination by the system noise. The experimental results clearly indicate that the low frequency excess noise over the two segments were correlated. This demonstrates that the low frequency noise was governed by a diffusion mechanism as predicted by the Thermal Fluctuation model.

It is interesting to compare our experimental results with quantum $1/f$ noise theory. Using a Hooge's parameter of 2×10^{-3} we obtained a fundamental noise power spectral density of $10^{-21}V^2/Hz$, for $I = 2\mu A$ and $f = 5$ Hz at 91.18 K, about four orders of magnitude smaller than our measured noise level.

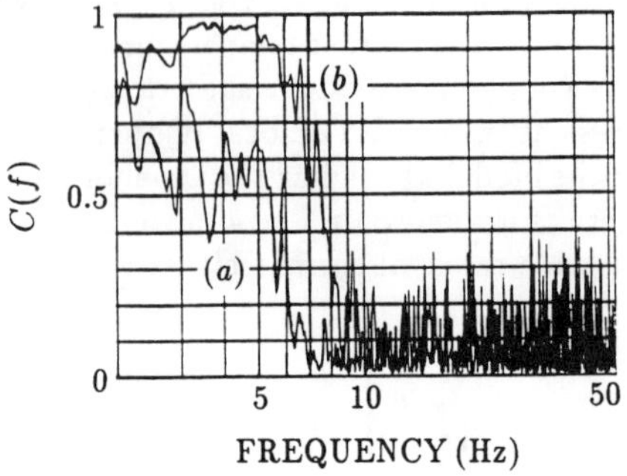

Fig. 5 The coherence factor versus frequency at $I = 2\mu A$ for $T = 91.65$ K (a), and 91.18 K (b) respectively.

CONCLUSION

We have conducted experimental studies on the low frequency noise in YBCO thin films near the transition temperature. Our results show that the voltage noise power spectra vary as I^2 and $(\partial R/\partial T)^2$. Moreover, the noise was found to be correlated over a distance of 300 μm. These phenomena can only be explained by the Temperature Fluctuation model. Also, an excellent quantitative agreement between the experimental and the computed cutoff frequency based on the Temperature Fluctuation model has been established. Our experimental results provide strong evidence that low frequency noise in YBCO thin films near the transition region originated from equilibrium local temperature fluctuations.

ACKNOWLEDGEMENT

This work is supported by NSF grant ECS-9102396.

References

[1] R.F. Voss and J. Clarke, Phys. Rev. **B13**, 556 (1976).

[2] J. Clarke and T.Y. Hsiang, Phys. Rev. **B13**, 4790 (1976).

[3] L.S. Fletcher, G.P. Peterson, and R. Schaup, J. Heat Transfer **113**, 274 (1991).

COMPARATIVE LOW-FREQUENCY NOISE STUDIES OF YBaCuO FILMS

I. A. Khrebtov, V. N. Leonov, A. D. Tkachenko
S. I. Vavilov State Optical Institute, St.Petersburg, 199034

A. V. Bobyl, V. Yu. Davydov, V. I. Kozub
Ioffe Physico-Technical Institute, St.Petersburg, 198021

ABSTRACT

Comparative noise studies of YBaCuO films produced by laser evaporation and magnetron sputtering on $SrTiO_3$, $LaAlO_3$, $NdGaO_3$, Al_2O_3, YSZ, MgO/$BaSrTiO_3$, Si/ZrO_2 substrates have been carried out. Noise spectra in the frequency range 1 Hz–1 MHz, temperature dependence of noise in temperature range 78–300 K, current, resistance and magnetic field dependences have been measured. Two components of excess $1/f$ noise have been distinguished. The possible origins of noise are discussed.

INTRODUCTION

Successful results in HTSC applications in low-current cryoelectronics are basically attributed to low-noise films manufactured from those materials.[1] The present work was aimed to comparative noise studies of high-quality YBaCuO films manufactured by various methods with substrates of different types. The films were previously selected according to their structural quality. The applied nature of this work is related to the developments of new bolometers of nytrogen cooling level which determined the choice of YBaCuO films deposited on substrates made of the most prospective materials for bolometer manufacturing.[2]

SPECIMENS AND EXPERIMENTAL TECHNIQUES

The films $YBa_2Cu_3O_{7-x}$ have been investigated. The films on $SrTiO_3$(100), Al_2O_3(1012), $NdGaO_3$(100) and YSZ(100) substrates have been manufactured by laser deposition method. The films on MgO(100) substrates with $BaSrTiO_3$ buffer layer, Si(100) substrates with ZrO_2 buffer layer and $LaAlO_3$(110) substrates have been produced by magnetron sputtering method. Investigations of phase composition and structure quality of films have been carried out by x-ray microanalysis, x-ray structure analysis, electron microscopy and Raman light scattering methods. While using those methods epitaxial films with high-structure quality have been chosen. Films on Al_2O_3, $NdGaO_3$, YSZ, $SrTiO_3$ and Si/ZrO_2 substrates had c axis oriented perpendicular to substrate plane. Films manufactured on $LaAlO_3$ substrates had c- and $b(a)$-axes oriented at 45° to substrate plane and having, as a result, a high macroscopic anisotropy of transport properties. Films on MgO substrates had c-axis lied in substrate plane. The film thicknesses are ~1500–2000 Å. Other parameters of test specimens of meander-type (No 1-3) and bridge-type (No 4-9) are presented in Table 1. Specimens were placed in vacuum cavity of metallic liquid nytrogen cryostat having temperature stability by 10^{-2} K. Noise measurements have been carried out with the help of four-probe method. The dependences of the noise on current, magnetic field and resistance in 78–300 K temperature range and at frequencies 1 Hz–1 MHz have been studied.

MEASUREMENT RESULTS AND THEIR DISCUSSION

It is seen at figure 1 that a typical temperature dependence of the normalized noise spectral density within $T = 300$ K to critical temperature T_c has the form of $S_V/V^2 \sim T^\beta$, where S_V is noise power density, V is voltage drop across specimen, $\beta = 1.3$–3. In the superconducting transition region with temperature and resistance decrease noise has a further decrease (figure 2). However with further temperature decrease noise is beginning to increase reaching the maximum at a certain resistance $R_m \leq 0.5\,R_n$ (R_n is normal state resistance). Position of this maximum does not coincide with $\mathrm{d}R/\mathrm{d}T$ maximum. With current and magnetic field increase this noise increases; it can be even higher than the noise at normal state. For films of higher structural quality and smaller resistivity ρ having more narrow superconducting transition, the noise maximum has been observed at smaller temperatures and has been more sensitive to magnetic field (figure 2).

Table 1. Parameters of YBaCuO films

No	Substrate material	A (mm^2)	$\rho(300)$ (Ω cm)	$T_c(R_n/2)$ (K)	ΔT (K)	α $R(300)$	α $R_n/2$
1	$SrTiO_3$	40×0.025	90	88	1.5	1.3	0.48
2	$SrTiO_3$	90×0.05	70	90	1.6	33	1.4
3	$MgO/BaSrTiO_3$	10×0.1	600	85.1	2.2	0.54	2.3
4	Al_2O_3	0.5×0.035	900	85.5	1.4	1.4	0.12
5	$LaAlO_3$	0.5×0.025	2200	86	4.0	12	0.6
6	Si/ZrO_2	0.5×0.035	920	88.3	4.0	7.2	0.31
7	YSZ	0.01×0.007	1660	85.5	5.4	9.2	0.18
8	$NdGaO_3$	0.014×0.004	960	90.3	1.5	3.4	0.12
9	Si/ZrO_2	0.5×0.05	5000	88.5	9	1570	220

For bridges No 5 and No 7 fine structure of noise temperature curves is observed at transition edge resistance, i.e. the presence of a few noise maxima with amplitudes depending on temperature, current and magnetic field.[3] The noise voltage $S_v^{1/2}$ had linear dependence on current at room temperature, at normal state and at transition tail. However in the last case that dependence was weaker.

Noise spectra for all films from $T = 300$ K to T_c are close to $S_v \sim f^{-\gamma}$ behavior where $\gamma \simeq 1$. The frequency dependence of noise is more complex at the transition tail where noise maxima are observed. For microbridges[3] coefficient γ has varied in the range 0,5–2. Flicker noise of microbridges on $NdGaO_3$ substrate (No 8) has extended to the frequencies 10–20 kHz. At the higher frequencies noise has been frequency independent and can be described as $S_v \sim (I\,dR/dT)^2$, where I is current. Noise level was close to thermal noise level stipulated by fluctuations of heat flow to substrate.

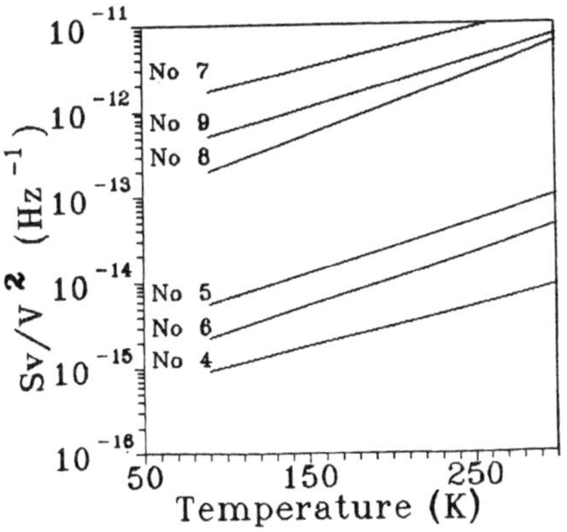

Fig.1. Normalized noise spectral density of YBaCuO films versus temperature at 12 Hz.

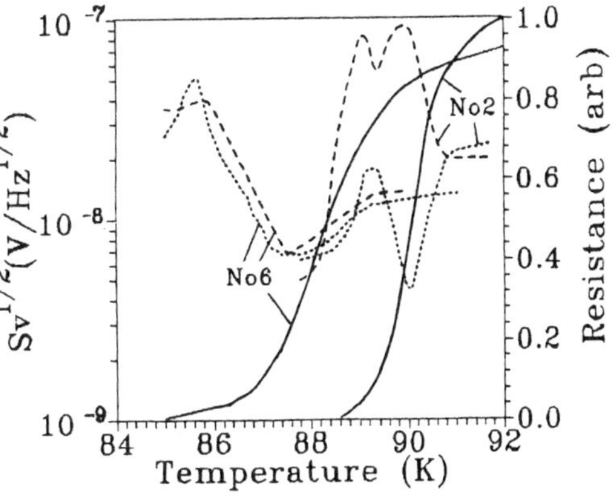

Fig.2. Temperature dependences of the resistance (——) and noise voltage at 12 Hz with $B = 0$ (···) and $B = 34$ mT (– – –). No 2—$I = 1$ mA, $R_n = 3.5\,k\Omega$; No 6—$I = 1$ mA, $R_n = 270\,\Omega$.

Hooge's coefficient α was used to compare noise level for different films (see Table). For calculation of α we have assumed the carrier density to be $n = 5 \times 10^{21}$ cm^{-3}. It is seen that at $T = 300$ K for high-quality films $\alpha = 0.5$–1.4. In transition range at $R_n/2$ where bolometers usually operate coefficient α decreases by more than the order of magnitude. For films with defect structure (No 9) α is much larger. Probably in course of deposition of YBaCuO film the defects of intergrain boundary type arose due to microcracks in buffer layer ZrO_2. As a result resistivity ρ has increased, superconducting transition has broadened and noise has increased as well. We believe that component of excess noise in normal state is stipulated by structural fluctuation processes in the vicinity of extended structural defects like intergrain boundaries.[4] In this case HTSC film can be regarded as an irregular-structure material with slow-relaxing excitations of the atomic nature exhibiting exponentially-broad relaxation time distribution and thus leading to $S_v \sim f^{-1}$ dependence.[5]

At the tail of transition region where an infinite superconducting cluster is formed and the resistance is formed by vortices flow one deals with another noise component related to the vortices motion. Such a noise should obviously depend on the current and on external magnetic field. Two situations are possible. On the one hand, there are weak Josephson links at the boundaries between grains providing a motion of Josephson vortices. With the increase of current and magnetic field these links are suppressed, i.e. the number of fluctuators decreases and the noise stipulated by these mechanisms decreases as well. That mechanism will be obviously more vividly manifested in the granular films with larger ρ that we observed. In epitaxial high-quality structure films when the sizes of microcrystals are larger and the number of weak links is smaller, the noise is stipulated by the fluctuations of a number of moving Abrikosov vortices due to its trapping and detrapping by pinning centers.

We can assume that for spectral fluctuation density of a number of vortices in the specimen N the following estimation can be obtained:

$$(N,N)_\omega \equiv S_\omega \sim N \frac{\tau}{1+(\omega\tau)^2} \frac{\tau_f \tau_t}{(\tau_f+\tau_t)^2} \quad (1)$$

where $\tau^{-1} = \tau_f^{-1} + \tau_t^{-1}$, τ_f and τ_t are time periods of the vortices in free and pinning states. $\tau_f \sim (n_c \sigma V_v)^{-1}$, where n_c and σ are the concentration and cross-section of pinning centers, respectively, and V_v is characteristic vortice motion speed. $\tau_t = \tau_0 e^{(E_B/T)}$, where E_B is the pinning center activation energy. The noise maximum corresponds to the situation when τ_f and τ_t are comparable. The fine structure of temperature noise dependence is probably related to the existence of percolation superconducting cluster provided by local fluctuations of T_c values. In this case the noise maximum is due to a some narrow critical region in such structure providing the increased noise level. The maximum correspond to a temperature of formation of such a spot where its size is minimal.

SUMMARY

Epitaxial YBaCuO films with optimum stoichiometric composition have Hooge'coefficient $\alpha = 0.5$–12 at $T = 300$ K. For the best specimens in the temperature region where dR/dT has maximum $\alpha = 0.12$–2.3. At the transition tail ($R < 0.5R_n$) a noise maximum is observed. That noise for some films was larger than in normal state. It is supposed that at normal state as well as at the beginning of superconducting transition noise is produced by some structural defects with internal degree of freedom situated supposedly near grain boundaries while at the tail of transition noise is related to vortices motion. Noise quality of the films studied appear to provide a possibility to produce the bolometers of high quality.[2].

REFERENCES

1. R. C. Lacoe et al, IEEE Trans. Magn. **27**, 2832 (1991).
2. I. A. Khrebtov, Supercond. Phys. Chem. Technol. **5**, 558 (1992).
3. V. N. Leonov and I. A. Khrebtov, Supercond. Phys. Chem. Technol. **4**, 1760 (1991).
4. Y. Song, A. Misra, P. P. Crooker, and J. R. Gaines, Phys. Rev. Lett. **66**, 825 (1991).
5. Yu. M. Galperin, V. L. Gurevich, V. I. Kozub, Fiz. Tverd. Tela. **31**, 155 (1989).

INFLUENCE OF THE OXYGEN CONTENT ON THE 1/f NOISE PROPERTIES OF $YBa_2Cu_3O_{7-x}$ CERAMICS

A. N. Lavrov, V. E. Fedorov
Institute of Inorganic Chemistry, Russian Academy of Sciences, 3, Acad. Lavrentiev pr., Novosibirsk, Russia

A. G. Cherevko, M. P. Tigunov
Institute of Communication, 86, Kirov st., Novosibirsk, Russia

ABSTRACT

Measurements of the noise properties were carried out on the $YBa_2Cu_3O_{7-x}$ ceramics with different oxygen contents. The obtained value of the Hooge parameter enhanced by two orders of magnitude with increasing of the oxygen content from 6.25 to 6.38. This change of the oxygen stoichiometry induced also the phase transition from the tetragonal to orthorhombic. The enhancement of the noise magnitude is proposed to be due to the formation of the orthorhombic twin boundaries.

INTRODUCTION

The copper oxide high T_c superconducting materials, both ceramics and single crystals, have an extremely large magnitude of 1/f noise spectral density in the normal state[1,2]. Excess noise would limit the ultimate sensitivity of many potential high T_c based devices such as DS-SQUIDs, infrared detectors. To date the dominant noise source has not been unambiguously identified because a wide range of various structural defects exists in this materials. A quantity and structure of defects in the $YBa_2Cu_3O_{7-x}$ compound can be essentially modified by changing of the oxygen stoichiometry and post-preparation low temperature aging[3,4]. That allows to obtain an additional information on the origin on the anomalous noise properties.

In this paper we present a study of the influence of the oxygen content on the noise properties of $YBa_2Cu_3O_{7-x}$ ceramic samples.

EXPERIMENTAL METHODS

The starting material was the ceramic with the $YBa_2Cu_3O_{6.96}$ composition prepared by conventional sinter-processing procedures. The stable low resistance contacts for the electrical measurements were obtained by annealing of the samples with attached silver pads. Then the oxygen stoichiometry of the samples was reduced by annealing in an appropriate air-helium mixture and fixed by subsequent quenching into liquid nitrogen. The oxygen content was determined from chemical titration measurements.

Noise measurements were carried out by the six-probe cross-correlation method with two current and four potential contacts. A battery generated DC-current (1-300 mA) was passed through the sample and ballast resistor. The noise signals taken from both pairs of potential contacts were amplified by two independent channels and after that the correlative components of the signals were extracted. That allowed to exclude the contact noise contribution. In order to improve the accuracy 2000 measurements were done for every set of parameters with subsequent averaging.

The temperature dependence of the noise spectral density was measured in the wide frequency band ($1-10^3$ Hz) with a resolution of 0.5 Ohm in units of equivalent heat noise resistively. The frequency dependence of the noise power was measured from 0.5 Hz to 2 kHz by an CK4-72 signal analyzer. The noise power obtained with switched off current was used as the background noise and subtracted from the data.

RESULTS AND DISCUSSION

The noise measurements were carried out at different bias currents. For all samples the obtained magnitude of the $1/f^n$ noise power depends on the DC-voltage across the sample V as $S_V \propto V^2$ (or equivalently $S_V \propto I^2$), see Fig. 1, evidencing that the noise signal originates from resistivity fluctuations.

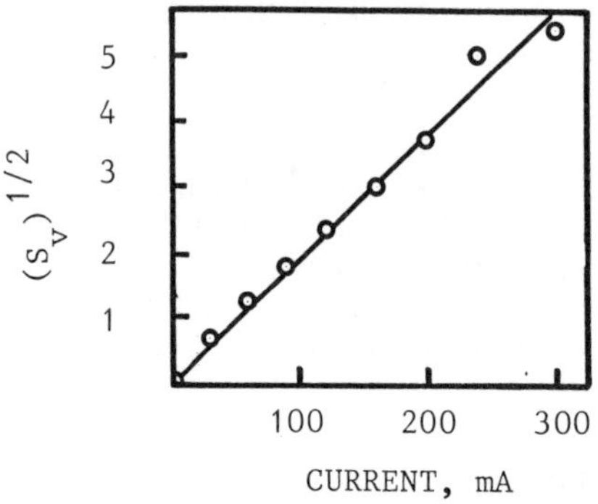

Fig.1. Current dependence of noise voltage $(S_V)^{1/2}$ for the $YBa_2Cu_3O_{6.38}$ sample.

Measurements revealed that both the magnitude and the temperature dependence of the excess noise are strongly affected by the oxygen content of the samples. The magnitude of the excess noise in the tetragonal phase $YBa_2Cu_3O_{6.25}$ ceramic does not depend essentially on temperature, see Fig. 2. The estimated value of the Hooge parameter for this sample averaged over the frequency band ($1-10^3$ Hz) was $\alpha_H=(0.9 \pm 0.1)\cdot 10^4$.

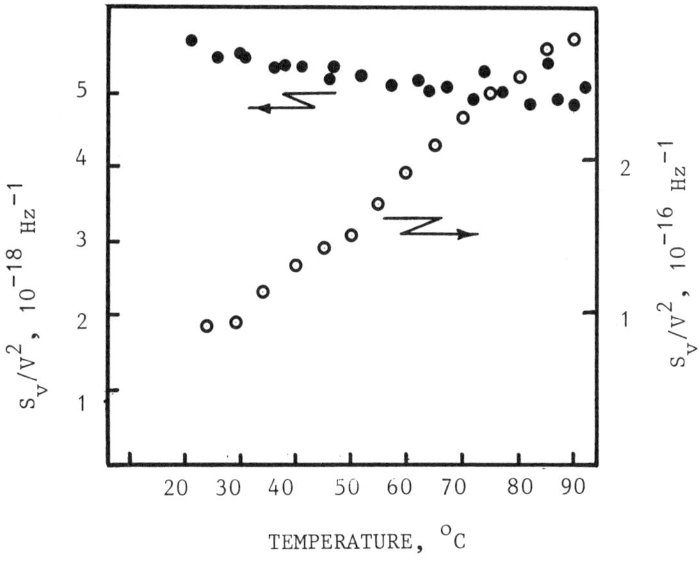

Fig.2. Temperature dependence of the normalized excess noise for $YBa_2Cu_3O_{6.25}$ (solid circles) and $YBa_2Cu_3O_{6.38}$ (open circles) samples.

The Hooge parameter for the orthorhombic sample $YBa_2Cu_3O_{6.38}$ was found to be much higher and temperature dependent, Fig. 2. It increases with temperature from $\alpha_H=(4.5 \pm 0.3)\cdot 10^5$ at T=300K to $\alpha_H=(1.3 \pm 0.1)\cdot 10^6$ at T=363K. The obtained data of the Hooge parameter are several orders of magnitude larger than for conventional metals and coincide with values reported for the copper-oxide ceramics.

Measurements of the frequency dependence of noise power spectral density have shown that properties of the $YBa_2Cu_3O_{6.25}$ sample differ from those of the oxygen-rich samples. While the parameter $n = \partial \ln S_v / \partial \ln f$ in oxygen-rich samples was close to 1.0 (n=1.0-1.1) [1,2], a higher value of n was observed for the $YBa_2Cu_3O_{6.25}$ sample, Fig. 3. The value of $S_v(f)$ was measured for 600 frequency points at fixed temperatures. The S_v versus f dependence for $YBa_2Cu_3O_{6.25}$ sample can be well described in the range $0.5-2\cdot 10^3$ Hz by the same fit for all the temperatures: $S_v \propto 1/f^n$, $n=1.27 \pm 0.06$.

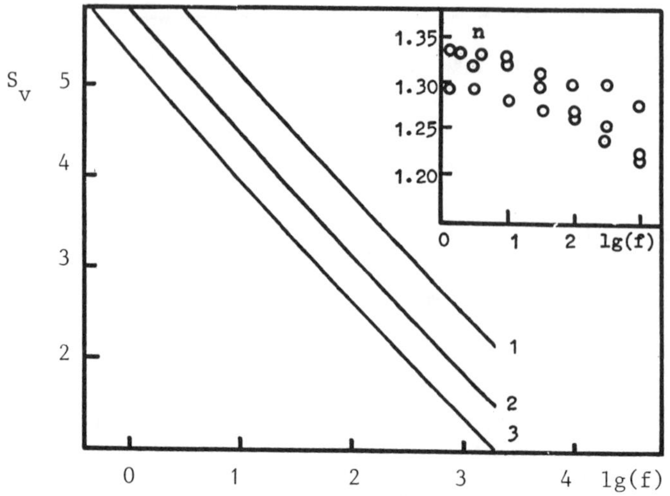

Fig.3. Frequency dependences of the noise power spectral density for the $YBa_2Cu_3O_{6.25}$ sample measured at temperatures 295K (1), 333K (2), 363K (3). The inset shows the frequency dependence of the parameter $n= \partial \ln S_V(f)/ \partial \ln f$.

All the studied samples originate from the same pellet, thus they have a similar granular and impurity structures. Nevertheless their noise spectral densities differ by a factor of 10^2. We propose that the effects induced by the change of the structural type are responsible for the difference in the noise properties. Orthorhombic samples have an additional set of the structural formations - ordered orthorhombic domains separated by the domain or twin boundaries. The enhanced noise in orthorhombic samples may be due to the resistivity fluctuations generated at the twin boundaries that coincides with results of Ref. 1.

At the temperatures above 290-300K the increased mobility of the oxygen ions allow the oxygen rearrangement processes to occur [3,4]. The degree of ordering was shown to be temperature dependent [3,4]. Thus the temperature dependence of the noise power observed in the $YBa_2Cu_3O_{6.38}$ sample may be attributed to the evolution of the twin structure with temperature caused by the oxygen rearrangement.

1. Y.Song, A.Misra, P.P.Crooker, J.R.Gaines, Phys.Rev.-Lett. 66,825(1991).
2. J.P.Zheng, Q.Y.Ying, S.Y.Dong, H.S.Kwok, S.H.Liou, J.Appl.Phys. 69,553(1991).
3. J.D.Jorgensen, S.Pei, P.Lightfoot, H.Shi, A.P.Paulikas, B.W.Veal, Physica C167,571(1990).
4. I.N.Kuropyatnik, A.N.Lavrov, Physica C197,47(1992).

LOW-FREQUENCY NOISE, ELECTRIC AND MAGNETIC CHARACTERISTICS OF Bi-BASED THICK FILMS IN THE SUPERCONDUCTING TEMPERATURE REGION

V. Palenskis, A. Stadalnikas, J. Juodviršis
Vilnius University, Vilnius 2054, Lithuania

B. Vengalis
Semiconductor Physics Institute, Vilnius 2600, Lithuania

ABSTRACT

The low-frequency noise characteristics in Bi-based superconductor thick films have been measured as a function of frequency, temperature, bias current, and applied magnetic field. These results have been compared with the resistance, temperature coefficient of the resistance dependences on temperature, and also with ones of Y-based thick film superconductor. Our investigation show that the peak of the voltage noise is not related with the peaks both of the temperature coefficient of the resistance and the voltage-to-flux transfer function. We suppose that voltage fluctuations are taking place at grain boundaries (weak Josephson links) between superconducting and normal grains, and that depend on the relative filling of the film by the superconducting grains. We represent a model that explains the resistance fluctuation peak in the superconducting transition temperature region.

INTRODUCTION

The investigations of the fundamental fluctuation processes in high-temperature superconducting materials are very useful on the standpoint to obtain an additional information on superconductive transition mechanism. In many cases it is believed that voltage fluctuation peak in the transition temperature region arises due to flux Φ fluctuations and that there is an ordinary relation between spectral densities: $S_U=S_\Phi(dU/d\Phi)^2$. On the other hand, there are also many studies on the ground of conductance fluctuations. Our earlier investigations[1-3] on the base of $Y Ba_2 Cu_3 O_{7-x}$ thick films show that voltage fluctuations can be explained by the grain boundary (Josephson weak link) resistance fluctuations. In this paper we represent the analogical data for superconducting $Bi_2 Sr_2 Ca Cu_2 O_{8-x}$ thick films and give a model for explanation of the resistance fluctuation peak in the superconducting transition temperature region.

EXPERIMENTAL RESULTS

The thick $Bi_2 Sr_2 Ca Cu_2 O_{8-x}$ (BSCCO) films were fabricated onto the single

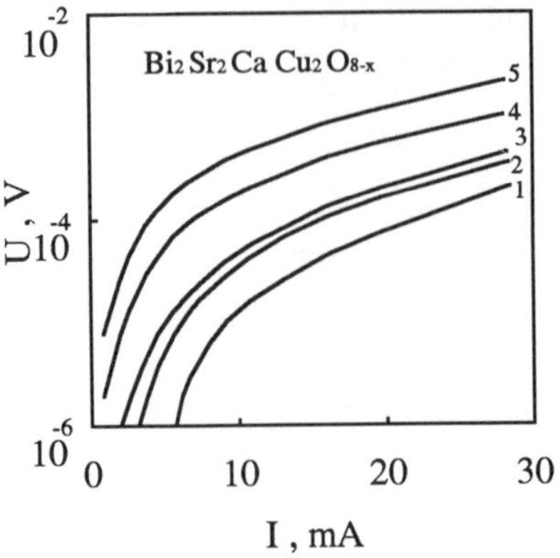

Fig. 1. Current-voltage characteristics of Bi-based thick films at various temperatures T, K: 1-77.3; 2-78.6; 3-79.1; 4-80.8; 5-82.2.

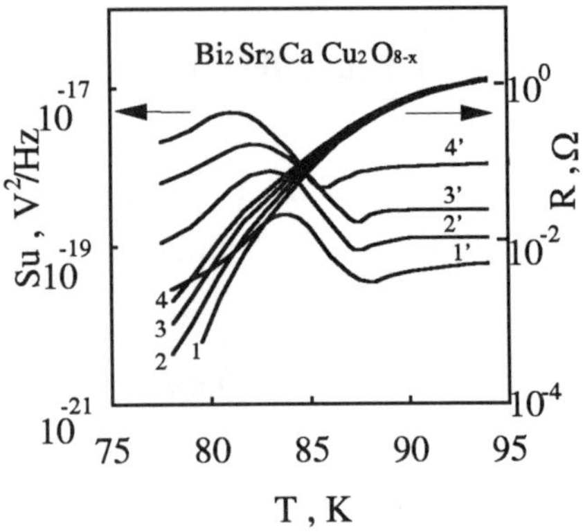

Fig. 2. The voltage noise spectral density Su (1'-4') and the resistance R (1-4) dependences on temperature at various dc currents I, mA: 1,1'-2.4; 2,2'-4.1; 3,3'-6.8; 4,4'-9.9.

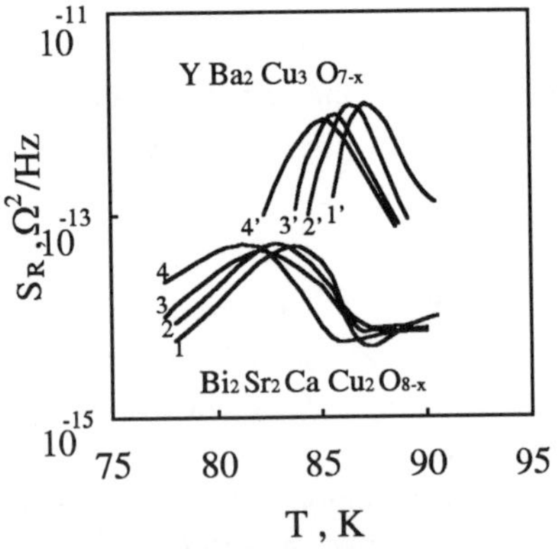

Fig. 3. The resistance fluctuation spectral density S_R dependences on temperature at various dc currents I, mA: for Bi-based films 1-2.4; 2-4.1; 3-6.8; 4-9.9; for Y-based films 1'-2.9; 2'-5.1; 3'-6.9; 4'-10.1.

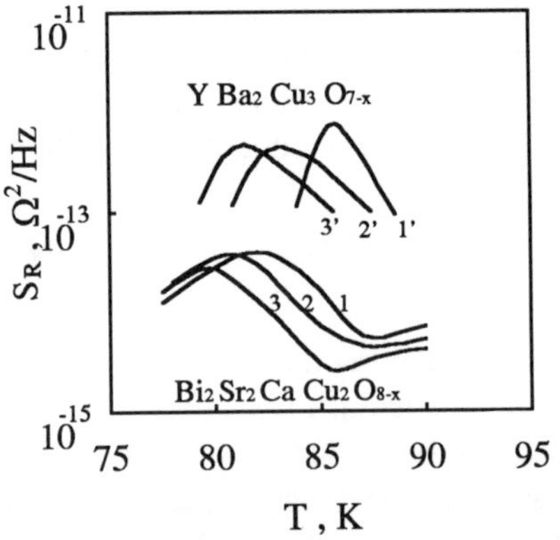

Fig. 4. The resistance fluctuation spectral density S_R dependences on temperature at various magnitudes of the magnetic induction B, mT: for Bi-based films (I=6.8 mA) 1-0; 2-1.33; 3-2.67; for Y-based films (I=8 mA) 1'-0; 2'-0.67; 3'-1.33.

crystal MgO (100) substrate by using superconducting BSCCO (2212) ceramic powders . That superconducting powders have been alloyed on the substrate at temperature 1340 K and annealed during 3 h at temperature 1110 K . We obtain

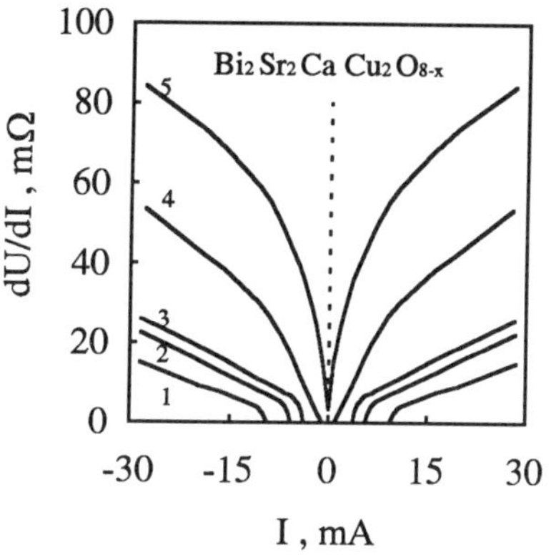

Fig. 5. The differential resistance dU/dI dependences on dc current at various temperatures T, K: 1-77.3; 2-78.6; 3-79.1; 4-80.8; 5-82.2.

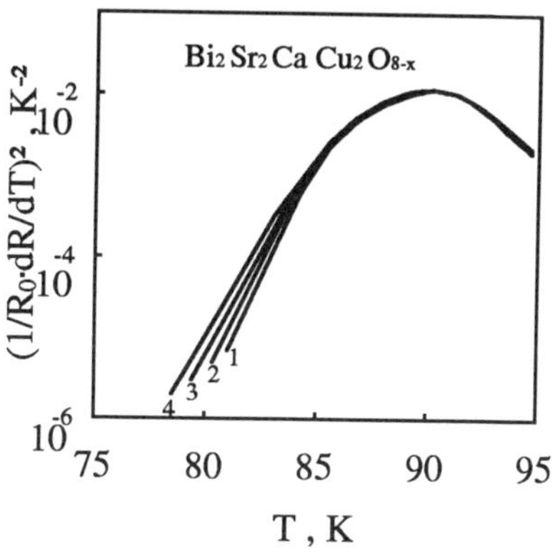

Fig. 6. The square of the temperature coefficient of the resistance dependences on temperature at various dc I, mA: 1-2.4; 2-4.1; 3-6.8; 4-9.9.

superconducting BSCCO films about 20 μm thickness. The contact are prepared by silver film evaporation. All measurements performed by four-probe technique in termostat placed in solenoid and screened to prevent Earth magnetic field influence. Noise spectra have been measured from 20 Hz to 1 kHz, they have 1/f type frequency dependence. In this paper all voltage fluctuation spectral density Su data are given at 30 Hz.

The current-voltage characteristics of BSCCO thick films at various temperatures in the transition region are presented in fig.1, they, in general, show SNS (superconductor-normal material-superconductor) type behavior. The voltage fluctuation spectral density Su and the resistance R dependences on temperature at different dc currents I are shown in fig.2. The square dependence of Su on I suggests that the noise arises from the resistance fluctuation, which spectral density S_R can be expressed by $S_R=Su/I^2$. The resistance fluctuation spectral density S_R dependance on temperature both at different dc currents and at different magnetic fields are presented in fig.3 and 4, respectively. For comparison in these figures there are shown analogical data for $Y B_2 Cu_3 O_{7-x}$ thick films[3]. Investigations both of differential resistance dU/dI (fig.5) and temperature coefficient of resistance $(1/R_0)(dR/dT)$ (fig.6) and comparison these results with fluctuation measurements did not show direct mutual correlations, i.e. low-frequency noise is not produced by thermal fluctuations.

DISCUSSION

In earlier our papers[1-3] we suggested that the fluctuation peak originates from the resistance fluctuation and not from magnetic flux noise in the film, i.e. low-frequency noise in the superconducting transition temperature region is caused by the

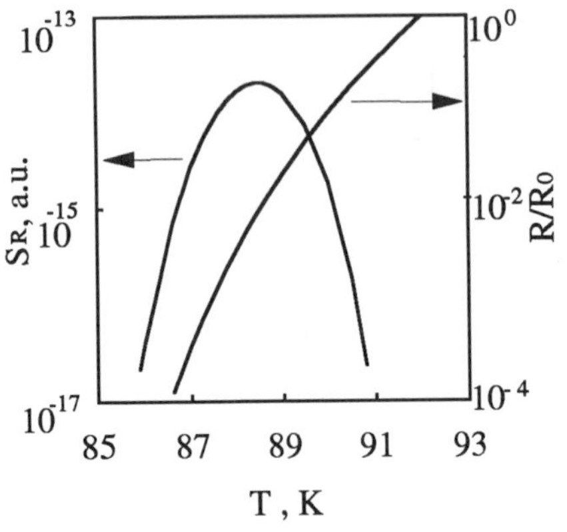

Fig. 7. The temperature dependence of the resistance and the resistance fluctuations (in arbitrary units) in the transition region.

superconducting transition temperature region is caused by the switching of S-N-S, S-N-N, S-N-I-S and like them Josephson junction due to charge carrier trapping processes in the boundary grain regions. In the superconductive transition temperature region the resistance temperature dependence may be aproximated by such relation:

$$R/R_0=[(T-T_c)/(T_{on}-T_c)]^n , \qquad (1)$$

where $T_c \leq T \leq T_{on}$; R_0 is the resistance at the onset of the transition temperature $T=T_{on}$; T_c is the temperature at which $R=0$. On the other hand, the formation and transition of elementary Josephson weak links to completely sperconducting state may be evaluated as:

$$n_s(T)/n_{so}=[(T_{on}-T)/(T_{on}-T_{cs})]^m, \qquad (2)$$

where $T_{cs}<T_c<T_{on}$, n_{so} is the general number of elementary Josephson weak links in the sample; $n_s(T)$ is the number of junctions which are completely in the superconducting state; T_{cs} is the temperature at which sample is completely in superconducting state.

If we consider that switching processes of Josephson weak links in the transition region due to charge trapping are independent, we can express the resistance fluctuations as:

$$S_R/R^2=(C/f)[n_s(T)/n_{so}] \text{ or } S_R=(CR^2/f)[n_s(T)/n_{so}] , \qquad (3)$$

where C is the parameter.

The temperature dependence both of the resistance R/R_0 (1) and the resistance fluctuations (3) (in arbitrary units) are shown in fig.7.

It is obvious that such model can in principle to explain the obtain results.

REFERENCES

1. V. Palenskis, Z. Šoblickas, R. Simanavičius and B. Vengalis, Lithuania Phys. J. 30, 567 (1990).
2. V. Palenskis, Z. Šoblickas, R. Simanavičius and B. Vengalis, Pis'ma v ZhTPh 16, 27 (1990).
3. V. Palenskis, Z. Šoblickas, A. Lukauskas and A. Stadalnikas, In Proc. "Noise in Physical Systems and 1/f Fluctuations". Ed. by T. Musha et al. (Ohsma, 1991), p.43.

VOLTAGE NOISE OF $YBa_2Cu_3O_{7-\delta}$ FILMS IN THE VORTEX-LIQUID PHASE

P. J. M. Wöltgens, C. Dekker,* S. W. A. Gielkens, and H. W. de Wijn

Faculty of Physics and Astronomy, and Debye Institute, University of Utrecht, P.O. Box 80.000, 3508 TA Utrecht, The Netherlands

ABSTRACT

We have measured the voltage noise in the vortex-liquid phase of $YBa_2Cu_3O_{7-\delta}$ films at high magnetic fields. The voltage-noise spectral density S_V is found to be of the $1/f$-type, and to vanish critically at the superconducting phase transition according to $S_V \propto (T - T_g)^x$, where T_g is the vortex-glass transition temperature and $x = 1.8 \pm 0.3$. A model is presented which explains the experimental observations. The model is based on a distribution of critically divergent lifetimes of vortex-glass domains.

The present paper reports on noise experiments in films of the high-T_c superconductor $YBa_2Cu_3O_{7-\delta}$ in the resistive state at high magnetic fields. This resistive state is accepted to be a liquid of vortices carrying the magnetic flux lines. Whereas previous noise studies in $YBa_2Cu_3O_{7-\delta}$ have focused on the regime near zero magnetic field, this work studies the regime at high magnetic fields. The principal new result is that in a high magnetic field the voltage noise vanishes at the superconducting phase transition according to a critical power law.

The magnetic field-temperature (H-T) phase diagram of high-T_c superconductors like $YBa_2Cu_3O_{7-\delta}$ consists of several phases (Fig. 1): (*i*) the normal phase at high temperatures and fields ($H > H_{c2}$); (*ii*) the superconducting Meissner phase at low fields ($H < H_{c1}$); (*iii*) the superconducting vortex-glass phase[1] at higher fields, where the vortices are frozen into a glassy state; and (*iv*) the vortex-liquid phase in between the vortex-glass transition and the upper critical field H_{c2}. In the vortex liquid, the motion of the vortices causes the resistance to be finite. We have examined the voltage noise in this phase with the aim to study the dynamics of vortex lines. The sample was a c-axis-up 3000-Å $YBa_2Cu_3O_{7-\delta}$ film, which was laser ablated onto a $SrTiO_3$ substrate. The film was photolithographically patterned to four-probe patterns with a central stripe of $150 \times 20\ \mu m^2$. A gold layer was deposited onto the contact pads. Subsequent annealing yielded a contact resistance of much less than 1 Ω. In zero magnetic field the film exhibited zero resistance below 91.2 K.

The voltage noise spectral density was obtained while biasing the superconductor with a constant dc current I. The noise signal was passed through a low-noise

*Present address: Department of Applied Physics, Delft University of Technology, P.O. Box 5046, 2600 GA Delft, The Netherlands.

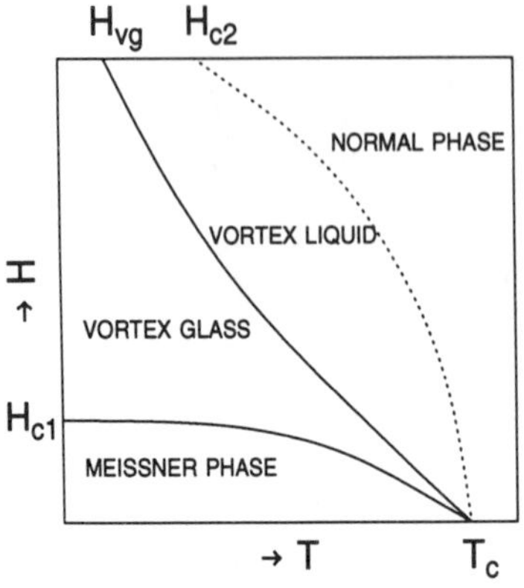

Figure 1: The *H*-*T* phase diagram of $YBa_2Cu_3O_{7-\delta}$.

1:100 transformer and subsequently amplified by a low-noise preamplifier. The resulting signal was fed to a fast-Fourier transform spectrum analyzer. The noise spectrum was corrected for background noise and the system response,[2] to obtain the excess ($I > 0$) voltage-noise spectral density S_V. We measured S_V as a function of frequency f, current I, temperature T, and magnetic field H (up to 5 T along the c axis).

The low-frequency (< 2 kHz) noise is found to depend on the frequency essentially as $1/f^n$, with $n = 0.96 \pm 0.06$, independent of I, T, and H. A typical noise spectrum is shown in Fig. 2. At all temperatures and fields, we find that S_V depends on the current approximately as $S_V \propto I^2$. In the normal state this points to resistance fluctuations as the noise source. In the vortex-liquid phase, the I^2 dependence of S_V is not fully understood. Presumably it is related to the complicated crossover from ohmic to power-law behavior upon approaching the vortex-glass transition temperature T_g ($T_g = 76.3$ K at 5 T, $T_g = 84.3$ K at 2 T).

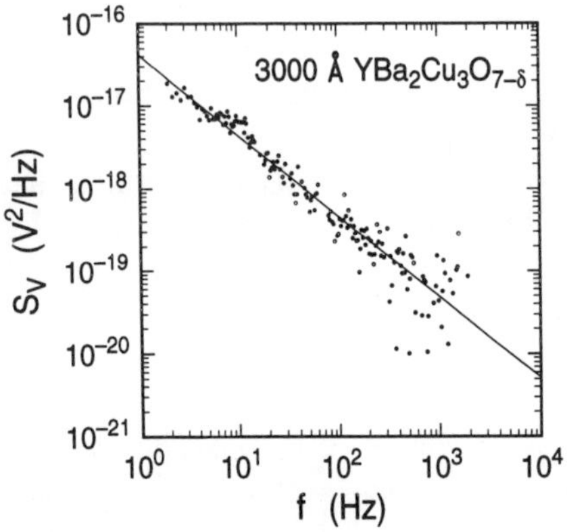

Figure 2: Typical excess noise spectrum, corrected for background noise and the system response; $H = 2$ T, $T = 90.11$ K, $I = 2$ mA. The solid line denotes a $1/f^{0.98}$ dependence.

In Fig. 3, we present S_V as a function of the reduced temperature $(T - T_g)/T_g$. With decreasing temperature, S_V vanishes at T_g. Quite remarkably, this temperature dependence can be described in terms of a critical power law

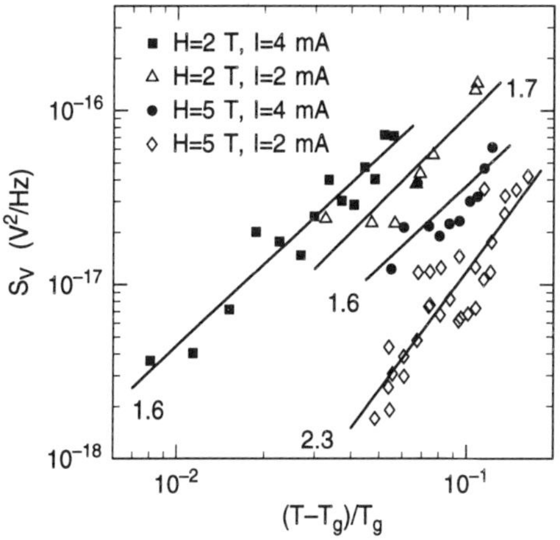

Figure 3: Excess noise spectral density S_V vs $(T-T_{\rm g})/T_{\rm g}$. The number entries are the slopes derived from fits to Eq. (1).

$$S_V \propto (T-T_{\rm g})^x \ , \tag{1}$$

with $x = 1.8 \pm 0.3$ (Fig. 3). Supplementary nonlinear current-voltage (I-V) curves were used to determine the vortex-glass phase transition temperature $T_{\rm g}$ and the critical vortex-glass exponents z and ν. These measurements were carried out according to the method used in Ref. 3. The I-V-curves were subjected to a critical scaling analysis[4] to extract values for $T_{\rm g}$, z and ν. For the critical exponents we find $z = 4.8$ and $\nu = 1.7$, in good agreement with earlier studies.[3,4] It should be pointed out that the observed noise is not simply the normal-state resistance noise extrapolated to the vortex-fluid state. This becomes evident in a plot of S_V/V^2 vs T, where we can see the temperature dependence of S_V/V^2 change from slowly decreasing with decreasing temperature above $T_{\rm c}$, to strongly diverging at the glass transition temperature below $T_{\rm c}$.

We model the observed $1/f$ noise to result from a superposition of many random processes with different characteristic times τ. These characteristic times are thought to be the lifetimes of bundles of correlated vortices. The lifetime of a vortex bundle scales with its size l as $\tau \propto l^z$. The average size of these vortex bundles is the vortex-glas correlation length ξ, which is known to diverge critically at $T_{\rm g}$ as $\xi \propto 1/(T-T_{\rm g})^\nu$.

For a distribution function $D(l)$ of the size of the vortex-glass domains, the spectral density of the voltage fluctuations is given by

$$S_V(\omega) \propto \int f(\omega, l) D(l)\, dl \ , \tag{2}$$

where $f(\omega, l)$ is the Lorentzian spectrum associated with the lifetime of a vortex bundle of size l. We have $f(\omega, l) \propto Cl^z/[1+(\omega Cl^z)^2]$, with C the proportionality constant in $\tau = Cl^z$. For $\tilde{l}(\omega) = (\omega C)^{-1/z}$ this Lorentzian reaches its maximum $f(\omega, \tilde{l}) = 1/2\omega$. If we assume the distribution function $D(l)$ to vary only slowly with l within the width of the Lorentzian distribution, Eq. (2) can be approximated by[5]

$$S_V(\omega) \propto D(\tilde{l}) \int \frac{Cl^z}{1+(\omega Cl^z)^2}\, dl \propto D(\tilde{l})/\omega^{1+1/z} . \tag{3}$$

Secondly, we make the reasonable assumption that the distribution $D(l)$ does not change much when scaling the length scale l to the vortex-glass correlation length ξ. We thus rewrite the distribution as

$$D(l) = D_0 \rho(l/\xi) , \tag{4}$$

where ρ is a normalized form function, and $D_0 = 1/\xi$ is a normalization constant determined by $\int_0^\infty D(l)dl = 1$. Note that, upon approaching T_g, the form of $D(l)$ does not change much, whereas the length scale l for which $D(l)$ reaches a maximum critically shifts upwards, while the value of this maximum critically vanishes. Combining Eqs. (3) and (4) and substituting $\xi \propto 1/(T-T_g)^\nu$, we finally arrive at

$$S_V(\omega, T) \propto \frac{(T-T_g)^\nu}{\omega^{1+1/z}} \rho(\tilde{l}/\xi) . \tag{5}$$

We thus expect that S_V vanishes upon approaching T_g from above according to $S_V \propto (T-T_g)^\nu$, with $\nu = 1.7$. This is in good agreement with the experimental result that S_V depends critically on the temperature with a critical exponent $x = 1.8 \pm 0.3$. From Eq. (5) a frequency dependence $S_V \propto 1/f^{1+1/z}$ would be expected, i.e., $S_V \propto 1/f^{1.2}$ for $z = 4.8$. The minor difference with the observed $1/f$ dependence can be accounted for by a slight length dependence of $D(l)$.[5]

In summary, we have observed that in the vortex-liquid phase (*i*) S_V has a $1/f$ character and (*ii*) S_V diverges critically upon approaching T_g. We have explained the $1/f$ character and the temperature dependence of the voltage fluctuations in the vortex fluid phase by the use of a model, based on a distribution of lifetimes of vortex glass domains with a critically diverging average lifetime.

We are indebted to the late Dr. W. Eidelloth of IBM Research at Yorktown Heights for providing the high-quality $YBa_2Cu_3O_{7-\delta}$ films. This research was in part supported by the Netherlands science foundations FOM and NWO.

1. M. P. A. Fisher, Phys. Rev. Lett. **62**, 1415 (1989); D. S. Fisher, M. P. A. Fisher, and D. A. Huse, Phys. Rev. B **43**, 130 (1991).
2. F. Hofman, R. J. J. Zijlstra, and J. C. M. Henning, Solid-State Electron. **31**, 279 (1988).
3. R. H. Koch, V. Foglietti, W. J. Gallagher, G. Koren, A. Gupta, and M. P. A. Fisher, Phys. Rev. Lett. **63**, 1511 (1989).
4. R. H. Koch, V. Foglietti, M. P. A. Fisher, Phys. Rev. Lett. **64**, 2586 (1990).
5. This is similar to the approach in P. Dutta and P. M. Horn, Rev. Mod. Phys. **53**, 497 (1981).

V. LIQUID CONDUCTORS

Fluctuations of Light Intensity Scattered by Aqueous LiCl Solution

Toshimitsu MUSHA, Koji *TAKADA and Keisuke NAKAGAWA

Department of Electrical Engineering, Science University of Tokyo
Kagurazaka, Shinjuku-ku, Tokyo, JAPAN 162
*Department of Applied Electronics, Tokyo Institute of Technology
Nagatsuta, Midoriku, Yokohama, JAPAN 227

Musha, et al. found that laser light intensity scattered by a single crystal quartz was subject to $1/f$ fluctuations.[1] This was attributed to fluctuations in energy partition among phonon modes which were almost in thermal equilibrium. Similar experiment was extended to light scattering by water[2] and ionic solutions, LiCl, KCl and NaCl, and $1/f$-like spectra were also observed. The fractional fluctuation of scattered light intensity decreases rapidly when the ionic molar ratio to water molecules becomes larger than 0.2. It is concluded that light scattering in water is attributed to cluster structure.

1. Introduction

The fractional fluctuations of the light intensity of scattered light is, in principle, proportional to the number of scatterers. In case of the Brillouin scattering in quartz, we found that the fractional spectral level is inversely proportional to the number of phonon modes involved in the scattering.

We repeated this experiment with water and $1/f$-like spectra were also observed which, however, leveled off at low frequencies. What is the mechanism of the scattering we have observed? The following two possibilities have been investigated.

(1) *Propagating scatterers*: If the scattering is caused by propagating density waves like phonon modes, the fractional fluctuations will be inversely proportional to the number of wave modes which are involved in the scattering. This possibility can be checked through dependence of fractional fluctuations on the single- and multi-mode or multi-line light incidence.

(2) *Non-propagating scatterers*: If the scattering is caused by non-propagating localized scatterers, the fractional fluctuations depend on the number of scatterers and insensitive to the single- and multi-mode or multi-line incidence.

2. Experiment

The experimental setup is shown in Fig.1. The source laser is an argon ion laser (Lexel Model 95-4) and the incident laser power was controlled from 10 mW to 200 mW with an ND filter, and the principal wavelengths are 514.5 nm, 488.0 nm and 496.5 nm. The power was stabilized by optical feedback to less than ± 0.2%/hr and the frequency was stabilized to less than ± 75 MHz/2hrs. Water as a specimen was filled in a quartz vessel of $10 \times 10 \times 40$ mm^3. The laser mode was single longitudinal mode with transverse mode TEM_{00} with line width 3 MHz and a beam diameter was 1.6 mm or multi-mode in which frequency spreads over a single line or multi-line in which frequency spreads over the three principal lines. An image of a laser beam passing in the specimen was generated with a lens on a slit and it was detected by a photomultiplier. The solid angle of the detected light beam was 0.82 steradian and the scattering volume was 2.9×10^{-11} m^3. A similar optical path was made by a beam splitter making a small angle to each other. Light intensity was measured by means of photon counting. The ratio of the photon counts of these two photomultipliers gives a fractional fluctuation of the scattering light intensity as was described in ref(1).

3. Results

When purified water was used the observed data were not always stable, and some bright spots were observed in a laser beam passing the specimen. They are probably small dust particles. When heavy water D_2O which was purchased in a glass capsule and used as specimen, no bright spots were observed and the data were reproducible. The number of possible phonon modes is determined by a spread of wave vectors of phonon modes or optical modes whichever is larger. In the multi-mode operation, the longitudinal mode interval is 150 MHz within bandwidth 5 GHz, and the phonon number is determined by the bandwidth of the related phonon modes. No definite conclusion can be obtained from this experiment. In the multi-line operation, however, the number of phonon modes is determined by a spread of the optical modes. However, no significant difference was observed in the spectral level. Therefore, it is concluded that the scattering should be attributed to non-propagating scatterers.

To investigate the property of the scattering, aqueous ionic solutions such as LiCl, NaCl and KCl were examined. They are soluble up to 46, 26 and 26 wt%, respectively. We had specially interesting result with LiCl solution because its saturation concentration is extremely large. The power spectral levels for three different ionic concentrations are shown in Fig.2, where the spectra are $1/f$-like and they level off at frequency below 0.01 Hz.

The spectral levels for these three ionic solutions at 10 Hz are shown in Fig.3, where the horizontal axis indicates the ionic molar ratio to water molecules in a given volume. With heavy water D_2O the $1/f$-like spectral level at 10 Hz is 10^{-5}, and hence for ionic molar ratios smaller than 0.1, the spectral level is an order of magnitude larger as compared with water. When the molar ratio of Li^+ to water molecules becomes larger than 0.2, the spectral level decreases rapidly with this ratio.

4. Discussion

Light scattering is caused by local fluctuations of refractive index. Water molecules form cluster structures, and the cluster structure will have a different refractive index from that of amorphous molecular configuration. Therefore, it is likely that clusters scatter the incident light. Li ions have small diameter and the electric fields generate larger clusters. According to computer simulation[3], the cluster in water consists of 4~5 water molecules. A Li^+ ion in a dilute solution attracts 6 water molecules while it attracts 4 water molecules in a concentrated solution.[4,5] Therefore, a dilute ionic solution decreases the number density of clusters, which raises the fractional fluctuations of scattered light intensity. When the molar ratio of Li^+ is larger than 0.2, there are more Li ions than water molecule-clusters. Clusters are mainly generated by Li^+ and the mean cluster size is smaller and hence the number density of clusters increases with increase of the ion concentration. This will be the reason for the spectral behavior in Fig.3. Although further investigation is needed, we conclude based on the observations so far that the spectrum is superposition of Lorentzian spectra and the origin of the light scattering is attributed to cluster structure in water.

The shoulder frequency of the $1/f$-like spectrum increases when the temperature of water is raised. This is probably due to convection flows of water due to temperature inhomogeneity.

---

[1] T. Musha, G. Borbely and M. Shoji, Phys. Rev. Lett., **64** (1990) 2394

[2] T. Musha and G. Borbely, Jpn J. Appl. Phys., **31** (1992) L370

[3] H. A. Frank and W. Y. Wen, Proc. Faraday Soc., **24** (1957) 133.

[4] R. Letz, A. T. Halgler and H. A. Scherage, J. Phys. Chem., **78** (1974) 1531.

[5] A. T. Hagler, H. A. Scherage and G. Némethy, J. Phys. Chem., **76** (1972) 3229.

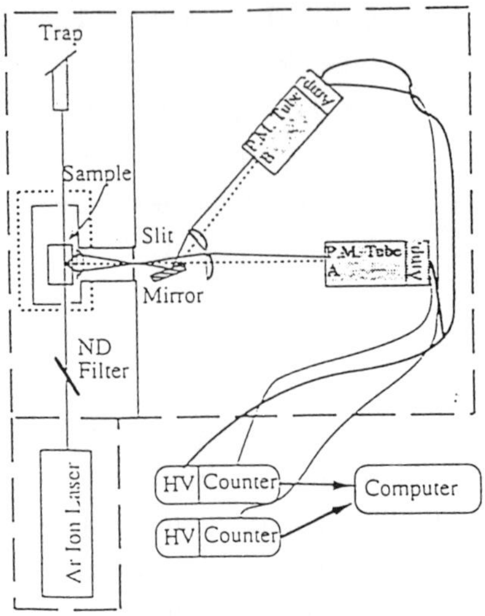

Fig.1 Experimental setup

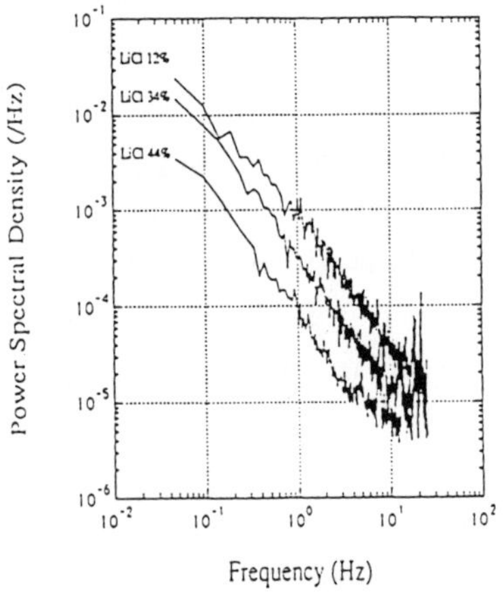

Fig.2 The fractional power spectra of scattered light ntensity for three different concentration of LiCl solution.

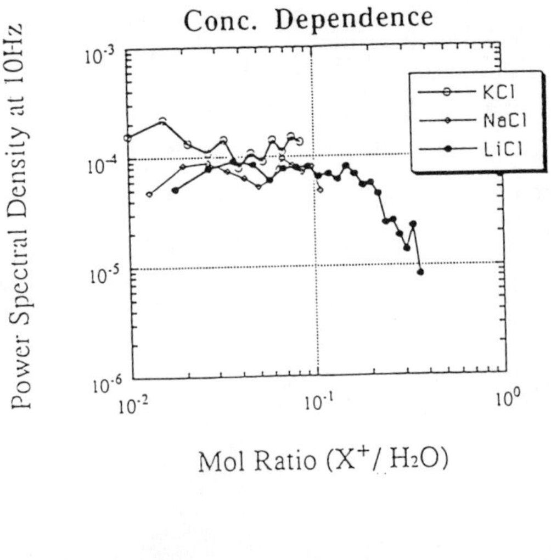

Fig.3 Dependence of the spectral level at 10 Hz on the ionic molar ratio to water molecules.

ANALYSIS OF $1/F^{\beta}$ FLUCTUATIONS IN AN INTERPHASE REGION AS A POSSIBLE TOOL FOR CHARACTERIZATION OF HIGHLY DILUTED SOLUTIONS

V.M.Uritsky, N.I.Muzalevskaya, E..Korolyov and G.P.Timoshinov
Nonequilibrium Systems Laboratory
Department of Science and Technology of SOFID Co.
196 140 Russia, St.Petersburg, Pulkovskoe shosse 86-56

ABSTRACT

A technique for testing water solutions of organic substances with low and tending to zero concentrations of solutes is proposed. The technique consists in measuring and interpreting $1/f^{\beta}$fluctuations of electrical potential of a double layer at the interface liquid to be tested - metal electrode. The observed activation of the fluctuations can determine to the large extent macroscopical properties of highly diluted solutions. Effect of weak low frequency magnetic field on fluctuations in such liquids is also studied.

INTRODUCTION

Difficultes inherent to existing methods of study of diluted solutions with concentrations of solutes below 10^{-5} - 10^{-6} M are caused to a large extent by high level of statistical noises in such solutions. However, the $1/f^{\beta}$ part of these very noises is known to be very sensitive to inhomogeneities in media under testing on micro- and macroscopic scales, and, consequently, can provide information about presence of molecules of a solute in a solvent. In the present pilot study this approach was tested experimentally on water solutions of histamine. Estimation of $1/f^{\beta}$ fluctuations in test samples was made on the basis of the method developed by us earlier[1] that consisted in analysing fluctuations of electrical potential at the interface metal - test liquid.

EXPERIMENTAL TECHNIQUE, RESULTS AND DISCUSSION

Pure water and water solutions of histamine in concentrations 10^{-7}M and 10^{-32} M were tested. All samples were provided by INSERM U200, France. Test samples were poured into a cell of volume 40 ml with a small area measuring electrode and an auxiliary electrode, both made of Pt. $1/f^{\beta}$ fluctuations of electric surface potential were registered at f<10Hz as an excess noise of the cell's asymmetry potential. Computer analysis of fluctuations included calculation of Fourier power spectrum S(f) and a number of its numerical parameters. Fluctuation response of 10^{-32}M histamine to weak low frequency electromagnetic disturbance was also studied.

Fig.1 shows smoothed S_0-histograms, S_0 being the value of S(f) at the lowest frequency of analysis 0.1Hz. Each curve was obtained on the basis of 12-14 separate spectral measurements, averaged over 20-minutes intervals, and belonging to several different samples and days of observation. Mean value of S_0 for 10^{-7}M histamine was larger than for pure water, increased was also confidence interval of S_0, which fact implicitly pointed to increase in intensity of fluctuations below 0.1Hz. Distribution curve $N(S_0)$ for 10^{-7}M histamine stretched towards larger S_0. Mean value of β for 10^{-7}M histamine was larger than that for pure water (1.32) and amounted to 1.57, the difference being statistically significant. It means that not only intensity of $1/f^{\beta}$ fluctuations in diluted solution was increased in comparison with pure water, but that the frequency makeup of fluctuation spectra was modified. Shift to larger β may mean reorganization of fluctuations of solution's volume properties in favour of spectral components with lower frequencies.

The obtained data for 10^{-7}M histamine could be the result both of direct influence of molecules of the solute on $1/f^{\beta}$ fluctuations, and of modification of the solvent itself in the precess of dilution. The existence of the latter mechanism is confirmed by the tests with the samples of 10^{-32}M histamine obtained by step-by-step dilution[2]. The differences of stochastic

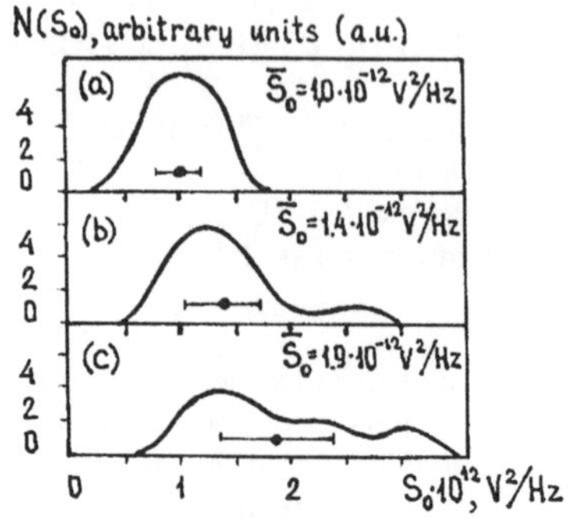

Fig.1. Distribution $N(S_0)$ for pure water (a) and for water solutions of histamine in concentrations 10^{-32}M (b) and 10^{-7}M (c). Segments below each curve correspond to confidence intervals (p=.05).

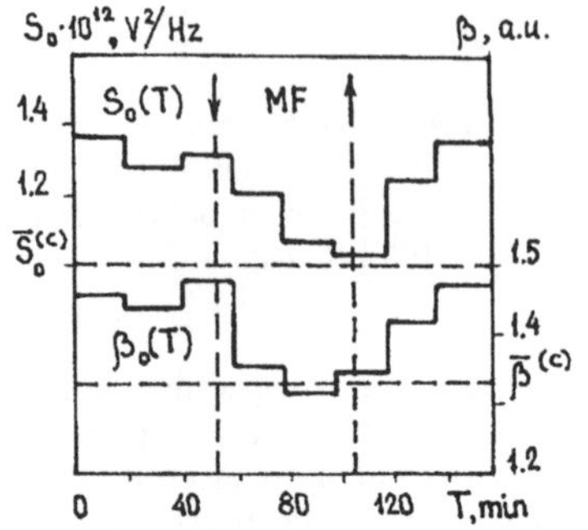

Fig.2. Effect of magnetic field on fluctuation parameters S_0 and for 10^{-32}M histamine. Arrows mark beginning and termination of exposition. $\bar{S}_0^{(c)}$ and $\bar{\beta}^{(c)}$ are mean values for pure water (control).

parameters of 10^{-32}M histamine and 10^{-7}M histamine from those of pure water were the same, though for 10^{-32}M histamine they were less pronounced (Fig.1b; $\bar{\beta}$ =1.45). There were practically no molecules of solute in 10^{-32}M histamine, so it was low frequency random molecular dynamics of the solvent itself that may have undergone modification in the process of dilution, and the modification appeared as the increase in intensity of fluctuations with large correlation times. Random molecular dynamics of the solvent modified in this way may be largely responsible for macroscopic properties of such virtual solutions, in particular, for their biological activity[2,3]. Solutions with concentrations of solutes tending to zero are likely to represent a special class of substances which may be called stochastically modified liquids (SML).

To check up a hypothesis concerning electromagnetic origin of modification in SML[3,4] we studied effect of sinusoidal magnetic field with frequency in the range of $1/f^{\beta}$ fluctuations (H=2.5A/m, f=0.5Hz) on 10^{-32}M histamine. After exposition to the field for 60 min., fluctuation properties of the sample became identical with those of pure water, but then after 40-60 min. were restored (Fig.2). Magnetic field in ref.[3] was 10^4 times stronger and cancelled biological activity of SML irreversibly, whereas our weak magnetic field acted as information disturbance[4] causing reversible change in the stochastic electromagnetic level to which it appeared to be adressed. Critical sensitivity to magnetic fields suggest that distinctive properties of SML are determined by their inner stochastic electromagnetic organization.

The obtained preliminary results demonstrate efficiency and high informative potential of the fluctuation testing for study of solutions with ultra-low concentrations of solutes. Super sensitive devices for detection of vanishing amounts of dissolved substances by regestering changes in stochastic propereties of solvents can be constructed on the bases of the proposed method.

We thank Dr. J.Benveniste (INSERM U200) for supplying the test samples and for valuable discussions of this work.

REFERENCES

1. N.I.Muzalevskaya, V.M.Uritsky, E.V.Korolyov and G.P.Timoshinov, Proc.12th Int.Conf.on Noise in Physical Systems and 1/f Fluctuations (Tokyo, Ohmsha, 1991), p.465.

2. E.Davenas, F.Bauvais, J.Amara, et al., Nature, 333, 816 (1988).

3. L.Hadji, B.Arnoux, J.Benveniste, FASEB J., 5, 1583 (1991).

4. N.I.Muzalevskaya, in: Informational Interactions in Biology (Tbilisy, Izd.TGU, 1990), p.28. (in Russian).

VI. QUANTUM $1/f$ NOISE

QUANTUM 1/f NOISE IN RADIOACTIVITY

K.Gopala and M. Athiba Azhar
Department of Studies in Physics, Manasagangothri,
Mysore - 570 006, India.

ABSTRACT

Radioactivity results mainly in the emission of charged particles or gamma photons. Charged particle emission can be considered to be an electric current that has fluctuations due to random emission. The fluctuations are generally poissonian exhibiting shot noise. But, Handel's quantum 1/f noise theory predicts that, in addition, 1/f noise should also exist. Experiments with α particles and γ photon emission have not shown the existence of this quantum 1/f noise. However, experiments with ß particles have shown the existence of such a noise.

INTRODUCTION

Fluctuations in the otherwise expected smooth behavior of a physical quantity is considered to be noise. Thus the fluctuations in current flow or in voltage constitute noise. Thermal noise and shot noise have uniform spectral density, where as 1/f noise has a spectral density inversely proportional to the frequency. 1/f noise has been observed in a variety of systems: Semiconductor devices[1], Music[2], Height of floods in the Nile[3], Earthquake cycles[4], Thunderstorms[4], Biological systems[5,6], Traffic flow[7], etc[8].

1/f noise in semiconductor devices and other electronic devices has received by far the greatest attention[9-19]. Several models have been proposed for 1/f noise and are shown to hold good in some cases.

In an attempt to understand the basic principles underlying the existence of 1/f noise in particle emissions, Handel has proposed a quantum theory of 1/f noise[20,21], according to which infraquanta generated in the emission process (any emission process can be considered to be a scattering process in one way or the other) will render some of the particles to have slightly less energy than the others. Interference between these two sets of particles leads to 1/f noise.

A series of articles have appeared on the theory of quantum 1/f noise. References 20 to 30 are a fair representation of these.

Several objections [31-34] have been raised against Handel's theory. However, Handel's detailed theory based on second quantisation [28] seems to answer many of the criticisms. Van Vliet's paper of reference 30 is a good recent review on 1/f noise. Some of her theoretical conclusions need to be experimentally verified.

THEORETICAL BACKGROUND

Measurement of radioactive current as such is difficult because it is very weak. But one can easily measure the counts M_T^i for an interval of time T repeatedly. From such data one can calculate the Allan variance given by

$$A(T) = [1/2\,(N-1)] \sum_{i=1}^{N-1} \left| M_T^i - M_T^{i+1} \right|^2 \qquad (1)$$

where N is the number of measurements. This Allan variance is related to the noise spectral density of the flux fluctuations $S_m(\omega)$ through a transform equation[35]

$$A(T) = \int_0^\infty (S_m(\omega)/\omega^2)\ \sin^4(\omega T/2)d\omega \qquad (2)$$

where ω is the angular frequency.

It is known that for the poisson shot noise

$$S_m(\omega) = 2m_o \qquad (3)$$

where m_o is the average count rate.

Hence from eqn. (2) one finds for the poisson shot noise

$$A(T) = m_oT = \langle M_T \rangle \qquad (4)$$

which is the mean count. Thus we see that in case of poisson shot noise the Allan variance is simply the normal variance.

But, for 1/f noise, the spectral density can be written as

$$S_m(\omega) = 2\pi C/\omega \qquad (5)$$

where C is a constant. Using eqn. (2) again, we get

$$A(T) = 2CT^2 \ln 2 \qquad (6)$$

If the noise is composed of both shot noise and 1/f noise, then one can write

$$A(T) = m_oT + 2CT^2 \ln 2 \qquad (7)$$

We define a quantity called Relative Allan variance as

$$R(T) = A(T) / \langle M_T \rangle^2 \qquad (8)$$

Since $\langle M_T \rangle = m_oT$, using eqn. (7), we get

$$R(T) = (1/m_oT) + 2C' \ln 2 \qquad (9)$$

where $C' = C/m_o^2$ is the characteristic strength of the normalised excess noise $S_m(f)/m_o^2$. The constant $2C'\ln 2$ is called the Flicker floor F as it arises due to the Flicker noise (i.e., 1/f noise). We see that a log-log plot of R(T) versus $1/m_oT$ would deviate from a straight line and reach a constant value F from which C' can be determined.

Handel's theory predicts that $C' = 2Z^2\alpha\zeta A$, where Z is the charge of the emitted particle, α is the fine structure constant, ζ is the coherence factor,

$$A = (2\pi/3)(\Delta v/c)^2 = (2\pi/3)\ [1-(1+(E/m_z c^2))^{-2}] \qquad (10)$$

Δv = velocity change of the particles in the emission process, c is the velocity of light in vacuum, m_zc^2 is the rest mass energy of the particle and E is the kinetic energy of the emitted particle. Hence,

$$F = 4Z^2\alpha A\zeta \ln 2 \qquad (11)$$

which shows that F/A is a constant.

EXPERIMENTAL WORK

The very first experimental work on quantum 1/f noise was done by Gong et.al[36]., with α particles of ^{241}Am. They found a Flicker floor of $\simeq 10^{-7}$. Subsequent study[37] from the same group confirmed the earlier result. But Kennet and Prestwitch[38] made an exhaustive study and found that such a Flicker floor does not exist for ^{241}Am α particles. Another paper from the same group[39] confirmed the negative result. So also did the paper of Jones et.al [40]. We at Mysore [41] studied α decay statistics of ^{210}Po and again found that no Flicker floor exists. Now it is well acknowledged[30] that α decay does not exhibit 1/f noise.

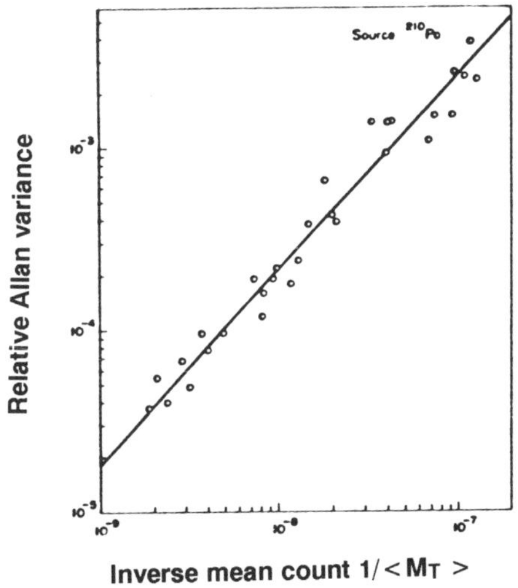

Fig. 1: A plot of Relative Allan variance versus inverse mean count for α particle counting of ^{210}Po

We extended our experimental study to β emission and found for the first time that β decay statistics of ^{204}Tl and ^{90}Y do exhibit 1/f noise [42,43] with a Flicker floor of $\simeq 10^{-5}$. β particles were detected in a plastic scintillator coupled to a photomultiplier. Pulses from the photomultiplier were passed through a pre-amplifier and a pulse amplifier and then analysed in a Multichannel Analyser. The β spectra were taken for 5 minutes, consecutively, for a large number of times. It was repeated for 10 min., 20min., etc. The calibration was done by the Compton electron scattering of γ rays from ^{137}Cs which gives a clear edge at 477 keV. The gain was checked periodically, and for large counting periods it was checked before and after the counting period. If there was any shift in the location of the Compton edge that data was rejected. Analysis was done for a series of selected energy channels.

While calculating the Relative Allan variance R(T) the number of consecutive measurements used for calculation becomes important. It is found that for ^{204}Tl β counting, for a 10 min., period one should take at least 35 count measurements to get a consistent value of R(T) whereas for 1000 min., period just 10 count measurements would suffice.

Figure 2 shows the variation of R(T) with inverse mean count for 387 keV β emission of ^{204}Tl. We see that R(T) deviates from poisson distribution for mean counts greater than 12500. The Flicker floor F is 2.1×10^{-5}.

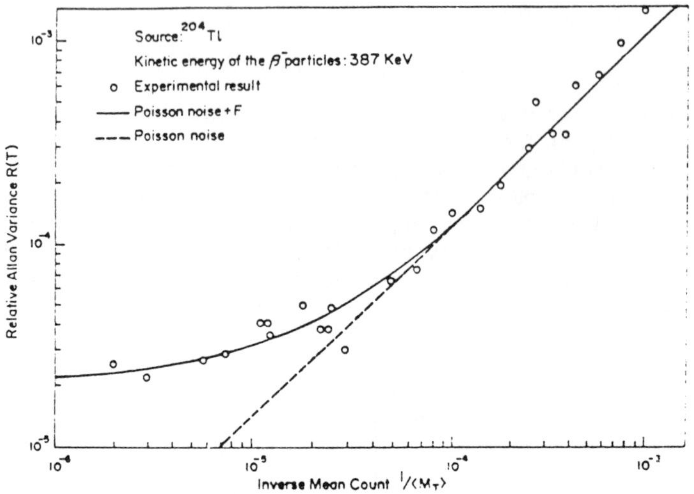

Fig.2: A plot of Relative Allan variance versus inverse mean count for 387 keV ß emission of ^{204}Tl.*

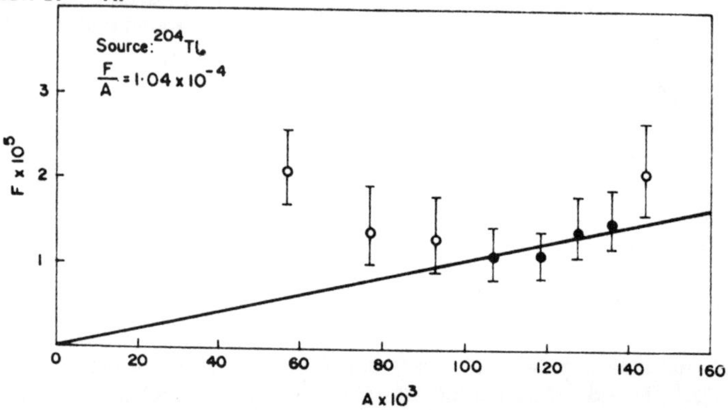

Fig.3: A plot of Flicker floor versus A for ^{204}Tl. Only filled circles are considered for drawing the straight line.

We have determined F for different energies of the ß spectrum. Figure 3 shows the variation F with A, for ^{204}Tl. A straight line has been drawn to pass through the origin, because it is the theoretical expectation. The slope of the straight line, of course, gives the mean F/A. The F/A values determined thus for ^{204}Tl and ^{90}Y are 1.04×10^{-4} and 1.43×10^{-4}, respectively.

CONCLUSIONS

One can notice that in Figs. 3 and 4, there is a general trend for the points to follow a bowl shaped curve. This trend has also been found for ^{90}Sr - ^{90}Y ß particles in the energy region below 546 keV. This is not in accordance with the existing theory.

Further experiments have to confirm this observation.

The objection raised for our finding of the existence of Flicker floor in case of ß decay was that it might have occurred due to instrumental instabilities. But we did the same type of experiment[44] for the γ emission of ^{137}Cs with almost the same experimental set up except for the detector and could not find a Flicker floor. Hence we do not believe that instrumental instabilities are the cause for the observed Flicker floor in ß -decay.

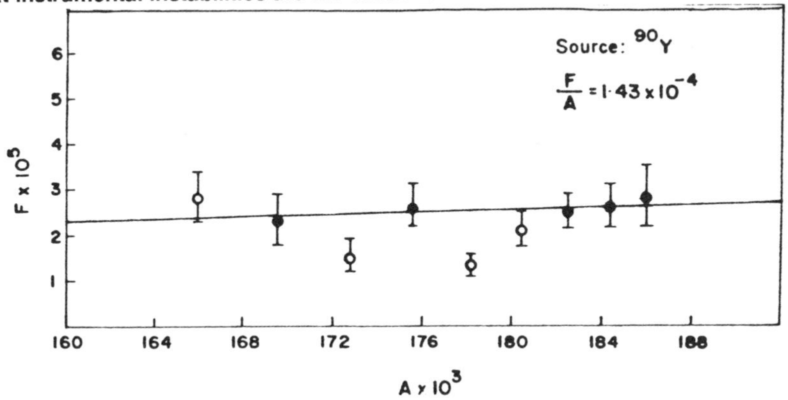

Fig.4: A plot of Flicker floor versus A for ^{90}Y. Only filled circles are considered for drawing the straight line.

Now it is fairly certain that 1/f noise is not present in α and γ decay but is present in ß-decay.

The Relative Allan variance is a measure of the "error" i.e., the average difference between consecutive measurements . Eqns. 4 & 8 show that, in case of shot noise, this error decreases indefinitely with increasing mean count. But when 1/f noise is present, this error can never be lower than F.

REFERENCES

1. A. Van der Ziel, Proc. IEEE., 76, 233 (1988)
2. R.F. Voss and J. Clarke, J. Acoust. Soc. Am 68, 258 (1978).
3. M. Gardner,Scientific American 238,16 (Apr. 1978)
4. S. Machlup, Proc.6th Int. Conf. on Noise in Physical Systems, Gaithersburg (Ed. by P.H.E. Meijer, National Bureau of Standards, Washington, DC, 157 (1981).
5. F.N. Hooge, Physica 83B, 9 (1976).
6. M. S. Keshner, Proc. IEEE 70, 212 (1982).
7. T. Musha. H. Higuchi, Proc. of the Symp. on 1/f fluctuations, Ed. by T. Musha (Tokyo Institute of technology, Tokyo, 187, 1977).
8. D. Wolf, Noise in Physical Systems, (Springer Verlag, New York, NY 1978).
9. A.L.Mc Whorter, Semiconductor Surface Physics (Univ. of Pennsylvania Philadel-phia, 1957).
10. D.A. Bell, Proc. Roy. Soc. 72, 27 (1958).
11. F.N. Hooge, Phys. Lett. A29, 139 (1969).
12. A. Van der Ziel, Proc. IEEE 58, 1178 (1970).
13. F.N. Hooge, Physica 60, 130 (1972).
14. T.H. Bell Jr., J. Appl. Phys. 45, 1902 (1974).
15. R.F. Voss, and J. Clarke, Phys. Rev. B13, 556 (1976).

16. M.Mikulinski and S. Fishman, Proc. Symp. on 1/f fluctuations, Tokyo, 7 (1977).
17. A. Van der Ziel, Adv. Electron Phys. 49, 225 (1979).
18. P. Dutta and P.M. Horn, Rev. Mod. Phys. 53, 497 (1980).
19. F. N. Hooge, T.G.M. Kleinpenning , and L.K.J. Vandamme, Rep. Prog. Phys. 44, 479 (1981).
20. P.H. Handel, Phys. Rev. Lett. 34, 1492 (1975).
21. P.H. Handel, Phys. Rev. A22, 745 (1980).
22. C.M. Van Vliet, Physica 1134, 261 (1982).
23. P.H. Handel, "Starting points of the quantum 1/f noise approach", unpublished .
24. P.H. Handel, "Keldysh-Schwinger method calculations of 1/f low frequency current fluctuations", unpublished.
25. J. Kilmer, C.M. Van Vliet, G. Bosman, A. Van der Ziel and P.H. Handel, Phys. Stat. Sol. 121, 429 (1984).
26. P.H. Handel and T. Musha, J. Phys. B:Condensed Matter, 70, 515 (1988).
27. P.H. Handel and Q. Peng, IX Int. Conf. on Noise in Physical Systems, Montreal (1987).
28. P.H. Handel, Proc. of the IV Symp. on Quantum 1/f noise and other low frequency fluctuations in electronic devices, Minneapolis, May 10-11 (1990) Ed. by P.H. Handel (Univ. of Missouri, Publication office, St. Louis, M063121) p.55 and P.107.
29. P.H. Handel , "Answer to objections against my theory" , unpublished.
30. C.M. Van Vliet, Solid State Electronics, 34, 1 (1991).
31. A.M. Tremblay, Ph. D. Thesis, Massachusetts Institute of Technology, 1978.
32. L.B. Kiss and P. Heszler, J. Phys. C19, 2631 (1986).
33. Th. M. Nieuwenhuizen, D. Frenkel and N.G. Van Kampen, Phys. Rev. A35, 2750 (1987).
34. N.G. Van Kampen, Proc. 9th Int. Conf. on Noise in Physical Systems, Ed. by C.M. Van Vliet, (World Scientific Publishers, Singapore, 1987) p. 3.
35. C.M. Van Vliet, and P.H. Handel, Physica A (Amsterdam) 113A, 261 (1982).
36. J. Gong, C.M. Van Vliet W.H. Ellis, G. Bosman and P.H. Handel, Noise in Physical Systems and 1/f noise, M. Savelli, G. Lecoy and J.P. Nougier (eds) (Elsevier Publishers, New York, 1983) p. 381.
37. G.S. Kousik et.al., Can. J. Phys. 65, 365 (1987).
38. T.J. Kennett and W.V. Prestwitch, Phys. Rev. A40, 4630 (1989).
39. W.V. Prestwitch, T.J. Kennett and G.T. Pepper, Phys. Rev. A34, 5132 (1986).
40. B.K. Jones, M.G. Berry and G. Hughes, Can. J. Phys. 67, 1022 (1989).
41. M. Athiba Azhar and K. Gopala, Phys. Rev. A39, 5311 (1989).
42. M. Athiba Azhar and K. Gopala, Phys. Rev. A39, 4137 (1989).
43. M. Athiba Azhar and K. Gopala, Phys. Rev. A43, 1044 (1991).
44. K. Gopala and M. Athiba Azhar, Phys. Rev. A37, 2173 (1988).

1/f NOISE FROM THE "UNIVERSAL" DIELECTRIC RESPONSE.

Agafonov V. M., Antohin A. Y.
Moscow Institute of Physics and Technology
Mailing address: Nagornaya 8-31, Dolgoprudniy, Moscow Region, 141700, Russia.

Handel's quantum theory [1] predicts 1/f noise in any physical system whenever the cross section for the quasielastic scattering of charged particles exhibits a low-frequency divergence due to infraquanta of any nature emission. Usually following the original Handel's paper[1] the 1/f noise of electromagnetic origin is considered. Handel and Musha[2] calculated the noise from the piezo-electric coupling, and a mechanism of 'phonon bremstrahlung' was suggested.

In the paper presented we discuss the possibility of infrared divergence in the quasielastic cross section from the charged particle-dielectric polarization interaction in crystal, whose dielectric response function is "universal"[3,4]. Numerous measurements showed that when a step voltage U is applied to a capacitor at the moment t=0, the polarization varies in the following way:

$$dp/dt = U/(ht^{n}), \quad t>0 \tag{1}$$

where n is a number less but very close to 1. h is a constant. It was shown in a lot of experiments that eq.(1) is valid at least at time intervals of many days while the sensitivity of equipment remains sufficient. In Fourier space the image part of the dielectric function may be written (at low frequencies) as:

$$\varepsilon''(\omega) = c\omega^{n-1} \tag{2}$$

Following the Handel's theory let us consider a single electron interacting with crystal according to the formula:

$$\hat{H}_{int} = \hat{H}_{im} + \hat{H}_{die} \tag{3}$$

where $\hat{H}_{im}$ represents the interaction of the electron with scattering centers, for instance impurities, and the second term H_{die} describes the interaction with polarization in crystal. Now we denote initial state of the system as $n,k_0>$, and the ultimate as $m,k_1>$, where n,m are the quantum states of the crystal, and k_0, k_1 - wave vectors of the electron. According to gold Fermi rule the probability of such transition is given by:

$$w_{nm}(k_0,k_1)=2\pi\, T_{nm}(k_0,k_1)\,^2\delta(E_0+E_n-E_1-E_m) \tag{4}$$

where

$$T_{nm}(k_0,k_1)=\langle m,k_1\, H\, \psi^+_{n,0}\rangle \tag{5}$$

$$\psi^+_{n,0}\rangle = n,k_0\rangle + \frac{1}{E_n+E_0-\hat{H}_0+i\gamma}\hat{H}_{int}\,\psi^+_{n,0}\rangle \tag{6}$$

In the first order of excitation theory all terms except those linear on H_{die} may be omitted. After some calculation we find for the probability of the electron scattering in the solid angle $d\Omega_1$ with energy quantum ω emission :

$$\frac{d\sigma_1(\omega,\Omega_1)}{d\omega} = \frac{\sigma_0(\Omega_1)V}{(2\pi)^3}\int dq w_1(q,\omega)\left[\frac{1}{qv_0-\omega-i\gamma} - \frac{1}{qv_1-\omega+i\gamma}\right]^2 \tag{7}$$

$$w_1(q,\omega)=2e^2\varepsilon''(q,\omega)/(q^2V) \tag{8}$$

The cross section $d\sigma_1(\omega,\Omega_1)$ may be subdivided in two parts:

$$(d\sigma_1/d\omega) = (d\sigma_1/d\omega)_{Ch}+(d\sigma_1/d\omega)_{sc} \tag{9}$$

with the first term describing the process when the quantum of energy is generated long before or after the scattering on the impurity potential (similar to Cherenkov's irradiation). It corresponds to the case when $v_0=v_1$ and quantum 1/f noise in the cross section of inelastic scattering must be considered. It may be omitted if the scattering on the impurities is more intensive than inelastic processes and determines the mobility of charge carriers.. For the second term in (9) after integration over solid angle $d\Omega_q$ and substitution $x=q\,v\,/\omega$ the following expression is obtained:

$$(d\sigma_1(\omega,\Omega_1)/d\omega)_{sc}= \sigma_0(\Omega_1)\varepsilon''(\omega)g(\phi)/\omega \tag{10}$$

$$g(\phi)=\int dx\,\mathrm{Re}\,\frac{1}{\sqrt{2(1-\cos\phi)-x^2\sin^2\phi}}\times$$

$$\times \mathrm{Ln}\left[\frac{1-x^2\cos\phi+[2(1-\cos\phi)-x^2\sin^2\phi]^{1/2}}{1-x^2\cos\phi-[2(1-\cos\phi)-x^2\sin^2\phi]^{1/2}}\right] \tag{11}$$

Following the original Handel's theory infrared divergence in the cross section corresponds to the noise with the spectral density:

$$\frac{S_{\Delta\sigma}}{\sigma^2} = \frac{2g(\phi)ce^2}{N\,\omega^{2-n}} \qquad n \approx 1 \tag{12}$$

where $g(\phi)$ is averaged over different scattering processes. It is worth mentioning that the same result may be obtained from classical EM field theory analogously to section 5 in paper [1].

The "universal" dielectric response in matter is rarely discussed in the literature. Nevertheless the similar effect is known not only in dielectrics but in all systems that involve matter. For instance the mechanical properties often show the same behavior with the frequency. The aim of this paper was to point out close connection between two universal low-frequency phenomena: 1/f noise and universal long time response to external influence on matter.

REFERENCES

1. P. H. Handel, Phys. Rev. A22, 745 (1980).
2. P. H. Handel, T. Musha, Proc. of Int. Conf. Noise in Phys. Syst. and 1/f Noise.(Elsevier Science Publications, 1983), p. 101.
3. Sv. Westerlung. Physica scripta. v.43, p. 174, 1991.
4. A. K. Jousher. Nature. v.267, p.673, 1977.

"(1/ω)" NOISE AND THE DYNAMICAL CASIMIR EFFECT

E. Sassaroli, Y.N. Srivastava and A. Widom
Northeastern University, Boston, MA 02115
and
University of Perugia and INFN, Perugia, Italy

ABSTRACT

The dynamical Casimir effect for a frequency modulated electromagnetic oscillator is described in terms of time reflections of the oscillator mode. From an experimental viewpoint, time reflections appear as radiated photon noise. The radiation noise temperature due to a modulation pulse is computed as a function of frequency ω. The noise temperature is shown to exhibit a "$(1/\omega)$" singularity in the limit $\omega \to 0$.

1. INTRODUCTION

The so-called Casimir effect describes non-classical electromagnetic forces between atoms, molecules and condensed matter objects : Van der Waals forces and force of attraction between two parallel perfectly conducting plates, for example[1,2].

On the other hand, the terminology dynamical Casimir effect shall be used when condensed matter objects are accelerated so that they can actually radiate photons[3]. Our purpose is to show how the dynamical Casimir effect serves as a source of "$1/\omega$" noise[4,5,6]. To be more specific , what we want to show is how a pulse can cause an electromagnetic signal moving backward in time to be reflected forward in time[7]. This appears in laboratory situations as an excess noise after the pulse and the spectrum shows a "$(1/\omega)$" noise behavior as the frequency ω approaches zero.

2. PHOTON RADIATION AND "1/ω" NOISE

To illustrate our point we shall consider a single photon oscillator with a time varying frequency modulation $v(t)$. The Hamiltonian is

$$H(t) = (1/2)P^2 + (1/2)[\omega^2 + v(t)]Q^2, \tag{1}$$

One may write $Q(t) = q(t) + q^*(t)$, with the equation of motion

$$[(d/dt)^2 + \omega^2 + v(t)]q(t) = 0, \tag{2a}$$

and the backward in time signal at frequency ω has the form

$$q_{in}(t) = e^{i\omega t}. \tag{2b}$$

The free oscillator propagator (Feynman-Stükelberg boundary condition)

$$D_o(t) = (i/2\omega)e^{-i\omega|t|}, \tag{3}$$

determines the complex amplitude $q(t)$ via Eqs.(2) and (3) as

$$q(t) = q_{in}(t) - \int_{-\infty}^{\infty} ds D_o(t-s)v(s)q(s). \tag{4}$$

Thus, for a given modulation pulse $v(t)$ there is a reflection forward in time and an attenuated signal backward in time

$$q(t \to \infty) = e^{i\omega t} + \rho(\omega)e^{-i\omega t}, \quad q(t \to -\infty) = \xi(\omega)e^{i\omega t}, \tag{5}$$

where

$$\xi(\omega) = 1 - i(1/2\omega)\int_{-\infty}^{\infty} dt e^{-i\omega t}v(t)q(t), \quad \rho(\omega) = -i(1/2\omega)\int_{-\infty}^{\infty} dt e^{i\omega t}v(t)q(t). \tag{6}$$

We here define the noise temperature $T_n(\omega)$ via the Boltzmann factor for the probability that a signal initially moving backward in time is reflected forward in time by the modulation pulse $v(t)$, i.e. $|\rho(\omega)|^2 = e^{-\hbar\omega/kT_n(\omega)}$ with

$$|\rho(\omega)|^2 + |\xi(\omega)|^2 = 1. \tag{7}$$

With the Hamiltonian of Eq.(1), the Heisenberg operators $Q(t)$ and $P(t)$ can be used to define creation $a^\dagger(t)$ and destruction $a(t)$ operators, i.e.

$$Q(t) = \sqrt{\hbar/2\omega}[a(t)e^{-i\omega t} + a^\dagger(t)e^{i\omega t}], \quad P(t) = -i\sqrt{\hbar\omega/2}[a(t)e^{-i\omega t} - a^\dagger(t)e^{i\omega t}]. \tag{8}$$

In the limit of times long before and long after the pulse[8]

$$a(t \to -\infty) = a_{in}, \quad a(t \to +\infty) = a_{out}. \tag{9}$$

The linear equations of motion dictate that

$$a_{out} = Ua_{in} + Va_{in}^\dagger, \tag{10}$$

where

$$|U|^2 - |V|^2 = 1, \tag{11a}$$

is required by the equal time commutation relation

$$[a(t), a^\dagger(t)] = 1. \tag{11b}$$

If long before the pulse the oscillator was at zero temperature,

$$a_{in}|0>=0, \tag{12a}$$

then long after the pulse the mean number of quanta in the oscillator

$$\bar{N} =<0|a^{\dagger}_{out}a_{out}|0> \tag{12b}$$

obeys

$$\bar{N} = |V|^2. \tag{12c}$$

The coefficients U and V in Eq.(10) are related to ρ and ξ of Eqs.(5) via

$$|U|^2 = |1/\xi|^2, \qquad |V|^2 = |\rho/\xi|^2, \tag{13}$$

so that Eq.(11a) is equivalent to Eq.(7), i.e.

$$|V|^2 = [|\rho|^2/(1-|\rho|^2)]. \tag{14}$$

Thus the noise temperature is determined by the Planck mean number of photons after the pulse

$$\bar{N}(\omega) = 1/(e^{\hbar\omega/kT_n(\omega)} - 1). \tag{15}$$

as determined by Eqs.(12c) and (14). The noise produced by the pulse is a clear manifestation of measuring signals due (in part) to time reflected photon propagation.

A pulse shape which produces a noise temperature that can be calculated in analytic form is given by

$$v(t) = [\Omega/cosh(\pi t/\tau)]^2, \tag{16}$$

where Ω determines the strength of the pulse and τ describes the pulse duration (see Fig.1). The reflection probability is given by

$$|\rho(\omega)|^2 = [cos^2\theta/(sinh^2(\omega\tau) + cos^2\theta)], \tag{17}$$

where $\theta = \sqrt{(\pi/2)^2 + \Omega^2\tau^2}$.

In Fig.2 a logarithmic plot of noise temperature as a function of frequency is exhibited, where $T_o = (\hbar/\tau k)$ and the pulse strength $\Omega = (2/\tau)$ for purposes of illustration. As $\omega \to 0$ there is a clear "$1/\omega$" rise in the noise temperature

$$T_n(\omega \to 0) = (\hbar/k\tau_c^2)(1/\omega), \tag{18}$$

where τ_c is the "characteristic scattering time" of the pulse.

Eq.(18) is our central result concerning "$(1/\omega)$" noise. For higher frequency, the noise temperature obeys $T_n(\omega \to \infty) \to (\hbar/2k\tau)$, characteristic of this particular pulse shape.

In the more general case, for a pulse of finite duration, the behavior of the transmission amplitude as $\omega \to 0$ is given by

$$|\xi(\omega)|^2 \to (\omega\tau_c)^2, \tag{19}$$

apart from exceptional values of the pulse strength. Eq.(19) leads directly to the "$1/\omega$" noise in Eq.(18).

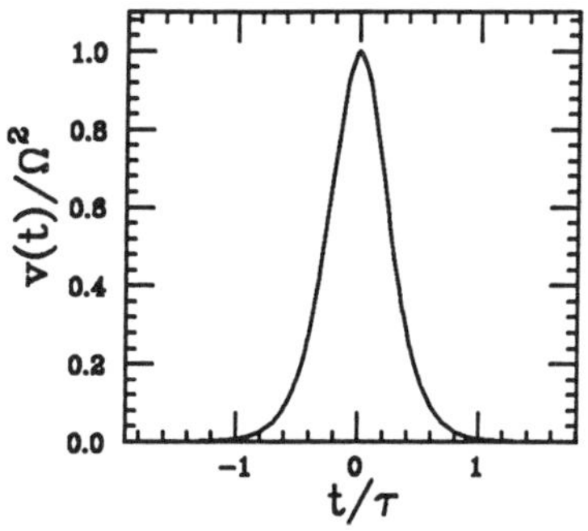

Fig.1. Modulation pulse $v(t)$ vs t.

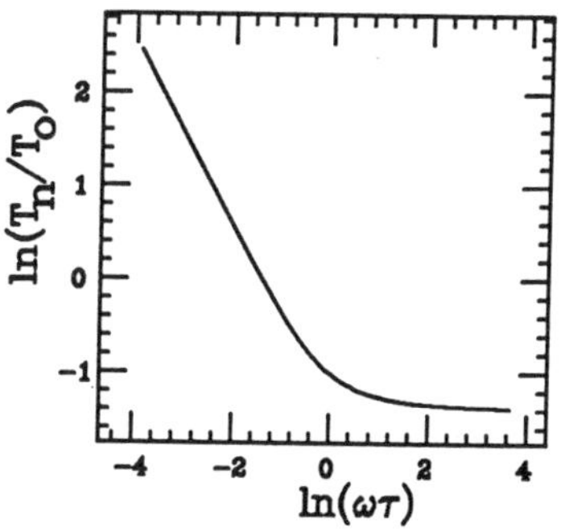

Fig.2. A plot of $ln(T_n/T_o)$ vs $ln(\omega\tau)$.

3. CONCLUSIONS

There are many physical systems that show the dynamical Casimir Effect. Examples include: (i) soft photon radiation noise in electronic devices[4,5,6], (ii) electromagnetic radiation from cavities whose geometry changes with time, e.g. light emission from bubble cavitation in a fluid[9].

The dynamical Casimir effect in theory allows for sources of radiation exploiting merely the modulation of the vacuum.

REFERENCES

1. G. Plunien, B. Müller, and W. Greiner, Physics Reports 134, 87 (1986).
2. Yu. S. Barash and V. L. Ginzburg, Sov. Phys. Usp. 143, 467 (1984).
3. J. Schwinger, Pro. Natl. Acad. Sci. 89, 4091 (1992).
4. P. H. Handel, Phys. Rev. Lett. 34, 1492 (1975); Phys. Rev. A22, 745 (1980).
5. C. M. Van Vliet, Solid-State Eletronics A34, 1 (1991).
6. Y. N. Srivastava and A. Widom, Physics Reports 148, 1 (1987).
7. R. P. Feynman, Phys. Rev. 74, 939 (1948); 80, 440 (1950).
8. S. A. Fulling, Aspects of Quantum Field Theory in Curved Space-Time (Cambridge University Press, 1989), p. 36, p. 138.
9. B. Barber and S. J. Putterman, Nature 352, 318 (1991).

THE NATURE OF FUNDAMENTAL 1/f NOISE

P.H. Handel
Department of Physics, University of Missouri-St. Louis
St. Louis, MO 63121, USA

ABSTRACT

Fundamental 1/f noise is shown to arise from nonlinearity and homogeneity, in particular from the nonlinear interaction of charged particles with their own field in QED.

I. INTRODUCTION

The universal presence of fluctuations with a spectral density proportional to 1/f both in collision-dominated solid state devices and in ballistic devices, or in vacuum tubes and beams of charged particles has been considered to be one of the most peculiar aspects of electrophysics, ever since people learned to amplify electrical signals in the first quarter of our century. The plot thickened after Hooge[1] pointed out in 1969 that the known inverse proportionality of 1/f noise with the volume or the number of carriers of the investigated sample can be generalized as a 1/N dependence with a universal proportionality constant of 2 10^{-3} independent of the nature of the sample. With N denoting the total number of current carriers in the sample, Hooge empirically claimed the simple expression 2 10^{-3}/fN would give the spectral density of fractional current, voltage, resistance and carrier mobility fluctuations in any condensed matter sample, including metals and electrolytes. The claim was initially successful, and the 1/f fluctuations were interpreted as mobility fluctuations. Later, his claim was found to be wrong. The coefficient was found to be dependent on the sample, as people had known before.

Hooge's empirical relation came three years after the magnetohydrodynamic turbulence model of 1/f noise[2,3] had provided for the first time a universal physical 1/f spectrum, derived directly from Maxwell's equations and classical Newtonian fluid mechanics without identifying the instabilities needed to trigger the turbulence at arbitrarily low bias, and therefore without a possibility of predicting the magnitude of 1/f noise in any device. Nevertheless, the success of the basic idea of the turbulence model in obtaining a universal 1/f spectrum of the chaotic current fluctuations, and in separating the inner dynamical equilibrium of turbulent chaos from the various instabilities which may have triggered the turbulence, deeply influenced the perception of 1/f noise and encouraged those who, like Hooge, were looking for a general unified explanation of the ubiquitous 1/f noise in 1966.

The search (1966-1974) for zero-threshold instabilities which could generate the turbulence was unsuccessful; all new instabilities found by this author had a finite bias threshold. Therefore he included the fundamental quantum unrest into the theory, instead of any particular type of instability. This quantization resulted in the creation of the conventional quantum 1/f theory[4-7] in 1975. The new theory turned out to be nothing but plain quantum electrodynamics (QED), showing for the first time that the most important infrared radiative corrections to the cross sections and process rates of quantum mechanics had been overlooked by Bloch, Nordsieck, Dirac, Schwinger, Feynman, and the other founders of QED, because they were time-dependent quantum fluctuations which were usually ignored in QED, being considered to be physically irrelevant. The spectral density of conventional quantum 1/f noise fractional fluctuations, derived below in Sec.II, is given by the simple formula $2\alpha A/fN$ which has the empirical Hooge form, with the coefficient 2 10^{-3} of Hooge replaced by $2\alpha A$. Here $\alpha=e^2/\hbar c=1/137$ is Sommerfeld's fine structure constant, and $A=2(\Delta \mathbf{v})^2/3\pi c^2$ is the square of the velocity change $\Delta \mathbf{v}$ in in the scattering process considered, in units of the speed of light c, multiplied by $2/3\pi$. The quantity $\alpha A/f$ is the well-known probability of bremsstrahlung emission into the unit frequency interval at f, per scattered particle. Van Vliet[8] has extended the proof to a level equivalent to all orders of perturbation theory in the Van Hove weak interaction limit, thereby confirming the $1/f^{1-\alpha A}$ form as the exact shape of the spectrum, which includes all infrared radiative corrections.

The conventional quantum 1/f theory was criticized initially because it yielded in general Hooge coefficients lower than the value $2\ 10^{-3}$ observed in most macroscopic samples. This weakness soon became a strength, when the focus of the scientific community shifted to mesoscopic samples and ultrasmall electronic devices, in which the observed Hooge parameters turned out to be close to the values predicted by the quantum 1/f theory, and much smaller (e.g. 10^{-8}), than Hooge's value of $2\ 10^{-3}$. The latter obtained a new partial lease on life for the limit of large samples and devices from the quantum 1/f theory[9,10] which yields $2\alpha/\pi = 4.6\ 10^{-3}$, in the coherent state limit. This limit, known as coherent quantum 1/f theory, is closely related to the conventional quantum 1/f theory, and also represents plain QED, this time for the case of large samples. The intermediary sizes are covered by a physical interpolation[10,20]. The coherent quantum 1/f effect has recently been derived from a well-known QED propagator (Sec. III). The whole quantum 1/f theory was verified experimentally by van der Ziel[11] and Tacano[11] both in collision dominated and ballistic devices and systems.

Fundamental 1/f noise in general is defined as any true 1/f fluctuation, independent of any particular parameters, which did not appear accidentally, through the approximation of a power law with some closeness to the 1/f dependence in a limited frequency interval, e.g., by a fortuitous superposition of Lorentzian spectra in a certain frequency interval. In Sec. IV we show that in any chaotic system the coexistence of nonlinearity and homogeneity guarantees a 1/f spectral density which we consider fundamental from a conceptual or epistemological point of view.

Finally, we show how coherent and conventional quantum 1/f noise can be combined in a heuristic interpolation formula with a clear physical basis, and how this combination can be considered from an ontological or constructive point of view as the most fundamental form of 1/f noise, while being a particular case of our universal sufficient criterion. The quantum 1/f theory and the universal sufficient criterion are basic physics, chapters of quantum mechanics or QED, with no additional hypotheses or free parameters of any kind, but are often misunderstood as models or obscure hypothetical theories, in part because of early misinterpretations by critics.

II. CONVENTIONAL QUANTUM 1/F EFFECT

This effect is present in any cross section or process rate involving charged particles or current carriers. The physical origin of quantum 1/f noise is easy to understand. Consider for example Coulomb scattering of current carriers, e.g., electrons on a center of force. The scattered electrons reaching a detector at a given angle away from the direction of the incident beam are described by DeBroglie waves of a frequency corresponding to their energy. However, some of the electrons have lost energy in the scattering process, due to the emission of Bremsstrahlung. Therefore, part of the outgoing DeBroglie waves is shifted to slightly lower frequencies. When we calculate the probability density in the scattered beam, we obtain also cross terms, linear both in the part scattered with and without bremsstrahlung. These cross terms oscillate with the same frequency as the frequency of the emitted bremsstrahlung photons. The emission of photons at all frequencies results therefore in probability density fluctuations at all frequencies. The corresponding current density fluctuations are obtained by multiplying the probability density fluctuations by the velocity of the scattered current carriers. Finally, these current fluctuations present in the scattered beam will be noticed at the detector as low frequency current fluctuations, and will be interpreted as fundamental cross section fluctuations in the scattering cross section of the scatterer. While incoming carriers may have been Poisson distributed, the scattered beam will exhibit super-Poissonian statistics, or bunching, due to this new effect which we may call quantum 1/f effect. The quantum 1/f effect is thus a many-body or collective effect, at least a two-particle effect, best described through the two-particle wave function and two-particle correlation function.

Let us estimate the magnitude of the quantum 1/f effect semiclassically by starting with the classical (Larmor) formula $2q^2\mathbf{a}^2/3c^3$ for the power radiated by a particle of charge q and acceleration **a**. The acceleration can be approximated by a delta function $\mathbf{a}(t) = \Delta\mathbf{v}\delta(t)$ whose Fourier transform $\Delta\mathbf{v}$ is constant and is the change in the velocity vector of the particle during the almost instantaneous scattering process. The one-sided spectral density of the emitted Bremsstrahlung power $4q^2(\Delta\mathbf{v})^2/3c^3$ is therefore also constant. The number $4q^2(\Delta\mathbf{v})^2/3hfc^3$ of emitted photons per unit frequency interval is obtained by dividing with the energy hf of one photon. The probability amplitude of photon emission $[4q(\Delta\mathbf{v})^2/3hfc^3]^{1/2}e^{i\gamma}$ is given by the square root of this photon number spectrum, including also a phase factor $e^{i\gamma}$. Let ψ be a representative Schrödinger catalogue wave function of the scattered outgoing charged particles, which is a single-particle function, normalized to the actual scattered particle concentration. The beat term in the probability density $\rho=|\psi|^2$ is linear both in this Bremsstrahlung amplitude and in the non-Bremsstrahlung amplitude. Its spectral density will therefore be given by the product of the squared probability amplitude of photon emission (proportional to 1/f) with the squared non-Bremsstrahlung amplitude which is independent of f. The resulting spectral density of fractional probability density fluctuations is obtained by dividing with $|\psi|^4$ and is therefore

$$|\psi|^{-4}S_{|\psi|^2}(f) = 8q^2(\Delta\mathbf{v})^2/3hfNc^3 = 2\alpha A/fN = j^{-2}S_j(f), \tag{1}$$

where $\alpha = e^2/\hbar c = 1/137$ is the fine structure constant and $\alpha A = 4q^2(\Delta\mathbf{v})^2/3hc^3$ is known as the infrared exponent in quantum field theory, and is known as the quantum 1/f noise coefficient, or Hooge constant, in electrophysics.

The spectral density of current density fluctuations is obtained by multiplying the probability density fluctuation spectrum with the squared velocity of the outgoing particles. When we calculate the spectral density of fractional fluctuations in the scattered current j, the outgoing velocity simplifies, and therefore Eq. (1) also gives the spectrum of current fluctuations $S_j(f)$, as indicated above. The quantum 1/f noise contribution of each carrier is independent, and therefore the quantum 1/f noise from N carriers is N times larger; however, the current j will also be N times larger, and therefore in Eq. (1) a factor N was included in the denominator for the case in which the cross section fluctuation is observed on N carriers simultaneously.

The fundamental fluctuations of cross sections and process rates are reflected in various kinetic coefficients in condensed matter, such as the mobility μ and the diffusion constant D, the surface and bulk recombination speeds s, and recombination times τ, the rate of tunneling j_t and the thermal diffusivity in semiconductors. Therefore, the spectral density of fractional fluctuations in all these coefficients is given also by Eq. (1).

When we apply Eq. (1) to a certain device, we first need to find out which are the cross sections σ or process rates which limit the current I through the device, or which determine any other device parameter P, and then we have to determine both the velocity change $\Delta\mathbf{v}$ of the scattered carriers and the number N of carriers simultaneously used to test each of these cross sections or rates. Then Eq. (1) provides the spectral density of quantum 1/f cross section or rate fluctuations. These spectral densities are multiplied by the squared partial derivative $(\partial I/\partial\sigma)^2$ of the current, or of the device parameter P of interest, to obtain the spectral density of fractional device noise contributions from the cross sections and rates considered. After doing this with all cross sections and process rates, we add the results and bring (factor out) the fine structure constant α as a common factor in front. This yields excellent agreement with the experiment[11] in a large variety of samples, devices and physical systems.

Eq. (1) was derived in second quantization, using the commutation rules for boson field operators. For fermions one repeats the calculation replacing in the derivation the commutators of field operators by anticommutators, which yields[7]

$$\rho^{-2}S_\rho(f) = j^{-2}S_j(f) = \sigma^{-2}S_\sigma(f) = 2\alpha A/f(N-1). \tag{2}$$

This causes no difficulties, since $N \geq 2$ for particle correlations to be defined, and is practically the same as Eq. (10), since usually $N >> 1$. Eqs. (1) and (2) suggest a new notion of physical cross sections and process rates which contain 1/f noise, and express a fundamental law of physics, important in most high-technology applications[7].

We conclude that the conventional quantum 1/f effect can be explained in terms of interference beats between the part of the outgoing DeBroglie waves scattered without bremsstrahlung energy losses above the detection limit (given in turn by the reciprocal duration T of the 1/f noise measurement) on one hand, and the various parts scattered with bremsstrahlung energy losses; but there is more to it than that: exchange between identical particles is also important. This, of course, is just one way to describe the reaction of the emitted bremsstrahlung back on the scattered current. This reaction, itself an expression of the nonlinearity introduced by the coupling of the charged-particle field to the electromagnetic field, thus reveals itself as the cause of the quantum 1/f effect, and implies that the effect can not be obtained with an independent boson model. The effect, just like the classical turbulence-generated 1/f noise[2,3], is a result of the scale-invariant nonlinearity of the equations of motion describing the coupled system of matter and field. Ultimately, therefore, this nonlinearity is the source of the 1/f spectrum in both the classical and quantum form of the theory. We can say that the quantum 1/f effect is an infrared divergence phenomenon, this divergence being the result of the same nonlinearity. The quantum 1/f effect is, in fact, the first time-dependent infrared radiative correction. Finally, it is also deterministic in the sense of a well determined wave function, once the initial phases γ of all field oscillators are given. In quantum mechanical correspondence with its classical turbulence analog[2,3], the new effect is therefore a quantum manifestation of classical chaos which we can take as the definition of a certain type of quantum chaos.

III. DERIVATION OF COHERENT QUANTUM 1/F NOISE

The present derivation is based on the well-known new propagator $G_s(x'-x)$ derived relativistically[12,13] in 1975 in a new picture required by the infinite range of the Coulomb potential. The corresponding nonrelativistic form[14] was provided by Zhang and Handel:

$$-i\langle\Phi_o|T\psi_{s'}(x')\psi_s^\dagger(x)|\Phi_o\rangle \equiv \delta_{ss'}\, G_s(x'-x)$$

$$= (i/V)\sum_{\mathbf{p}}\{\exp i[\mathbf{p}(\mathbf{r}-\mathbf{r}')-\mathbf{p}^2(t-t')/2m]/\hbar\}n_{\mathbf{p},s}$$

$$\times\{-i\mathbf{p}(\mathbf{r}-\mathbf{r}')/\hbar+i(m^2c^2+\mathbf{p}^2)^{1/2}(t-t')(c/\hbar)\}^{\alpha/\pi}. \quad (3)$$

Here $\alpha=e^2/\hbar c=1/137$ is Sommerfeld's fine structure constant, $n_{\mathbf{p},s}$ the number of electrons in the state of momentum $\mathbf{p}$ and spin s, m the rest mass of the fermions, $\delta_{ss'}$ the Kronecker symbol, c the speed of light, $x=(\mathbf{r},t)$ any space-time point and V the volume of a normalization box. T is the time-ordering operator which orders the operators in the order of decreasing times from left to right and multiplies the result by $(-1)^P$, where P is the parity of the permutation required to achieve this order. For equal times, T normal-orders the operators, i.e., for t=t' the left-hand side of Eq. (3) is $i\langle\Phi_o|\psi_s^\dagger(x)\psi_{s'}(x')|\Phi_o\rangle$. The state Φ_o of the N electrons is described by a Slater determinant of single-particle orbitals.

The resulting spectral density coincides with the result $2\alpha/\pi fN$, first derived[8] directly from the coherent state of the electromagnetic field of a physical charged particle. The connection with the conventional quantum 1/f effect was suggested later[9].

To calculate the current autocorrelation function we need the density correlation function, which is also known as the two-particle correlation function. The two-particle correlation function is defined by

$$\langle\Phi_o|T\psi_s^\dagger(x)\psi_s(x)\psi_{s'}^\dagger(x')\psi_{s'}(x')|\Phi_o\rangle = \langle\Phi_o|\psi_s^\dagger(x)\psi_s(x)|\Phi_o\rangle\langle\Phi_o|\psi_{s'}^\dagger(x')\psi_{s'}(x')|\Phi_o\rangle$$

$$- \langle\Phi_o|T\psi_{s'}(x')\psi_s^\dagger(x)|\Phi_o\rangle\langle\Phi_o|T\psi_s(x)\psi_{s'}^\dagger(x')|\Phi_o\rangle. \quad (4)$$

The first term can be expressed in terms of the particle density of spin s, $n/2 = N/2V = \langle\Phi_o|\psi_s^\dagger(x)\psi_s(x)|\Phi_o\rangle$, while the second term can be expressed in terms of the Green function (1) in the form

$$A_{ss'}(x-x') \equiv \langle\Phi_o|\psi_s^\dagger(x)\psi_{s'}^\dagger(x')\psi_{s'}(x')\psi_s(x)|\Phi_o\rangle = (n/2)^2 + \delta_{ss'}\, G_s(x'-x)G_s(x-x'). \qquad (5)$$

The "relative" autocorrelation function $A(x-x')$ describing the normalized pair correlation independent of spin is obtained by dividing by n^2 and summing over s and s'

$$A(x-x') = 1 - (1/n^2)\sum_s G_s(x-x')G_s(x'-x)$$

$$= 1 - (1/N^2)\sum_s \sum_{\mathbf{pp'}}\{\exp i[(\mathbf{p}-\mathbf{p}')(\mathbf{r}-\mathbf{r}')-(\mathbf{p}^2-\mathbf{p}'^2)(t-t')/2m]/\hbar\}n_{\mathbf{p},s}n_{\mathbf{p}',s}$$
$$\times\{\mathbf{p}(\mathbf{r}-\mathbf{r}')/\hbar-(m^2c^2+\mathbf{p}^2)^{1/2}(t-t')(c/\hbar)\}^{\alpha/\pi}$$
$$\times\{\mathbf{p}'(\mathbf{r}-\mathbf{r}')/\hbar-(m^2c^2+\mathbf{p}'^2)^{1/2}(t-t')(c/\hbar)\}^{\alpha/\pi}. \qquad (6)$$

Here we have used Eq. (1). We now consider a beam of charged fermions, e.g., electrons, represented in momentum space by a sphere of radius p_F, centered on the momentum $\mathbf{p}_o$ which is the average momentum of the fermions. The energy and momentum differences between terms of different $\mathbf{p}$ are large, leading to rapid oscillations in space and time which contain only high-frequency quantum fluctuations. The low-frequency and low-wavenumber part A_l of this relative density autocorrelation function is given by the terms with $\mathbf{p}=\mathbf{p}'$

$$A_l(x-x') = 1 - (1/N^2)\sum_s \sum_{\mathbf{p}} n_{\mathbf{p},s}$$
$$\times\{\mathbf{p}(\mathbf{r}-\mathbf{r}')/\hbar-(m^2c^2+\mathbf{p}^2)^{1/2}(t-t')(c/\hbar)\}^{2\alpha/\pi} \qquad (7)$$

$$\approx 1 - (1/N)|\mathbf{p}_o(\mathbf{r}-\mathbf{r}')/\hbar-mc^2\tau/\hbar|^{2\alpha/\pi} \qquad \text{for } p_F \ll |p_{o3}-mc^2\tau/z|. \qquad (8)$$

Here we have used the mean value theorem, considering the $2\alpha/\pi$ power as a slowly varying function of $\mathbf{p}$ and neglecting $\mathbf{p}_o$ in the coefficient of $\tau \equiv t-t'$, with $z \equiv |\mathbf{r}-\mathbf{r}'|$. Using the identity[15], with arbitrarily small cutoff ω_o, we obtain from Eq. (8) with $\theta \equiv |\mathbf{p}_o(\mathbf{r}-\mathbf{r}')/\hbar-mc^2\tau/\hbar|$ the exact form

$$A_l(x-x') = 1 + [(2\alpha/\pi N)\int_{\omega_o}^{\infty}(mc^2/\hbar\omega)^{2\alpha/\pi}\cos(\theta\omega)d\omega/\omega]$$

$$\times\{\cos\alpha + (2\alpha/\pi)\sum_{n=0}^{\infty}(\theta\omega_o)^{2n-2\alpha/\pi}[(2n)!(2n-2\alpha/\pi)]^{-1}\}^{-1}. \qquad (9)$$

This indicates a $\omega^{-1-2\alpha/\pi}$ spectrum and a 1/N dependence of the spectrum of fractional fluctuations in density n and current j, if we neglect the curly bracket in the denominator which is very close to unity for very small ω_o. The fractional autocorrelation of current fluctuations δj is obtained by multiplying Eq. (5) on both sides with ep_o/m, and dividing by $(enp_o/m)^2$ which is the square of the average current density j, instead of just dividing by n^2. It is the same as the fractional autocorrelation for quantum density fluctuations. Then Eq. (9) for the coherent Quantum Electrodynamical chaos process in electric currents can be written also in the form

$$S_{\delta j/j}(k) \approx [2\alpha/\pi\omega N][mc^2/\hbar\omega]^{2\alpha/\pi} \approx 2\alpha/\pi\omega N = 0.00465/\omega N. \qquad (10)$$

Being observed in the presence of a constant applied field, these fundamental quantum current fluctuations are usually interpreted as mobility fluctuations[1]. Most of the conventional quantum 1/f fluctuations in physical cross sections and process rates are also mobility fluctuations, but some are also in the recombination speed or tunneling rate.

IV. SUFFICIENT CRITERION FOR FUNDAMENTAL 1/F NOISE

In spite of the practical success of our quantum 1/f theory in explaining electronic 1/f noise in most high tech devices, and in spite of the conceptual success of our earlier classical turbulence approach to 1/f noise, the question about the origin of nature's omnipresent 1/f spectra remained unanswered. During the last three decades, we have claimed repeatedly that nonlinearity is a general cause of 1/f noise. The present paper proves that nonlinearity always leads to a 1/f spectrum if homogeneity is also present in the equation(s) of motion. Specifically, if the system is described in terms of the dimensionless vector function $\mathbf{Y}(\mathbf{x},t)$ by the mth order nonlinear differential equation

$$\partial\mathbf{Y}/\partial t + \mathbf{F}(\mathbf{x}, \mathbf{Y}, \partial\mathbf{Y}/\partial x_1 ... \partial\mathbf{Y}/\partial x_n, \partial^2\mathbf{Y}/\partial x_1^2 ..... \partial^m\mathbf{Y}/\partial x_n^m) = 0 \tag{11}$$

a 1/f spectrum is obtained if the nonlinear function $\mathbf{F}$ satisfies the homogeneity condition

$$\mathbf{F}[\lambda\mathbf{x}, \mathbf{Y}, \partial\mathbf{Y}/(\lambda\partial x_1) ... \partial\mathbf{Y}/(\lambda\partial x_n), \partial^2\mathbf{Y}/(\lambda\partial x_1)^2 ..... \partial^m\mathbf{Y}/(\lambda\partial x_n)^m]$$
$$= \lambda^{-p}\mathbf{F}(\mathbf{x}, \mathbf{Y}, \partial\mathbf{Y}/\partial x_1 ... \partial\mathbf{Y}/\partial x_n, \partial^2\mathbf{Y}/\partial x_1^2 ..... \partial^m\mathbf{Y}/\partial x_n^m), \tag{12}$$

for any real number λ. The order of homogeneity is the number $-p$. Performing a Fourier transformation of Eq. (10) with respect to the vector $\mathbf{x}(x_1, x_2, ...x_n)$, we get in terms of the Fourier-transformed wavevector $\mathbf{k}$ the nonlinear integro-differential equation

$$\partial\mathbf{y}(\mathbf{k},t)/\partial t + \mathbf{G}[\mathbf{k}, \mathbf{y}(\mathbf{k},t), k_1\mathbf{y}(\mathbf{k},t) ... k_n\mathbf{y}(\mathbf{k},t), k_1^2\mathbf{y}(\mathbf{k},t) ..... k_n^m\mathbf{y}(\mathbf{k},t)] = 0, \tag{13}$$

where $\mathbf{y}(\mathbf{k},t)$ is the Fourier transform of $\mathbf{Y}(\mathbf{x},t)$. Due to Eq. (12), the nonlinear integro-differential operator $\mathbf{G}$ satisfies the relation

$$\mathbf{G}[\lambda\mathbf{k}, \mathbf{y}, \lambda k_1\mathbf{y} ... \lambda k_n\mathbf{y}, (\lambda k_1)^2\mathbf{y} ..... (\lambda k_n)^m\mathbf{y}]$$
$$= \lambda^p\mathbf{G}[\mathbf{k}, \mathbf{y}, k_1\mathbf{y} ... k_n\mathbf{y}, k_1^2\mathbf{y} ..... k_n^m\mathbf{y}], \tag{14}$$

where the integration differentials dk, dk', etc., are excepted from replacement with λdk, λdk', etc. Eq. (13) can thus be rewritten in the form

$$d\mathbf{y}/d(t/\lambda^p) + G[\lambda\mathbf{k}, \mathbf{y}, \lambda k_1\mathbf{y} ... \lambda k_n\mathbf{y}, (\lambda k_1)^2\mathbf{y} ..... (\lambda k_n)^m\mathbf{y}] = 0, \tag{15}$$

Taking $\lambda=1/k$, where $k=|\mathbf{k}|=(k_1^2+....+k_n^2)^{1/2}$, and setting $k^p t=z$, we notice that k has been eliminated from the dynamical equation, and only $\mathbf{k}/k$ is left. This means that there is no privileged scale left for the system in $\mathbf{x}$ or $\mathbf{k}$ space, other than the scale defined by the given time t, and expressed by the dependence on z. We call this property of the dynamical system "*sliding-scale invariance*".

In certain conditions, instabilities of a solution of Eq. (10) may generate chaos, or turbulence. In a sufficiently large system described by the local dynamical equation (10), in which the boundary conditions become immaterial, homogeneous, isotropic turbulence, (chaos) can be obtained, with a spectral density determined only by Eq. (10). The stationary autocorrelation function $A(\tau)$ is defined as an average scalar product, the average being over the turbulent ensemble

$$A(\tau) = <\mathbf{Y}(\mathbf{x},t)\mathbf{Y}(\mathbf{x},t+\tau)> = \int <\mathbf{y}(\mathbf{k},t)\mathbf{y}(\mathbf{k},t+\tau)> d^n k = \int u(k,z) d^n k. \tag{16}$$

Here we have introduced the scalar

$$u(k,z) = <\mathbf{y}(\mathbf{k},t)\mathbf{y}(\mathbf{k},t+\tau)> \tag{17}$$

of homogeneous, isotropic chaos (turbulence), which depends only on $|\mathbf{k}|$ and $z=k^p\tau$. All integrals are from minus infinity to plus infinity. The chain of integro-differential equations for the correlation functions of any order obeys the same sliding-scale invariance which we have noticed in the fundamental dynamical equation above. *Therefore, in isotropic, homogeneous, conditions, u can only depend on k and z.* Furthermore, the direct dependence on k must reflect this sliding-scale invariance, and is therefore of the form

$$u(k,z) = k^{-n}v(z). \tag{18}$$

Indeed, only this form insures that $u(k,z)d^n k$ and therefore also the corresponding integrals and multiple convolutions in k space have the necessary sliding-scale invariance.

According to the Wiener-Khintchine theorem, the spectral density is the Fourier-transform of A(t),

$$S_y(f) = \int e^{2\pi i f\tau} A(\tau) d\tau = (1/f) \int e^{2\pi i t'} \int k'^{-n} v(z) d^n k' dt' = C/f, \tag{19}$$

where we have set $f\tau=t'$, $k^n=fk'^n$, $z=k^n\tau=k'^n t'$, and the integral

$$C = \int e^{2\pi i t'} \int k'^{-n} v(z) d^n k' dt' = \int e^{2\pi i t'} \int k''^{-n} v(k''^n) d^n k'' dt' \tag{20}$$

is independent of f. We have defined the vector $\mathbf{k}''=t'^{1/n}\,\mathbf{k}$.

The general form of our criterion considers a system described in terms of the integro-differential system of equations

$$\Phi[t, \mathbf{x}, \mathbf{Y}, \partial\mathbf{Y}/\partial t, \partial\mathbf{Y}/\partial x_1 ... \partial\mathbf{Y}/\partial x_n, \partial^2\mathbf{Y}/\partial t^2, \partial^2\mathbf{Y}/\partial x_1^2 ..... \partial^m\mathbf{Y}/\partial x_n^m) = 0 \tag{21}$$

where the vector function Φ may be nonlinear in any of its arguments, with the partial derivative with respect to . *If a number θ exists such that Eq. (11) implies*

$$\Phi[\lambda^\theta t, \lambda\mathbf{x}, \mathbf{Y}, \partial\mathbf{Y}/\lambda^\theta\partial t, \partial\mathbf{Y}/\lambda\partial x_1 ... \partial\mathbf{Y}/\lambda\partial x_n, \partial^2\mathbf{Y}/\lambda^{2\theta}\partial t^2, \partial^2\mathbf{Y}/\lambda^2\partial x_1^2 ..... \partial^m\mathbf{Y}/\lambda^m\partial x_n^m) = 0 \tag{22}$$

for any real number λ, the power spectral density of any chaotic solution for the vector function ***Y*** *defined by Eq. (11) is proportional to 1/f.*

Here we have assumed that there are no boundary conditions associated with Eq. (22), or that any boundary conditions included would satisfy the same homogeneity conditions.

In conclusion, nonlinearity + homogeneity = 1/f noise, provided the system is chaotic. The ultimate cause of the ubiquitous 1/f noise in nature is the omnipresence of nonlinearities (no matter how weak) and homogeneity. The latter is finally related to rotational (or Lorentz) invariance and therefore to the isotropy of space (or space-time). All our four specific theories of 1/f chaos in nonlinear systems are just special cases to which this criterion is applicable. They include our magneto-plasma theory of turbulence for current carriers in intrinsic symmetric semiconductors[2] (1966), our similar theory for metals[3] (1971), the quantum 1/f theory[4-7] (pure QED, 1975, see below), and the spectral theory of Musha's highway traffic turbulence results[16] (1989). Applied to the motion of a nonlinearly interacting chain of atoms, it predicts no 1/f spectrum. Starting from a wrong defining equation of the chain, both our criterion and direct calculation allowed for 1/f noise in a special case[16], but the correct defining equation does not fulfill the criterion, and no 1/f spectrum is expected. However, 1/f fluctuations in phonon number, in frequency, and in phase are predicted by the criterion, are derived directly[17] with the quantum 1/f theory, and have been experimentally verified[17,18], in piezoelectric crystals.

V. APPLICATION TO QED: QUANTUM 1/F THEORY AS A SPECIAL CASE

The nonlinearity causing the 1/f spectrum of turbulence in both semiconductors and metals is caused by the reaction of the field generated by charged particles and their currents back on themselves. The same nonlinearity is present in quantum electrodynamics (QED), where it causes the infrared divergence, the infrared radiative corrections for cross sections and process rates, and the quantum 1/f effect. We shall prove this on the basis of our sufficient criterion for 1/f spectral density in chaotic systems.

Consider a beam of charged particles propagating in a well-defined direction which we shall call the x direction, so that the one-dimensional Schrödinger equation describes the longitudinal fluctuations in the concentration of particles. Considering the non-relativistic case which is encountered in most quantum 1/f noise applications, we write in second quantization the equation of motion for the Heisenberg field operators ψ of the in the form

$$i\hbar\partial\psi/\partial t = (1/2m)[-i\hbar\nabla - (e/c)\,A]^2\psi, \tag{23}$$

With the non-relativistic form $J = -i\hbar\psi^*\nabla\psi/m$ + hermitian conjugate, and with

$$A(x,y,z,t) = (\hbar/2cmi)\int\frac{[\psi^*\nabla\psi-\psi\nabla\psi^*]}{|x-x'|}dx' \tag{24}$$

we obtain

$$i\hbar\partial\psi/\partial t = (1/2m)\Big[-i\hbar\nabla - (e\hbar/2c^2mi)\int\frac{[\psi^*\nabla\psi-\psi\nabla\psi^*]}{|x-x'|}dx'\Big]^2\psi. \tag{25}$$

At very low frequencies or wave numbers the last term in rectangular brackets is dominant on the r.h.s., leading to

$$i\hbar\partial\psi/\partial t = (-1/2m)\Big[(e\hbar/2c^2m)\int\frac{[\psi^*\nabla\psi-\psi\nabla\psi^*]}{|x-x'|}dx'\Big]^2\psi. \tag{26}$$

For x replaced by λx, and x' replaced by $\lambda x'$, we obtain

$$i\hbar\partial\psi/\partial t = (-1/2m)\Big[(e\hbar/2c^2m)\int\frac{[\psi^*\nabla/\lambda\psi-\psi\nabla/\lambda\psi^*]}{\lambda|x-x'|}\lambda^3dx'\Big]^2\psi = \lambda^2H\psi = \lambda^{-p}H\psi. \tag{27}$$

This satisfies our homogeneity criterion with $p=-2$. Our sufficient criterion only requires homogeneity, with any value of the weight p, for the existence of a 1/f spectrum in chaos. Therefore, we expect a 1/f spectrum of quantum current-fluctuations, i.e., of cross sections and process rates in physics, as derived in detail earlier[1-3,12-14]. This is in agreement with the well-known, and experimentally verified, results of the Quantum 1/f Theory.

In conclusion, we realize that, both in classical and quantum mechanical nonlinear systems, the limiting behavior at low wave numbers is usually expressed by homogeneous functional dependences, leading to fundamental 1/f spectra on the basis of our criterion.

VI. DISCUSSION

The derivations of conventional and coherent quantum 1/f noise in Sec. II and III correspond to different physical situations. These two situations have been discussed on the first page of the 1966 turbulence paper, at the beginning of this long journey which led us from the classical hydromagnetic or plasma turbulence to quantum 1/f noise and the general sufficient criterion. The discussion of these two situations was repeated identically[10] for the quantized form of our turbulence theory, i.e., for the two related quantum 1/f effects in 1985. It shows us that conventional quantum 1/f noise is observed in small samples, for which most of the drift energy of the current carriers is included in the sum of their individual kinetic energies $mv^2/2$. For larger samples, and larger values[20] of the parameter s measuring this proportion numerically, most of the drift energy of the carriers is in their collective magnetic energy $LI^2/2$. The transition between the two situations is given by a physical interpolation formula[10], and is the focus a present research effort discussed in the paper by Handel and Zhang[20] in this volume.

Our criterion shows how homogeneity provides the ingredient leading from nonlinearity to 1/f noise. Physically, the homogeneity is required both by the physical requirement of dimensional homogeneity of terms in the equations of physics, and by the invariance of the three-dimensional space with respect to rotations, i.e., by the isotropy of space, which requires x_1, x_2, and x_3 to enter in the same way into the basic laws of nature. In general, we conclude that the ubiquity of the 1/f spectrum is caused by the omnipresence of nonlinearities, no matter how small, and by the simultaneous requirement of rotational and Lorentz invariance which shape the world of classical and relativistic physics respectively. In general, we conclude that ontologically, i.e., from the construction of our world with quarks and leptons, quantum 1/f noise theory gives the cause of fundamental 1/f noise, while epistemologically, i.e., in the world of general notions, the combination of nonlinearity and homogeneity required by our general sufficient criterion is the ultimate cause of all fundamental 1/f noise, including the ontlogically primordial quantum 1/f noise as a special case. Mathematically, this happens in all fundamental 1/f spectra on the basis of the idempotence of $1/f^{1-\varepsilon}$ with respect to convolutions in the limit $\varepsilon \to 0$, with $\varepsilon = \alpha A$ in the case of quantum 1/f noise. In practice, however, the idempotent property of 1/f does not allow us to distinguish which systems will show 1/f fluctuations, while the general sufficient criterion, first presented at the Symposium on 1/f Nose and Chaos In Tokyo, March 1991, allows us to easily recognize the systems which generate 1/f spectra, if their mathematical definition is given in terms of a dynamical system of nonlinear integro-differential equations, or in simpler terms.

We note that our sufficient criterion explains the ubiquity of 1/f noise through a homogeneity which can be established sometimes even without knowing the exact form of the dynamical equation(s) governing a nonlinear system. The derivation of the criterion shows that it is obviously connected with (actually based on) the idempotent property of the 1/f spectrum with respect to the convolution operation. Therefore the 1/f spectrum corresponds to an accumulation point in Hilbert space, as was first demonstrated[2,3] directly in 1966 and 1971. Due to the divergence of the integral of 1/f at f=0, this author reformulated this accumulation point property in dimensional analysis terms before submitting his paper[19] for publication in 1980; in this form, the argument is more elegant and avoids the divergence at f=0. The idea was rediscovered by Kawai in Japan and is presented by him in the present volume without reference to the 1980 paper, although he was informed about it in time.

The author acknowledges the support of the Air Force Office of Scientific Research and of the National Science Foundation.

REFERENCES

1. F.N. Hooge, Phys. Lett. A29, 139 (1969); Physica 83B, 19 (1976).

2. P.H. Handel: "Instabilities, Turbulence and Flicker-Noise in Semiconductors I, II and III", Zeitschrift für Naturforschung 21a, 561-593 (1966).

3. P.H. Handel: "Turbulence Theory for the Current Carriers in Solids and a Theory of 1/f Noise", Phys. Rev. A3, 2066 (1971).

4. P.H. Handel: "1/f Noise - an 'Infrared' Phenomenon", Phys. Rev. Letters 34, 1492 - 1494 (1975); "Nature of 1/f Phase Noise", Phys. Rev. Letters 34, 1495-1498 (1975).

5. P.H. Handel: "Quantum Approach to 1/f Noise", Phys. Rev. 22A, p. 745 (1980).

6. P.H. Handel and D. Wolf: "Characteristic Functional of Quantum 1/f Noise", Phys. Rev. A26, 3727-30 (1982).

7. P.H. Handel: "Starting Points of the Quantum 1/f Noise Approach", Submitted to Physical Review B; "Fundamental Quantum 1/f Fluctuation of Physical Cross Sections and Process Rates", submitted to Phys. Rev. Letters.

8. *C.M. Van Vliet: "Quantum Electrodynamical Theory of Infrared Effects in Condensed Matter. I. and II.", Physica A165, 101-155 (1990).

9. P.H. Handel, "Any Particle Represented by a Coherent State Exhibits 1/f Noise" in *"Noise in Physical Systems and 1/f Noise"*, (Proceedings of the VIIth International Conference on Noise in Physical Systems and 1/f Noise) edited by M. Savelli, G. Lecoy and J.P. Nougier (North - Holland, Amsterdam, 1983), p. 97.

10. P.H. Handel, "Coherent States Quantum 1/f Noise and the Quantum 1/f Effect" in *"Noise in Physical Systems and 1/f Noise"* (Proceedings of the VIIIth International Conference on Noise in Physical Systems and 1/f Noise) edited by A. D'Amico and P. Mazzetti, Elsevier, New York, 1986, p. 469.

11. A. van der Ziel, "Unified Presentation of 1/f Noise in Electronic Devices; Fundamental 1/f Noise Sources", Proc. IEEE 76, 233-258 (1988 review paper); "The Experimental Verification of Handel's Expressions for the Hooge Parameter", Solid-State Electronics 31, 1205-1209 (1988); "Semiclassical Derivation of Handel's Expression for the Hooge Parameter", J. Appl. Phys. 63, 2456-2455 (1988); 64, 903-906 (1988); A.N. Birbas et al., J. Appl. Phys. 64, 907-912 (1988); A. van der Ziel et. al., "Extensions of Handel's 1/f Noise Equations and their Semiclassical Theory", Phys. Rev. B 40, 1806-1809 (1989); M. Tacano, Proc. XI. Int. Conf. on Noise in Physical Systems and 1/f Fluctuations, T. Musha, S. Sato and Y. Mitsuaki Editors, Ohmsha Publ. Co., Tokyo 1991, pp. 167-170; M. Tacano, Proc. Fifth van der Ziel Conference "Quantum 1/f Noise and other Low Frequency Fluctuations" AIP Conference proceedings #282, P.H. Handel and A.L. Chung Editors, 1992; see also "Quantum 1/f Bibliography" by P.H. Handel, unpublished.

12. D. Zwanziger, Phys. Rev. D7, 1082 (1973); Phys. Rev. Lett. 30, 934 (1973); Phys. Rev. D11, 3481 and 3504 (1975).

13. T.W.B. Kibble, Phys. Rev. 173, 1527; 174, 1882; 175,1624 (1968); J. Math. Phys. 9, 315 (1968).

14. Y. Zhang and P. H. Handel, Proc. Fifth van der Ziel Conference "Quantum 1/f Noise and other Low Frequency Fluctuations" AIP Conference proceedings #282, P.H. Handel and A.L. Chung Editors, 1992.

15. J.S. Gradshteiyn and I.M. Ryzhik, *"Table of Integrals, Series and Products"* Sec. 3.761, No. 9 and No. 7, Academic Press, New York 1965.

16. P.H. Handel, Proc. XI. Int. Conf. on Noise in Physical Systems and 1/f Fluctuations, T. Musha, S. Sato and Y. Mitsuaki Editors, Ohmsha Publ. Co., Tokyo 1991, pp. 151-157.

17. F.L. Walls, P.H. Handel, R. Besson and J.J. Gagnepain: "A New Model relating Resonator Volume to 1/f Noise in BAW Quartz Resonators", Proc. 46. Annual Frequency Control Symposium, pp.327-333, 1992.

18. T. Musha, G. Borbely and M. Shoji, Phys. Rev. lett. 64, 2394 (1990).

19. P.H. Handel, T. Sherif, A. van der Ziel, K.M. van Vliet and E.R. Chenette: "Towards a More General Understanding of 1/f Noise", submitted Physics Letters early in 1980, rejected by T. Musha in the same year, even after I answered his objections. This unpublished manuscript received a wide distribution, even in Japan.

20. P.H. Handel and Y. Zhang in this volume.

GENERAL DISCUSSION OF COHERENT QUANTUM 1/F NOISE IN SMALLER SEMICONDUCTOR SAMPLES

Y. Zhang and P.H. Handel
Department of Physics, University of Missouri St. Louis, Mo 63121, USA

ABSTRACT

Various avenues of research into the connection between coherent and conventional quantum 1/f noise are discussed in terms of the relevant propagator.

INTRODUCTION

From the beginning of the theory of fundamental 1/f noise in semiconductors and metals two situations were distinguished[1]. The first, applicable to small semiconductor samples and very small (mesoscopic) metallic samples, has most of the energy excess $Nmv_d^2/2$ present in the stationary state carrying a finite current through the sample, (excess over the energy of the equilibrium state), contained in the sum of the individual kinetic energies of the N current carriers $\Sigma_i\, mv_i^2/2$. Here the velocities v_i of the carriers of mass m contain a small drift term v_d. The second, applicable in larger semiconductor or metal has most of that energy excess contained in the collective magnetic energy of the current carrying state, $\int(B^2/8\pi)d^3x = LI^2/2$. The ratio s of this magnetic energy to the kinetic energy excess is roughly equal[1,2] to the number of carriersN' per unit length of the sample, multiplied by the classical radius of the electron $r_o=e^2/mc^2$: $s = N'r_o$. This situation was considered already in Handel's classical magnetic turbulence theory[1,3].

In the first situation conventional quantum 1/f noise is applicable for fluctuations in physical scattering cross sections σ, in physical process rates Γ, and in mobility μ or diffusion coefficient D, (the latter two only if exclusively limited by σ or Γ)

$$\sigma^{-2}S_\sigma(f) = \Gamma^{-2}S_\Gamma(f) = \mu^{-2}S_\mu(f) = 2\alpha A/fN, \qquad (s<<1) \qquad (1)$$

because in this case the coherent, collective, term in the Hamiltonian is negligible. In the second case, however, the coherent quantum 1/f effect[4] is dominant

$$j^{-2}S_j(f) = \mu^{-2}S_\mu(f) = 2\alpha/\pi fN, \qquad (s>1) \qquad (2)$$

because the incoherent, kinetic, term can be neglected.

For the intermediary case, an interpolation formula was proposed[2]

$$j^{-2}S_j(f) = \mu^{-2}S_\mu(f) = (2\alpha/fN)\{A/(s+1) + s/\pi(s+1)\}, \qquad (3)$$

which is heuristic. The main purpose of this paper is to discuss various avenues to derive the correct form for the intermediary situation, and to consider initially the problem of coherent quantum 1/f noise in the $s\leq1$ case.

We start in Sec. II with a brief derivation of coherent quantum 1/f noise from a quantum-electrodynamical propagator. In Sec. III we discuss various lines of attack in the derivation of a propagator which would be applicable to small condensed matter samples and devices. Finally, in Sec. IV we discuss the difficulties present and the physics which has still to be incorporated.

II. COHERENT QUANTUM 1/F NOISE

The present derivation is based on the well-known new propagator $G_s(x'-x)$ derived relativistically[6,7] in 1975 in a new picture required by the infinite range of the Coulomb potential. The corresponding nonrelativistic form[8] was provided by Zhang and Handel:

$$-i\langle\Phi_o|T\psi_{s'}(x')\psi_s^{\dagger}(x)|\Phi_o\rangle \equiv \delta_{ss'}\,G_s(x'-x)$$

$$= (i/V)\sum_{\mathbf{p}}\{\exp i[\mathbf{p}(\mathbf{r}-\mathbf{r}')-\mathbf{p}^2(t-t')/2m]/\hbar\}n_{\mathbf{p},s}$$
$$\times\{-i\mathbf{p}(\mathbf{r}-\mathbf{r}')/\hbar+i(m^2c^2+\mathbf{p}^2)^{1/2}(t-t')(c/\hbar)\}^{\alpha/\pi}. \tag{4}$$

Here $\alpha=e^2/\hbar c=1/137$ is Sommerfeld's fine structure constant, $n_{\mathbf{p},s}$ the number of electrons in the state of momentum **p** and spin s, m the rest mass of the fermions, $\delta_{ss'}$ the Kronecker symbol, c the speed of light, x=(**r**,t) any space-time point and V the volume of a normalization box. T is the time-ordering operator which orders the operators in the order of decreasing times from left to right and multiplies the result by $(-1)^P$, where P is the parity of the permutation required to achieve this order. For equal times, T normal-orders the operators, i.e., for t=t' the left-hand side of Eq. (4) is $i\langle\Phi_o|\psi_s^{\dagger}(x)\psi_{s'}(x')|\Phi_o\rangle$. The state Φ_o of the N electrons is described by a Slater determinant of single-particle orbitals.

The resulting spectral density coincides with the result $2\alpha/\pi fN$, first derived[4] directly from the coherent state of the electromagnetic field of a physical charged particle. The connection with the conventional quantum 1/f effect was suggested later[2].

To calculate the current autocorrelation function we need the density correlation function, which is also known as the two-particle correlation function. The two-particle correlation function is defined by

$$\langle\Phi_o|T\psi_s^{\dagger}(x)\psi_s(x)\psi_{s'}^{\dagger}(x')\psi_{s'}(x')|\Phi_o\rangle = \langle\Phi_o|\psi_s^{\dagger}(x)\psi_s(x)|\Phi_o\rangle\langle\Phi_o|\psi_{s'}^{\dagger}(x')\psi_{s'}(x')|\Phi_o\rangle$$
$$-\langle\Phi_o|T\psi_{s'}(x')\psi_s^{\dagger}(x)|\Phi_o\rangle\langle\Phi_o|T\psi_s(x)\psi_{s'}^{\dagger}(x')|\Phi_o\rangle. \tag{5}$$

The first term can be expressed in terms of the particle density of spin s, $n/2 = N/2V = \langle\Phi_o|\psi_s^{\dagger}(x)\psi_s(x)|\Phi_o\rangle$, while the second term can be expressed in terms of the Green function (4) in the form

$$A_{ss'}(x-x') \equiv \langle\Phi_o|\psi_s^{\dagger}(x)\psi_{s'}^{\dagger}(x')\psi_{s'}(x')\psi_s(x)|\Phi_o\rangle = (n/2)^2+\delta_{ss'}\,G_s(x'-x)G_s(x-x'). \tag{6}$$

The "relative" autocorrelation function A(x-x') describing the normalized pair correlation independent of spin is obtained by dividing by n^2 and summing over s and s'

$$A(x-x') = 1-(1/n^2)\sum_s G_s(x-x')G_s(x'-x)$$

$$= 1-(1/N^2)\sum_s\ \sum_{\mathbf{p}\mathbf{p}'}\{\exp i[(\mathbf{p}-\mathbf{p}')(\mathbf{r}-\mathbf{r}')-(\mathbf{p}^2-\mathbf{p}'^2)(t-t')/2m]/\hbar\}n_{\mathbf{p},s}n_{\mathbf{p}',s}$$
$$\times\{\mathbf{p}(\mathbf{r}-\mathbf{r}')/\hbar-(m^2c^2+\mathbf{p}^2)^{1/2}(t-t')(c/\hbar)\}^{\alpha/\pi}$$
$$\times\{\mathbf{p}'(\mathbf{r}-\mathbf{r}')/\hbar-(m^2c^2+\mathbf{p}'^2)^{1/2}(t-t')(c/\hbar)\}^{\alpha/\pi}. \tag{7}$$

Here we have used Eq. (4). We now consider a beam of charged fermions, e.g., electrons, represented in momentum space by a sphere of radius p_F, centered on the momentum $\mathbf{p}_o$ which is the average momentum of the fermions. The energy and momentum differences between terms of different **p** are large, leading to rapid oscillations in space and time which contain only high-frequency quantum fluctuations. The low-frequency and low-wavenumber part A_l of this relative density autocorrelation function is given by the terms with **p**=**p'**

$$A_l(x-x') = 1-(1/N^2)\sum_s\sum_{\mathbf{p}} n_{\mathbf{p},s}$$
$$\times\{\mathbf{p}(\mathbf{r}-\mathbf{r}')/\hbar-(m^2c^2+\mathbf{p}^2)^{1/2}(t-t')(c/\hbar)\}^{2\alpha/\pi} \tag{8}$$

$$\approx 1 - (1/N)|\mathbf{p}_o(\mathbf{r}-\mathbf{r}')/\hbar - mc^2\tau/\hbar|^{2\alpha/\pi} \qquad \text{for } p_F << |p_{o3} - mc^2\tau/z|. \tag{9}$$

Here we have used the mean value theorem, considering the $2\alpha/\pi$ power as a slowly varying function of **p** and neglecting $\mathbf{p}_o$ in the coefficient of $\tau \equiv t-t'$, with $z \equiv |\mathbf{r}-\mathbf{r}'|$. Using the identity[9], with arbitrarily small cutoff ω_o, we obtain from Eq. (9) with $\theta \equiv |\mathbf{p}_o(\mathbf{r}-\mathbf{r}')/\hbar - mc^2\tau/\hbar|$ the exact form

$$A_l(x-x') = 1 + [(2\alpha/\pi N)\int_{\omega_o}^{\infty}(mc^2/\hbar\omega)^{2\alpha/\pi}\cos(\theta\omega)d\omega/\omega]$$

$$\times\{\cos\alpha + (2\alpha/\pi)\sum_{n=0}^{\infty}(\theta\omega_o)^{2n-2\alpha/\pi}[(2n)!(2n-2\alpha/\pi)]^{-1}\}^{-1}. \tag{10}$$

This indicates a $\omega^{-1-2\alpha/\pi}$ spectrum and a 1/N dependence of the spectrum of fractional fluctuations in density n and current j, if we neglect the curly bracket in the denominator which is very close to unity for very small ω_o. The fractional autocorrelation of current fluctuations δj is obtained by multiplying Eq. (6) on both sides with ep_o/m, and dividing by $(enp_o/m)^2$ which is the square of the average current density j, instead of just dividing by n^2. It is the same as the fractional autocorrelation for quantum density fluctuations. Then Eq. (10) for the coherent Quantum Electrodynamical chaos process in electric currents can be written also in the form

$$S_{\delta j/j}(k) \approx [2\alpha/\pi\omega N][mc^2/\hbar\omega]^{2\alpha/\pi} \approx 2\alpha/\pi\omega N = 0.00465/\omega N. \tag{11}$$

Being observed in the presence of a constant applied field, these fundamental quantum current fluctuations are usually interpreted as mobility fluctuations. Most of the conventional quantum 1/f fluctuations in physical cross sections and process rates are also mobility fluctuations, but some are also in the recombination speed or tunneling rate.

III. PROPAGATOR FOR SMALL SAMPLES OR DEVICES

For a finite sample or device Eq. (4) should be replaced by a propagator which approaches the classical free particle propagator of the Schrödinger equation when the transversal sample size, or the number of particles per unit length of the sample, approach zero. This would cause the coherent quantum 1/f effect to become very small compared with the conventional quantum 1/f noise present in the beam due to the particular way in which the beam was generated. A formula like Eq. (3) would then express the fact that conventional quantum 1/f is always presen, but is masked in larger samples by the coherent quantum 1/f effect. However, a formula with a size-dependent infrared exponent intermediary between the coherent and conventional limits of α/π and αA, present both in the coefficient and the would express the same transition in a slightly different, physically more meaningful form:

$$j^{-2}S_j(f) = \mu^{-2}S_\mu(f) = 2\beta/f^{1-\beta}N, \qquad \text{with} \qquad \beta = \alpha A/(s+1) + \alpha s/\pi(s+1). \tag{12}$$

So far we have not derived an expression equivalent to Eq. (12) in any way. However, the physical unity of coherent and conventional quantum 1/f effects speaks in favor of a more sophisticated relation than (3). This same physical content can be expressed in a slightly different way by noting that Eq. (4) is equivalent to a energy momentum relation which is not sharp, allowing for quantum fluctuations of the rest mass of the charged particle, or of any other particle with infrared divergent coupling to a group of massless infraquanta. Describing these quantum fluctuations of the rest mass μ with the help of a distribution function $\rho(\mu)$ peaked at the measured rest mass m, we could attempt to write Eq. (4) in the form

$$-i\langle\Phi_o|T\psi_{s'}(x')\psi_s^{\dagger}(x)|\Phi_o\rangle \equiv \delta_{ss'}\, G_s(x'-x)$$

$$= (i/V)\sum_{\mathbf{p}}\{\exp i[\mathbf{p}(\mathbf{r}-\mathbf{r}')-\mathbf{p}^2(t-t')/2m]/\hbar\}n_{\mathbf{p},s}\times\{-i\mathbf{p}(\mathbf{r}-\mathbf{r}')/\hbar+i(m^2c^2+\mathbf{p}^2)^{1/2}(t-t')(c/\hbar)\}^{\alpha/\pi}.$$

$$= (i/V)\int d\mu\rho(\mu)\sum_{\mathbf{p}}\{\exp i[\mathbf{p}(\mathbf{r}-\mathbf{r}')-\mathbf{p}^2(t-t')/2m]/\hbar\}n_{\mathbf{p},s}. \qquad (13)$$

The distribution function $\rho(\mu)$ could be used to transform various classical results calculated simply with the Schrödinger propagator into the corresponding quantum 1/f results.

At the present time both lines of attack of this problem, the one based on Eq. (12) and the one presented in Eq. (13) are actively pursued in our investigations. We hope this discussion will stimulate quantum 1/f thought everywhere.

REFERENCES

1. P.H. Handel: "Instabilities, Turbulence and Flicker-Noise in Semiconductors I, II and III", Zeitschrift für Naturforschung 21a, 561-593 (1966).

2. P.H. Handel, "Coherent States Quantum 1/f Noise and the Quantum 1/f Effect" in *"Noise in Physical Systems and 1/f Noise"* (Proceedings of the VIIIth International Conference on Noise in Physical Systems and 1/f Noise) edited by A. D'Amico and P. Mazzetti, Elsevier, New York, 1986, p. 469.

3. P.H. Handel: "Turbulence Theory for the Current Carriers in Solids and a Theory of 1/f Noise", Phys. Rev. A3, 2066 (1971).

4. P.H. Handel, "Any Particle Represented by a Coherent State Exhibits 1/f Noise" in *"Noise in Physical Systems and 1/f Noise"*, (Proceedings of the VIIth International Conference on Noise in Physical Systems and 1/f Noise) edited by M. Savelli, G. Lecoy and J.P. Nougier (North - Holland, Amsterdam, 1983), p. 97.

5. P.H. Handel and E. Bernardi, Proc. XI. Int. Conf. on Noise in Physical Systems and 1/f Fluctuations, T. Musha, S. Sato and Y. Mitsuaki Editors, Ohmsha Publ. Co., Tokyo 1991, pp. 159-162.

6. D. Zwanziger, Phys. Rev. D7, 1082 (1973); Phys. Rev. Lett. 30, 934 (1973); Phys. Rev. D11, 3481 and 3504 (1975).

7. T.W.B. Kibble, Phys. Rev. 173, 1527; 174, 1882; 175,1624 (1968); J. Math. Phys. 9, 315 (1968).

8. Y. Zhang and P. H. Handel, Proc. Fifth van der Ziel Conference "Quantum 1/f Noise and other Low Frequency Fluctuations" AIP Conference proceedings #282, P.H. Handel and A.L. Chung Editors, 1992.

9. J.S. Gradshteiyn and I.M. Ryzhik, *"Table of Integrals, Series and Products"* Sec. 3.761, No. 9 and No. 7, Academic Press, New York 1965.

GRAPHICAL REPRESENTATION OF QUANTUM 1/f MOBILITY FLUCTUATION SPECTRA IN SILICON

P.H. Handel
Department of Physics, University of Missouri-St. Louis MO 63121, USA
and T.H. Chung
Department of Electrical Engineering, Stanford University, Stanford, CA 94305

We briefly present here the mobility fluctuation values of the quantum 1/f alpha parameter for impurity scattering and for intervalley scattering, calculated with the new cross-correlation formula[1].

The earlier graphical representations[2] were erroneously drawn with the value of the degeneration concentration N applicable always to the temperature of 300K. Therefore the data at temperatures different from 300K have to be multiplied by the factor $(T/300)^{3/2}$ in the two graphs[2] presented in 1991. The corresponding results we have at this stage of the numerical calculation are based on the equations given earlier[2] and are graphed below.

These values replace the results obtained initially[3,4] without the cross correlation formula[1]. In the first graph the values of $\alpha_{impurity}$ are given for three values of the impurity concentration. Everywhere na stand for the new, o for the old α parameter values[3].

The second graph brings the $\alpha_{intervalley}$ values for scattering *along* the direction of the electron energy minima with umklapp (g scattering), and for scattering along a wave vector $\Delta\mathbf{k}$ not in the direction of any energy minimum (f scattering). In both cases $\Delta\mathbf{k}$ links two energy minima.

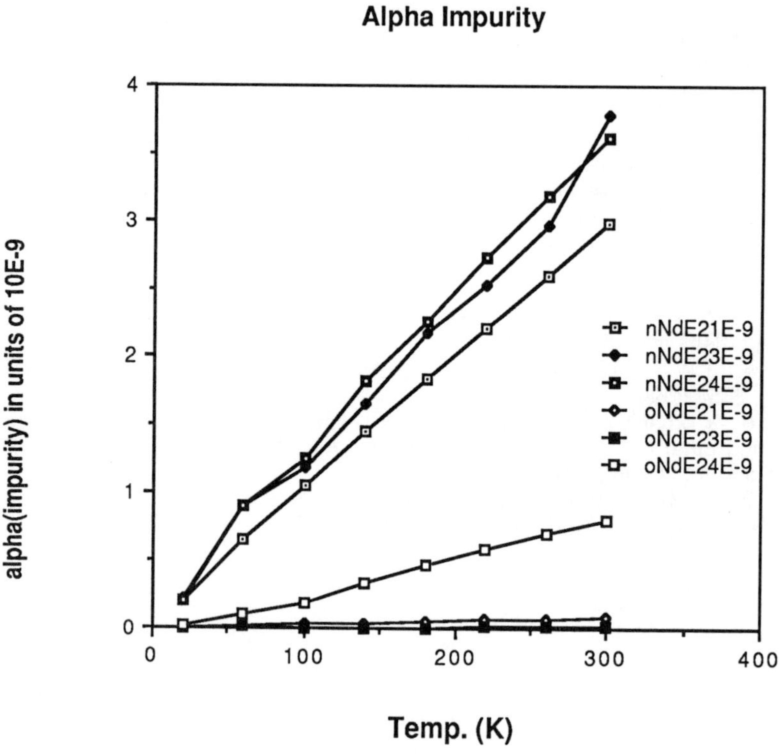

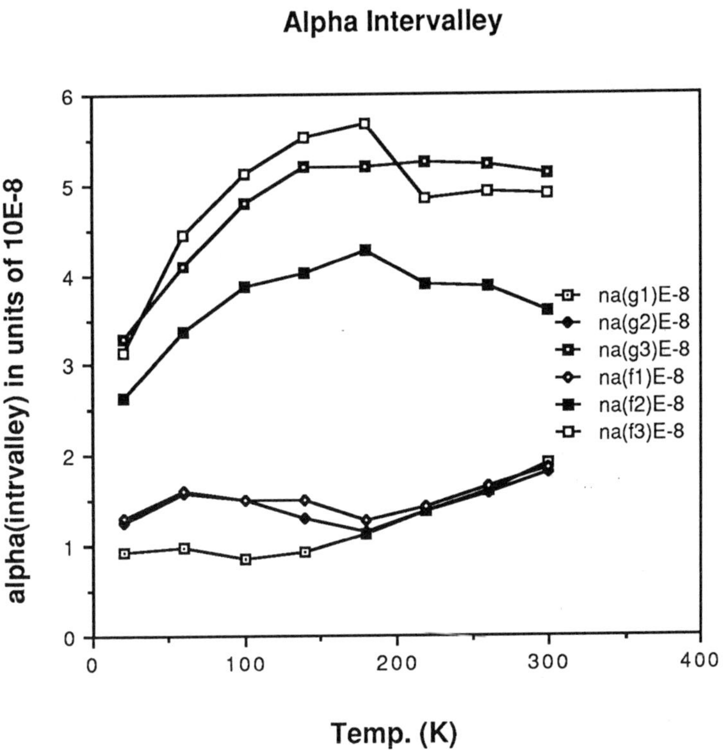

There are three g-type alphas α_{g1}, α_{g2} and α_{g3} (from LA, TA and LO phonons repectively) and three f-type contributions α_{f1}, α_{f2} and α_{f3} (from TA, LA and TO phonons). Their values are given in the graph above and are in general a few times larger than the old values.

The various quantum 1/f contributions α_i derived here can be approximately superposed to yield the resultant quantum 1/f coefficient α_H according to the rule (where μ_i are the partial mobilities)

$$\alpha_H = \Sigma_i(\mu/\mu_i)^2 \alpha_i.$$

The support of the Air Force Office of Scientific Research, and of the National Science Foundation is thankfully acknowledged.

REFERENCES

1. P.H. Handel: "Starting Points of the Quantum 1/f Noise Approach", p.1-26, Subm. to Phys. Rev. B, February 1988.

2. P.H. Handel, Proc. XI. Int. Conf. on Noise in Physical Systems and 1/f Fluctuations, T. Musha, S. Sato and Y. Mitsuaki Editors, Ohmsha Publ. Co., Tokyo 1991, pp. 151-157.

3. G.S. Kousik, C.M. Van Vliet, G. Bosman and P.H. Handel, Advances in Physics **34**, 663-702 (1985).

4. G.S. Kousik, C.M. Van Vliet, G. Bosman, and Horng-Jye Luo, Phys. stat. sol. **154**, 713-726 (1989).

VII. SEMICONDUCTOR DEVICES

NOISE IN CONFINED STRUCTURES OF DOPED SEMICONDUCTORS

V.Bareikis, R.Katilius, A.Matulionis
Semiconductor Physics Institute, Vilnius 2600, Lithuania

ABSTRACT

Hot electron noise properties in bulk semiconductors and semiconductor structures are reviewed. Sensitivity of microwave noise to doping is demonstrated. An emphasis is made on high electric field region where the effect of doping on kinetic properties was earlier assumed to be weak. Possibilities are stressed of microwave noise measurement technique as a method of diagnostics of kinetics of nonequilibrium electrons and energy levels in conduction band, especially if combined with selective doping, hydrogenation and electron gas confinement.

INTRODUCTION

Confined structures are among the most modern objects of solid state physics[1]. Doping of semiconductor structures, being inevitable for existence of free electrons, influences electron transport and noise properties important for high-speed low-noise microelectronics[2].

Effect of doping on electron transport has been thoroughly studied and well understood at low applied electric fields. Less attention was paid to high fields: electron scattering by ionized impurities was known to become negligible for high energy electrons.

However, during the last years, especially due to nonequilibrium noise spectroscopy applications[3], it became clear that role of doping at highly-nonequilibrium conditions can also be very important[4]. Indeed, due to sensitivity of hot-electron noise to details of scattering mechanisms and band structure, noise measurements at high electric fields opened new possibilities of diagnostics of confined electron gas[5,6].

The paper discusses recent results of noise investigation of GaAs-based structures at high electric fields, with an emphasize on doping effects. Such effects as impurity-resonant-scattering-induced noise[4,7], influence of impurity and defect passivation by hydrogen[6,8], noise in δ-doped structures[9], doping-dependent intersubband transfer noise in modulation-doped heterostructures[5,10] are reviewed. The main results have been deduced from noise measurements performed by microwave technique[3].

Understanding of these physical phenomena is important while seeking to increase the speed of operation and to reduce the noise level in microwave devices (speed-noise trade-off problems).

INFLUENCE OF RESONANT IMPURITY LEVELS

In compound semiconductors, impurities form energy levels located either in the forbidden gap or above the conduction band edge (resonant levels). There is no conventional experimental technique to investigate resonant impurity levels (unless the density of states in the conduction band is abnormally low). Fortunately, their contribution to nonequilibrium noise in GaAs is reported to be important[11].

Fig.1 shows two regions of electric field where spectral density of current fluctuations depends on doping[12]. Such a dependence at low fields can be interpreted in terms of ionized impurity scattering, which tends to decrease when electron energy increases. As a result, the dependence on doping disappears at intermediate fields (cf. points for the undoped and doped material in Fig.1). A similar behaviour has also been obtained by Monte Carlo simulation for a model which accounts for electron scattering by phonons and ionized impurities (Fig.1, curves 1 and 2)[12].

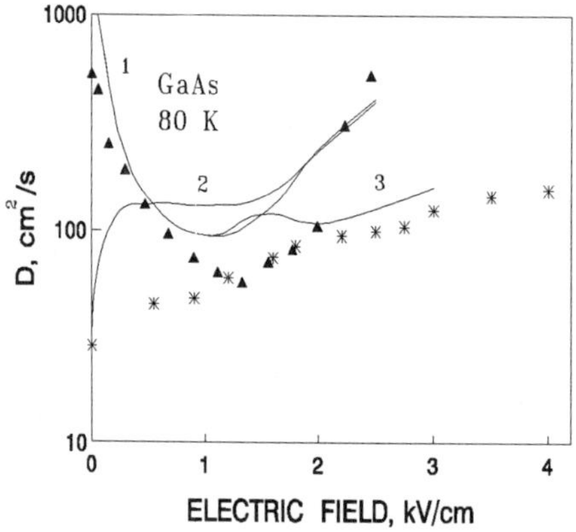

Fig.1. A comparison of experimental results (points) on spectral density of current fluctuations (normalized to the diffusion coefficient value at zero field) and diffusion coefficient calculated by Monte Carlo technique (curves) at 80 K[12]. Electron density: ▲ - $9 \cdot 10^{14}$ cm^{-3}, * - $3 \cdot 10^{17}$cm^{-3}; ion density: 1- $5 \cdot 10^{14}$ cm^{-3}, 2 - $5 \cdot 10^{16}$ cm^{-3}, 3 - $5 \cdot 10^{14}$ cm^{-3} plus resonant impurity scattering.

In contrast to this model, experimental results of Fig.1 demonstrate a strong dependence on doping at high electric fields. This behavior cannot be accounted for by ionized impurity scattering, and the frequency used excludes generation-recombination noise. The observed data can be interpreted as arising from resonance impurity scattering. This can be seen from a satisfactory fit of the the experimental results for the doped sample at high fields and Monte Carlo calculations performed for a model which accounts for resonance impurity scattering of electrons in addition to scattering by phonons and ionized impurities (Fig.1)[12].

Recent observation of resonant impurity levels in GaAs by optoelectronic modulation spectroscopy[13] confirms existence of the levels with the resonance energy close enough to that obtained by the noise spectroscopy[4,7].

IMPROVEMENT OF RESOLUTION BY HYDROGENATION

The effect of doping at high electric fields can be partially masked by intervalley noise. To suppress the intervalley noise and reveal more details

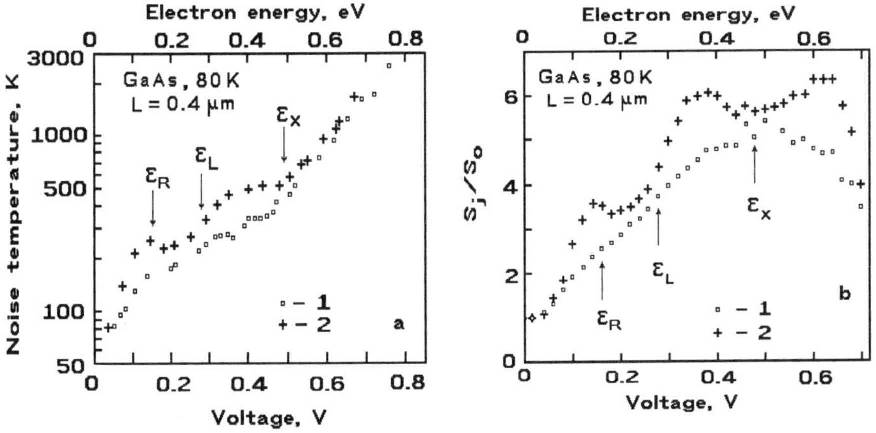

Fig.2. Dependence of noise temperature (a) and spectral density of current fluctuations (b) on voltage (lower scale) and top electron energy (upper scale): 1 - before and 2 - after hydrogenation ($10^{17}cm^{-2}$ dose)[6].

induced by doping, short enough samples (L $\leq$ 0.5 μm) were used[11].

The main advantage of short samples is a possibility to resolve noise sources which need different time and space to manifest themselves (L-E spectroscopy[3]). As a result of shortening, a wider range of electric fields is opened for manifestation of the noise features caused by resonant impurity scattering. Moreover, top energy electrons (which succeed to suffer no energy loss before reaching certain threshold energy) become important in short samples[6]. This has enabled determination of resonant level position in the conduction band at room temperature from noise spectra[4,7].

The spectral resolution at low temperatures has been improved by sample treatment with low energy protons (hydrogenation)[6]. The effect of hydrogenation on hot electron

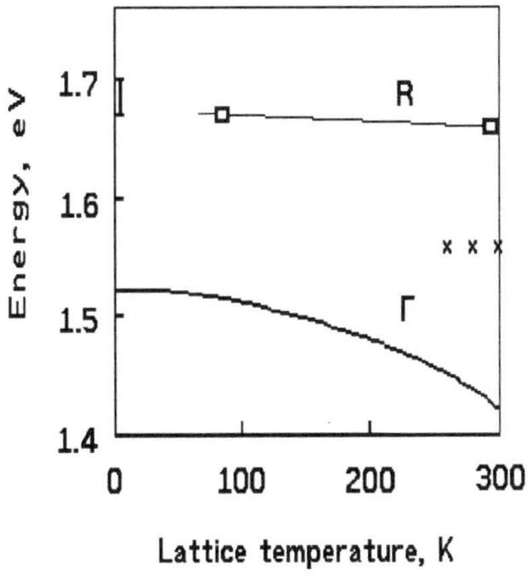

Fig.3. Temperature dependence of Γ and R level position: I[14], x[13], □[6,7,12].

noise temperature and spectral density of current fluctuations is illustrated by Fig.2. After hydrogenation a reasonably good resolution was obtained. Shoulders and maxima in Fig.2 were ascribed to interaction of top electrons with intrinsic (L, X) and resonant impurity (R) levels. The noise L-E spectroscopy data on hydrogenated GaAs are sufficient to determine energy position of the levels in the conduction band.

Fig.3 compares data on the resonant impurity level position obtained from the noise measurements at different lattice temperatures[6,7] with that extrapolated from data on highly compressed GaAs[14] and optoelectronic modulation spectra[13]. A weak dependence on the lattice temperature of the R level position suggests these levels to originate from an isovalent impurity (e.g. nitrogen).

NOISE IN δ-DOPED STRUCTURES

New features of microwave noise spectrum have been observed when impurities were concentrated in a restricted region of the sample[9,15]. GaAs structures under investigation contained two heavily-doped planes separated by a thin layer of undoped GaAs. Though the planar transport was characterized by a low mobility at low fields, a rather high drift velocity was obtained at high electric fields[16].

Data on noise temperature and spectral density of current fluctuations for undoped, uniformly doped and δ-doped GaAs are compared in Fig.4[15]. Low noise for the δ-doped GaAs at low fields is in consistence with the low value of electron mobility in the δ-doped planes. The curves for doped and δ-doped - samples cross at intermediate fields, and a powerful noise source with the

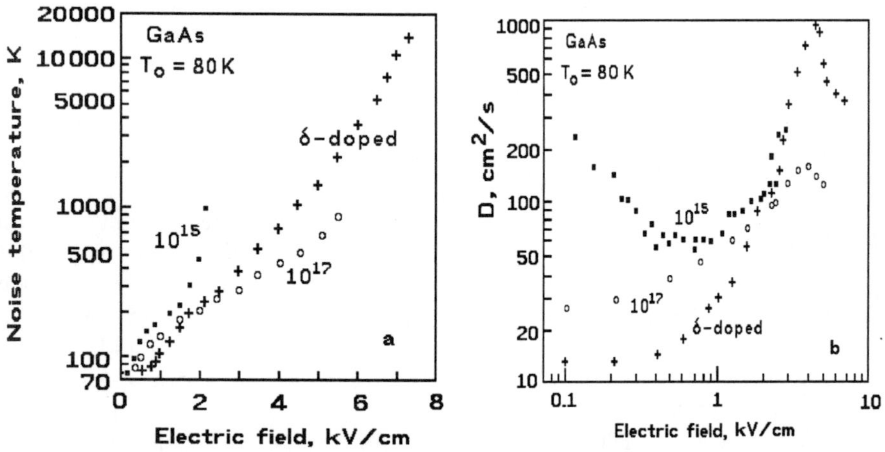

Fig.4. Dependence of (a) noise temperature and (b) spectral density of current fluctuations (normalized to the value of diffusion coefficient at zero field) for undoped, doped and δ-doped GaAs.

maximum spectral density at 4 kV/cm appears in the δ-doped material[9,15].

This excess noise was interpreted as arising from real space transfer: heating of electrons induces their transfer into the undoped material, and random back and forth transitions of the electrons give rise to the excess longitudinal current fluctuations due to difference of electron drift velocities in doped and undoped material.

NOISE IN MODULATION-DOPED HETEROSTRUCTURES

Another type of structures with interesting noise characteristics is a modulation-doped heterostructure[5,9,10,17-20]. The main advantage of such structures is spatial separation of electrons from their native donors. When an undoped spacer is grown between the doped layer and the quantum well, a direct influence of impurity scattering on noise is weak. But the doping changes the shape of the quantum well, and, for sufficient doping, the intersubband separation energy can exceed the mean thermal energy. Electron heating can transfer electrons into the nearest empty subband resulting a maximum on noise vs field dependence.

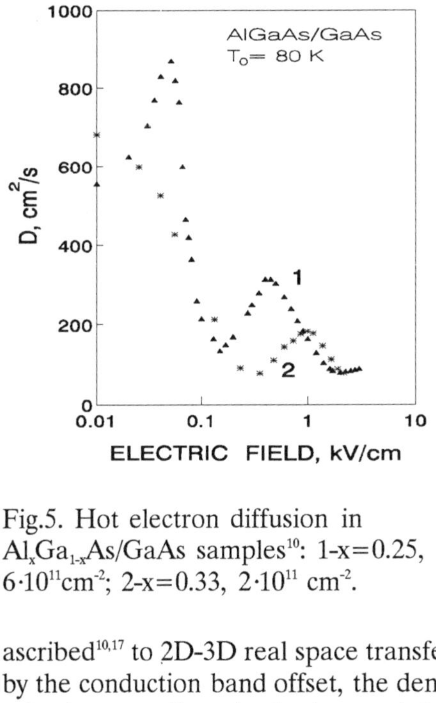

Fig.5. Hot electron diffusion in $Al_xGa_{1-x}As/GaAs$ samples[10]: 1-x=0.25, $6\cdot10^{11}cm^{-2}$; 2-x=0.33, $2\cdot10^{11}$ cm^{-2}.

The maximum, presumably of this origin, has been observed at 50 V/cm for the more heavily doped heterostructure (Curve 1, Fig.5)[5,10]. Extra correlation due to electron-electron collisions[3] can also be partially responsible for the maximum. No effect of mole composition is expected for low and intermediate fields, where the electron heating is moderate.

Position and magnitude of the maximum at high fields depend on Al mole ratio in AlGaAs[10]. It has been ascribed[10,17] to 2D-3D real space transfer[18] over/through the heterobarrier shaped by the conduction band offset, the density of two-dimensional electrons, and the selective as well as the background doping.

CONCLUSIONS

The results presented demonstrate an expansion of noise spectroscopy into a new region: physics of confined doped semiconductor structures. The microwave noise technique proves to be a powerful method of diagnostics of hot electron kinetics and energy levels in conduction band.

REFERENCES

1. T.Ando, A.Fowler, F.Stern, *Rev.Mod.Phys.* 54, 7 (1982).
2. *High-Speed Semiconductor Devices*, ed. S.M.Sze (Wiley, N.Y. 1990) p.621.
3. V.Bareikis, R.Katilius, J.Pozhela, S.Gantsevich, V.Gurevich, in *Spectroscopy of Nonequilibrium Electrons and Phonons* (Elsevier, Amsterdam 1992) p.237.
4. V.Bareikis, J.Liberis, A.Matulionis, R.Miliušytė, J.Požela, P.Sakalas, *Solid-State Electr.* 32, 1647 (1989).
5. V.Aninkevičius, V.Bareikis, J.Liberis, A.Matulionis, P.Sakalas, *Solid-State Electronics* (in press)
6. V.Bareikis, J.Liberis, A.Matulionis, P.Sakalas, M.Capizzi, submitted to *Semicond. Sci. & Technology.*
7. V.Bareikis, J.Liberis, A.Matulionis, R.Miliušytė, J.Požela, P.Sakalas, in *Proc. 20th Int. Conf. Phys. Semicond.* (World Scientific, Singapore 1990) p.2479.
8. P.Sakalas, V.Bareikis, J.Liberis, A.Matulionis, in *Proc. 6th Sci. Conf. Fluctuation Phenomena in Phys. Systems* (University Press, Vilnius 1991) p.30.
9. V.Bareikis, V.Aninkevičius, J.Liberis, T.Lideikis, A.Matulionis, P.Sakalas, G.Treideris, R.Katilius, in *Proc.Int. Semicond. Device. Res. Symp.* (University of Virginia, Charlottesville 1991) p.469.
10. V.Aninkevičius, V.Bareikis, J.Liberis, A.Matulionis, M.R.Leys, P.S.Kop'ev, in *Proc. 8th Vilnius Symp. Ultrafast Phenomena in Semiconductors* (in press).
11. V.Bareikis, R.Katilius, in *Proc. 10th Int. Conf. Noise in Phys. Systems* (Academiai Kiado, Budapest 1990) p.189
12. A.Matulionis, P.Sakalas, T.Smertina, V.Aninkevičius, J.Liberis, V.Bareikis, in *Ann. Repts Semiconductor Physics Inst.*, (Vilnius 1993) p.39.
13. M.Di Marco, G.Swanson, *J.Electrochem.Soc.* 139, 262 (1992); G.Swanson, Private communication.
14. D.J.Wolford, in *Proc. 18th Int. Conf. Phys. Semicond.* (World Scientific, Singapore 1987) p.1115.
15. J.Liberis, V.Aninkevičius, V.Bareikis, T.Lideikis, A.Matulionis, G.Treideris, in *Proc. 6th Sci. Conf. Fluctuation Phenomena in Phys. Systems* (University Press, Vilnius 1991) p.35.
16. V.Balynas, A.Krotkus, T.Lideikis, A.Stanionis, S.Treideris, *Electronics Lett.* 27, 2 (1991).
17. V.Aninkevičius, V.Bareikis, J.Liberis, A.Matulionis, P.S.Kop'ev, in *Proc.11th Int. Conf. Noise in Phys. Systems* (Ohmsha, Tokyo 1991) p.183.
18. J.Gest, H.Fawaz, H.Kabbaj, J.Zimmermann, ibid., p.291.
19. C.F.Whiteside, G.Bosman, H.Morkoç, *IEEE Trans.* ED-34, 2530 (1987).
20. A.D. Van Rheenen, G.Bosman, H.Morkoç, in *Proc. 9th Int. Conf. Noise in Phys. Systems* (World Scientific, Singapore, 1987) p.150.

ON THE ANALYSIS OF LOW-FREQUENCY NOISE IN BIPOLAR TRANSISTORS

Theo Kleinpenning
Eindhoven University of Technology, Dept. EE, P.O. Box 513,
Eindhoven, The Netherlands

ABSTRACT

Until recently, the 1/f noise sources in bipolar transistors were generally accepted as being located between emitter and collector and between emitter and base. However, in modern (sub)micrometer transistors the influence of the internal base and emitter series resistances on both the I-V characteristics and the l.f. noise at higher currents becomes important. In this paper expressions are presented for the l.f. noise in transistors where the influence of the internal parasitic series resistances has been taken into account. These expressions are compared with the common expressions in the literature. How to locate the l.f. noise sources is demonstrated by analyzing the new expressions both at low and high currents and at different values of the external resistances.

INTRODUCTION

Up to the present a number of papers has been published on low-frequency (l.f.) noise, especially 1/f noise, in bipolar transistors. A review of the literature over the period 1985-1989 has been given at the 10th ICNF[1]. The l.f. noise sources were generally accepted as being located between emitter and collector and between emitter and base. These sources are represented by two noise current generators: S_{I_b} between emitter and base, and S_{I_c} between emitter and collector[1-4]. Both generators have 1/f noise and shot noise. Recent developments in transistor technology have led to a significant lateral downscaling of bipolar devices. In such down-scaled transistors, deviations from the expected current dependence of the 1/f noise and white noise in both the emitter current and the base current are observed. Kleinpenning[1] suggested that these deviations may be ascribed to fluctuations in the parasitic internal series resistances. These resistances can produce 1/f noise, Nyquist noise, generation-recombination noise, and, possibly, random telegraph signal (RTS) noise.

The purpose of this paper is to present new formulas for the l.f. noise in bipolar transistors, taking into account the contributions from the series resistances. The presented formulas will be compared with the common formulas in the literature. In addition, how to locate the l.f. noise sources in transistors will be shown by analyzing the formulas both at high and low currents and at different values of the external resistances.

LOW-FREQUENCY NOISE SOURCES

Bipolar transistor noise is usually measured with the transistor in a circuit as shown in Fig. 1. The internal series resistances are given by r_e, r_b, and r_c, respectively. Three external wire-wound or metal film resistors R_E, R_B, and R_C can be inserted. Such resistors have only 4kTR-noise. Here we consider a pnp transistor with the emitter-base forward-biased and the base-collector reverse-biased. In the circuit of Fig. 1 we have the following l.f. noise sources

$$S_{I_b} = 2qI_B + S_{I_b}^{1/f}, \quad S_{I_c} = 2qI_C + S_{I_c}^{1/f}$$

$$S_{V_{r_x}} = 4kTr_x + I_x^2 S_{r_x} \quad \text{with } x = b,e,c \tag{1}$$

$$S_{V_{R_X}} = 4kTR_X \quad \text{with } X = B,E,C$$

The spectral current noise density in the base and collector currents at constant voltage drop across the emitter-base junction, S_{I_b} and S_{I_c}, is the sum of shot noise and 1/f noise. The internal series resistances have Nyquist noise and resistance noise S_r (1/f, g-r and/or RTS noise). Expressions for $S_{I_b}^{1/f}$ and $S_{I_c}^{1/f}$ can be found in Refs. 1, 4

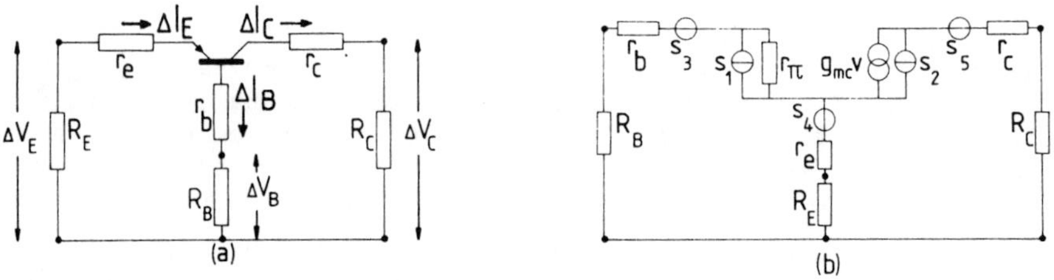

Fig. 1. (a) General circuit for l.f. noise measurements.
(b) Equivalent circuit, including noise sources: $S_1 = S_{I_b}$, $S_2 = S_{I_c}$, and S_3, S_4, S_5 are related to $S_{V_{r_x}}$.

and 5. Usually one observes $S_{I_b}^{1/f} \sim I_B$ and $S_{I_c}^{1/f} \sim I_C$.

GENERAL FORMULAS

For the fluctuations in the emitter, base, and collector currents we have

$$\Delta I_E = \Delta I_B + \Delta I_C \tag{2}$$

$$\Delta I_B = \Delta I_b + \Delta V_{eb}/r_\pi \tag{3}$$

$$\Delta I_C = \Delta I_c + g_{mc}\Delta V_{eb} \tag{4}$$

Here $g_{mc} = dI_C/dV_{eb}$ is the collector transconductance, $r_\pi = dV_{eb}/dI_B$ the input resistance, ΔI_b and ΔI_c the spontaneous fluctuations in the base and collector currents at constant V_{eb}, with V_{eb} the voltage drop across the emitter-base junction, excluding the voltage drops across the series resistances r_e and r_b. The fluctuations in V_{eb} are given by

$$\Delta V_{eb} = -(R_B+r_b)\Delta I_B - I_B\Delta r_b - (R_E+r_e)\Delta I_E - I_E\Delta r_e + \Delta V^N_{eb} \quad (5)$$

with ΔV^N_{eb} the fluctuations in V_{eb} due to the Nyquist noise in R_B, r_b, R_E, and r_e. Hence we have

$$S_{V^N_{eb}} = 4kT(R_B+r_b+R_E+r_e) \quad (6)$$

In Eq. (4) we have assumed I_C to be independent of V_{bc}, so the Early effect has been neglected.

Combining Eqs. (2-5), the fluctuations in the currents I_E, I_C, and I_B can be expressed in terms of the spontaneous noise sources ΔI_b, ΔI_c, Δr_b, Δr_e, and ΔV^N_{eb}. We then obtain

$$\Delta I_E.Z = [r_\pi - \beta(r_b+R_B)]\Delta I_b + [r_\pi+r_b+R_B]\Delta I_c - (\beta+1)[I_B\Delta r_b + I_E\Delta r_e - \Delta V^N_{eb}] \quad (7)$$

$$\Delta I_C.Z = -\beta[r_b+R_B+r_e+R_E]\Delta I_b + [r_\pi+r_b+R_B+r_e+R_E]\Delta I_c - \beta[I_B\Delta r_b + I_E\Delta r_e - \Delta V^N_{eb}] \quad (8)$$

$$\Delta I_B.Z = [r_\pi + \beta(r_e+R_E)]\Delta I_b - [r_e+R_E]\Delta I_c - [I_B\Delta r_b + I_E\Delta r_e - \Delta V^N_{eb}] \quad (9)$$

with $\beta = g_{mc}r_\pi$ is the current-amplification factor and

$$Z = R_B + r_b + r_\pi + (\beta+1)(R_E+r_e) \quad (10)$$

For the fluctuations in the voltage across R_E, R_C, and R_B we have

$$\Delta V_E = \Delta V^N_E + R_E\Delta I_E, \quad \Delta V_C = \Delta V^N_C + R_C\Delta I_C, \quad \Delta V_B = \Delta V^N_B + R_B\Delta I_B \quad (11)$$

Here ΔV^N_E, ΔV^N_C, and ΔV^N_B are the Nyquist fluctuations in the external resistors R_E, R_C, and R_B, respectively. Using Eqs. (6-11) and going over to the spectra, we then have

$$S_{V_X} = 4kTR_X + R_X^2 S_{\Delta I_X} + 2R_X S_{\Delta V^N_X, \Delta I_X} \quad (12)$$

with X = E, B, and C, respectively. The last term in Eq. (12) results from the correlation between the term ΔV_X^N and the term ΔV_{eb}^N in the relations for ΔI_X, which gives rise to

$$S_{\Delta V_C^N, \Delta V_{eb}^N} = 0, \quad S_{\Delta V_B^N, \Delta V_{eb}^N} = -4kTR_B, \quad S_{\Delta V_E^N, \Delta V_{eb}^N} = -4kTR_E \tag{13}$$

The minus signs stem from the fact that positive Nyquist voltage fluctuations in R_B and R_E lead to negative fluctuations in ΔV_{eb}^N. Taking the spontaneous noise sources, $\Delta I_b, \Delta I_c, \Delta r_b, \Delta r_e$, and ΔV_{eb}^N, to be uncorrelated, we obtain

$$(Z/R_E)^2 S_{V_E} = [r_\pi - \beta(r_b + R_B)]^2 S_{I_b} + [r_\pi + r_b + R_B]^2 S_{I_c} + (\beta+1)^2 S_{V_r} + $$
$$+ 4kT[(\beta+1)^2(R_B + R_E) - 2(\beta+1)Z + Z^2/R_E] \tag{14}$$

$$(Z/R_C)^2 S_{V_C} = \beta^2[r_b + R_B + r_e + R_E]^2 S_{I_b} + [r_\pi + r_b + R_B + r_e + R_E]^2 S_{I_c} + $$
$$+ \beta^2 S_{V_r} + 4kT[\beta^2(R_B + R_E) + Z^2/R_C] \tag{15}$$

$$(Z/R_B)^2 S_{V_B} = [r_\pi + \beta(r_e + R_E)]^2 S_{I_b} + [r_e + R_E]^2 S_{I_c} + S_{V_r} + $$
$$+ 4kT[R_B + R_E - 2Z + Z^2/R_B] \tag{16}$$

with

$$S_{V_r} = I_B^2 S_{r_b} + I_E^2 S_{r_e} + 4kT(r_b + r_e) \tag{17}$$

By considering Eqs. (7-10) and (14-17) the importance of the internal series resistances can be seen. The influence of these resistances is twofold. They can produce l.f. noise and, by means of feedback, they can influence the magnitude of the noise originating from the sources S_{I_b} and S_{I_c}.

By short-circuiting the emitter, the collector, and the base to ground ($R_E = R_C = R_B = 0$) and with the help of Eqs. (7-10), we can calculate the spectral current noise densities of I_E, I_C, and I_B, respectively. We then find

$$Z_o^2 \cdot S_{I_E} = (r_\pi - \beta r_b)^2 S_{I_b} + (r_\pi + r_b)^2 S_{I_c} + (\beta+1)^2 S_{V_r} \tag{18}$$

$$Z_o^2 \cdot S_{I_C} = \beta^2 (r_e + r_b)^2 S_{I_b} + (r_\pi + r_b + r_e)^2 S_{I_c} + \beta^2 S_{V_r} \tag{19}$$

$$Z_o^2 \cdot S_{I_B} = (r_\pi + \beta r_e)^2 S_{I_b} + r_e^2 S_{I_c} + S_{V_r} \tag{20}$$

with

$$Z_o = Z(R_B=R_E=0) = r_b+r_\pi+(\beta+1)r_e \qquad (21)$$

Note that for $r_e=r_b=0$ we obtain $S_{I_E} = S_{I_b}+S_{I_c}$, $S_{I_C}=S_{I_c}$ and $S_{I_B}=S_{I_b}$. It should be noted that the cross-correlation spectral densities of the currents I_B, I_C and I_E can easily be obtained from Eqs. (7-10).

COMPARISON WITH LITERATURE

The most recent expressions for the current noise spectral densities are given by Van der Ziel et al.[3]. They did not take into account noise contributions from the series resistances ($S_{V_r} = 0$). They considered the base and collector current noise for both $r_e = 0$ and $r_e >> r_b$, $r_\pi/(\beta+1)$, with the condition $r_\pi >> r_b$. For these situations Eqs. (19-20) lead to the same equations as found by Van der Ziel et al.. Here we have to remark that Van der Ziel et al. took into account a current noise between base and collector. However, in practical cases the base-collector junction is reverse-biased, so that the base-collector current is slight and its noise can be neglected.

With respect to voltage noise spectral densities S_{V_E}, S_{V_C}, and S_{V_B}, in the literature only expressions can be found where $S_{r_b}=S_{r_e}=0$ and $r_e = 0$. These expressions can be derived from Eqs. (14-16) making the appropriate approximations. In the common-emitter configuration one mostly measures the voltage noise S_{V_C}. With $S_{r_b}=S_{r_e}=0$ and $r_e+R_E=0$, Eq. (15) leads to the well-known relation[6]

$$S_{V_C}/R_C^2 = S_{I_c} + \left[\frac{\beta(r_b+R_B)}{r_\pi+r_b+R_B}\right]^2 \left[S_{I_b} + \frac{4kT}{r_b+R_B}\right] + \frac{4kT}{R_C} \qquad (22)$$

In the common-collector configuration S_{V_E} is mostly measured. With $R_E >> R_B+r_b+r_\pi$, $r_e = 0$ and $S_{r_e} = S_{r_b} = 0$ we obtain the well-known relation[7,8]

$$S_{V_E} = \left[\frac{r_\pi}{\beta+1} - \frac{\beta}{\beta+1}(r_b+R_B)\right]^2 S_{I_b} + \left[\frac{r_\pi+r_b+R_B}{\beta+1}\right]^2 S_{I_c}+4kT(r_b+R_B) \qquad (23)$$

USUAL MEASURING METHODS

With the help of the new expressions presented here, one can determine the l.f. noise sources in transistors, as well as the magnitude of the series resistances r_b and r_e. Some examples can be found in the literature, e.g. see Refs. (1,2,4,6,7,8,9). Here we shall present some measuring methods based on Eqs. (14-16).

Common-collector configuration ($R_C = 0$)

For $R_B >>> r_b + r_\pi + (\beta+1)(R_E+r_e)$ we have $Z \simeq R_B$ and thus

$$S_{V_E}/R_E^2 = \beta^2 S_{I_b} + S_{I_c} + 4kT/R_E \tag{24}$$

For the same value of R_B and with $R_E = 0$ we obtain

$$S_{V_B} = [r_\pi + \beta r_e]^2 S_{I_b} + r_e^2 S_{I_c} + S_{V_r} \tag{25}$$

For $(\beta+1)R_E >> R_B+r_b+r_\pi+(\beta+1)r_e$ and $R_B = 0$ we have $Z \simeq (\beta+1)R_E$ and thus

$$S_{V_E} = \left[\frac{r_\pi-\beta r_b}{\beta+1}\right]^2 S_{I_b} + \left[\frac{r_\pi+r_b}{\beta+1}\right]^2 S_{I_c} + S_{V_r} \tag{26}$$

At low currents, where $r_\pi >> \beta r_b$, βr_e and where S_{V_r} can be neglected, Eqs. (25,26) reduce to

$$S_{V_B} = r_\pi^2 S_{I_b} + r_e^2 S_{I_c} \text{ and } S_{V_E} = g_{me}^{-2}[S_{I_b}+S_{I_c}] \tag{27}$$

with $g_{me} = (\beta+1)/r_\pi$.

By measuring the 1/f noise in S_{V_E} versus I_E under the conditions of Eq. (26), and for the situation that $S_{I_c} \sim I_C$, $S_{I_b} \sim I_B$, and $I_B \sim I_C \sim I_E$, we have the results as sketched in Fig. 2. The noise in the base current can be eliminated completely in the case where $r_b = r_\pi/\beta = kT/qI_C$. Here we can have the opportunity to determine the

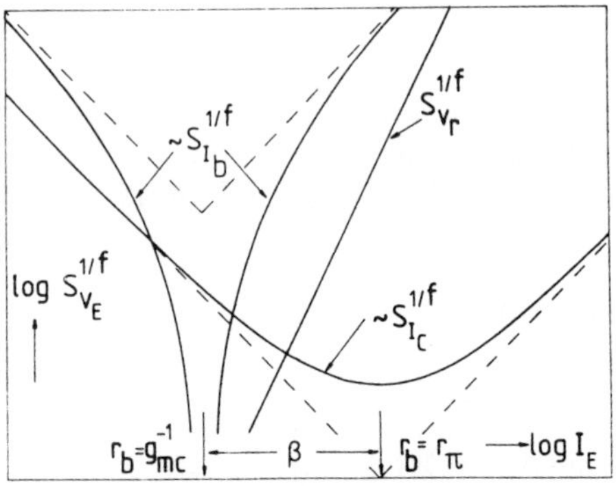

Fig. 2. $S_{V_E}^{1/f}$ versus I_E according to Eq. (26); $Z \simeq (\beta+1)R_E$ and $R_B = 0$.

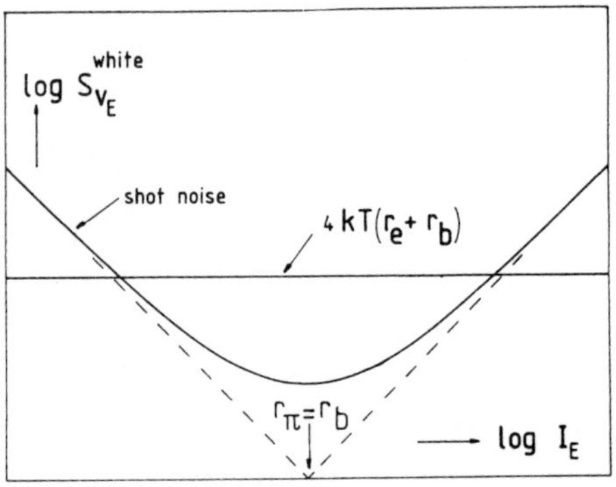

Fig. 3. $S_{V_E}^{white}$ versus I_E according to Eq. (26); $Z \simeq (\beta+1)R_E$ and $R_B = 0$.

value of r_b. If S_{I_c} is dominant, then we find a minimum at $r_b = r_\pi = kT/qI_B$. At higher currents it is possible to find a strong increase in the noise due to fluctuations in r_b and r_e. By measuring the white noise in S_{V_E} versus I_E, then at low currents we have $S_{V_E} = 2(kT)^2/qI_E$, at intermediate currents $S_{V_E} = 4kT(r_e+r_b)$, and at high currents $S_{V_E} = [r_b/(\beta+1)]^2 2qI_E$. From these results we can determine r_e+r_b and r_b (see Fig. 3). According to Eq. (25) the 1/f noise in S_{V_B} versus I_E is as sketched in Fig. 4. If the noise in S_{I_b} dominates, we find a minimum at $r_e = r_\pi/\beta = kT/qI_C$. In this case r_e can be determined.

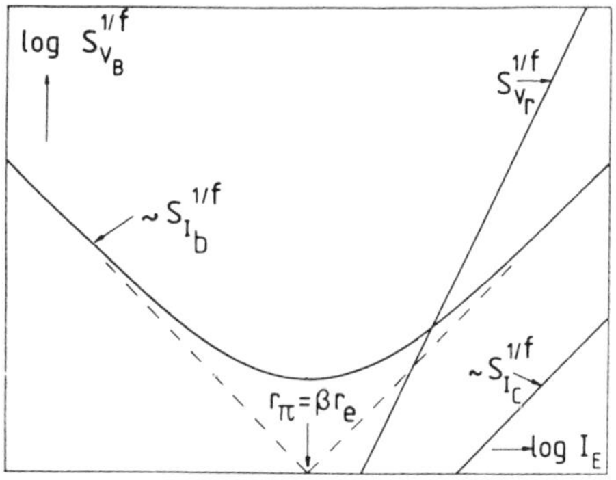

Fig. 4. $S_{V_B}^{1/f}$ versus I_E according to Eq. (25); $Z = R_B$ and $R_E = 0$

Common-emitter configuration ($R_E = 0$)

For $R_B \ggg r_b + r_\pi + (\beta+1)r_e$, thus $Z \simeq R_B$, we obtain

$$S_{V_C}/R_C^2 = \beta^2 S_{I_b} + S_{I_c} + 4kT/R_C \tag{28}$$

Here S_{V_B} is equal to Eq. (25). Measuring of S_{V_C} and S_{V_B} gives the same information as measuring S_{V_E} and S_{V_B} according to Eqs. (24) and (25).

REFERENCES

1. T.G.M. Kleinpenning, Noise in Physical Systems (Akadémiai Kiadó, Budapest, 1990), p. 443.
2. A. van der Ziel, Proc. IEEE 76, 233 (1988).
3. A. van der Ziel, X. Zhang, and A.H. Pawlikiewicz, IEEE-ED 33, 1371 (1986).
4. T.G.M. Kleinpenning, IEEE-ED 39, 1501 (1992).
5. T.G.M. Kleinpenning, Physica 154B, 27 (1988).
6. J. Kilmer, A. van der Ziel, and G. Bosman, Solid-State Electron. 26, 71 (1983).
7. T.G.M. Kleinpenning, Physica 138B, 244 (1986).
8. J.L. Plumb and E.R. Chenette, IEEE-ED 10, 304 (1963).
9. T.G.M. Kleinpenning and A.J. Holden, IEEE-ED 40, nr. 6 (1993).

RANDOM TELEGRAPH FLUCTUATIONS IN $GaAs/Al_{0.4}Ga_{0.6}As$ RESONANT TUNNELING DIODES

Sze-Him Ng and Charles Surya
Northeastern University
Department of Electrical and Computer Engineering
409 Dana Research Center
Boston, MA, 02115
Elliott R. Brown and Paul A. Maki
Lincoln Laboratory, MIT
Lexington, MA 02173

ABSTRACT

We have characterized discrete conductance switching noise in Resonant Tunneling Diodes (RTDs) fabricated on $GaAs/Al_{0.4}Ga_{0.6}As$ material system. Temperature dependence of the high and low resistive states time constants indicated that the noise arises from thermal activation of electrons to localized states in the energy barrier. Both time constants are found to decrease with increasing negative bias at the emitter indicating that the noise is caused by hopping conduction of electrons from the energy barrier. The magnitude of the discrete conductance fluctuation is studied in detail. A model has been postulated in which the capture of an electron by a trap in the energy barrier causes fluctuations in the transmission coefficient of the electron due to the modulation of the local barrier potential. The computed step height based on the proposed model is compared to the experimental results. Excellent agreement between theory and experiment is observed.

INTRODUCTION

Much interest has arisen over the physics and applications of resonant tunneling diode (RTD) in recent years. This is due to the potential of the device in low-power, high-speed applications. The device demonstrates promising results in microwave oscillation, detection and mixing up to THz range [1, 2]. In addition, RTDs are found to have significant potential in digital applications. It is also shown that a low-power, high-speed switching device [3] can be realized by a circuit involving an RTD structure and a heterojuction bipolar transistor. Low frequency excess noise has significant effects on the operation and reliability of the device. The up-conversion of the low frequency noise affects the performance of the device as microwave oscillator and mixer [4]. Also, excess current in the valleys of I-V curves is attributed in part to trap-assisted tunneling process.

In this paper, we investigate the origin and the physical processes underlying low frequency noise in RTDs. Previous studies which relied exclusively on spectral measurements failed to fully characterize the noise mechanism [5]–[8]. We report on time-domain measurements to monitor the activities of individual traps. From the characteristics of the Random Telegraph Noise (RTN) we deduced the trapping kinetics and the underlying mechanism for the low frequency noise.

EXPERIMENT AND RESULTS

The RTDs we studied were MBE grown at Lincoln Laboratory. The structure of the devices and the experimental techniques are reported in recently published papers [8, 9]. Measurements of the voltage fluctuations in the time-domain were performed over a temperature range between 20 – 70 K, and the data acquisition time ranging from 80 ms to 400 s [9]. While RTN were observed in most devices, the switching patterns measured on different samples under similar experimental conditions show large disparity in the time scales of transitions. Furthurmore, switching times differing by two orders of magnitdes were observed on the same device at the same temperature at two different biases. Figure 1 shows a complete time-capture frame of RTN measured at 60 K under a bias voltage of −0.4 V. Step sizes ranging from about 5 to 20 μV were observed from the plot. The 5-μV switching events are caused by the capture and emission of carriers by multiple traps, as is demonstrated by the consecutive 5-μV decrements in the data. In contrast, the 20-μV switching events, which are being enumerated, alternate between decrements and increments, consistent with a two-state process involving a single trap. This phenomenon was observed in several other devices fabricated from the same MBE wafer. In view of the uncertainty in the number of traps involved in the RTN with smaller step heights, we focus in this paper on the 20-μV switching sequences. It was also observed that the switching rate increased significantly with the device temperature. From the histograms of the duration the device spent at the high and low resistive states, the corresponding time constants of the RTN at different temperatures and biases were obtained through exponential curve fitting. Figure 2 shows the Arrhenius plots of the high and low resistive state time constants, τ_h and τ_l, from which the activation energies for τ_h and τ_l were found to be 81 meV and 51 meV respectively.

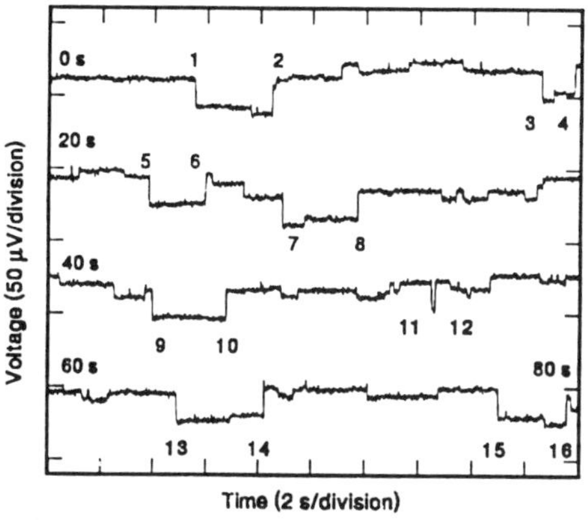

Fig. 1. A complete time capture frame of RTN in RTD at $V = -0.4$ V and $T = 60$ K. All discrete switches with step height $\sim$ 20-μV are enumerated.

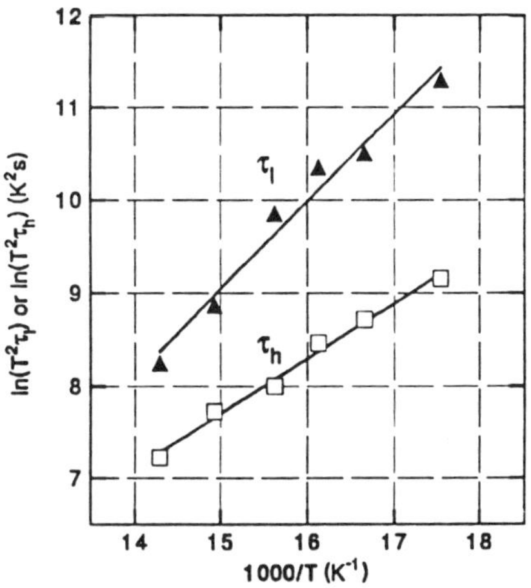

Fig. 2. Arrhenius plots of τ_l and τ_h.

Strong bias dependence on the magnitudes of the time constants were observed as shown in Fig. 3a, where both τ_h and τ_l were found to decrease exponentially with increasing negative bias at the emitter.

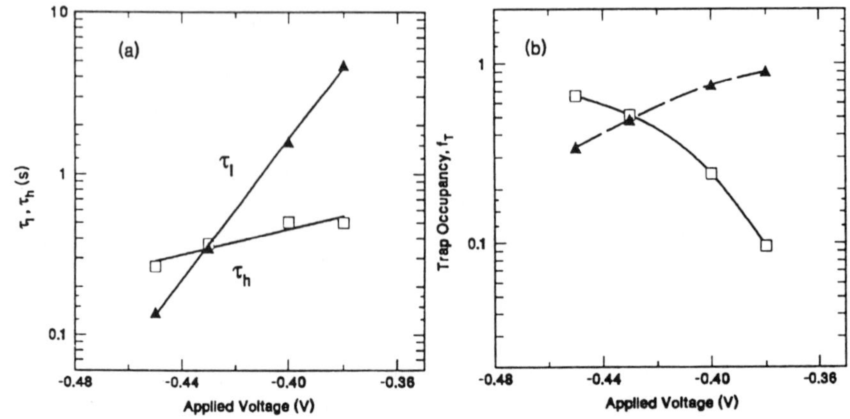

Fig. 3. (a) The bias dependence of τ_l (solid triangle) and τ_h (open square); (b) trap occupancy f_T for $\tau_l \equiv \tau_c$ (open square), and $\tau_h \equiv \tau_c$ (solid triangle).

PROPOSED LOW FREQUENCY NOISE MODEL

The observation of RTN in RTDs presents a strong evidence for the defect origin of low frequency noise in these devices. Furthermore, our study of the temperature and bias dependence of time constants reveals the kinetics of traps. To complete the model of noise based on electron trapping, we need to explain the modulation mechanism whereby the microscopic process causes fluctuations in the device conductance. Following Machlup's analyses of RTN statistics, the current noise power spectral density resulting from a single trap is shown to be

$$\begin{aligned} S_I(f) &= 4(\Delta I_0)^2 \frac{\tau_e \tau_c}{(\tau_e + \tau_c)^2} \frac{\tau}{1 + 4\pi f^2 \tau^2}, \\ &= 4(\Delta I_0)^2 f_T (1 - f_T) \frac{\tau}{1 + 4\pi f^2 \tau^2}. \end{aligned} \tag{1}$$

To establish the model, we consider the discrete current change, ΔI_0, caused by carrier trapping of a single defect, and compare such theoretical results with our measured switching steps in RTN. We consider the modulation of the Coulomb potential arising from the excitation of a single trap in the barrier. Such an event may cause fluctuations in the thermionic emission current by modulating the effective barrier height as well as the transmission probability of the electrons, and thus the tunneling current due to changes in the local band profile. To investigate these effects we calculate the three-dimensional Coulomb potential, ϕ_s, of a single ionized trap by modeling the trapped charge as a delta function of strength $Q = \pm q$. Due to the layered structure of RTDs, the computation of ϕ_s needs to account for the abruptly varying dielectric constants. In addition, since a high concentration of electrons ($> 10^{18}\mathrm{cm}^{-3}$) reside in the emitter accumulation region adjacent to the first barrier, considerable screening of the Coulomb potential is expected and needs to be included in the calculations. Therefore, Poisson's equation can be solved in cylindrical coordinates:

$$\nabla^2\phi_s(r,z) \equiv \frac{1}{r}\frac{\partial \phi_s}{\partial r} + \frac{\partial^2 \phi_s}{\partial r^2} + \frac{1}{\epsilon_S}\frac{\partial}{\partial z}\left(\epsilon_S \frac{\partial \phi_s}{\partial z}\right) = -\frac{Q}{\epsilon_S}\frac{\delta(r,z)}{2\pi r} + \frac{\phi_s}{L_B^2}, \quad (2)$$

where z and r are the perpendicular and radial distances with respect to the interfacial plane, and L_B is the Debye length. The second term on the right-side of Eq. (2) accounts for screening of the Coulomb potential by electrons. The presence of an ionized trap results in a modified potential, $U_s(r)$. The tunneling current density J_t is given by

$$J_t = \frac{qm^* k_B T}{2\pi^2 \hbar^3}\int_0^\infty dE_z T_r(E_z) \ln\left[\frac{1+\exp\left(\frac{E_F-E_z}{k_BT}\right)}{1+\exp\left(\frac{E_F-E_z-qV}{k_BT}\right)}\right], \quad (3)$$

where the transmission coefficient, T_r, of the electrons through the device is evaluated by solving the time-independent Schrödinger equation following the quantum mechanical wave impedance method [10] based on a transmission line analogy for a one-dimensional potential $U(z)$. Fluctuations in current density due to ϕ_s are computed by taking the differences of the current densities calculated for U_s and that calculated for U, giving $\Delta J_t(r)$. The total current fluctuation is obtained by integrating $\Delta J_t(r)$ over the device area.

DISCUSSION

We compare the model with the experimental results. First, the capture of an electron in the barrier results in the increase in the device resistance. Consequently, according to the model, τ_l corresponds to capture time constant and τ_h corresponds to emission time constant of the trap. To examine this experimentally, we examine the 20-μV switchings in the RTN data which shows that both τ_l and τ_h vary exponentially with the applied voltage. The trap occupancy, f_T, is related to the capture and emission time constants by $f_T = \tau_e/(\tau_e + \tau_c)$. Since it is not known a priori whether τ_l or τ_h corresponds to τ_c, we plot the two possible cases in Fig. 3b where the open circles and open triangles represent the cases where $\tau_l \equiv \tau_c$ and $\tau_h \equiv \tau_c$ respectively.
As a larger negative bias is applied to the emitter, the Fermi level in the emitter region is raised leading to an increase in the probability of the occupancy of the trap. From the figure it is clear that τ_l corresponds to the capture time constant, which is consistent with the proposed model.
The calculated current fluctuation, ΔI_0, arising from the capture of an electron by a trap within the first energy barrier gives rise to a voltage fluctuation $\Delta V = \Delta I_0/g_D$, where g_D is the dynamic conductance. The results are shown in Fig. 4, where the solid triangles are the experimental data for the trap that exhibits step size of approximately 20-μV, the solid line and dashed line correspond to calculated step size for a positively charged and neutral trap

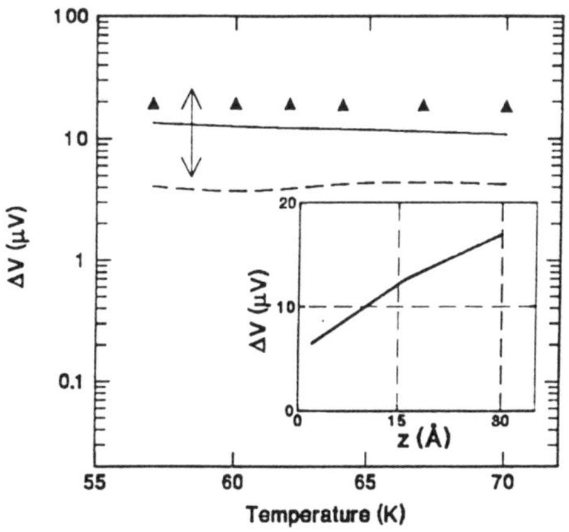

Fig. 4. Comparison between experimental and calculated RTN step sizes.

located at the center of first barrier from the emitter respectively, and the double arrow indicates the range of step sizes observed in the experiment. The inset indicates the magnitudes of the step heights calculated for a positively charged trap located at different positions within the first barrier from the emitter for $T = 60$ K. Clearly, the theoretical estimates of ΔV agree well with the experimental data, providing strong support for the model.

In Fig. 3a, both τ_c and τ_e are shown to decrease with increasing negative bias at the emitter. For traps that capture electrons from the emitter and subsequently releasing them back to the emitter, one would expect that the application of a more negative voltage would lower the capture barrier but raise the emission barrier. Since the time constants τ_c and τ_e vary exponentially with the capture and the emission barriers, respectively, τ_c should decrease, and τ_e should increase as is typically seen in RTN measured from MOSFETs [11, 12]. Experimental data in Fig. 3b, however, indicate otherwise. An alternative view is that of hopping conduction via localized states, in which carriers are captured from the emitter, but, instead of being emitted back to the emitter, are released into the quasi-bound states in the quantum well. An increase in the negative bias raises the emitter Fermi level accompanied by a reduction in the capture barrier. Similarly, the emission energy barrier is also decreased. Consequently, both the capture and the emission time constants decrease with increasing negative bias. The bias dependences of our RTN data clearly indicate hopping conduction as the underlying mechanism for the RTN.

The RTN technique can be utilized to characterize the traps. The capture rate of the trap can be expressed as

$$\frac{1}{\tau_c} = c_n n = \sigma_n \overline{v_n} n, \tag{4}$$

where $\overline{v_n}$ is the average thermal velocity of the carriers, n is the carrier density, and σ_n is the capture cross section which can be expressed as

$$\sigma_n = \sigma_{n0} \exp\left(-\frac{E_{ca}}{k_B T}\right), \tag{5}$$

in which σ_{n0} is the capture cross section at infinite temperature. Our calculations show that $\sigma_{n0} \approx 1.4 \times 10^{-18}\text{cm}^2$. To evaluate the density of traps, we estimate that over the range of sampling rates, from 51.2 Hz to 256 kHz, used in measuring random switching patterns and over the temperature range of 57–70 K the total number of active traps observed from the RTN data is of the order of 10. With a device area of 1.26×10^{-7} cm^2, the density of active traps is estimated to be $\sim 10^9$ cm^{-2}eV^{-1}.

CONCLUSION

Detailed studies of RTN in GaAs/AlGaAs resonant tunneling diodes over a wide range of temperatures and biases have been conducted. Based on the analyses of the experimental data, we presented a noise model that accounts for the observed low frequency fluctuations in the device. We showed that the low frequency noise arises from defect assisted hopping of electrons through the energy barrier. This process gives rise to fluctuation in the transmission coefficient of the tunnel barrier. We have also shown that detailed trap parameter can be extracted from the RTN data. In our studies of the RTN, we focused on one of the traps that produces a discrete switching of 20-μV.

We found that the capture cross section for this trap is of the order 1.4×10^{-18}cm^2 and the trap density of the device is estimated to be about 10^9cm^{-2}eV^{-1}.

References

[1] T. C. L. G. Sollner, P. E. Tannenwald, D. D. Peck, and W. D. Goodhue, Appl. Phys. Lett., vol. 45, pp. 1319–1321 (1984).

[2] E. R. Brown, W. D. Goodhue, and T. C. L. G. Sollner, J. Appl. Phys., vol. 64, pp. 1519–1529 (1988).

[3] C. E. Chang, P. M. Asbeck, K. C. Wang, and E. R. Brown, IEEE Trans. Electron Devices, vol. 40, pp. 685–691 (1993).

[4] H. J. Siweris and B. Schiek, IEEE Trans. Microwave Theory Tech., vol. 33, pp. 233–242 (1985).

[5] M. H. Weichold, S. S. Villareal, and R. A. Lux, Appl. Phys. Lett., vol. 55, p. 1969 (1989).

[6] X. M. Li, M. J. Deen, S. P. Stapleton, R. H. S. Hardy, and O. Berolo, Cryogenics, vol. 30, p. 1140 (1990).

[7] Y. Lin, A. D. van Rheenan, and S. Y. Chou, Appl. Phys. Lett., vol. 59, p. 1105 (1991).

[8] S. Ng and C. Surya, Solid-State Electron., vol. 35, p. 1213 (1992).

[9] S. Ng, C. Surya, E.R. Brown, and P.A. Maki, Appl. Phys. Lett., May (1993).

[10] A. N. Khondker, M. R. Khan and A. F. M. Anwar, J. Appl. Phys., vol. 63, p. 5191 (1988).

[11] M. J. Kirton and M. J. Uren, Advances in Physics, vol. 38, pp. 367–469 (1989).

[12] M. Schulz and A. Karmann, Physica Scripta, vol. T35, pp. 273-280 (1991).

INVITED

LOW - FREQUENCY NOISE MEASUREMENTS AS A CHARACTERIZATION AND TESTING TOOL IN MICROELECTRONICS

Zeynep Çelik-Butler
Southern Methodist University, Department of Electrical Engineering, Dallas, TX 75275 USA

ABSTRACT

An overview of low-frequency noise (LFN) techniques is done where a brief description of experimental techniques and results is given about methodologies developed for characterization and testing in microelectronic systems. Three such evaluation techniques are discussed as examples. The first one involves the characterization of electromigration parameters and prediction of the life-times in Very - Large - Scale - Integrated (VLSI) circuit metallization layers and vias. By performing LFN measurements in accelerated electromigration (EM) conditions such as elevated temperatures and at stressing current densities, one can evaluate the EM parameters like activation energy and current exponent. Then, these parameters are used to establish a correlation between the life-time of the metallization layer and the LFN the film exhibits, which in turn can be used as a predictor of life-time. Secondly, LFN measurements can be utilized to characterize $YBa_2Cu_3O_{7-\delta}$ thin films and to determine the quality of the film related to its composition, crystallinity, Resistance-Temperature (R-T) characteristics, and the critical current. The third, and perhaps the most widely reported application of the LFN characterization techniques is the measurement of semiconductor - insulator interface state densities, specifically in HgCdTe and Si Metal - Insulator - Semiconductor (MIS) structures. This can be done by varying the bias conditions of the structure, temperature or both.

INTRODUCTION

Although there is no single Low-Frequency Noise (LFN) technique that is universally used for characterization and testing purposes, all such methods involve noise measurements at the lower ranges of the frequency spectrum either as a function of temperature or bias conditions to extract parameters that are inherent to the mechanism(s) proven to be the origin of these fluctuations. Therefore, first there should be profound evidence to the fact that the characterized mechanism indeed causes these low-frequency fluctuations. Second, the proposed LFN method should be easy to implement and should surpass other existing techniques in at least one aspect. There have been few applications of noise measurements as a characterization and testing tool. In the next three sections, I am going to describe noise measurement techniques and its applications as a characterization method for three different systems. These are the detection of electromigration in thin metal films, characterization of interface states in MIS structures, and establishing a correlation between noise and electrical and material characteristics of $YBa_2Cu_3O_{7-\delta}$ thin films. The list of papers sited in this overview is not intended to be a comprehensive list but merely a representative of the different investigations in these areas.

ELECTROMIGRATION CHARACTERIZATION

Currently over a quarter of all failures in the VLSI circuits is due to the metallization layers. Moreover, the vast majority of metallization failures is from electromigration damage. Most common techniques used to characterize electromigration parameters in thin metal films and to predict their life-times are time consuming and destructive. Low - frequency noise measurements, on the other hand, provide information about the electromigration parameters and were shown to be good indicators of their life-time. Moreover, if the measurement time is low, the amount of damage induced in the film is negligible.

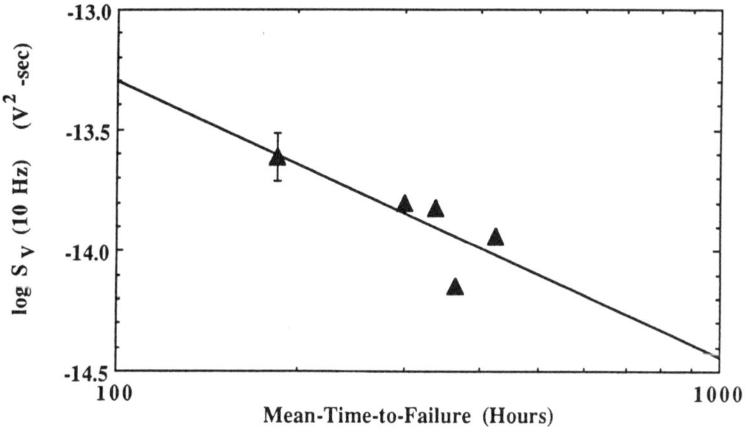

Figure 1. Voltage noise power spectral density at 10 Hz vs MTF for 5 wafers. The fitted line depicts $S_V \propto 1/MTF$.

A method that utilizes LFN measurements for electromigration studies has been first suggested by Celasco[1] et al. in 1976. Later, similar methods have been developed by Koch et al.[2], Diligenti et al.[3,4], Cottle et al.[5], van der Ziel et al.[6], and Çelik-Butler et al.[7,8]. A model relating observed low-frequency noise to electromigration mechanisms was developed based on electron mobility fluctuations due to scattering from vacancies migrating along the grain boundaries. Through this model, the voltage power spectral density measured across the film can be written as[8]:

$$S_V \propto \frac{SLV^2}{f^{SL+1}} \operatorname{cosec}\left[\frac{\pi(SL+1)}{2}\right] \exp(-E_a/kT) \tag{1}$$

where L is the mean distance of travel by the vacancy before it reaches a vacancy sink or meets an interstitial where it is annihilated. V is the applied voltage, E_a is the electromigration activation energy, T is the temperature, and k is the Boltzmann constant. S is F/kT where F is the driving force for electromigration which can be expressed as qZ^*_bE in terms of the electric field E, the effective charge number Z^*_b for grain boundary electromigration and the elementary charge q. The predictions of the above expression have been verified experimentally as far as the bias and temperature dependences of noise magnitude S_V and the frequency exponent γ of the $1/f^{\gamma}$ spectrum (γ=SL+1) are concerned[8]. Moreover, the activation energies computed using noise measurements were found to be in the same range for grain boundary vacancy migration activation energies measured using other techniques[9].

In addition to the characterization of electromigration parameters such as electromigration activation energy, low-frequency noise measurements can also be used to predict the life - times of VLSI metallization layers due to electromigration failures. Two such investigations have been reported showing that there is, in deed, an inverse relationship between the initial noise exhibited by the structure and the mean - time - to - failure (MTF). Cottle et al. showed such a correlation exists between noise and fabrication processes, and life-times for TiW / Al interconnections. Çelik-Butler and Ye proved the same for W / Al -Cu multilayered metallizations. Figure 1 depicts the voltage noise power spectral density at 10 Hz vs MTF for the multilayered metallizations. Noise measurements were performed at T=150 °C with the samples biased with $J=4x10^6$ A/cm^2. Life-time tests were done at T=200 °C (T_{stripe}=216 °C) and $J=2x10^6$ A/cm^2. The least-squares fitted line shows an inverse relationship between these parameters.

There are problems still remaining to be investigated in relating low-frequency noise to electromigration failure, especially in vias between multilayered metallization layers and contacts since it has been shown that some of the failures are due to voids in these regions.

CHARACTERIZATION OF HIGH T_c SUPERCONDUCTING FILMS

Contrary to the low-frequency noise of thin metal films and MIS structures, there is no established and widely accepted model for the noise behavior of high T_c superconducting thin films. . The reported noise magnitudes as well as the spectral shape, for $YBa_2Cu_3O_{7-\delta}$ for example, differ vastly from one sample to sample, making it impossible to evaluate a consistent value for Hooge's constant α in the expression [10]:

$$S_V = \frac{\alpha V^2}{N_C f} \tag{2}$$

Here, V is the voltage, and N_C is the total number of charge carriers between the voltage probes. Moreover, there are contradicting reports[11-14] on agreement between specific data and the generally accepted 1/f noise models, such as the equilibrium temperature fluctuations model[15], Hooge's model[10], or Dutta's thermal activation model[16]. There also has been an attempt to explain the noise behavior of $YBa_2Cu_3O_{7-\delta}$ thin films in the normal state through a McWhorter-like tunneling mechanism[17]. If the low-frequency noise technique is to be used as a characterization tool, first a detailed investigation has to be done on the origin of these fluctuations.

Some preliminary investigations have been reported relating the 1/f noise magnitude to the crystal structure and orientation. R. D. Black et al.[12] showed that 1/f noise magnitude measured on c - axis - normal epitaxial $YBa_2Cu_3O_{7-\delta}$ films on $SrTiO_3$ substrate was two to three orders of magnitude lower that that measured on granular $YBa_2Cu_3O_{7-\delta}$ films on ZrO_2 buffer layers with different substrates. It was also found that as the R - T characteristics deteriorated, the 1/f noise component increased in these films[17,18]. The 1/f noise magnitude increased with time as the film degraded when it was left exposed to air. Figure 2 shows the normalized noise power spectral density at 500 Hz measured on $YBa_2Cu_3O_{7-\delta}$ film sputtered on MgO. The increased S_V/V^2 was accompanied by a 15% resistance increase with time[18].

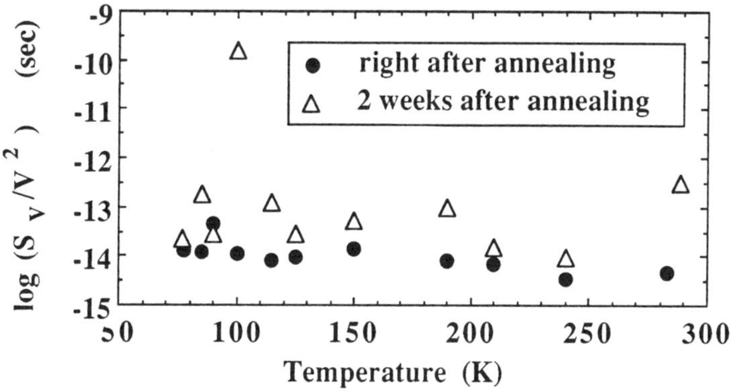

Figure 2. Degradation of the YBCO film with time.

1/f noise measurements promise to be good indicators of film quality in high T_c thin film superconductors. However, a thorough study linking the 1/f noise to the crystal structure, R-T characteristics and film deposition techniques on different substrates is still lacking.

CHARACTERIZATION OF INTERFACE STATES

Perhaps the most widely reported application of 1/f noise measurements as a characterization method is the determination of dielectric traps in a semiconductor - dielectric system. It is widely accepted now in the noise community that the major contributor to noise in surface conduction devices is the trapping centers at the semiconductor - dielectric interface and in the bulk dielectric whether this is due to carrier number or carrier mobility fluctuations or both. Therefore in theory, measurement of this noise should yield information about the magnitude and energy distribution of these traps. Different experimental techniques have already been developed where MISFETs, MIS capacitors, ungated oxide - semiconductor structures, made of Si and HgCdTe have been probed to extract interface trap densities. There have been also variations in the type of models used. There are several advantages of using noise techniques to measure interface trap densities: first, capability to probe into band gap energies close to band - edges and to energies even beyond the semiconductor band-gap; second, ability to probe into slow interface states which reside within the dielectric.

One of the first investigations was done on ungated diffused resistors with thermally grown oxide where 1/f noise measurements at cryogenic temperatures were used in conjunction with McWhorter's number fluctuation theory to extract $Si\text{-}SiO_2$ interface traps density around the valence band edge[19]. Later, scattering effect of the interface traps was also taken into account. R. Jayaraman et al.[20] and Hung et al.[21] independently developed theories where correlated mobility fluctuations due to traps as well as the number fluctuations are considered to determine the trap distribution. Jayaraman and Sodini[20] demonstrated that 1/f noise data taken on MOSFETs as a function of gate voltage and frequency can be used to extract oxide trap density as a function of distance from the Si-SiO_2 interface and energy. A thermal activation theory was utilized by Surya and Hsiang[22] to compute the interface states in Si MOSFETs. Here, the noise was assumed to be due to capture and emission of carriers by oxide traps through thermal activation.

For determination of HgCdTe - ZnS interface states, so far two different studies have been published. 1/f noise measurements have been performed at 77 K on HgCdTe MISFET's with varying band gaps, where the HgCdTe - ZnS interface states have been extracted as a function of band - gap energy[23,24]. Figure 3 shows the trap density for a p-substrate 240 meV HgCdTe found through the investigation of the drain voltage noise power spectral density dependence on gate and drain bias. MIS capacitors operated in pulsed charge integrating mode were also utilized to probe into HgCdTe - ZnS interface trap density[25].

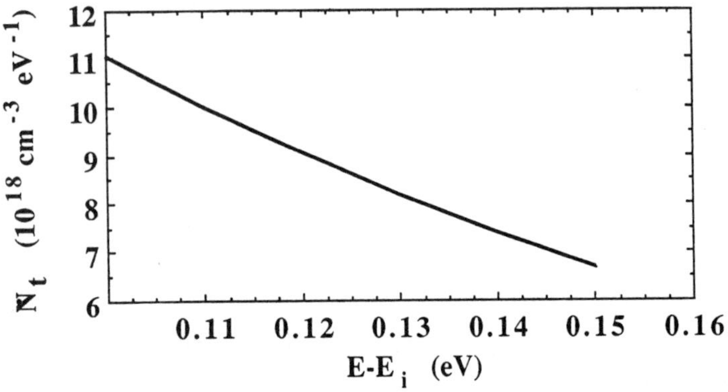

Figure 3. Extracted Zn trap density vs energy from a HgCdTe MISFET.

Although there have been some variations due to different surface treatments, interface state values obtained through noise measurements are in general agreement with those obtained by conventional techniques such as conductance and capacitance measurements.

CONCLUSION

Low - frequency noise measurements promise to be good characterization tools that have found various applications in microelectronics where the parameters involved in the mechanism(s) causing these fluctuations can be identified and investigated. Some of these parameters are interface states in surface conduction devices, electromigration parameters in stressed metallization layers, and crystallinity and crystal orientation in high T_c superconductors.

ACKNOWLEDGEMENTS

Author's work summarized here was supported by Texas Instruments Inc., THECB Texas Advanced Technology Program under Grants No. THECB ATP003613-004, THECB ATP003613-008, THECB ATP003613-005 and the National Science Foundation under Grants No. ECS-9116209, ECS-8908533.

REFERENCES

1. M. Celasco, F. Fiorillo, P. Mazetti, Phys. Rev. Lett., **36**, 38 (1976).
2. R. H. Koch, J. R. Lloyd, and Cronin, Phys. Rev. Lett., **55**, 2487 (1985).
3. A. Diligenti, B. Neri, P. E. Bagnoli, A. Barsanti, and M. Rizzo, IEEE Electron Device Lett., **EDL-6**, 606 (1985).
4. B. Neri, A. Diligenti, and P. E. Bagnoli, IEEE Tran. Electron Devices, **ED-34**, 2317 (1987).
5. J. Cottle, N. S. Klonaris, and Mark Bordelon, IEEE Electron Device Lett., **11**, 523 (1990).
6. A. van der Ziel, T. M. Chen, P. Fang, Solid - St. Electronics, **33**, 1025 (1990).
7. Z. Çelik-Butler and M. Ye, Solid - St. Electronics, **35**, 1209 (1992).
8. W. Yang, and Z. Çelik-Butler, Solid - St. Electronics, **34**, 911 (1991).
9. Z. Çelik-Butler, W. Yang, H. H. Hoang, W. R. Hunter, Solid-St. Electronics, **34**, 185 (1991).
10. F.N. Hooge, Phys. Lett., **29A**, 139 (1969).
11. P. Rosenthal, R.H. Hammond, and M.R. Beasley, IEEE Trans. Magn., **25**, 973 (1989).
12. R.D. Black, L.G. Turner, A. Mogro-Campero, and T.C. McGee, Appl. Phys. Lett., **55**, 2233 (1989).
13. J.H. Lee, S.C. Lee, Z.G. Kim, Phys. Rev., **B40**, 6806 (1989).
14. Z. Çelik-Butler, W. Yang, D. P Butler, Appl. Phys. Lett., **60**, 246 (1992).
15. R.F. Voss, and J. Clarke, Phys. Rev., **B13**, 556 (1976).
16. P. Dutta, P. Dimon, and P.M. Horn, Phys. Rev. Lett., **43**, 646 (1979).
17. Y. Song, A. Misra, P.P. Crooker, and J.R. Gaines, Phys. Rev. Lett., **66**, 825 (1991).
18. S. M. Alamgir, Z. Çelik-Butler, W. Yang, J. Wang, and D. P. Butler, Proc. 1990 Applied Superconductivity Conf., 45, September 24-28, 1990, Snowmass Village, CO.
19. Z. Çelik-Butler, and T. Y. Hsiang, IEEE Tran. Electron Devices, **ED-35**, 1651 (1988).
20. R. Jayaraman, and C. G. Sodini, IEEE Tran. Electron Devices, **ED-36**, 1773 (1989).
21. K. K. Hung, P. K. Ko, C. Hu, Y. C. Cheng, IEEE Tran. Electron Devices, **ED-37**, 654 (1990).
22. C. Surya, T. Y. Hsiang, Solid-St. Electronics, **31**, 959 (1988).
23. R. A. Schiebel, Solid-St. Electronics, **32**, 1003 (1989).
24. Z. Çelik-Butler, S. M. Alamgir, and S. R. Borrello, Solid-St. Electronics, **33**, 585 (1990).
25. W. He, and Z. Çelik-Butler, 1993 Proc. International Conf. Noise in Physical Systems and 1/f Fluctuations. (This issue).

NOISE SPECTROSCOPY, DIAGNOSTICS AND RELIABILITY OF ELECTRONIC DEVICES

J. Sikula
Technical University, Zizkova 17; Brno, CZECH Republic
A. Touboul,
IXL (URA 846-CNRS), University.of Bordeaux, 33405 Talence, FRANCE

ABSTRACT

Application of low frequency (LF) noise analysis to defect characterization, diagnostic and reliability of electronic devices is based on the evaluation of the G-R and 1/f noise.After some remarks on the parameters shifts induced by non-stationary processes, specific examples on electromigration, p-n junction devices and FET's are detailed.

INTRODUCTION

Application of low frequency (LF) noise analysis to defect characterization, diagnostic and reliability of electronic devices is based on the evaluation of the G-R and 1/f noise. It is now established that there are basically two kinds of LF 1/f noise in semiconductor devices. The first type, sometimes called fundamental 1/f noise belongs to the family of theoretical sources of noise along with thermal, shot and G-R noises. The other kind is related to defects in the device structure. Strictly speaking, its spectral density is of the $1/f^{\gamma}$ type, where $\gamma \simeq 1$. We call it excess 1/f noise and it is just this kind of noise that we will deal with henceforth.

In the early 60's, it has been accepted that a device which exhibits excessive noise level represents a major reliability risk. Hence a device generating important noise level will prematurely be "going out ot specs" than another sample the noise of which is found in a "normal" range. This approach supports the trend of performing screening tests to detect and eliminate samples which may present potential failures.

It has been suggested long ago that the excess 1/f noise can be used for assessment of the device technology, to reveal defects, briefly, to non-destructive device testing[1]. It was mainly due to the fact that the excess noise is by orders of magnitude more sensitive to the presence of defects than other parameters, such as an electrical resistance.

Noise analysis provides, in general, information about interface traps, volume traps in MOSFET's and also in the depletion layer of P-N junctions. At the same time, these methods can provide information on Silicon and III-V FET's. Information about lack of ideality in various devices can be obtained from the frequency, voltage, current and temperature dependences of the spectral density. The next steps were made to suggest that 1/f noise can be used to assess the quality and to predict the reliability of electronic devices. In the past the reliability prediction was mainly based on evaluation of time-average values of the first moments of the realizations of the measurable quantities.

Use of noise for quality assessment and reliability prediction is based on analogous quantities, but it is the second moment of the realization that is evaluated. And this is also the reason why this method is more sensitive. It is, therefore, much more suitable for control of technology processes and for the prediction of the device life time. Hence, noise measurements become an effective screening tool for reliability evaluation instead of the usual time-consuming and destructive life tests generally performed under accelerating ageing conditions.

NOISE SPECTROSCOPY AND DEFECT CHARACTERIZATION

Any kind of defects, structural and chemical, located in the active region of a semiconductor device will generate excess noise. The resulting low frequency excess noise can be either stationary or non-stationary.

For stationary processes, it is possible to use the noise spectroscopy to determine some parameters of localized states in the band gap. In this way, energy levels and capture cross-sections can be determined. This is applicable both to crystalline and polycrystalline materials.

Noise spectroscopy is used to determine the parameters of deep levels, known as DX centers in GaAs based heterostructures[2] (fig.1).

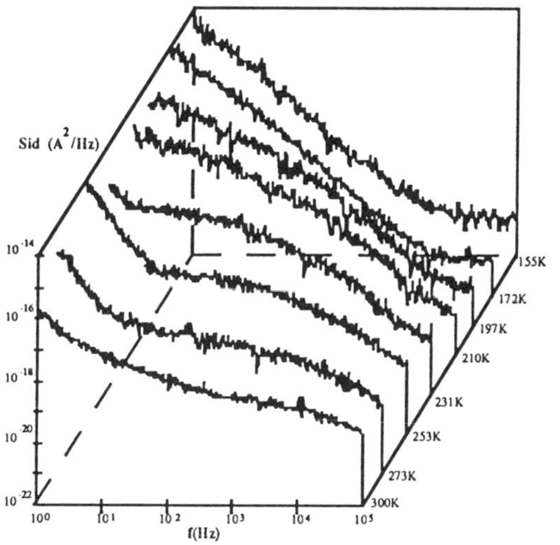

Figure 1 : G-R. Noise Spectra at different temperatures in AlGaAs/GaAs HEMT's.

Simeon et al.[3] applied a combined low-frequency and RTS analysis to Silicon MOSFET's and showed that the noise is most probably related to the presence of deep level G-R centers in the depletion region of the MOST. In their model the excess noise amplitude was proportional to the defectiveness of the SOI film.

Madenach and Werner[4] applied the same approach to ion-implanted polysilicon thin films and were able to find out the defect parameters.

The noise spectroscopy can be used also to identify single traps. Schulz and Pappas[5] determined parameters of individual defects in MOS interface from the RTS noise.Defects that are caused by radiation generate sources of noise, whose spectral density is, as a rule, proportional to the square root of the radiation dose. This was shown on copper films by Pelz et al.[6].

On the other hand, it has been shown by Vandamme and Oosterhoff[7] that annealing decreases the concentration of structure defects and the noise spectral density.

NON-STATIONARY PROCESSES AND PARAMETRIC DEGRADATION

Non-stationary noise may be generated when the defects in the device active region are unstable. From the point of view of parametric degradation and reliability only those defects whose positions change as a consequence of the temperature, thermal and concentration gradients, mechanical and electrical stress are important. In other words, degradation takes place only when the electrical field intensity of other external driving forces exceed a certain value necessary to activate the motion of these defects, which may result in parametric degradation or even a failure.

In the millihertz region the non-stationary components of noise can manifest their presence in the measured quantities as follows :
i) a slow drift, or a trend, in the first moment of the realization in the time domain,
ii) a higher value of the exponent γ in the $1/f^{\gamma}$ plot in the frequency domain.

The processes associated with the above mentioned non-stationary phenomena bring about changes in operational characetristics of the device in question and may eventually result in a parametric failure. Any means suitable to detect such processes is therefore of high importance for reliability prediction.

ELECTROMIGRATION

When dealing with the use of noise in this field, it should be emphasized that conventional testing techniques respond to the presence of mostly immobile defects, whereas noise measurements, particularly in the millihertz region, provide information on defects that are mobile, that is, about migrating atoms, moving dislocation lines, etc. A typical phenomenon of this kind is electromigration, which became one of the most important problems in VLSI integrated circuits.

The fundamental mechanisms of the electromigration phenomena, which influence substantially the reliability of semiconductor devices, consists in transport of aluminium in conducting tracks. Assuming that the excess noise is generated by random processes associated with diffusion mechanisms, a thermal activation of the noise is empirically modelled either for $1/f^2$ noise or in the most general case for $1/f^{\gamma}$ noise. The activation energy derived from excess noise dependence on temperature can be expressed on the basis of the Black's equation by using two different temperatures corrresponding respectively to noise measurement and to electromigration test.

Bagnoli[8] showed that activation energy as determined from the Arhenius plot of the noise spectral density has nearly the same value as when determined by other methods. It has been found to range from 0.6 to 1.2 eV, according to the passivation and impurity concentration and is close to that derived from the grain boundary diffusion in Al films.

It must be precised that this attempt still requires work to precise the origin of discrepancies in the experimental values of the activation energies obtained by both procedures. Sun et al.[9] and Celik-Butler[10] have found a relation between the mean time to failure and the narrow band noise voltage (fig. 2).

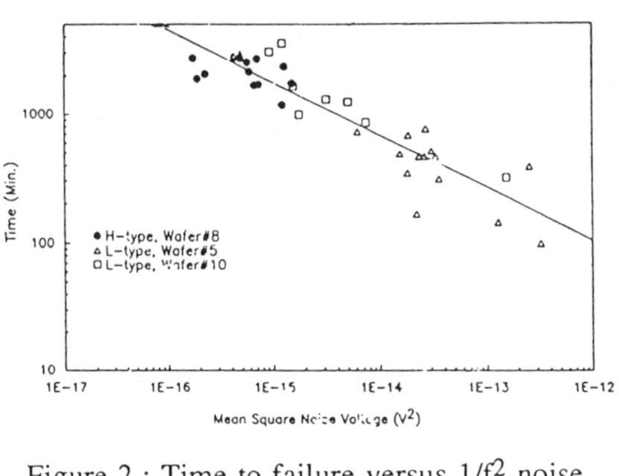

Figure 2 : Time to failure versus $1/f^2$ noise voltage.

Jones et al.[11] showed that excess noise is usually observed in good Al tracks at high current densities and temperatures, whereas electromigration damaged tracks show excess noise at low currents and temperatures.

Touboul et al.[12] studied electromigration in Al and damages in NiCr stripes by means of the LF noise and showed that a weak degradation of electrical parameters corresponds to a great increase in the excess noise level (fig. 3).

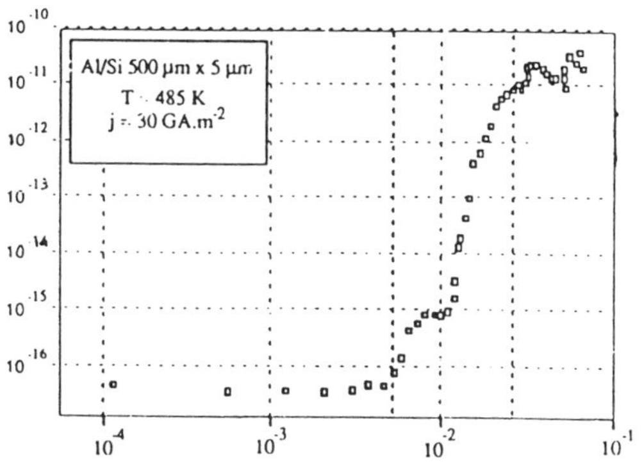

Figure 3 : Noise spectral intensity (arb. units) as a function of the relative increase of resistance ΔR/R.

They have found out that the noise power increases by four orders of magnitude whereas the resistance increases by two per cent, when the sample is stressed. Moreover, they have found noise of the $1/f^{\gamma}$ spectrum type, with γ ranging from 1 to 2.6.

A quantitative explanation why resistance noise measurements are more sensitive than resistance measurements only, was given by Vandamme[13] by modelling degradation of conducting carbon thin films by holes and kinks and the consequent generation of the 1/f noise.

In what concerns the higher value of the exponent γ we suggest an explanation of this phenomenon, which stems from the non-stationarity of the noise generation. Presence of non-stationary processes is manifested in the frequency domain through an $1/f^{\gamma}$ component, where γ is greater than unity, sometimes it holds $\gamma \simeq 2$.

The exponent γ appears to be stress-dependent : an increase of γ is often noticed when increasing the current density or the stripe temperature. To obtain a pure, or near-pure 1/f noise, one has to substract from the entire realization in the time domain the higher-exponent noise component which is usually identified as a time-drift (Vasina et al.[14]). It is clearly established that a $1/f^2$ law in electromigration test stems from erroneous causes and generally has deterministic origins : it appears important to discriminate $1/f^2$ noise arising from vacancy fluctuations or else from the resistance value drift of the sample under test. On the other hand, existence of such a trend in the millihertz region can be used to advantage as an indicator of slow ion motion, e.g., electromigration. Then a $1/f^2$ noise is an accurate predictor of failure times. This component of the noise carries most information concerning the non-stationary microscopic processes which in turn are responsible for low device reliability.

In spite of the potential of the noise measurements, it is not possible, at present, to foresee the effect of a prolonged stress and to make a precise estimate of the lifetime of the sample[15]. It still seems controversial to conclude on the slope of the noise spectra effectively produced by electromigration and hence on the valididty of the MTF derived from its expressions.

RELIABILITY INDICATORS IN P-N JUNCTION DEVICES

The concept of a reliability indicator was introduced by Jensen and Moltoft[16] in 1986 and now represents one of the major features of most reliability evaluation procedures.

Use of noise for study of the degradation process in semiconductor devices needs to define measurable quantities, which could provide information not only about the defects but also about their localization. The noise parameters have to be completed by transport parameters. For P-N junction devices it has been verified that the reliability indicators are the following :

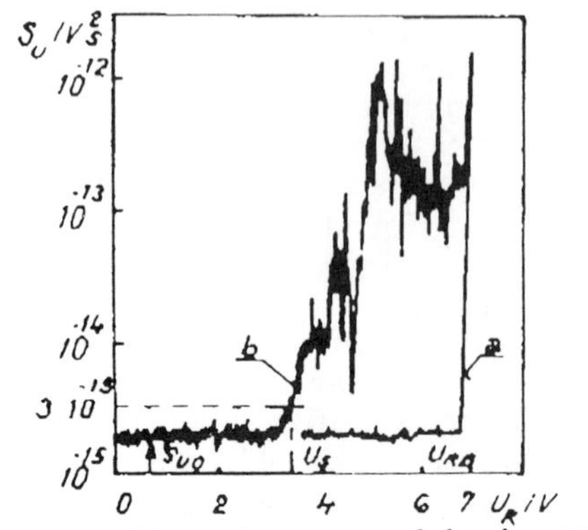

Figure 4 : Noise voltage spectral density versus the diode reverse voltage plots :
a) defectless diode - b) diode with defects.

i) maximum value of the noise voltage spectral density S_{UM} in forward direction across a load resistance R_L,
ii) the reverse voltage U at which the noise voltage spectral density across a given load resistance equals twice that of the background noise (fig.4),
iii) the burst noise, where the structure defects modulate the current.
These indicators were used by Sikula et al. for reliability prediction of LED's[17] and bipolar IC's[18].

RELIABILITY INDICATORS IN FET's

SILICON MOS TRANSISTORS

Two illustrative approaches will be underlined for the applications of noise to evaluate damages. Let us consider first the important contribution to define radiation hardness test on the basis of preirradiation noise measurements and its correlation with the post irradiation oxide-trapped charge density[19,20,21] (fig. 5).

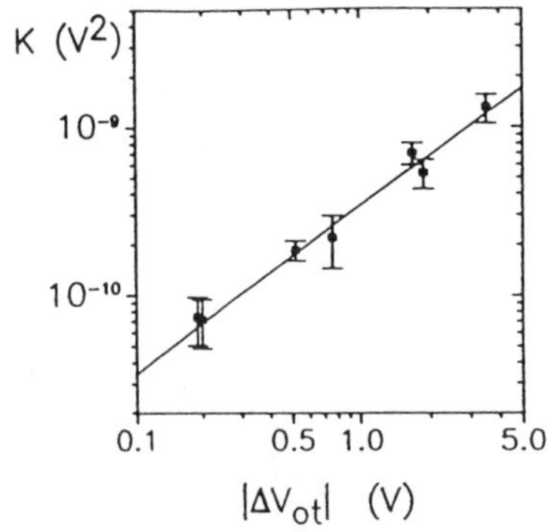

Figure 5 : Noise level K versus ΔV_{ot}

Further investigation in that direction have revealed the contribution of the channel access resistances. Their noise can in some cases screen the intrinsic channel noise and hence restricts the field of application of this method to a well-defined technological MOS process[22].
Another promising application is the analysis of the gate voltage and drain current noises for scaled-down MOS transistors. For small gate area, extensive work deals with the R.T.S. components in the total noise.

Once more, 1/f behaviours may be distinguished : those from the RTS fluctuations and the "residual" channel excess noise. Specific processing of the drain current noise will enable one to extract the RTS fluctuations and give access to the analysis of an individual trapping center[23]. Nevertheless, no precise trend can be drawn to predict the LF noise of S-D MOS transistors and it remains of great interest for VLSI integrated circuits reliability.

Excess noise measurements performed by Jones and Xu[24] on CMOS logic integrated circuits have shown that the presence and size of the noise is a sensitive indicator of the quality and hence reliability of the device.

III-V DEVICES

This part will be restricted to the impact of technology on the noise of HEMT's. Noise has been successfully applied to quantify the effect of DX centers through the correlation between the transconductance frequency dispersion and the fluctuation of deep levels occupancy[25]. The evolution of the noise with bias and temperature reveals individual G-R mechanisms which identify these DX centers. Another application is the gate current noise analysis to detect failures. This approach appears to be more sensitive than the drain current noise analysis, even if its experimental implementation is not straightforward. It is completed by coherence measurements between these two noise currents (from gate and drain)[26].

CONCLUSION

Electrical noise can be used as an indicator to characterize the quality, device reliability, and for prediction of electronic device failures. Noise spectral density measurements are by several orders of magnitude more sensitive than other measurements of various effects of physical or chemical nature.

It is a more sensitive indicator than a resistance change or - in PN junctions - the reverse current, the breakdown voltage, the parameter change in other devices, such as the threshold voltage shifts in MOS devices. The further step will be a reasonable prediction of devices life expectancy on the basis of their excess noise analysis.

REFERENCES

1 A. Van der Ziel et al, Electronics, 95, (1966)
2 N. Labat et al. Qual. and Rel. Eng., 8, 301, 1992
3 E. Simeon et al, Applied Surf Sc, 63, 285, (1992)
4 A.J. Madenach et al, Phs. Rev. B, 38, 1958, (1988)
5 M. Schulz et al, Proc. of ICNF 91 (Kyoto) Edit. T. Musha, 265, (1991)
6 J. Pelz et al, Phys. Rev., B, 38, 10371 (1988)
7 L.K.J. Vandamme et al, J. Appl. Phys. 59, 3169, (1986)
8 P.E. Bagnoli et al, J. Appl. Phys. 63, 1448, (1988)
9 M.I. Sun, Proc. of ICNF 89 (Budapest), Edit. Ambrozy, 519, (1990)
10 Z. Celik-Butler, Sol. St. El, 35, 1209, (1992)
11 B.K. Jones et al, Microelec. Rel. 31, 351, (1991)
12 A. Touboul et al,Proc. ICNF 91 (Kyoto), Edit. T. Musha, 249,(1991)
13 L.K.J. Vandamme, Proc. of ESREF 92, Edit. VDE, 377 (1992)
14 P. Vasina et al, Proc. of "Fluct. in Solids" - Brno, 69, (1992)
15 A. Scorzoni et al, Mat. Sc. Repots, 7, 143, (1991)
16 F. Jensen et al, Qual. and Rel. Eng. Int., 2, 39, (1986)
17 J. Sikula et al, Proc. of ICNF 89, (Budapest) Edit Ambrozy, 479 (1990)
18 J. Sikula et al, Proc. of ESREF 92, (Germany), Edit. VDE, 415, (1992)
19 J.H. Scofield et al, IEEE Trans. on Nucl. Sc, NS-36, 1946 (1989)
20 D.M. Fleetwood et al, Phys. Rev. Letters, 64, 579, (1990)
21 J.H. Scofield et al, IEEE Trans. on Nucl SC, NS-38, 1567, (1991)
22 D. Rigaud et al, Proc. of RADECS 93, To be published
23 G. Ghibaudo et al, Proc. of ICNF 91, (Kyoto) Edit Musha, 229, (1991)
24 B.K. Jones et al, Microelec. Rel., 31, 351 (1991)
25 C. Canali et al, IEEE Proc. G. 138, (1991)
26 L.K.J. Vandamme, IEEE Trans. on El. Dev., ED 35, 1071, (1988)

RTS NOISE IN INTEGRATED CIRCUITS

Z. CHOBOLA, P. VASINA, J. SIKULA
DEPARTMENT OF PHYSICS
TECHNICAL UNIVERSITY OF BRNO, ZIZKOVA 17, 602 00
CZECH REPUBLIC

ABSTRACT

The objective of this paper is twofold. First, the model theory of the RTS noise developed by Sikulova[1] is applied to an analysis of the RTS noise in bipolar Schottky IC's. This theory makes it possible to evaluate quantities characteristic of the processes of carrier capture, emission and recombination. Second, a new non-invasive method of potential spectroscopy of IC's is presented, which makes exclusive use of the IC external pins. An attempt is made to localize the sources of the observed RTS noise. Experiment results are in a good agreement with the mentioned theoretical model.

INTRODUCTION

In the model of the RTS noise[1] two processes are taken into consideration:

i) The primary process X(t), which consists in carrier transitions between a trap and the bands. This process is a three-state one and is inaccessible for direct measurement.

ii) The secondary process Y(t) which consists in the device current modulation and has two states, α and β. The distributions of the α and β state occupation times are

$$g_\alpha = \frac{1}{\tau_\alpha} \exp(-t/\tau_\alpha),$$

$$g_\beta = \frac{1}{\tau_\alpha} \exp(-t/\tau_1) + \frac{1}{\tau_\beta} \exp(-t/\tau_2),$$

where

$$\tau_\alpha = (1 + c_n n_1 + c_p p_s)^{-1}, \qquad \tau_\beta = \tau_\alpha \left(\frac{n_1}{n_s} + \frac{p_s}{p_1} \right).$$

The dispersion of the occupation time distribution is related to the occupation times themselves, as follows from the equations:

$$D(T_\alpha) = \tau_\alpha^2, \quad D(T_\beta) = a_1 \tau_1^2 + a_2 \tau_2^2 + a_1 a_2 (\tau_1 - \tau_2)^2.$$

The model[1] of the RTS noise is neither material nor device specific. It is based on general features of the processes of carrier capture, emission and recombination in structure where there is a trap in the gap. In this model, an important role is played by the relative position of the Fermi level and that of the trap.

The objective of this paper consists, therefore, in application of this theory to bipolar Schottky IC's which exhibit RTS noise in a certain region of the applied DC voltage and in a temperature range.

EXPERIMENT

Fig. 1 shows the circuit diagram of the IC under test. Only the pin No 16 and the input pin are connected. A ramp voltage U_F source is connected in series with a load resistor R_L and feeds the mentioned pins, the pin No 16 being positive. The noise voltage U_N across the load resistor is analyzed both in frequency and the time domain. Fig.2 shows a typical noise voltage spectral density versus U_F plot. Several peaks are seen in the diagram. The sequence of the peaks corresponds to the sequence of forward currents of various junctions in the IC.

The RTS noise was observed in the voltage range around 2 volts the corresponding peaks are shown in Fig.2.

Fig.3 shows the probability density of the on-time of the impulses. Two time constants are apparent, namely τ_1 = 13 μs, τ_2 = 80 μs. In the case where the time constant differ substantially, as is our case, the mixed system can be easily divided into two clean statistical ensembles with different statistical parameters.

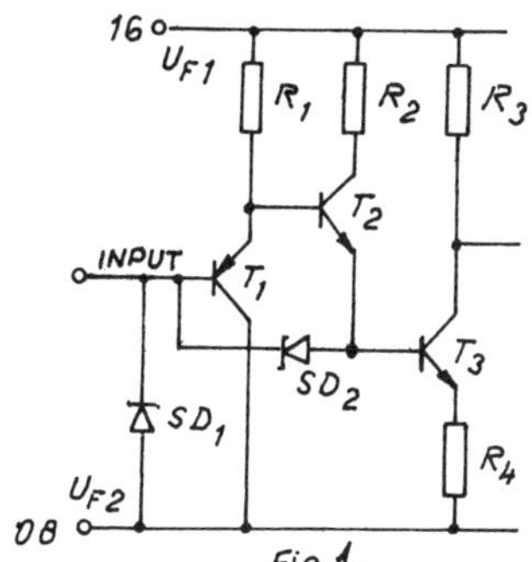

Fig.1.

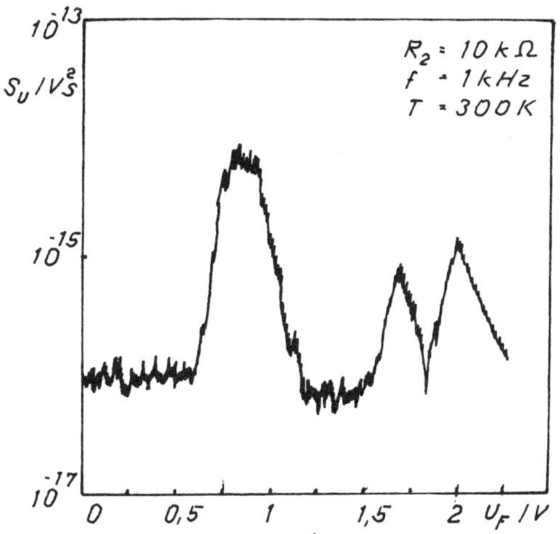

Fig. 2.

If the two parameters are close to one another, in other words, if $c_n n$ is approximately equal $c_p p_1$, which means physically that the transition probability density for the electron capture is approximately the same as the transition probability density for a hole emission, then the two processes are inseparable. Two different microscopic processes play the role here, namely: Either an electron from the conductivity band or another electron from the valence band are emitted and captured by the trap.

It follows that for a clean ensemble the dispersion of the occupation time is equal to the square of the average value of this quantity, wheras for a mixed ensemble a more complicated formula holds, (see equ.(8) in paper[1]), in which the probabilities of both above mentioned processes play the role.

The absolute probability densities Π_α , Π_β are independent of the capture cross sections. Therefore, the number of variables is lowered. Moreover these quantities are easily measureble from the occupation times,

$$\Pi_\alpha = \frac{T_\alpha}{T}, \quad \Pi_\beta = 1 - \Pi_\alpha ,$$

where T_α is the total time during which the system is in the state α and T is the total measurement time. The quantity Π_α as a function of the applied voltage for the sample No 85/03-16 is in Fig.4. We can see that for the voltage of 1,94 V the absolute probability density distribution is 1/2. In this case the Fermi level coincides with the trap level, provided that the degeneration is neglected.

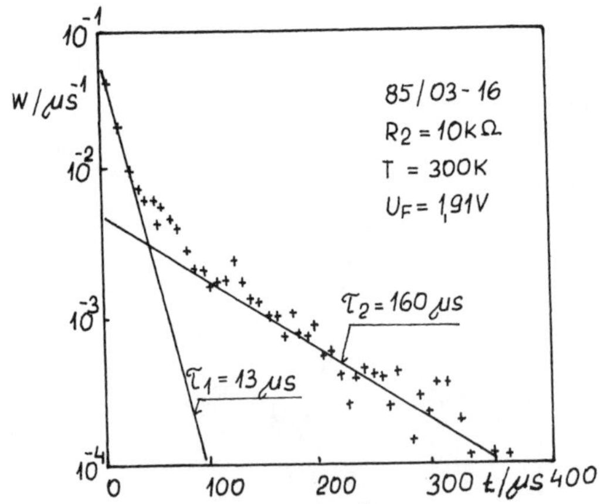

Fig.3.

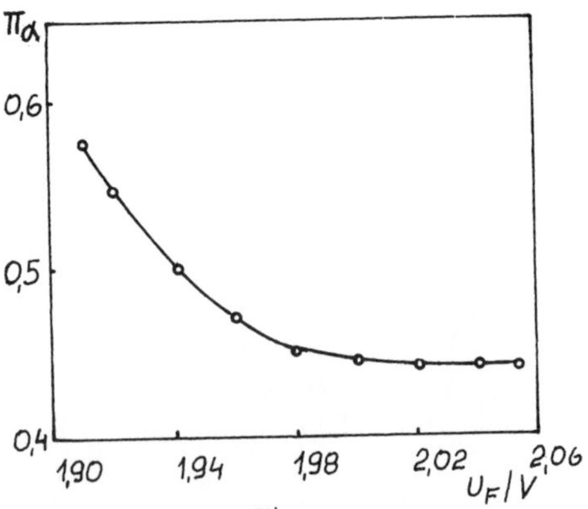

Fig.4.

If we assume that the trap is of the acceptor-like type, then using equ. 3 in paper[1], we may draw the probability density $\Pi_\alpha = \Pi_0$ as a function of the variable

$$x = \frac{E_t - E_{En}}{k\,T}$$

In this diagram, fig.5 our results are represented by crosses, results obtained by Schultz[3] for level I by open circles, for level II by triangles. It is seen that for the RTS noise to be generated it is necessary that x is within ± 3kT, or, the Fermi level is very close to the trap level.

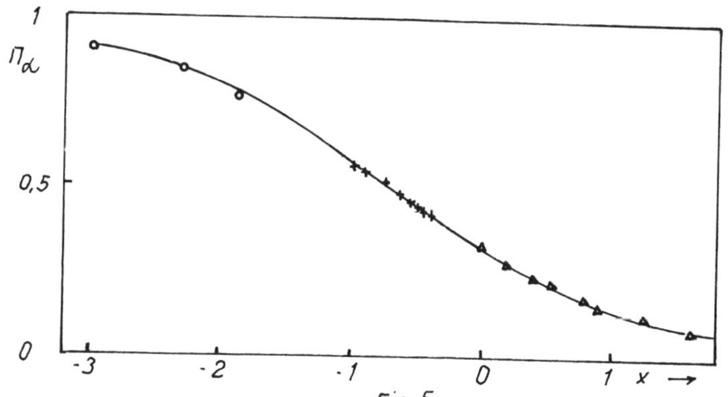

Fig.5.

CONCLUSIONS

Application of the the model[1] to the RTS noise in IC provides a better insight into the microscopic processes that take place structures with PN junctions.

REFERENCE

1. M. Sikulova, J. Sikula, Model of the RTS Noise in semiconductor devices in these Proceedings.
2. S.T. Hsu, R.J. Whittier, C.A. Mead: Physical Model for Burst Noise in Semiconductor Devices. Solid State Electronics, 13 (1970), p.1055 - 1071.
3. M. Schulz, A. Karmann, Single, Individual Traps in MOSFETs. Physica Scripta. Vol.T35 (1991), p.273-280.

NOISE CHARACTERIZATION AND MODELLING OF POLYSILICON EMITTER BIPOLAR JUNCTION TRANSISTORS AT MICROWAVE FREQUENCIES

M.J. Deen and J.J. Ilowski
Department P813, Mailstop 014, Northern Telecom Limited
185 Corkstown Road, Nepean, Ontario, Canada K2H 8V4

ABSTRACT

Microwave noise parameters of polysilicon emitter bipolar junction transistors with emitter areas A_E ranging from 3.2 μm^2 to 144 μm^2, collector current densities J_C from $\sim 7.8x10^{-5} mA/\mu m^2$ to $\sim 0.31 mA/\mu m^2$, and for frequencies between *1* and *5.4GHz*, have been measured and modelled from a $0.8\mu m$ *BiCMOS* process. Our results show that the minimum noise figure NF_{MIN} at the typical working J_C of $\sim 0.16 mA/\mu m^2$ and frequency of 1GHz varied as $A_E^{-0.26}$, the real $R_{S,OPT}$ and imaginary $X_{S,OPT}$ parts of the optimal source impedance varied as $A_E^{-0.7}$ and A_E^{-1} respectively, the noise resistance R_N varied as A_E^{-1}, and the noise voltage V_N varied as $A_E^{-0.41}$. These noise parameter variations with A_E are in first order agreement with a simple noise model of the bipolar transistor, and with the experimental finding of the variation of the base resistance $R_B \sim A_E^{-0.92}$ and the emitter junction capacitance $C_{JE} \sim A_E^{-1}$. The detailed modelling of NF_{MIN}, NF_{50} (referenced to a 50Ω input), and R_N as a function of J_C and frequency showed good agreement with the experimental results.

INTRODUCTION

Polysilicon emitter (poly-emitter) bipolar junction transistors (BJTs) are being increasingly used in high frequency telecommunication applications in the GHz frequency range. Because they can be fabricated on the same wafer as CMOS transistors, they allow very high frequency performance (BJT) as well as dense signal processing, memory, and other relevant circuits (MOS) to be integrated on the same chip. These applications require accurate models for the simulation of BJT high frequency noise. In addition, it is useful to investigate the effects of emitter area scaling on the device noise parameters.

EXPERIMENTS

The devices were characterized on wafer using a HP8510C vector network analyzer and a HP 8970 noise figure meter. The measurements were controlled by an ATN NP5B computer-aided noise parameter test set which included a solid-state source impedance tuner and analysis software. The calibration procedure measured system losses and the reflection coefficients of the system components, and the software performed vector error correction. Noise figures were measured at 16 different impedance states, and the 4 noise parameters NF_{MIN}, $R_{S,OPT}$, $X_{S,OPT}$ and R_N were determined from curve fitting.

The devices studied had nominal emitter areas varying from $3.2\mu m$ to 144 μm^2, and they were characterized by both d.c. and microwave measurements. The d.c current gain β of these devices was typically between 110 and 140 over all current ranges used, with a supply voltage V_{CC} of 10V. The unity gain-bandwidth product f_T was in the range of 5 to 9GHz over a typical operating collector current I_C range, and f_T measurements followed the usual variation with I_C. The collector-emitter breakdown voltages BV_{CEO} were greater than 14V. These transistors were designed to have high $BV_{CEO} \cdot f_T$ products in excess of $100V \cdot GHz$. Note that as f_T increases due to shorter collector lengths, BV_{CEO} decreases, and so the product $BV_{CEO} \cdot f_T$ can be regarded as a useful figure of merit of these devices.

Using s-parameter measurements, we determined the base and emitter resistances and f_T at each bias current value. Other parameter values for modelling were determined from d.c and high frequency measurements using the TECAP parameter extraction program.

NOISE MODEL, RESULTS AND DISCUSSIONS

The small signal noise equivalent circuit used for the poly-emitter bipolar transistor is similar to that previously described[1,2], and is shown in Fig. 1. While this equivalent circuit only considers the base resistance R_B and the base-emitter junction capacitance C_{JE} (it neglects other device capacitances and parasitic resistances), and the current gain α neglects the phase shift factor due to the collector region delay time, we have found that it provides good agreement with measured device noise characteristics over the frequency range up to 6 GHz, and over a wide J_C range. Using this model and appropriate assumptions[1,3], the following expressions for the noise parameters are derived.

$$NF = 1 + \frac{R_B}{R_S} + \frac{r_e}{2R_S} + \left(\frac{\alpha_O}{|\alpha|^2} - 1\right) \cdot \frac{(R_S + R_B + r_e)^2 + X_S^2}{2r_eR_S}$$

$$+ \frac{\alpha_O}{|\alpha|^2} \cdot \frac{r_e}{2R_S} \cdot \{(\omega C_{JE}X_S)^2 - 2\omega C_{JE}X_S + (\omega C_{JE})^2(R_S + R_B)^2\} \tag{1}$$

$$X_{S,OPT} = \omega C_{JE} r_e^2 \left(1 - \frac{|\alpha|^2}{\alpha_O} + \omega^2 C_{JE}^2 r_e^2\right)^{-1} \tag{2}$$

$$R_{S,OPT}^2 = R_B^2 - X_{S,OPT}^2 + r_e(2R_B + r_e)\left(1 - \frac{|\alpha|^2}{\alpha_O} + \omega^2 C_{JE}^2 r_e^2\right)^{-1} \tag{3}$$

$$R_N = \frac{R_B}{|\alpha|} + \frac{r_e}{2}\left(1 + \left\{\frac{R_B}{r_e}\right\}^2 \cdot \left\{\frac{f}{f_e}\right\}^2\right) \tag{4}$$

where $\alpha^2 = \alpha_O^2 \cdot \left(1 + \left(\frac{f}{f_b}\right)^2\right)^{-1}$, $f_b = (2\pi\tau_b)^{-1}$, $f_e = (2\pi r_e C_{JE})^{-1}$, r_e is the emitter dynamic resistance defined as $(k \cdot T/q \cdot I_e)$, and α_O is the low frequency value of α. To compare calculations using the model equations listed above to experiments, all parameter values were determined from either high frequency measurements, or from d.c. measurements.

Figure 2 shows the measured minimum noise figure NF_{MIN} and the associated gain for a BJT with $A_E = 48\mu m^2$ at a collector current of 5mA, which is typical of the current biases used in telecommunication circuit applications. Also shown in Fig. 2 is the computed NF_{MIN} using equation (1), and as shown, the calculated and experimental results are in good agreement. Figure 3 compares the computed to measured NF_{MIN} values for the same device, as a function of collector current, and the agreement between measured and calculated values is good at both frequencies. Also shown in this figure is the *gain* as a function of I_C.

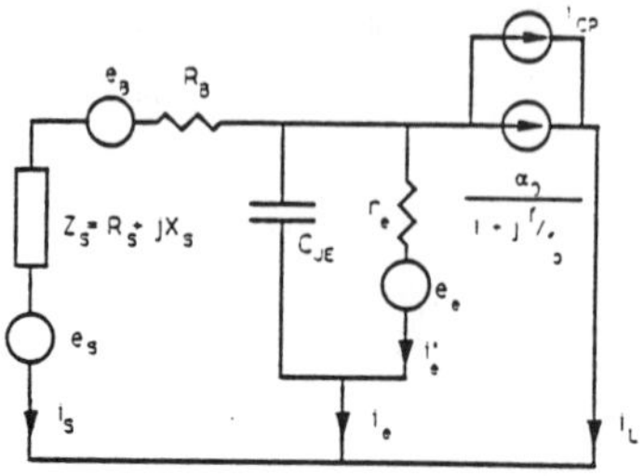

$$\overline{i_{CP}^2} = 2kT\cdot(\alpha_0 - |\alpha|^2)/r_e \qquad \overline{e_B^2} = 4kTR_B$$

$$\overline{e_e^2} = 2kTr_e \qquad \overline{e_S^2} = 4kTR_S$$

Figure 1: Simple small-signal T-equivalent circuit for the poly-emitter bipolar transistor used in the noise calculations.

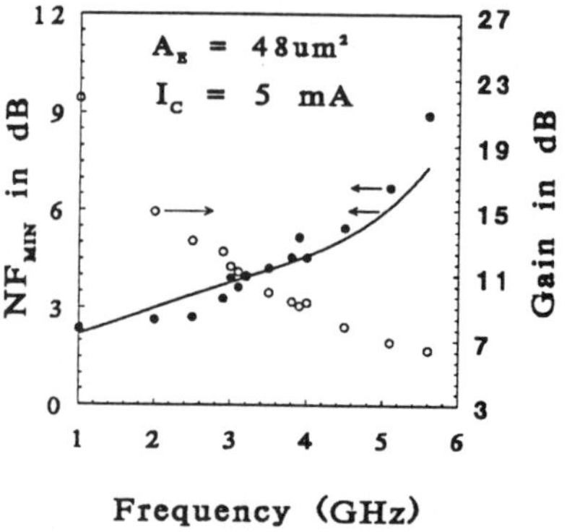

Figure 2: NF_{MIN} and Gain versus Frequency. The symbols are the experimental data and the solid line represents the calculated NF_{MIN} values.

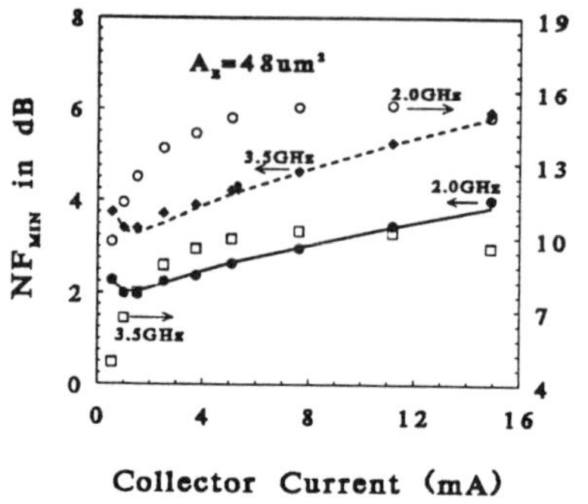

Figure 3: NF_{MIN} and Gain versus I_C. The symbols are the experimental data and the solid lines represent the calculated NF_{MIN} values.

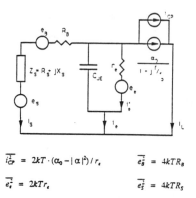

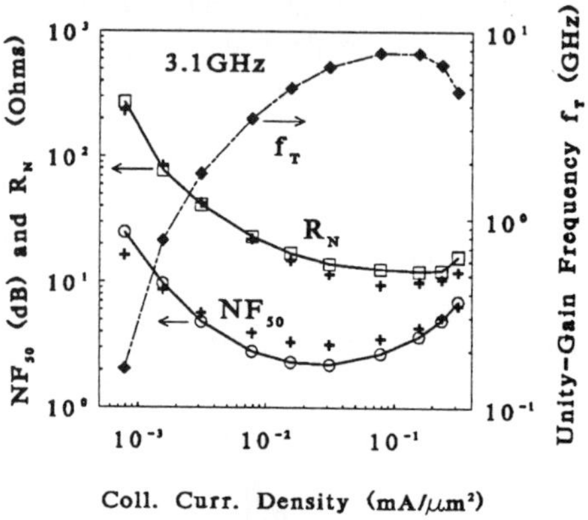

Figure 4: NF_{50}, R_N and f_T versus J_C. The symbols are the experimental data and the solid lines represent the calculated values.

Figure 4 shows the variation of NF_{50} and R_N with J_C for the largest device of area $144\mu m^2$. Note that for variations in J_C of more than 2 decades, good agreement between model calculations and experimental results is obtained. The f_T's are also shown in the figure, and the approximate relations f_B~$2.7f_T$ and f_e~$3.8f_T$ are used to describe their base-emitter voltage bias dependence, and $R_B \sim (10 \pm 1)\Omega$ for the model calculations.

To investigate the scaling dependence of the noise parameters, Fig. 5 shows the variation of $(NF_{MIN} - 1)$, and figure 6, the variation of R_N, $R_{S,OPT}$, and $X_{S,OPT}$ as a function of nominal emitter area and at a collector current density of $0.16 mA/\mu m^{-2}$. In Fig. 5, the $(NF_{MIN} - 1)$ variation with A_E shows an almost linear dependence on a log-log plot for all devices except the smallest one. The reason for the deviation for this device is because pad impedances are quite important, and their contribution to the noise parameters must be removed since they result in an apparent

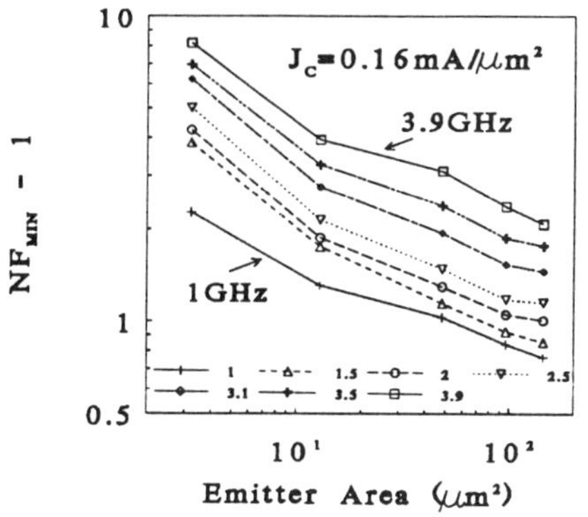

Figure 5: Variation of $NF_{MIN} - 1$ (as a number) with A_E (μm²), at a J_C of 0.16mA/μm².

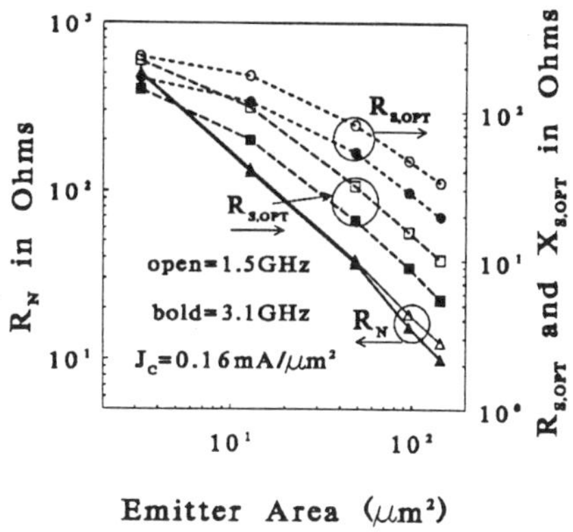

Figure 6: R_N, $R_{S,OPT}$ and $X_{S,OPT}$ variation with A_E (μm²), at a J_C of 0.16mA/μm².

increase in the device NF_{MIN}. This conclusion is supported by the $R_{S,OPT}$ and $X_{S,OPT}$ data in Fig. 6 which shows a similar deviation for the 3.2μm² device. In table 1, the area dependence of these parameters is listed at 2 frequencies of 1.5 and 3.1GHz. For ($NF_{MIN} - 1$), the variation of δ with frequency is not significant, indicating that at this J_C, R_B is the dominant noise source and the area dependence of ($NF_{MIN} - 1$) is determined mainly by the R_B/R_S variation with A_E. Since $R_B \sim A_E^{-0.92}$ from measurements of several devices at current densities typical of working devices, and $R_{S,OPT} \sim A_E^{-0.7}$ then the variation shown in the table is expected. Similar conclusions can be obtained by observing the emitter area dependence of other noise parameters in equations (2) to (4).

Parameter	δ at 1 GHz	δ at 3.1 GHz
$NF_{MIN} - 1$	0.23	0.26
R_N (Ω)	0.97	1.06

Parameter	δ at 1 GHz	δ at 3.1 GHz
R_S (Ω)	0.68	0.73
X_S (Ω)	0.97	1.01
V_N ($V/\sqrt{Hx}$)	0.41	0.47

Table 1: Listing of the area dependence δ in $A_E^{-\delta}$ of the noise parameters. Only the 4 largest devices were used in the fitting, and the noise voltage V_N was calculated from NF_{MIN} and R_S.

CONCLUSION

We have experimentally investigated in detail the emitter area, collector current density and frequency dependence of four important noise related parameters (NF_{MIN}, R_S, X_S and R_N) of poly-emitter bipolar junction transistors that allow for accurate noise modelling. We obtained good agreement between model calculations and experiments of all four parameters under the various operating conditions tested, and we explicitly gave the variation of these parameters with emitter area at a current density typical of the working current in the devices.

REFERENCES

1. R.J. Hawkins *Solid-State Electronics*, Vol 20, pp. 191-196 (March 1977).
2. H. Fukui *IEEE Transactions on Electron Devices*, Vol 13, pp. 329-341 (March 1966).
3. M.J. Deen and J.J. Ilowski *IEE Electronics Letters*, Vol 29(8), pp. 676-7 (April 1993).

SPATIAL ANALYSIS OF VOLTAGE FLUCTUATIONS IN SEMICONDUCTOR n^+nn^+ STRUCTURES

Tomás González, Daniel Pardo
Departamento de Física Aplicada, Universidad de Salamanca,
Plaza de la Merced s/n, 37008 Salamanca, Spain

Luca Varani, Lino Reggiani
Dipartimento di Fisica ed Istituto Nazionale di Fisica della Materia,
Universitá di Modena, Via Campi 213/A, 41100 Modena, Italy

ABSTRACT

We present a novel Monte Carlo method to investigate electronic noise at different spatial points in one-dimensional semiconductor structures by employing voltage-noise operation. This method provides a spatial map of voltage fluctuations in the structures. The results obtained for n^+nn^+ structures of Si and GaAs show that at equilibrium most of the low-frequency noise originates from the n region, while at increasing voltages an important contribution comes from the hot electrons which penetrate the drain region.

INTRODUCTION

When trying to optimize the performances of electronic devices, the analyses of fluctuations are specially interesting, since they provide specific information about the transport processes which control their performances. This is particularly important when dealing with non-equilibrium phenomena in submicron structures. In this paper we present an original Monte Carlo method which provides a spatial analysis of voltage fluctuations in one-dimensional semiconductor structures. To this end voltage-noise operation[1,2] is employed, which means that constant current conditions through the structure are imposed, and the spectral density of voltage fluctuations of the open circuit, $S_V(x,f)$, is determined as a function of different positions x inside the structure as measured from one of the contacts. The calculations make use of an ensemble Monte Carlo simulation coupled with a one-dimensional Poisson solver. In this way fluctuations in carrier velocity and electric field are self-consistently accounted for, and no approximations related to the statistical properties of the microscopic noise sources are introduced, as usually done by employing more traditional methods such as the impedance-field[3] or the transfer-impedance[4].

THEORETICAL ANALYSIS

The theory underlying the present method is based on the following. In a one-dimensional structure of length L, the total current, $I(t)$, is given by[2]:

$$I(t) = I_c(t) - \frac{\varepsilon_0 \varepsilon_r A}{L} \frac{d}{dt} \Delta V(L,t) \tag{1}$$

where ε_0 is the free space permittivity, ε_r the relative static dielectric constant of the material, A the cross-sectional area, $\Delta V(L,t)$ the instantaneous voltage drop between the terminals, and $I_c(t)$ the conduction current defined by:

$$I_c(t) = -\frac{e}{L}\sum_{i=1}^{N_T} v_i(t) \qquad (2)$$

with e the absolute value of the electron charge, N_T the total number of carriers inside the structure (here taken constant since periodic boundary conditions are considered[5]), and $v_i(t)$ the instantaneous velocity along the field direction of the ith particle.

By imposing that the total current is constant in time, $I(t)=I_0$, from Eq. (1) we obtain:

$$\frac{d}{dt}\Delta V^I(L,t) = \frac{L}{\varepsilon_0\varepsilon_r A}\left[I_c(t) - I_0\right] \qquad (3)$$

where the superscript I is to recall the use of voltage-noise operation. From this expression the instantaneous voltage drop between the terminals can be calculated with the following procedure:

(i) Starting from the stationary operation point in the device corresponding to I_0, $[\Delta V^I(L,0),I_0]$, one solves the Poisson equation, simulates one time-step Δt (typically 10 fs for Si and 2.5 fs for GaAs) and gets the conduction current $I_c(\Delta t)$.

(ii) Once $I_c(\Delta t)$ is evaluated, Eq. (3) is integrated by employing a finite differences scheme. The new instantaneous voltage drop between the terminals, $\Delta V^I(L,\Delta t)$, is thus calculated as:

$$\Delta V^I(L,\Delta t) = \Delta V^I(L,0) + \frac{L}{\varepsilon_0\varepsilon_r A}\left[I_c(\Delta t) - I_0\right]\Delta t \qquad (4)$$

(iii) With the new value $\Delta V^I(L,\Delta t)$ one solves the Poisson equation and obtains the value of the voltage drop at each position x of the structure as measured from the first terminal $\Delta V^I(x,\Delta t)=V^I(x,\Delta t)-V^I(0,\Delta t)$.

(iv) A successive time step is then simulated to obtain the new value of I_c, and the process is iterated by repeating from point (ii). The number of time steps simulated must be enough to get a sufficient resolution in the calculation of the autocorrelation function of voltage fluctuations, $C_V(x,t)=\overline{\delta\Delta V^I(x,0)\delta\Delta V^I(x,t)}$, where the bar indicates time average and $\delta\Delta V^I(x,t)=\Delta V^I(x,t)-\overline{\Delta V^I(x)}$. By Fourier transformation, the corresponding spectral density as a function of frequency, $S_V(x,f)$, is obtained.

RESULTS

As application we present the results for the case of a Si n^+nn^+ structure at T=300 K with $n^+=10^{17}$ cm^{-3}, $n=10^{16}$ cm^{-3} and length 0.20-0.20-0.20 μm, respectively. The microscopic model is the same of Ref. 5. The time evolution of the autocorrelation functions at equilibrium [Fig. 1(a)] depends on the contribution to the voltage fluctuations which comes from each region in the structure through the value of their resistance and doping. Accordingly, at the shortest times the n^+ regions introduce an oscillatory behavior associated with the plasma time that is suppressed by the dielectric relaxation at increasing times[1]. In the n region the evolution is mainly determined by the dielectric relaxation through a contribution which decreases exponentially with time. Fig. 1(b) shows the corresponding spectral density, where the different influence of each region is clearly emphasized. The n region, due to its larger resistance, introduces most of the noise at low frequencies. When the frequency is higher, its contribution decreases, while that of the n^+ regions increases, reaching its maximun value for the associated plasma frequency (around 1300 GHz).

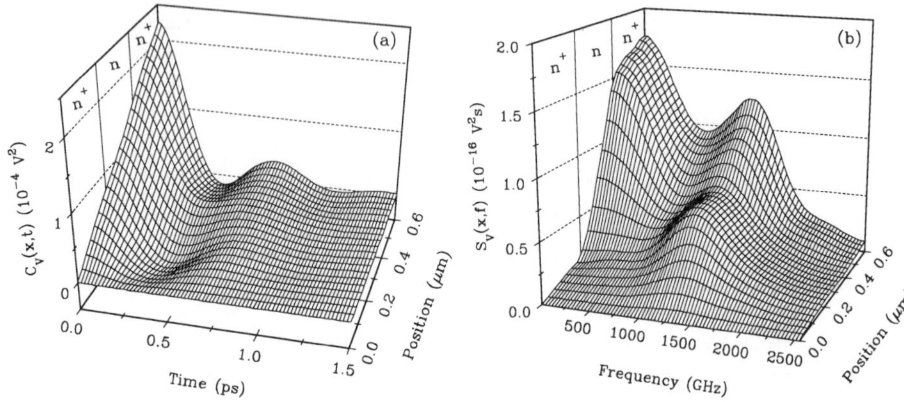

Fig. 1. (a) Autocorrelation function of voltage fluctuations as a function of time and position and (b) spectral density as a function of frequency and position in a Si n^+nn^+ structure at thermal equilibrium with T=300 K, $n^+=10^{17}$ cm^{-3}, $n=10^{16}$ cm^{-3} and length 0.20-0.20-0.20 μm, respectively.

Under far-from-equilibrium conditions the onset of hot-carriers effects in the structure produces important changes in the behavior of voltage fluctuations, mainly at the lowest frequencies. Fig. 2 shows the spectral density of voltage fluctuations for an average voltage $\Delta V^T(L)$=0.4 V. The low-frequency noise increases considerably due to the appearance of hot carriers, leading to a low local mobility in the high-field region of the device. This effect persists even at the beginning of the second n^+ region, and makes the low-frequency noise penetrate this zone. This penetration of the noise sources into the drain region is more pronounced at increasing voltages, as it is shown in Fig. 3, where $S_V(x,0)$ is presented for several average voltages.

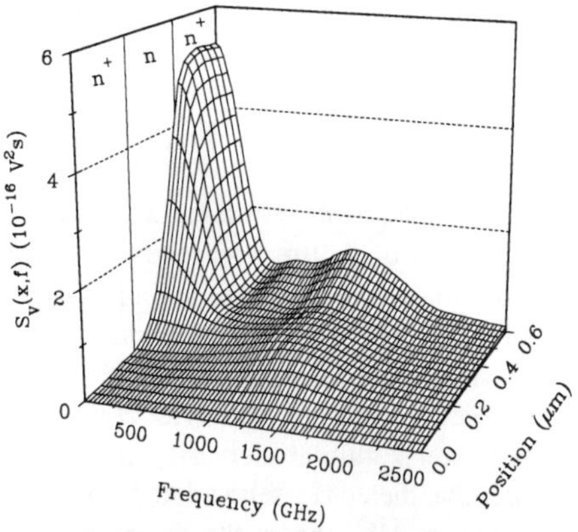

Fig. 2. Spectral density of voltage fluctuations around an average voltage of 0.4 V as a function of frequency and position in the same structure of Fig. 1.

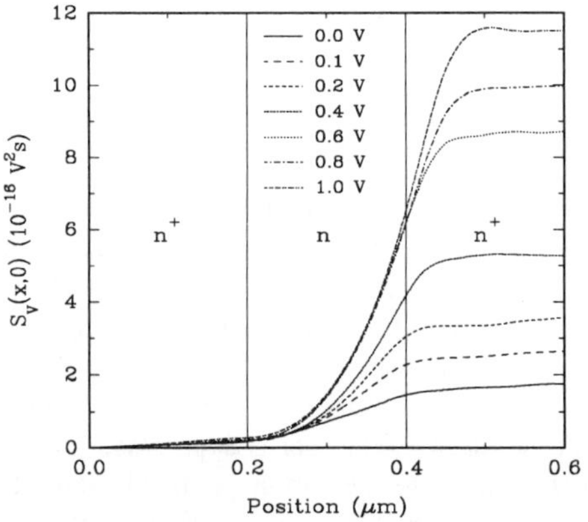

Fig. 3. Low-frequency value of the spectral density of voltage fluctuations around several average voltages as a function of position in the same structure of Fig. 1.

To compare the behavior of the voltage spectral density between Si and GaAs devices we have made the calculations for the case of a GaAs n^+nn^+ structure at T=300 K with $n^+=10^{17}$ cm^{-3}, $n=10^{16}$ cm^{-3} and length 0.15-0.25-0.50 μm, respectively. The microscopic model is the same of Ref. 6. The fluctuations are around an average voltage $\Delta V^{I}(L)$=0.4 V. In this material the peak associated to the plasma frequency of the n^+ regions appears at a higher frequency (around 3000 GHz), as corresponds to its lower effective mass. Furthermore, in the low-frequency region the contribution of the drain is even more important than in the case of Si. This is due to the presence in this region of carriers in the upper satellite valleys, with higher effective mass, which involves a deeper penetration of hot carriers into the drain before they can thermalize, making this region highly resistive and thus an important source of noise.

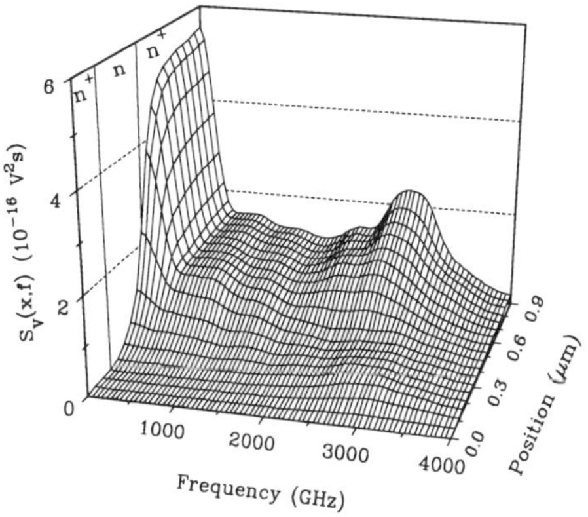

Fig. 4. Spectral density of voltage fluctuations for an average voltage of 0.4 V as a function of frequency and position in a GaAs n^+nn^+ structure with T=300 K, $n^+=10^{17}$ cm^{-3}, $n=10^{16}$ cm^{-3} and length 0.15-0.25-0.50 μm, respectively.

CONCLUSIONS

We present a new Monte Carlo technique able to provide a spatial analysis of voltage fluctuations in one-dimensional n^+nn^+ semiconductor structures. When applied to the cases of Si and GaAs we find that, in addition to the n region, the drain region can be an important source of noise at high average voltages, due to the onset of hot-carriers effects.

This work is partially supported by the SA-14/14/92 project from the Conserjería de Cultura de la Junta de Castilla y León, and by the Italian Ministero dell' Università e della Ricerca Scientifica e Tecnologica (MURST).

REFERENCES

1. J. Zimmermann and E. Constant, Solid-State Electron. **23**, 915 (1980).
2. L. Reggiani, T. Khun and L. Varani, Appl. Phys. **A54**, 411 (1992).
3. W. Shockley, J. A. Copeland and P. James in *Quantum Theory of Atoms, Molecules and the Solid State*, Ed. P. O. Löwdin (Academic, New York, 1966), p. 537.
4. K. M. van Vliet, A. Friedmann, R. J. J. Zijlstra, A. Gisolf and A. van der Ziel, J. Appl. Phys. **46**, 1804 (1975).
5. L. Varani, T. Kuhn, L. Reggiani and Y. Perlès, Solid-State Electron. **36**, 251 (1993).
6. T. González, J. E. Velázquez, P. M. Gutiérrez and D. Pardo, Appl. Phys. Lett. **60**, 613 (1992).

NOISE AND SUBMILLIMETER WAVE GENERATION IN InP DIODES

V. Gruzinskis, V. Mitin,
Department of Electrical and Computer Engineering
Wayne State University, Detroit, MI 48202
E. Starikov, and P. Shiktorov
Semiconductor Physics institute, Vilnius 2600, Lithuania

ABSTRACT

High frequency (350 to 750 GHz) generation in submicron InP diodes is investigated by modified hydrodynamic (MH) and Monte Carlo particle (MCP) techniques. The noise power spectral density P_n in the diode loaded by resistor R and generation spectra P_g in series resonant LR circuit are calculated by MCP technique. It is shown that at the biases above the generation threshold the P_n has a peak at the frequency f_{max} which corresponds to the highest generation frequency at the given R. The excess noise arise in the frequency region where the real part of diode impedance has negative values. The P_g broadening at high frequencies is the result of interaction between the self-oscillations at frequency f_{max} and LR circuit driven oscillations.

In recent years, the chief progress in increasing the operating frequency of millimeter and submillimeter wave systems has been through development of Transferred Electron Devices (TED). A recent report [1] describes an InP TED operating at 272 GHz frequency. High frequency generation (HFG) in usual TED arises due to the Gunn effect, which utilizes the negative differential conductivity (NDC) of the steady-state velocity-field characteristic resulting from the carrier transfer into the upper valleys with larger effective mass. Generation frequency in TED depends on the length of n-region. Shorter length of n-region results in higher generation frequencies. On the other hand, the steady-state NDC in shorter diodes becomes weaker and disappears at near-micrometer lengths due to electron injection from n^+-regions. Therefore, the usual Gunn effect in short samples is impossible. But this is not a sufficient reason to say that short diodes can not be used for the microwave generation. Electrons in short diodes drift from the source to the drain contact under highly nonstationary conditions [2]. Under such conditions a high frequency generation (HFG) mechanism different from the classical Gunn effect can arise [3]. The HFG in short diodes arises due to heating and transit time delay, which causes the velocity overshoot in real space. In this paper we present the results of HFG investigation in submicron InP diodes by modified hydrodynamic (MH) [4] and Monte Carlo particle (MCP) techniques.

In our simulation the parameters of n^+-n-n^+ InP are chosen as follows: the length of n-region is 0.25 μm, the doping concentration in n and n^+ regions is $2 \cdot 10^{17}$ cm^{-3} and $3 \cdot 10^{18}$ cm^{-3}, respectively. These parameters yield highest available frequency with reasonable value of conversion efficiency at the lattice temperature T=300 K [4]. Current-voltage relation calculated by MCP technique is shown in Fig.1. There is no region with static NDC on this current-voltage relation. However, the linear analysis carried out by MH method in current driven regime shows the presence of dynamic NDC above the bias U_d=1.1 V. In particular, at U_d=3 V (see Fig.2) the NDC (i.e, negative values of real part of impedance (ReZ)) exists in the frequency range from 350 GHz to 1050 GHz with maximum value $3.6 \cdot 10^{-10}$ $\Omega \cdot m^2$ at frequency 520 GHz. The MCP simulation in LR circuit yields the conversion efficiency at 520 GHz as high as 1.85 % (see Fig.3 (dots)). The load resistance R corresponding to maximum efficiency is 10^{-10} $\Omega \cdot m^2$, and average voltage drop on the diode is U_d=3 V. The MCP (dots) and MH (solid line) calculations agree qualitatively with each other (see Fig.3). The same agreement is obtained for the generation power dependence on frequency (see Fig.4 where MCP (dots) and MH (solid line) results are presented).

The diode noise power spectral densities P_n are calculated analyzing voltage-voltage correlations on a load resistor R. The noise is simulated by MCP technique in the diode loaded by $R=10^{-10}\ \Omega \cdot m^2$. The P_n for two different U_d shown in Fig.5. At U_d=1 V the ReZ is positive in the entire frequency range. Therefore, the P_n has a usual Lorenzian shape. The high frequency noise in that case can be related to the shot noise due to the carrier flight from source to drain. At U_d=3 V the noise power spectral density P_n has a maximum in the frequency range where ReZ is negative (compare Figs.2 and 5). This is due to the self-excitation of oscillations in the circuit. The most powerful generation arises at the frequency where the condition ReZ+R=0 is satisfied. To check this assumption the dependence of P_n on the length of n-region is calculated at U_d=3 V. Other parameters of the structure are left unchanged. In Fig.6 the P_n are shown for three lengths of n-region. Same for the ReZ is shown in Fig.7 (calculated by MH technique). The comparison of Fig.6 and Fig.7 shows that the frequency of P_n maxima coincide well with upper frequency limits of the negative ReZ. The negative values of ReZ increase with increasing the length of n-region (see Fig.7). As a consequence, the noise power increases as well (see Fig.6). Therefore, the P_n peaks occur due to the self-excitation of generation.

The generated power spectral density $P_g(f)$ in LR circuit at certain values of L is calculated in the same way as $P_n(f)$. The parameters of structure and circuit are same as for the power and efficiency calculations (Figs.3 and 4). The calculations are performed for three values of L_1=10, L_2=6 and $L_3=4\cdot 10^{-23}$ Hm^2. The spectral densities $P_g(f)$ are presented in Fig.8 (MCP calculations). The central maxima frequencies of $P_g(f)$ for L_1, L_2, and L_3 are f_1=517, f_2=622, and f_3=718 GHz, respectively. The narrowest peak is at frequency f_1. At f_2 the peak is wider. Finally, at f_3 we have three well defined maxima approximately of the same order in magnitude. Broadening of spectrum at frequency f_3 is a result of interaction between the self-oscillations and the circuit driven oscillations. As is shown by the simulation in LR circuit the frequency of self-oscillations (i.e., the frequency of $P_n(f)$ peak) is the highest generation frequency at the given value of R (compare Figs.4 and 5).

On the basis of obtained results it have to be outlined that enhanced noise arises in the frequency region where the real part of diode impedance is negative. Noise power spectral density at the biases above the generation threshold has a peak at frequency which corresponds to the microwave power generation upper frequency limit at the given value of load resistance. Broadening of microwave power generation spectrum at high frequencies is the result of interaction between the self-oscillations and the circuit driven oscillations. Practically important result is that InP submicron diodes can be used as the efficient sources of microwave power generation up to the frequencies of about 750 GHz. The diodes with crossection area 100 μm^2 can generate microwave power 100 mW at 500 GHz and 20 mW at 750 GHz.

REFERENCES

1. A. Rydberg, IEEE Electron Dev. Lett., **11**, 439 (1990).
2. M.P. Shaw, V. Mitin, E. Scholl, and H. Grubin. The Physics of instabilities in Solid State Electron Devices. Plenum Press, New York, 1992, 467 p.
3. M.R. Friscourt, P.A. Rolland, A. Cappy, E. Constant, and G. Salmer, IEEE Trans. Electron Devices, **ED-30**, 223 (1983).
4. V. Gruzhinskis, E. Starikov, P. Shiktorov, L. Reggiani, M. Saraniti, and L. Varani, Appl. Phys. Lett., **61**, 1456 (1992).

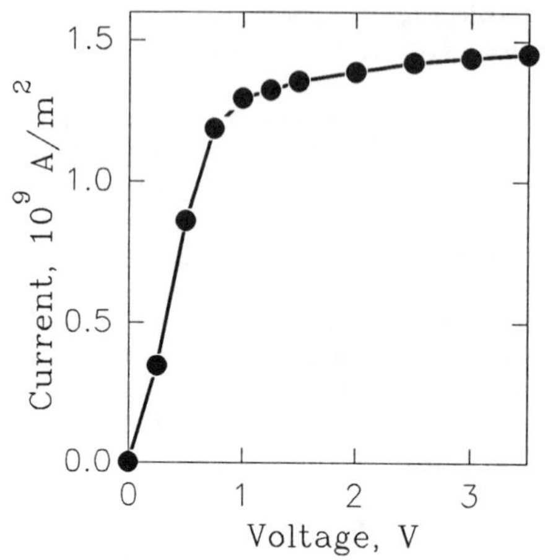

Fig.1. Current–voltage relation.

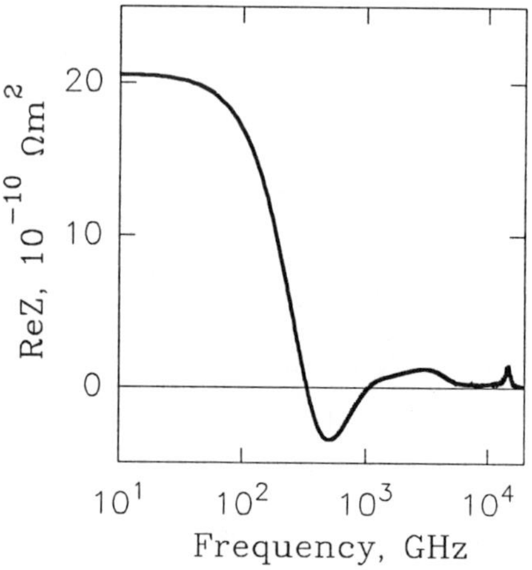

Fig.2. Real part of impedance spectrum at U_d=3 V.

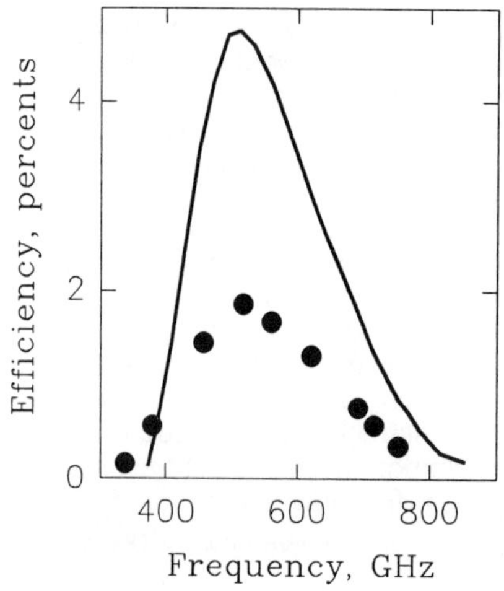

Fig.3. Generation efficiency vs frequency at U_d=3 V.

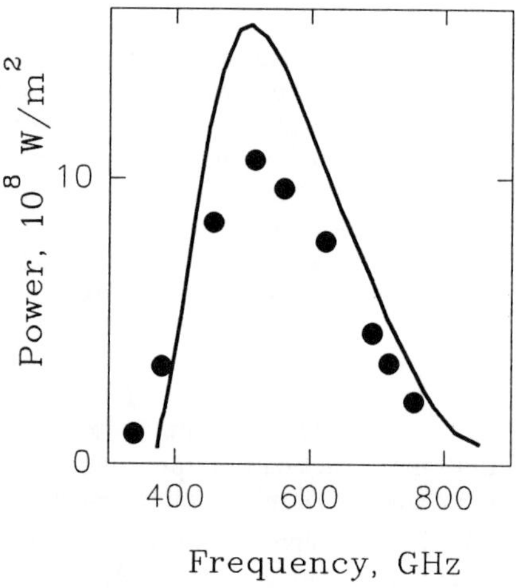

Fig.4. Generation power vs frequency at U_d=3 V.

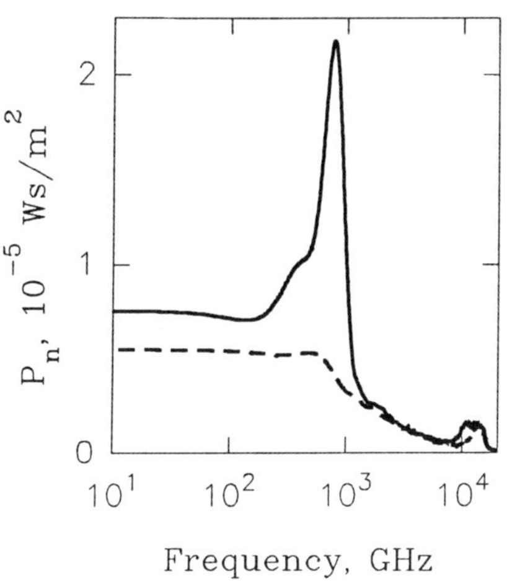

Fig.5. Noise power density spectra at U_d=1 V (dashes) and 3 V (solid line).

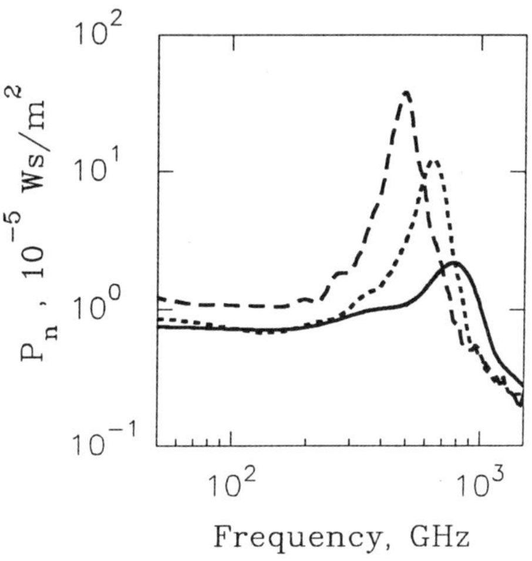

Fig.6. Noise power density spectra at U_d=3 V and n−region lengths: 0.25 (solid), 0.4 (dots), and 0.6 (dashes) micrometers.

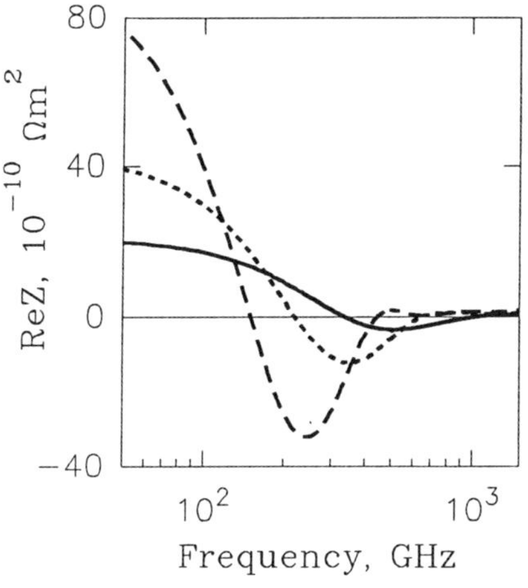

Fig.7. Real part of impedance spectra. Notation and conditions same as in Fig.6.

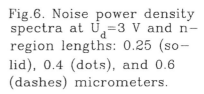

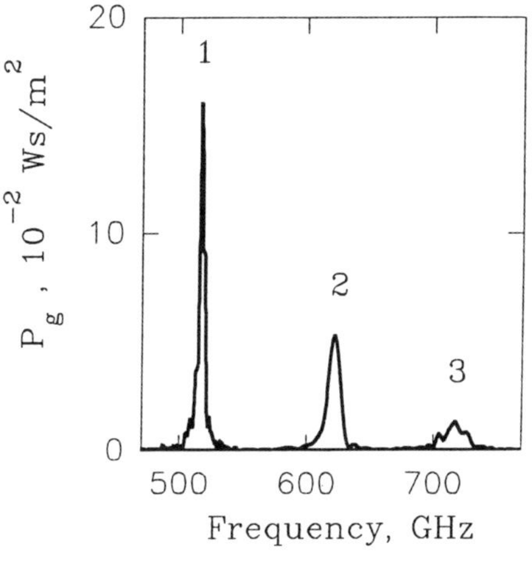

Fig.8. Generation power density spectra at frequencies: 1−517 GHz, 2−622, and 3−718.

TEMPERATURE DEPENDENCE OF $1/f$ NOISE IN AlGaAs/InGaAs HEMT

S.Hashiguchi, H.Horiuchi, M.Ohki
Department of Electronics, Yamanashi University, Kofu 400, Japan

M.Yajima
Sankyo Seiki Mfg., Simosuwa, Nagano, Japan

M.Tacano
Kyocera Ltd., Yamashina, Kyoto, Japan

ABSTRACT

Temperature dependence of Hooge's $1/f$ noise parameter α in intrinsic InGaAs was measured using heterostructure. It was estimated as $3 \times 10^{-3} \exp(-T/150)$. The bulges in the noise spectra were attributed to the generation-recombination noise caused by the DX centers in the doped AlGaAs layer and by the electron traps in the intrinsic InGaAs.

INTRODUCTION

In $1/f$ noise spectra of the current in GaAs layer of HEMT there have been observed several bulges[1]. These have been attributed to the defects in GaAs bulk. We measured the temperature dependence of the drain current noise of AlGaAs/InGaAs HEMT in order to determine the temperature dependence of Hooge's $1/f$ noise α and to identify the origins of the bulges in the noise spectra in intrinsic in undoped InGaAs.

MEASUREMENT

The sample, AlGaAs/InGaAs super HEMT(Fujitsu), has a MBE grown intrinsic InGaAs layer between an intrinsic GaAs buffer layer and a Si-doped n-AlGaAs layer. The doping density in the n-AlGaAs layer is $10^{24}\mathrm{cm}^{-3}$. The gate length is 0.15μm, the gate width is 200μm, and the gate is 0.5μm apart from each of the drain and the source electrodes. The electron mobility is $4000\mathrm{cm}^2/\mathrm{V{\cdot}s}$ at 290K and $12000\mathrm{cm}^2/\mathrm{V{\cdot}s}$ at 77K, respectively.

The sample was biased at $V_{\mathrm{GS}} = 0$ and $V_{\mathrm{DS}} = 2$V, and the noise spectra of the drain current were measured for the frequency from 0.05Hz to 100kHz between 77K and 290K.

RESULTS AND DISCUSSIONS

Figure 1 shows the temperature dependence of the spectrum $S_{\mathrm{I}}(f)$ of the drain current noise. The HEMT under measurement was operating at the drain voltage of $V_{\mathrm{DS}} = 2$V, the gate voltage of $V_{\mathrm{GS}} = 0$V. The drain current I_{D} was 18.2mA at 290K and 14.6mA at 77K, respectively. The levels of S_{I} and the relative fluctuation S_{I}/I^2 decrease slowly with the rise of temperature.

Figure 2 shows the temperature dependence of the frequency-weighted spectrum $fS_I(f)$. There are three separate bulges observed on the spectra, one is at 300Hz at 77K, another is at 70kHz at 288K, and the other is at 2Hz at 288K. The bulge at 300K at 77K shifts to the right with the rise of temperature, with losing its height. The bulge at 70kHz at 288K emerges on the left at 159K, also shifts to the right with the rise of temperature. The height of the bulge reaches its peak about 220K. The smallest bulge at 2Hz at 288K emerges on the left at 260K and shifts to the right.

Figure 3 shows Arrhenius plots of the center frequencies of the three bulges. The activation energy for each bulge is 0.005eV for the higher frequency bulge (f_1), 0.3eV for the middle frequency bulge (f_2), and 0.52eV for the lower frequency bulge (f_3). The activation energy 0.3eV corresponds to the electron emission of the deep level (DX centers) in AlGaAs, and 0.53eV to the electron traps in InGaAs ($In_{0.15}Ga_{0.85}As$)[2].

Hooge's α is estimated by using the total carrier number N of 10^6, which is 20% of the product of the doping density $10^{24}m^{-3}$ in the AlGaAs layer and the volume 5×10^6 (200μwide, 1.15μm long, and 20nm thick) of the AlGaAs layer between the gate and the drain electrodes. This value of the total carrier number is a rough estimation of the electrons under the gate electrode where the current density is higher than any other portion in the sample.

Figure 4 shows the temperature dependence of Hooge's α. It is of the order of 10^{-3}, and slightly decreases with the rise of temperature. The hump at about 180K is the effect of the bulge at f_2, and the rise at 280K is the effect of the bulge at f_1. The solid line is a rough estimation of the temperature dependence of Hooge's α when the effects of the bulges are compensated. This solid line is expressed by the relation

$$\alpha = 3 \times 10^{-3} \exp(-T/150). \qquad (1)$$

CONCLUSIONS

Hooge's α in intrinsic InGaAs is about 10^{-3} and slowly decreases with rising temperature. Electron traps in InGaAs are responsible to the generation-recombination noise which causes remarkable bulges in the $1/f$ spectrum.

REFERENCES

1. S.Kugler, IEEE Trans. on ED, ED-35, 623 (1988)
2. A.Mircea, Phys. Rev. B16, 3665 (1977)

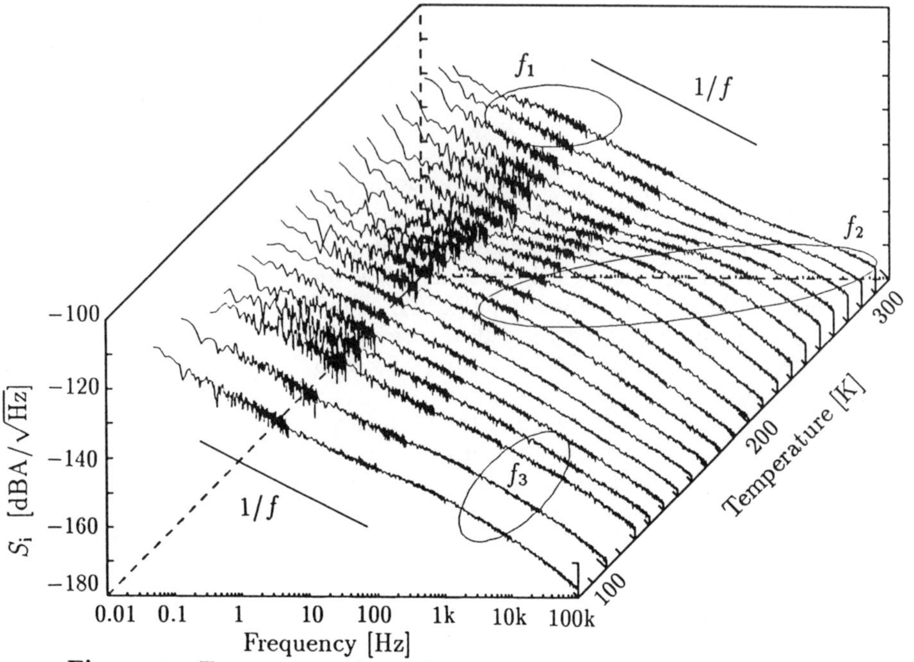

Figure 1 Temperature dependence of the noise spectrum in HEMT

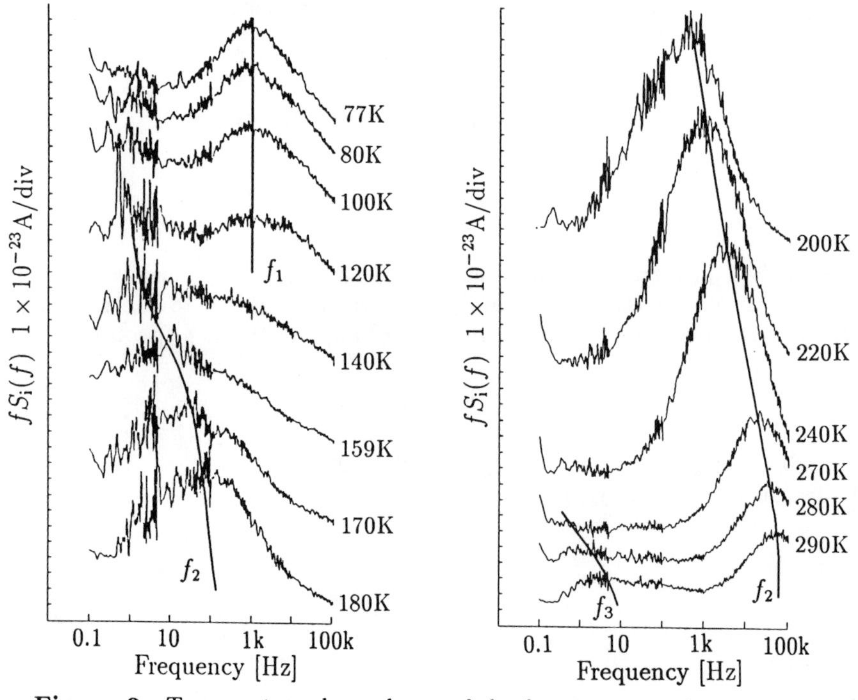

Figure 2 Temperature dependence of the frequency-weighted spectrum

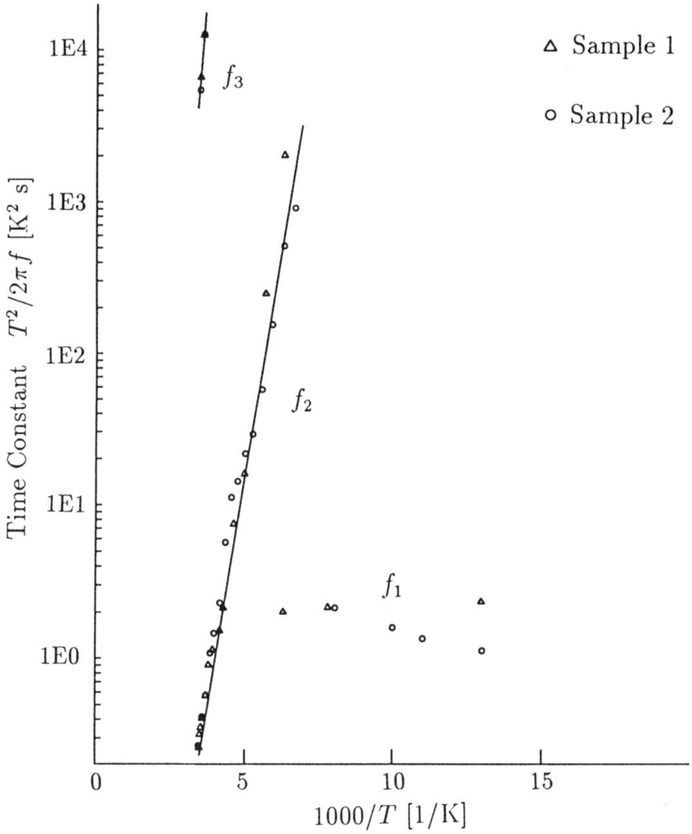

Figure 3 Arrhenius plots of the center peak frequencies of the three bulges

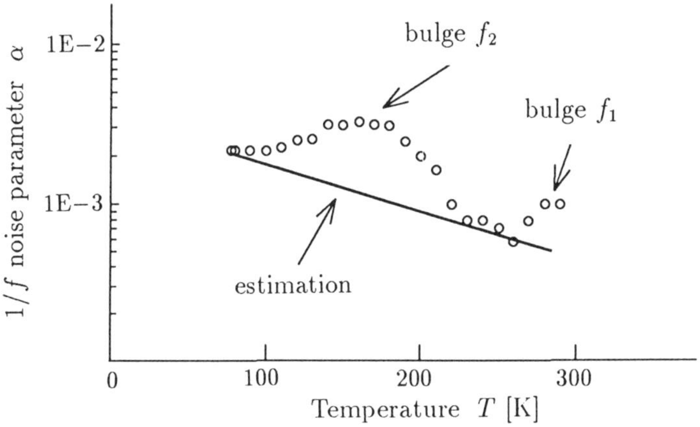

Figure 4 Temperature dependence of Hooge's α

NOISE INVESTIGATIONS OF RADIATION INDUCED DEFECTS IN MOS/SIMOX TRANSISTORS

A. Ionescu, A. Chovet, S. Cristoloveanu
LPCS, ENSERG, 23, Rue des Martyrs, BP 257, 38016 Grenoble Cedex, France

P. Jarron, E. Heijne
CERN, Geneva, Switzerland

ABSTRACT

The 1/f noise is investigated in n- and p-channel MOSFET's/SIMOX in linear and saturation regions of operation, prior and after γ irradiation (up to 10 Mrad). Low-frequency (10 Hz - 100 kHz) noise investigations are correlated to static and dynamic evaluations of radiation induced defects. It is found that in n-channels the excess of 1/f noise induced by irradiation correlates with the oxide trapped charge. An underlying not simple correlation is observed for p-channel devices since the dominant degradation is interface state creation. Sidewall damage is found to be responsible for the most important increase in the normalized noise, for n-channels, in the saturation region of operation.

INTRODUCTION

Noise measurements are very useful to obtain complementary data and a deeper insight into the physical mechanisms responsible for MOSFET degradation after irradiation. Because the radiation induced degradation is very complex and strongly depends on device bias and geometry, an accurate investigation can be made only by correlating multiple characterization techniques. In this work, we compare low-frequency (10 Hz - 100 kHz) noise measurements with: (i) static characterizations, (ii) charge pumping evaluations, and (iii) Zerbst-type analysis of drain current transients, in order to investigate the effects of γ-irradiation on MOS/SIMOX transistors.

SAMPLES

The experimental devices were n- and p-long channel (L=15µm) enhancement MOSFET's. The SIMOX wafers were synthesized by 200 keV multiple oxygen implants into Si at 600°C followed by annealing at 1300°C. Two structures of MOSFET's were used: edgeless (W=70µm) and edged (W=170µm). The thicknesses of the gate oxide, buried oxide and Si film were 23, 380 and 150 nm, respectively. The transistors have been irradiated up to 1, 3 and 10 Mrad with a calibrated ^{60}Co source at room temperature. During irradiation biases close to usual working were applied: I_{DS}=50 µA (corresponding to $|V_{Gf}-V_{Tf}|$=250 mV), V_{Gb}=-4V for n-channel and V_{Gb}=+4V, for p-channel, edged devices, and $V_D=V_S$=0V, V_{Gf}=+4V, V_{Gb}=-4V for n-channel and V_{Gf}=-4V, V_{Gb}=+4V for p-channel, edgeless devices (f, b = front, back interfaces, D=drain, S=source, G=gate, V_T=threshold voltage). The substrate doping level for all transistors was $1x10^{17}$ cm^{-3}.

STATIC CHARACTERIZATION, CHARGE PUMPING AND ZERBST-TYPE TRANSIENTS

Static characterization has been performed in order to evaluate the variations of the threshold voltage and carrier mobility with γ radiation dose. All parameters were extracted with the opposite interface in accumulation in order to avoid interface coupling effects. A "rebound" in the threshold voltage was observed after 1Mrad dose for n-channel devices and it was considered as a consequence of interface states creation (acceptor-like in n-channels). No rebound was observed for p-channel transistors. The interface states were extracted from charge pumping experiments, thus eliminating all errors due to potential fluctuations when using weak-inversion slope method.The oxide trap charge was calculated using:

$$\Delta N_{ot}[cm^{-2}] = \pm 2\,\Phi_F\,\Delta N_{it}[eV^{-1}cm^{-2}] - C_{ox}\,\Delta V_T/q \qquad (1)$$

where "+" is for n-channels and "-" is for p-channels. Field effect mobilities were deduced from the maximum of the transconductance curves. It was found that mobility degradation correlates with the density of interface states, Fig.1. In Fig.2 are presented the deduced oxide trapped charge and interface states densities. Due to the positive bias of the front gate, the creation of both interface states and oxide traped charge appears to be important in n-channel devices. Dominant interface state creation was observed for front p-channels where the negative gate bias reduces the oxide charge trapping, a balance between trapping of holes and electrons being also possible. An original point was to use Zerbst-type drain current transients in order to extract the generation lifetime (which is a measure of crystal and edge defects) and the surface generation velocity (related to interface states). More than one order of magnitude degradation of generation lifetime (from 13 μs, prior to irradiation, to 0.4 μs after 10Mrad, for edged devices) is observed which indicates the creation of deep centers in Si film and important defects on the edges. Surface generation velocity was increased by about two orders of magnitude, corresponding to the interface state density increase.

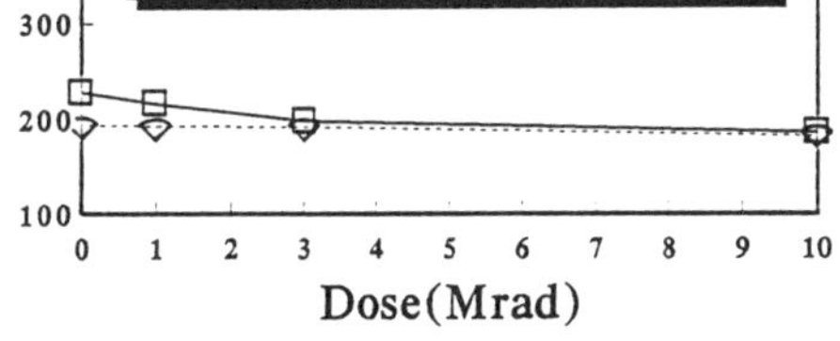

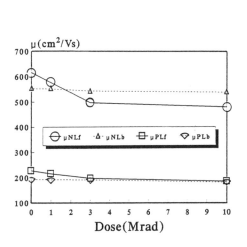

Fig.1 Front (f) and back (b) effective mobilities versus radiation dose ("N", "P"=n-, p-channel, "L"=linear (edged) device geometry).

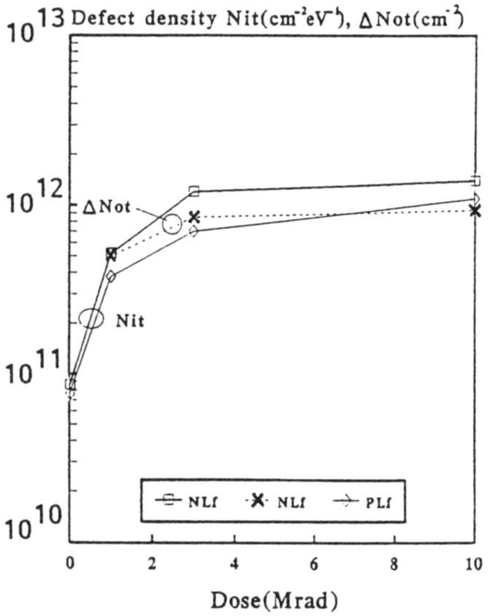

Fig.2 Defect densities, ΔN_{it}, ΔN_{ot}, versus γ radiation dose.

NOISE INVESTIGATIONS

1/f noise measurements were performed both in the linear and the saturation regions of transistor operation (Figs. 3 and 4), using a digital spectrum analyser. The investigated frequency range was 10 Hz to 100 kHz. In order to remove the dependence of the noise on gate and drain biasing, the following normalized noise power[1,2] is deduced from the drain voltage or current noise power spectral densities (S_{VD} or S_{ID}):

in the linear region: $K_{lin} = S_{VD} [(V_G - V_T) / V_D]^2 f^{\gamma}$ (2)

and in the saturation region:

$$K_{sat} = S_{ID}/I_D^2 \, L/L_{eff} \, V_{Dsat}^2/2 \, \{\ln [(qV_{Dsat}/2nkT) + 1]\}^{-1} f^{\gamma} \quad (3)$$

where V_G, V_D are the gate and drain biases, $V_{Dsat} = V_G - V_T$, V_T is the threshold voltage, f is the frequency, γ is a positive exponent (~1), L_{eff} is the effective channel length, and $n=(C_{ox} + C_{it} + C_d)/C_{ox}$; C_{ox}, C_{it} and C_d being the oxide, interface state and depletion capacitances. For the measured devices, we observed γ in the range 0.825 - 1.05. It is worth noting that γ is increased after irradiation. Prior to irradiation, a 1/f dependence is observed in the range 10 Hz - 300 Hz in linear operation, and 10 Hz - 2 kHz in saturation. The normalized noise is found to be smaller in p-channel devices.

After irradiation, all investigations were performed by adjusting the gate voltage in order to keep the value V_G-V_T constant. Our results, Figs. 5 and 6, shows that the 1/f noise of n-MOSFET's is less affected by irradiation than that of p-MOSFET's, and correlates classically[3,4] with the trapped oxide charge, N_{ot}, rather than with interface charge, N_{it}. The higher 1/f noise increase in saturation, for n-MOSFET's, is related to the contribution of edges where an important amount of oxide trapped charge was revealed by static characterization. Indeed, due to the change of the channel conductance of the main transistor in saturation, the noise contribution of the edge transistors becomes important.The case of p-channel transistors is more complex, even if radiation-induced interface states are dominant for negative gate biases. Their normalized noise is substantially increased but data shows that a simple linear correlation with interface states is not suitable. Moreover, it is difficult to know exactly the influence of the balance between possible both hole and electron trapping. However, there is a non-negligible contribution of interface states for the p-channel noise. Fig.6 shows the noise power density for p-channels, prior and after 10 Mrad radiation dose.

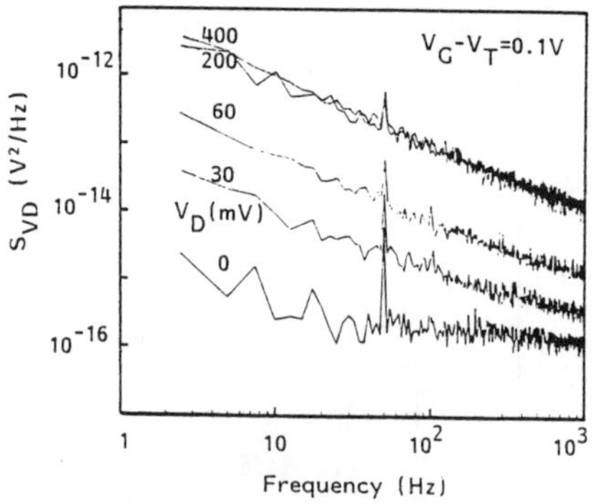

Fig.3 Power spectral density, S_{VD}, of an unirradiated n-channel MOSFET.

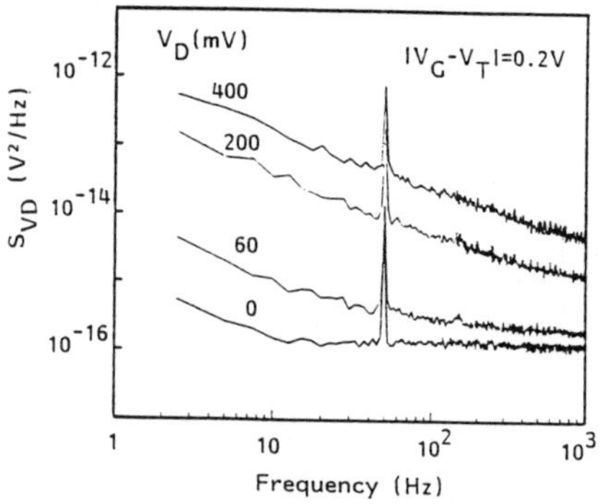

Fig.4 Power spectral density, S_{VD}, of an unirradiated p-channel MOSFET.

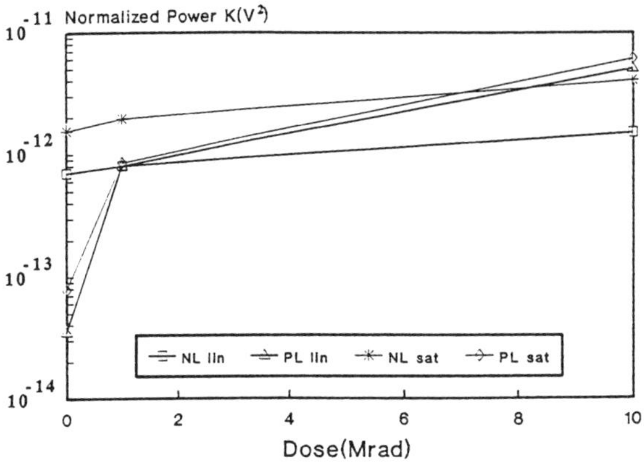

Fig. 5 Normalized noise, K, as a function of total dose for n- and p-channel devices.

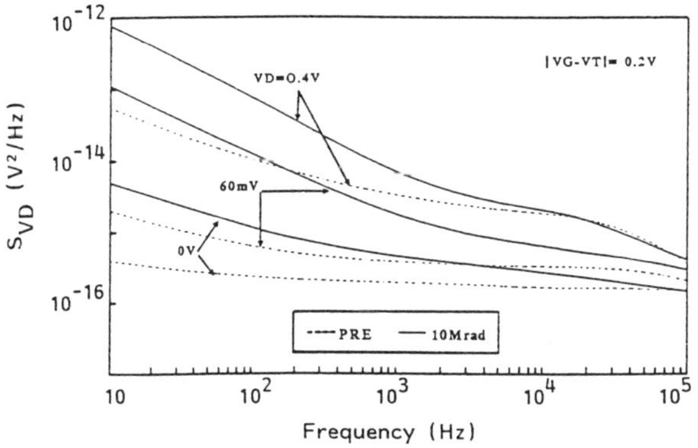

Fig. 6 Power spectral density for a p-channel MOSFET, in linear (V_{DS}=60mV) and saturation (V_{DS}=0.4V) operation, before and after 10 Mrad radiation dose.

CONCLUSION

1/f noise investigations shows that in n-channel enhancement-mode MOSFET's, the noise increase is mainly due to oxide charge trapping. The role of interface states cannot be excluded even if their faster time constants make their influence less likely. In p-channel there is an underlying, not yet understood, relation between noise and interface states. The device bias and geometry (edge contribution) were found important for radiation degradation process. For positive gate bias, the role of edge defects becomes important in terms of saturation noise. However, due to the rather reduced 1/f noise increase after irradiation it appears that SIMOX is a very good candidate for radiation-hardened technologies.

REFERENCES

1. J. Scofield, T. Doerr and D. Fleetwood, IEEE Trans. Nucl. Sc., 36 (1989), 1946
2. Z. Fang, A. Chovet, Q. Zhu, J. Zhao, Solid-State El., 34 (1991), 327.
3. T. Meisenheimer, D. Fleetwood, IEEE Trans. Nucl. Sc., 37 (1990), 1696.
4. T. Meisenheimer, D. Fleetwood, M. Shaneyfelt, L. Riewe, IEEE Trans. Nucl. Sc., 38 (1991), 1297.

THE 1/f NOISE IN GaAs FIELD-EFFECT TRANSISTORS

M.A. Abdala and B.K. Jones
School of Physics and Materials, Lancaster University,
Lancaster. LA1 4YB, United Kingdom

ABSTRACT

The 1/f noise has been measured in the channels of GaAs FETs biased in the ohmic regime. The devices have been studied using many different techniques so that there is a good understanding of their trap and defect properties and their d.c. operation. The significant results are that the 1/f noise is extremely variable between specimens and with temperature. There is evidence that traps with a comparable time constant within the device are related in some way to the 1/f noise.

INTRODUCTION

The resistance fluctuation, 1/f noise, found in many materials and electronic devices is still not well understood and there are considerable differences between the results of different experiments. The experiments reported here are on well characterised samples with many variables so that the analysis can allow for several possible effects. The noise in compound semiconductors is large compared with that in silicon or metals since there can be more defects and the surface is not well passivated. In particular the presence of deep levels produces generation-recombination (g-r) noise which must be allowed for in any 1/f noise measurement. The results reported here were made on samples which have been extremely well characterised in the ohmic channel bias regime ($V_{DS} << V_p$). The device d.c. operation has been characterised, including the substrate-channel interface. Many different experiments have been performed to understand the properties, densities and locations of the numerous traps. The experiments include the analysis of the noise and the separation of the 1/f and g-r noise contributions, mutual conductance dispersion, channel conductance deep level transient spectroscopy (DLTS) with various excitations and modifications and substrate current oscillations[1,2,3,4,5,6,7]

EXPERIMENT

The samples were commercial designs made from an epitaxial layer of GaAs on a Cr doped HB semi-insulting substrate with an undoped buffer-layer. The device is on a mesa and the gate length is $\sim 0.8\ \mu$m long. A significant feature is that there are significant access regions of ungated channel between the gate and source or drain. Their resistance does not vary with gate voltage and the surface is not perfectly passivated so that this device component may add to any apparent contact noise.

All measurements have been made in the ohmic regime in the temperature range 77-420 K. The noise measurements were conventional, between 1 Hz and 25 kHz, and analysed by fitting to a sum of 1/f noise, white noise and one or more Lorentzian g-r spectra. The location, densities, cross-sections and activation energies of the traps found by the noise agree with the other measurements and we concentrate here on the 1/f noise results. Although the spectra at first sight appear to be dominated by the 1/f noise it is found essential to separate the many g-r components to obtain informative data.

RESULTS

The significant feature of the results is the variability of the 1/f noise between specimens. The relative concentrations of deep levels also varies considerably. This variability of both noise components is within batches as well as between batches with different processing conditions where changes might be expected. An example is shown in figure 1.

In these samples the channel carrier mobility is approximately temperature independent and at a very low value because of the large carrier density and impurity scattering. For this reason it is not possible with any accuracy to use the temperature dependence of the noise to consider the relative applicability of the various bulk models such as those due to Hooge in its original form, $\frac{S_R}{R^2} = \frac{\alpha}{Nf}$, or in its modified form $\frac{S_r}{R^2} = \left(\frac{\mu}{\mu_l}\right)^2 \frac{\alpha}{Nf}$. Here S_R/R^2 is the normalised spectral density, N is the total number of carriers in the sample and μ and μ_l are the total mobility and the lattice mobility. The constant α is normally taken as 2 $\times$ 10^{-3}. For samples from both batches the noise shows a rapid increase with increasing temperature at the higher temperatures as found by others[8]. At low temperatures the noise varies greatly between specimens. Values of α using the original equation are in the range 3 to 50 $\times$ 10^{-6}.

In these ohmic samples a valuable method of analysis is to consider the slope of the log (S_R/R^2) – log R curve since this will vary between + 2 and - 2 in uniform samples depending on the location and type of the noise[9]. A change of slope may be expected at the point where the resistances of the gated and ungated sections of the channel are equal.

Two examples of this variation are shown normalised to the values at $V_G = 0$ in Figure 2. These are taken at temperatures when there is also a g-r peak present from a known trap. The dependence for that noise is also shown. It can be seen that this variation is very temperature dependent and has a different form for different traps.

In general there seems to be no simple pattern to the values of the slopes. This is to be expected since the other measurements suggest that most of the traps are in the substrate and hence have a very complex behaviour which appears as if they were inhomogeneously distributed. The significant feature is that the variation in many cases is very similar between the 1/f noise and the g-r noise as it is for trap NH3 in Figure 2(a). This suggests that the traps contribute to both the g-r and 1/f noise in the same manner as the bias is varied. This could be possible if the g-r noise was produced by traps in a uniform region of the sample but the 1/f noise derived from a region where the g-r characteristic times are distributed.

REFERENCES

1. M.A. Abdala and B.K. Jones, Solid-State Electron. 35 1713-9 (1992).
2. G. Jin and B.K. Jones, Semicond. Sci. Technol. 5 395-403 (1990).
3. G. Jin and B.K. Jones, Semicond. Sci. Technol. 3 1083-93 (1988).
4. M.A. Abdala and B.K. Jones, Noise in Physical systems 1991, Ohmsha Limited, pps. 187-90.
5. N. Sengouga and B.K. Jones, IEEE Trans. ED 46 471-9 (1993).

6. N. Sengouga and B.K. Jones, Solid-State Electron. 36 229-36 (1993).
7. M.A. Abdala and B.K. Jones, Solid-State Electron. 36 237-45 (1993).
8. L. Ren and F.N. Hooge, Physica B176 209-12 (1992).
9. B.K. Jones, J. Phys. D14 471-90 (1981).

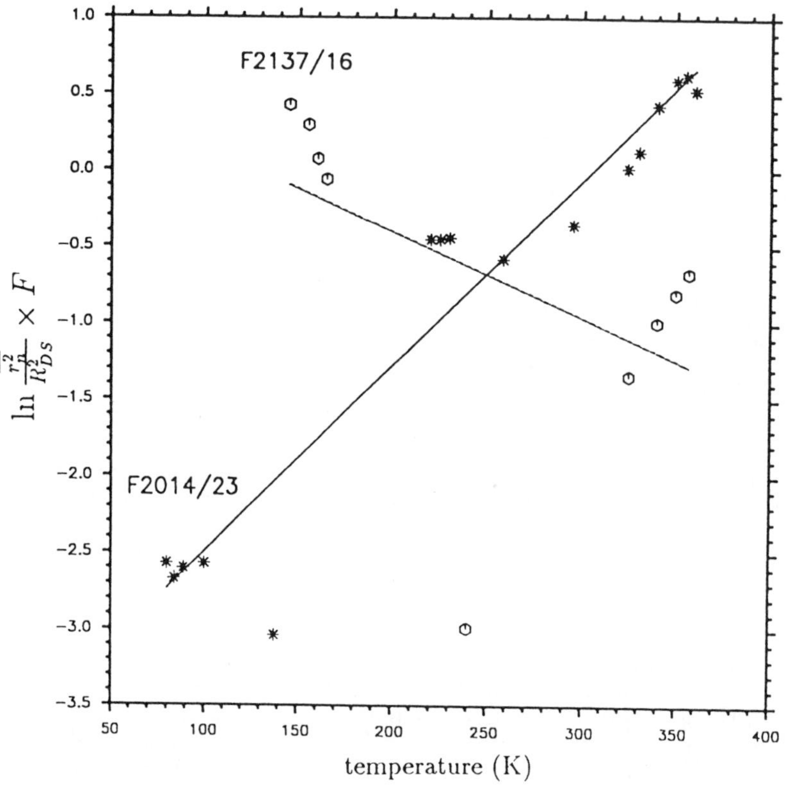

Fig. 1.
The 1/f resistance fluctuation of two specimens as a function of temperature at $V_{GS} = V_B = 0$ and $V_{DS} = 0.15$V. The constant F is 1.5×10^{12}.

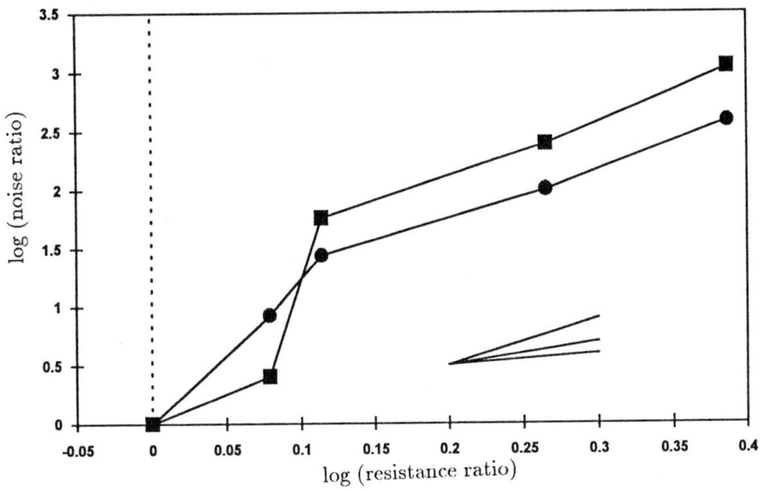

(a)

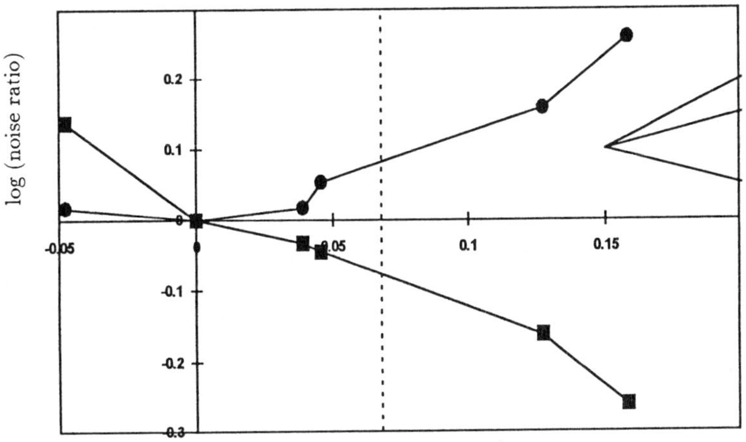

(b)

Fig. 2.

The plots of log (noise) against log (resistance) normalised to the values at $V_{GS} = V_{B_{sub}} = 0$.

● for 1/f noise ■ for g-r noise. The vertical broken line indicates the value where $R_{ch} \sim (R_D + R_S)$. The open symbols represent the values for $V_G = 0$ and $V_{sub} = -10V$. Lines are shown at various slopes.

(a) at T = 220 K near trap NH3, slopes 1, 2, 4.

(b) at T = 150 K near trap NH4, slopes -1, 1, 2.

THE EXCESS NOISE IN SILICON BIPOLAR TRANSISTORS

B.K. Jones and R.C.J. Smets
School of Physics and Materials, Lancaster University,
Lancaster. LA1 4YB, United Kingdom

ABSTRACT

The excess noise of low noise silicon bipolar transistors has been measured with base current and temperature as variables. A generation-recombination component has been found with an activation energy of 0.087 eV. The intensity of this noise, and that of the 1/f noise, varies roughly as the fourth power of the non-ideal base current.

INTRODUCTION

Although the excess noise in silicon bipolar transistors has been extensively studied there are still many unresolved problems. These include the location and mechanism of the 1/f noise and most aspects of burst noise. We report here measurements of the 1/f and generation-recombination noise observed in a bipolar transistor and relate it to the model for the device operation. Generation-recombination (g-r) noise has not often been reported in silicon bipolar devices.

EXPERIMENT

The devices studied were commercial NPN transistors type BC109. The experiments were carried out in the temperature range 77-340 K over a wide range of bias. The d.c. characteristics were measured in both the normal bias configuration ($V_C > V_B > V_E$) and the inverted configuration ($V_C < V_B < V_E$). The Gummel plots were fitted in the conventional manner with good consistency. The base current was fitted to the sum of ideal (I'_B) and non-ideal (I''_B) components

$$I_B = I'_B + I''_B = I'_{BO} \exp \frac{eV}{kT} + I''_{BO} \exp \frac{eV}{n\,kT} \tag{1}$$

The non-ideallity factor n was about 1.7 but tended to decrease as the temperature was raised and was higher in the normal configuration than the inverted bias arrangement. The gain was several hundred. The ideal current takes part in the transistor action while the non-ideal component is due to generation in the base-emitter junction. The temperature coefficient of the non-ideal component gave an activation energy of about 0.6 eV.
The collector current noise measurements were conventional with a collector load R_C and were analysed assuming that the noise was the sum of 1/f, g-r and white components.

$$S_V = R_C^2\, S_I = \frac{A}{f} + \frac{B\,\tau_1}{1+(f\tau)^2} + D \tag{2}$$

so that the parameters A, B τ, and D were obtained. Under most conditions D was not large and could usually be neglected while B was very small except for the specimen described here. This small g-r component can be overlooked unless the analysis is made in great detail[1]. In this specimen the g-r noise was detectable but not dominant. From the time trace there was no sign of burst noise. The d.c. characteristics of this specimen were typical.

RESULTS

The l/f noise was analysed in several ways. The most significant feature was that the l/f noise intensity A varied approximately as $(I_B)^{2.8}$ and $(I''_B)^{3.8}$ at all temperatures. This is unexpected since detailed investigations on similar specimens[2] showed a clear $(I''_B)^2$ dependence and this has been found on other devices[3,4].

The g-r noise was also analysed in several ways. The characteristic time (τ) of the process had an activation energy of 86.9 meV measured at constant current and making an Arrhenius plot of $ln(\tau T^2)$ against $1/T$. This energy was found over a range of currents. The value of τ varied accurately as $1/I_B$ at each temperature. There does not seem to be a simple explanation for this behaviour. The value of the activation energy is larger than that of likely common donors and too small for deep levels.

The intensity of the g-r noise ($B\,\tau$) was found to vary as $(I_B)^{1.8}$ and $(I''_B)^{3.0}$. This variation could again suggest that it is not related to the non-ideal base current but could be related to the total base current or the collector current. The latter is not likely because τ varies with I_B when the collector voltage is kept constant.

REFERENCES

1. M.A. Abdala and B.K. Jones, Noise in Physical Systems, (ed. T. Musha, S. Sato and M. Yamamoto) Ohmsha, 1991 p. 187-90.
2. C.T. Green and B.K. Jones, J. Phys. D18 2269-75 (1985).
3. D.A. Kozlowski and B.K. Jones, IEE Colloq. on Optical Detectors, Digest No. 1990/014, 13/1-16 (1990).
4. J. Sikula et al., Phys. Stat. Sol. A84 693-6 (1984).

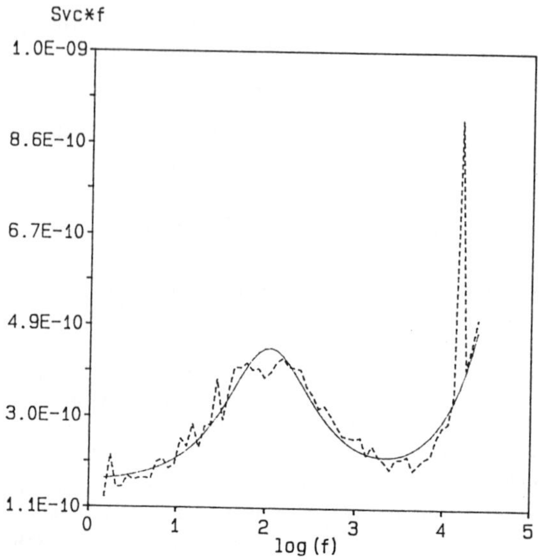

Fig. 1

Typical data fitted to the function of S_V where S_V is given by equation (2). The broken line is the data and the solid line the curve fit.

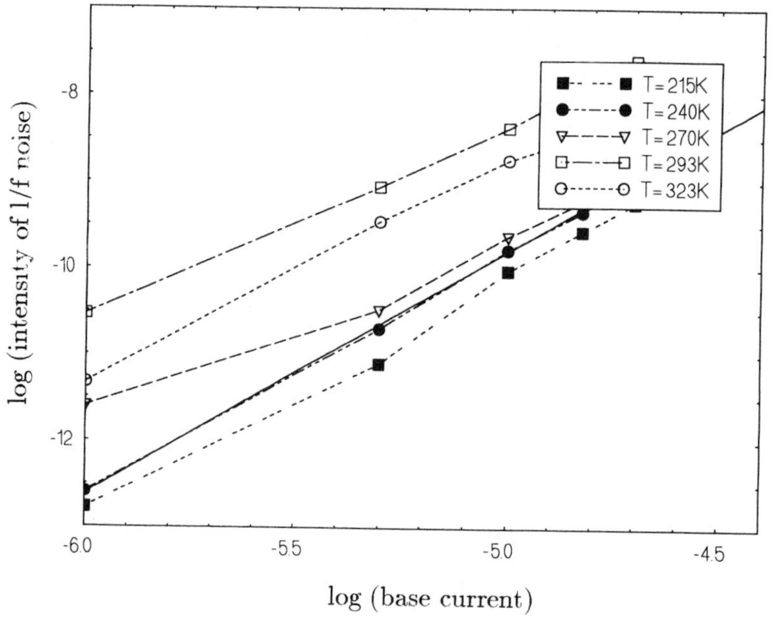

Fig. 2

The 1/f noise intensity against total base current.

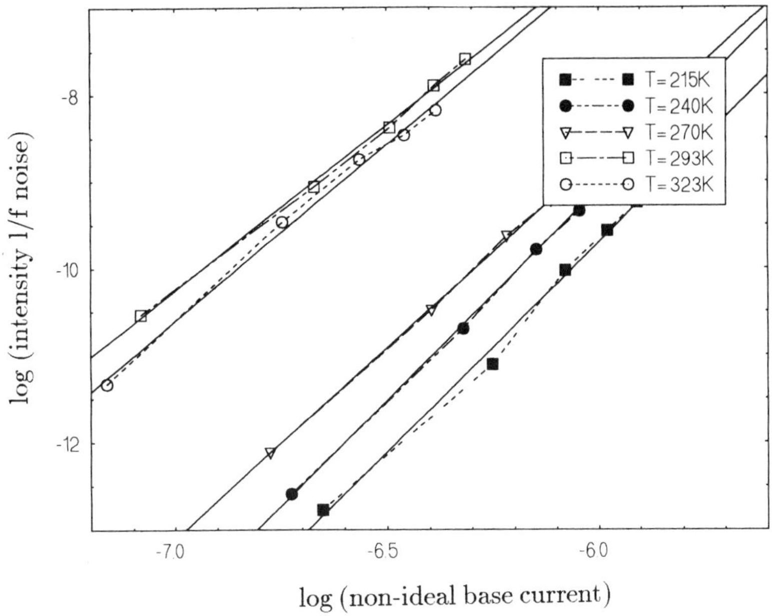

Fig. 3

The 1/f noise intensity against the non-ideal base current.

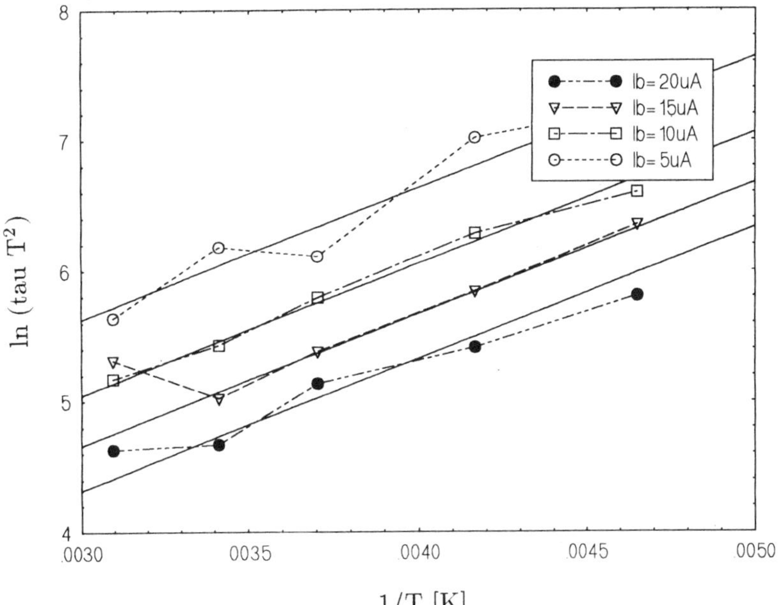

Fig. 4

The Arrhenius plot of the g-r noise characteristic time constant.

1/f NOISE IN NPN GaAs/AlGaAs HBT's

Theo Kleinpenning
Eindhoven University of Technology, Dept. EE, P.O. Box 513,
Eindhoven, The Netherlands

Anthony Holden
GEC-Marconi Materials Technology Ltd.
Caswell, Towcester, Northants, U.K.

ABSTRACT

1/f Noise experiments are performed for npn GaAs/AlGaAs HBT's as a function of forward bias at room temperature. The experimental data are discussed with the help of new expressions for the 1/f noise where the influence of internal series resistances has been taken into account. At low forward currents the 1/f noise is determined by spontaneous fluctuations in the base and collector currents, where the collector current noise exceeds the base current noise. At higher forward currents the series resistances and their 1/f noise become important.

INTRODUCTION

It is well known that the 1/f noise in GaAs HBT's and FET's can limit the bandwidth and stability of circuit operation at high speed. Heterojunction bipolar transistors (HBT's) featuring very low 1/f noise are of considerable interest for high frequency applications.

Until now only a few papers have been published on 1/f noise in HBT's and their results do not agree. Moreover, the experimental data are often analyzed with the help of inaccurate expressions for the 1/f noise in bipolar transistors.

In this paper we describe 1/f noise studies of npn GaAs/AlGaAs HBT's. The transistors are made by GEC-Marconi. They have a single 2.5 x 5 μm^2 emitter with 2 bases and 1 collector, the cut-off frequency is 30 GHz. The internal series resistances of the emitter, base and collector are roughly 40 Ω, 80 Ω and 30 Ω, respectively. The experimental noise results are discussed in terms of new expressions for the 1/f noise, including series resistances effects. The magnitude of the 1/f noise sources in our transistors is compared with 1/f noise data in the literature.

EXPERIMENTAL RESULTS

Noise measurements were made putting the HBT in a circuit as shown in Fig. 1. The internal series resistances are given by r_e, r_b, r_c, the external resistances by R_E, R_B, R_C. In order to locate and to identify the 1/f noise sources, measurements have been carried out for both the common-collector (CC) ($R_E >> \beta/g_{mc}$, $R_B = R_C = 0$) and the common-emitter (CE) configuration ($R_B >> \beta/g_{mc}$, $R_E = 0$, $R_C \approx 1\ k\Omega$). The noise measurements were performed in the current ranges $1\ \mu A < I_B < 100\ \mu A$ and $1\ \mu A < I_C < 1\ mA$. In all devices 1/f noise was observed. Mostly the 1/f noise was observed over at least three decades of frequency, generally in the range of 1 Hz up to 10 kHz.

From the I-V characteristics of the collector and base current versus V_{BE} at $V_{CB} = 3$ V, the ideality factor for I_C was found to be 1.3 and for I_B it was 2.0. At high forward currents deviations of the exponential behaviour were found due to r_e and r_b. The current gain β was found to be proportional to $I_C^{0.4}$ with $\beta \simeq 1$ at $I_C = 1\ \mu A$.

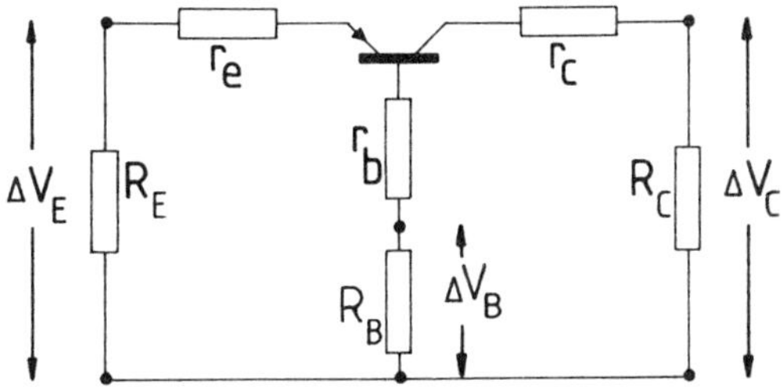

Fig. 1. General circuit for 1/f noise measurements

In Fig. 2 typical 1/f noise plots are presented of S_{V_B}, S_{V_C} and S_{V_E} versus the collector current I_C. The spectral density S_{V_E} is measured in the CC-configuration, the other two densities, S_{V_B} and S_{V_C}, in the CE-configuration. According to Kleinpenning[1,2] the following expressions have to be used

$$g_{me}^2 S_{V_E} \approx S_{I_b} + S_{I_c} + g_{me}^2 \left[I_B^2 S_{r_b} + I_E^2 S_{r_e} \right] \qquad (1)$$

$$S_{V_C}/R_C^2 \approx \beta^2 S_{I_b} + S_{I_c} \qquad (2)$$

$$S_{V_B}/r_\pi^2 \approx (1 + \beta r_e/r_\pi)^2 S_{I_b} \qquad (3)$$

Here S_{I_b} and S_{I_c} are the 1/f current noise densities in the base and collector currents, and S_{r_b} and S_{r_e} the 1/f resistance noise in r_b and r_e.

Interpreting the experimental data of Fig. 2 in terms of Eqs. (1)-(3) leads to the following results:

- At $I_C = 1\ \mu A$, where $\beta \approx 1$, we have $S_{I_c} \approx S_{I_b} \approx 5 \times 10^{-21}\ A^2/Hz$ at $f = 1$ Hz.
- At $I_C = 10\ \mu A$, where $\beta \approx 4$, we have $S_{I_c} \approx 1.6 \times 10^{-18}\ A^2/Hz$ and $S_{I_b} \approx 4 \times 10^{-19}\ A^2/Hz$ at $f = 1$ Hz.
- Above $I_C = 100\ \mu A$ we observe a change in the current dependences of both S_{V_E} and S_{V_B}. The change with respect to S_{V_B} can be ascribed to the fact that in Eq. (3) the term $\beta r_e/r_\pi$ becomes dominant.

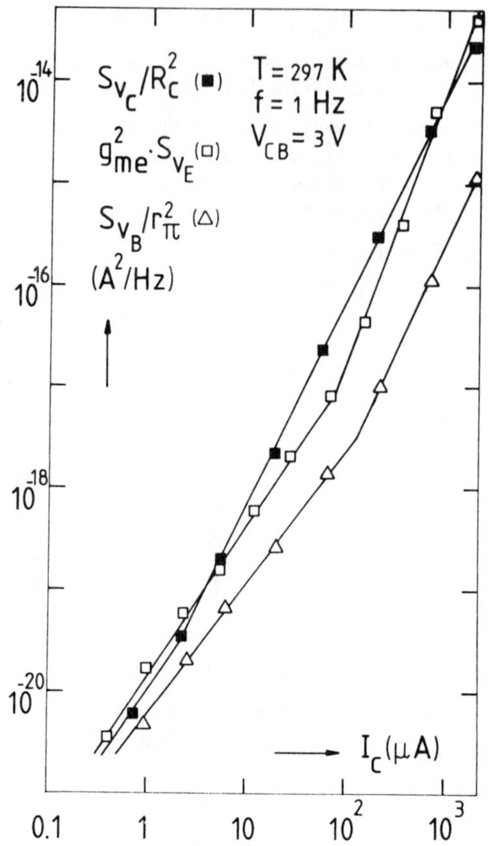

Fig. 2. Typical plots of the 1/f noise versus collector current. For definitions see Eqs. (1)-(3).

The change in the slope of S_{V_E} versus I_C can be ascribed to the 1/f noise in the series resistances.

In view of the above we conclude that at lower currents, i.e. 1 μA < I_C < 100 μA and I_B < 25 μA, the 1/f noise density $g_{me}^2 S_{V_E}$ is practically determined by S_{I_c} and the density S_{V_B}/r_π^2 by S_{I_b}. We then find $S_{I_c} \sim I_C^{1.4}$ and $S_{I_b} \sim I_B^{1.5}$. The steep increase at higher currents is due to the noise in the series resistances (for S_{V_E}) and due to the feedback of the emitter series resistance (for S_{V_B}). Therefore in Fig. 3 we have plotted our results of $g_{me}^2 S_{V_E}$ versus I_C together with some data in the literature of S_{I_c} versus I_C. In Fig. 4 we have plotted S_{V_B}/r_π^2 versus I_B and for comparison some literature data of S_{I_b} versus I_B. Comparison of our data with data from the literature shows a corresponding magnitude of the 1/f noise. There is only a difference in contributions from the parasitic series resistances at higher currents.

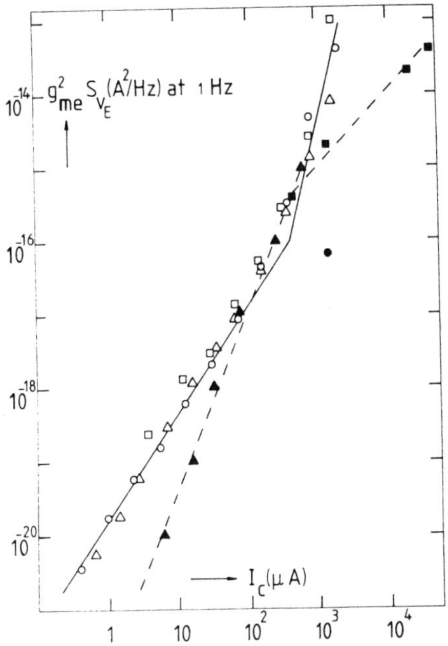

Fig. 3. 1/f Noise density in terms of $g_{me}^2 S_{V_E}$ versus I_C (o,Δ,□). For comparison, S_{I_C} versus I_C measured by: Tutt[3] (■); Jue[4] (▲); Hayama[5] (•)

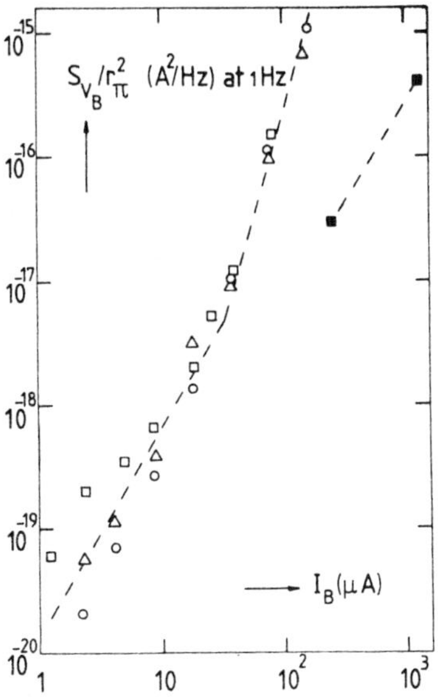

Fig. 4. 1/f Noise density S_{V_B}/r_π^2 versus I_B (o,Δ,□). For comparison, S_{I_b} versus I_B measured by Tutt[3] (■).

CONCLUSIONS

At lower forward currents the 1/f noise is determined by spontaneous fluctuations in both the collector and the base current, with $S_{I_b} \sim I_B^{1.5}$ and $S_{I_c} \sim I_C^{1.4}$. At higher forward currents the parasitic series resistances play an important role. These resistances influence the noise of a transistor in a biasing circuit due to feedback effects and they have 1/f resistance fluctuations. Comparison of our data with the data from the literature shows a corresponding magnitude of the 1/f noise both in the base and in the collector current. There is only a difference in contributions from the series resistances.

REFERENCES

1. T.G.M. Kleinpenning, IEEE-ED 39, 1501 (1992).
2. T.G.M. Kleinpenning and A.J. Holden, IEEE-ED 40 (1993) vol. 6.
3. M.N. Tutt, D. Pavlidis, and B. Bayraktaroglu, Inst. Phys. Conf. Ser., no. 106, IOP Publ. Ltd., 1990, pp. 701-706.
4. S.C. Jue, D.J. Day, A. Margittai, and M. Svilans, IEEE-ED 36, 1020 (1989).
5. N. Hayama, S.R. Le Sage, M. Madihian, and K. Honjo, IEEE MTT Int. Microwave Symp. Dig., vol. II, 1988, pp. 679-682, IEEE Cat.no. 88CH2489-3.

CHARACTERISATION OF MODFET's WITH 1/f NOISE

H. A. W. Markus and T. G. M. Kleinpenning
Eindhoven University of Technology, Eindhoven, The Netherlands

ABSTRACT

We present the results of a comparative 1/f noise study on experimental MODFET's of two fabrication processes. The two processes differ in the way the gate is etched. The first process uses etching fluids, the second plasma etching. The last process may lead to considerable damage in the semiconductor layer under the gate. We measured both the 1/f noise in the gate source voltage and in the drain source voltage. We compared the results of both processes with the results of commercial available transistors. We found that the 1/f noise in the gate source voltage is increased clearly by the damages caused by the plasma etching, whereas the 1/f noise in the drain source voltage shows no apparent difference between the two production processes

Subsequently we present some methods to determine the series resistance and the threshold voltage of MODFET's with the help of 1/f noise measurements.

INTRODUCTION

There are two processes to etch the gate area of a MODFET. In this paper we present noise results obtained from MODFET's made by different technologies. In one process the gate area is etched using etching fluids. In the other process this is done by plasma etching. In the latter process the high energy of the ions in the plasma may lead to considerable damage in the semiconductor layer directly under the etched area.

1/f Noise can increase significantly when the crystalline structure of a device is damaged. Therefore we can use 1/f noise measurements to characterise the MODFET's made with the two production processes. Differences in the 1/f noise performance can be caused by the production processes. These processes can influence the quality of the crystalline structure of the MODFET's and so the 1/f noise.

EXPERIMENTAL RESULTS

The experimental MODFET's are made by IMEC and have gate lengths of 0.3 μm and 0.7 μm. The commercial transistors are made by Fujitsu and have a gate length of 0.25 μm. We measured both the I_G-V_G characteristics and the noise in the gate source voltage of the Schottky barrier under forward bias. At high currents the Schottky barrier dominates the conduction, but at low currents we found deviations which can be interpreted in terms of a leakage resistance parallel to the barrier. For the total current we can write:

$$I_G = I_0[\exp(qV_G/\eta kT) - 1] + V_G/R_p \qquad (1)$$

Here I_0 denotes the saturation current, η the ideality factor and R_p the parallel resistance. For all MODFET's we found the 1/f noise in the gate current S_{I_G} to be proportional to I_G^2 in the region where the conduction is dominated by the barrier. The experimentally obtained values for S_{I_G}/I_G^2 in this region are $1.6 \cdot 10^{-9}$ Hz^{-1} for the Fujitsu FHX31, $9.0 \cdot 10^{-9}$ Hz^{-1} and $1.1 \cdot 10^{-8}$ Hz^{-1} for the fluid etched experimental IMEC transistors, and $5.0 \cdot 10^{-8}$ Hz^{-1} and $5.0 \cdot 10^{-7}$ Hz^{-1} for the plasma etched experimental IMEC transistors.

At low gate currents, where the parallel leakage resistance dominates the conduction, we found a significantly higher 1/f noise in the IMEC transistors with 0.3 μm gate length. In this current region we found S_{I_G} also to be proportional to I_G^2. Here we calculate the current noise stemming from such a parallel ohmic leakage $S_{I_{Rp}}$ and of the barrier $S_{I_{BAR}}$ as a function of the total gate source current. We have to consider two current regions. For sufficiently low currents, where $\partial V/\partial I \gg R_p$ and $V/I \gg R_p$, the current through R_p dominates the total current and therefore we find with Hooge's relation[2] $S_{I_{Rp}} \propto I_{Rp}^2 \approx I^2$. For high currents where $\partial V/\partial I \ll R_p$ and $V/I \ll R_p$, the barrier current dominates the total current and therefore we can make the approximations:

$$S_{I_{Rp}} \propto I_{Rp}^2 = \left(\frac{V}{R_p}\right)^2 \approx \left\{\frac{\eta kT}{qR_p}\ln\left(\frac{I}{I_0}\right)\right\}^2 \propto \{\ln(I)\}^2 \quad \text{and} \quad S_{I_{BAR}} \propto I_{BAR}^2 \propto I^2 \qquad (2)$$

We ascribe the higher 1/f noise level at low currents in the IMEC transistors with a gate length of 0.3 μm to a strongly fluctuating parallel resistance. However at high currents we found $S_{I_G} = S_{I_{Rp}} + S_{I_{BAR}}$ to be proportional to I^2. So at high currents we conclude that the Schottky barrier dominates S_{I_G}.

We measured both the I_{DS}-V_{DS} characteristics with different gate voltages and the noise in the drain source voltage of the channel in the ohmic region. We found the noise in the drain source voltage to be proportional to I_{DS}^2. With Hooge's relation[2] $S_{V_{DS}} = \alpha_{CH} V_{DS}^2/fN$ we can calculate α_{CH}. For the channel we found α_{CH} values of $5 \cdot 10^{-5}$ for both the Fujitsu FHX35 and the Fujitsu FHC31. For the plasma etched transistors we found α_{CH} values of 10^{-3} and $2 \cdot 10^{-3}$, and for the fluid etched transistors $2 \cdot 10^{-3}$ and $5 \cdot 10^{-3}$ respectively.

CHARACTERISATION METHODS

Peransin et al.[1] proposed a method for determining the physical location of the dominant 1/f noise source in MODFET's by considering the relative 1/f noise in the drain current S_{I_D}/I_D^2 as a function of the effective gate voltage. For the same purpose we present a more convenient method considering $S_{V_{DS}}/I_D^2$ as a function of the effective gate voltage. We have

$$S_{V_{DS}}/I_D^2 = S_{R_{CH}} + S_{R_S} \tag{3}$$

where $S_{R_{CH}}$ represents the 1/f noise in the channel resistance and S_{R_S} the 1/f noise in the series resistance. The total resistance between source and drain is $R_{TOT} = R_{CH} + R_S$. For the 1/f noise we use Hooge's relation[2], hence

$$S_{R_{CH}} = \frac{\alpha_{CH} R_{CH}^2}{f N_{CH}} = \frac{\alpha_{CH} q \mu R_{CH}^3}{f L_G^2} \propto V_G^{-3} \quad \text{and} \quad S_{R_S} = \frac{\alpha_S R_S^2}{f N_S} \propto V_G^0 \tag{4}$$

where N_{CH} is the number of carriers in the channel, α_{CH} Hooge's parameter for the channel, L_G the gate length, μ the mobility, N_S the number of carriers in the series resistance, α_S Hooge's parameter for the series resistance and V_G the effective gate voltage, $V_G = V_{GS} - V_T$ with V_T the threshold voltage. In the $S_{V_{DS}} / I_D^2$ versus V_G plot there are two regions. In the region with slope -3 the channel determines the noise, in the region with slope 0, the noise stems from the series resistance. An example is given in Fig. 1.

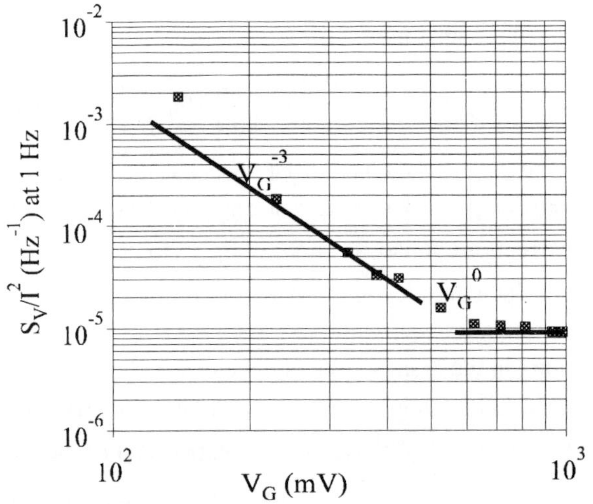

Fig. 1. $S_{V_{DS}}/I_D^2$ at 1 Hz versus the effective gate voltage V_G of an IMEC MODFET

From the I-V curves of the transistors we only could make a rough estimation of the values of the series resistance R_S and the threshold voltage V_T, due to the parasitic MESFET and the gate forward current at high positive V_{GS}. Here we present a convenient method for extracting these values from the 1/f spectra measured at low drain bias. Assuming the noise stems from the channel, we find with Eqs. (3,4):

$$\left(S_{V_{DS}}/I_D^2\right)^{1/3} = \left(\alpha_{CH} q \mu / f L_G^2\right)^{1/3} \cdot (R_{TOT} - R_S) \tag{5}$$

By plotting $(S_{V_{DS}}/I_D^2)^{1/3}$ versus R_{TOT} we obtain the series resistance from the intersection at $(S_{V_{DS}}/I_D^2)^{1/3} \to 0$. An example is given in Fig. 2. From the slope of the plot we calculate the parameter α_{CH}.

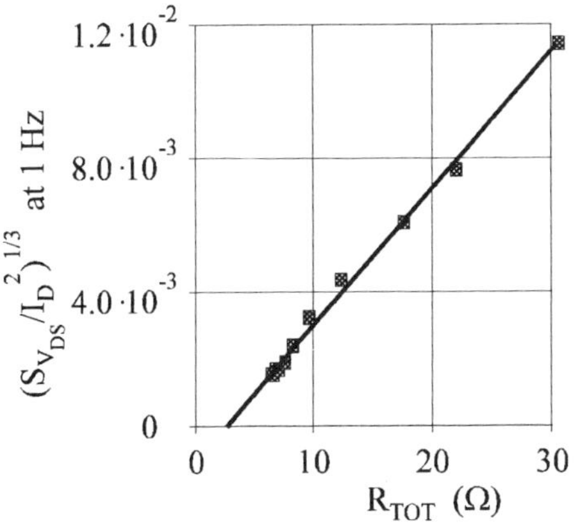

Fig. 2. Plot for determination of R_S
Fujitsu FHX31

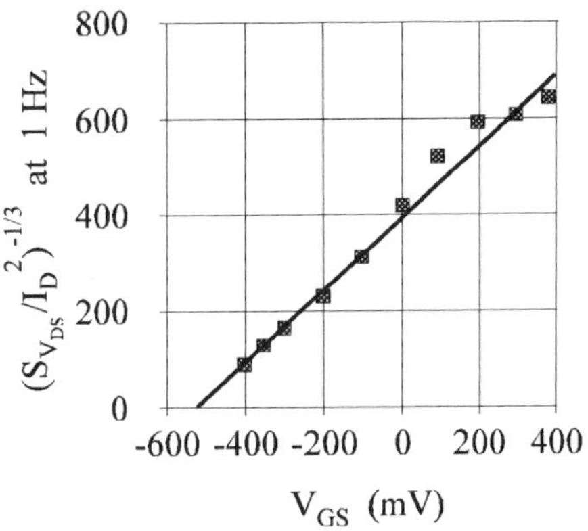

Fig. 3. Plot for determination of V_T
Fujitsu FHX31

According to Eqs. (3,4) the plot of $(S_{V_{DS}}/I_D^2)^{-1/3}$ versus the applied gate source voltage V_{GS} is linear and the intersection at $(S_{V_{DS}}/I_D^2)^{-1/3} \to 0$ gives the threshold voltage V_T. An example of this plot is given in Fig. 3.

CONCLUSIONS

The 1/f noise in the gate source voltage is clearly increased by the damages caused by the plasma etching, whereas the 1/f noise in the drain source voltage shows no apparent difference between the two production processes. All experimental transistors show far more 1/f noise in the gate source voltage and in the drain source voltage then the commercial available transistors.

In cases, where the method for determination of the physical location of the dominant 1/f noise source proposed by J. Peransin et al.[1] doesn't provide clear answers, the new method presented here ends in better interpretable results. The values of R_S and V_T obtained from the methods presented in this paper are consistent with the values estimated from the I-V curves.

REFERENCES

1. J. Peransin et al., IEEE-ED 37, 2250 (1990).
2. F. N. Hooge et al., Rep. Prog. Phys. 44, 479 (1980).

COHERENCE OF 1/f VOLTAGE FLUCTUATIONS AT GATE AND DRAIN IN MODFET'S

Theo Kleinpenning, Philippe Hervé and Bas Vermeulen
Eindhoven University of Technology, Dept. EE, P.O. Box 513,
Eindhoven, The Netherlands

ABSTRACT

The gate and drain voltage fluctuations and their coherence have been investigated on MODFET's (HEMT's). New expressions are presented for the coherence. These expressions are compared with experimental results. A fair agreement has been found between theory and experiment.

INTRODUCTION

Recently, Vandamme et al.[1] reported on the coherence γ_I^2 of 1/f fluctuations in the gate and drain current in MESFET's and MODFET's.

$$\gamma_I^2 = S_{I_G I_D}^2 / S_{I_G} . S_{I_D} \quad \text{with } 0 \leq \gamma_I^2 \leq 1 \tag{1}$$

Here $S_{I_G I_D}$ is the cross-power spectrum of gate- and drain-current fluctuations. Some investigated devices showed an absence of coherence, i.e. below the detection limit of 0.01. Other devices showed a coherence as high as 0.55. Vandamme et al. interpreted their results with the help of a fluctuating ohmic leakage conductance between gate electrode and channel. They considered several types of leakage paths. The calculated coherence depends on the type of leakage path. Without leakage conductance there should be no coherence.

In this paper we present calculations of the coherence of ideal MODFET's and MESFET's biased in the ohmic region. There is only a Schottky barrier in between gate and channel, there are no leakage paths. For such devices we found non-negligible values for the coherence. Our calculated results are compared with experimental results.

CALCULATIONS OF THE COHERENCE FOR OPEN CIRCUIT AND SHORT CIRCUIT

Consider a FET in the ohmic region with a homogeneous channel, a homogeneous distributed gate current (thus $V_{DS} < kT/q$) and $I_G << I_D$.

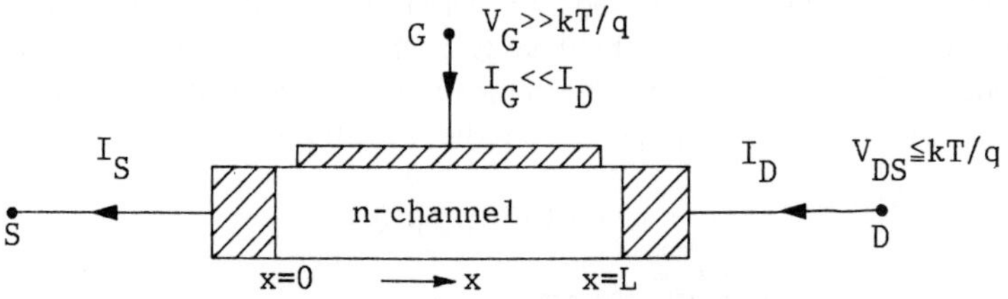

Fig. 1. Diagram of the MODFET (MESFET)

For the fluctuations in the gate current we have

$$\Delta I_G = \Delta I_G^* + \Delta I_G^{ind} = {}_0\int^L \Delta i_g^*(x)dx + {}_0\int^L [di_g(x)/dV_B(x)]\Delta V_B(x)dx \quad (2)$$

where ΔI_G^* is the spontaneous fluctuation and ΔI_G^{ind} the induced one. $V_B(x) = V_G(x) - V(x)$ is the voltage across the Schottky barrier at spot x, and $i_g(x)$ the gate current per unit length. For a homogeneous situation we have $di_g(x)/dV_B(x) = d(I_G/L)/dV_B = 1/(LZ_B)$.
The fluctuations in the channel current $\Delta I(x)$ are given by

$$\Delta I(x) = (I/\sigma)\Delta\sigma^*(x) + (I/\sigma)\Delta\sigma^{ind}(x) + A\sigma\Delta E(x) \quad (3)$$

with $\Delta\sigma^*$ the spontaneous fluctuations in the conductivity, $\Delta\sigma^{ind}(x)= (d\sigma/dV_B)\Delta V_B(x)$ the induced fluctuations, and $\Delta E(x)$ the electric field fluctuations. For the fluctuations $\Delta I(x)$ we can also write

$$\Delta I(x)=\Delta I_D+ {}_x\int^L \Delta i_g(y)dy = \Delta I_D + {}_x\int^L \Delta i_g^*(y)dy+\frac{1}{LZ_B}\,{}_x\int^L \Delta V_B(y)dy \quad (4)$$

where ΔI_D is the fluctuation in the drain current. With the help of Eqs. (2-4) the coherence can be calculated for a FET operating in the strong inversion region where $\sigma \sim V_G - V_T = V_G^* > 0$. Here V_T is the threshold voltage. Then we obtain $d\sigma/dV_B = d\sigma/dV_G^* = \sigma/V_G^*$. Furthermore, we make the approximation

$$V_G^*/Z_B \approx V_B/Z_B \approx (qV_B/kT)I_G \ll I_D \approx I$$

which implies that the Schottky barrier is forward-biased ($V_B > 0$). The open-circuit gate- and drain-voltage 1/f fluctuations can be calculated with the help of Eqs. (2-4) and taking $\Delta I_D = \Delta I_G = 0$. We obtain with Eqs. (3) and (4)

$$\Delta V_{DS} = {}_0\int^L \Delta E(x)dx \simeq \frac{I}{A}\,{}_0\int^L \Delta\rho^*(x)dx + \frac{\rho}{A}\,{}_0\int^L x\Delta i_g^*(x)dx - \frac{V_{DS}}{LV_G^*}\,{}_0\int^L \Delta V_B(x)dx \quad (5)$$

with $\rho = 1/\sigma$ and $V_{DS} = I\rho L/A$. From Eq. (2) we find

$${}_0\int^L \Delta V_B(x)dx = -LZ_B\,{}_0\int^L \Delta i_g^*(x)dx = -LZ_B\Delta I_G^* \quad (6)$$

Combining Eqs. (5) and (6) yields

$$\Delta V_{DS} \simeq (I/A)\,{}_0\int^L \Delta\rho^*(x)dx + (V_{DS}Z_B/V_G^*)\Delta I_G^* \quad (7)$$

The fluctuations in V_{GS} are given by

$$\Delta V_{GS} = \frac{1}{L}\,{}_0\int^L \left[\Delta V_B(x) + \Delta V(x)\right]dx = -Z_B\Delta I_G^* + {}_0\int^L \left(1 - \frac{x}{L}\right)\Delta E(x)dx \quad (8)$$

The last integral in Eq. (8) consists of two contributions. According to Eqs. (5) and (7) the first contribution is given by

$$\frac{I}{A}\int_0^L \left(1 - \frac{x}{L}\right)\Delta\rho^*(x)dx \tag{9}$$

and the second one is lower than $(V_{DS}Z_B/V_G^*)\Delta I_G^*$ due to the factor $(1 - x/L) \leq 1$. Since $V_{DS} << V_G^*$ the second contribution can be neglected with respect to the term $Z_B\Delta I_G^*$ in Eq. (8). Hence we have the approximation

$$\Delta V_{GS} = - Z_B\Delta I_G^* + \frac{I}{A}\int_0^L \left(1 - \frac{x}{L}\right)\Delta\rho^*(x)dx \tag{10}$$

For the open-circuit gate- and drain-voltage 1/f noise we then obtain with Eqs. (7) and (10) and $\Delta R^* = \int\Delta\rho^*(x)dx/A$

$$S_{V_{GS}} = (1/3)I^2S_{R^*} + Z_B^2S_{I_G^*} \tag{11}$$

$$S_{V_{DS}} = I^2S_{R^*} + (V_{DS}/V_G^*)^2Z_B^2S_{I_G^*} \tag{12}$$

$$\gamma_V^2 = \frac{S^2_{V_{DS}V_{GS}}}{S_{V_{DS}} . S_{V_{GS}}} = \frac{\left[(1/2)I^2S_{R^*} - (V_{DS}/V_G^*)Z_B^2S_{I_G^*}\right]^2}{S_{V_{DS}} . S_{V_{GS}}} \tag{13}$$

Here γ_V^2 is the coherence between gate- and drain-voltage 1/f noise, $S_{I_G^*}$ the spontaneous noise in the gate current and $S_R{}^*$ the spontaneous noise of the channel resistance.

EXPERIMENTAL RESULTS

We have performed coherence measurements on MODFET's from Fujitsu at 300 K. The measurements were carried out at low frequencies and with open-circuit gate and drain. At frequencies below 1 kHz the 1/f noise prevails both at the gate and at the drain. We have measured γ_V^2 as a function of V_{DS}. According to Eq. (13) we have

$$\gamma_V^2 = \frac{[(1/2)V_{DS} - V_o^2/V_G^*]^2}{[1 + V_o^2/V_G^{*2}][(1/3)V_{DS}^2 + V_o^2]} \approx \left[\frac{V_{DS} - 2V_o^2/V_G^*}{2V_o}\right]^2 \tag{14}$$

with

$$V_o = \left[Z_B^2S_{I_G^*}/(S_R{}^*/R^2)\right]^{\frac{1}{2}} \sim N_{ch}^{\frac{1}{2}} \sim V_G^{*\frac{1}{2}} \tag{15}$$

The approximation made in Eq. (14) is justified for $V_{DS} << V_o << V_G^*$.

For the MODFET (type FHX05LG) at f = 1 Hz we found experimentally $Z_B^2 S_{I_G^*} \approx 10^{-12}$ V^2/Hz to be almost independent of the gate current hence $V_o \sim [S_R^*/R^2]^{-\frac{1}{2}} \sim N_{ch}^{\frac{1}{2}} \sim V_G^{*\frac{1}{2}}$. The coherence measurements were carried out at $V_G^* \approx 1$ V and with $V_{DS} < kT/q$. The relative channel resistance noise at f = 1 Hz is found to be $S_R^*/R^2 \approx 10^{-10} Hz^{-1}$, hence $V_o \approx 0.1$ V. For $V_{DS} << 2V_o^2/V_G^*$ we calculate $|\gamma_V|$ to be $V_o/V_G^* \approx 0.1$. In Fig. 2 we have plotted both the calculated and the experimental results of $|\gamma_V|$ versus V_{DS} of the MODFET with $2V_o^2/V_G^* = 12$ mV.

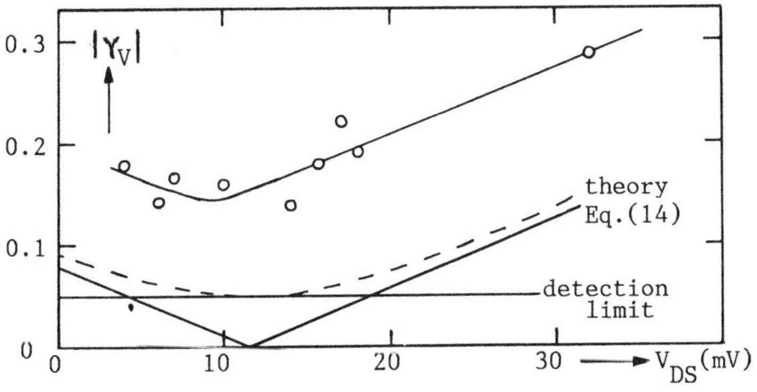

Fig. 2. Coherence $|\gamma_V|$ as a function of V_{DS} for MODFET FHX05LG

We have also measured γ_V as a function of V_{DS} ($\leq kT/q$) of two MODFET's of type FHX35LG. Here we found $V_o = 0.3$ V at $V_G^* = 0.7$V and thus $2V_o^2/V_G^* = 0.26$V. For both devices we found experimentally $\gamma_V \approx 0.1$ and theoretically $\gamma_V \approx V_o/V_G^* \approx 0.4$.

In Fig. 2 there is a fair agreement between theory and experiment. However, for the FHX35LG's we find a rather low value for γ_V^{exp}. This can be caused by imperfections in the MODFET. For example, if the 1/f noise in V_{DS} is determined by the internal series resistances, then Eq. (13) does not apply. Fluctuating leakage conductances between gate and channel, as observed by Vandamme et al.[1], make Eq. (13) also invalid. Nevertheless, it is obvious that ideal devices have non-negligible coherences.

REFERENCES

1. L.K.J. Vandamme, D. Rigaud and J.-M. Peransin, IEEE-ED 39, 2377, (1992).

CURRENT FLUCTUATIONS IN A M-I-M SYSTEM

B. Koktavy, J. Sikula and P. Vasina
Technical University of Brno, Zizkova 17, 602 00 Brno
Czech Republic, tel/fax: 0042 5 41211125

ABSTRACT

A Metal-Insulator-Metal (M-I-M) system, which is realized in capacitors, may be a source of current fluctuations provided that the voltage applied to this system is a slowly varying function of the time. We studied such a system using a ramp voltage with a slope ranging from 1 to 100 V/s. At higher electric fields the interface between the metal and the insulator makes up a source of fluctuations. In the time domain these fluctuations are realized by current pulses with random amplitude and random time gap between the consecutive pulses. The charge transferred is random, too, its value reaching as much as 10 pC. At a constant applied DC voltage the pulse occurrence is very low. Therefore, we carried out our measurements with the slowly varying DC voltage only. The average impulse rate was found to be directly proportional to the slope of this ramp voltage.

INTRODUCTION

In this paper we deal with current impulse noise that is generated in polyethylene terephtalate capacitors under the conditions of a ramp voltage applied across it.

The capacitors under study feature the following values: capacity 15 nF; nominal voltage 630 V; the dielectric thickness 15 micrometers; the relative permittivity ε_r = 3,3; the loss factor tgδ = 0,02 at a frequency of 1 MHz; the breakdown electric strength amounts to 580 MV/m. Furthermore, capacitors whose capacity was 10 nF and nominal voltage 1000 V were studied, too.

The capacitors were studied in a circuit which consisted of a series load resistor whose value was of the order of 1 kΩ. Thus produced RC network was joined to the ramp voltage supply whose ramp slope was adjustable.

The impulse noise voltage across the load resistor was amplified by a low-noise amplifier and subsequently fed into the analyzer whose output was displayed by a recorder. In this way, the time dependence of the noise voltage, the influence of the applied voltage magnitude and ramp rate on the noise and the average impulse rate were depicted.

1. TIME AND VOLTAGE DEPENDENCIES OF THE IMPULSE NOISE

When a sufficiently high DC voltage is applied an impulse current noise is produced in the circuit. The impulse noise voltage across the load resistor to be analyzed is directly proportional this current.

If the applied DC voltage is constant then the number of the voltage impulses is very low. The time separation between the adjacent impulses is determined by the electrical conductivity of the dielectric and may range between 10^4 to 10^6 s. The impulse rate can be substantially increased when a ramp voltage is used.

The shape of the impulse can be observed on the screen of a memory oscilloscope. A typical impulse is shown in Fig.1. The measured time dependence corresponds to the discharging of a capacitor C through a load resistor R_L. It depends on the time constant $\tau = R_L C$ and the amplifier bandwidth. Theoretical analysis of the impulse shape is given[1].

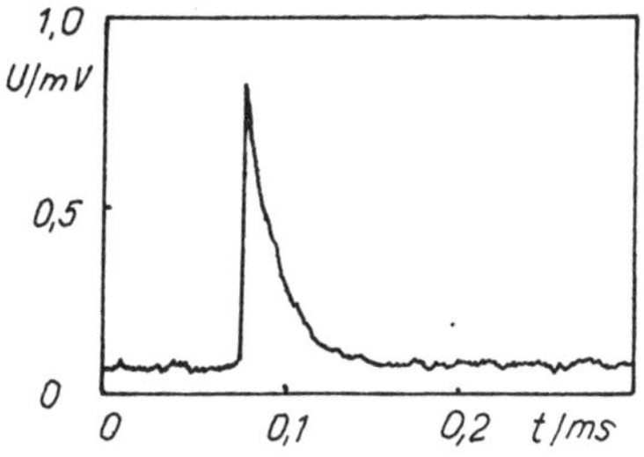

Figure 1 The shape of the impulse

The current impulse amplitude is random. Its magnitude corresponds to fluctuations of the capacitor charge which amounts much as 10 pC. The amplitude distribution function has been studied[2]. It has been found that the distribution function is exponential. In the mentioned paper a statistical analysis of the time between to neighbouring impulses is carried out and it has been found out that the distribution function of this quantity is exponential, too.

Let us suppose that the time constant of the noise voltage amplifier is sufficiently high. When we apply a ramp voltage we get a diagram of the noise voltage across the load resistor versus the instantaneous value of the ramp voltage. The results are represented in Fig.2. Here U_{NO} is the background noise voltage of the measurement setup, U_t is the threshold voltage of the current impulse generation.

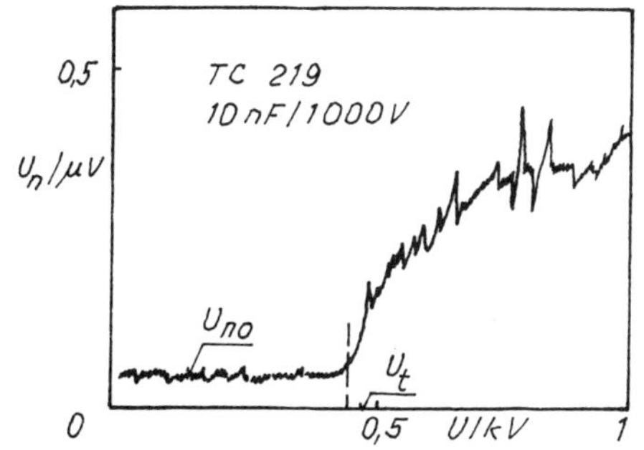

Figure 2 The noise voltage vs. the DC voltage plot (the first voltage application)

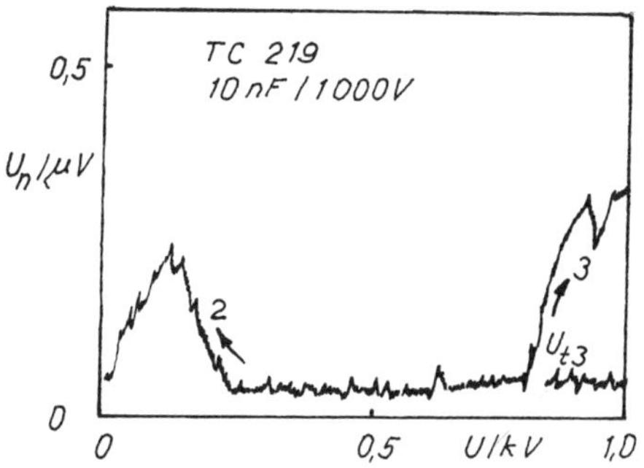

Figure 3 The noise voltage vs. the DC voltage plot (the second voltage application)

The noise voltage U_N is directly proportional to the amplitude and the average frequency of the current impulses. This suggests that one can take these two quantities as quality indicators of the capacitor under study.

When now the DC voltage is decreased we get a noise voltage which is shown in Fig.3, curve 2. When the ramp voltage is applied again then the threshold voltage is approximately doubled.

2. THE DEPENDENCE OF THE AVERAGE IMPULSE RATE ON THE DC CAPACITOR VOLTAGE

In the experiment we studied the dependence of the average impulse rate f_a on the ramp voltage slope v. The result is shown in Fig.4. It is seen that the average impulse rate f_a is directly proportional to the ramp voltage slope.

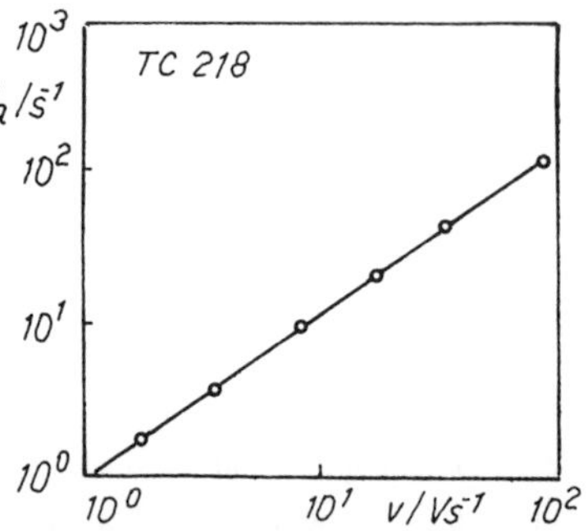

Figure 4 The average impulse rate versus the ramp voltage slope plot

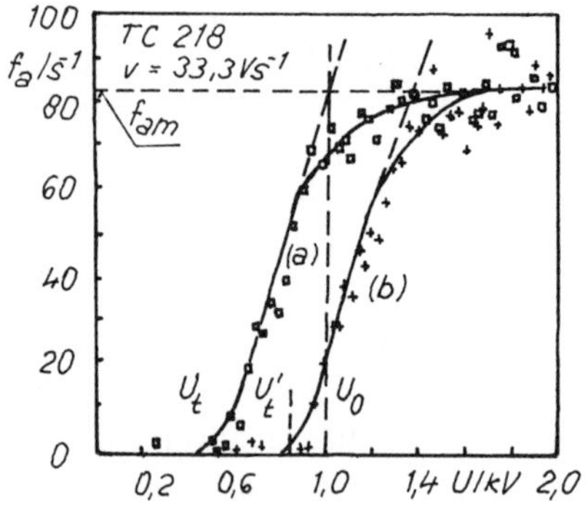

Figure 5 The average impulse rate versus the DC voltage plot

Furthermore, we studied the dependence of f_a on the voltage across the capacitor at a given ramp slope v. This is shown in Fig.5, where U_t is the threshold voltage at which the current impulses begin to appear.

We distinguish three regions:

a) $U < U_t$ no current noise region

b) $U_t < U < U_o$ transition region in which the average impulse rate f_a increases with growing DC voltage,

c) $U > U_o$ saturation region, where f_a is practically constant.

The boundaries between the particular regions depend on the state in which the specimen was before the measurement. If the first measurement is carried out with the ramp voltage growing from zero to a voltage $U > U_o$, (Fig.5, curve a), then the average impulse rate reaches a certain value f_{am}. At the second measurement the threshold voltage is approximately doubled (Fig.5., curve b). The average impulse rate is the same in the both cases.

3. DISCUSSION

A qualitative explanation of the obtained results is based on the assumption that the impulse noise is generated by partial discharges in cavities between the electrode and the insulating layer. A strong electric field in the cavity brings about a discharge as a consequence of the gas ionization.

The discharge transfers a charge on the surface of the cavity, which makes the electric filed intensity in the cavity decrease. The threshold voltage of the impulse generation depends on the cavity width and the gas ionization potential. Application of the Paschen's law to the air gives the threshold voltage U_t = 300 V at the cavity width of 10 micrometers and a pressure of 10^5 Pa. The values of the threshold voltage that we have found experimentally ranged from 100 V to 1000 V for various samples.

If the applied voltage $U > U_t$ is constant the time separation of the pulses is very large reaching as much as 10^6 s.

On the other hand, if a ramp voltage is used the partial discharges are generated in various cavities which have different threshold voltages. The average impulse rate is then directly proportional to the ramp slope.

Partial discharges in a M-I-M structure cause charging of the surface dielectric layer. As the electrical conductivity of the dielectric is usually very low, the electrical neutrality is set within 10^5 to 10^7 s.

Subsequent decreasing the DC voltage across the capacitor results in discharge generation between the cavity surface and the metallic electrode. The amplitude of the impulses is reversed with respect to that of the DC voltage growth. The surface of the dielectric is not discharged entirely, so that when a growing DC voltage is applied for the second time the discharge takes place at a voltage that is higher than in the first capacitor charging process.

4. CONCLUSION

Application of an electric field to the M-I-M system brings about random current impulses, from the statistical characteristics of which information on their nature may be yielded. These partial discharges result in electrical corrosion of the dielectric and make a source of the electrical degradation of the capacitors.

REFERENCES

1. B. Koktavy, in Proc. of the Symposium '92 on Fluctuations in Solids (Brno, 1992), p.75.
2. J. Macur, B. Koktavy, J. Sikula, P. Vasina, Noise in Physical Systems and 1/f Fluctuations (OMSHA 1991), p.313.
3. J. Sikula, A. Cermakova, M. Cermak, P. Vasina, in Proc. of the Int.Conf., Noise in Physical Systems (Washington 1981), p. 125.
4. J. Sikula, B. Koktavy, P. Vasina, P. Schauer, L. Strasky, in Proc. of 10th Int.Conf. on Noise in Physical Systems 1989, Budapest (Budapest, 1990), p.297.

1/f NOISE AS INDICATOR OF QUALITY OF POWER TRANSISTORS

A. KONCZAKOWSKA
Katedra Aparatury Pomiarowej
Wydział Elektroniki Politechniki Gdańskiej
Narutowicza 11/12, 80-952 Gdańsk, Poland

ABSTRACT

The methodology of investigations enabling the classification of power bipolar transistors into groups, depend on their expected quality, based on their 1/f noise is presented. It has been proposed to divide the low frequency noise generated by power transistors into a natural (essential) and a redundant noise. Transistors generating both a natural and a redundant noise are mismanufactured. The analysis of natural sources of noise and importance of a choice of measurement noise conditions were presented.

PROBLEM STATEMANT

The current interest in links between low frequency noise of semiconductors devices and their quality (reliability) is reflected in a number of papers published up to the present. In this paper the methodology enabling the prediction (individually) of a quality of power transistors is proposed. It means that power transistors are classified into groups dependently on their expected quality (high, good, low, poor) immediately after manufacturing. Transistors are classified on the basis of a intensity of low frequency noise (noise parameter X). To apply this methodology it is required to carry out:

* a preliminary study of noise in low frequency range,
* an experimantal verification.

Taking into account a quality of power transistor prediction as the aim it is proposed to devide the low frequency noise generated by power transistors into a natural (essential) noise and a redundant noise. **The natural noise** is connected with phenomena, which are involved with a process of acting of power transistors. This part of low frequency noise can be minimized by improving materials or technology. The essential noise exists always in power transistors, of course in whole lot of electronic components, and a mean essential noise can be evaluated by statistical methods. **The redundant noise** of power transistors is caused by imperfections of materials or/and of technology, therefore it is due to defects, which are produced during manufacturing. This part of noise does not exist if transistor is indeed manufactured in conformity with a technology.

Having made this partition, one can consider that if a power transistor generates only an essential noise it is a high quality one. Transistors generating both an essential and redundant noise in low frequency range are probably mis-manufactured.

The proposed preliminary study for a power transistors should consists of:

* an analysis of a natural noise,
* an analysis of links between typical defects and redundant noise,
* a choice of a noise parameter X for a quality prediction,

* a proposal of a classification algorithm which enables to classify transistors (individually) into the quality groups.

In the paper only the analysis of natural sources of noise and an importance of a choice of measurement noise conditions (frequency f, source resistance R_S, dc current I_B, I_C) of bipolar transistors were presented. Also a proposal of the classification algorithm were described. Presented investigation (a preliminary study of 1/f noise) should be extended on an experimental verification.

LOW FREQUENCY NOISE OF BIPOLAR POWER TRANSISTOR

Noise measurements were carried out for power transistors BDP 286 putting the devices in the common-emitter configuration at different values of the current I_C and source resistance R_S. The spectra were measured in the frequency range 10 Hz-150 kHz.

Low frequency equivalent circuit of the common-emitter configuration, showing the locations of natural noise sources (the thermal noise, shot and 1/f noise), is presented in Fig.1.

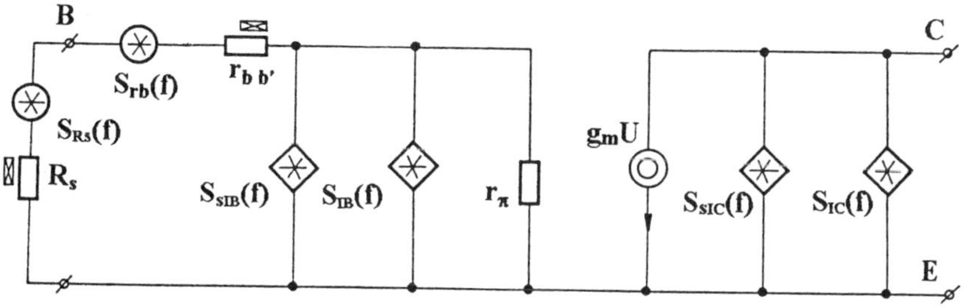

Fig. 1. Equivalent circuit of common-emitter configuration

In the low frequency range a total natural noise generated by transistor can be represented by equivalent input noise source $S_V(f)$ given by the relation [1,2]:

$$S_V(f) = S_{Rs}(f)+S_{rb}(f)+(r_{bb'}+R_S)^2\,[S_{IB}(f)+S_{sIB}(f)]+[(r_{bb'}+r_{\Pi}+R_S)^2/\beta_o^2][S_{IC}(f)+S_{sIC}(f)] \qquad (1)$$

Here S_{Rs}=4kTR_S is the Nyquist noise of R_S, S_{rb}=4kT$r_{bb'}$, the Nyquist noise of $r_{bb'}$, S_{IB} - the 1/f noise density of I_B [1.eqs.(6)-(8)], S_{IC} - the 1/f noise density of I_C [1,eq.(5)], S_{sIB}=2qI_B the shot noise of I_B, S_{sIC}=2qI_C the shot noise of I_C, r_{Π} - the input resistance, $r_{bb'}$ - the bulk resistance, R_S - the source resistance, β_o - the current amplification factor, and g_m is the transconductance.

The value of an optimal source resistance depends on measurement noise conditions and for a room temperature it is equal to:

$$R_{So}^2 = 2\beta_o r_{bb'} 0.026/I_E + \beta_o(0.026/I_E)^2 \qquad (2)$$

For the middle frequency range we have:

$$S_V = S_{Rs} + S_{rb} + (r_{bb'} + R_S)^2 S_{sIB} + [(r_{\Pi} + r_{bb'} + R_S)^2/\beta_o^2]\, S_{sIC} \qquad (3)$$

For this frequency range it is very easy to evaluate components of the relation (3), there are: S_{Rs}, S_{rb}, $S_{VB} = (r_{bb'} + R_S)^2 S_{sIB}$, $S_{VC} = [(r_{\Pi} + r_{bb'} + R_S)^2/\beta_o^2]\, S_{sIC}$.

In the Fig.2 the S_V and its components:S_{Rs}, S_{rb}, S_{VB}, S_{VC} versus source resistance at f = 100 kHz for the transistor BDP 286 Nr 1 at I_C=16mA (r_{Π}=107.8Ω, $r_{bb'}$=0.3Ω, β_o=69, I_B=0.23mA, R_{So}=13.49Ω) and at I_C=580mA (r_{Π}=2.58Ω, $r_{bb'}$=0.3Ω, β_o=60,

I_B=9.66mA, R_{So}=0.347Ω) were exemplified.

a b

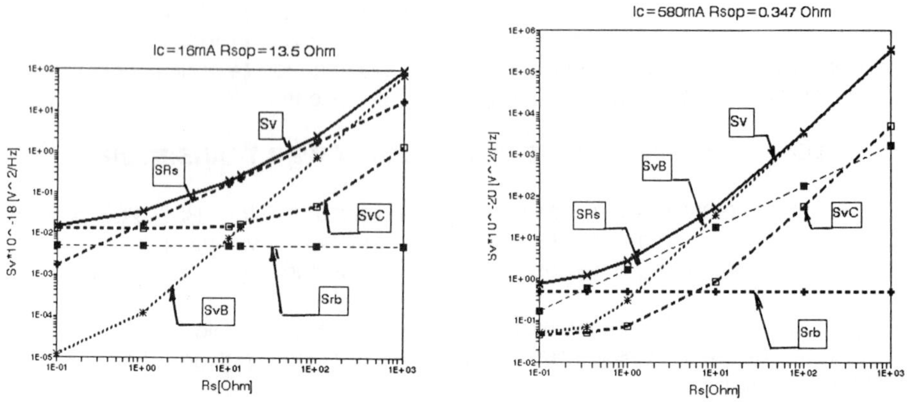

Fig. 2. The equivalent noise source S_V and its components for transistor BDP 286 Nr 1 at f = 100kHz a - I_C = 16 mA, b - I_C = 580 mA

It is easy to recognize that for I_C = 16mA (Fig.2a) at R_S = 0,1Ω (R_S<100R_{So}) the S_{VC} and S_{rb} dominate in the equivalent noise source, the S_{VB} can be neglected. For I_C = 580mA (Fig.2.b) at R_S = 100Ω (R_S>100R_{So}) the S_{VB} dominates in the S_V markedly. By a choice of measurement noise conditions one can evaluate for example a bulk resistance $r_{bb'}$, sources S_{sIB} or S_{sIC}.

For the lower range of frequency the 1/f noise can not be omited, and for example the S_{VIB}(f=300Hz) = $(r_{bb'}+ R_S)^2 S_{IB}$(f=300Hz) at I_C=580mA is dominated (see Fig.3). As we see the S_{VIB}=S_V is located parallely to S_{VB}, and in S_V other components can be neglected. By choice the following measured conditions: R_S=100Ω, I_C=580mA, f=300Hz it is possible to evaluate the 1/f noise source in the base. In Fig.4 the S_{IB}(f) for transistors Nr1, Nr2, Nr3 is exemplified.

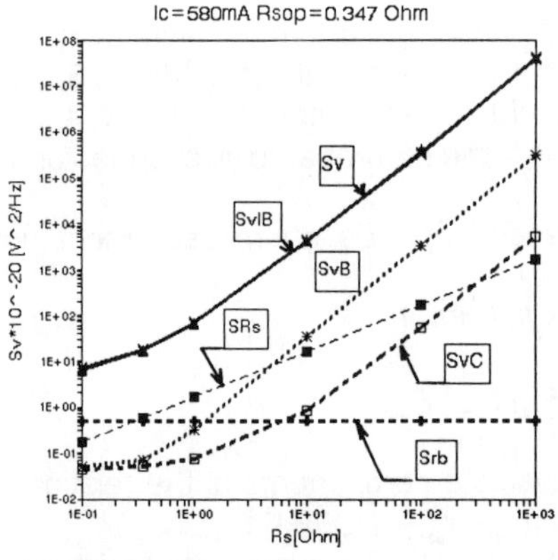

Fig. 3. The equivalent noise source S_V and its components f=300Hz, I_C=580mA

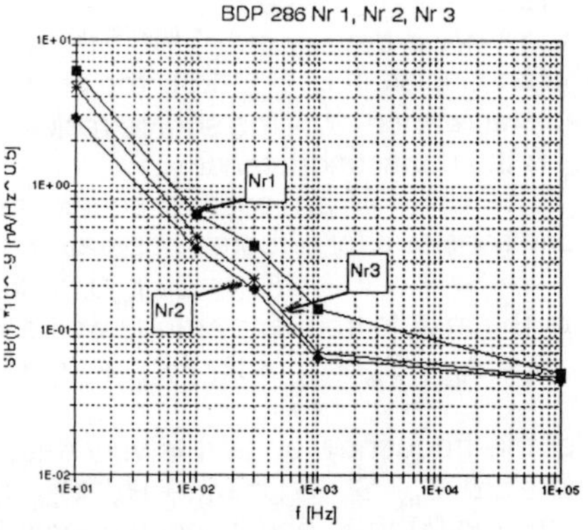

Fig. 4. The S_{IB} versus frequency for transistors Nr1, Nr2, Nr3

Which part of noise is a natural, which one is a redundant?

The natural 1/f noise can be statistically evaluated based on noise measurements of a few selected power transistors (without any defects) or of a large number of power transistors drawn from a population of manufactured transistors.

All transistors which generate a higher 1/f noise than a natural 1/f noise (evaluated statistically) should be recognized as a mismanufactured,

Following phenomena: crystal defects and contaminations, emmiter edge dislocations, electromigration, imperfection of chip bonding, radiation (external and intristic) have been indentified as the main sources of low frequency noise in bipolar transistors. In the majority these phenomena are reasons of failures of bipolar transistors. Influence of mentioned phenomena on an intensity of 1/f noise sources should be specified during a preliminary study of noise.

PROPOSAL OF CLASSIFICATION ALGORITHM

One of the major problem in the presented methodology is a critical analysis of a classification algorythm. In the [2] the following procedure has been proposed.

On the basis of measurements of noise parameter X (carried out for a random sample drawn from a population of manufactured transistoirs) two moments of X: the mean value $\overline{X}$ and variance S_x should be evaluated. Assuming that the random variable $\underline{X}$ follows the normal distribution in investigated sample (and in population as a whole) and that manufactured transistors will be brought into one of four reliability groups it is needed to calculate three values which define the border values for quality groups, namely X_{bm}:

$$X=\overline{X} + S_x z_{\alpha m} \tag{4}$$

where: $z_{\alpha m}$ - standardized normal random variable.

It has been assumed that the X_{b1} value defines a mean essential noise for investigated power transistors.

The rules of power transistors classification into four groups are as follows:

$X_j < X_{b1}$ - first group - high quality is expected,

$X_{b1} \leq X_j < X_{b2}$ - second group - good quality is expected,

$X_{b2} \leq X_j < X_{b3}$ - *third group - low quality is expected,*

$X_j \geq X_{b3}$ - fourth group - poor quality is expected.

These rules of power transistors classification should be experimentally verify.

CONCLUSION

In the individual prediction of quality of power transistor very important problem is a choice of measurement conditions of a noise parameter X . If conditions of measurement are not well fixed, the measured noise parameter X will not reflected quality of tested transistors.

REFERENCES

1. T.G.M. Kleinpenning, Noise in Physical Systems, (A.Ambróży, Budapest, Hungary, Akademiai Kiado, 1990) p.443.
2. A. Konczakowska, Zeszyty Naukowe Politechniki Gdanskiej, Elektronika Nr LXXVII, Gdansk, 1992, (in polish).

INVESTIGATIONS ON THE ORIGIN OF AlGaAs-GaAs HEMTS LF CHANNEL NOISE

N. Labat, N. Saysset, D. Ouro Bodi, A. Touboul, Y. Danto
IXL (URA 846-CNRS), University of Bordeaux, 33405 Talence, FRANCE
C. Tedesco
D.E.I., Universita Degli Studi di Padova, 35131 Padova, ITALY
A. Paccagnella
Istituto di Elettrotecnico, Universita di Cagliari, 09123 Cagliari, ITALY
C. Lanzieri
ALENIA Spa, Roma, ITALY

ABSTRACT

The influence of deep levels on AlGaAs/GaAs HEMTs operation has been evaluated from the transconductance frequency dispersion and the low frequency channel noise analysis. The LF noise level of damaged devices has been shown to increase after hot electron ageing tests.

INTRODUCTION

The purpose of this work is to analyse the drain excess noise of depletion mode AlGaAs-GaAs HEMT's designed for microwave power and low noise applications. The influence of deep levels has been extensively studied and they are identified as responsible for many degradations of the performances of FET devices. These traps, some of which are identified as donor-aluminium complexes (DX) mainly induce frequency and bias-dependent anomalies[1,2]. They show typical capture and emission frequencies below 1 MHz, even at 300 K.
As well as these traps, hot electron ageing tests induce damages in HEMT's operation and this work mainly deals with the influence of both deep levels and hot electrons on the LF channel noise behaviour of conventional AlGaAs-GaAs HEMT's.

DESCRIPTION OF THE DEVICES UNDER TEST

The devices under test are divided into three sets (A, B, HT).Their structure is composed by a n-doped AlGaAs layer grown on the undoped GaAs layer. AuGeNi ohmic contacts are processed on a n+ GaAs cap layer. They mainly differ on the recessed gate geometry and on the Al molecular fraction of the n-doped AlGaAs layer : respectively 0.3 and 0.27 for type A and B devices.

INFLUENCE OF THE DX CENTERS ON THE HEMT OPERATION

ANALYSIS OF THE CHANNEL NOISE

Low frequency channel noise measurements were performed in the ohmic regime (V_{DS} < 100 mV). For values of the gate bias V_{GS} close to the threshold voltage of the component, the channel noise presents a 1/f behaviour. The current noise spectral intensity is then proportional to the square of the drain-source current (Fig.1). Even for such low bias conditions, the theoretical thermal noise level is not reached for frequencies below 100kHz.
When increasing the gate bias, G-R noise contributions appear in the noise spectra. At room temperature, the G-R "plateau" value of type A devices is about two orders of magnitude above the channel noise level of type B HEMTs. Due to the V_{GS} dependence of the G-R components appearence in the noise spectra, the contributed traps are supposed to be located in the AlGaAs active layer.

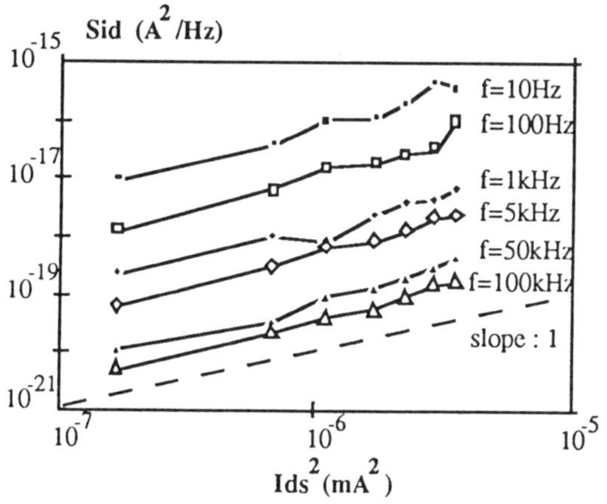

Figure 1 : Dependence of the 1/f channel noise current on I_{DS}(type B devices)

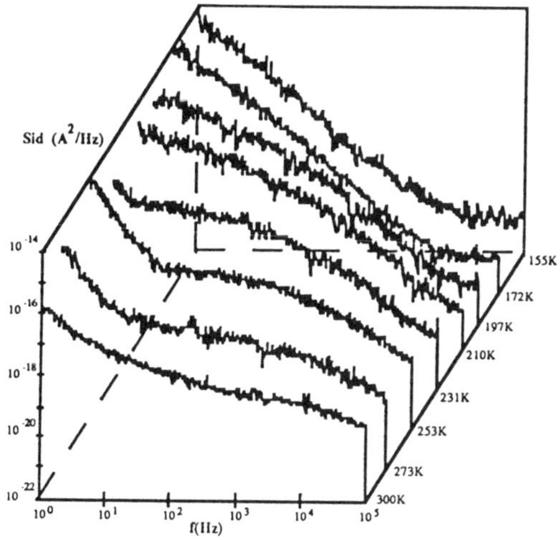

Figure 2 : Spectral intensity of the noise current vs. temperature (type B devices)

By assuming the influence of a single preeminent level and a non spatially-dependent time constant τ, the G-R current noise spectral intensity is expressed as[2,3] :

$$S_I(f) = \frac{4 I_{DS}^2 d \tau N_T f_T (1 - f_T)}{n_S^2 L W} \frac{1}{1 + (2 \pi f \tau)^2} \quad (1)$$

where $N_T f_T (1 - f_T)$ is the effective traps density, n_s the electron density in the 2DEG and d the space charge region width. A mean value of the active traps density is obtained : 2.5×10^{14} cm^{-3} and 8.6×10^{13} cm^{-3} for type A and B devices respectively. When decreasing the temperature to 150K, the 1/f excess noise is screened by the increasing amplitude of the G-R contribution (Fig. 2). The shift of the corner frequency f_c vs. temperature is reported on the Arrhenius plot of $\log(T^2/f_c)$.

The activation energy Ea and the capture cross-section σ are obtained from a least-squares linear regression (the electron effective mass of the L-valley in $Al_xGa_{1-x}As$ with respectively x=0.3 and 0.27 has been used to determine σ). The value of these parameters (reported in table I) clearly identify the DX center.

FREQUENCY DISPERSION OF THE TRANSCONDUCTANCE

The influence of these traps on the HEMTs electrical characteristics, as undesirable frequency-dependent effects has been investigated[4].

The transconductance $g_m(f)$ has been measured in the 0.1Hz-1MHz frequency range. Measurements were performed at V_{DS}=50mV and V_{GS}=0V, at various ambient temperatures. A sinusoïdal signal of 100 mV peak-to-peak amplitude was applied to the gate contact with source grounded and a 10 Ω resistance connected between the drain and the DC supply terminal. $g_m(f)$ is computed from I_{DS} measurements across the 10 Ω resistance with an impedance-gain phase analyzer and is normalized to the lowest frequency value.

Typical results are reported in figure 3. The increase of $g_m(f)$ at frequencies higher than 1 kHz is a typical feature of DX centers. At frequencies lower than the DX characteristic frequency, the electron trapped in these centers reduce the modulation of the 2DEG density induced by changing the applied gate voltage. Thus, transconductance values at low frequencies are noticeably smaller than for higher frequency values.

Activation energy and capture cross-section are obtained from the temperature dependence of the frequency f corresponding to g_m transitions (Table II). The low frequency decrease of $g_m(f)$ identifies traps with an activation energy of 0.48 eV. In fact, by driving the device toward pinch-off, the low frequency decrease of $g_m(f)$ is still evident, while the DX related increase of $g_m(f)$ at high frequencies fades.

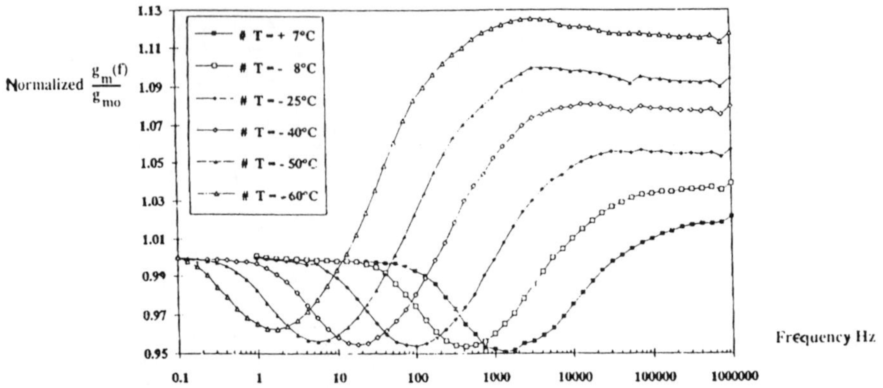

Figure 3 : Normalized $g_m(f)$ vs. frequency and temperature (type A devices)

Devices	Ea (eV)	σ (cm^2)
Type A	0.43 ± 0.02	5.5x10^{-15}
Type B	0.44 ± 0.02	8x10^{-15}

Table I : Characteristic parameters of the DX levels (from LF noise analysis)

Devices	Ea (eV)	σ (cm^2)
Type A	0.42	2x10^{-15}
Type B	0.40	2.9x10^{-15}

Table II : Characteristic parameters of the DX levels(from $g_m(f)$ dispersion)

HOT ELECTRONS DEGRADATION OF THE ELECTRICAL PARAMETERS

To create impact ionization effects, a set of devices have been biased at an operating point corresponding to a high ratio I_G/I_D (typically V_{DS}= 6V). The percentual decreases of the drain-source current and of the transconductance have already been correlated with the respective I_G/I_D ratio of the accelerated test[5]. Figure 4 gives the transconductance measured in the ohmic regime for two sets of devices : HT1 and HT10 (no ageing) - HT3 and HT14 (after 215 hours of life-tests performed at room temperature). The collapse of the transconductance is confirmed after the ageing.

Low frequency noise is measured from 1 Hz to 100 kHz at room temperature. The devices are biased in the linear regime (V_{DS} = 50 mV) and the gate voltage varies between the device threshold voltage V_T and zero.

As shown previously[2], the noise current spectral intensity as a function of the gate bias shows two different behaviours : for V_{GS} lower than the gm maximum bias conditions (figure 4), the drain current noise follows a 1/f law whereas for higher V_{GS} values, G-R components are superimposed on the 1/f noise. The governing equation of the 1/f noise is :

$$S_{I_D} = \frac{\alpha_H I_{DS}^2}{N\ f} \qquad (2)$$

where α_H is the Hooge parameter resulting and N the total electron number.

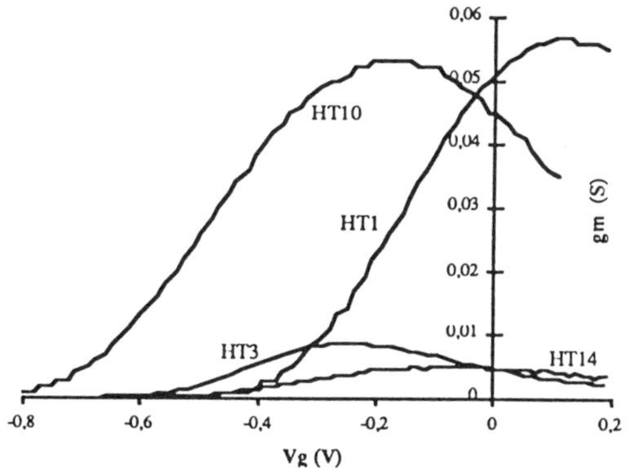

Figure 4 : Transconductance of HT1, HT10 (no ageing) and HT3, HT14 (after degradation) in the ohmic regime

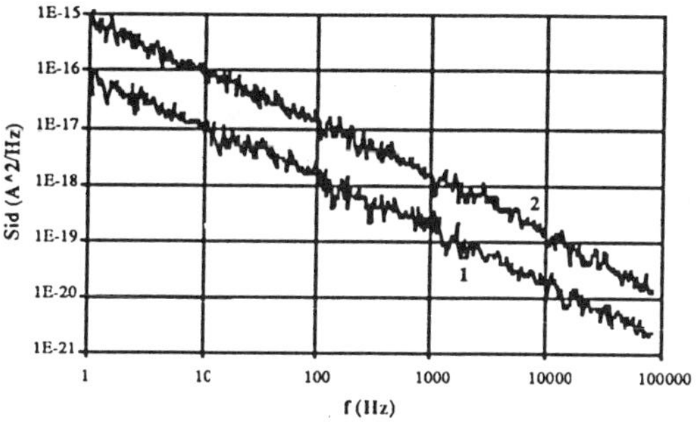

Figure 5 : Drain noise spectral intensity of HT1 (1 : virgin) and HT3 (2 : after degradation) devices for I_{DS} = 0.19 mA

The plot of the drain current spectral intensity for the same drain current (figure 5) clearly shows an increase of the drain 1/f noise of degraded devices (HT3) compared to the excess noise of the virgin ones (HT1). This difference has to be corrected according to the dispersion of the respective DC parameters (I_{DS} and g_m vs. V_{GS}) of each device. The "normalised" drain noise current appears to be only depending on the gate bias (in the linear regime), on the technological parameters of the device and on its intrinsic physical noise sources (interfaces regions - access regions - 2DEG).

CONCLUSIONS

DX centers have been characterised in conventional HEMTs according to the G-R LF noise dependence on temperature. Activation energy value of these traps is in very close agreement with the value determined from the temperature-dependent increase of the transconductance g_m in the high frequency range (f>10kHz at T=280K). However, the capture cross-section determined from the noise analysis is found to be higher. DX centers have then been found to induce a frequency dependence of the transconductance.

The drain excess noise of AlGaAs/GaAs HEMTs in the ohmic regime is mainly controlled by fluctuations occuring in the channel. After life tests at room temperature, an increase in the drain excess noise level has been correlated with degradation resulting from hot electrons effects. The identification of the physical origins of the noise increase will therefore need a separation between contributions deriving from either electron mobility or from density fluctuations in the access regions and in the gate controlled channel.

REFERENCES

[1] C. Canali et al., IEEE Proc.-G, 138, 1991
[2] N. Labat, et al., Qual. and Rel. Eng. Int., 8, 301, (1992)
[3] S. Kügler , J. Appl. Phys., 66, (1989)
[4] M. A. Abdala, B. K. Jones, Sol. St. Elect., 35, 1992
[5] C. Tedesco et al., Proc. of ESREF92, Edit. VDE, Germany, (1992)
[6] A.Van Der Ziel, Sol. St. Elect., 26, 385, (1983)
[7] D. Delagebeaudeuf and al., IEEE Trans. Elect. Dev., ED-29, 955, (1982)

SURFACE AND BULK 1/F NOISE IN SILICON BIPOLAR TRANSISTORS

G.Leontjev
Vilnius University, Vilnius 2054, Lithuania

ABSTRACT

Investigations of 1/f-noise in the bipolar transistors with field electrode (gate) revealed, that spectral, current and temperature characteristics of bulk and surface 1/f-noise are very similar. This result shows their common nature: both bulk and surface 1/f-noise are superposition of many sources of burst noise.

INTRODUCTION

Bipolar transistor with gate is suitable object for investigations of 1/f-noise. The earlier study of those transitions have shown, that the main origin of 1/f-noise in the bipolar transitions is surface recombination current [1]. Since this study some changes in view on 1/f-noise nature occurred and considerable progress in technology of the bipolar transitions producing has been achieved. Thus, investigations of 1/f-noise in samples made by modern technology, seem be actual.

RESULTS

The results of the experimental investigations of the characteristics of the 1/f-noise in the special made bipolar transistors with gate over emitter junction and part of base are presented. The devices were made by a planar diffusion process into silicon using boron and phosphorous as impurity dopants.

Fig. 1 shows the dependences of the base current I_B on the gate potential U_G. One can see the peak of base current at $U_G \sim -16$ V, which is related with equal concentrations of electrons and holes in the undergate surface. This situation corresponds to the maximum surface recombination velocity. Further decrease of potential U_G leads to rapid decrease of current I_B and appearance of inversion induced p-n junction under the gate. Independently on the surface state current $I_B \sim \exp(qU_{BE}/mkT)$ (Fig.2). At U_G=0 V base current consists of the diffusion component and in this case "nonideal" coefficient m=1. At U_G=-16 V contribution of the component of the surface recombination current to the base current is main and coefficient m=2. At U_G= -30 V base current consists of several nearly equal parts of diffusion and the base current, caused by maximum velocity of surface recombination at U_G ~ 45 V is more weakly expressed, then in the p-n-p type transistors.

To distinguish the components of 1/f noise associated with the surface recombination base current $I_B{}^S$ and the bulk base currents the dependence of 1/f noise on the potential U_G at a fixed bias of the emitter junctions (Fig. 3) has been measured. Bulk base current in our case consisted of the main emitter junctions current $I_B{}^V$ and the induced emitter current $I_B{}^{VI}$. The components of 1/f noise are marked as $I_{eq}{}^S$, $I_{eq}{}^V$ and $I_{eq}{}^{VI}$ respectively. From the measured dependences $I_B = f(U_G)$ and $I_{eq} = f(U_G)$ for a several values of U_{BE} the dependences $I_{eq}{}^S = f(I_B{}^S)$, $I_{eq}{}^V = F(I_B{}^V)$ and $I_{eq}{}^{VI} = f(I_B{}^{VI})$ were founded.

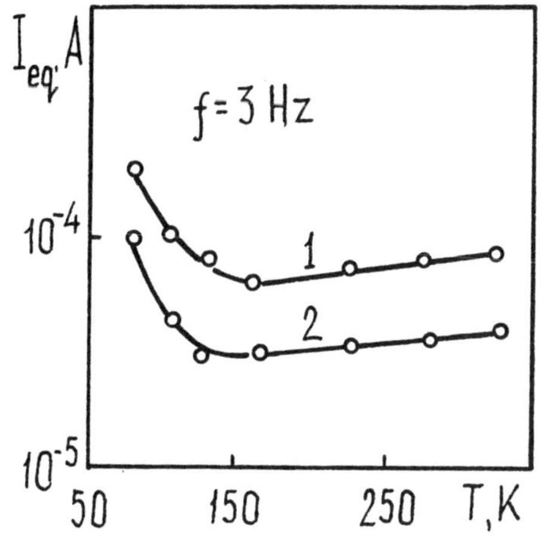

Fig. 4. Temperature dependence of noise currents $I_{eq}{}^V$ (curve 1) and $I_{eq}{}^S$ (curve 2).

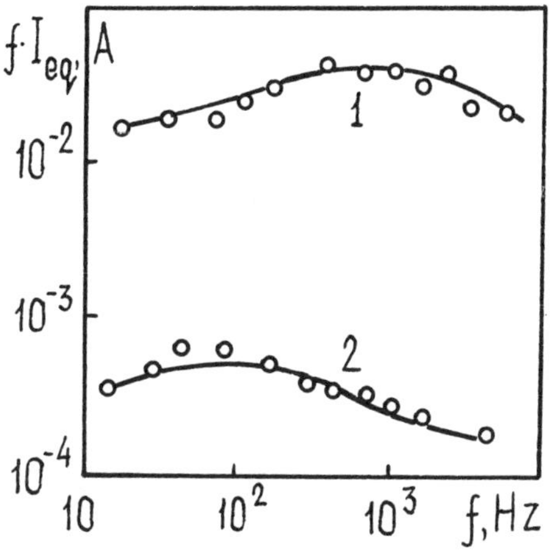

Fig. 5. Normalized noise spectral density at potential U_G = 0 V at base current I_B:1 - 10^{-5} A; 2 - 10^{-6} A.

Results of investigations have shown that 1/f-noise of $I_B{}^S$ component is much smaller then noise of $I_B{}^V$ and $I_B{}^{VI}$ components both in the n-p-n and p-n-p transistors. Beside, dependences of noise current I_{eq}, $I_{eq}{}^V$ and $I_{eq}{}^{VI}$ on recombination components of current in the main and induced emitter junction, and value of m is between 1 and 2.

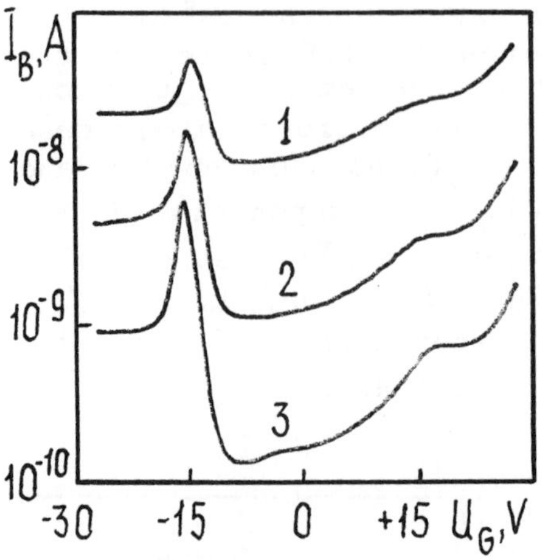

Fig. 1. Dependences of I_B on U_G for p-n-p transistor at some emitter-base voltages U_{BE}:1 - 0,5 V; 2 - 0,44 V; 3 - 0,38 V.

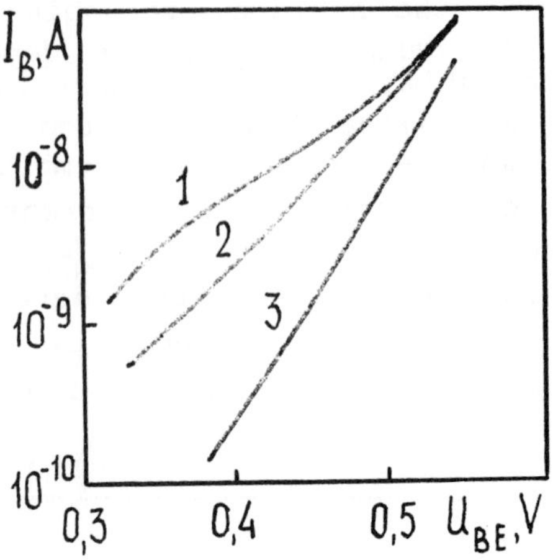

Fig. 2. Dependences of I_B on the emitter-base voltage at some potential U_G: 1 - -30 V; 2 - -16 V; 3 - 0 V.

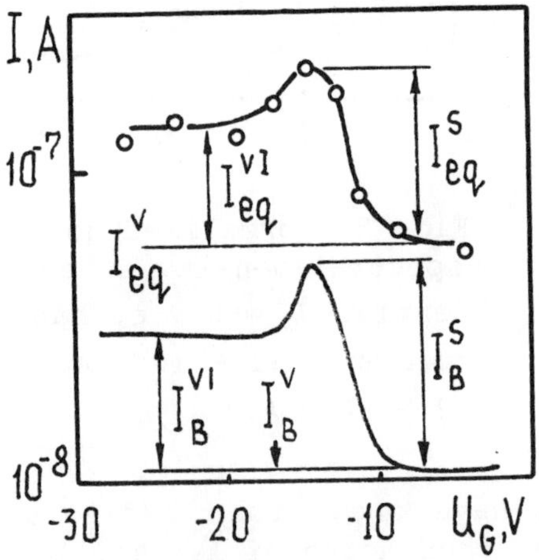

Fig. 3. Dependence of base current (curve 1) and noise current (curve 2) on gate potential.

The similar dependences of current I_B on potential U_G were observed in the n-p-n transistors. However, peak of corresponding currents are quite similar for many samples and proportional to ~ $I_B^{1,4}$.

Temperature dependences of noise currents I_{eq}^{S}, I_{eq}^{V} and I_{eq}^{VI} are alike too (Fig. 4).

Investigations of noise spectra in the frequency region 2Hz to 20 kHz have shown, that spectra change similar to 1/f, but have fine structure. Beside, observed maxima of the normalized spectra move to the high frequencies with the increase of the current I_B (Fig. 5).

DISCUSSIONS

From the present investigations, we can conclude that 1/f-noise is the sum of the big number of the burst noise sources, located in bulk and surface region. This conclusion is confirmed by oscilograms of 1/f-noise, which show that noise is not "smooth" and it is possible to pickout big number of pulses, of various amplitude and durations. According to the models of the noise of burst sources [2,3], there are defects in the emiter-junction through which flow current of increased density. Recharge of g-r centres, located in the defect region, modulates this current. Beside, charge carriers while interacting with g-r centres owercome recombination barrier [4]. This enables to explain required interval of relaxation times, necessary for the synthesis of 1/f spectrum.

Time of recombination of charge carries through the g-r centres decrease in proportion to concentration of charge carries, thats means in proportion to current. The above presented view is in good aqreement with behaviour of spectra presented in Fig. 5. Using characteristics of burst noise [3] we explained observed dependence of 1/f-noise on current I_B.

REFERENCES

1. S.T.Hsu, Sol.-State Electronics. 13, 843 (1970).
2. S.T.Hsu, R.J.Whittier, and C.A.Mead, Sol.-State Electronics. 12, 867 (1969).
3. C.B.Cook, Jr. and A.J.Brodersen, Sol.-State Electronics, 14, 1237 (1971).
4. M.K.Sheikman, A.J.Shik, Fizika i tekhnika poluprovodnikov (in Russian), 10, 209 (1976)

REDUCED SHOT NOISE IN A QUASI-ONE-DIMENSIONAL CHANNEL

F. Liefrink, R.W. Stok, and J.I. Dijkhuis

Faculty of Physics and Astronomy, and Debye Research Institute, University of Utrecht, P.O. Box 80000, 3508 TA Utrecht, The Netherlands

M.J.M. de Jong, H. van Houten, and C.T. Foxon

Philips Research Laboratories, 5600 JA Eindhoven, The Netherlands

ABSTRACT

We present a study on shot noise in a quasi-one-dimensional channel much longer than the electron mean free path. We find that the shot-noise intensity is suppressed below the full shot-noise level and compare this result with recent theoretical calculations predicting reduced shot noise.

Shot noise in nanostructure devices has attracted much attention in the last few years.[1–7] A primary issue has been the suppression of shot noise in quantum point contacts, conductors much shorter than the electron mean free path. Such devices exhibit a quantization of the conductance related to the nearly unit transmission probability of electrons residing in a one-dimensional (1D) subband.[8] The zero-frequency, zero-temperature equivalent current-noise spectral density S_I of a two-terminal 1D conductor under conditions of a small voltage V between the two terminals, has been calculated to equal[1]

$$S_I = 2e\,|V|\,\frac{2e^2}{h}\sum_{n=1}^{\infty} T_n(1 - T_n), \tag{1}$$

with T_n the transmission probability for an electron in the nth 1D subband. One thus expects the shot noise intensity to vanish at the conductance plateaus, where all occupied subbands have $T_n \approx 1$. The only experimental indication for a reduction of shot noise in the ballistic transport regime to occur thus far may be the observation by Li *et al.*[9] of a white contribution to the noise spectral density, with an intensity that is reduced compared to $2eI$.

Recently, the attention has been turned to shot noise in conductors much longer than the electron mean free path. Calculations starting from the general equation (1) and based on results obtained from the random-matrix theory of quantum transport, have shown that the shot-noise power in a disordered phase-coherent conductor, much longer than the electron mean free path, is expected to be one-third of the classical value of a Poisson process.[10,11] It was pointed out that this reduction is a consequence of the presence of noiseless open quantum channels.

A similar prediction for the suppression of shot-noise was obtained by Nagaev[12]

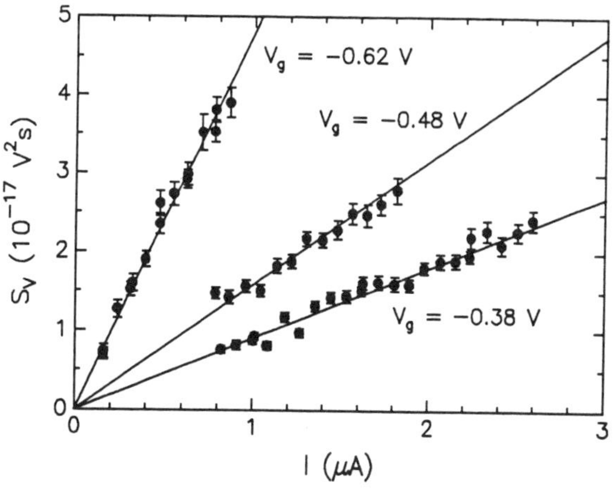

Figure 1: S_V vs I for three different values of V_g.

for metal contacts with a small electron mean free path. He used the Boltzmann-Langevin equation to determine the fluctuations of the electron distribution function and found also a one-third reduction of the shot-noise power. However, in contrast to the interpretation in Refs. 10 and 11, here the reduction was attributed to a nonequilibrium electron distribution in the center of the contact.

In this paper we present the first experimental evidence for the suppression of shot noise in a quasi-one-dimensional conductor much longer than the electron mean free path. In some range of parameters, indeed a reduction by a factor of 1/3 was found, however also other values emerged. Thus far, there is no satisfying explanation for this phenomenon.

The conductor is defined by means of a split-gate lateral depletion technique in the degenerate two-dimensional electron gas (2DEG) of a GaAs/AlGaAs heterostructure.[8] The length of the device is 16.7 μm, much longer than the electron mean free path $\ell_e \approx$ 2 μm at 4.2 K. The resistance R of the channel can be varied by controlling the electron density (2.6×10^{15} m^{-2} or less) and width (500 nm or less) of the channel by means of a gate voltage V_g applied to the split-gate on top of the structure (see lower part of Fig. 2). All experiments discussed in this paper were carried out in a four-terminal configuration, in order to eliminate spurious contributions to the noise produced at the current contacts and voltage probes. The noise spectral density was measured in the frequency range of 250 mHz to 100 kHz.

At frequencies below 1 kHz the noise intensity is completely dominated by a $1/f$ contribution, which will be discussed elsewhere. At higher frequencies a white-noise spectrum appears with an intensity dependent of I. In order to demonstrate that these fluctuations indeed can be associated to shot noise, we measured the intensity of the white-noise level of the equivalent voltage-noise spectral density $S_V \equiv R^2 S_I$ as a function of the bias current I. In Fig. 1 we have plotted S_V vs I for three different

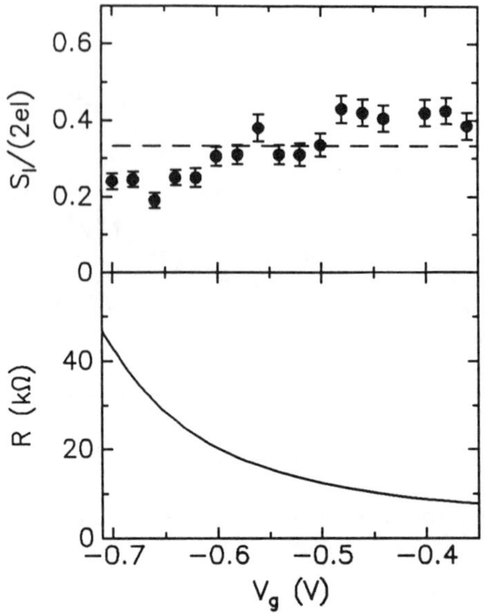

Figure 2: Upper part: $S_I/2eI$ vs V_g. The dashed line indicates the theoretical prediction of a one-third reduction of the shot-noise level. Lower part: R vs V_g in the same gate-voltage range.

values of V_g for $T = 4.2$ K, showing a clear proportionality of S_V with I. We checked that R is ohmic and therefore conclude that the white-noise level originates from shot noise. Indeed, the constant γ defined by the relation $S_I = 2\gamma eI$ (with e the elementary charge) is reduced below the classical value of 1 but depends on V_g. In the upper part of Fig. 2 we present $S_I/2eI$ vs V_g in the gate-voltage range of -0.35 to -0.7 V. From this figure, γ appears to equal approximately 0.4 at low $|V_g|$ down to 0.2 at high $|V_g|$. We conclude that the shot noise in our conductor is suppressed below the full shot noise level, but that the reduction factor deviates from the theoretically predicted value of 1/3, indicated by the dashed line in Fig. 2.

We also studied the temperature dependence of the shot-noise level (not shown here) and found, that γ strongly decreases at higher temperatures ($T \gtrsim 10$ K) for all gate voltages examined. Below 10 K the shot-noise level is approximately independent of temperature. A similar reduction of γ is predicted in Refs. 10 and 11 and attributed to a decrease of the ineleastic scattering length with temperature.

In conclusion, we reported the first experimental results on the suppression of shot noise in a quasi-one-dimensional channel, much longer than the electron mean free path. This reduction is in accordance with recent theoretical calculations concerning reduced shot-noise. However, these theories do not account for the details of our experimental observations, and the origin of the observations remains therefore unclear.

1. G.B. Lesovik, JETP Lett. **49**, 592 (1989).
2. M. Büttiker, Phys. Rev. Lett. **65**, 2901 (1990).
3. R. Landauer and Th. Martin, Physica **175**, 167 (1991).
4. M. Büttiker, Physica **175**, 199 (1991).
5. C.W.J. Beenakker and H. van Houten, Phys. Rev. B **43**, 12066 (1991).
6. Th. Martin and R. Landauer, Phys. Rev. B **45**, 1742 (1992).
7. M. Büttiker, Phys. Rev. B **45**, 3807 (1992).
8. For a review of quantum transport in semiconductor nanostructures, see: C.W.J. Beenakker and H. van Houten, Solid State Physics **44**, 1 (1991).
9. Y.P. Li, D.C. Tsui, J.J. Heremans, J.A. Simmons, and G.W. Weimann, Appl. Phys. Lett. **57**, 774 (1990).
10. C.W.J. Beenakker and M. Büttiker, Phys. Rev. B. **45**, 1889 (1992).
11. M.J.M. de Jong and C.W.J. Beenakker, Phys. Rev. B **46**, 13400 (1992).
12. K.E. Nagaev, Phys. Lett. A **169**, 103 (1992).

1/f NOISE IN A QUASI-ONE-DIMENSIONAL CHANNEL

F. Liefrink, A. van Die, R.W. Stok, and J.I. Dijkhuis

Faculty of Physics and Astronomy, and Debye Research Institute, University of Utrecht, P.O. Box 80000, 3508 TA Utrecht, The Netherlands

A.A.M. Staring, H. van Houten, and C.T. Foxon

Philips Research Laboratories, 5600 JA Eindhoven, The Netherlands

ABSTRACT

We perform a noise-spectroscopy study on electron transport through a narrow split-gate GaAs/AlGaAs channel. The noise is found to be dominated by a $1/f$ contribution, which is interpreted to originate from time-dependent fluctuations in the electrostatic potential in the channel due to electron trapping and detrapping processes with a broad distribution of time constants.

The improvements in the techniques of lithography have driven down the size of semiconductor devices to submicron length scales. Although the time-averaged electron transport properties of such devices has been investigated thoroughly,[1] fluctuations have received less attention. We present a noise-spectroscopy study in a quasi-one-dimensional channel, that reveals new phenomena in electron transport, which can not be obtained otherwise. We demonstrate that the the noise spectrum is dominated by a $1/f$ contribution and identify the mechanism. We calculate the temperature dependence based on a model developed for the analysis of our earlier measurements in quantum point contacts[2] and find perfect agreement.

We used a GaAs/$Al_{0.33}Ga_{0.67}As$ heterostructure, supplied by Philips Research Laboratories. The structure consists of a GaAs substrate on top of which a thick layer of high purity GaAs has been deposited, followed by a 20-nm undoped AlGaAs layer, a 40-nm silicon doped AlGaAs layer, and a 20-nm GaAs cap layer. At the interface of GaAs and AlGaAs a two-dimensional electron gas (2DEG) is formed. The 2DEG mobility equals 23 m^2/Vs at 4.2 K. Its density amounts to 2.6×10^{15} m^{-2}, independent of temperature. A narrow channel of length $L = 16.7$ μm is defined by biasing two half gates, separated by a slit of 500 nm, located on top of the structure. The gate voltage V_g controls the channel width W, the electron density n, and consequently the channel resistance R.

The experiments were performed in a four-terminal configuration with the sample biased with a dc current I. The excess noise $S_V(f)$ was measured in the voltage $V \equiv IR$ across the channel. Generally $S_V(f)$ appeared to consist of a $1/f$ contribution, with an intensity that goes quadratically with I, and a white noise contribution, with an intensity linear in I. In Fig. 1 a typical noise spectrum is plotted. The solid line in the figure represents a fit to the sum of a $1/f$- and a frequency-independent contribution.

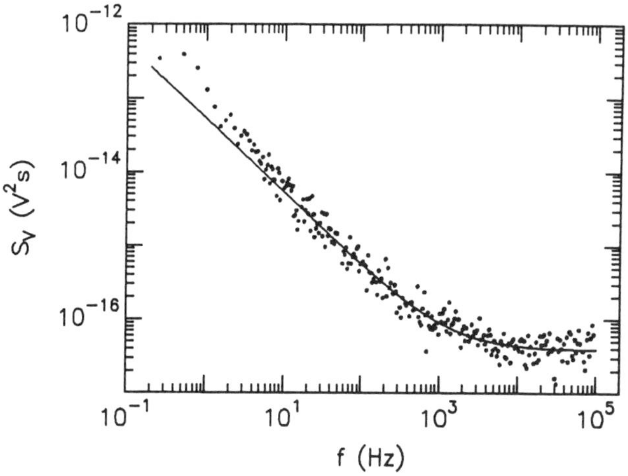

Figure 1: A typical example of the equivalent noise spectral density for T = 4.2 K, $V_g = -0.62$ V, and $I = 0.78$ μA. The solid curve represents a fit by a $1/f$ contribution combined with a white noise contribution.

We interpret the white noise part of the spectrum as an example for reduced shot noise and will discuss this elsewhere. We measured the temperature dependence of the relative $1/f$-noise intensity $S_V(f)/V^2$ at 1 Hz for three values of V_g. The results are plotted in Fig. 2. For low temperatures the noise intensity increases approximately linearly with temperature, but with a slope dependent on V_g. For $V_g = -0.62$ and -0.73 V, the experimental data start to deviate from the straight line at higher temperatures ($T \gtrsim$ 15 K). However, no significant deviation from linearity is observed for $V_g = -0.38$ V, at least in the temperature range examined.

In order to construct a tractable model to account for the experimental results, we assume that R obeys the Drude formula $R = (L/W)(1/ne\mu)$, and thus neglect contributions of the Sharvin resistance and weak localization. These can indeed be shown to give only very small corrections. Furthermore, we write for the reciprocal channel mobility $\mu^{-1} = \mu_0^{-1} + \alpha T$, with μ_0 the zero-temperature mobility and αT referring to the reduction of the mobility by acoustic-phonon scattering. We checked R and observed it to vary linearly with T. We note that $\alpha\mu_0$ varies roughly linearly with V_g, and that $\alpha\mu_0 T \ll 1$ for all temperatures and gate voltages examined.

We contend that the resistance noise is caused by fluctuations in the electrostatic potential ε_0 in the channel originating from traps that induce thermally activated trapping and detrapping processes of electrons. The noise can be quantitativily understood if a sufficiently broad range of activation energies of these traps is present near or in the channel. The mechanism is as follows. Whenever electron capture takes place, the electrostatic potential near the trap rises because of the Coulomb potential of the electron. As soon as the electron is released, the electrostatic potential is restored. Raising the electrostatic potential lowers n and consequently R. The connection between $S_V(f)/V^2$ and the time-dependent fluctuations $S_{\varepsilon_0}(f)$ in ε_0 is straightforwardly given by

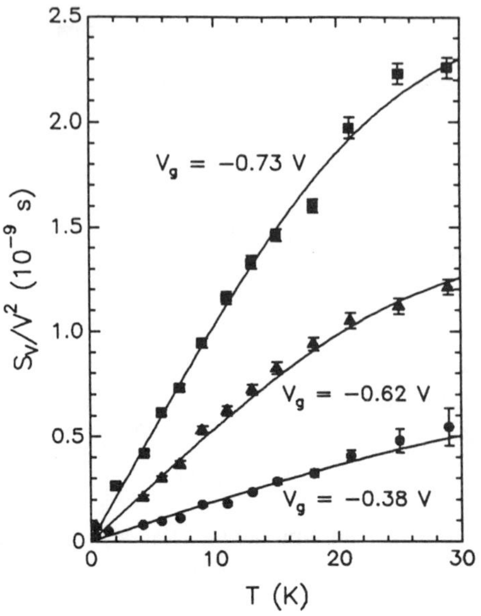

Figure 2: The $1/f$-noise spectral density at 1 Hz vs T for V_g = −0.38 V (circles), −0.62 V (triangles), and −0.73 V (squares). Solid curves are fits to the data by Eq. (2).

$$\frac{S_V(f)}{V^2} = \frac{1}{R^2}\left(\frac{\partial R}{\partial \varepsilon_0}\right)^2 S_{\varepsilon_0}(f). \tag{1}$$

In order to evaluate Eq. (1) we insert the expressions for R and μ as mentioned above, and use the definition of the Fermi energy ε_F. Following Ref. 3, we assume that μ_0 is proportional to $n^{3/2}$ and use the observation for our channel that $\alpha\mu_0$ is proportional to n. We then arrive at the result

$$\frac{S_V(f)}{V^2} = \left[1 - \exp\left(-\frac{nh^2}{4\pi mkT}\right)\right]^2 \left(\frac{2\pi m}{nh^2}\right)^2 \left(\frac{3\alpha\mu_0 T + 5}{\alpha\mu_0 T + 1}\right)^2 S_{\varepsilon_0}(f). \tag{2}$$

A remarkable result from the present analysis is that $S_V(f)/V^2$ is independent of both W and L. Furthermore, Eq. (2) shows that the $S_V(f)/V^2$ strongly increases with decreasing n.

At low temperatures ($kT < nh^2/4\pi m$), the measurements show that $S_V(f)/V^2$ is proportional to T. From Eq. (2) we then may conclude that $S_{\varepsilon_0}(f)$ must be linear in T. This linear increase with T can generally be produced by a set of electron traps with a sufficiently flat distribution $g(E_a)$ of activation energies E_a. If we denote the thermally activated time at a single site with $\tau = \tau_0 \exp(E_a/kT)$ ($1/\tau_0$ an attempt frequency), it can be shown,[4] that

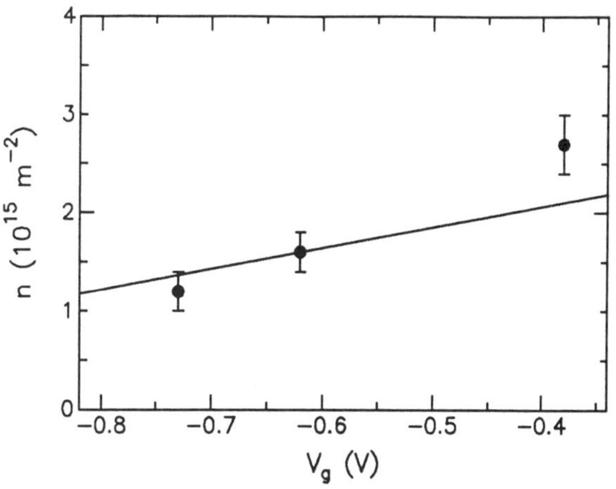

Figure 3: The electron density in the channel vs V_g. Data points indicate results from measurements of $S_V(f)/V^2$ vs T. The solid line represents the result as obtained from high magnetic-field measurements of R.

$$S_{\varepsilon_0}(f) \propto \int_0^\infty \frac{\tau}{1+4\pi^2 f^2 \tau^2} g(E_a) dE_a \propto \frac{kT}{f}. \tag{3}$$

The solid curves in Fig. 2 are fits by Eq. (2) to the data, taking values of $\alpha\mu_0$ as obtained from the temperature dependence of R and a linear temperature dependence of S_{ε_0}. From the fits we extract that $S_{\varepsilon_0}(f) = (3 \times 10^{-10}\ \text{meV}^2\text{K}^{-1}) \times T/f$, independent of V_g within the experimental error, and we find values for n as plotted vs V_g in Fig. 3. In this figure also the V_g dependence of n is shown as obtained from high magnetic-field measurements of R. Clearly, the results for the two routes to measure n vs V_g agree reasonably well. This corroborates the validity of our model for the $1/f$ noise in a narrow GaAs/AlGaAs channel.

In conclusion, we have shown that the resistance of a narrow GaAs/AlGaAs channel is sensitive to fluctuations in the electrostatic potential. The fluctuations were found to be $1/f$ like and interpreted to originate from thermally activated processes having a flat distribution of activation energies. From measurements of the temperature dependence of the $1/f$-noise intensity, the electron density is observed to increase linearly with gate voltage, consistent with high magnetic-field measurements of the resistance.

1. For a review of quantum transport in semiconductor nanostructures, see: C.W.J. Beenakker and H. van Houten, Solid State Physics **44**, 1 (1991).
2. C. Dekker, A.J. Scholten, F. Liefrink, R. Eppenga, H. van Houten and C.T. Foxon, Phys. Rev. Lett. **66**, 2148 (1991).
3. H.Z. Zheng, H.P. Wei, D.C. Tsui, and G. Weimann, Phys. Rev. B **34**, 5635 (1986).
4. P. Dutta and P.M. Horn, Rev. Mod. Phys. **53**, 497 (1981).

Noise Modeling for *W*-Band Pseudomorphic *InGaAs* HEMT's

K.-W. Liu, A. F. M. Anwar and Roger D. Caroll*
Electrical and Systems Engineering Department
The University of Connecticut
Storrs, CT 06269-3157
*United Technologies Research Center
East Hartford, CT

ABSTRACT

A model to determine the noise properties for *AlGaAs/InGaAs/GaAs* Pseudomorphic HEMTs is presented. The analysis is based on a self-consistent solution of Schrödinger and Poisson's equations. Pseudomorphic HEMTs (PHEMTs) have a lower noise figure and temperature as compared to normal *AlGaAs/GaAs* HEMTs. Minimum noise temperature in pseudomorphic HEMTs decreases with increasing quantum well (QW) width. Noise temperature in general increases with increasing gate length.

Introduction

In this paper a noise model is presented which is based on the self-consistent solution of Schrödinger and Poisson's equations. A velocity-electric field characteristic[1,2] is used that approximates the experimental curve more closely and makes the calculations analytic in nature. The calculated results are compared with experimental data[3,4] which shows excellent agreement.

Model

The self-consistent noise model presented by Anwar et al.[1] is extended to calculate noise in PHEMTs. Schrödinger and Poisson's equations are solved self-consistently to calculate the average distance of the electron cloud x_{av}, from the first heterointerface and the position of the Fermi level E_F with respect to the tip of the conduction band[1]. Using the functional relationship of x_{av} and E_F with two-dimensional electron gas (2DEG) concentration n_s, the d.c. small-signal parameters are evaluated and used eventually to calculate noise figure and temperature. By accounting for different noise sources[2] and matching the optimized external source impedance to the transistor, the minimum noise figure F_{min}, minimum noise temperature T_{min} and noise conductance g_n are calculated as[2] :

$$F_{min} = 1 + 2 \cdot g_n(R_c + \sqrt{R_c^2 + \frac{r_n}{g_n}}) \tag{1}$$

$$T_{min} = 2 \cdot T \cdot g_n \cdot (R_c + R_{s,opt}) \qquad (K) \tag{2}$$

where

$$g_n = g_m \cdot (\frac{f}{f_T})^2 \cdot (R + P - 2C\sqrt{PR}) \tag{3}$$

$$R_c = R_s + R_g + R_i \tag{4}$$

where r_n is the noise resistance, R_c is the correlation resistance, R_s and R_g are the source and drain resistances and $R_i = \frac{L_g}{v_s C_{gs}}$, is the gate charging resistance. L_g represents the length of the gate and v_s is carrier saturation velocity in the conducting channel. P, R and C represents the noise coefficients and $R_{s,opt}$ is the optimal external source resistance defined by Anwar et al.[1]. g_m and C_{gs} are the device transconductance and gate capacitance, respectively. T is the operating temperature in ^0K. In some cases, the external source impedance is not matched to the optimal source impedance $Z_{s,opt}$. This mismatch results in a modified expression for noise temperature :

$$T_n = T_{min} + \frac{T \cdot g_n}{R_{sou}} \cdot [(R_{sou} - R_{s,opt})^2 + (X_{sou} - X_{s,opt})^2] \tag{5}$$

where $Z_{sou} = R_{sou} + jX_{sou}$ is the external source impedance presented to the transistor.

Results and Discussion

In Fig.1, the calculated minimum noise figure, F_{min} (dB), is plotted as a function of drain-source current for a 0.1×200 μm^2 pseudomorphic $InGaAs$ HEMT with V_{ds}=1.06 V at 94 GHz. On the same plot experimental data[2] are shown and reflects a good fit. Using this noise model, the quantum well (QW) width and gate length dependance on the noise properties of pseudomorphic $InGaAs$ HEMT operating at W-band is studied.

In Fig.2, minimum noise figure F_{min} is plotted as a function of frequnecy for a 0.35×200 μm^2 PHEMT at 300K with V_d = 2.0 V and I_{ds} = 10 mA. On the same graph, the calculated results are compared with experimental data (diamond)[3] and show an excellent agreement.

In Fig.3, The calculated minimum noise temperature is compared with room temperature experimental data for 0.15×100 μm^2 $AlGaAs/InGaAs/GaAs$ PHEMT[4] at 30 GHz. v_s and μ_0 are assumed to be 1.2×10^7 cm/sec and 9,000 $cm^2/V-sec$ for PHEMTs at room temperature. As observed the agreement is excellent. We believe that the use of a self-consistent calculation to determine the effective channel

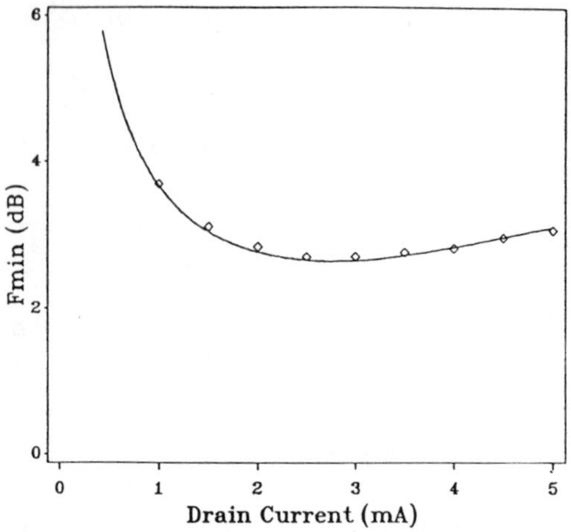

Figure 1: Minimum noise figure, F_{min} (dB) is plotted as a function of saturation drain current and compared to the experimental data (diamond)[2].

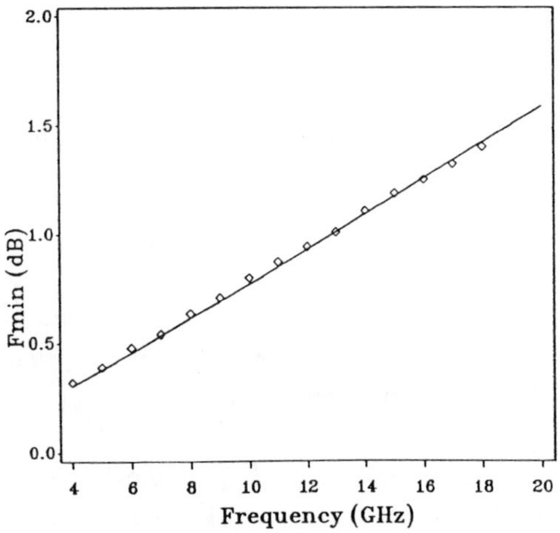

Figure 2: Minimum noise figure F_{min} is plotted as a function of frequency for a $0.35\times200\ \mu m^2$ PHEMT at 300K. The diamonds represent the experimental data[3].

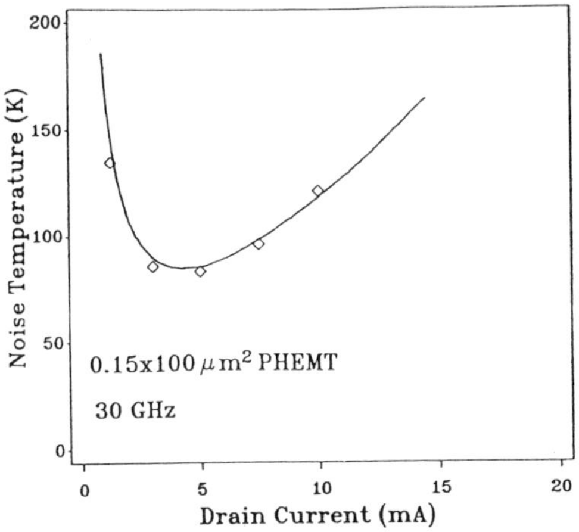

Figure 3: Noise temperature is plotted as a function of drain current for PHEMTs (solid line). The experimental data is shown (diamond)[4].

width as a function of the 2DEG concentration and velocity-electric field characteristic that approximates the experimental results more closely enabled such an agreement.

Conclusion

A self-consistent model is presented to model noise performance for pseudomorphic *InGaAs* HEMTs. An improved velocity-electric field (v_d-$\mathcal{E}$) characteristic is used to make the calculation analytic in nature. The calculated F_{min} and T_{min} are compared to experimental data[3,4] and show excellent agreement.

Reference

1. A. F. M. Anwar and Kuo-Wei Liu, IEEE Trans. Electron Device, June, 1993.

2. K. L. Tan, R. M. Dia, Dwight C. Streit, A. C. Han, Tien Q. Trinh, J. R. Velebir, P. H. Liu, Tzuenshyan Lin, H. C. Yen, M. Sholley, and L. Shaw, IEEE Electron Device Letters, Vol. 11, p. 303, 1990.

3. Robert Plana, Laurent Escotte, Olivier Llopis, Hicham Amine, Thierry Parra, Michel Gayral and Jacques Graffeuil, IEEE Trans. Microwave Theory and Technol., vol. MTT-40, no. 5, pp. 852, May 1993.

4. Kazukiyo Joshin, Yutaka Mimiho, Souji Ohmura and Yasutake Hiracchi, IEEE Trans. Electron Devices, vol. ED-39, no. 3, pp. 515-519, 1992.

ON NOISE SUPPRESSION IN DOUBLE-BARRIER DEVICES AND IN QUANTUM POINT CONTACTS

M. Macucci and B. Pellegrini
Dipartimento di Ingegneria dell'Informazione: Elettronica, Informatica, Telecomunicazioni Università degli Studi di Pisa Via Diotisalvi, 2, I-56126 Pisa, Italy

ABSTRACT

We investigate the origin of shot noise suppression in Double-Barrier Resonant-Tunneling Structures and in Quantum Point Contacts. We find out that the suppression factor for DBRTSs can be derived, obtaining the same result, from different approaches. We also show how the current pulse representation can be applied to an intuitive derivation of shot noise in QPCs

INTRODUCTION

Widespread interest has been recently received by the problem of shot noise suppression in Double-Barrier Resonant-Tunneling Structures (DBRTSs) and in Quantum Point Contacts (QPCs). This is the consequence of experimental findings[1,2] exhibiting a significant reduction in the noise power spectral density in such devices with respect to the full shot noise level. Several explanations of this phenomenon have been proposed, based on the role played by the exclusion principle[3] as a consequence of the fermionic nature of the electrons, on the statistics of the charge stored between the barriers[4] and on the scattering matrix approach to charge transport[5].

For quantum point contacts an expression of the noise spectral density at zero frequency has been obtained by Lesovik[6] and by Büttiker[7] using a second quantized representation of the current operator. Beenakker and Van Houten[8] have obtained a semi-classical derivation of the same result.

NOISE IN DOUBLE-BARRIER DEVICES

In this paper we discuss the shot noise reduction in DBRTSs and QPCs, and investigate the differences in the origin of such reduction.

We first examine the problem of excess noise in DBRTSs: the theories in Ref. 4 and 5 produce results with a different frequency dependence for the shot noise suppression factor η, but agree on the value of η for zero frequency:

$$\eta(0) = 1 - \frac{1}{2}T_{res}, \tag{1}$$

where T_{res} is the transmission coefficient of the double-barrier structure at resonance. T_{res} can be expressed as a function of the transmission and reflection coefficients T_1, T_2 of the single barriers. By summing the amplitude scattering series[9], including the direct term and all the possible multiply reflected ones, we obtain, at resonance, the expression:

$$T_{res} = \left(\frac{\sqrt{T_1}\sqrt{T_2}}{1-\sqrt{1-T_1}\sqrt{1-T_2}}\right)^2 \tag{2}$$

For T_1, T_2 much smaller than unity, as in the practical examples of resonant tunneling devices, we have, as a function of the ratio α between T_1 and T_2,

$$T_{res} \approx \frac{4\alpha}{(1+\alpha)^2}, \tag{3}$$

therefore

$$\eta = \frac{1+\alpha^2}{(1+\alpha)^2}. \tag{4}$$

Looking at the problem from a macroscopic, circuital point of view, we can consider the case of two distinct tunneling junctions connected by a wire: each of them can be represented with a current noise generator with power spectral density $S_I = 2qI$ in parallel with a resistor (R_1, R_2)whose value is inversely proportional, through the constant β, to the transmission coefficient of the corresponding barrier (see Fig. 1). I is the bias current and q is the electron charge. If the average time spent by an electron between the two junctions is much longer than the observation time, the two generators can be considered as independent. The contribution to the total noise current from the generator corresponding to the left barrier is

$$S_1 = \left(\frac{R_1}{R_1+R_2}\right)^2 S_I, \tag{5}$$

while S_2, for the other generator, is given by the same expression with R_1 and R_2 interchanged. Being the generators uncorrelated, the power spectral densities are additive, therefore, once again,

$$\eta = \frac{S_1+S_2}{S_I} = \left(\frac{\frac{\beta}{T_2}}{\frac{\beta}{T_1}+\frac{\beta}{T_2}}\right)^2 + \left(\frac{\frac{\beta}{T_1}}{\frac{\beta}{T_1}+\frac{\beta}{T_2}}\right)^2 = \frac{1+\alpha^2}{(1+\alpha)^2}. \tag{6}$$

The methods in Ref. 3-5 do not take into account the displacement current, but only the conduction current. The approach in Ref. 10 represents an improvement from this point of view, including also the contribution of the displacement current. It also allows an explanation of the increase with respect to the full shot noise level observed in some cases. This is explained as a consequence of electrons going back into the emitter contact. In the hypothesis of completely uncorrelated input and output pulses and of no re-emission back into the left contact, the shot noise reduction factor is given[10] by:

$$\eta = 1 + 2\phi(\phi - 1), \tag{7}$$

where ϕ is the ratio between the charge carried by one of the two pulses and the electron charge. The derivation of the value of this ratio is, in general, non-trivial, but in the particular case of symmetrical barriers $\phi = 0.5$ from evident symmetry considerations. Therefore also with this approach the shot noise is reduced by one half for symmetrical barriers.

From the above discussion we obtain quite singular a result: very different approaches, starting from apparently different hypotheses, lead to exactly identical results. The calculation in Ref. 3, for example, does not make any, at least explicit, assumption about pulse independence, while the derivation in Eq.(5-6) relies on such independence, but does not require require the effect of the exclusion principle. Still, they yield the very same result for the shot noise suppression factor.

Let us apply the current pulse representation in a more detailed example. If we consider a random process, representative of the current through the DBRTS, whose realizations are made up of pairs of current pulses corresponding to the crossing of the two barriers by an electron, we can compute its autocorrelation function and, from it, the associated power spectral density. If the escape probability is constant in time, as it happens for the emission probability out of a quantum well, the time the electron spends between the barriers is well represented by a random variable with exponential distribution and mean value σ. The main constraint that must be imposed for this process to be representative of the current in a DBRTS is that the time integral of the two current pulses corresponding to the same electron (one pulse for entering the well region and one for leaving it) must equal q, the electron charge. In Fig. 2 we report the autocorrelation function for the current fluctuations, in the hypothesis of rectangular pulses with an identical duration (symmetrical barrier) of 2×10^{-13} s separated by an exponentially distributed time with average value $\sigma = 4.2 \times 10^{-12}$ s. The time constants we have chosen differ only by an order of magnitude for graphic representation convenience.

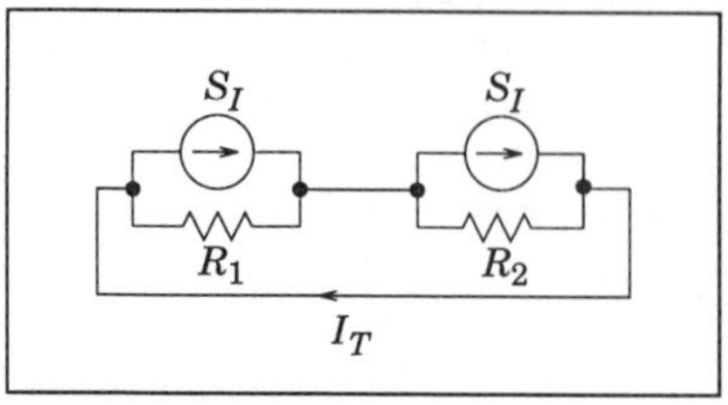

Fig. 1 Equivalent noise circuit of two tunnel junctions in series.

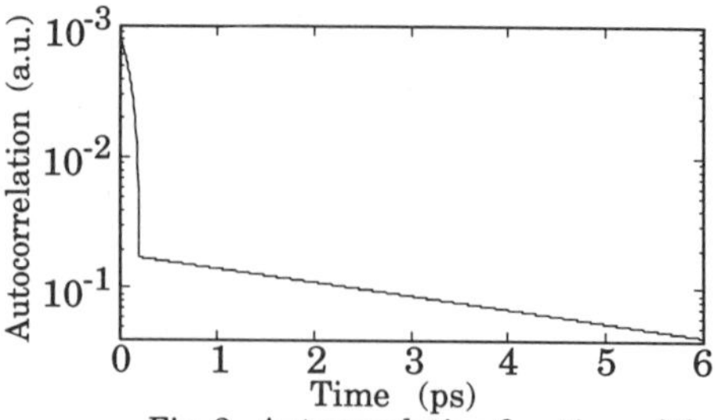

Fig. 2 Autocorrelation function of the current pulses in a DBRTS.

There are a linear part (which appears logarithmic because of the logarithmic scale on the ordinate axis), corresponding to the autocorrelation of each single pulse and a long exponential tail due to the correlations between the input and output pulses. The power spectral density at zero frequency is given by four times the integral of the autocorrelation function[11]. It is easy to show that the contribution from the linear part of the autocorrelation function yields exactly one half of the full shot noise level, while the exponential part would yield the other half. Therefore, if the observation time is much longer than σ, we should recover the full shot noise level, notwithstanding the fact that the separation between the two pulses due to the same electron is a random variable. If, instead, there were some mechanism, such as slow time constants associated with the capacitance and the differential resistance of the tunnel junctions, capable of extending the dwell time of the electron in the DBRTS, the measured noise would be one half of the full shot. This is true as long as the "correlation time" between the input and output pulses is longer than the observation time, i.e., in the case of an FFT signal analyzer, the length of each time record.

Unfortunately, Ref. 1 is the only experimental reference on this topic: more data are needed in order to decide which of the hypotheses we have discussed holds true in DBRTSs.

NOISE IN QUANTUM POINT CONTACTS

In quantum point contacts the number of current carrying states is limited (the current itself is limited by the number of available conduction channels), thus the exclusion principle and the fact that electrons are fermions play an important role, leading to a suppression of the shot noise which can be total.

We shall now show how the current pulse representation allows a simple and intuitive explanation of this phenomenon. A similar demonstration has been given, with orthogonal wave-packets, by Martin and Landauer[12].

When all the current carrying states are occupied, a new electron is prevented from entering the constriction: for transmission $T = 1$, the current carried by each state is represented by an orderly sequence of contiguous pulses, therefore there are no fluctuations and noise is totally suppressed. If there is a region with transmission coefficient less than unity, each electron has a probability T of traversing it and a probability $1 - T$ of being reflected. The current is in this case represented by a process randomly high or low with probabilities T and $1 - T$ during each time slot corresponding to the transit of one electron. Let us evaluate the autocorrelation $r(\tau)$ of the current fluctuations:

$$r(\tau) = \langle \delta i(t) \delta i(t + \tau) \rangle = \langle i(t) i(t + \tau) \rangle - \langle i(t) \rangle^2. \tag{8}$$

If we define the length of each current pulse as $t_p = q/I_p$, where I_p is the instantaneous current associated with the pulse, for $0 < \tau < t_p$ we have

$$r(\tau) = I_p^2 T \left(1 - \frac{\tau}{t_p}\right) + \frac{\tau}{t_p} I_p^2 T^2 - I_p^2 T^2 = I_p^2 T(1-T)\left(1 - \frac{\tau}{t_p}\right), \tag{9}$$

where the first term in the right hand side is due to the overlap of the same pulse, the second term comes from the overlap between different pulses and the third corresponds to the square of the average current. Beyond t_p the autocorrelation vanishes because different pulses are independent. Therefore $r(\tau)$ has a triangular shape and the power spectral density is given by four times its area,

$$S_I(O) = 4\frac{1}{2} I_p^2 T(1-T)\frac{q}{I_p} = 2qI(1-T), \tag{10}$$

in agreement with the results derived in Ref. 6, 7, 12. Within this picture it is also easy to understand why, for very small values of T, the full shot noise value is recovered: occurring at a low rate, the current pulses are not significantly affected by the non-overlap rule, and the Poisson process limit is reached.

REFERENCES

1. Y. P. Li, A. Zaslavsky, D. C. Tsui, M. Santos, and M. Shayegan, Phys. Rev. B **41**, 8388 (1990).
2. Y. P. Li, D. C. Tsui, J. J. Heremans, J. A. Simmons, G. W. Weimann, Appl. Phys. Lett. **57**, 774 (1990).
3. L. Y. Cheng and C. S. Ting, Phys. Rev. B **43**, 4534 (1991).
4. L. Y. Cheng and C. S. Ting, Phys. Rev. B **46**, 4714 (1992).
5. M. Büttiker, Phys. Rev. B **45**, 3807 (1992).
6. G. B. Lesovik, Pis'ma Zh. Eksp. Teor. Fiz. **49**, 515 (1989) [JETP Lett. **49**, 594 (1989)].
7. M. Büttiker, Phys. Rev. Lett. **65**, 2901 (1990).
8. C. W. J. Beenakker and H. van Houten, Phys. Rev. B **43**, 12066 (1991).
9. M. Johnson and Anna Grincwajg, Appl. Phys. Lett. **51**, 1729 (1987).
10. B. Pellegrini and M. Macucci, "Splitting of Electron Current Pulses and Noise in Double-Barrier Heterostructures," (paper submitted to this same conference).
11. A. van der Ziel, "Noise in Measurements," (Wiley, New York, 1976) pp.30-31.
12. Th. Martin and R. Landauer, Phys. Rev. B **45**, 1742 (1992).

LOW-FREQUENCY NOISE SOURCES IN POLYSILICON EMITTER BIPOLAR TRANSISTORS : INFLUENCE OF HOT-ELECTRON-INDUCED DEGRADATION

A Mounib*, F. Balestra, N. Mathieu, J. Brini, G. Ghibaudo, A. Chovet,
Laboratoire de Physique des Composants à Semiconducteurs (URA-CNRS)
ENSERG/INPG, 23 rue des Martyrs, BP 257, 38016 Grenoble, France

*On leave from Institut d'Electronique, Université de Blida, Algeria

A. Chantre, A. Nouailhat
CNET/CNS, BP 98, 38243 Meylan, France

ABSTRACT

The noise properties of polysilicon emitter bipolar transistors are studied. The influences of the various chemical treatments and annealing temperatures, prior and after polysilicon deposition, on the noise magnitude are shown. The impact of hot-electron-induced degradation and post-stress recovery on the base and collector current fluctuations are also investigated in order to determine the main noise sources of these devices.

INTRODUCTION

A number of papers have been published on low-frequency noise in bipolar junction transistors (BJT's). The main advantages of these devices are associated with their high speed operation and low noise properties. Up to now, the 1/f noise in bipolar transistors has been discussed in terms of mobility fluctuations [1,2,3] and also in terms of carrier number fluctuations [2,3]. Moreover, there is no agreement on the location of the noise sources [1-6]. In particular, polysilicon emitter BJT's, which have a very specific structure, seem to present a different noise behavior from the conventional bipolar transistors [5-7]. Indeed, the noise in these structures has been mainly attributed to the fluctuations associated with carrier interactions with the silicon/polysilicon interface [5-7]. On the other hand, aging experiments on conventional BJT's have shown that interface states created after degradation at the $Si\text{-}SiO_2$ interfaces can play an important role on the noise magnitude [8].

This paper deals with CMOS compatible self-aligned etched-polysilicon emitter BJT's for BiCMOS technology (NPN Devices) [10]. We show the dependences of the low frequency (1/f) collector and base current spectral densities, obtained for different technological variations and various geometries. In addition, the influence on the noise properties of a reverse bias of the emitter-base junction, which induces a significant degradation of the current-voltage characteristics, is underlined. The effect of a forward bias recovery on the gain and the noise magnitude of the BJT's is also analysed. These various investigations allow us to determine the physical nature and the location of the noise sources.

RESULTS AND DISCUSSION

The influence of variations in the technological process has been studied for different chemical treatments (RCA, HF) given to the silicon surface prior to polysilicon deposition, and annealing temperatures after polysilicon deposition (1030° C - 1100° C). The structures with an HF etch have a very thin and non-uniform oxide layer averaging $\simeq$ 0.4 nm at the interface, which may be discontinuous. The RCA clean has been observed to produce 1-1.4 nm of relatively uniform oxide layer at the interface [9]. The RTA annealing leads to the break-up of the interfacial layer, this effect occuring more rapidly at higher temperatures [9]. These technological treatments have been shown to have a significant impact on the gain of the BJT's. For these different treatments, we have studied the variations of the base SI_b and collector SI_c current spectral densities as a function of the base I_b and collector I_c current respectively, for a large emitter surface.

In figure 1 are shown the dependences of SI_b as a function of the base current for the same device structure (1000x10 μm^2) with various technological processes. A RCA clean with a RTA at 1060° C (type 1), for which an important interfacial oxide is expected, and an HF etch with a RTA at 1100° C (type 2), for which no oxide layer is expected, are used. For these transistors, the same dependences for SI_b as I_b for low current and I_b^2 for strong current appear. However, the noise is between 0.5 and 1 order of magnitude lower for the type 2 BJT's whatever the base current is. On the other hand, the same collector current spectral density SI_c, with a variation as I_c, is observed at low I_c for both devices (Figure 2). However, for strong collector current, a difference is obtained, with a variation as $I_c^{3.5}$ for the transistor of type 1 and a dependence as I_c^2 for the transistor of type 2 (Figure 2).

These results point out that the influence of the interfacial oxide treatment is higher for the base current noise. They can be correlated with the strong influence of the polysilicon/Si interface which is observed for the base current, and emphasize in particular the fundamental difference for the transport of electrons and holes accross this oxide layer.

For small geometry BJT's (2x1 μm^2 or 5x2 μm^2), significant differences compared with large surface have been obtained. Variations as $I_b^{1.5}$ and as $I_c^{1.5}$ have been observed whatever the current is in the device, for the base and collector current spectral density respectively (see for instance the initial curve of figure 5). Nevertheless, as for large geometry BJT's, the influence of the interfacial oxide is higher for the base current noise.

The influence of a reverse bias in the emitter-base junction on the noise behavior, which has been shown to induce a significant degradation of the gain of polysilicon emitter BJT's [11], is analysed in detail. Indeed, hot-carriers are created at the periphery of the emitter-base junction, where a large electric field exists due to the presence of heavy dopings. These hot-carriers have a strong impact on the creation of interface states at the $Si\text{-}SiO_2$ interface and on the trapping in this oxide. An important aging can result from these effects, which is underlined by an increase of the base current of the devices. Therefore, it is very useful to study the influence of these degradations on the noise behavior of the BJT's.

A reverse bias in the emitter-base junction has been applied for a sufficient time in order to observe a significant degradation of the base current at low V_{be}. After this aging experiment, the collector current spectral density is unchanged at low current, and is slightly modified at high current (Figure 3). On the other hand, unlike SI_c, a strong variation ($\simeq$2 decades) appears in the base current noise whatever I_b is (Figure 4). This very interesting result, obtained whatever the geometry is, shows that the defects created at the $Si\text{-}SiO_2$ interface at the periphery of the emitter-base junction have a fundamental impact on the fluctuations in the base current and a very low influence on the fluctuations of the collector current.

The noise in the base current can therefore primarily be attributed to dynamic exchanges of carriers with the $Si\text{-}SiO_2$ interfaces at the periphery of the emitter-base junction, and with the interfacial oxide layer at the silicon/polysilicon interface which was also shown to be important. Nevertheless, the noise in the collector current seems, at low I_c, to be independent of the $Si\text{-}SiO_2$ interface degradation and of the oxide layer at the polysilicon/Si interface. Therefore, at low I_c, it can be concluded that this collector current noise can be mainly attributed to fluctuations in carrier mobility. On the other hand, at high I_c, the influence of the interfacial oxide and of the degradation of the periphery of the emitter-base junction shows that an additionnal mechanism probably associated with carrier number fluctuations has to be taken into account.

The effect of a forward bias on the base current and noise behaviors after a reverse bias aging is also analysed. After forward biasing the emitter-base junction for a sufficient time, the forward and reverse base currents are slightly reduced and increased, respectively. Moreover, the base current noise is strongly lowered after this experiment (Figure 5, for a 5x2 μm^2 device). These phenomena are attributed to (1) the injection of holes, during degradation at high reverse V_{be}, in the oxide at the periphery of emitter-base junction, and (2) the partial detrapping of these holes by injection of electrons in this oxide during a conventional operation of the BJT's with a forward V_{be}. These results underline the impact of a dynamic operation for which both forward and reverse biases are applied.

CONCLUSION

The noise properties of polysilicon emitter bipolar transistors have been studied. The influences of the various chemical treatments and annealing temperatures, which are used in order to control the interfacial oxide prior and after polysilicon deposition, on the noise magnitude have been shown. The defects created at the Si-SiO_2 interface at the periphery of the emitter-base junction after aging have been observed to affect substantially the base current noise and to have a low influence on the collector current noise of the devices. The main source of base current fluctuations can be attributed to dynamic exchanges of carriers with the Si-SiO_2 interface at the periphery of the emitter-base junction, the effect of the interfacial oxide layer being less significant. The main source of collector current fluctuations can be attributed to carrier mobility at low current. At high current, an additional mechanism associated with carrier number fluctuations due to the presence of the oxide layers seems also to be important.

REFERENCES

1. T.G.M. Kleinpenning, 10th Int. Conf. on Noise in Physical Systems, 1989, Budapest, Hungary, Proceedings p. 443.
2. A. Van der Ziel et al, IEEE Trans. Electron Dev., ED-33, p. 1371 (1986).
3. A. H. Pawlikiewicz and A. Van der Ziel, IEEE Trans. Electron Dev., ED-34, p. 2009 (1987).
4. T.G.M. Kleinpening, IEEE Trans. Electron Dev., ED-39, p. 1501 (1992).
5. T.L. Crandell and T.M. Chen, 11th Int. Conf. Noise in Physical Systems and 1/f fluctuations, 1991, Kyoto, Japan, Proceedings p. 209.
6. W.S. Lau et al, Jpn. J. Appl. Phys., 31, p. L 1021 (1992).
7. N. Siabi-Shahrivar et al, ESSDERC'90, Nottingham, U.K., 1990, Proceedings p. 341.
8. C. J. Sun et al, IEEE Trans. Electron Dev., ED-39, p. 2178 (1992).
9. P. Ashburn, Design and realization of bipolar transistors, John Wiley & Sons (1988).
10. A. Nouailhat et al, Electron. Lett., 24, p. 1581 (1988).
11. J.D. Burnett and C. Hu, IEEE Trans. Electron Dev., ED-35, p. 2238 (1988).

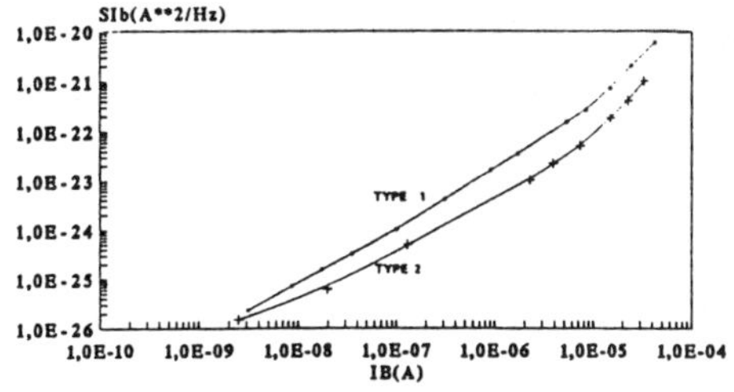

Fig. 1 : Base current spectral density versus I_b for various technological processes (type 1 : RCA clean, RTA 1060° C; type 2 : HF etch, RTA 1100° C). f=10 Hz, V_{ce} = 2V.

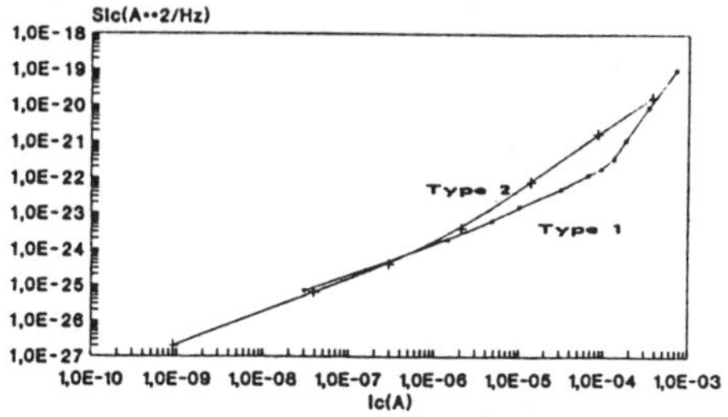

Fig. 2 : Collector current spectral density versus I_c for various technological processes (type 1 : RCA clean, RTA 1060° C; type 2 : HF etch, RTA 1100° C). f=10 Hz, V_{ce} = 2V.

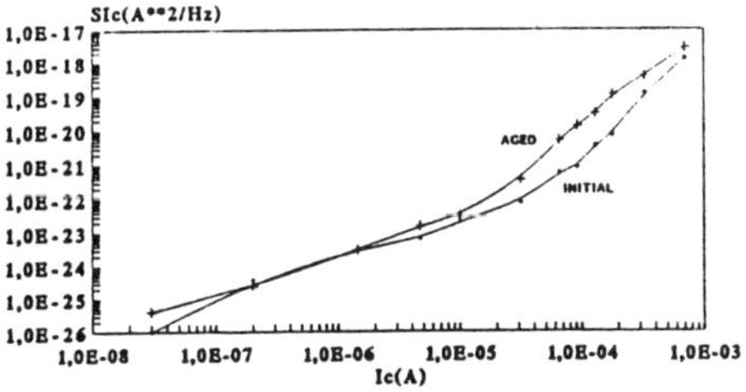

Fig. 3 : Collector current noise as a function of I_c before and after aging with strong reverse V_{be}. f = 1 Hz, V_{ce} = 2V.

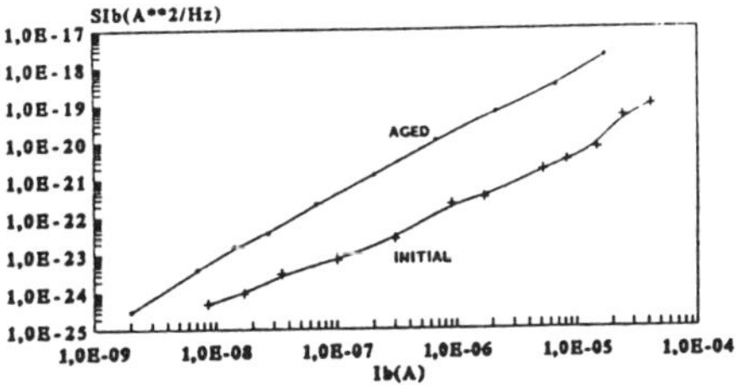

Fig. 4 : Base current noise as a function of I_b before and after aging with strong reverse V_{be}. f = 1 Hz, V_{ce} = 2V.

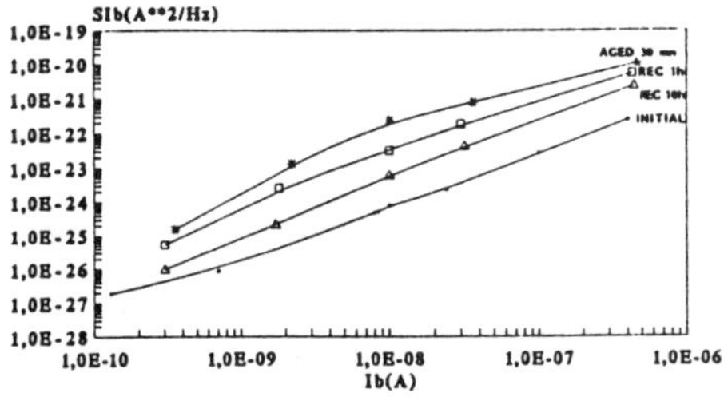

Fig. 5 : Base current noise versus I_b after aging with a reverse V_{be} of 6 V during 30 mn, and several forward bias recoveries (V_{be} = 0.8 V, V_{ce} = 2 V) during 1 and 10 h.

BIAS DEPENDENCE OF LOW-FREQUENCY NOISE AND NOISE CORRELATIONS IN LATERAL BIPOLAR TRANSISTORS

A. Nathan
Electrical and Computer Engineering
University of Waterloo, Waterloo, Ontario, Canada N2L 3G1

ABSTRACT

The low frequency noise behaviour in dual-collector lateral npn and pnp bipolar transistors is presented for various bias conditions in both forward active and saturation regimes. The correlation in collector output noise is very high in medium injection and degrades at high injection, depicting a behaviour analogous to the forward current gain of the device. The high degree of coherence makes such devices potentially useful in high resolution magnetic field sensing applications.

INTRODUCTION

The low frequency noise behaviour, and in particular, the correlation in output noise has been investigated for a wide variety of multi-terminal devices. Although these device structures can be employed for the detection of magnetic fields, the minimum detectable field is ultimately limited by the intrinsic device noise behaviour. For example, in dual-drain MOSFETs the output noise is independent in the linear region but a negative coherence is observed when the device is operated in the saturation region[1]. In modulation doped AlAs/GaAs superlattice dual-drain devices, the coherence is a strong function of frequency due to the predominance of different noise mechanisms at different frequency ranges[2]. With dual-collector transistors fabricated in MOS or bipolar technologies, the degree of coherence between output collectors turns out to be as high as 99.999%, independent of the frequency. As a result, the low frequency noise, which is predominantly 1/f, is virtually eliminated in the differential output which is at least five orders of magnitude less than the single-ended counterpart. This feature makes such structures extremely desirable for resolution of low field strengths[3] ($B < \mu T$).

In this paper, the low frequency noise behaviour in lateral npn and pnp dual-collector structures is studied for various bias conditions. In particular, we investigate the dependence of the noise correlation between output collectors for transistors operated at different injection levels in forward active as well as in the saturation regimes.

DEVICE THEORY AND MEASUREMENT TECHNIQUE

The samples considered here are dual-collector lateral npn and pnp transistors fabricated in-house using a bipolar process. A cross section of the fabricated structures is shown in Figs. 1 and 2. In principle, the structures are sensitive to magnetic fields parallel to the chip surface, although the pnp transistor suffers from inherent drawbacks from the viewpoint of circuit integration because of its collectors in the substrate. However, the correspondence between the correlation behaviour and injection level observed in this structure will equally be valid for other similar structures. The P+ emitter of the pnp transistor is surrounded by an N+ ring to supress lateral injection. With an appropriate N+ ring voltage, such configurations have been found to considerably increase the device sensitivity to magnetic fields[4].

The forward current gain, β_F of the npn transistor was approximately constant at 110 for a wide range of base currents ranging from 10 to 46 μA. However, with the pnp transistor the current gain was strongly dependent on bias; the maximum β_F was 51 at a base current of 3 μA and degraded to 30 at 33 μA. The coherence Γ was evaluated based on measurement of the single-ended power spectral (S_{xx}, S_{yy}) and cross-spectral (S_{xy}) densities, viz., $\Gamma(jf) = S_{xy}(jf)/[S_{xx}(f)\ S_{yy}(f)]$. The measurements were performed with appropriate shielded enclosures and by using low noise circuit components and amplifiers. The coherence was evaluated using a HP3561 dynamic signal analyzer. The detailed measurement procedure is given in Ref. 5.

RESULTS AND DISCUSSION

The behaviour of output collector noise was investigated with the transistors biased at different injection levels and operation modes. No noticeable difference was observed between the single-ended spectra, S_{xx} and S_{yy}. A near ideal 1/f noise characteristic was observed for both npn and pnp devices for frequencies up to 10 kHz and 1 kHz, respectively. In forward active operation under medium injection, the single-ended noise current power spectral densities (PSDs) increased with increasing collector current. The differential noise current PSDs were consistently at least 50 dB lower than the single-ended counterparts, implying a correlation in noise very close to unity between the output collector terminals. But, as the level of injection was increased, a degradation in the degree of coherence was observed. This is illustrated in Figs. 3 and 4 which show a correspondence in correlation behaviour with the forward current gain for the respective npn and pnp transistors. For both devices, we observe that the degree of coherence is the highest at moderate injection levels when the devices operate at the regime of maximum current gain.

The exact cause of degradation in coherence with bias is uncertain, although this may be caused by the same effects responsible for the reduction in current gain with bias, i.e., the predominance of generation-recombination current in the space charge layer at low injection and emitter crowding effects at relatively large base currents at high injection levels. It is well known that in bipolar transistors, the low frequency noise originates primarily from the emitter-base junction and includes both surface and bulk effects. In common-emitter configuration, the low frequency fluctuations are amplified through to both collectors and, by virtue of device symmetry, these fluctuations are simultaneously felt by both collectors. At the limits of very low and very large base currents (leading to degradation in β_F), there is reduced spatial coherence in these noise sources. In particular, as the level of injection increases, there is appreciable potential drop in the base region which possibly lessens the degree of spatial coherence of these fluctuations at the emitter-base junction, consequently, degrading the correlation in collector outputs. In fact, in the limit of very high injection, the output collector noise becomes statistically independent[6]. The above predictions are consistent with observations of temperature dependence of noise behaviour performed for lateral npn structures,[5,7] over the range 250 K $\leq T <$ 373 K. Although the single-ended noise increases with increasing temperature, the correlation in output noise turns out to be temperature independent; the spatial correlation in fluctuations at the emitter-base space-charge layer appears to be affected only by the magnitude of the base current and not explicitly by variations in temperature.

In contrast to the dependence of the output noise on bias current in forward active, the single-ended noise current PSDs under saturation conditions decreased with increasing collector (or base) currents, and was strongly frequency dependent.

Figure 5 illustrates the noise behaviour for the pnp transistor for various collector currents as the transistor moves from forward active (I_C ~ 23, 44, and 165 μA) into saturation (I_C ~ 239, 405 μA). The dip in the spectra beyond 20 kHz is due to the degradation in the transistor's a.c. current gain. For increasing values of collector current (Fig. 5), the values of the base current increased correspondingly and the transistor current gain decreased from its maximum value to 30. The corresponding behaviour of coherence is shown in Fig. 6. In the forward active regime, the degree of coherence is close to unity but it degrades with increasing frequency and when operated in saturation; the degree of coherence being the highest at the edge of saturation. The exact causes of the pronounced degradation at higher frequencies and when saturated, is presently unclear. It may be possibly due to effects of capacitive coupling between collectors which increases as the transistor is driven harder into saturation.

CONCLUSIONS

The low frequency noise behaviour in lateral dual-collector npn and pnp transistors has been investigated for various bias conditions in both forward active and saturation regimes. With the transistor in forward active, the single-ended noise current PSD increases with increasing collector (or base) currents, with the degree of coherence being the highest when the forward current gain of the transistor is maximum. The coherence degrades at high injection levels, which can be attributed to potential drops in the base region which become appreciable at relatively large base currents. In the saturation regime, however, the dependence of the output noise on bias is reversed. The single-ended noise current PSD decreases with increasing collector (or base) current and there is correspondingly a degradation in the coherence. The degree of coherence is maximum only at the edge of saturation and is highly frequency dependent.

ACKNOWLEDGEMENTS

This work was funded by the Natural Sciences and Engineering Research Council of Canada (NSERC) and the Federal (Micronet) and Provincial (ITRC) Centres of Excellence in Microelectronics.

REFERENCES

1. D. R. Briglio, A. Nathan, H. Baltes, in Noise in Physical Systems, Ed. C. M. Van Vliet, (World Scientific Publishing, Singapore, 1987), p. 453.
2. A. Nathan, H. P. Baltes, R. Castagnetti, Y. Sugiyama, D. R. Briglio, Sensors and Actuators **A21-A23**, 776 (1990).
3. A. Nathan, H. P. Baltes, D. R. Briglio, M. Doan, IEEE Trans. Electron Devices **ED-36**, 1073 (1989).
4. L. Ristic, H. P. Baltes, T. Smy, I. Filanovsky, IEEE Electron Device Letts. **EDL-8**, 395 (1987).
5. W. Kung, A. Nathan, IEEE Trans. Electron Devices **ED-40**, 910 (1993).
6. A. Chovet, Ch. S. Roumenin, D. Dimopoulos, N. Mathieu, Sensors and Actuators **A21-A23**, 790 (1990).
7. W. Kung, A. Nathan, A. Salim, T. Manku, in Noise in Physical Systems and 1/f Fluctuations, Eds. T. Musha, S. Sato, Y. Yamamoto, (Ohmsha Ltd, 1991), p. 217.

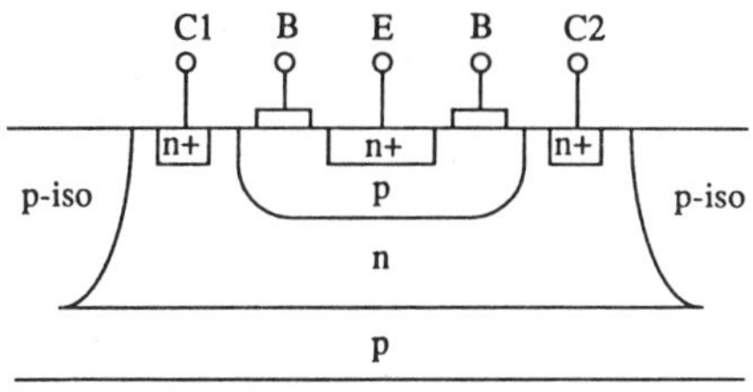

Fig. 1 Cross section of lateral dual-collector npn transistor.

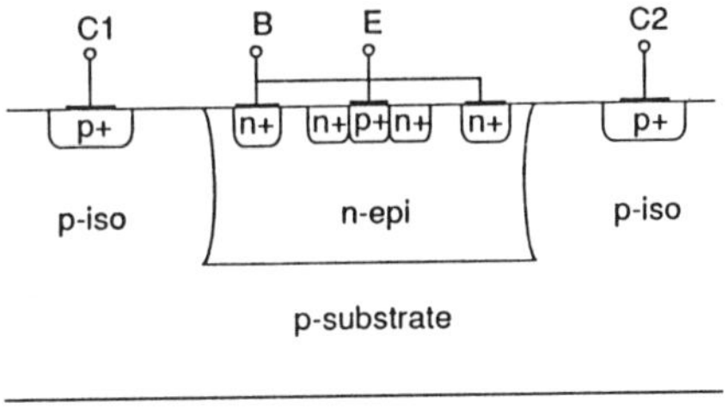

Fig. 2 Cross section of dual-collector pnp transistor.

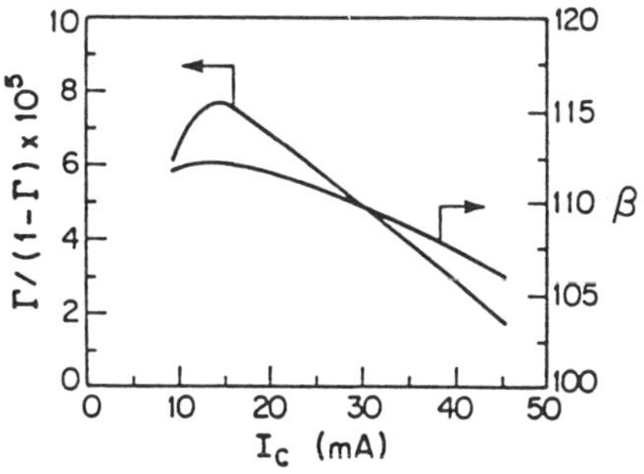

Fig. 3 The coherence and current gain behaviour as a function of bias currents for the npn transistor.

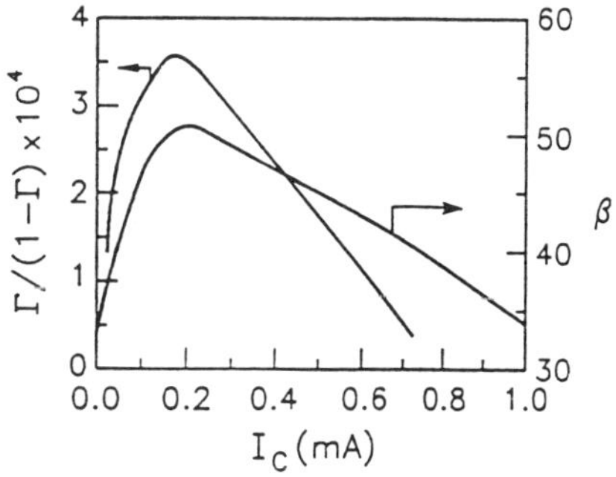

Fig. 4 As in Fig. 3, but for the pnp transistor.

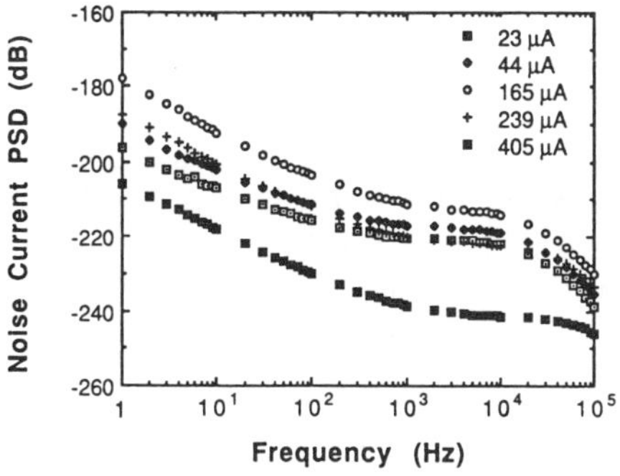

Fig. 5 The single-ended output noise for the pnp in forward active and saturation.

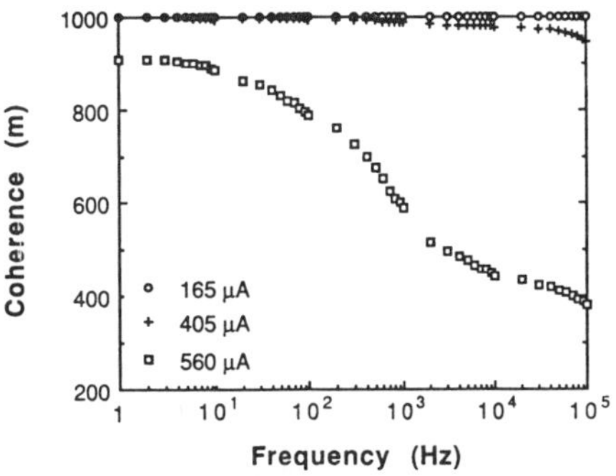

Fig. 6 The coherence in collector outputs as a function of frequency in forward active and saturation.

LOW-FREQUENCY NOISE MEASUREMENTS IN GaAlAs/GaAs HETEROJUNCTION BIPOLAR TRANSISTORS

F. Pascal, S. Jarrix, C. Delseny, G. Lecoy
Centre d'Electronique de Montpellier, CNRS URA 391, Place E. Bataillon,
Université de Montpellier II, 34095 Montpellier Cedex 5, France

C. Dubon-Chevallier, J. Dangla, H. Wang
CNET, 196 Avenue H. Ravera, 92220 Bagneux, France

ABSTRACT

Low-frequency noise measurements on GaAlAs/GaAs Heterojunction Bipolar Transistors are reported. Noise spectra exhibit excess noise composed of 1/f noise and several generation-recombination (g-r) levels. To try to localise and identify noise sources, results from various measurement modes and at different working temperatures are investigated.

INTRODUCTION

Until now, the low-frequency noise measurements on Heterojunction Bipolar Transistors (HBTs) published[1,2,3,4] suggest that the excess noise is mainly due to the emitter-base junction of the transistor. The aim of this paper is to try to locate and to identify the noise sources. First, classical DC characteristics on GaAlAs/GaAs HBTs are presented. Second, several experiments in different set-up configurations for noise measurements at room temperature are related. Finally, temperature evolutions of the spectra are shown and interpreted in terms of trap levels responsible for the g-r noise.

DEVICE CHARACTERISTICS

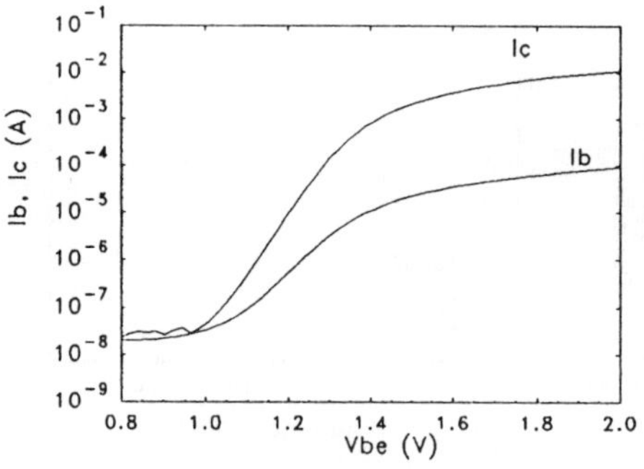

Fig. 1. Gummel-Plot

Npn GaAlAs/GaAs HBTs supplied by CNET Bagneux are unpassivated and processed with a non-self-aligned double mesa technology. Base is C-doped ($4\ 10^{19}$ cm^{-3}). Emitter area is $2x20\mu m^2$. Gummel Plots (Fig. 1.) show a typical non-constant variation of the static gain, β, with bias. Values concerning the collector current are about $1.54\ 10^{-21}$A for the saturation current and 1.28 for the ideality factor n at 300K. This n value significantly deviates from the ideality factor of 1 usually observed for Be-doped HBTs. Static current gain values reach 100 and above as reported for these C-doped transistors [5].

NOISE MEASUREMENTS

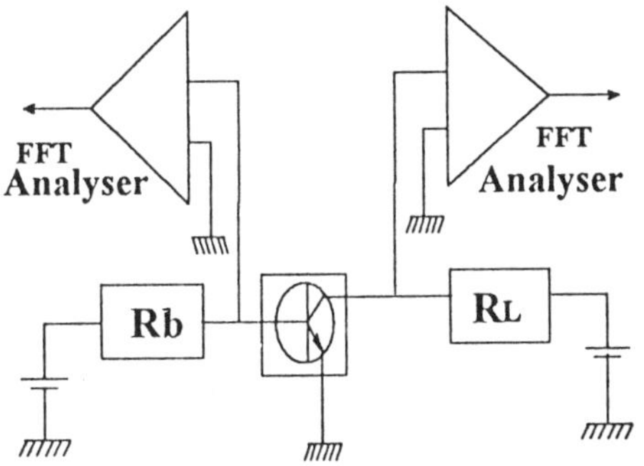

Fig. 2. Measurement Set-up

Noise measurements were performed in the normal region, at Vce equal to 1.5V in the common emitter configuration for frequencies ranging from 1Hz to 100 kHz, and extended to 1MHz in some cases. Noise voltage is measured at the output or at the input of the device through the bias collector resistance RL or the bias base resistance Rb, respectively, as shown on figure 2. Expressions for the spectral densities at the input and at the output are calculated with a simplified equivalent circuit, and by neglecting series resistances. If $r\pi$ is the dynamic input resistance of the transistor the spectral density Svi at the input is:

$$S_{V_i} = \left(\frac{Rb\, r_\pi}{Rb + r_\pi}\right)^2 \left(\frac{4kT}{Rb} + S_{I_b}\right) \qquad (1)$$

If Hfe is the dynamic current gain of the transistor the spectral density Svo at the output is given by:

$$S_{V_o} = 4kT\, R_L + R_L^2\, S_{Ic} + Hfe^2 \left(\frac{R_L}{Rb + r_\pi}\right)^2 \left(4kT\, Rb + Rb^2\, S_{I_b}\right) \qquad (2)$$

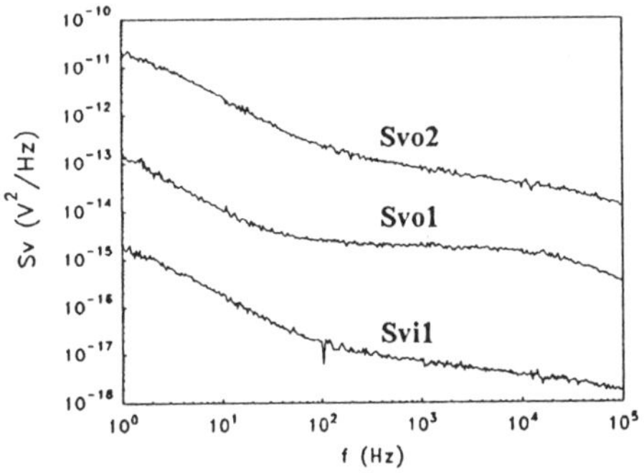

Fig. 3. Spectral densities for Ib=20μA at 300K

A series of noise measurements were undertaken at room temperature with Rb1=50 Ω at the input of the transistor. First RL is set to 50 Ω, and then the output is short-circuited by a capacitor. In both cases, for the same bias, spectral densities Svi1 are the same, indicating no interaction of the output onto the input in our measurement frequency range and for RL values up to 50 Ω. An example of the spectral density Svi1 is given figure 3.

At the output, two cases are studied:

- with Rb1<<$r\pi$, (here Rb1=50 Ω), the spectral density Svo1 can be expressed by:

$$Svo1 \approx 4kT\, R_L + R_L^2 S_{Ic} \qquad (3)$$

- with Rb2>>rπ (here Rb2=320 kΩ): as RL=Rb1, the spectral density Svo2 at the output and Svi1 at the input can be related by:

$$Svo2 \approx (Hfe)^2 \, (Svi1 - 4kT\,Rb1) \tag{4}$$

Expression (4) is in good agreement with our experimental data. Thus, noise measurements performed at the output with Rb>>rπ image the noise from the input of the transistor.

Whatever the base bias current Ib, all spectra measured at the output exhibit excess noise. The same behaviour on GaAlAs/GaAs HBTs has been recently reported by different authors [3,4]. The white noise is nearly reached for low base current values at 1MHz for Svi1 and Svo2. All spectra can be decomposed into a 1/f component and at least two Lorentzians by using the following expression:

$$Sv = \frac{b}{f} + \sum_{i=1}^{n}\left(\frac{Ci}{1+(\omega\tau i)^2}\right) + Sv_{wh} \tag{5}$$

with Svwh the theoretical white noise calculated with equations (1) and (2). An example is given on figure 4.

All measurements at room temperature were performed in the frequency range 1Hz-1MHz with the base current Ib varying from 5 to 80 μA. All b coefficients used for the curve-fit are quasi-quadratic with Ib and vary at the output as $Ib^{2.05}$ for Svo2, $Ib^{2.10}$ for Svo1, and at the input as $Ib^{1.9}$ when Ib>10 μA. This suggests the base or the base-emitter junction to be responsible for the 1/f noise.

Three Lorentzians are needed to fit Svo2 and Svi1, and only two for Svo1. All g-r time constants extracted from the fits vary experimentally with base current[3,6] as Ib^{-r}. They are nearly identical for Svi1 and Svo2. With Rb1, Svo1 has two very close levels, as shown by the plateau on figure 3. In all cases, the coefficient r of the last level is the same and equal to 0.89.

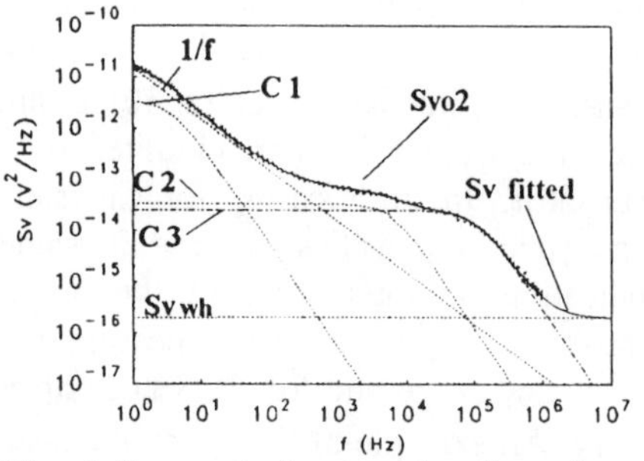

Fig. 4. Spectral density showing the various components for Ib=20μA at T=300K

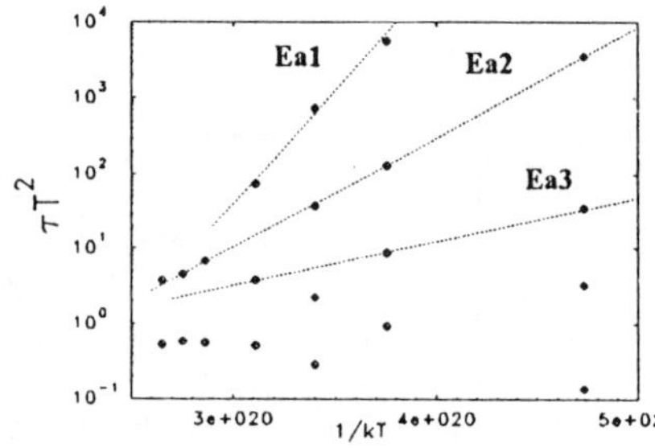

Fig. 5. Arrhenius Plot for Ib=5μA

To characterise these g-r levels and to determine the activation energies E_t of the associated traps, noise measurements at the output with Rb2 were performed in the temperature range 300K-77K. The g-r time constants can be expressed in terms of Fermi-Dirac statistics [7,8], by:

$$\frac{1}{\tau} = \sigma v_{th} \, Nc \, \exp\frac{(E_f - Ec)}{kT}\left[B \exp\left(\frac{E_t - E_f}{kT}\right) + 1\right] \tag{6}$$

with σ capture cross-section, B a spin degeneracy factor, E_f Fermi level. The thermal velocity v_{th} is proportional to $T^{1/2}$, the effective density of states Nc is proportional to $T^{3/2}$. By plotting $\ln(T^2\tau)$ versus 1/kT (Arrhenius plot) the slope will give E_t. All spectral densities Svo2 were simulated with equation 5. From 283K to 253K, curves are fitted with two Lorentzians having cut-off frequencies about a decade apart. At 233K the first level is masked by a new "bump" which appears. Below 233K, the curves are smeared out and at least four levels are needed. By following the evolution of each Lorentzian versus temperature, activation energies were extracted (Fig. 5). Ea1=420 ± 40 meV for the first trap level, Ea2=210 ± 20 meV for the second trap level, Ea3=85 ± 10meV for the third. The second value is close to that found by Costa et al.[3]. The third value has been found elsewhere for a shallow level [1].There is too much inaccuracy on the fourth trap to extract the activation energy value

CONCLUSION

Our noise measurements on GaAlAs/GaAs HBTs show that the 1/f noise observed at room temperature can be associated with the emitter-base junction. Studies at low temperatures reveal the presence of two or more g-r components, suggesting that different traps are active in different temperature regimes. When close to 77K, noise spectra exhibit a broad Lorentzian with a 1/f-like slope, due to the numerous relaxation time constants. Activation energies of certain traps could be extracted. The DX center may be a possible trap candidate as proposed by Costa et al.[3].

REFERENCES

1. V. K. Raman, C. R. Viswanathan, M. E. Kim, IEEE Trans. on Microwave Theory and Tech., Vol. 39, No. 6, p.1054, (1991)
2. S. C. Jue, D. J. Day, A. Margittai, M. Svilans, IEEE Trans. on Electron. Devices, Vol. 36, No.6, p.1020, (1989)
3. D. Costa, J. S.Jr. Harris, IEEE Trans. on Electron. Devices, Vol. 39, No. 10, p.2383, (1992)
4. N. Hayama, K. Honjo, IEEE Trans. on Electron. Devices, Vol. 39, No. 9, p.2180, (1992)
5. J. L. Benchimol, F. Alexandre, C. Dubon-Chevallier, F. Heliot, R. Bourguiba, J. Dangla, B. Sermage, Electron. Lett., Vol. 28, No. 14, p. 1344, (1992)
6. R. Plana, Thèse, Université Paul Sabatier, Toulouse, France (1993)
7 A. D. Van Rheenan, G. Bosman, R. Zijlstra, Solid-State Electron, Vol 30, No. 3, p 259, (1987)
8 Y. Dai, Solid-State Electron., Vol. 32, No. 6, p. 439, (1989)

SPLITTING OF ELECTRON CURRENT PULSES AND NOISE IN DOUBLE–BARRIER HETEROSTRUCTURES

B. Pellegrini and M. Macucci
Dipartimento di Ingegneria dell'Informazione: Elettronica, Informatica e Telecomunicazioni, Università degli Studi di Pisa, Via Diotisalvi 2, 56126 Pisa, Italy

ABSTRACT

The aim of this paper is to show (i) that the total output current pulses of the double–barrier resonant–tunneling structures due to single electrons are split into two parts as a consequence of the displacement current contribution and of the temporary storage in the double barrier well, and (ii) that this phenomenon can either increase or reduce the shot noise in comparison to the full shot noise. The noise increase is due to the electrons which, after being stored in the well, return to the emitter, whereas the noise reduction is associated to those which cross the whole structure.

SPLITTING OF ELECTRON CURRENT PULSES

Present microelectronic technologies allow the fabrication of double–barrier resonant tunneling structures (DBRTS), whose properties are being intensively investigated.

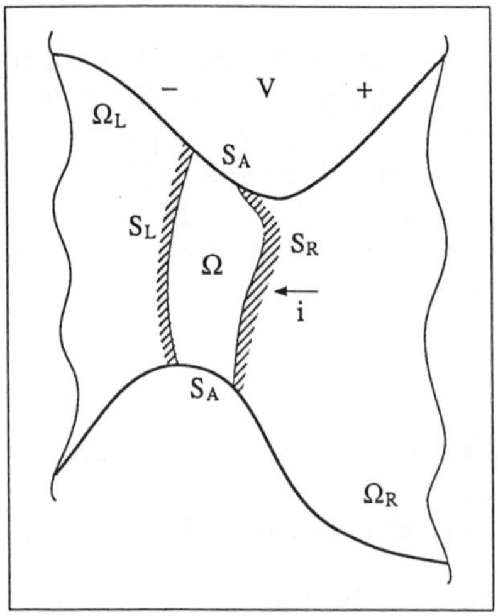

Fig. 1 – Two electrode device.

The aim of this paper is to show that the output current pulses of such structures due to single electrons are split into two parts as a consequence of the displacement current and of the temporary storage in the double barrier well, and that this phenomenon can either increase or reduce the shot noise in comparison to the full shot noise.

To the first end, we exploit a recent quantum electrokinematics theorem[1] which takes into account the displacement current due to the motion of each electron, and which allows a direct evaluation of the total output current due to the electron just from its probability density.

Let us apply it to a two electrode device (Fig. 1) of volume Ω with a surface which is constituted of an external part S_A and of the inner surfaces S_L and S_R of the two electrodes (for instance, the n^+ layers of the DBRTS) between which the voltage V is kept constant.

According to this theorem the total current i(t) entering Ω across S_R at time t becomes $i(t) = \sum_{\mu=1}^{N} i_\mu(t)$, where i_μ is the current pulse due to the μ–th of the N(t) electrons interacting with Ω at t; i_μ, in its turn, can be expressed as a function of the charge density $\rho_\mu(\mathbf{r},t)$ only of μ–th electron itself.

To this purpose let us use Ω_L (Ω_R) to indicate the volume of the region on the left (right) of S_L (S_R) (Fig. 1) which is enclosed by S_L (S_R) itself, by the external surface of the device and of any other parts of the closed loop in which the device current flows, and by any cross section S_C of the loop characterized by $\rho_\mu=0$ at least during the dwell time interval $[t'_\mu, t''_\mu=t'_\mu+\sigma_\mu]$ that the electron spends in Ω.

Then the theorem, for $S_A << S_R$ and / or for low frequencies, yields[1]

$$i_\mu = \int_{(\Omega_L+\Omega)} \frac{\partial \rho_\mu(\mathbf{r},t)}{\partial t} d^3\mathbf{r} - \int_\Omega \Phi(\mathbf{r}) \frac{\partial \rho_\mu(\mathbf{r},t)}{\partial t} d^3\mathbf{r}, \tag{1}$$

where the function $\Phi(\mathbf{r})$ is defined by the following equation and boundary conditions: $\nabla \cdot [\varepsilon(\mathbf{r}) \nabla \Phi(\mathbf{r}, t)] = 0$, $\Phi(\mathbf{r}) = 1$ on S_R, $\Phi(\mathbf{r}) = 0$ on S_L and S_A, $\varepsilon(\mathbf{r})$ being the static permittivity.

Equation (1), unlike other methods which only consider the contribution of the first integral, i.e., the conduction current, takes into account the displacement current by means of the potential function Φ, and it is a useful tool for analyzing the shape of the current pulses $i_\mu(t)$ due to single electrons.

Let the μ–th electron be in Ω_L with certainty for $t < t'_\mu$ and let it be in Ω_R or, again, in Ω_L for $t > t''_\mu$.

In such cases ρ_μ is null in Ω and its integral over Ω_L is equal to $-q$ or to 0, so that, from (1), we get $i_\mu(t)= 0$ for both $t < t'_\mu$ and $t > t''_\mu$ (Fig. 2).

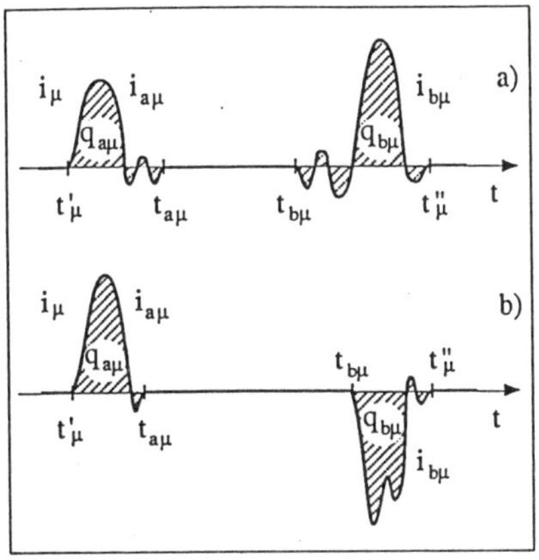

Fig. 2 – Current–pulse splitting.

By definition, on the contrary, ρ_μ has to be different from zero somewhere in Ω for $t'_\mu < t < t''_\mu$, so that, according to (1), i_μ can be different from zero during the dwell time.

The dwell time, in its turn, can be divided into three intervals[2] (Figs. 2a and 2b). In the two edge intervals $[t'_\mu, t_{a\mu}]$ and $[t_{b\mu}, t''_\mu]$, the electron transits across the barriers of the DBRTS and, as a consequence, we have $i_\mu(t)\neq 0$ because $\rho_\mu(\mathbf{r},t)$ is a function of time somewhere in Ω.

In the remaining intermediate interval $[t_{a\mu}, t_{b\mu}]$ the electron is trapped in a (quasi) stationary state, so that ρ_μ becomes time independent and, in virtue of (1), i_μ becomes equal to zero, i.e., the current pulse is split into two pulses of the type shown in Fig. 2a (Fig. 2b) if the electron crosses the whole structure (comes back to the injection electrode).

The ratio between the average values of the transit times $t_{a\mu} - t'_\mu$ and $t''_\mu - t_{b\mu}$ across the barriers and of the dwell time σ_μ in the whole structure is of the order of the greatest between the tunneling probabilities T_L and T_R of the left and right barriers, respectively,[2] at the resonance energy, so that the average separation between the two current pulses is much greater than their duration.

On the other hand, the dwell time, that has a mean value of the order of 10^{-10}s, [3] can have any random value σ_μ for each μ–th electron. Therefore, even for a complete phase coherence of the electron wave function, i.e., in absence of any inelastic scattering in the whole tunneling process across the DBRTS, the occurrence of its second current pulse could be considered as an event independent from the first one.

Moreover, the average time interval $q/2\bar{i}$ between consecutive split pulses (that carry an average charge q/2, q being the electron charge) is smaller than 10^{-13} s for the average current $\bar{i}$ greater than 0.9 μA,[4] so that, for instance, we can have hundreds of split pulses of other electrons between the ones of the same electron.

Therefore, on the basis of such considerations, it seems reasonable to consider the split current pulses of the same electron as uncorrelated pulses, also when the tunneling process is coherent.[3]

A fortiori, we have that the current pulses are certainly independent events when the electron is scattered inelastically in the well and it loses any phase memory before escaping from the well itself, i.e., when the tunnneling mechanism across the structure is completely incoherent (sequential).[3]

For instance, for an inelastic scattering time $\tau_i < 4\text{x}10^{-12}$s and a dwell time $\sigma > 4\text{x}10^{-10}$s, we have hundreds of randomizing events, leading to independence of the two current pulses of the same electron.

Therefore, the temporary storage of the electron in Ω splits its current pulse into two distinct and uncorrelated parts $i_{a\mu}(t-t_{a\mu})$ and $i_{b\mu}(t-t_{b\mu})$ that start at the times $t=t_{a\mu}$ and at $t=t_{b\mu}$, respectively, so that the current across the structure becomes

$$i(t) = \sum_{\mu=1}^{N}\left[i_{a\mu}\left(t - t_{a\mu}\right) + i_{b\mu}\left(t - t_{b\mu}\right)\right] = \sum_{j=1}^{N_I} i_j\left(t - t_j\right), \tag{2}$$

where $j = a\mu$, $b\mu$ and $N_I(t)$ is the number of the split current pulses that occur at t.

In their turn, according to (1), the pulses $i_{a\mu}$ and $i_{b\mu}$ carry fractions $q_{a\mu}=\varphi_\mu q$ and $q_{b\mu}=(1-\varphi_\mu)q$ of the electron charge q, respectively, with $0<\varphi_\mu<1$ if the electron crosses the whole structure, while they carry opposite fractions $q_{a\mu}=-q_{b\mu}=\varphi_\mu q$ if the electron comes back to the emitter electrode after the storage in the well (Fig. 2).

From (1), in the resonant state, we also obtain that $\varphi_\mu=\varphi$ is equal for all the electrons.

SHOT NOISE

The particular transport mechanism of electrons in the DBRTS's we have discussed affects their current noise, whose computation is the second aim of the paper.

If, according to the previous considerations, the current pulses $i_j(t-t_j)$ in (2) can be considered as independent events occurring at random at the average rate $\langle n \rangle$, from Carson's theorem [5] we obtain the spectral power denity S_i of the current i at low frequency in the form

$$S_i = S_u = 2\langle n\rangle\langle q_j^2\rangle = 2q\langle i\rangle\{1 + 2\varphi[\varphi(1+\vartheta) - 1]\}, \tag{3}$$

where $\langle i\rangle = q\langle n_f\rangle$ is the average current, ϑ is the ratio between the average rate $\langle n_b\rangle$ of the electrons which from the well come back to the emitter and the average rate $\langle n_f\rangle$ of those which reach the collector, forthermore $q_j = q_{a\mu}$, $q_{b\mu}$.

If the pulses $i_j(t - t_j)$, occurring at the rate n(t), do not obey the Poisson statistics, we can compute S_i according to Milatz's theorem,[5] or, ultimately, according to the extension of Burgess's variance theorem to the autocorrelation function[5,6], in the form

$$S_i = 2\langle i_j\rangle^2\left(\int_{-\infty}^{\infty}\langle\Delta N_I(t)\Delta N_I(t+s)\rangle ds - \langle N_I\rangle\right) + S_u, \tag{4}$$

that, in the case of Poisson statistics for which $\int_{-\infty}^{\infty}\langle\Delta N_I(t)\Delta N_I(t+s)\rangle ds = \langle N_I\rangle$, again gives (3).

The computation of $\langle\Delta N_I(t)\Delta N_I(t+s)\rangle$ is beyond the scope of the present work.

Nevertheless, in a first approximation that the split pulses are independent, from (3) we obtain results which agree with the experimental findings.[4]

For example, from (3) we obtain that only for the values $\varphi=1/2$ and $\vartheta=0$ the noise can be reduced down to a minimum value of one half of the full shot noise.[4]

On the contrary, for $\vartheta > 1/\varphi - 1$, the noise becomes greater than the full shot noise.

In conclusion, the noise of DBRTS, in agreement with the experimental findings[4], can be both greater and smaller than the full shot noise.

The noise increase is due to the electrons which, after their storage within the well, return to the emitter, whereas the noise reduction is associated to those which cross the whole structure.

REFERENCES

1. B. Pellegrini, "Extension of the electrokinematics theorem to the electromagnetic field and quantum mechanics", Nuovo Cimento D (Accepted for the publication)
2. R. Ricco, and M.Y. Azbel, Phys. Rev. B29, 1970 (1984).
3. M. Jonson and A. Grincwaig, Appl. Phys. Lett. 51, 1729 (1987).
4. Y.P. Li, A. Zaslavsky, D.C. Tsui, M. Santos, and M. Shayegan, Phys. Rev. B 41, 8388 (1990).
5. A. van der Ziel, "Noise in solid state devices and circuits", (J. Wiley & Sons, New York 1986), pp.14–18.
6. A. van der Ziel and K. M. van Vliet, Physica 113B, 15 (1982).

MODELING RANDOM TELEGRAPH NOISE BY MEANS OF THE SINGLE-ELECTRON TRANSISTOR

R. S. Popovic
Landis & Gyr, Corp. Research and Development
6300 Zug, Switzerland

ABSTRACT

A trap in the gate oxide of a MOSFET is modeled by the single-electron tunneling (SET) transistor. The island of the SET transistor simulates the trap center itself and the two SET transistor's junctions in parallel represent the connection between the trap center and the channel of the MOSFET. In the certain ranges of the gate voltage, the island of the SET transistor may exchange electrons with the channel. This gives rise to random telegraph signal (RTS) noise in the MOSFET. The results of the Monte Carlo simulations of the trap potential reveal a dependence of the duty cycle of RTSs on the gate voltage.

INTRODUCTION

Random telegraph signal (RTS) noise is interesting both in its own right and for its connection to 1/f noise in MOSFETs [1, 2]. The smaller the electronic device, the higher the impact of RTS noise on its operation. In sub-micrometer MOSFETs, RTS noise is already a serious concern. In nanoscopic devices, such as the single-electron transistor, RTS noise may be the major obstacle for their application [3].

In a small MOSFET, RTS noise appears as discrete value fluctuations in the channel resistance of up to 1% in magnitude. It has been found that the switching events are caused by the capture and emission of individual electrons by traps in the gate oxide [1, 2]. A change in the charge of a trap affects both the channel charge (due to the neutrality condition) and the channel mobility (since a trap may act as a scaterer). In mesoscopic devices, there is another cause of RTS noise: the universal conductance fluctuations (see [5] and references therein). This mechanism of RTS noise is beyond the scope of this paper.

In the present paper, I propose a model of a trap in the gate oxide of a MOSFET based on the single-electron tunneling transistor [4]. Inherent to the model is the concept of the Coulomb blockade, which proved crucial for the detailed understanding of the traps and the related RTS noise in MOSFETs.

MODEL AND ANALYSIS

Fig. 1 shows the transformation of the single-electron tunneling (SET) transistor into the model of a trap in the gate oxide of a MOSFET. The SET transistor (Fig. 1 (a)) consists of two tunnel junctions (j1 and j2) connected in series. Each junction is characterized by its tunnel resistance Ri and capacitance Ci (i=1, 2). The two junctions enclose a small "island" electrode (il), the electrostatic potential of which can be influenced by the gate voltage Vg through the capacitance Cgi. We can incorporate the SET transistor into a MOSFET (Fig. 1 (b)) so that the island il simulates the trap center (tc) itself, the junctions j1 and j2 in parallel represent the tunnel connection between the trap center and the channel (ch) of the MOSFET, and the input capacitance Cgi models the capacitance between the gate and the trap center. The tunneling rate of electrons between the trap and the channel is characterized by the tunnel resistance Rt=(R1 || R2). The total capacitance of the trap is given by C=C1+C2+Cgi.

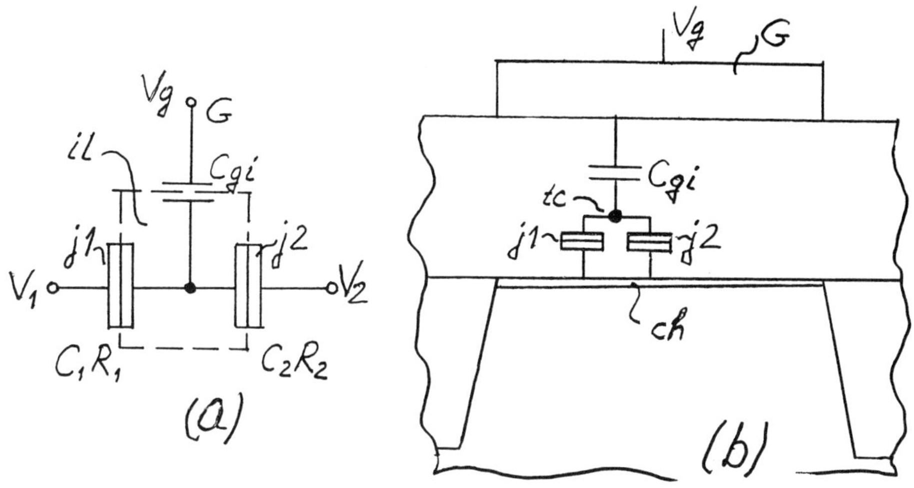

Fig. 1 The transformation of the single-electron tunneling transistor (SET transistor) into the model of a trap in the gate oxide of a MOSFET. (a) The circuit diagram of the SET transistor; (b) The SET transistor incorporated into a MOSFET.

In a rigorous treatment, one should certainly take into account the quantization of the trap energy. However, in order to demonstrate the main features of the model, we shall apply here the so called "orthodox theory" of single-electron tunneling and neglect the quantization.

The total capacitance of a trap defines the Coulomb energy Ec associated with the charging of the trap by an electron:

$$Ec = q^2 / 2C \qquad (1)$$

Here q denotes the elementary charge. If the Coulomb energy is large relative to the average energy of thermal fluctuations,

$$Ec \gg kT \qquad (2)$$

in usual notation, the Coulomb blockade shows up: If

$$q(n-1/2) < Cgi\ Vg < q(n+1/2) \qquad (3)$$

where n denotes the number of excess electrons on the trap, the state with n electrons on the trap is stable. Then there is no charge fluctuation at the trap and no RTS noise. But if

$$\text{Cgi Vg} \simeq q(n \pm 1/2) \tag{4}$$

the trap may exchange one electron back and forth with the channel, and the traped charge fluctuates. The discrete fluctuations in the trap charge affect the channel charge and the channel mobility. This gives rise to RTS noise in the MOSFET. The better the equality (4) is fulfilled, the closer is the average duty cycle of the RTSs to unity. Thereby the "sharpness" of the equality (4) is measured relative to the quantity kTC/q. This explains the experimentally observed fact that the existence of the random telegraph signals in MOSFETs and its duty cycle strongly depend on the gate voltage and temperature.

We can simulate the thermal noise voltage at the trap center (tc) with the aid of the Monte Carlo method: we let the individual charge carriers transit each of the junctions at random and observe the resulting voltage fluctuations (Vi) at the island. Some details on the simulation and complementary results are given elsewhere [6].

RESULTS AND CONCLUSIONS

Fig. 2 shows the results of the Monte Carlo simulation of a trap using the described model. The parameters of the model used in the simulations

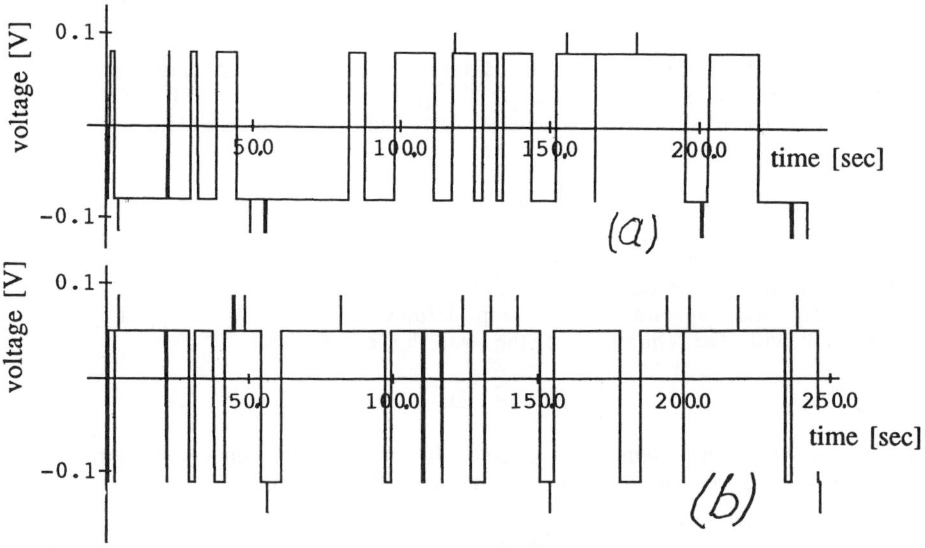

Fig. 2 The results of the Monte Carlo simulation of a trap using the present model. Shown is the electrical potential of the trap as a function of time. (a) Vg=-0.8 V; (b) Vg=-1.1 V. Note the dependence of the duty cycle of RTS noise on the gate voltage.

are as follows: $C=10^{-18}$ F (the capacitance of a sphere with the radius of about 1 nm, situated in the oxide, 1 nm away from the silicon-oxide interface), $Cgi=10^{-19}$ F, $Rt=10^{18}$ ohm, and T=300 K. Shown is the electrical potential of the trap as a function of time for two different gate voltages: one (a) corresponding exactly to equation (4) (with n=0), and the other (b) when equation (4) is only approximately fulfilled. The plots reveal clearly the expected duty cycle dependence on the gate voltage. Further simulations demonstrated also the experimentally observed fact that a certain switching pattern appears only over a limited range of gate voltage and temperature.

The noise spectra, obtained by applying the Fourier transform on the noise signals such as those in Fig. 2, are Lorentzian [6]. This also corroborates the experimental observations of the RTS noise [1].

Therefore, the single-electron tunneling (SET) transistor is a suitable means to model a trap in the gate oxide of a MOSFET as a cause of random telegraph signal (RTS) noise. A salient inherent feature of the model is the fact that the existence and the appearance of the RTS noise is a periodic function of the gate voltage, see equations (3) and (4). This is due to the Coulomb blockade effect.

REFERENCES:

1. K. S. Rals et al., Phys. Rev. Lett. 52, 228-231 (1984).
2. M. J. Uren et al., Appl. Phys. Lett. 47, 1195-1197 (1985).
3. G. Zimmerli et al., Appl. Phys. Lett. 61, 237-239 (1992).
4. H. Grabert and M. H. Devoret (eds.), Single charge tunneling, Plenum Press, New York 1992.
5. G. L. Timp and R. E. Howard, Proc. IEEE, 79, 1188-1207 (1991)
6. R. S. Popovic, Thermal noise in ultrasmall tunnel junctions (in this Proceedings).

THERMAL NOISE IN ULTRASMALL TUNNEL JUNCTIONS

R. S. Popovic
Landis & Gyr, Corp. Research and Development
6300 Zug, Switzerland

ABSTRACT

The thermal noise voltage at the island of a single-electron tunneling transistor is simulated using the Monte Carlo method. The appearance of the time dependence of the noise voltage changes dramatically as the thermal energy kT approaches the Coulomb energy $Ec=q^2/2C$. If kT <= Ec, the noise voltage either degenerates into the random telegraph signal; or, due to the Coulomb blockade, the fluctuations cease altogether. Surprisingly, the noise spectra stay Lorentzian at all temperatures. Moreover, the thermodynamic noise relations apply even if the noise voltage degenerates into the random telegraph signal.

INTRODUCTION

An ultrasmall tunnel junction is a semiconductor or metal-insulator-metal tunnel junction with nanometer dimensions and a capacitance of $C=10^{-15}$ F or less. In such a junction, the electrostatic energy related to the transit of a single electron across the junction, $Ec=q^2/2C$ (the Coulomb energy), is comparable to or larger then the average energy of thermal fluctuations, kT. This gives rise to the appearance of a remarkable effect called the Coulomb blockade. Ultrasmall tunnel junctions are basic building blocs of several novel devices, such as the single-electron transistor, supersensitive electrometer, current standard, etc.

The subject of ultrasmall junctions has been attracting a lot of interest [1]. However, it seams that the issue of thermal noise in such junctions has not been addressed yet. Is there any influence of the Coulomb blockade on noise? How the noise spectrum looks like when just one single electron jumps back and forth across the junction? Do the thermodynamic noise relations still apply in this case? These are the main questions that we shall try to answer in this paper.

ANALYSIS

As a representative example, we shall treat the single-electron tunneling (SET) transistor, shown in Fig. 1. The SET transistor consists of two tunnel junctions (j1 and j2) connected in series between the source (s) and the drain (d). Each junction is characterized by its tunnel resistance Ri and capacitance Ci, i=1,2. The two junctions enclose a small "island" electrode (il), the electrostatic potential of which can be influenced by the gate voltage Vg through the capacitance Cgi. The total capacitance of the island is given by C=C1+C2+Cgi. We are interested only in thermal noise and therefore assume the thermal equilibrium conditions. Thus we keep the source and the drain at equal potentials, i.e. V1=V2.

For simplicity, we describe the current-voltage characteristics of each of the junctions by the diode-type equation

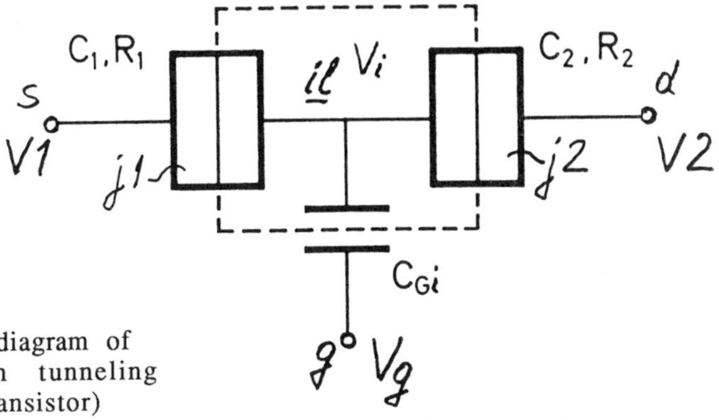

Fig. 1 The circuit diagram of the single-electron tunneling transistor (SET transistor)

$$I = Io\ (\exp(qV/kT) - 1) \quad (1)$$

in usual notation. Such a rectifying characteristic has, for instance, a very asymmetrical metal-insulator-metal junction. The choice of a definite current-voltage characteristic has no influence on the generality of the final conclusions, but it makes the analyses much simpler.

We analyze thermal noise in the circuit of Fig. 1 with the aid of the Monte Carlo method. Briefly, we let the individual charge carriers transit the junctions at random and observe the resulting voltage fluctuations (Vi) at the island. First, we have to calculate the probability that one single electron transits a junction in given direction. We do so for each of the two junctions and each of the two directions. These probabilities depend on the instantaneous value of the voltage Vi. Than, according to this probabilities, we select at random one of the possible four transit events. Finally, we update the voltage Vi, and repeat the procedure.

The probability that an electron transits a junction in a given direction is proportional to the corresponding average current. If an electron travels in the easy (forward) direction of the junction j1, the average current for this event is given by the corresponding effective voltage-current characteristics

$$I1 = Io\ \exp[\ q\ (V1\text{-}Vi\text{-}q/2C)\ /\ kT\ ]. \quad (2)$$

Here the term -q/2C is the voltage equivalent of the increase in the potential barrier height during the transit of the electron ($Ec=q^2/2C$). This increase in the barrier height is called the Coulomb barrier and may give rise to the Coulomb blockade. For the electron transit in the reverse direction, the effective voltage-current characteristics stays as in the macroscopic case, viz. I2 = Io. Similar equations hold also for the electron transits through the junction j2.

The probability that during the sample time period Dt a carrier transits the junction j1 in the easy direction is now given by

$$P1 = (\ Dt\ /\ (q\ /\ I\ )\ )\ (I1\ /\ I) \quad (3)$$

where I denotes the arithmetic sum of all four average currents. Similar equations hold also for the other three probabilities.

RESULTS AND CONCLUSIONS

A few typical results of the Monte Carlo simulations are shown in Figs. 2 and 3.

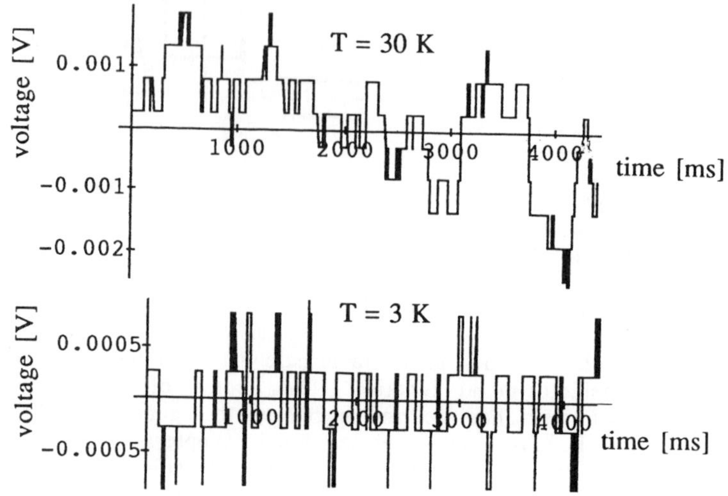

Fig. 2 Two examples of the noise voltage at the island of the SET transistor as function of time. The parameters: C1=C2=Cgi=10^{-16} F and thus C=3*10^{-16} F, Io=10^{-18}A, Vg=0.8 mV, and V1=V2=0 V. Note that for the chosen capacitances the Coulomb energy equals the thermal energy at the temperature T=3.1 K.

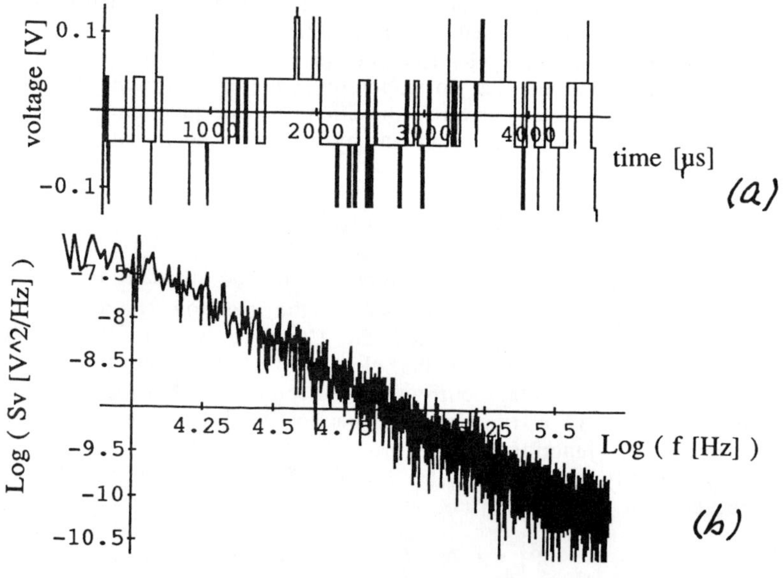

Fig. 3 Noise voltage at the island of the SET transistor as function of time (a) and the corresponding noise spectrum (b). The parameters: C1=C2=0.5·10^{-18} F, Cgi=10^{-18} F, Io=10^{-15}A, T=300 K, Vg=-0.08 V, and V1=V2=0 V.

The appearance of the noise voltage as function of time changes dramatically as the thermal energy kT varies relative to the Coulomb energy $Ec=q^2/2C$. At high temperatures, when kT >> Ec, the picture is almost as usual, with the exception that the voltage fluctuates only in finite increments. The magnitude of the increments equals q/C. As kT approaches Ec, the voltage excursions contain less increments, and the voltage steps become very clear. If kT <= Ec, the noise voltage either degenerates into the random telegraph signal (RTS), see Fig. 2; or, due to the Coulomb blockade, the fluctuations cease altogether. What exactly happens in this case depends very much on the gate voltage [2]. Briefly, the fluctuations are possible only if the relation

$$Cgi\ Vg \simeq q(n \pm 1/2) \tag{4}$$

is fulfilled. Here n denotes the number of excess electrons at the island of the SET transistor. But note that if kT <= Ec, the charge at the island fluctuates only for 1q. Equation (4) is fulfilled for different combinations of Vg and n. Therefore, we expect that the appearance of the RTS is a periodic function of the gate voltage.

By applying the Fourier transform on the noise signals in the time domain, such as those in Fig. 2, we obtain the corresponding noise power spectra in the frequency domain. One example of the two representations of a fluctuation signal is shown in Fig. 3.

The noise spectra reveal a surprizing property: they stay qualitatively similar and Lorentzian at all temperatures. Moreover, the noise voltage spectral density at sufficiently low frequencies equals the value given by the Johson-Niquist equation, viz.

$$Sv = 4\ k\ T\ R. \tag{5}$$

Here R denotes the dynamic resistance of the two junctions, at V=0, connected in parallel. In accordance with these facts, the mean square total fluctuation of the voltage is also given by the equation valid for macroscopic systems, namely

$$< V^2 > = kT / C. \tag{6}$$

Therefore, the thermodynamic noise relations (5) and (6) apply even if the capacitance of the system is so small, that the noise voltage degenerates into the random telegraph signal.

REFERENCES:

1. H. Grabert and M. H. Devoret (eds.), Single charge tunneling, Plenum Press, New York 1992.
2. R. S. Popovic, Modeling random telegraph noise by means of the single electron transistor (in this Proceedings).

SMALL-SIGNAL AND NOISE CHARACTERISTICS OF SUBMILLIMETER DIODE GENERATORS

V. Gružinskis, E. Starikov, P. Shiktorov
Semiconductor Physics Institute, A. Goštauto 11, 2600 Vilnius, Lithuania

L. Reggiani, M. Saraniti, L. Varani
Dipartimento di Fisica ed Istituto Nazionale di Fisica della Materia,
Università di Modena, Via Campi 213/A, 41100 Modena, Italy

ABSTRACT

By using a closed hydrodynamic model coupled with the field impedance method we evaluate the small-signal and noise spectra of submillimeter n^+nn^+ diode generators. Calculations performed for the case of InP with applied fields near the switching on of self-oscillations evidence a sharp peak of the current spectral-density at the generation frequency.

INTRODUCTION

The small-signal and noise spectra play a fundamental role in characterizing carrier transport of semiconductor devices. The frequency behavior of these quantities reflects both dynamic and relaxation processes of the microscopic mechanisms, hence, it can be used to investigate the physical phenomena responsible for the performances of the device of interest. The aim of this communication is to present an original calculation of the small-signal and noise characteristics of near micron n^+nn^+ InP diodes. The current and voltage spectral-densities are calculated within a hydrodynamic model which is validated by comparing the results with a direct simulation performed with an ensemble Monte Carlo method.

THEORY

To calculate the current and voltage noise in the framework of a hydrodynamic model we have recently developed[1] it is convenient to use the general procedure based on the impedance field[2] or transfer impedance[3] methods. For one-dimensional geometry the voltage spectral-density per unit surface, $S_U(f)$, caused by the fluctuations of the carrier velocities for spatially uncorrelated sources, is given by[2]:

$$S_U(f) = e^2 \int_0^L n(z)|\nabla Z(z,f)|^2 S_v(z,f) dz \qquad (1)$$

where e is the electron charge, L the length of the device between the probing electrodes taken along the z direction, f the frequency, $n(z)$ the local carrier concentration, $|\nabla Z(z,f)|$ the absolute value of the local impedance-field per unit surface, $S_v(z,f)$ the local spectral density of velocity fluctuations. According with Ref. [2,3] the impedance field can be represented as the ratio of the Fourier components of the local electric field to the total current flowing through the device. Integration of the impedance field over the spatial coordinate z gives the small-signal impedance of the whole device:

$$Z(f) = \int_0^L \nabla Z(z,f)dz \tag{2}$$

Thus, for the noise analysis under current driven operation, we need to evaluate the three quantities: $n(z)$, $\nabla Z(z,f)$ and $S_v(z,f)$. Conversely, from the Langevin theorem[4], the current spectral-density $S_j(f)$ can be evaluated as:

$$S_j(f) = S_U(f)|Y(f)|^2 \tag{3}$$

where we have introduced the small-signal admittance $Y(f) = Z^{-1}(f)$. For the calculation of $S_v(z,f)$ it is assumed that the spatial dependence is determined by the local instantaneous energy $\epsilon(z)$ only, and the energy-field dependence is the same as in the homogeneous steady-state. Under these assumptions, $S_v(z,f) \equiv S_v^0(\epsilon,f)$ can be expressed analytically through the parameters of the hydrodynamic model[1]. Following the notation of Ref. [1], it is:

$$S_v^0(\epsilon,f) = \frac{4\phi_+\nu_+}{\nu_+^2 + (2\pi f)^2} + \frac{4\phi_-\nu_-}{\nu_-^2 + (2\pi f)^2} \tag{4a}$$

for the case of two real eigenvalues $\nu_\pm$ of the spectrum of the hot-carrier relaxation rates, and as:

$$S_v^0(\epsilon,f) = 4 < \delta v^2 > \frac{\nu[\nu_R^2 + \omega_0^2 + (2\pi f)^2] + B\omega_0[\nu_R^2 + \omega_0^2 - (2\pi f)^2]}{[\nu_R^2 + (\omega_0 - (2\pi f))^2][\nu_R^2 + (\omega_0 + (2\pi f))^2]} \tag{4b}$$

for the case of two complex conjugate eigenvalues $\nu_R \pm i\omega_0$ of the spectrum of the hot-carrier relaxation rates. Here $\phi_\pm$ and $< \delta v^2 >$ (dimension velocity square) and B (dimensionless) are the coeffcients of the matrix expression with the above eigenvalues. The impedance field is calculated under current driven operation by using the transient response of the local electric field $E(z,t)$ to a delta like perturbation of the total current. To this end, the steady-state inhomogeneuos distributions $n_0(z)$, $v_0(z)$, $\epsilon_0(z)$ and $E_0(z)$ corresponding to the total current density $j_o = en_0v_0$ are first calculated. Then, a small perturbation of the local electric field δE_0 is added to $E_0(z)$ at every point and the transient response of the resulting electric field $E(z,t) = E_0(z) + \delta E(z,t)$ is calculated at the same j_o. The transient response is thus used to obtain the local response functions:

$$K_E(z,t) = \frac{1}{\epsilon_r\epsilon_0}\frac{\delta E(z,t)}{\delta E_0} \tag{5}$$

where ϵ_r is the relative static dielectric constant, and ϵ_0 the vacuum permittivity. Finally, the impedance field is calculated by Fourier transforming as:

$$\nabla Z(z,f) = \int_0^\infty K_E(z,t)exp(-i2\pi ft)\, dt \tag{6}$$

Such an approach allows the impedance field to be calculated at once for the whole frequency range of interest.

RESULTS

Calculations are applied to the case of n^+nn^+ InP diodes at room temperature with doping levels of $n = 2 \times 10^{16}$ and $n^+ = 1 \times 10^{18}\ cm^{-3}$. The cathode, n region and anode lengths are respectively 0.05, 1.0 and 0.1 μm. Abrupt homojunctions are assumed. Under voltage driven operation this diode is unstable in the restricted region of applied voltages $2 < U_0 < 5\ V$ while it is stable outside this region. The hydrodynamic calculations are performed under the current driven operation at $j_0 = 1.03 \times 10^9\ Am^{-2}$ which corresponds to an applied voltage $U_0 = 5.0\ V$. The results of the small-signal calculations are presented in Figs. 1 and 2. Figure 1 shows the real and imaginary parts of $Z(f)$. $Re[Z(f)]$ is negative in the frequency region $90 < f < 235\ GHz$. Amplification of microwave power is possible in this region. $Im[Z(f)]$ is negative everywhere. This provides a stability of the unloaded diode. The real and imaginary part of the admittance are presented in Fig. 2. The resonant behavior of $Y(f)$ near $f = 230\ GHz$ is evident. This frequency corresponds to the minimum time required for the accumulation layer to transit across the diode. The spectral density of voltage fluctuations obtained by the various approaches are compared in Fig. 3. The dashed curve corresponds to the hydrodynamic calculation performed using Eq. (1). The solid curve is the result of a direct simulation by the Monte Carlo method. Dots corresponds to $S_U(f) = S_j(f)|Z(f)|^2$, where $S_j(f)$ is calculated directly from the Monte Carlo method under the voltage driven operation. This $S_j(f)$ is reported in Fig. 4 by the dots. Here, dashed and solid lines correspond to the spectral density of the current fluctuations calculated from Eq. (3) using the $S_U(f)$ obtained from the hydrodynamic and Monte carlo approaches, respectively. One can observe full quantitative and reasonable qualitative agreement among the three various procedures. This agreement validates the present calculation which has the merit of providing reliable results within a simple and low-cost computer enviroment. In this way, transport and noise modelling can be simultaneuosly carried out for device engineering and physical analysis under extreme conditions. This work is partially supported by the Commission of European Comunity (CEC) through the contract CIPA3510PL921499.

REFERENCES

1. V. Gružinskis, E. Starikov, P. Shiktorov, L. Reggiani, M. Saraniti and L. Varani, Semicond. Science Technol., to be published (July, 1993).
2. W. Shockley, J.A. Copeland, R.P. James in: Quantum theory of atoms, molecules and solid state, Ed. P.O. Lowdin (Academic Press, New York, 1966) p. 537.
3. K.M. van Vliet, A Friedman, R.J.J. Zijlstra, A. Gisolf and A. van der Ziel, J. Appl. Phys., **46**, 1804 (1975).
4. J.P. Nougier, Methodes de calcul du bruit de fond, (Montpellier, 1975) unpublished.

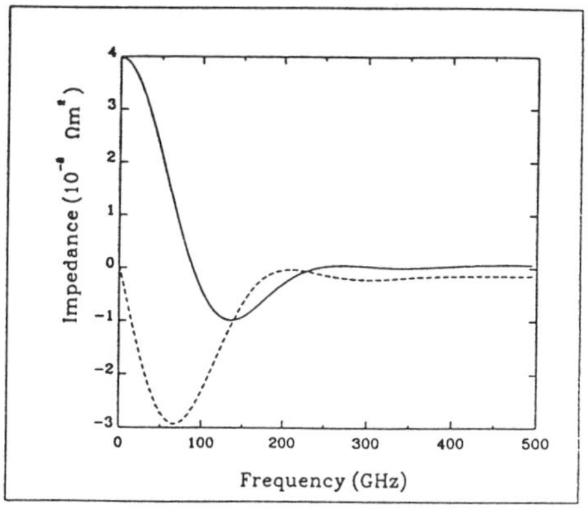

Fig. 1 - Frequency dependence of the real (solid line) and imaginary (dashed line) parts of $Z(f)$ calculated with the hydrodynamic model at room temperature with $U_0 = 5\ V$. Parameters of the n^+nn^+ InP structure are: $n = 2 \times 10^{16}$ and $n^+ = 1 \times 10^{18}\ cm^{-3}$. The cathode, n region and anode lengths are respectively 0.05, 1.0 and 0.1 μm.

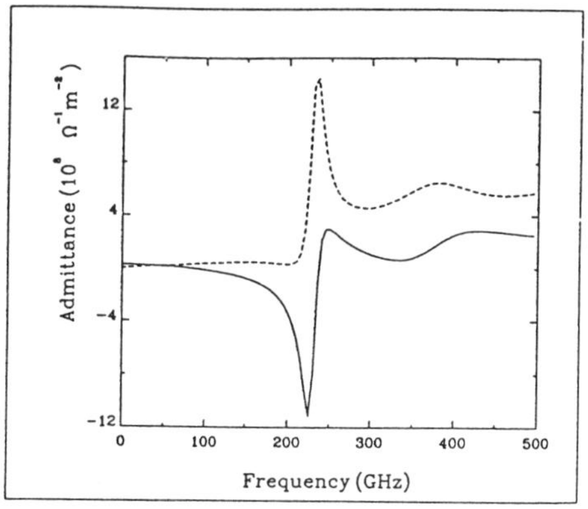

Fig. 2 - Frequency dependence of the real (solid line) and imaginary (dashed line) parts of $Y(f)$ calculated from the data of Fig. 1.

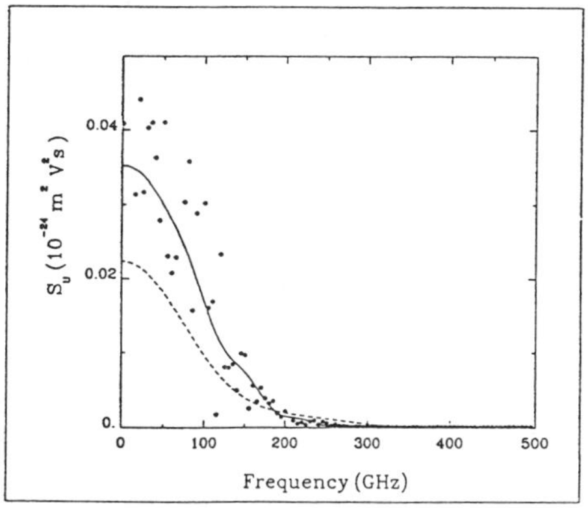

Fig. 3 - Frequency dependence of $S_U(f)$. Dashed line reports the results obtained from the hydrodynamic model, solid line from Monte Carlo simulations, dots from Eq. (3) substituting for $S_j(f)$ the results of the Monte Carlo simulation.

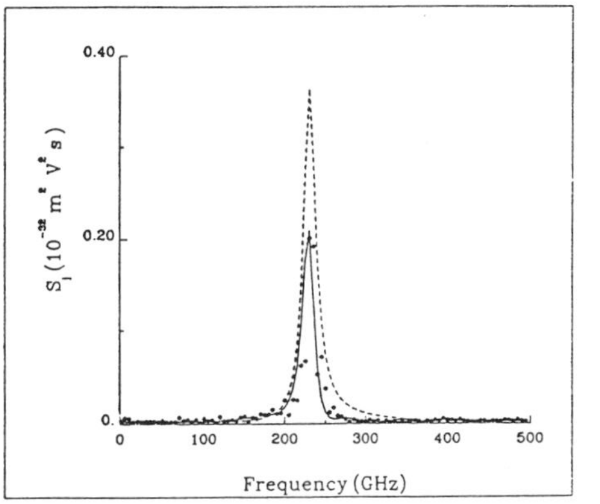

Fig. 4 - Frequency dependence of $S_j(f)$. Dots report the results obtained from Monte Carlo simulations, solid and dashed lines from the analogous curves of Fig. 3 using Eq.(3).

MODEL OF THE RTS NOISE IN SEMICONDUCTOR DEVICES

M. Sikulova and J. Sikula,
Technical University of Brno, Zizkova 17, 60200 Brno,
Czech Republic, tel/fax 00 425 74 23 57

ABSTRACT

The RTS noise is assumed to be induced by quantum transitions of charge carriers from trap energy levels to both the conductivity and the valence bands. The charge state of the trap controls the channel conductivity in MOSFET's. Formulas for the time constants describing the RTS noise in a general case of both n-type and p-type semiconductors have been derived. From the statistics of the time duration additional information on the generation-recombination process is yielded. It is shown that microscopic processes related to quantum transitions are through the channel current modulation transformed into measurable quantities.

INTRODUCTION

The RTS noise was observed in many semiconductor devices, bipolar as well as unipolar. Recently, attention has been focused mainly on the source-drain current noise in micron or submicron sized FET's.

The RTS noise provides information on the charge carrier trapping processes caused by single defects. The crucial role in submicron MOSFET's is played by the interface between silicon and its thermally grown oxide. The degree of perfection corresponds to 1 through 100 defects per square micrometer[1].

From experimental studies it is known that:

i) Individual defects at the $Si{:}SiO_2$ interface create traps with a broad energy level interval in the forbidden band. Traps with deep energy levels act as g-r centers whereas the shallow ones as capture and emission centers.

ii) The local carrier density n_S for the trap level position at the interface is given by

$$n_s = n_0 \exp\left[-e\,\psi_s/kT\right] \quad . \tag{1}$$

Here ψ_s is the surface potential, n_0 is the carrier density in the bulk.

iii) For the transition of an electron from the channel to the trap to be possible, an additional energy ΔE is required. Coulomb barrier changes the energy level of the neutral trap and its capture cross section. This results in a wide range of the time constants for the capture and emission which is of the order from 10^{-3} to 10^{+3} s.

QUANTUM TRANSITIONS AND TRANSITION PROBABILITY INTENSITIES

The theories[1-4] are based on the assumption that the quantum transitions of charge carriers take place between a trap energy level and one band only. This is correct for shallow traps and thermodynamic equilibrium. If the system is not in the thermodynamic equilibrium than generation and/or recombination processes appear and the simple models do not provide satisfactory results. Moreover, in a general case it is to be distinguished between two stochastic processes:

The primary process $X(t)$, whose states are described by the numbers of electrons in the conductivity band and the numbers of holes in the valence band and the charge states of the trap. The states of the primary process cannot be measured. Measurable quantity is the current modulation, which we denote by the secondary process $Y(t)$.

THE PRIMARY PROCESS $X(t)$

To describe our model we suppose that a single defect acts as an acceptor-like center with an energy level E_t and that the quasi-Fermi levels near the location of the trap are E_{Fns} and E_{Fps}. Then for an n-type channel the primary process $X(t)$ is described in Table I., where the processes of generation and recombination are described separately.

Table I *The primary and secondary states*

	recombination			generation		
$X(t)$	0	1	2	0	1	2
trap level	1	0	0	1	0	0
conduction band	$n-1$	n	$n-1$	n	n	$n+1$
valence band	p	p	$p-1$	$p+1$	p	$p+1$
$Y(t)$	α	β	β	α	β	β

In what follows we describe the case, in which the following transitions are possible: capturing, emission and recombination.

At $X(t) = 0$ one electron is captured by the neutral center and the number of electrons in the conductivity band is decreased by one. The number of the holes in the valence band remains constant. The state $X(t) = 1$ corresponds to the emission of an electron from the center to the conductivity band whereas the state $X(t) = 2$ to the transition of an electron from the center to the valence band. In this way the recombination process was realized.

The transition probability density μ_{01} from the state $X(t) = 0$ to the state $X(t) = 1$ is given, using Shockley-Read notation, by $\mu_{01} = c_n n_1$, in which case the electron is emitted from the trap to the conductivity band.

Similarly, the transition probability density μ_{02} from the state $X(t) = 0$ to the state

$X(t) = 2$ is $\mu_{02} = c_p p_s$. In this case the electron is emitted from the trap to the valence band or a hole is captured by the trap.

The primary process is described by a matrix of the transition probability densities for the capture, emission and recombination or generation[5]:

recombination *generation*

$$\mu_{ij} = \left\| \begin{matrix} -c_n n_1 - c_p p_s & c_n n_1 & c_p p_s \\ c_n n_s & -c_n n_s & 0 \\ c_p p_1 & 0 & -c_p p_1 \end{matrix} \right\| \qquad \mu_{ij} = \left\| \begin{matrix} -c_n n_1 - c_p p_s & c_p p_s & c_n n_1 \\ c_p p_1 & -c_p p_1 & 0 \\ c_n n_s & 0 & -c_n n_s \end{matrix} \right\| \qquad (2)$$

for $i,j \in (0, 1, 2)$.

As was shown[5], the eigenvalues of this matrix are reciprocal values of the time constants of the primary process.

The absolute probability distribution Π_i of the $X(t)$ states for a stationary case and thermodynamic equilibrium is given by[5]

$$\Pi_0 = (1 + \frac{n_1}{n_s} + \frac{p}{p_1})^{-1} = (1 + 2\, e^{x})^{-1}, \qquad \Pi_1 = \Pi_2 = (2 + e^{-x})^{-1}, \qquad (3)$$

where

$$n_s = N_c \exp\{(E_{Fns} - E_c)/kT\} \qquad p_1 = N_v \exp\{(E_v - E_t)/kT\}$$

$$n_1 = N_c \exp\{(E_t - E_c)/kT\} \qquad n_1/n_s = \exp\{(E_t - E_{Fns})/kT\} = e^{x} \qquad (4)$$

$$p_s = N_v \exp\{(E_v - E_{Fps})/kT\} \qquad p_1/p_s = \exp\{(E_{Fps} - E_t)/kT\} = e^{-y}$$

THE SECONDARY PROCESS $Y(t)$

It consists in the current modulation and has two states, α and β. The corresponding pulse time durations are τ_α and τ_β. The absolute probability distribution Π_α is given by the absolute probability distribution of the primary process: $\Pi_\alpha = \Pi_0$ and is of the Fermi-Dirac type (see Fig.1). The absolute probability distribution $\Pi_\beta = 1 - \Pi_\alpha$.

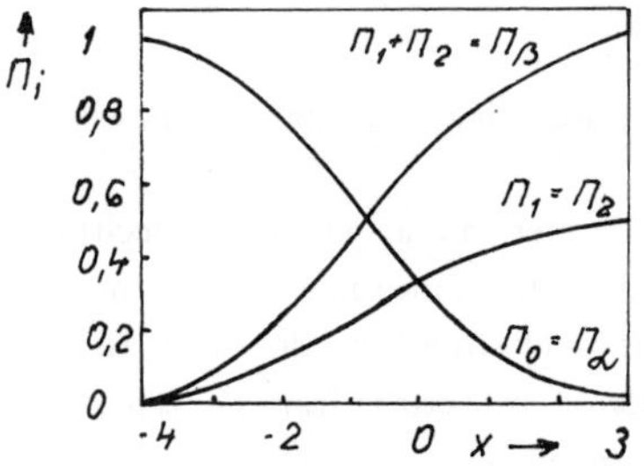

Figure 1 *The absolute probability distribution as a function of trap energy level position*

The probability density of the occupation times in the state α and β is

$$g_\alpha = \frac{1}{\tau_\alpha} \exp(-t/\tau_\alpha) \quad , \qquad (5)$$

$$g_\beta = \frac{c_n^2 n_s n_1}{c_n n_1 + c_p p_s} \exp(-t/\tau_1) + \frac{c_p^2 p_s p_1}{c_n n_1 + c_p p_s} \exp(-t/\tau_2) \quad , \tag{6}$$

where

$$\tau_\alpha = (c_n n_1 + c_p p_s)^{-1} , \quad \tau_\beta = \tau_\alpha \left(\frac{n_1}{n_s} + \frac{p_s}{p_1}\right) , \quad \tau_1 = 1/c_n n_s , \quad \tau_2 = 1/c_p p_1 . \tag{7}$$

An important information can be gained from the dispersion of the occupation time. For the α and β state it holds

$$D\{T_\alpha\} = \tau_\alpha^2 , \qquad D\{T_\beta\} = a_1\tau_1^2 + a_2\tau_2^2 + a_1 a_2(\tau_1 - \tau_2)^2 + a_1 a_2 (\tau_1 - \tau_2)^2, \tag{8}$$

where $a_1 = (1 + \frac{c_p p_s}{c_n n_1})^{-1}$; $a_2 = 1 - a_1$. The ratio of these times is

$$\frac{\tau_\alpha}{\tau_\beta} \approx \frac{2n_s}{n_1} = 2 \exp[(E_{Fns} - E_t/kT] . \tag{9}$$

In Table II. a comparison of our model with other ones[1-4,6] is made.

Table II *Time constants for three and two states model*

3 states model	$\tau_\alpha = 1/(c_n n_1 + c_p p_s)$	$\tau_\beta = 1/(c_n n_s + c_p p_s(n_s/n_1)) + 1/(c_p p_1 + c_n n_1(p_1/p_s))$
Ref. 4, 6	$\tau_e = 1/c_n n_1$	$\tau_c = 1/c_n n$

CONCLUSION

The RTS noise gives information on quantum transitions of the carriers between the trap level and bands. Moreover, it is possible to distinguish between the two or three state primary processes by measuring the probability density and the dispersion of the occupation time in the β - state.

REFERENCES

1. M. Schulz, Surf.Sci. 132, 422 (1983).
2. E. Simoen, U. Magnusson and C. Claeys, Appl.Surf.Science, 63 (1993), 285.
3. M.J. Kirton et all, Semicond.Sci.Technol 4, 1116 (1989).
4. M. Schulz, J.Appl.Phys. (to appear 1993).
5. J. Sikula, M. Sikulova, P. Vasina and B. Koktavy: Burst Noise in Diodes, in Proc. of the 6-th Symp. on Noise in Physical Systems, p.100, Nat. Bureau of Stand. Washington D.C. (1981).
6. S.T. Hsu, R.J- Whittier and C.A. Mead, Solid.State Electron. 13, 1055 (1970)

PECULIARITIES OF HOT ELECTRON FLUCTUATIONS IN SUBMICRON SEMICONDUCTOR STRUCTURES: LIMITATION AND SUPPRESSION

V.N.Sokolov, V.A.Kochelap, and N.A.Zakhleniuk
Institute for Semiconductor Physics, Academy of Sciences of Ukraine
Pr. Nauka 45, Kiev-28 252650, Ukraine

ABSTRACT

We present theoretical investigations of fluctuations of hot electrons in submicron active regions, where the dimension 2d of the region is comparable to the electron energy relaxation length L_ϵ. In the low-frequency limit, we find an exact solution of the Langevin equation for space-dependent electron fluctuations. The numerical calculations of spectral densities of fluctuations of the electron temperature and current are presented. The new physical phenomenon is reported: the fluctuations depend on the sample thickness, with $2d < L_\epsilon$ a suppression of fluctuations arises in the range of fluctuation frequencies $\omega\tau_\epsilon \ll 1$, τ_ϵ is the electron energy relaxation time.

INTRODUCTION

The latest achievements in modern semiconductor technology have made possible the investigations of submicron structures with different thicknesses and properties of conducting channels, their interfaces, etc. The main attention in studying of such structures is usually given to the quantum phenomena. Among them the quantum size effects take up of particular place. These cause the lowering of the electron gas dimensionality, i.e. the formation of 2D, 1D, and 0D electron systems. Under these conditions the electron transport shows the set of new behaviors. Such effects take place, in general, at low temperatures. With increasing of temperature, the quantum effects vanish, because the phonon scattering prohibits long range coherence arising from the wave nature of electrons. However, the quality and physical dimensions of active layers are such that the kinetic lengths characterizing nonequilibrium transport processes still remain comparable to or exceed these dimensions. In this way using the submicron structures, we enter with increasing of temperature into the region of classic size effects.

The classic size effects have been extensively investigated for metal and semiconductor films. Some of them have resulted in the improvement of conductive layers characteristics in comparison with those of the bulk samples. The progress in modern submicron technology enables to investigate the size effects at a qualitatively new level in order to deepen understanding of the physical processes and to control the electrophysical characteristics of semiconductor structures.

One of such possibilities of substantial alteration and control of fluctuation processes under size effect conditions is discussed in this report. We shall demonstrate the new phenomenon of limitation of hot electron fluctuations and suppression of current noise. It can be noted that the phenomenon studied here is completely different from the noise suppression under ballistic and diffusive quantum (vertical) transport in semiconductor structures.

The previous kinetic theories of hot electron fluctuations have been developed without taking into account the influence of crystal boundaries. In this case all statistical average characteristics of the electron gas are spatially homogeneous and only volume stochastic sources of fluctuations are essential. The works on the theory of hot electron fluctuations taking into consideration the influence of crystal boundaries are, practically, absent in the literature.

THE PROBLEM, APPROACH, AND CRITERIA

In this report the fluctuations of spatially inhomogeneous hot electron gas are theoretically investigated under size effect related with the characteristic length of the electron energy relaxation.

We consider the sample of a small extent with the thickness of $2d$ in the y-direction (the smallest size of the sample), which is comparable to the characteristic diffusion length $L_\epsilon = \sqrt{D\tau_\epsilon}$, D is diffusion coefficient. External d.c. heating electric field $E = E_x$ and electric current $j = j_x$ are in the x-direction. All kinetic parameters of electron transport considered here are supposed to depend only on y-coordinate. The electron gas remains homogeneous in the xz plane independently on the heating. We study the situation which is typical for hot electron plasma, when the following criteria take place:

$$\tau_{\vec{p}} \ll \tau_{ee} \ll \tau_\epsilon, \quad l_{\vec{p}}, l_D \ll 2d, L_\epsilon,$$

where $\tau_{\vec{p}}$, $l_{\vec{p}}$ are characteristic time and length of momentum relaxation, respectively; τ_{ee} is the time of electron-electron collisions, l_D is Debye length. The first set of these criteria means that the electron distribution function has a Maxwellian form with electron temperature T being dependent on y (electron temperature is the hydrodynamical parameter of hot electron gas).

Using the electron temperature approximation, the transport stochastic Boltzmann-Langevin equation has been obtained taking into account the hot electron size effect. The problem has following peculiarities: (i) the internal transverse, in the y-direction, d.c. and a.c. (fluctuative) electric fields arise in the sample; (ii) the fluctuations under the size effect are space-dependent; (iii) there are the additional surface sources of fluctuations besides of the volume ones. These peculiarities are taken into account in the calculations.

It is shown that all fluctuating quantities can be expressed only through the electron temperature fluctuation $\delta T(y,t)$. The fluctuation $\delta T(y,t)$ is defined by the stochastic equation of energy balance that follows from the Boltzmann-Langevin equation. The boundary conditions include the surface noise sources

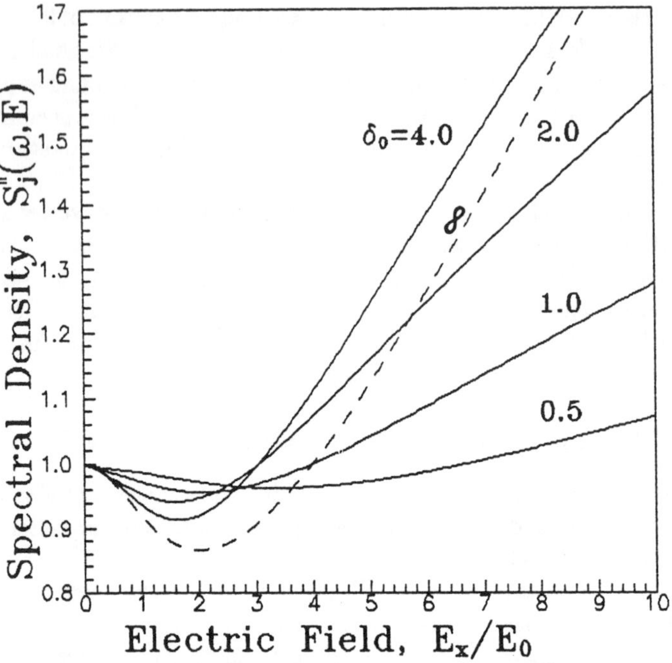

Fig.1. Field dependence of current fluctuations

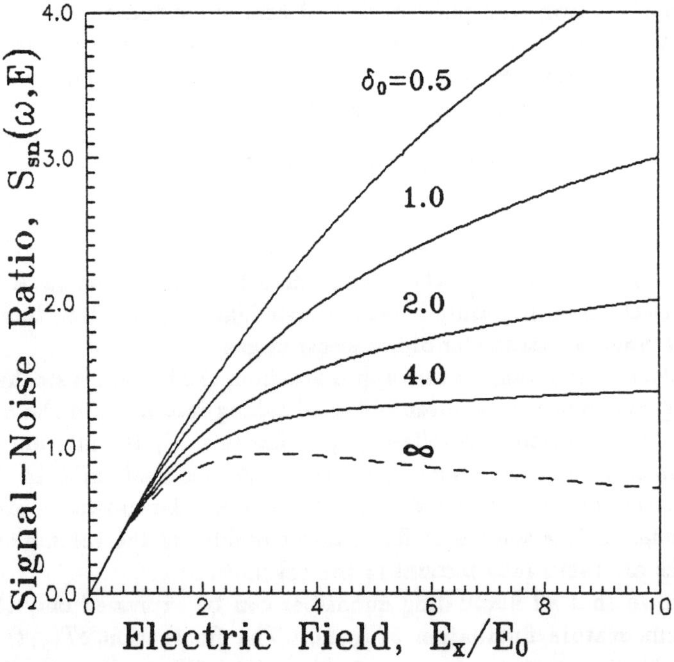

Fig.2. Field dependence of signal-noise ratio

and energy relaxation velocities $S^{\pm} = S(y = \pm d)$. If the surface rates of energy relaxation are very high $S^{\pm} \gg D/L_e$, the boundary conditions have the simple form: $T(y = \pm d) = T_0$, $\delta T(y = \pm d, t) = 0$, T_0 is the lattice temperature. We found the analytical solutions of the discussed equations in the low-frequency limit $\omega\tau_e \ll 1$ using the known solution of associated steady state problem, i.e. the electron temperature distribution $T(y)$.

RESULTS

The general expressions have been obtained for: spectral densities of electron temperature and electron current fluctuations, low-signal conductivity, noise temperature, and signal-noise ratio. The field and size dependences of these quantities have been numerically calculated and investigated. The electron interaction with acoustic and optical phonons has been taken into account.

In Fig.1 the dependence of longitudinal spectral density of current fluctuations, $S_j^{\parallel}(\omega, E)$, on E_s/E_0 (here $E_0 = k_0 T_0/eL_e$ is characteristic field of electron heating) for different sample thicknesses $\delta_0 = d/L_e$ is shown.

Fig.2 demonstrates field dependence of signal-to-noise ratio $S_n(\omega, E)$ for the same conditions as in Fig.1. In both cases considered in Fig.1,2 interaction only with optical phonons has been taken into account.

In the bulk samples at high electric fields the current density is saturated and the spectral density $S_j^{\parallel}(\omega, E)$ is proportional to E_x. Then $S_n^{\infty}(\omega, E) \sim E_x^{-1/2}$, i.e. the signal-to-noise ratio decreases with increasing of E_s (dashed line in Fig.2). One can follow the tendency of behavior of the signal-to-noise ratio in the example of extremely thin specimen, where Ohm's law takes place at high electric fields and the limitation of the noise at the level of the equilibrium fluctuations is realised at the same time. As a result the enhancement of signal-to-noise ratio parameter $S_n(\omega, E)$ can be attained.

CONCLUSION

Thus, there is the set of new interesting features of the fluctuation phenomena for the size effect considered. Among them the suppression of the low-frequency current noise and the significant increase of the signal-to-noise ratio for submicron structures with respect to the volume samples take place. These results can be extremely of interest for some modern device applications. Above we have presented the results for the case of extremely high scattering of the electron energy at the surfaces, when the effects are maximum. But the results obtained are not critically dependent upon this assumption. The effect of limitation of hot electron fluctuations has to be observed at the arbitrary finite rate of the surface energy scattering. Note that the presence of additional energy loss mechanisms at the sample surfaces does not lead to decreasing of integral intensity of fluctuations. Only the redistribution of the fluctuations occurs over the spectrum frequencies, i.e the decreasing for low frequencies and increasing in the range of $\omega \approx D/(2d)^2$.

1/f NOISE AND THERMAL NOISE OF A GaAs/$Al_{0.4}Ga_{0.6}As$ SUPERLATTICE

L.K.J. Vandamme*, S. Kibeya, B. Orsal and R. Alabedra
* Eindhoven University of Technology, Dept. EE, P.O. Box 513, Eindhoven, The Netherlands
Université de Montpellier II, CEM, 34095 Montpellier, France

ABSTRACT

The current fluctuations have been investigated in different GaAs/$Al_{0.4}Ga_{0.6}As$ superlattice structures. New expressions are presented for the observed 1/f and white noise. These expressions are compared with experimental results. A fair agreement has been found between theory and experiment.

INTRODUCTION

Three MBE grown multiquantum wells have been investigated with a thickness of 0.5; 0.2 and 2 μm and consisting of 25, 10 and 20 periods of GaAs wells and $Al_{0.4}Ga_{0.6}As$ barriers with thickness of 100 - 100 Å and 500 - 500 Å, respectively. These structures are designed to selectively enhance impact ionization by electrons in Capasso avalanche photodiodes.

The 1/f current noise shows the proportionality $S_{I_{1/f}} \propto I^{3/2}$ and the thermal noise $S_{I_{th}} = (2/3)4kT/R_{dp}$ with R_{dp} the current dependent dynamic resistance of the superlattice structure. The experimentally observed trends agree with calculations based on an injection diode model applied to a superlattice period.

MODEL FOR CURRENT TRANSPORT IN MULTIQUANTUM WELLS

The current voltage characteristic of a n^+nn^+ current injection diode depends on the space-dependent carrier distribution in the short n region. Here the GaAs interfaces are considered as electron reservoirs comparable to a n^+ contact in a n^+nn^+ structure and the AlGaAs barrier plays the role of the n-material. In such a structure the current can be space charge limited.

CALCULATIONS FOR THE CURRENT AND CURRENT FLUCTUATIONS IN MULTIQUANTUM WELLS

An analysis of the dynamic resistance as function of current shows $R_d \propto I^{-1/3}$ (see fig. (1)). If we assume a constant mobility $\mu = \mu_o$ in the $Al_xGa_{1-x}As$ barrier we obtain for the Mott-Gurney law $I = 9\varepsilon\mu_o SV^2_{ext}/(8\ell^3)$ with $\varepsilon = \varepsilon_o\varepsilon_r$ the permitivity and ℓ the length of the low doped barrier. If we assume a saturated velocity in the barrier then $I = 2\varepsilon Sv_{sat}V_{ext}/\ell^2$ and for the assumption of balistic transport in the barrier with a velocity given by $1/2\ mv^2 = qV_{ext}$ we expect $J = (4/9)(2q/m^*)^{\frac{1}{2}}\varepsilon V_{ext}^{3/2}/\ell^2$.[1]

If the current-voltage across one period of the multiquantum well is given by $I = I_o(V/V_o)^{\upsilon}$ with $2 \geq \upsilon \geq 1$, I_o and V_o characteristic values and V the voltage drop across one period, then we have for p periods

$$I = I_o(V_t/pV_o)^{\upsilon} \tag{1}$$

with V_t the applied voltage across p periods in series. The dynamic R_d and static resistances R_s then become for one single period

$$R_d = dV/dI = 1/\upsilon\ V/I = 1/\upsilon\ R_s \propto I^{(1-\upsilon)/\upsilon} \propto I^{-\gamma} \tag{2}$$

with $\gamma = (\upsilon-1)/\upsilon$. The static resistance for p periods R_{sp} becomes $R_{sp} = pR_s$ and for the dynamic resistance of p periods holds $R_{dp}=pR_d$, while the dependence on current remains as for a single period. Experimental results show $1/3 \leq \gamma < 1/2$ or $\upsilon \approx 3/2$ for the 500 - 500 Å structure and $\upsilon = 2$ for the 100 - 100 Å structure.

Starting from Poisson equation $\delta F/\delta x = \delta\rho/\varepsilon_o\varepsilon_r$ we calculate I vs V assuming that $q\delta n = \delta\rho$ and the excess charge of mobile carriers is not compensated by ionized impurities and the mobility is given by

$$\mu = \mu_o(F/F_o)^{-m} \tag{3}$$

with F_o a critical field strength and $0 < m < 1$ we find

$$I = \frac{(3-m)^{(2-m)} S.\varepsilon\mu_o F_o^{\ m}}{(2-m)^{(3-m)} L^{(3-m)}} V^{(2-m)} = I_o\left(\frac{V}{V_o}\right)^{\upsilon} \tag{4}$$

Hence
$$\upsilon = 2-m = 1/1-\gamma \tag{5}$$

To explain our experimental results we need $3/2 < \upsilon \leq 2$ or $1/3 < \gamma < 1/2$ which means $0 \leq m \leq 1/2$.

The 1/f and thermal noise are calculated with the Langevin method in a one dimensional treatment resulting in $I(x,t) = \overline{I} + \Delta I(x,t)$ with $\overline{\Delta I(x,t)} = 0$ and $\overline{\Delta I(x,t).\Delta I(x',t)} \propto (x-x')$. We use for the cross correlation spectrum for the thermal noise $S_{\Delta I}(x,x',f) = 4kTSq\mu n(x)\delta(x-x')$.

The current spectrum is obtained after integration over a barrier length ℓ by

$$S_I(f) = \frac{1}{\ell^2} \int_0^\ell \int_0^\ell S_{\Delta I}(x,x',f)dx.dx' = \frac{4kTSq\mu}{\ell^2} \int_0^\ell n(x)dx \qquad (6)$$

Assuming no mobility degradation ($m = 0$, $\upsilon = 2$ and $\gamma = \frac{1}{2}$) we find for the concentration

$$n(x) = \frac{3\varepsilon V}{4q\ell^{3/2}x^{1/2}} \qquad (7)$$

The thermal noise for one period in the MQW structure then becomes with eqs. (6) and (7)

$$S_I = 4kTS \cdot \frac{3\varepsilon\mu V}{2\ \ell^{3/2}} = (2/3)\ \frac{4kT}{R_d} \qquad (8)$$

Hence for $m = 0$ holds $I \propto V^2$ and $S_I \propto I^{\frac{1}{2}} \propto 1/R_d$ which agrees with the observed experimental results in fig. (2).
For a MQW with p periods in series the thermal noise S_I in eq. (8) must be divided by p which results in

$$S_{I_{MQW}} = (2/3)\ \frac{4kT}{p\ R_d} = (2/3)\ \frac{4kT}{R_{dp}} \qquad (9)$$

with R_d and R_{dp} the dynamic resistance of one and p periods respectively. The calculated thermal noise from eq. (9) agrees with the experimentally observed noise.

The 1/f noise is calculated by applying the empirical relation [2,3] and using the cross spectral density for the 1/f noise source $S_{\Delta I}(x,x',f) = (\alpha\ I^2/fSn(x))\delta(x-x')$. The current spectrum of one period is found after integration of $\int dx/n(x)$ using eq. (7). Assuming $m = 0$ ($\upsilon = 2$) we find

$$S_I = \frac{8\alpha q\ell I^2}{q\varepsilon SVf} \propto V^3 \propto I^{3/2} \qquad (10)$$

Substituting the value of the dynamic or static resistance leads to

$$S_I = \frac{\alpha q\mu_o R_s I^2}{\ell^2 f} = \frac{2\alpha q\mu_o R_d I^2}{\ell^2 f} \propto I^{3/2} \qquad (11)$$

The expected voltage fluctuations $S_v = R_d^2 S_I$ are proportional to $I^{\frac{1}{2}}$ for $\upsilon = 2$. The 1/f current noise of the complete MQW is found by dividing S_I for one period by the number of p periods in series.
Hence for the complete MQW the 1/f current noise becomes

$$S_I = \frac{\alpha q\mu_o R_{sp} I^2}{(p\ell)^2\ f} = \frac{2\alpha q\mu R_{dp} I^2}{(p\ell)^2\ f} \propto I^{3/2} \qquad (12)$$

Fig. 3 shows the experimental results for the 1/f noise. The calculated α values from the experimentaly obtained results [4] using the abovementioned equations based on mobility fluctuations in the AlGaAs barriers are of the order of $\alpha = 10^{-5}/\mu$ with μ in cm²/Vs.

EXPERIMENTAL RESULTS

We have confined ourselves to a current range of 10^{-6} A to 10^{-3} A where the dynamic resistance shows a proportionality of about $R_{dp} \propto I^{-1/3}$. The results for two different structures are presented in fig. 1.

We have performed noise measurements at room temperature in a frequency range of 1 Hz to 1 MHz. At frequencies below 20 kHz the 1/f noise prevails, above the white (thermal) noise is dominant. We have measured the noise spectra as a function of the average current passing perpendicular to the multi quantum well structures.

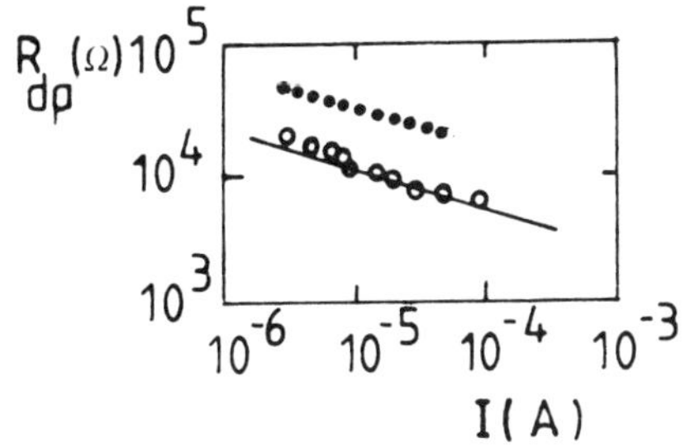

Fig. 1: The dynamic resistance R_{dp} versus current of a superlattice structure showing $R_{dp} \propto I^{-1/3}$. Dots are from 20 periods of 500 Å GaAs wells and 500 Å $Al_{0.4}Ga_{0.6}As$ barriers; circles are from 25 periods of 100 Å - 100 Å.

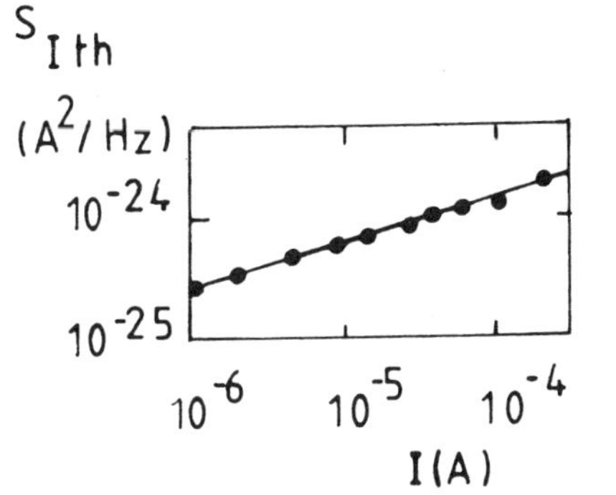

Fig. 2: Thermal noise at f = 100 kHz versus current of a superlattice with 20 periods of 500 Å - 500 Å, showing $S_{I_{th}} \propto I^{1/3}$

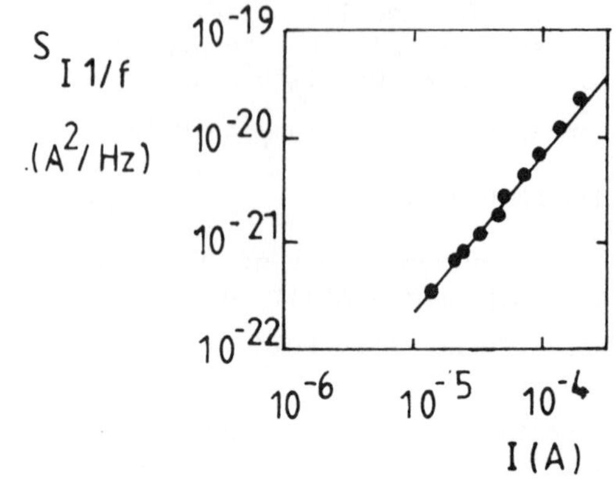

Fig. 3: The 1/f current noise at f = 3 Hz versus current of a superlattice with 20 periods of 500 Å - 500 Å showing $S_{I_{1/f}} \propto I^{3/2}$.

REFERENCES

1. K. Hess, "Advanced Theory of Semiconductor Devices", Prentice-Hall International, Englewood Cliffs, New Jersey, pp. 138-141, 1988.
2. F.N. Hooge, T.G.M. Kleinpenning and L.K.J. Vandamme; Rep. Prog. Phys., vol. 44, pp. 479-532, 1981.
3. T.G.M. Kleinpenning, Physica vol. 94B+C, pp. 141-151, 1978.
4. S. Kibeya, "Etude électrique et analyse du bruit électronique des photo-détecteurs à multipuits quantiques GaAs/GaAlAs". Thèse de doctorat Université Montpellier II, April 1992.

NUMBER AND CURRENT FLUCTUATIONS IN SUBMICRON SEMICONDUCTOR STRUCTURES

Luca Varani, Lino Reggiani
Dipartimento di Fisica ed Istituto Nazionale di Fisica della Materia,
Università di Modena, Via Campi 213/A, 41100 Modena, Italy

Tilmann Kuhn
Institut für Theoretische Physik, Universität Stuttgart,
Pfaffenwaldring 57, 7000 Stuttgart 80, Germany

Patrice Houlet, Jean Claude Vaissiére, Jean Pierre Nougier
Centre d'Electronique de Montpellier, Université des Sciences et
Techniques du Languedoc, 34095 Montpellier Cedex 5, France

Tomás González, Daniel Pardo
Departamento de Física Aplicada, Universidad de Salamanca,
Plaza de la Merced s/n, 37008 Salamanca, Spain

ABSTRACT

We present an analysis of number and current fluctuations in homogeneous n-Si resistors of submicron dimensions by employing analytical and Monte Carlo calculations. Results show the presence of long-time tails in the ballistic regime, which vanish when the length of the sample becomes comparable with the carrier mean-free-path. In the diffusive regime different time-scales related to diffusion, drift and dielectric relaxation are shown to characterize the behaviour of fluctuations.

INTRODUCTION

The recent trend in scaling down the dimensions of a device deep into the submicron size has emphasized the importance of a better understanding of fluctuations mainly because: (i) Fluctuations increase with decreasing system size; (ii) The electric field may reach quite large values, thus provoking the onset of hot-carrier phenomena[1]; (iii) Contacts play a dominant role[2]. The aim of this contribution is to present analytical and Monte Carlo calculations of the autocorrelation functions of number and current fluctuations in homegenous n-Si resistors when going from the ballistic to the diffusive regime.

BALLISTIC REGIME

The system we consider is a one-dimensional resistor of length L, terminated by ideal contacts i.e.: (i) When arriving at a contact the carrier is thermalized immediately, thus any correlation is destroyed; (ii) The contacts always remain at thermal equilibrium, hence emitting carriers according to the Maxwell-Boltzmann distribution

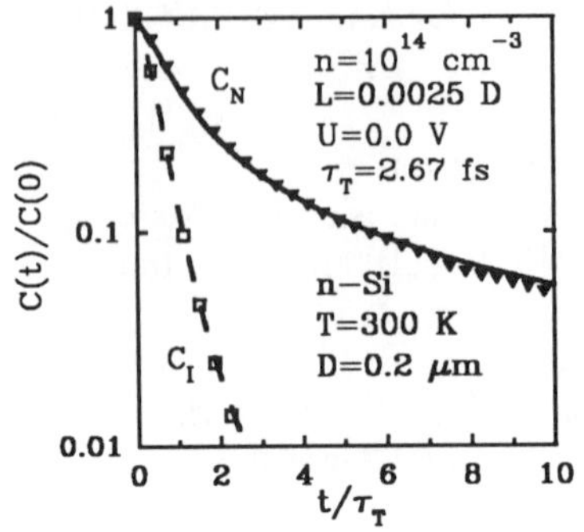

Fig. 1 - Normalized autocorrelation functions of number and current fluctuations at equilibrium for a Si ballistic resistor at $T=300$ K for the reported parameters. The lines are obtained from the analytical formula [Eqs. (1) and (2)] and the symbols represent the results of Monte Carlo calculations.

if the system can be considered as classical. In this case, at equilibrium, the autocorrelation functions of number and current fluctuations C_N and C_I are found to be given by[3]:

$$C_N(t) = N\left\{\text{erf}\left(\frac{\tau_T}{t}\right) - \frac{t}{\sqrt{\pi}\tau_T}\left[1-\exp\left(-\frac{\tau_T^2}{t^2}\right)\right]\right\} \tag{1}$$

$$C_I(t) = \frac{Ne^2}{2\tau_T^2}\left\{\text{erf}\left(\frac{\tau_T}{t}\right) - \frac{2t}{\sqrt{\pi}\tau_T}\left[1-\exp\left(-\frac{\tau_T^2}{t^2}\right)\right]\right\} \tag{2}$$

where e is the absolute value of the electronic charge, N the number of carriers in the sample and τ_T the average transit-time, here equal to $L\sqrt{m/2k_BT}$, k_B being the Boltzmann constant, T the lattice temperature and m the effective mass. We notice that both C_N and C_I exhibit a nonexponential behavior, which, in the limit of $t \to \infty$, is proportional to τ_T/t and $(\tau_T/t)^3$, respectively. The long-time tail of C_N is related to carriers injected with small velocity. The above long-time tail is not present in C_I since for the calculation of the current the carriers are weighted by their velocity. These functions are shown in Fig. 1, where, together with the analytical curves of Eqs. (1) and (2), we have reported the Monte Carlo results obtained with the same microscopic model of Ref. [4]. These last have been calculated by considering the transport properties of a small slice, with dimension L (much smaller than the carrier mean free path), of a homogeneous resistor of length D. We remark the excellent agreement between the analytical and the Monte Carlo results, giving us confidence in the microscopic model used in the simulation.

FROM BALLISTIC TO DIFFUSIVE REGIME

Taking advantage of the above agreement, we have investigated situations where analytical calculations are not available. To this end, we have increased the length of the sample to study the transition from the ballistic to the diffusive regime. This is shown in Fig. 2 for the case of number fluctuations. Here we observe that, by gradually increasing the length of the sample, C_N decays more slowly at short times. This is attributed to the fact that very fast carriers have a large probability to undergo

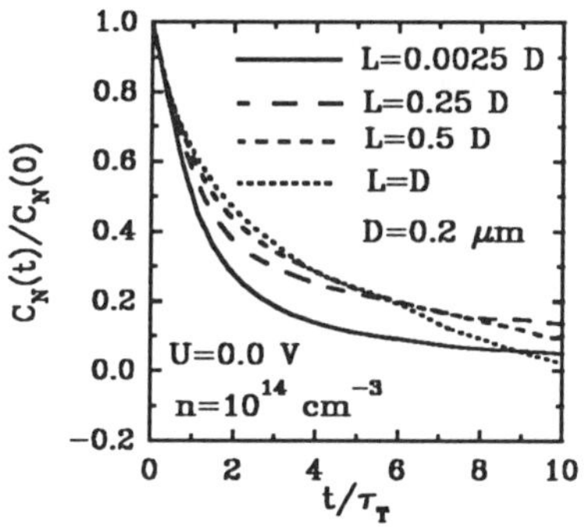

Fig. 2 - Normalized autocorrelation functions of number fluctuations at equilibrium for the reported parameters and lengths of the sample.

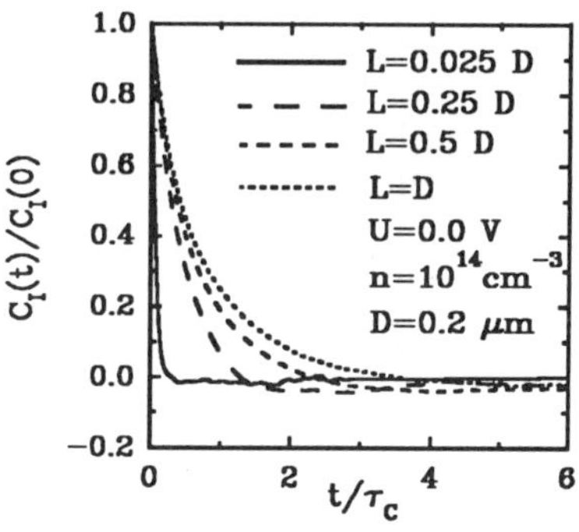

Fig. 3 - Normalized autocorrelation functions of current fluctuations at equilibrium for the reported parameters and lengths of the sample.

scatterings and therefore, once leaving one contact they reach the opposite one at longer times. A second effect is the suppression of the long-time tail (in units of t / τ_T), which is related to the decreasing importance played by carriers entering the sample with small velocities. Indeed, because of scatterings they can return more or less quickly to the contact from which they originate, or increase their velocity from the energy gained by the bath. The results for C_I are reported in Fig. 3 where the time t has been divided by the collision time $\tau_c = \mu m / e$, where μ is the ohmic mobility. Here, by increasing the length of the sample, C_I decays more slowly since the characteristic time for current fluctuations goes from the transit-time to the collision time, finally reaching the expected exponential shape with a time constant given by the collision time when the regime is completely diffusive.

DIFFUSIVE REGIME

In the presence of scattering processes and/or of an applied electric field, Eqs. (1) and (2) are no longer valid. We conjecture that the characterization of the number fluctuations by means of a characteristic time is still possible. Plausible candidates are: the diffusion time $\tau_D = L^2 / D(E)$, the drift time $\tau_{vd} = L / v_d(E)$ or the differential dielectric relaxation time $\tau_\varepsilon = \varepsilon / en\mu'(E)$. Here D is the longitudinal diffusion coefficient, v_d the average drift velocity, ε the dielectric constant of the material, n the carrier concentration, μ' the differential mobility and E the electric field. On the basis of this conjecture we introduce the characteristic time τ, given by the parallel of the above three times as: $\tau^{-1} = \tau_D^{-1} + \tau_{vd}^{-1} + \tau_\varepsilon^{-1}$. Then we have investigated the number fluctuations in the diffusive regime at increasing field strengths. Even in this case the function C_N is found to exhibit a non-exponential shape. Anyway, from the decay of the correlation function we have extracted a characteristic time which has been compared with the analytical one. This latter has been determined through phenomenological expressions for D, v_d and μ' as functions of E^5. In Fig. 4 we have

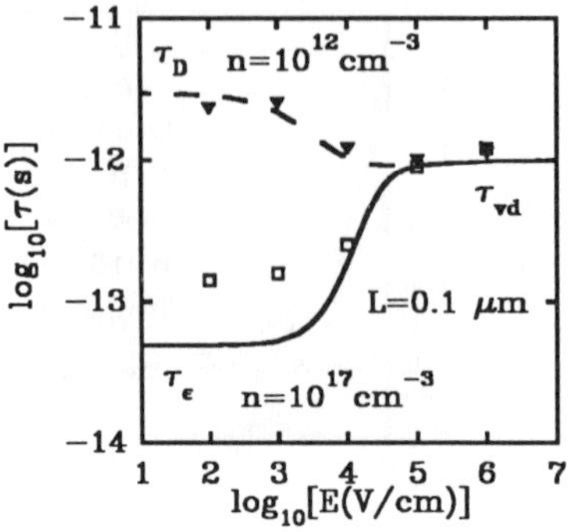

Fig. 4 - Characteristic times associated with fluctuations of *N(t)* as a function of the electric field in the diffusive regime for the reported parameters. The dashed line and the closed triangles refer to calculations for a low carrier concentration of 10^{12} cm^{-3}; the continuous line and the open squares to calculations for a high carrier concentration of 10^{17} cm^{-3}.

reported the results obtained from the Monte Carlo (symbols) and the analytical calculations (lines) for two different carrier concentrations. The good agreement found between the Monte Carlo results and the analytical expression for τ supports our conjecture. Furthermore, we remark that, at low density, the characteristic time goes from the large value of τ_D to the short of τ_{vd} On the contrary, at high density, it goes from the short value of τ_ε to the large of τ_{vd}.

CONCLUSIONS

The main results of the calculations are summarized as follows: (i) Excellent agreement has been found between the analytical expressions for the autocorrelation functions of number and current fluctuations in the ballistic regime and the corresponding Monte Carlo calculations. (ii) In the ballistic regime we have evidenced a long-time tail in number fluctuations related to the dominant role played by carriers with small velocity. (iii) In passing from the ballistic to the diffusive regime the long-time tail vanishes due to the presence of scatterings. (iv) In the diffusive regime different time-scales related to diffusion, drift and dielectric relaxation are shown to characterize the behaviour of number fluctuations.

This work has been partially supported by the CEC ERASMUS ICP91-F-1018 project, by the Italian Ministero dell' Università e della Ricerca Scientifica e Tecnologica (MURST), by the Special Action TIC92-0229-E from the Comisión Interministerial de Ciencia Y Tecnología (CICYT) and by the contract ERBCHBICT920162 from the CEC *"Human Capital and Mobility"* Programme.

REFERENCES

1. L. Reggiani, *Hot Electron Transport in Semiconductors*, Topics in Applied Physics Vol. 58 (Springer-Verlag, Berlin, 1985).
2. W. R. Frensley, Rev. Mod. Phys. **62**, 745 (1990).
3. L. Varani, L. Reggiani, T. Kuhn, P. Houlet, T. González and D. Pardo, to be published.
4. P. Lugli, L. Reggiani and J. J. Niez, Phys. Rev. **B40**, 12382 (1989).
5. M. A. Omar and L. Reggiani, Solid State Electron. **30**, 692 (1987).

$1/f$ CYCLE SLIPPING BEHAVIOR IN PHASE–LOCKED LOOPS WITH TIME DELAYS

W. Wischert
Institut für Theoretische Physik und Synergetik, Universität Stuttgart,
D–7000 Stuttgart 80, Fed. Rep. of Germany

M. Olivier and J. Groslambert
Laboratoire de Physique et Métrologie des Oscillateurs du CNRS,
32, av. de l'Observatoire, F–25000 Besançon, France

Abstract

In this paper we report on delay–induced instabilities observed in first order phase–locked loops which lead to 2π–periodic jumps of the phase error signal., a so–called cycle slipping motion. Spectral measurements of this behavior yield a $1/f$ spectrum.

1. Introduction

Phase-locked loops (PLL's) under the influence of external modulation can show a variety of dynamical instabilities when driven in their nonlinear regime[1]. When taking into account the finite propagation time of the signal in the nonlinear feedback loop, time delay effects may become important giving rise to chaotic instabilities even in first order PLL's.[2] Such delay–induced instabilities, first predicted in physics by Ikeda[3] found a widespread application especially in optical bistable devices[4,5,6] but also in many other scientific disciplines like physiological control systems[7] and economics[8]. In applications where a synchronous operation of the components is required, instabilities leading to a cycle slipping motion are an annoying feature since they influence the quality of the device under consideration. It is therefore desirable to elucidate this phenomenon both from an experimental and a theoretical point of view. To this end we first present an experimental method how to detect these phase jumps and how to analyse their spectral characteristics. We then show that a simple delay differential equation modelling the dynamical behavior of the loop can exhibit the same phenomena.

2. Experimental Results

The experimental setup of the delayed synchronization of a first order PLL is shown in Fig. 1. The experiments were done at a reference frequency of 135 MHz with a sinusoidal phase detector (HP 10514A Mixer). The analog delay line ($\tau = 15\mu sec$) models time delays which could arise in the feedback loop. The low–pass filter shown in Fig. 1 serves us only to eliminate higher frequency components. Its bandwidth will be assumed to be sufficiently high as not to influence the order of the loop.

Experimentally, phase jumps in multiples of $\pm 2\pi$ cannot be detected directly at the output of a phase detector since it is insensitive to signals which are shifted by 2π. Therefore we have realized a system using digital frequency dividers ($N = 10$) which reduces the phase jumps to multiples of $\frac{2\pi}{10}$. At an additional phase detector these jumps could then be detected. As a control parameter we have chosen the sensitivity of the local oscillator which depends linearly from the open loop gain K. When $K > 104$

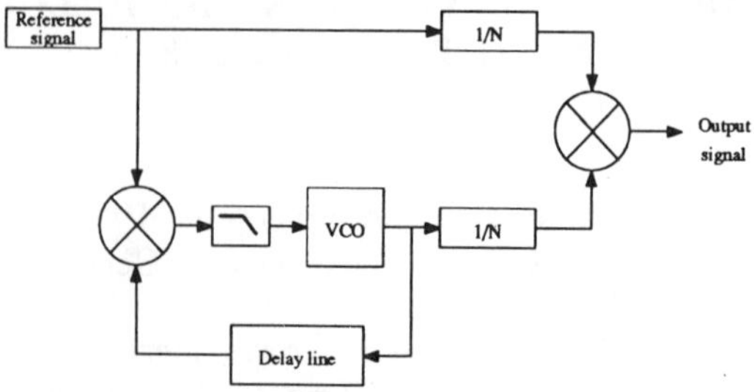

Figure 1: Experimental Setup of a first order PLL with analog delay lines.

kHz an oscillatory instability of the phase error signal occurs. Although the output signal is oscillatory in this regime, the loop still "synchronizes" the phase difference of the two oscillators. This can be seen by altering the angular frequency deviation $\Delta\omega$ which leads to an additional dc component of the phase error signal.

When the control parameter exceeds $K > 211$ kHz the loop desynchronizes and a cycle slipping motion of the phase error signal occurs which means that the signal leaves its basin of attraction and begins to jump in an irregular way between different attractors shifted by $|\Delta\phi| = 2\pi$ (Fig. 2). A spectral analysis of this cycle slipping behavior leads to a $1/f$ spectral distribution (Fig. 3). The specific spectral characteristics

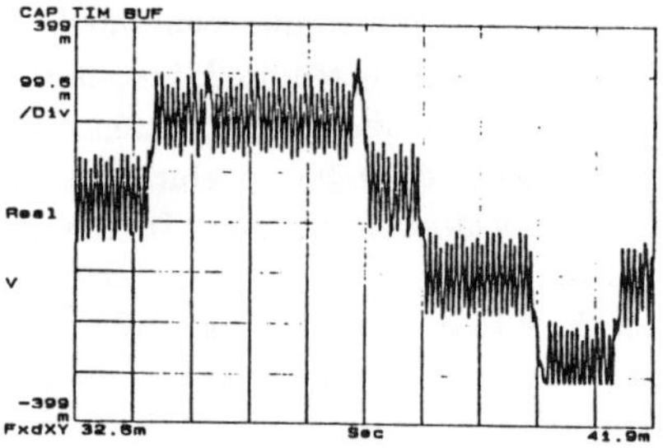

Figure 2: Cycle slipping of a first order PLL with time delay.

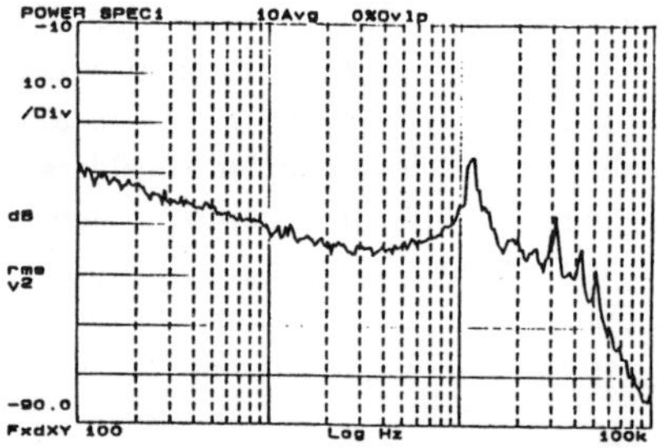

Figure 3: Spectral distribution of the cycle slipping motion

of the cycle slipping process is due to an intrinsic deterministic effect of the nonlinear dynamics of the loop and doesn't result from a fluctuating environment since white or $1/f^2$ noise introduced as a perturbation into the loop leaves the spectrum unchanged. It would be of great interest to know if the $1/f$ spectrum is related to an intermittent behavior of the delay system. Recent investigations of an intermittency route to chaos in delay systems[9] as well as studies on the chaotic behavior of Josephson junctions[10] leading to a similar cycle slipping process of the quantum phase difference across the junction seem to confirm this conjecture.

Experiments done with first order PLL's in the absence of time delays point out that the cycle slipping motion originating from this device is also associated with a

$1/f^{\alpha}$ spectrum with $\alpha \approx 0.8$ (see Figures 4 and 5). However, in this case the PLL has

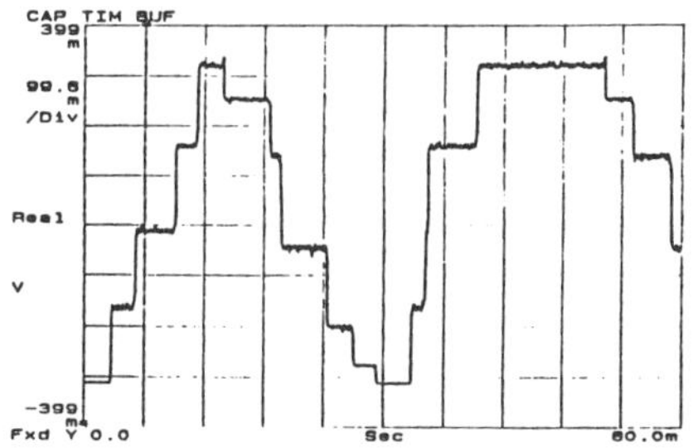

Figure 4: Cycle slipping of an ordinary first order PLL

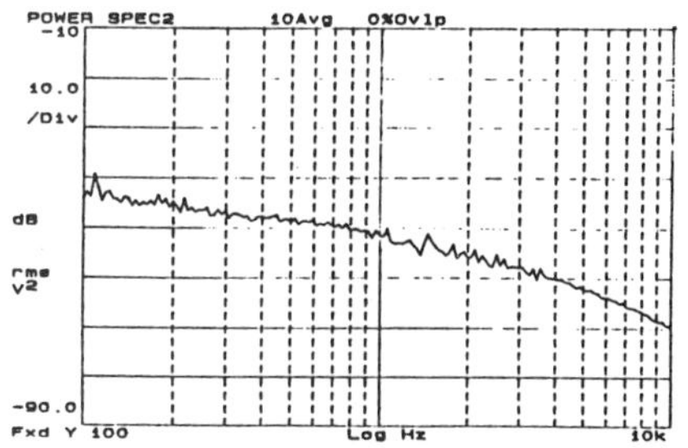

Figure 5: Spectral distribution of the cycle slipping motion

to be operated in the vicinity of its out–of–lock condition and random perturbation are required in order to induce the cycle slipping motion.

These results suggest that the $1/f$ spectrum is a very general feature although its origin can be quite different. Especially in the case where the loop has to synchronize over a wide frequency range (high open loop gain), even small time delays can induce instabilities of the above kind which will considerably influence the quality of the device.

3. Theoretical Considerations

The dynamical behavior of the phase error signal $\phi(t)$ of a first order PLL with time delay can be described by the delay differential equation[2]

$$\frac{d}{dt}\phi(t) + K\sin[\phi(t-\tau)] = \Delta\omega \tag{1}$$

where τ denotes the delay time and $\Delta\omega$ the angular frequency deviation between the two oscillators. The experimentally observed oscillatory instability is easily derived from a linear stability analysis of eq. (1). It can be shown that a Hopf bifurcation occurs when the control parameters K and τ exceed $K\tau \geq \pi/2$. The corresponding angular frequency of the oscillation is then given by $\Omega_1 = \pi/(2\tau)$. As the control parameters are increased further, more and more modes become unstable at the threshold values $K\tau = (n-\frac{1}{2}\pi)\,, n = 1, 2, \ldots$ with corresponding angular frequencies at $\Omega_n = (2n-1)\Omega_1$. When the amplitude of the signal crosses a separatrix between different attractors separated by $|\Delta\phi| = 2\pi$ a cycle slipping motion can be observed in the numerical simulations (Fig. 6). In Fig. 7 the corresponding attractors are reconstructed in the phase plane by plotting $\phi(t-\tau)$ versus $\phi(t)$. Although the values at the onset of the jump process differ from the measured values the spectral behavior yields the same $1/f$ characteristics (Fig. 8a). It should be emphasized that the spectral distribution changes qualitatively if the sinusoidally modulated phase error signal taken from the output of the first phase detector is analyzed. In this case the $1/f$ spectrum is destroyed (Fig. 8b). The reason is, that due to the insensitivity of the phase detector the output signal does no longer contain phase jumps.

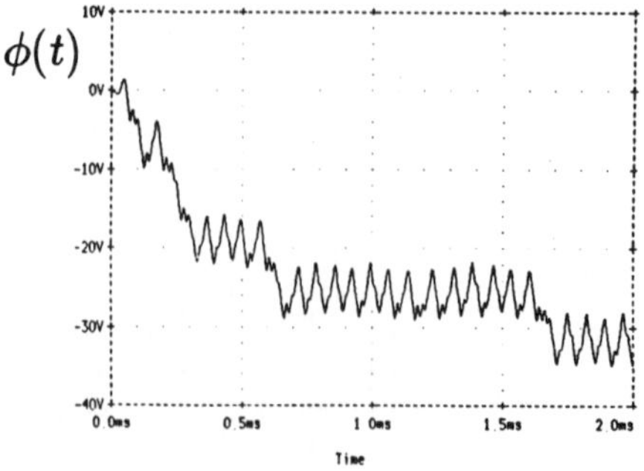

Figure 6: Numerical simulations of the cycle slipping motion

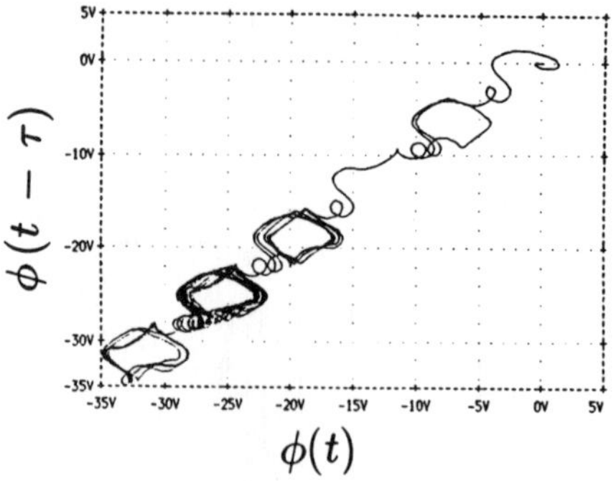

Figure 7: Reconstruction of the corresponding attractors in the phase plane from a time series

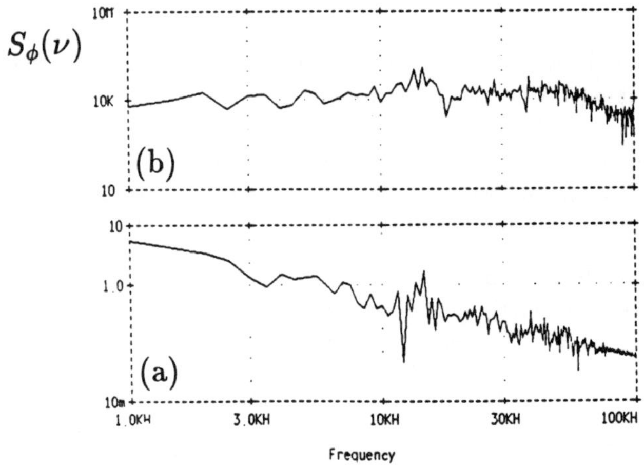

Figure 8: Spectral distribution of the phase error signal from eq. (1) (a); sinusoidally modulated signal as it could be observed at the output of a phase detector (b).

References

1. T. Endo and L.O. Chua, IEEE Trans. Circuits Syst., **35**, 987 (1988) and 255 (1989)
2. W. Wischert, M. Olivier, and J. Groslambert, Proc. 1992 IEEE Frequency Control Symposium, Hershey (1992)
3. K. Ikeda, Opt. Comm., **30**, 257 (1979)
4. H. M. Gibbs, F. A. Hopf, D. L. Kaplan, and R. L. Shoemaker, Phys. Rev. Lett.,**46**, 474 (1981)
5. R. Vallée and C. Delisle, Phys. Rev. **A31**, 2390 (1985)
6. M. Le Berre, E. Ressayre, and A. Tallet, Phys. Rev. **A41**, 6635 (1990)
7. A. Longtin, J. G. Milton, J. E. Bos, M. C. Mackey, Phys. Rev. **A41**, 6992 (1990)
8. P. Chen, System Dynamics Review **4**, 81 (1988)
9. M. Le Berre, E. Ressayre, and A. Tallet, J. Opt. Soc. Am. **B5**, 1051 (1988)
10. W. J. Yeh and Y. H. Kao, Appl. Phys. Lett., **42**, 299 (1983)

VIII. MOSFET

1/*f* NOISE AND OXIDE TRAPS IN MOSFETS

Daniel M. Fleetwood and Timothy L. Meisenheimer
Sandia National Laboratories, Dept. 1332
Albuquerque, New Mexico 87185-5800 USA

John H. Scofield
Oberlin College, Physics Department
Oberlin, OH 44074 USA

ABSTRACT

Comparative studies of the 1/*f* noise and radiation response of MOS transistors strongly suggest that similar point defects are responsible for their 1/*f* noise and radiation-induced oxide-trap charge buildup. A prime candidate for this common defect has been identified in radiation effects studies as an E' (E-prime) center, which is a trivalent Si center in SiO_2 associated with a simple oxygen vacancy. Thus, methods to reduce the number of vacancies in the SiO_2 layer of MOS technologies can dramatically reduce the 1/*f* noise of irradiated or *unirradiated* MOS devices. MOSFETs built with radiation-hardened process techniques show 1/*f* noise levels approaching the low levels of JFETs. 1/*f* noise measurements can also be used to *predict* the oxide-trap charge buildup in MOS electronics in radiation environments, such as those encountered in many military, nuclear reactor, high-energy accelerator, and satellite and spacecraft environments. Using "radiation-hardened" circuits and devices in these applications can dramatically improve the performance of analog MOS electronics.

INTRODUCTION

MOSFETs historically have exhibited large 1/*f* noise magnitudes because of carrier-defect interactions that cause the number of channel carriers and their mobilities to fluctuate.[1] Uncertainty in the type and location of defects that lead to the observed noise have made it difficult to optimize MOSFET processing to reduce the level of 1/*f* noise.[2] This has limited one's options when designing devices or circuits (high-precision analog electronics, preamplifiers, etc.) for low-noise applications at frequencies below ~10-100 kHz.

Over the last five years we have performed detailed comparisons of the low-frequency 1/*f* noise of MOSFETs manufactured with radiation-hardened and non-radiation-hardened processing. We find that the same techniques which reduce the amount of MOSFET radiation-induced oxide-trap charge can also proportionally reduce the magnitude of the low-frequency 1/*f* noise of both unirradiated and irradiated devices. MOSFETs built with radiation-hardened device technologies show noise levels up to a factor of 10 or more lower than standard commercial MOSFETs of comparable dimensions. Our quietest MOSFETs show noise magnitudes that approach the low noise levels of JFETs. We have also found that 1/*f* noise measurements provide the first nondestructive test for the radiation hardness of MOS devices, which is of great significance to military, space, reactor, and high-energy particle accelerator electronics.

EXPERIMENTAL RESULTS AND DISCUSSION

Two primary types of defects affect the electrical response of MOSFETs: oxide-trap charge and interface traps. Figure 1 schematically illustrates one method to separate their effects on MOSFET electrical response using ionizing radiation.[3,4] Shown are current-voltage traces before and after the device is irradiated. The shift in the (extrapolated) midgap voltage, ΔV_{mg}, has been shown to be approximately equal to the portion of the MOSFET threshold-voltage shift due to oxide-trap charge, ΔV_{ot}.[3-5] The difference between the net threshold-voltage shift, ΔV_{th}, and ΔV_{ot} is approximately equal to the threshold-voltage shift due to interface traps, ΔV_{it}. Note that ΔV_{ot} and ΔV_{it} are proportional to the number of radiation-induced trapped-charge precursor defects in the oxide and at the Si/SiO_2 interface, respectively. Thus, comparing the $1/f$ noise of *unirradiated* MOSFETs to ΔV_{ot} and ΔV_{it} for irradiated devices allows one to evaluate the relative importance of defects in the oxide and defects at the interface on MOSFET $1/f$ noise.

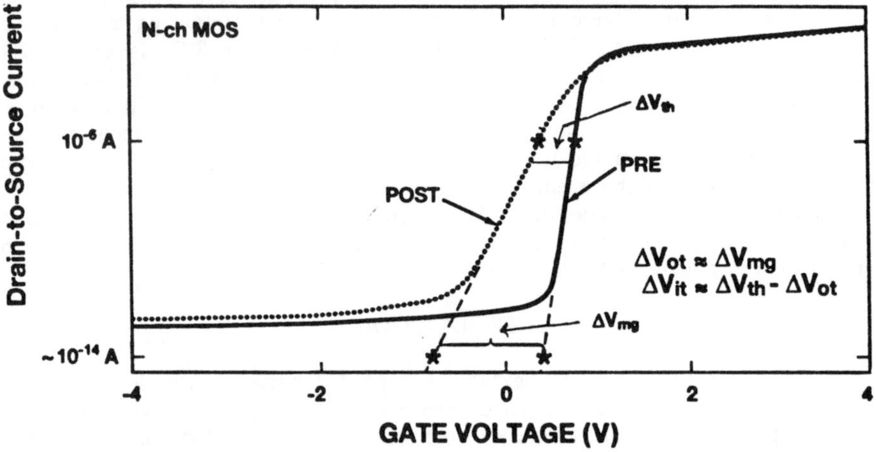

Fig. 1. Schematic illustration of a method to use current-voltage measurements before and after exposure of a MOSFET to ionizing radiation to infer relative densities of radiation-induced oxide-trap charge and interface traps (after Refs. 3 and 4).

Figures 2(a) and (b) show such a comparison for n-channel MOSFETs made in a single process lot at Sandia National Laboratories.[6,7] On the y-axis in each figure is the normalized noise magnitude, K, where

$$K \equiv S_V (V_G - V_{th})^2 f (V_D)^{-2} , \qquad (1)$$

S_V is the excess voltage-noise power spectral density measured at room temperature under constant current bias in the linear regime of transistor operation,[6] V_G and V_D are the gate and drain voltages during the noise measurements, and f is the frequency. For these measurements, which were performed on unirradiated devices, $S_V \sim f^{-1.0 \pm 0.1}$ for 0.2 Hz < f < 100 Hz.[6] On the x-axis in Figs. 2(a) and (b) are ΔV_{ot} and ΔV_{it}, respectively, inferred via the method of Fig. 1, following 100 krad(SiO_2) Co-60 irradiation at an oxide electric field of ~3 MV/cm for identical devices from the same wafers. Values of K varied by less than ~30% and values of

ΔV_{ot} and ΔV_{it} varied by less than ~10% for nominally identical devices from the same wafer. Results are shown for seven different wafers from the same lot which received gate-oxide processing ranging from moderately radiation-hardened (A) to commercial-like non-radiation-hardened technology (E). All other steps were the same, so differences in gate-oxide processing could be isolated from the effects of other potential differences in device processing on the noise. [Two wafers were measured from splits A and D, which showed identical ΔV_{ot} in Fig. 2(a) but slightly different ΔV_{it} on the expanded scale of Fig. 2(b).]

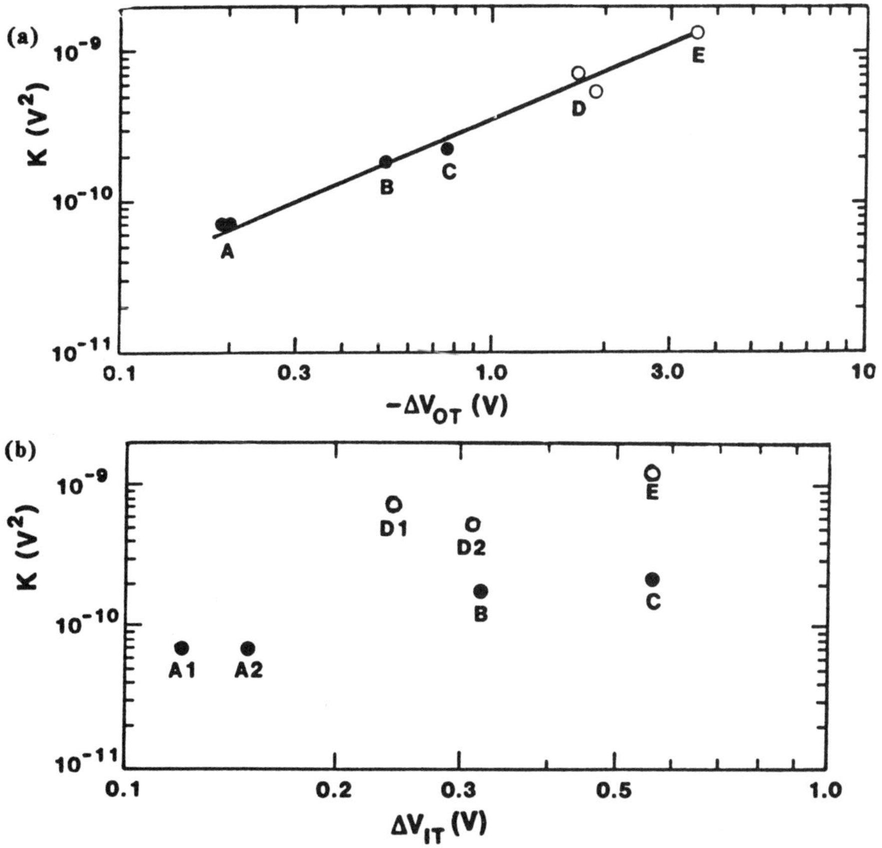

Fig. 2. K(10 Hz) vs. (a) ΔV_{ot} and (b) ΔV_{it} for MOSFETs built in the same process lot but with five different gate oxidation treatments (A-E). Solid symbols are hardened gate oxides, and open symbols are commercial gate oxides (after Refs. 6 and 7).

In Fig. 2(a) a one-to-one correlation is observed between the $1/f$ noise of *unirradiated* MOSFETs and ΔV_{ot} after irradiation. The noise of devices with the smallest ΔV_{ot} is more than 10-times lower than for those with the highest ΔV_{ot}. A similar correlation is not observed between K and ΔV_{it} in Fig. 2(b). In particular, wafers C and E show the same ΔV_{it} but noise levels that differ by nearly a factor of 10. Figures 2(a) and (b) show that the low-frequency (e. g., less than ~1 kHz) $1/f$ noise is more strongly related to oxide-trap charge precursor defects (density proportional to ΔV_{ot}) than to interface-trap precursor defects (density proportional to ΔV_{it}). We note that many previous studies have attributed the low-frequency $1/f$ noise of MOSFETs to "interface states." As we have discussed in detail in Refs. 8 and 9, interface- and oxide-trap charge effects may have been inadequately separated and/or imprecise nomenclature used in much of this work. This can lead to great confusion in trying to compare the results of different studies, and reinforces the need to critically assess the nature of the defects responsible for MOSFET noise in each case.[9] (High-frequency $1/f$ noise above ~1 kHz in the *subthreshold* regime of device operation appears truly to be caused by interface traps,[10] but near-interfacial oxide traps, i. e. "border traps," appear to cause the vast majority of MOSFET $1/f$ noise in the literature.[8,9])

In recent work, we have shown that the striking correlation between the low-frequency $1/f$ noise and oxide traps observed in Fig. 2(a) can be explained quite naturally using a simple number fluctuation model,[7,11-13] with the result that

$$K \approx kT\,(A\,\kappa_g\,f_y\,D\,\sigma\,E_g\,\varepsilon_{ox})^{-1}\,(-\Delta V_{ot})\;, \tag{2}$$

where k is the Boltzmann constant, T is the absolute temperature, A is the gate area of the device, κ_g is the number of electron-hole pairs generated per unit dose in the oxide of the MOSFET during the irradiations of Fig. 2,[7] f_y is the probability that an individual electron-hole pair escapes recombination,[7] D is the SiO_2 radiation dose, σ is the effective trap capture cross-section (assumed for simplicity to be equal for the $1/f$ noise and radiation-induced hole-trapping results reported here[7]), E_g is the SiO_2 band gap, and ε_{ox} is the SiO_2 dielectric constant.[7,13] The physical basis for Eq. (2) is discussed in detail in Refs. 7 and 13. Finally, we should point out that the proportionality between the *preirradiation* $1/f$ noise level, K, and the *postirradiation* value of ΔV_{ot} in Eq. (2) suggests that noise measurements can be used to screen MOS devices that are to be used in high-radiation environments such as those experienced in space, military, nuclear reactor, and high-energy particle accelerator applications. The advantages of using $1/f$ noise as a nondestructive screen of MOS radiation hardness are discussed in detail in Ref. 13.

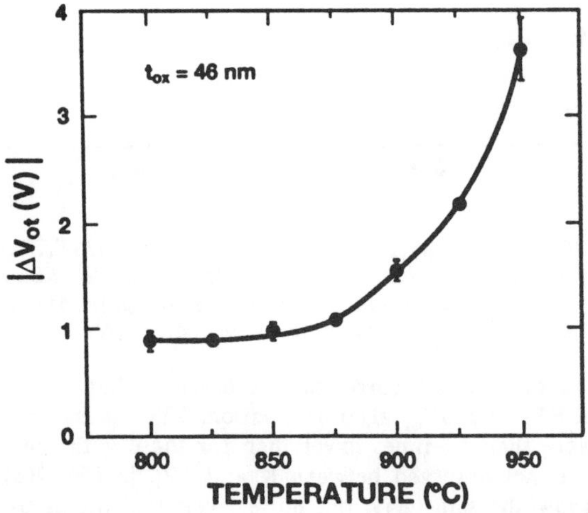

Fig. 3. Magnitude of ΔV_{ot} as a function of post-gate-oxidation N_2 anneal temperature for MOS devices. (After Ref. 14.)

In previous studies, it has been demonstrated that values of ΔV_{ot} are proportional to the number of oxygen vacancies near (i. e., within ~5 nm of) the Si/SiO_2 interface.[14-16] Figure 3 illustrates one way to reduce ΔV_{ot}, and thus the number of vacancies and the low-frequency $1/f$ noise. Here we show ΔV_{ot} as a function of the highest temperature a MOSFET experiences during processing after the gate oxide is grown. All of these post-gate anneals were performed in a nitrogen environment.[17] Clearly, any post-gate N_2 anneal above ~875°C results in a large increase in ΔV_{ot} and, per Fig. 2(a), the $1/f$ noise of the device. In previous work on radiation effects on MOS devices and circuits, many other process steps have been identified which can affect the radiation hardness of MOSFETs.[18] Thus, work on MOS radiation effects can be adapted to decrease the noise of *unirradiated* MOS devices. We have also shown that the noise of *irradiated* MOS devices correlates with the radiation-induced oxide-trap charge (and usually not with the interface-trap charge),[19,20] so reducing the number of vacancies in a MOS oxide can greatly reduce the noise of both irradiated and unirradiated devices.

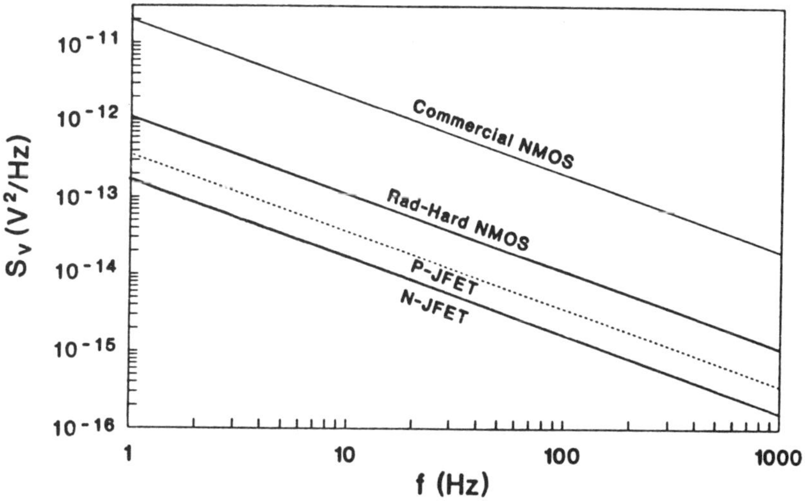

Fig. 4. Comparison of commercial and radiation-hardened MOSFET noise to n- and p-JFET noise for devices of similar dimensions. The commercial MOSFET and n- and p-JFET data are from Ref. 21. The hardened MOSFET is from this work.

In Fig. 4 we compare the noise of MOSFETs made in a radiation-hardened process with that of standard commercial MOSFETs of similar dimensions, as well as with n- and p-JFETs.[13,21] The radiation-hardened MOSFET is from our work; the other curves are from the literature.[21] The $1/f$ noise in the radiation-hardened MOSFET is more than a factor of 10 below the noise of the commercial MOSFET, and is only 3-6 times above that of the quiet JFET devices. This shows that understanding and controlling the defects in the oxide of a MOSFET can greatly reduce its $1/f$ noise. The option of designing circuits and systems with low-noise MOSFETs should greatly enhance one's flexibility in low-noise applications.

In summary, we have demonstrated a strong correlation between $1/f$ noise and radiation-induced oxide-trap charge trapping in MOSFETs. This allows process techniques which have already been developed to reduce the oxide-trap charge buildup in irradiated MOS devices to be used to reduce the $1/f$ noise of unirradiated or irradiated MOS devices. Noise measurements can provide a practical screen to assess the radiation hardness of MOSFETs without the need for destructive radiation testing. We have demonstrated that comparative studies of the $1/f$ noise and radiation response of MOSFETs provide (1) insight into the origin of the noise, (2) benefits for the design of transistors with lower noise, and (3) assistance for fielding systems with greater radiation tolerance.

We thank P. S. Winokur for stimulating discussions, and T. P. Doerr for experimental assistance. This work performed at Sandia National Laboratories and at Oberlin College was supported by the U. S. Department of Energy under Contract No. DE-AC04-76DP00789 and the Defense Nuclear Agency through its hardness assurance program.

References

1. A. van der Ziel, Adv. Electron. Electron Phys. 49, 225 (1979).
2. M. J. Kirton and M. J. Uren, Adv. Phys. 38, 367 (1989).
3. P. S. Winokur, J. R. Schwank, P. J. McWhorter, P. V. Dressendorfer, and D. C. Turpin, IEEE Trans. Nucl. Sci. 31, 1453 (1984).
4. P. J. McWhorter and P. S. Winokur, Appl. Phys. Lett. 48, 133 (1986).
5. D. M. Fleetwood, Appl. Phys. Lett. 60, 2883 (1992).
6. John H. Scofield, T. P. Doerr, and D. M. Fleetwood, IEEE Trans. Nucl. Sci. 36, 1946 (1989).
7. D. M. Fleetwood and John H. Scofield, Phys. Rev. Lett. 64, 579 (1990).
8. D. M. Fleetwood, IEEE Trans. Nucl. Sci. 39, 269 (1992).
9. D. M. Fleetwood, P. S. Winokur, R. A. Reber, Jr., T. L. Meisenheimer, J. R. Schwank, M. R. Shaneyfelt, and L. C. Riewe, accepted for publication, J. Appl. Phys. (May 15, 1993).
10. M. H. Tsai and T. P. Ma, IEEE Trans. Nucl. Sci. 39, 2178 (1992).
11. S. Christensson, I. Lundstrom, and C. Svennson, Solid State Electron. 11, 797 (1968); 11, 813 (1968).
12. G. Blasquez and A. Boukabache, in *Noise in Physical Systems and 1/f Noise*, edited by M. Savelli, G. Lecoy, and J.-P. Nougier (Elsevier, Amsterdam, 1983), pp. 303-306.
13. John H. Scofield and D. M. Fleetwood, IEEE Trans. Nucl. Sci. 38, 1567 (1991).
14. F. J. Feigl, W. B. Fowler, and K. L. Yip, Solid State Commun. 14, 225 (1974).
15. P. M. Lenahan, W. L. Warren, P. V. Dressendorfer, and R. E. Mikawa, Z. Phys. Chem. 151, 235 (1987).
16. W. L. Warren, E. H. Poindexter, U. Offenberg, and W. Muller-Warmuth, J. Electrochem. Soc. 139, 872 (1992).
17. J. R. Schwank and D. M. Fleetwood, Appl. Phys. Lett. 53, 770 (1988).
18. For example, T. P. Ma and P. V. Dressendorfer, *Ionizing Radiation Effects in MOS Devices and Circuits* (Wiley, New York, 1989).
19. T. L. Meisenheimer and D. M. Fleetwood, IEEE Trans. Nucl. Sci. 37, 1696 (1990).
20. T. L. Meisenheimer, D. M. Fleetwood, M. R. Shaneyfelt, and L. C. Riewe, IEEE Trans. Nucl. Sci. 38, 1297 (1991).
21. W. Buttler, H. Vogt, G. Lutz, P. F. Manfredi, and V. Speziali, IEEE Trans. Nucl. Sci. 38, 69 (1991).

1/f NOISE IN MOS TRANSISTORS DUE TO NUMBER OR MOBILITY FLUCTUATIONS

L.K.J. Vandamme and Xiaosong Li
Electrical Engineering Department, Eindhoven University of Technology
P.O. Box 513, 5600 MB Eindhoven, The Netherlands

D. Rigaud
University of Montpellier II, CEM, 34095 Montpellier, France

ABSTRACT

Some experimental studies on 1/f noise in MOS transistors are reviewed. Arguments are given for the two schools of thought about the origin of 1/f noise. The consequences of models based on number or mobility fluctuations on the device geometry and on the bias dependence of the 1/f noise are discussed. The correlation or lack of correlation between degradation effects by hot carriers or by irradiation on one hand and the 1/f noise on the other hand is considered in terms of a ΔN or $\Delta\mu$.

INTRODUCTION

A generally accepted model explaining the 1/f noise in all p- and n-channel MOS transistors is still lacking. The increase of 1/f noise through degradation by hot carriers or irradiation is often used as a proof for the surface and hence a ΔN origin of the 1/f noise. However the majority of results obtained not on MOS transistors but on homogeneous semiconductors or constriction dominated contacts can be described by an empirical relation[1, 2]

$$\frac{S_G}{G^2} = \frac{\alpha}{Nf} \tag{1}$$

where α is not a constant[2] but a volume and device length independent[4] 1/f noise parameter between 10^{-7} and 10^{-2}. N is the total number of free charge carriers in a homogeneous sample with perfect contacts or it is a well defined reduced number of free charge carriers in samples submitted to nonuniform fields[5] as is often the case in contacts without interface problems. Experimental results obeying the empirical relation imply that the 1/f noise source is homogeneously distributed in the bulk of the device which indicates a bulk origin for the 1/f noise.This relation has been applied successfully for p-n diodes and bipolar transistors[6].

The MOS transistor is by excellence an interface dominated device. The 1/f noise of n-channel devices has been ascribed successfully by carrier number fluctuations ΔN, which are caused by tunnelling of free-charge carriers into oxide traps close to the Si-SiO_2 interface[7]. Classical arguments in favour of the McWhorter model are the observed proportionality between trap density and 1/f noise[8-10]. Recent arguments for the ΔN origin of 1/f noise in MOS transistors are the evolution of dc and 1/f noise characteristics through degradation by hot electrons[11, 12] or by ionizing irradiation[13-19]. The p-MOS transistor has often a channel at a larger distance from the interface and is sometimes easier to interpret in terms of $\Delta\mu$ than in terms of ΔN. Here we discuss both points of view and explain why the 1/f noise in a MOS transistor is still a problem that gives rise to much controversy.

GEOMETRY AND BIAS DEPENDENCE IN VIEW OF ΔN OR Δμ

The 1/f noise parameter α is used in our analysis as a figure of merit and not to suggest a $\Delta\mu$ origin of the 1/f noise. Its value is gate length independent for both models[4]. For the sake of simplicity we assume no series resistance or edge current problems. This results in generally accepted dependence between the 1/f noise in MOSTs and the channel area, the dependence on oxide thickness is still under discussion. We start from empirical relation and do not suggest $\Delta\mu$ fluctuations. From eq. (1) $N = C_{ox}V_G^*Wl/q$ and the simple current-voltage equations for MOS transistors we find the calculated 1/f noise for MOSTs biased above threshold voltage[19]:
(i) below saturation ($V < V_s$, $I < I_s$) with V_s the drain source saturation voltage and I_s the saturation current

$$\frac{S_I}{I^2} = \frac{\alpha q \mu R}{l^2} \propto \frac{1}{Wl} \qquad (2)$$

$$S_I = \alpha q \mu^2 C_{ox} V_G^* V^2 \frac{W}{l^3} \propto \frac{W}{l^3} \qquad (3)$$

with μ the mobility, V_G^* the effective gate voltage, R the channel resistance, C_{ox} the oxide capacitance per unit area, W and l the channel width and length respectively. In the ohmic region $V < V_G^*/10$ holds, $S_{I/I2} = S_V/V^2 = S_R/R^2S_G/G^2$
(ii) in saturation holds for the l and W dependence

$$S_{I_s} \cong \alpha q \mu^2 C_{ox} V_G^{*3} \frac{W}{l^3 f} \propto \frac{W}{l^3} \qquad (4)$$

$$S_{I_s} \cong \frac{\alpha q \mu^{1/2} I_s^{3/2}}{W^{1/2} C_{ox}^{1/2} l^{3/2} f} \propto \frac{1}{W^{1/2} l^{3/2}} \qquad (5)$$

or in equivalent input noise voltage with $S_{V_{eq}} = S_{I_s}/g_m^2$

$$S_{V_{eq}} \cong \frac{\alpha q V_G^*}{W l C_{ox} f} \propto \frac{1}{Wl} \qquad (6)$$

$$S_{V_{eq}} = \frac{\alpha q I_s^{1/2}}{(WC_{ox})^{3/2} (l\mu)^{1/2} f} \propto \frac{1}{W^{3/2} l^{1/2}} \qquad (7)$$

The 1/f noise of devices with different channel area and W/l ratios have been compared with the proportionalities given in eqs. (2 - 7). If the devices do not suffer from important series resistance contributions[17, 18, 20-23] or channel edge currents, the electrical dimensions l and W are used, and the devices are biased at fixed effective gate voltages, no strong deviations between calculated and experimentally observed lW dependence are observed[19]. In Fig.1 experimental results in support of eq. (4) are presented.

To discriminate between Δn and $\Delta\mu$ models the 1/f noise must be studied as a function of gate voltage becomes both models predict the same dependence on l and W. The $\Delta\mu$ model predicts a gate voltage independent α value ($10^{-7} < \alpha < 10^{-3}$) if there is: (i) no appreciable mobility degradation with increasing bias voltages[10, 24, 25], (ii) a uniform noise source under the SiO_2 can be assumed[5] and (iii) the Si-SiO_2 interface is flat and there are no spatial nonuniformities in the oxide charge[26]. These conditions are important because: (i)mobility degradation by scattering mechanism other than lattice scattering reduces the α parameter[1, 27], (ii) reduced crystal quality results in high α values[28, 29], and (iii) at low V_G^* spatial field fluctuations at the interface induce cavities in the inversion layers. Si-SiO_2 interface roughness and spatial field fluctuations results in a "Swiss-cheese" channel. This results in an increased interface surface and trap number. This leads to current constrictions and a marked increase in 1/f noise at lower V_G^*. By overlooking the nonuniform current density in the channel an increasing apparent α value[26] is obtained with decreasing V_G^*. The Swiss-cheese model explains easily[26] an increase of resistance with 10% and an increase in 1/f noise by a factor 10 by taking into account an inhomogeneous current density in the same way as in Ref. 30. Increasing cavity holes with decreasing V_G^* is a good alternative to explain a dependence of $\alpha \propto V_G^{*-1}$ often interpreted as a proof for ΔN[26].

If the ΔN model holds we find for α

$$\alpha = \frac{q x_0 D_0 kT(x_0/x_2)}{\epsilon_0 \epsilon_r} \frac{t_{ox}}{V_G^*} \propto \frac{t_{ox}}{V_G^*} \tag{8}$$

with x_0 the characteristic decay length of the electron wave function (≈ 1Å), x_0D_0($\approx 10^{10}$cm^2(eV)$^{-1}$) trap density, x_2(≈ 30Å) the largest trapping distance resulting in a 1/f spectrum over 13 decades in frequency and ϵ_r the relative dielectric constant of the gate oxide. The proportionality from eq. (8) of $\alpha = \alpha_r E_c/(V_G^*/t_{ox})$ should be a proof for the validity of the ΔN model in MOS transistors. α_r is a reference value at a field strength of $V_G^*/t_{ox} = E_c$ a critical field strength. The typical ΔN-type dependence of α on V_G^* and t_{ox} has its consequences for the bias and oxide thickness dependence of the 1/f noise in eqs. (2)-(7). The 1/f noise is often expressed in simulation oriented equations. If the ΔN-model holds ($\alpha \propto t_{ox}/V_G^*$) we find for MOSTs biased in
(i)saturation

$$S_{I_s} = \frac{K^* I_s}{f} \tag{9a}$$

$$K^* = \frac{\alpha_r E_c t_{ox} q\mu}{l^2} \tag{9b}$$

$$S_{I_s} = \frac{B g_m^2}{Wlf} \tag{10a}$$

$$B = \frac{\alpha_r E_c t_{ox} q}{\epsilon_0 \epsilon_r} \tag{10b}$$

$$S_{V_{eq}} = \frac{A V_G^*}{C_{ox} W l f} = \frac{B}{W l f} \tag{11a}$$

$$A = \frac{B C_{ox}}{V_G^*} \tag{11b}$$

(ii)in the ohmic region holds

$$S_V = \frac{K V^2}{f^\gamma V_G^{*2}} \qquad \text{with } \gamma \approx 1 \tag{12a}$$

Eq. (12a) is often used to discuss 1/f noise and radiation hardness of devices[13-18]. In terms of α (not suggesting $\Delta\mu$-model) we find

$$K = \frac{\alpha q t_{ox} V_G^*}{W l \epsilon_0 \epsilon_r} \tag{12b}$$

In view of the ΔN model the value of K is given by

$$K = \frac{\alpha_r E_c t_{ox}^2 q}{\epsilon_0 \epsilon_r W l} = \frac{q^2 t_{ox}^2 x_0 D_0 k T (x_0/x_2)}{W l (\epsilon_0 \epsilon_r)^2} \tag{12c}$$

An overwhelming number of publications especially for N-channels showed $\alpha \propto V_G^{*-1}$ and its consequences in bias dependence of the 1/f noise see for example refs. (13-19), (8-10). The straightforward interpretation gives support for the ΔN model. Invoking increasing inhomogeneous current density in the inversion layer at decreasing V_G^* to explain $\alpha \propto V_G^{*-1}$ trends is also a possible in terms of the $\Delta\mu$-model.

A new 1/f noise model[31] based on $\Delta\mu$ and two-dimensional device simulator results for MOSTs in saturation and deep saturation showed almost no difference with the ΔN results. Their experimental results agree with both ΔN and $\Delta\mu$ interpretations.

The temperature dependence as predicted by ΔN eq. 8, $\propto$ T, has not been observed at T = 300K and 77K[32, 33].

HOT CARRIER DEGRADATION, A PROOF FOR ΔN?

The 1/f noise has been used as a more powerful tool than dc characteristics for evaluating the quality of a MOSFET and investigating hot-carrier degradation in devices[11, 12, 21, 34-39]. An LDD structure is often used in small MOSFETs and the series resistance attached to a channel becomes more pronounced to infect the dc characteristics as well as the 1/f noise. It was shown[21] that a series resistance with an acceptable value can be the dominant term in the 1/f noise behaviour of an edgeless MOSFET.

Another study[20] exhibited how to distinguish the 1/f noise from the series resistance and the channel part in a MOSFET biased in the ohmic region. Especially in a short channel LDD device, not only the value of the series resistance R_s but also its contribution to the 1/f noise of the device S_{Rs} increases significantly. The quantity of S_{Rs} can reach 20% of the total 1/f noise in a device. Fig. 2 shows an LDD device as a conventional MOSFET in series with two series resistors R_{Dd} at the drain side and R_{sS} at the source side[40-43]. Investigations[42-46] on an LDD MOSFET biased in the nonohmic region showed that the series resistance R_{Dd} at the drain side behaves differently from that at the source side R_{sS}. The assumption $R_{Dd} = R_{sS} = F(V_{GS})$ is only valid when a MOSFET is biased in the ohmic region. Above the ohmic region, it was found[42-46] that $R_{Dd} > R_{sS}$ and R_{Dd} is a function of V_{GS} and V_{DS} (see Fig. 3), although R_{sS} is still only a function of V_{GS}. When V_{GS} is constant, R_{Dd} increases strongly as V_{DS} increase towards the effective gate voltage V_G^*. Consequently, the internal drain source voltage is clamped, hence, the channel current is kept constant. This is used successfully to explain the experimental results of S_I as a function of V_{DS}/V_G^* in an LDD MOSFET which are at variance with those in a conventional MOSFET[44] shown in eq. (3).

Now a days hot carrier degradation of submicron MOSTs and the evolution of 1/f noise are often considered as a direct prove for the number fluctuation origin of the 1/f noise. It is thought that the 1/f noise increases after hot-carrier stressing due to an increase in the trap density in the oxide layer[11, 12, 39]. But the consequence of hot-carrier degradation in a MOSFET is far more complicated. Generally speaking, the shift of the threshold voltage, the changing slope in the subthreshold region, and the decrease in the channel current are considered as degradation phenomenon due to an increase of traps. It was also found after a short time of stressing, the threshold voltage and the slope in the subthreshold region do not change[47], but the series resistance increases especially at the drain side. Another degradation phenomenon is the reduction in effective channel length[48, 49] which implies an increase of the series resistance.

It was observed that 1/f noise level increased a lot in the reversed mode but hardly changed in the normal mode after hot-carrier stressing when a MOSFET is biased in saturation[11, 12]. This can be explained by the analysis used in[44] without using ΔN model. Before stressing, a MOSFET is symmetric, therefore there is no difference between 1/f noise levels in the normal mode and in the reversed mode. A post-stressed MOST biased in the ohmic region has nonsymmetric resistance $R_{sS} < R_{Dd}$ with about the same dependence on the biasing conditions. Hence, under the same external voltages, no matter in a normal mode or in a reversed mode, the internal biasing conditions and 1/f noise remains[11, 12] at least at low V_G^*. In saturation, our results showed that the series resistance in the drain side R_{Dd} increases significantly with V_{DS} (see Fig. 3) and the voltage drop V_{Dd} across R_{Dd} increases with R_{Dd} (see Fig. 4). The series resistance at the drain side is R_{Dd} in the normal mode and R_{sS} in the reversed mode. The hot carrier stressing increases the value of R_{Dd} but hardly R_{sS} in the ohmic region[47]. The fact that $R_{sS} < R_{Dd}$ results in more clamping and smaller channel voltage and current in the normal mode than in the reversed mode. Therefore we can expect that the 1/f noise is larger in the reversed mode than that in the normal mode as reported[11, 12]. The comparison of 1/f noise level, in a conventional MOSFET and an LDD device[50] or in the same device before and after hot-carrier stressing, should be done under the same internal biasing conditions. Otherwise the analysis leads to a wrong conclusions.

CORRELATION BETWEEN 1/f NOISE AND RADIATION HARDNESS A ΔN ARGUMENT?

A MOS transistor through ionizing irradiation shows a gradual shift in threshold voltage mainly due to an increase in positive oxide charge. The radiation hardness of a technology is expressed in threshold voltage shift per Krad (e.g. 3mV/Krad) and its estimate requires a destructive testing. A strong correlation was observed between the preirradiation 1/f noise of MOS transistors and their post irradiation threshold voltage shift[13-18] which makes 1/f noise a useful nondestructive radiation hardness test. This is not a surprising result because a too high 1/f noise is often a good indicator for a poor technology[13-16] or a poor crystal quality[28, 29].

Ionizing radiation causes not only bulk oxide charging but also an increase in interface-state density. The increase in interface state density causes a considerable decrease in channel mobility and a decline in g_m[51]. This makes explanations in terms of ΔN or $\Delta\mu$ possible.

Threshold voltage shift ΔV_T is attributed to an increase in oxide charge ΔV_{ot} and to a smaller extent to a change in the charge at the Si-SiO_2 interface traps ΔV_{it}, and is given by $\Delta V_T = \Delta V_{ot} + \Delta V_{it}$, the latter contribution can be observed with the midgap method[52]. The radiation-induced threshold voltage shifts depend on radiation dose (D) oxide thickness t_{ox} and bias conditions the latter can be removed by an expression that characterizes the inherent capability of the oxide to be charged[14]

$$-\Delta V_{ot} = \{\frac{qk_g D t_{ox}^2}{\epsilon_0 \epsilon_{ox}}\} f_y f_{ot} \tag{13}$$

where K_g is the number of electron-hole pairs produced per unit dose and per volume f_y is the probability that an electron-hole pair escapes recombination, and f_{ot} is the oxide charge-trapping efficiency, which is the ratio of the number of trapped holes to the number of electron-hole pairs created. It is expected that an oxide technology with a high f_{ot} will result in a noisy device. A comparable equation can be defined for ΔV_{it}. The 1/f noise observed in the ohmic region is given by eq.(12) where K is proportional to t_{ox}^2. Because ΔV_{ot} in eq. (13) is also proportional to t_{ox}^2 we can expect that keeping other parameters constant $K \propto \Delta V_{ot}$, which is a support for the ΔN model if also $KWl/t_{ox}^2 \propto \Delta V_{ot} \propto f_y f_{ot}$ over a large range in t_{ox}. All n-MOSTs in Refs. (13-18) obey $KWL \propto |\Delta V_{ot}|$ although no conduction has been observed with ΔV_{it}. Fig. 5 shows the results from Montpellier[17, 18] together with results obtained in USA[13-16].

For the $\Delta\mu$-model we expect $K \propto t_{ox}$ and increasing K and ΔV_{ot} should go hand in hand but not linearly.

The effective gate voltage should be kept constant at a low enough V_G^* in order to compare the 1/f noise through irradiation[17, 18]. N-channel devices are sensitive for leakage resistance through irradiation. A dramatic increase in noise is telling nothing about a ΔN or $\Delta\mu$ origin[17, 18].

CONCLUSIONS

Geometry dependence is well understood except the t_{ox} dependence.

Bias dependence points to a straight forward interpretation in terms of ΔN with $\alpha \propto V_G^{*-1}$, but $\Delta\mu$ is also possible, see Refs. 26, 30 31

Due to series resistance complications in hot carrier degraded MOSTs, the interpretation in terms of ΔN is not the only one.

$K \propto |\Delta V_{ot}|$ is in agreement with ΔN. If a rough interface between Si and SiO_2 goes hand in hand with a reduced radiation hardness then the Swiss-cheese model is also able to explain the observed trends in terms of $\Delta\mu$.

The temperature independent 1/f noise between 77 and 300 K, and the irradiation independent behaviour of the 1/f noise up to 300 Krad are more difficult to understand in terms of ΔN.

REFERENCES

1. F.N. Hooge, T.G.M. Kleinpenning and L.K.J. Vandamme, Rep. Prog. Phys. 44, 479 (1981).
2. L.K.J. Vandamme, Noise in Physical Systems (Elsevier, Amsterdam, 1983) 183.
3. R. Clevers, Physica B154 214 (1989).
4. L.K.J. Vandamme, Noise in Physical Systems (Akadémiai Kiadó, Budapest, 1990) 491.
5. L.K.J. Vandamme; IEEE Trans. Elec. Dev. ED-33, 1833 (1985).
6. T.G.M. Kleinpenning, Noise in Physical Systems (Akadémiai Kiadó, Budapest, 1990) 443.
7. A.L. McWhorter, Semiconductor Surface Physics, Ed. by R.H. Kingston (University of Pennsylvania Press 1957) 207.
8. F.M. Klaassen, IEEE Trans. Electron Devices, ED-18, 887 (1971).
9. H.E. Maes and S.H. Usmani, J. Appl. Phys. 54, 1937 (1983).
10. L.K.J. Vandamme and R.G.M. Penning de Vries, Solid-State Electr. 28, 1049 (1985).
11. M. Stegheer, Solid-State Electr. 27, 1055 (1984).
12. Z.H. Fang, S. Christoloveanu and A. Chovet, IEEE Trans. Electron Device Letters, EDL-7 371 (1986).
13. J.H. Scofield, T.P. Doerr and D.M. Fleetwood, IEEE Trans. Nucl. Sc. 36, 1946 (1989).
14. J.H. Scofield and D.M. Fleetwood, IEEE Trans. Nucl. Sc. 38, 1567 (1991).
15. D.M. Fleetwood and J.H. Scofield, Phys. Rev. Lett. 64, 579 (1990).
16. J.H. Scofield, M. Trawick, P. Klimecky and D.M. Fleetwood, Appl. Phys. Lett. 58, 2782 (1991).
17. A. Hoffmann, Thesis University of Montpellier April 1993.
18. A. Hoffmann, M. Valenza, D. Rigaud and L.K.J. Vandamme. This conference.
19. J. Jennen, ERASMUS Report: Université de Montpellier - Eindhoven University of Technology, April 1993.
20. X. Li and L.K.J. Vandamme, Solid-State Electronics, 35 1471 (1992).
21. X. Li and L.K.J. Vandamme, Solid-State Electronics, 35 1477 (1992).
22. X. Li and L.K.J. Vandamme, This conference.
23. X. Li and L.K.J. Vandamme, Solid-State Electronics, accepted for publication.
24. L.K.J. Vandamme, Solid-State Electronics, 23 317 (1980).
25. L.K.J. Vandamme, H.M.M. de Werd, Solid-State Electronics, 23 325 (1980).
26. L.K.J. Vandamme and X. Li, to be published.
27. F.N. Hooge and L.K.J. Vandamme, Phys. Lett. 66A 315 (1978).
28. L.K.J. Vandamme and S. Oosterhoff, J. Appl. Phys., 59 3169 (1985).
29. L.K.J. Vandamme, Solid State Phenomena 1 & 2 153 (1988). Ion Implantation in Semiconductors, Eds. D. Stievenard & J.C. Bourgoin, Trans Tech Publications Ltd. (1988) Aedermannsdorf Switzerland.
30. L.K.J. Vandamme and A. van Kemenade, Conf. Proc. ESREF92 3rd European Symp. on reliability of electron devices, failure physics and analysis, oct. 1992 Germany (VDE-Verlag Berlin 1992) 419.
31. Y. Zhu, M.J. Deen and T.G.M. Kleinpenning, J. Appl. Phys. 72, 5990 (1992).
32. E. Simoen and C. Claeys, Nucl. Instr. and Meth. A327 523 (1993).
33. L.K.J. Vandamme and A.H. de Kuijper, 5th Int. Conf. on Noise, Ed. D. Wolf (Springer Berlin 1978) 152.
34. C. Surya and T.Y. Hsinag, Solid-St. Electron., 31, 959 (1988).
35. H. Wong and Y.C. Cheng, IEEE Electron Devices, 37, 1743 (1990).
36. T.L. Meisenheimer and D.M. Fleetwood, IEEE Nuclear Science, 37, 1696 (1990).
37. Z. Celik-Butler and T.Y. Hsiang, Solid-St. Electron. 30, 419 (1987).
38. R. Jayaraman and C.G. Sodini, IEEE Electron Devices, 36, 1773 (1989).
39. C. Cheng and C. Surya, Solid-State Electronics, 36, 475, (1993).
40. H.G. de Graaff and F.M. Klaassen, Compact transistor modelling for circuit design, Springer, Berlin (1990).
41. C.C. McAndrew and P.A. Layman, IEEE Trans. Electron Devices, ED-39, 2298 (1992).
42. J.A.M. Otten and F.M. Klaassen, Proceedings ESSDERC 1992, 703 (1992).
43. F.J. Lai and J.Y. Sun, IEEE Trans. Electron Devices, Ed-32, 2803 (1985).
44. X. Li and L.K.J. Vandamme, accepted by Solide-State Electronics, (1993).

45. J.A.M. Otten and F.M. Klaassen, Microelectronic Engineering, 15, 555 (1991).
46. G.S. Huang and C.Y. Wu, IEEE Trans. Electron Devices, ED-34, 1311 (1987).
47. A. Bhattacharyya and S.N. Shabde, IEEE Trans. Electron Devices, 35, 1156, 1988.
48. B.S. Doyle and K.R. Misty, IEEE Trans. Electron Devices, 40, 152, (1993).
49. R. Woltjer, A. Hamada and E. Takeda, IEEE Trans. Electron Devices, 40, 392, (1993).
50. C.Y.H. Tsai and J. Gong, IEEE Trans. Electron Devices, ED-35, 2373 (1988).
51. H. Gesch, J.P. Leburton and G.E. Dorda, IEEE Trans. Electron Devices, ED-29, 915(1982).
52. P.S. Winokur, J.R..Schwank, P.J. McWhorter, P.V. Dressendorfer and D.C Turpin, IEEE Trans. Nuc. Sci. NS-31, 1453 (1984).

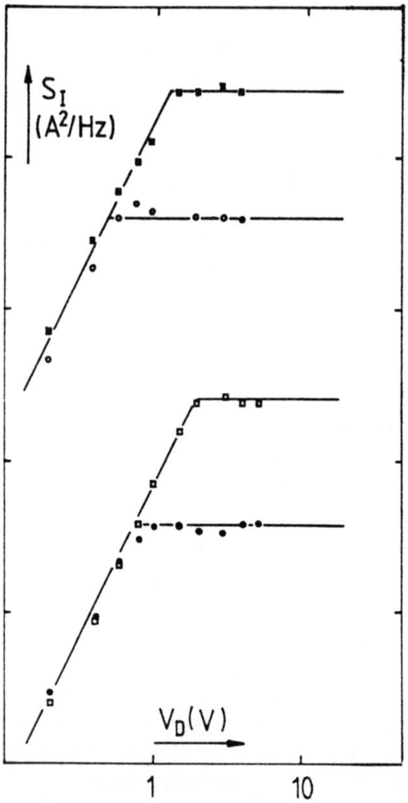

Fig. 1 Current noise S_I versus drain source voltage with V_G^* as a parameter for two different n-channel devices: ■ ○ for W=20μm, l= 5μm and □ ● for W=20μm, l= 20μm. V_G^* =1.2 V(●), 2.5V(□), 1.0V(○), 2.3V(■)

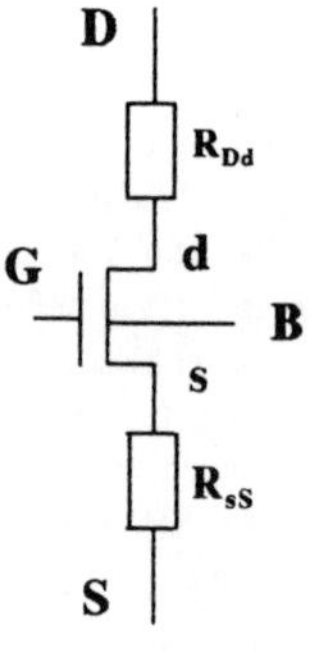

Fig. 2 An LDD MOSFET consists of a MOST and series resistors.

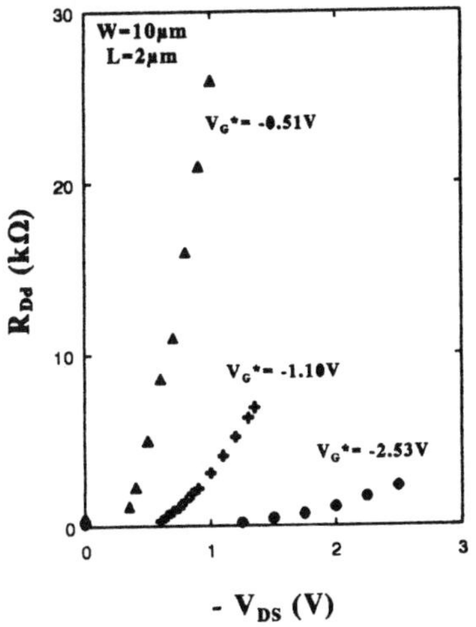

Fig. 3 The series resistance R_{Dd} vs. V_{DS}.

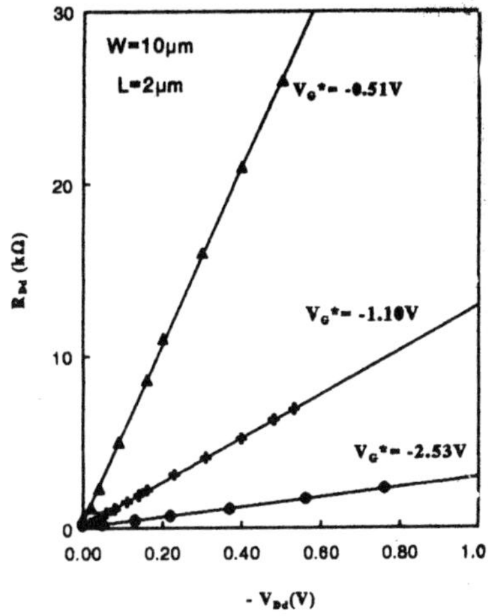

Fig. 4 The series resistance R_{Dd} vs. V_{Dd}.

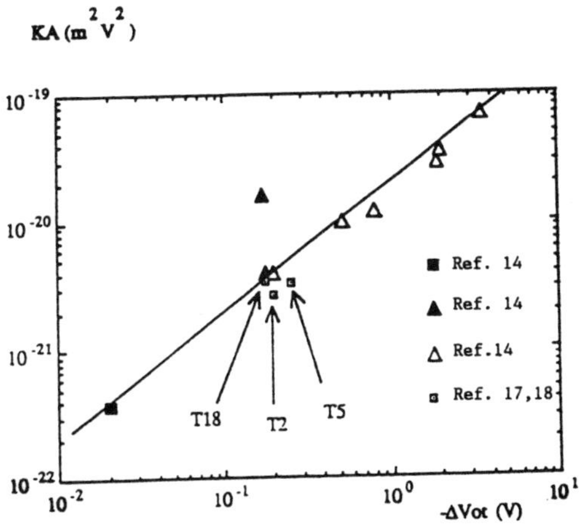

Fig. 5 The preirradiation 1/f noise of several devices (Ref. 14, 17, 18) versus the post irradiation threshold voltage shift.

DC, Small-Signal Parameters and Noise Performance for *SiGe/Si* FETs

A. F. M. Anwar, K.-W. Liu, M. M. Jahan and V. P. Kesan*
Department of Electrical and Systems Engineering
The University of Connecticut
Storrs, CT 06269-3157
*IBM Research Division, T. J. Watson Research Center
Yorktown Heights, NY 10598

ABSTRACT

An analytical model to evaluate d.c. small signal parameters and noise performance for the *SiGe/Si* based FETs is presented, that is based on a self-consistent solution of Schrödinger and Poisson's equations and an improved velocity-electric field (v_d - $\mathcal{E}$) characteristics. The presence of a self-consistent calculation provides a better insight in the dependence of the device parameters on the QW properties. Moreover, the inclusion of a modified velocity-electric field characteristic enables us to calculate small-signal parameters that are in excellent agreement with experimental data both at 300K and 77K, respectively. The theoretical calculation of noise properties for *SiGe/Si* based FETs are presented.

Introduction

In this paper, we report a consistent model to evaluate the devices performance of *SiGe/Si* FETs by using an improved velocity-field characteristics and a self-consistent solution of Schrödinger and Poisson's equations[1]. The model will aid in the understanding of the behavior of the QW formed in *Si* in a *SiGe/Si* system and will enable us to calculate d.c., small-signal parameters which showed an excellent agreement with experimental data. Based on the calculated results, we present the noise properties for *SiGe/Si* based FETs.

Model

Schr"odinger and Poisson's equations are solved self-consistently[1] for an electron QW formed in *Si* layer. The average distance of the electron cloud from the first heterointerface x_{av} and the position of the Fermi level with respect to the tip of the conduction band at the first interface are determined.

The evaluation of d.c. small signal parameters for such a device follows the treatment presented by Anwar et al.[2] where an improved velocity-electric field characteristic is used. Using the definition of reduced potential the transconductance g_m can be written as

$$g_m = \frac{G_0[1 - n\frac{p_s}{p}\cosh(\frac{pL_2}{2d})]}{1 + \frac{\gamma}{p|V_T|} - n[\frac{1}{2}ln(\frac{s+p_s}{p}) + \frac{p_s}{p^2}(\frac{s}{2} + \frac{\gamma}{|V_T|})]\cosh(\frac{pL_2}{2d})} \tag{1}$$

where $p_s = \sqrt{s^2 - p^2}$, s and p are the reduced potential at the source and drain, respectively[2], L_2 is the length of the saturation region of the channel, ε_{SiGe} is the permittivity of $SiGe$, Z is the gate width. The parameter $n=\sqrt{v_d^2/(v_s^2 - v_d^2)}$ and demands some discussion. In the saturation region when $v_d=v_s$ the present v_d-$\mathcal{E}$ characteristic gives indeterministic results. To avoid this situation v_d is assumed to be 99.9% (or better) of v_s. The drain resistance r_d and gate capacitance C_{gs} can be calculated by the above method.

The minimum noise figure, F_{min} can be written as :

$$F_{min} = 1 + 2 \cdot g_n \cdot (R_c + \sqrt{R_c^2 + \frac{r_n}{g_n}}) \tag{2}$$

where r_n is noise resistance defined by Anwar et al.[2], $Z_c = R_c + \mathrm{j}X_c$, is the correlation impedance, R_s and R_g are the source and drain impedance, $R_i = \frac{L_g}{v_s C_{gs}}$, is the gate charging resistance. L_g represents the length of the gate.

Results and Discussions

In Fig.1, the calculated transconductances as a function of applied gate voltage for a 0.25×200 μm^2 n-channel MODFET with a 106Å Si QW are compared with experimental data[3] at both room temperature and 77K. The mobility and saturation velocity used in the calculation are $\mu_0 = 1500\ cm^2V^{-1}sec^{-1}$ (6000) and $v_s = 1.0 \times 10^7 cmsec^{-1}$ (1.15×10^7) at 300K (77K). The results show an excellent agreement at both room temperature and at 77K. The fit at low gate voltage is only possible due to the incorporation of the self-consistent calculation. The plots are obtained using the calculated threshold voltage of -1.1 V at 300K and -1.0 V at 77K, respectively.

In Fig.2, the calculated current-voltage (I-V) characteristic for a $SiGe/Si$ p-channel FET is presented. The self-consistent calculation resulted in an excellent fit with experimental data[4]. For the p-channel FET $\mu_0 = 110 cm^2V^{-1}sec^{-1}$, $v_s = 3.9 \times 10^6 cmsec^{-1}$ at 300K are used. At room temperature, the drain, source and parasitic resistances are 4Ω, 6.5Ω and 5500Ω, respectively.

In Fig.3, the calculated minimum noise figures, F_{min} (dB), are plotted as a function of the drain-source current for n-channel MODFETs and p-channel MOSFETs with temperature as a parameter. The calculations are performanced at 15 GHz. It is observed that the minimum noise figure of n-channel MODFETs are smaller than that of p-channel MOSFETs. f_T for n-channel devices are much greater than p-channel FETs. This explains the lower F_{min} for n-channel devices as compared to p-channel FETs.

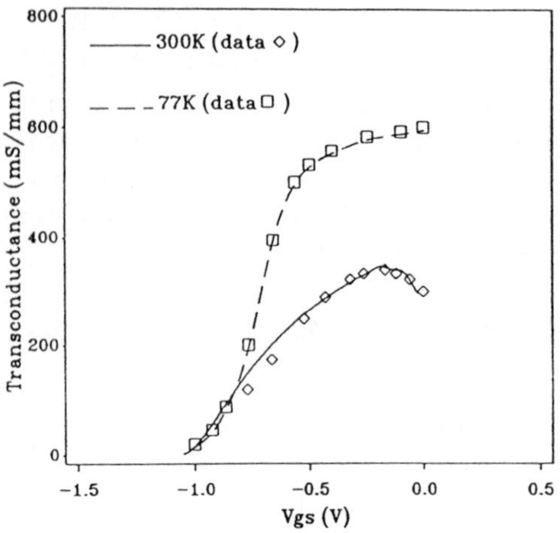

Figure 1: Comparison of the theoretical calculated transconductance with experimental data[3].

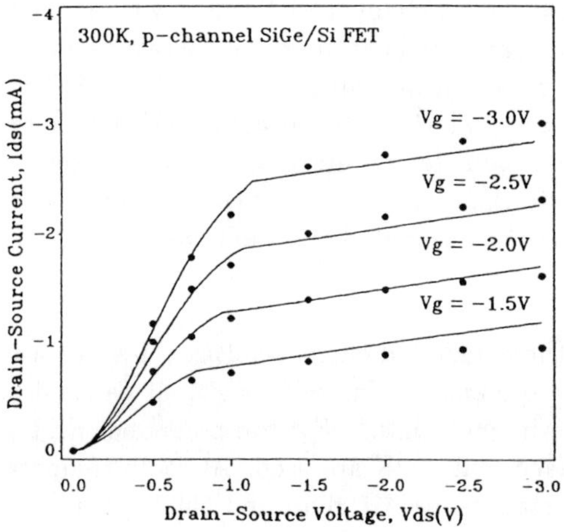

Figure 2: The current-voltage (I-V) characteristics for the p-channel FET[4] are plotted for 300K. Dots represent experimental data and the solid lines represent calculated results.

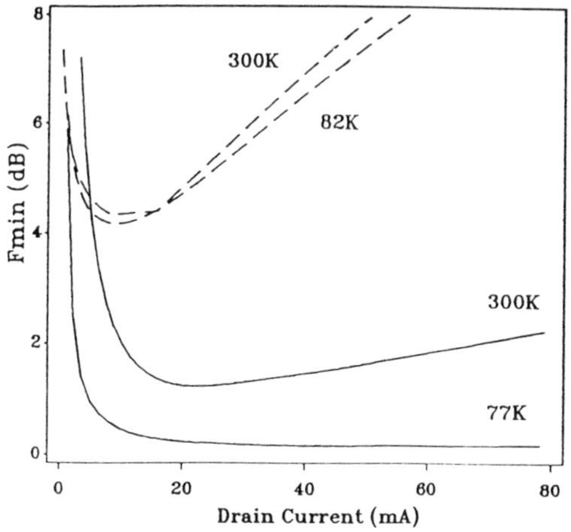

Figure 3: Minimum noise figures F_{min} are plotted as a function of drain current for $SiGe/Si$ n-channel MODFETs (solid lines) and p-channel MOSFETs (dashed lines) with temperature as a parameter (the drain-source current in p-channel FETs is scaled up by 5).

Conclusion

A self-consistent analysis is presented to model d.c. small-signal parameters noise performance for SiGe based FETs. The use of a modified velocity-electric field characteristic, that is in agreement with experimental results, Made the calculations analytically tractable. moreover, calculated results are in excellent agreement with the experimental data. This model will prove extremely useful in optimizing noise performance for SiGe based FETs.

References

1. A. F. M. Anwar, K. W. Liu and R. D. Carroll, to be published in Jour. of Appl. Phys., Aug. 1993.

2. A. F. M. Anwar, and K. W. Liu, to be published in IEEE. Trans. Electron Dev., June 1993.

3. V. P. Kesan, S. Subbanna, P. J. Restle, M. J. Tejwanl, IEDM 91, p. 2.2.1-2.2.4, 1991.

4. K. Ismail, B. S. Meyerson, S. Rishton, J. Chu, S. Nelson and J. Nocera, IEEE Electron Dev. Lett., vol. 13, no. 5, p. 229-231, May, 1992.

MODELING AND CHARACTERIZATION OF FLICKER NOISE IN CMOS TRANSISTORS FROM SUBTHRESHOLD TO STRONG INVERSION

Jimmin Chang and C. R. Viswanathan
Electrical Engineering Department, UCLA
University of California,
Los Angeles, CA 90024, U.S.A.

ABSTRACT

Flicker noise behavior of n- and p-channel silicon MOSFET's operating from subthreshold to strong inversion at room to low temperatures will be described. It is found that, for various bias and temperature conditions, input referred noise in n-channel devices show minimal gate bias dependence while p-channel transistors show gate voltage dependence.

INTRODUCTION

When the MOSFET gate voltage is below the threshold voltage, the semiconductor surface is in weak inversion and the drain current is dominated by diffusion rather than drift. It can be shown that in the subthreshold region [1],

$$I_D = \mu\left(\frac{W}{L}\right)\frac{aC_{ox}}{2\beta^2}\left[\frac{n_i}{N}\right]^2\left(1-e^{-\beta V_D}\right)e^{\beta\psi_s}\left(\beta\psi_s\right)^{-\frac{1}{2}}$$

where β=kT/q, $a=\sqrt{2}\,(\varepsilon_s/L_D)/C_{ox}$, N is the substrate doping and ψ_s is the surface potential. The subthreshold region of operation is particularly important for low voltage, low-power applications, such as in focal plane arrays, where noise directly affect memory errors and logic voltage swings. The subthreshold characteristics also describe how the device switches on and off. Even though noise in strong inversion has been studied extensively, very little is known in the noise performance of MOSFET's in weak inversion. In this paper, extensive characterization of the noise performance of both the n-channel and p-channel MOSFET's in the strong inversion to subthreshold regions of operations from room temperature to 50 K, is reported. A quasi one dimensional flicker noise model is then used to explain the measured data.

THEORY

In the current driven model [2], the drain voltage is calculated for a given drain current. The channel of the transistor is divided into a number of elementary sections of length ∂y, and the voltage drop along the channel can be calculated to be

$$V(y) = \int_0^y \frac{I_D}{W\mu(y)\,Q_i(y)}\partial y$$

where $\mu(y)$ and $Q_i(y)$ are the mobility and free carrier density along the channel. The carrier density fluctuation noise along the channel is calculated according to McWhorter's noise mode, given by

$$S_{Q_i} = \frac{q^2 N_T(E_F)\,kT}{\gamma f W}\delta(y)$$

where γ is McWhorter's tunneling parameter and $N_T(E_F)kT$ is the effective interface trap density chosen to be 6.5×10^{15} cm^{-3}. The mobility fluctuation noise is given by

$$S_\mu = \frac{q\alpha_H \mu^2}{fWQ_i}\delta(y)$$

where α_H is Hooge's parameter, chosen to be 8.9×10^{-6}.

EXPERIMENTAL

The samples used are from a p-well CMOS process with a nominal oxide thickness of 500 Å. The p-well doping is in the range of 1.5×10^{16} cm^{-3} while the n-type substrate is about 2×10^{15} cm^{-3}. Figure 1 shows the input referred noise spectrum of a 100×10 μm^2 n-channel MOSFET at room temperature. It can be seen that input referred noise does not change significantly as the gate bias decreases from strong inversion to subthreshold region. This suggests that carrier density fluctuation noise might be the dominating

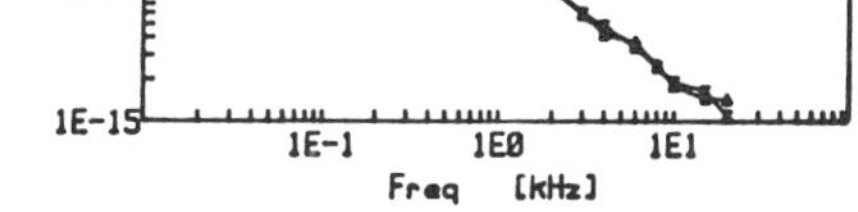

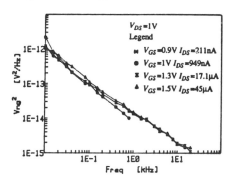

Fig.1 Input referred noise of n-channel device.

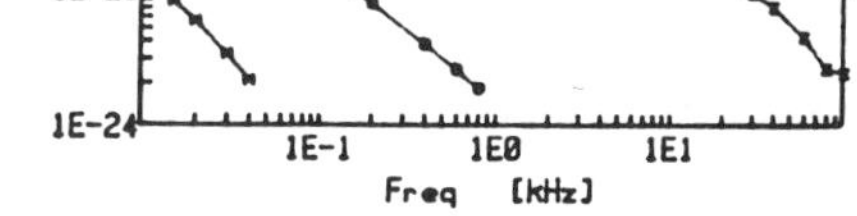

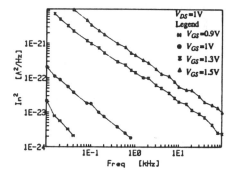

Fig.2 Output current noise of n-channel device.

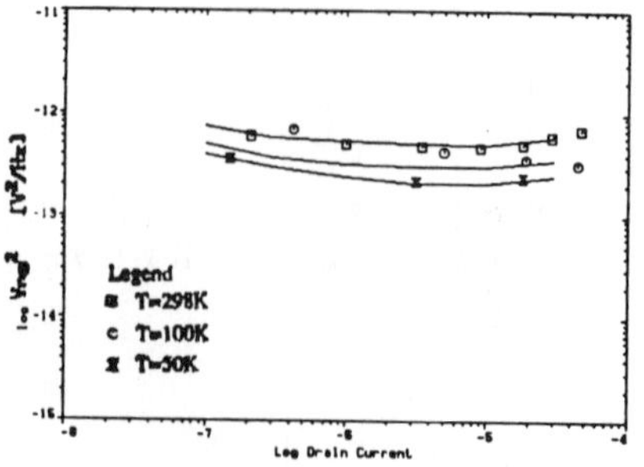

Fig. 3 Gate bias and Temperature dependence of n-MOS input referred noise

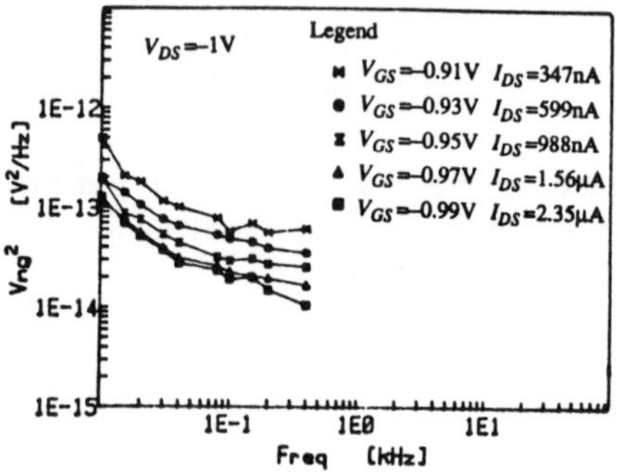

Fig. 4 Input referred noise of p-channel device.

noise source. The same experimental results expressed in output current noise is shown in Figure 2. Since the output current noise is obtained by multiplying the input-referred noise by the small-signal transconductance ($\partial I_D/\partial V_G$), and since from equation (1), $\partial I_D/\partial V_G$ decreases exponentially with V_G in the subthreshold region, the output current noise thus increases exponentially with gate voltage.

The temperature and bias dependence of the n-channel device input referred noise is shown in Figure 3. Only data down to 50 K is presented because freeze-out effects start to become dominant below this temperature, and the formation of the depletion region under the gate of the device has large time constants. In the figure, the symbols represent the spot noise data extracted at 20 Hz, and the solid lines show the theoretical carrier density fluctuation noise simulation using the current driven noise model. As can be seen, relatively small gate bias and temperature dependence is observed in the n-channel transistors.

Subthreshold noise characteristics in p-channel devices behave very differently from that of the n-channel FET's. Figure 4 shows the input-referred noise of a 100×5 μm^2 p-channel MOSFET at room temperature. The input referred noise decreases in magnitude

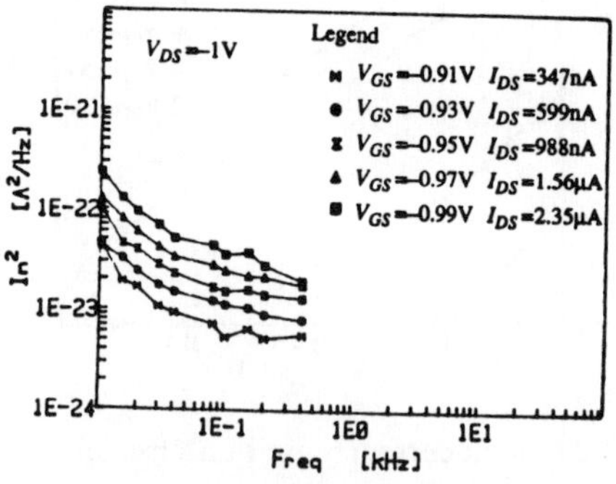

Fig. 5 Output current noise of p-channel device.

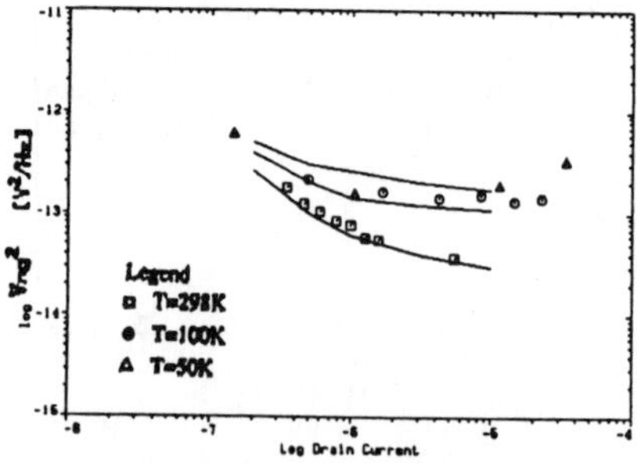

Fig. 6 Gate bias and temperature dependence of p-MOS input referred noise.

as the gate voltage increases from subthreshold into strong inversion. The same experimental results expressed in output current noise spectra is shown in Figure 5. Because of the exponential decrease of I_D versus V_G, the magnitude of the output current noise now increases from subthreshold to strong inversion. The strong gate bias dependence suggests that p-channel noise may be dominated by mobility fluctuation. Figure 6 shows how the gate bias and temperature dependence of the p-channel noise measurements compare to the theoretical mobility fluctuation noise model. As can be seen, there is significant gate bias and temperature dependence in the p-channel noise, as compared to the n-channel noise, suggesting that different noise mechanism is involved in the p-channel devices.

SUMMARY

Flicker noise behavior of CMOS transistors from subthreshold to strong inversion under various temperatures is presented. The results suggest that input referred 1/f noise in n-channel devices show relatively little gate bias and temperature dependence, and can be explained by carrier-density fluctuation. On the other hand, input-referred noise in p-channel devices show gate bias and temperature dependence that can be caused by mobility fluctuation.

REFERENCES

1. S.M. Sze, Physics of Semiconductor Devices (John Wiley & Sons, N.Y., 1981), p.446.
2. C.R. Viswanathan, Proc. Custom Integrated Circuit Conference, p. 199-204, 1982.

RADIATION EFFECTS ON RADIATION HARDENED LDD CMOS TRANSISTORS

A.HOFFMANN, M VALENZA, D. RIGAUD and L.K.J. VANDAMME*
Centre d'Electronique de Montpellier (CNRS URA 391)
Université Montpellier II
34095 Montpellier Cedex 5,France
*Eindhoven University of Technology, Dep. El. Eng.,
5600 MB Eindhoven, The Netherlands

ABSTRACT

Here we report on noise measurements from MOSTs through irradiations performed in a Co-60 gamma cell up to a total dose of 300 krad(SiO_2). We found that the noise in MOS transistors can be associated with channel and series resistance due to LDD structure. No change in 1/f noise level is observed when measurements are performed under constant effective gate voltage after irradiation. Nevertheless an increase in noise can reveal particular defects due to radiation damages.

INTRODUCTION

Conduction and noise measurements have been performed on n- and p-channel MOSTs before and after several steps of gamma-irradiation doses. Threshold voltage, mobility and 1/f noise levels have been investigated in order to correlate 1/f noise and radiation effects. There is still debate about the correlation of trap densities, trap efficiency and 1/f noise before and after irradiation[1-3].

Irradiation by 3 MeV electrons in MBE grown n GaAs layers created g-r noise by induced traps but no significant change in the 1/f noise parameter α at room temperature[4].

For Si MOS transistors significant change in 1/f noise has been reported after gamma-irradiation for large total dose under constant effective gate voltage or for the same external biasing conditions as before irradiation. In the later case the threshold voltage shift induces an increase of noise.

In this paper threshold voltage shifts are taken into account and the MOSTs are biased under the same effective gate voltage in order to compare. From our measurements [5] we conclude that the channel noise and series resistance noise remain about constant.

EXPERIMENTAL RESULTS AND DISCUSSION

Measurements have been performed on n- and p-channel LDD MOS transistors with several gate dimensions. For all transistors the gate oxide thickness is 19 nm. A Co-60 gamma cell was used for irradiations up to 300 krad(SiO_2) with a dose rate of 50 rad(SiO_2)/s.Threshold voltage shifts $|\Delta V_T|$ are about 3 mV/krad(SiO_2) but for large total doses a saturation effect appears.

The mobilities at low electric field have been investigated through irradiations and typical results for a large 25 µm / 25 µm n- and p-channel device are shown in fig.1 and 2. All p-channels show a slight decrease in mobility with increasing dose (10%) and n-channels have an opposite trend, which also has been observed in ref.6. Because a calculated mobility from a conductance measurement or a $\Delta G/\Delta V_G$ is always proportionnal to $\mu \propto$ L/W both trends can be explained in terms of a variation in channel width with increasing irradiation. Increasing irradiation goes hand in hand with increasing positive oxide charge in the birds beak at the rim of the channel. This provokes an increase in width for n-channels and sometimes even a leakage current path in parallel and for p-channels a reduction in width is observed. In addition no evolution of the subthreshold swing after irradiation is observed [5]. We conclude that interface state density remains below than 10^{11} cm^{-2} and mobility is kept constant up to 300 krad(SiO_2).

The 1/f noise has been observed between 10 Hz and 100 kHz in the ohmic region. From the observed current or voltage spectra the fluctuations in the resistance between source and drain have been calculated $S_R = S_V / I^2 = (S_I / I^2) R^2$.

The noise spectral density in the resistance consists of a channel and series resistance contribution in LDD MOSTs[7,8], each having a different dependence on the effective gate voltage:

$$S_R = S_{Rch} + S_{Rs}$$

with $R_{ch} = L / W C_{ox} V_G^* \mu$ the channel resistance and R_s the series resistance.

For low effective gate voltages $V_G^* = V_G - V_T$ we expect $S_R \equiv S_{Rch}$ and $S_{Rch}=\alpha_{ch} q\mu R_{ch}^3 / L^2 f \propto (L / W^3) / V_G^{*m}$ with often m=4 for n-channel devices which means $\alpha_{ch} \propto V_G^{*-1}$. If α_{ch} remains constant or the noise becomes dominated by series resistance a levelling off in S_R versus V_G^* can be expected with $S_R \propto V_G^{*-m}$ with $m \leq 3$.

For the series resistance we expect the following proportionality $R_s \propto s / W\mu V_G^*$ $\propto V_G^{*-1/2}$,where s is in order of 0,1 µm. We assume a weak reduction in mobility under the spacers with applied bias due to fringefields at the rim of the gate at the source and drain side and due to high fields in the low doped drain which results in $\mu \propto V_G^{*-1/2}$. For high effective gate voltage $S_R \equiv S_{Rs} = \alpha_s q\mu_s R_s^3/s^2 f$ and $S_R \propto V_G^{*-2}$ is the result if α_s remains constant with V_G^*.
If $S_R = S_{Rs}$ and the series resistance is located at the metal-semiconductor interface and independent of V_G^* then $S_R \propto V_G^{*0}$ can be expected[9].

In fig.3 the result of an unirradiated n-channel MOST with a width length ratio of W/L = 25 µm / 0,85 µm are shown; above V_G^*=2 V a levelling off indicates series resistance problems; $S_V \propto V_G^{*-2}$ for V_{DS} = 100 mV indicates that $\alpha_{ch} \propto V_G^{*-1}$ which is a classical result for number fluctuations.

Fig.4 shows the results through irradiation of a p-channel with a W/L = 25µm/25µm in a S_R versus V_G^* plot . The typical slopes -4 and -2 are indicated showing series resistance problems at least in the noise contribution for V_G^*>1 V. The 1/f noise in the channel and in the series resistance do not change with irradiation for this p-channel device.

Fig.5 shows the results through irradiation on a n-channel with W/L = 25 µm / 0,85 µm indicating no appreciable change in 1/f noise over the whole range of applied V_G^*.

All spectra values presented in figs.3, 4 and 5 are at 10 Hz and stem from 1/f spectra for 1 Hz < f < 10^4 Hz. Some devices showed a generation-recombination noise and these devices are not discussed here.

Noise measurements have been also carried out in the subthreshold conduction. Thermal noise in generally obtained without excess noise. Irradiation does not affect the low frequency noise. Nevertheless for some devices an increase of the white noise has been noticed. After investigations this increase has been associated with leakage paths located at the source-substrate or drain-substrate junctions and induced by irradiations. In the same way abnormal increases of the 1/f noise allow to detect radiation damages in the gate-channel insulation.

CONCLUSIONS

1/f noise in p- and n-channel MOSTs has been investigated before and after gamma-irradiations. Under same effective gate voltage no change in noise has been observed up to a 300 krad(SiO_2) total dose. Nevertheless 1/f noise before irradiation and the threshold voltage shift after irradiation agrees with previous published results[3].

The behaviour of 1/f noise levels versus V_G^* shows that this noise is due to number fluctuation . From this behaviour it has also pointed out the predominant part of the series resistance due to the LDD structure for high effective gate voltage.

REFERENCES

1 J.H. Scofield, T.P. Doerr and D.M. Fleetwood, IEEE Trans.Nucl.Sc., 36, 1946 (1989)
2 T.L.Meisenheimer, D.M. Fleetwood, M.R.Shaneyfelt and L.C.Riewe, IEEE Trans. Nucl. Sc., 38 , 1297 (1991)
3 J.H. Scofield and D.M. Fleetwood, IEEE Trans.Nucl.Sc., 38, 1567 (1991)
4 L.Ren, P.Baucour, F.N.Hooge, L.H.Luthjens and M.R. Leijs, J. Appl. Phys, 73, 2180 (1993)
5 A. Hoffmann, Thesis, University of Montpellier, April 1993
6 G.J. Dunn, B.J. Gross and C.G. Sodini, IEEE Trans. El. Dev., 39, 677 (1992)
7 X.Li and L.K.J. Vandamme, Solid. State. Elect., 35, 1471 (1992)
8 A. Bhattacharyya, S.N. Shabde, IEEE Trans. El. Dev., 35, 1156 (1988)
9 J.M.Peransin, P.Vignaud, D.Rigaud and L.K.J. Vandamme, IEEE Trans.El.Dev., 37, 2250 (1990)

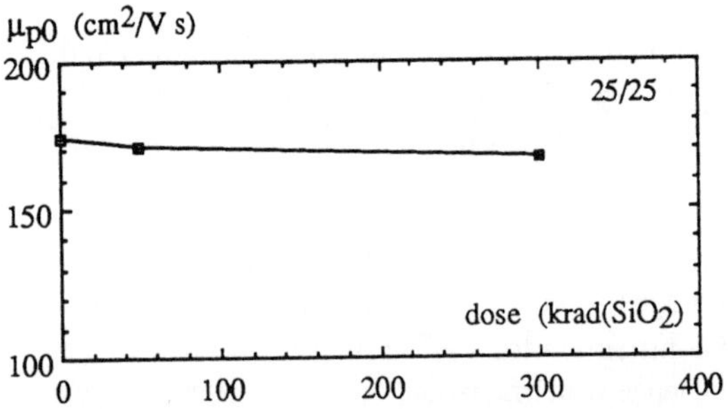

Fig.1 Hole mobility at low V_G^* versus radiation dose

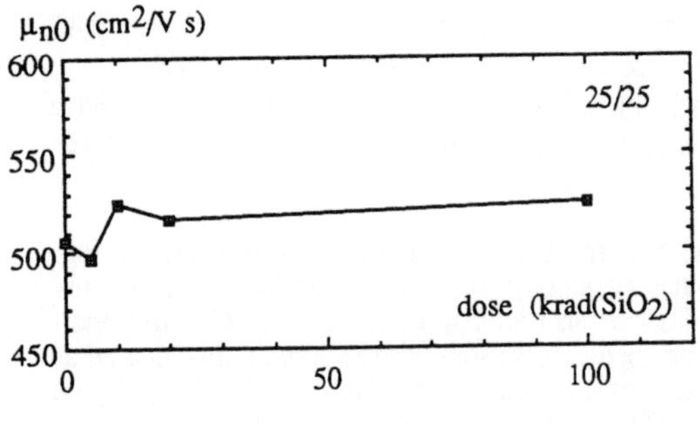

Fig.2 Electron mobility at low V_G^* versus radiation dose

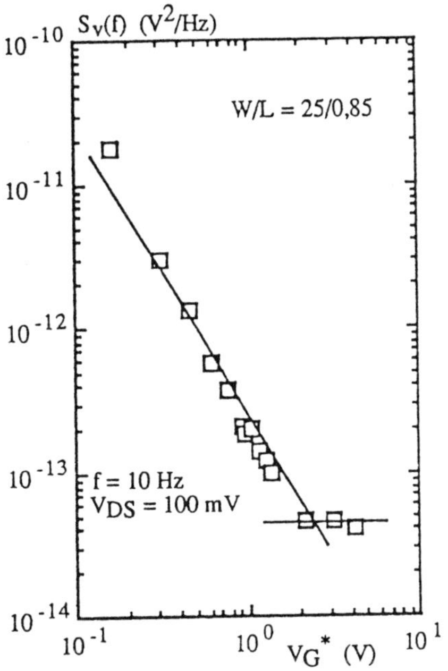

Fig.3. Voltage noise versus effective gate voltage of a short channel n-MOST before irradiation.

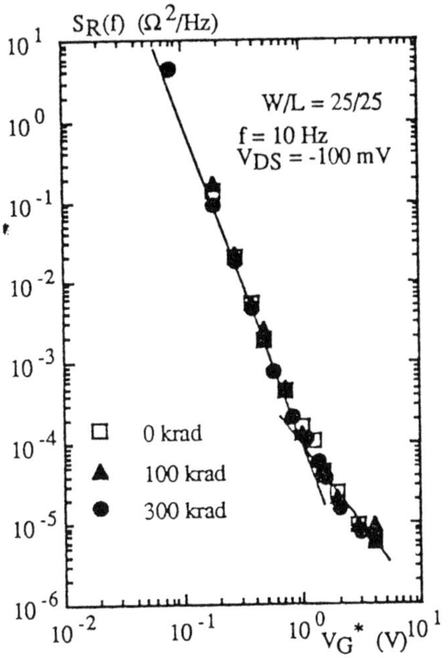

Fig.4. The resistance fluctuations S_R versus V_G^* through irradiation of a large p-MOS transistor.

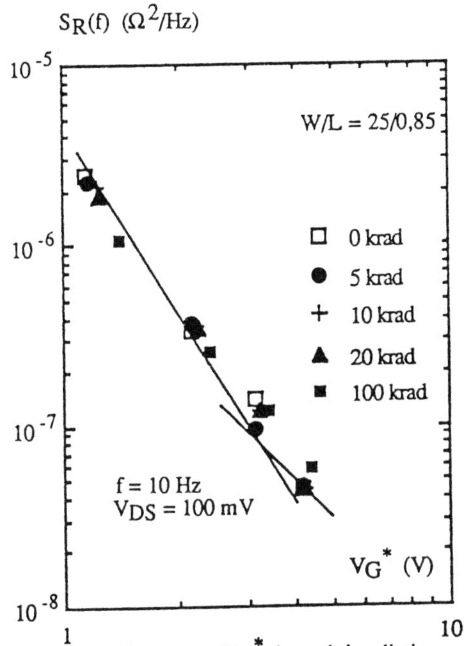

Fig.5. S_R versus V_G^* through irradiation of a short channel n-MOST. No evolution in 1/f noise.

LOW-FREQUENCY NOISE IN SILICON ON INSULATOR MOSFET's : EXPERIMENTAL AND NUMERICAL SIMULATION RESULTS

J. Jomaah, F. Balestra, G. Ghibaudo
Laboratoire de Physique des Composants à Semiconducteurs (URA–CNRS)
ENSERG/INPG, 23 rue des Martyrs, 38016 Grenoble, France

ABSTRACT

The low-frequency noise in SOI MOSFET's is studied experimentally and by numerical simulations. The behaviors of devices with completely depleted and partially depleted silicon film are investigated for various substrate biases. The importance of volume inversion in thin Si film is underlined. Moreover, the variations of the current noise around the kink is analysed for thin and thick film devices.

1. INTRODUCTION

A number of papers have been published on low frequency noise in MOS transistors. However, up to now, the noise behavior in silicon on insulator (SOI) MOSFET's, which have a very specific structure and very interesting electrical properties, has not been investigated in detail. It is worth noting that the leading technology in the field of SOI was in the 70's the SOS technology, and is presently the SIMOX technology, which presents much better electrical properties associated in particular with the strong decrease of defect density in the Si film. In these structures, a supplementary interface (Si film/buried insulator) can play an important role. In particular, for thin completely depleted Si film, a strong coupling of the front and back interface occurs, and the electrical properties of the devices depend on both interfaces and on the volume. Moreover, by biasing the front gate and the substrate (back gate) of the SOI transistors, a strong inversion can take place through the entire thickness of the silicon film [1]. On the other hand, for thick partially depleted Si film, a floating potential situation in the silicon film is created by the presence of the buried insulator, leading to an increase of the drain current in saturation for sufficiently large drain voltage. This kink effect is induced by a forward biasing of the source/thin Si film diode by the impact ionization current. The kink effect is suppressed when the floating potential situation vanishes, i.e. for a completely depleted thin silicon film.

Some noise investigations have been carried out for thick silicon on sapphire and partially depleted SIMOX MOSFET's [2,3] in order to observe the influence of the kink effect. An excess noise has been found around the kink, similar to that obtained for bulk Si MOSFET's in the liquid helium temperature range due to the frozen-out substrate [4-5]. On the other hand, the influence of the edges of the devices have been studied in thick film SIMOX MOSFET's [6]. Furthermore, the noise in thick and thin film depletion mode SIMOX MOS transistors has also been analysed [7]. However, no detail investigation has been performed for the noise in thin film enhancement type transistors, which represent the main devices in SOI applications. Therefore, in this study, the low frequency (1/f) noise for ultra thin film enhancement mode (EM) SIMOX MOSFET's is analysed for different coupling conditions depending on the back gate bias, in order to sweep the whole surface regimes at the back interface (inversion, depletion, accumulation). Moreover, the noise magnitude is studied as a function of the drain voltage for thin and thick film devices. In particular, the influence of the back gate on the noise, which can suppress the kink for thick film transistors by an inversion of the back interface, is considered. In addition, the effect of an accumulation of the back interface in thin film devices, which can induce a kink effect, is also analysed, and compared with the case of thick Si film.

2. RESULTS AND DISCUSSION

2.1. Experimental and numerical simulation results in linear operation

In figure 1 is shown the experimental dependence of the normalized drain current spectral density of a N-channel thin film EM SIMOX MOSFET as a function of the front gate voltage V_{g1} for various back gate voltages V_{g2}. For an inverted or a depleted back surface (V_{g2}=20V or 0V respectively), the contribution for the noise of the Si film/buried oxide interface should be maximum compared with an accumulated back surface (V_{g2}=-20V). However, unlike this foreseeable behavior, the noise with an accumulation at the back interface presents the maximum magnitude (Fig. 1). A similar feature has been obtained for P-channel devices. It is worth noticing that the drain current noise has been plotted as a function of V_{g1} in order to magnify the strong inversion region, but a similar behavior has been obtained by plotting the curves versus I_d. In order to understand this original behavior, a numerical simulator for SOI structures (ISIS I [8]) has been used. Some modifications of this program have been achieved in order to simulate the carrier number fluctuations by dynamic exchanges with the oxide traps, which represent the main contribution of noise in MOS transistors. These fluctuations are taken into account by varying the charges at both interfaces of the SOI structure and by simulating the corresponding change ΔI_d in drain current. With this method, the normalized drain current spectral density S_{I_d}/I_d^2 in very strong inversion does not decrease by inverting the back surface of the transistor (see dotted lines in Fig. 2), unlike the experiment. Therefore, we have tried to include in the model the supplementary fluctuations of the mobility induced by those of the interface charge [9,10]. The same results than the previous simulations were obtained by adding this physical phenomenon. It is worth noting that the noise associated with the mobility fluctuations alone, which is proportional to the inversion charge, cannot explain the experimental properties. Nevertheless, due to specific conduction in thin SOI film, we have taken into account in the model an other effect associated with the screening of the Coulomb scattering for carrier mobility.

The general expression for the variations of the mobility due to Coulomb scattering effects is of the form :

$$\frac{1}{\mu(x)} = \frac{1}{\mu_0} + \alpha(x)\, Q_{ss} \qquad (1)$$

where $\mu(x)$ is the effective mobility at a distance x from the interface, μ_0 is the maximum effective mobility, Q_{ss} is the interface charge which induces fluctuations in drain current (the same Q_{ss} is considered at both interfaces), and $\alpha(x)$ is a Coulomb scattering parameter depending on the distance from the interface, and taking into account the screening phenomenon, given by :

$$\alpha(x) = \alpha_0\, e^{-x/\lambda} \qquad (2) \qquad \text{and} \qquad \lambda = \frac{\epsilon_{Si}}{q^2}\frac{kT}{Q_i} \qquad (3)$$

where α_0 is the maximum Coulomb scattering parameter, λ is the screening length, ϵ_{si} is the silicon permittivity, q the electron charge, kT the thermal energy, and Q_i the inversion charge density calculated by numerical simulation.

These different equations (1-3) have been included in the ISIS I simulator.

Fig. 2 (full lines) shows the simulation results obtained in this case. A good agreement with the experimental behavior is underlined, with an increase of the noise by accumulating the back surface. In fact, the current noise is significantly changed for an inverted back surface compared with the results without screened mobility fluctuations (dotted lines in fig. 2), and is less affected for an accumulated back surface. This phenomenon is due to the location of the conduction channel. In the case of a back surface in accumulation, a strong electric field is present in the thin Si film and the channel is mainly confined close to the Si/SiO_2 front surface, where the oxide charge fluctuations may strongly influence the mobility fluctuations. On the other hand, for a strongly inverted back surface, an important part of the carrier transport is carried out in the Si volume far from both interfaces, where the oxide charge fluctuations are screened which reduces significantly the noise of the MOSFET's.

2.2. Experimental results in non-linear operation

The noise magnitude has also been studied as a function of the applied drain voltage. Fig. 3 is a plot of the experimental dependence of the normalized drain current spectral density of a thick film EM SIMOX MOSFET's as a function of I_d for various back gate voltages. For a zero V_{g2}, a strong increase of the noise is obtained around the kink. These supplementary fluctuations can be attributed to those of the substrate current which induces a change in the threshold voltage around the kink. An interesting result is also pointed out in this figure, i.e. the suppression of the excess noise associated with the application of a back gate voltage which induces an inversion of the back surface. In this situation, the back surface is strongly inverted which suppresses the floating potential in the Si film and consequently the kink effect.

In the case of a thin Si film (Fig. 4), no excess noise is observed whatever the drain voltage is. This behavior related to the suppression of the kink effect is very interesting for SOI applications. On the other hand, by accumulating the back surface (V_{g2}=-10V), a kink is created in the $I_d(V_d)$ characteristics and an important excess noise is underlined around the kink (Fig. 4). This behavior is due to the presence of a floating potential in the thin Si film with an accumulation of the back surface, and a similar noise feature than in thick film is underlined, which shows that this phenomenon is independent of the Si film thickness.

3. CONCLUSION

The low-frequency noise in SOI MOSFET's has been studied experimentally and by numerical simulations. The behaviors of devices with completely depleted and partially depleted silicon film have been investigated for various substrate biases. The importance of volume inversion in thin Si film has been underlined. In this mode of operation, a significant improvement of the noise magnitude has been shown. Moreover, the variations of the current noise around the kink has been analysed for thin and thick film devices. An appropriate substrate bias has been shown to induce or to suppress the excess noise around the kink.

REFERENCES

1. F. Balestra, S. Cristoloveanu, M. Benachir, J. Brini, T. Elewa, IEEE Electron Dev. Lett., EDL-8, p. 410 (1987).
2. W. Fichtner and E. Hochmair, Electron. Lett., 13, p. 675 (1977).
3. J. Chen, P. Fang, P.K. Ko, C. Hu, R. Solomon, T.-Y. Chan, C.G. Sodini, Proc. 1990 IEEE SOS/SOI Technology Conference, Key West, Florida, p. 40 (1990).
4. F. Balestra, L. Audaire, C. Lucas, Solid-St. Electron., 30, p. 321 (1987).
5. I.M. Hafez, G. Ghibaudo, F. Balestra, Solid-St. Electron., 33, p. 1525 (1990).
6. T. Elewa, B. Kleveland, S. Cristoloveanu, B. Boukriss, A. Chovet, IEEE

Trans. Electron Dev., ED-39, p. 874 (1992).

7. T. Elewa, B. Boukriss, H. Haddara, A. Chovet, S. Cristoloveanu, IEEE Trans. Electron Dev., ED-38, p. 323 (1991).
8. F. Balestra, J. Brini, P. Gentil, Solid-St. Electron., 28, p. 1031 (1985).
9. K.K Hung, P.K. Ko, C. Hu, Y.C. Cheng, IEEE Trans. Electron Dev., ED-37, p. 654 (1990).
10. G. Ghibaudo, O. Roux, Ch. Nguyen-Duc, F. Balestra, J. Brini, Phys. Stat. Sol. (a), 124, p. 571 (1991).

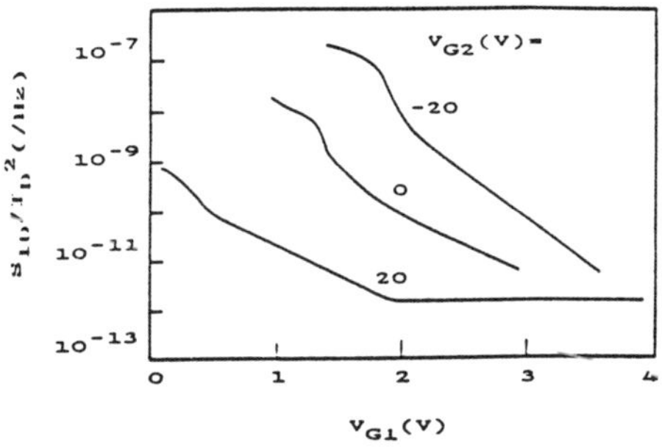

Figure 1. Experimental normalized drain current spectral density as a function of front gate voltage for various back gate voltages (N channel thin film (t_{si}=80 nm) EM SIMOX MOSFET's; f=10 Hz, V_d=50 mV).

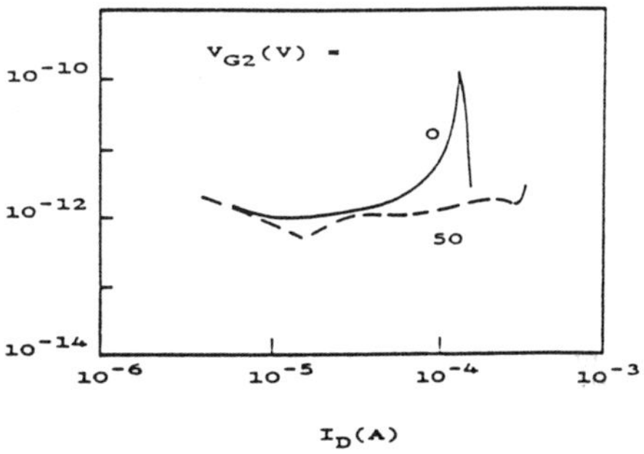

Figure 3. Experimental normalized drain current noise as a function of I_d by varying the applied drain voltage, for various back gate voltages (N channel thick film (t_{si}=150 nm) EM SIMOX MOSFET's; V_{g1}=3V, f=10 Hz).

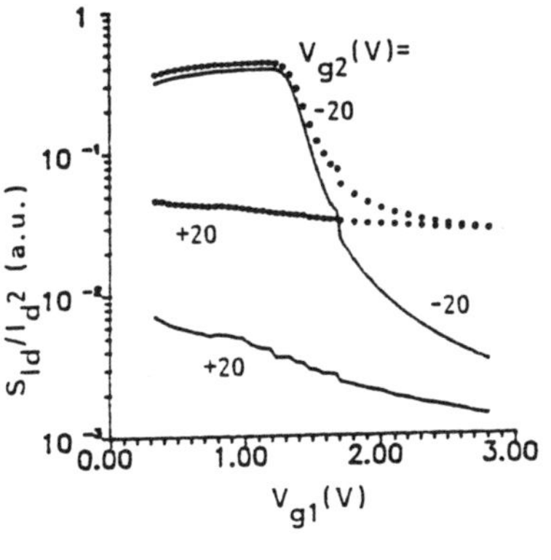

Figure 2. Simulated normalized drain current noise as a function of front gate voltage for various back gate voltages (N channel thin film (t_{si}=80 nm) EM SIMOX MOSFET's; Q_{ss}=5 10^{10}q/cm^2, μ_0=700 cm^2/Vs; dotted lines : α_0=0, full lines : α_0=12000 Vs/C).

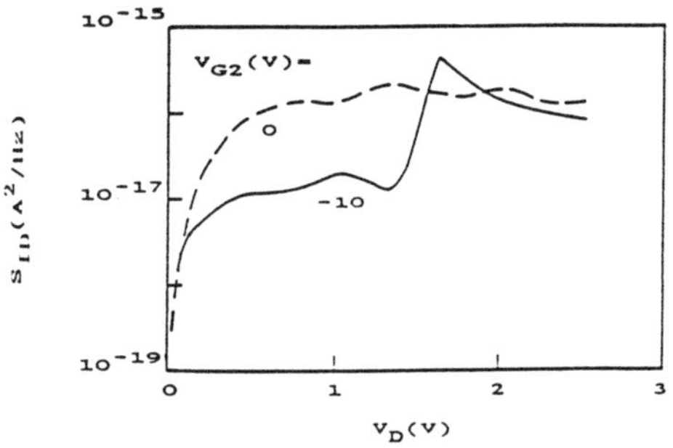

Figure 4. Experimental drain current noise as a function of drain voltage, for various back gate voltages (N channel thin film (t_{si}=80 nm) EM SIMOX MOSFET's; V_{g1}=1.7V, f=10 Hz).

A STUDY OF 1/f NOISE IN LDD MOSFETs

Xiaosong Li and L.K.J. Vandamme
Electrical Engineering Department, Eindhoven University of Technology
Postbus 513, 5600 MB Eindhoven, the Netherlands
Tel. #31-40-473242, FAX #31-40-448375

ABSTRACT

1/f Noise of the drain current S_I versus drain source voltage was studied in an LDD MOSFET. Analysis of the dc characteristics shows that the series resistance R_{Dd}, introduced by the LDD structure on the drain side, increases with the external drain source voltage V_{DS}. Consequently, the internal drain source voltage V_{ds} and the channel current I_D are reduced. This is used successfully to explain why the experimental results of S_I as a function of V_{DS}/V_G^* in an LDD MOSFET are at variance with those in a conventional MOSFET. Here a model and a procedure are proposed from which the internal drain source voltage V_{ds} and series resistance are calculated from measuring results. After recalculation of the experimentally observed S_I in terms of channel current fluctuations, we find that S_{Ich} versus normalized internal drain-source voltage V_{ds}/V_G^* shows the same trend as in a conventional MOSFET. The role of R_{Dd} becomes more important as the value of V_{DS}/V_G^* increases.

INTRODUCTION

Here we report on the contribution of nonlinear series resistance to the observed 1/f current noise in LDD MOSFETs. A shielded probe station has been constructed for measuring noise on wafers. We investigated p-channel LDD MOSFETs made with a 0.7μm technology.

In a conventional long-channel MOSFET, it was found that the 1/f noise power spectrum of the drain current S_{Ich} is proportional to $(V_{DS}/V_G^*)^2$ below saturation. V_{DS} is the external drain-source voltage and $V_G^* = V_{GS} - V_T$ the effective gate voltage. In saturation, S_{Ich} also saturates. This has been, for example, explained in Ref. 1. In an LDD device, below current saturation we observed that the 1/f noise current spectrum is proportional to $(V_{DS}/V_G^*)^{1+\delta}$ with $0 < \delta < 1$ depending on V_G^*, V_{DS}, and effective channel length l (see Fig. 1 and 2). For high values of V_G^* and l, $\delta \rightarrow 1$, but $\delta \rightarrow 0$ for $V_{DS} \rightarrow V_G^*$. The deviation of an LDD MOSFET from a long-channel device is due to the voltage dependent series resistance of the LDD structure.

EXPERIMENTAL RESULTS AND ANALYSIS

An LDD device can be seen as a conventional MOSFET in series with two resistors R_{Dd} and R_{sS}, one in the drain side and another in the source side (see Fig. 3a). The series resistance decreases channel current and it shares the external drain-source voltage with the channel[2-4]. Our analysis shows an increase in series resistance in the drain with increasing V_{DS} as already observed in Refs.[5, 6]. The channel current and the voltage over the conventional channel part are clamped.

In Fig. 3a, we assume the square-law model for the current versus the internal drain-source voltage V_{ds} is still valid in the channel part and neglect the second-order effects. Then the drain current I_D can be written as

$$I_D = \frac{\beta(V_G^* V_{ds} - \frac{1}{2}V_{ds}^2)}{1 + \beta R_{sS} V_{ds}} \quad \text{for } V_{ds} \leq V_g = V_G^* - I_D R_{sS} \quad (1)$$

$$V_{ds} = V_{DS} - I_D(R_{sS} + R_{Dd}) \quad (2)$$

$$\beta = \frac{\mu_{00} W C_{ox}}{l(1 + \theta V_g + \theta_c V_{ds})} \quad (3)$$

where W is effective channel width, μ_{00} the low field mobility, C_{ox} the gate oxide capacitance per unit area, θ the mobility degradation coefficient due to the normal electrical field, and θ_c the mobility reduction coefficient of the velocity saturation due to the lateral field. The internal effective gate voltage is $V_g = V_G^* - I_D R_{sS}$. Following the same procedure as in Ref. 7, with a set of devices having the same channel width but differing in channel length, μ_{00} and l are obtained. The total resistance of a MOSFET is $R_{tot} = R_{ch} + R_{sS} + R_{Dd}$, where R_{ch} is the channel resistance. R_{sS} is a function of V_G^* only and R_{Dd} a function of V_G^* and V_{DS}. The influence of V_{DS} on R_{Dd} is negligible in the ohmic region. Hence,we can assume $R_{Dd} = R_{sS}$. The dependence of R_{sS} on V_G^* is obtained from the intersections of $R_{tot} \sim l$ with V_G^* as a parameter at V_{DS} =-50mV. Combining eq. (1) and I_D versus V_{DS} we obtained V_{ds} with V_G^* as a parameter. Hence, $R_{Dd}(V_G^*, V_{DS})$ is known.

Fig. 3b shows the noise equivalent circuit of an LDD MOSFET. S_I is the measured current noise and S_{Ich} the noise from the conventional channel only. We get

$$S_I = \frac{S_{I_{ch}} + g_D^2 S_{V_s}}{(1 + R_{sS} g_m + R_{sS} g_D + R_{Dd} g_D)^2} \quad (4)$$

where g_m and g_D are internal transconductance and channel conductance, given by

$$g_m = \frac{\partial I_D}{\partial V_g} = \beta V_{ds} \quad (5)$$

$$g_D = \frac{\partial I_D}{\partial V_{ds}} = \frac{\beta(V_G^* - V_{ds} - \frac{1}{2}\beta R_{sS} V_{ds}^2)}{(1 + \beta R_{sS} V_{ds})^2} \quad \text{for } V_{ds} < V_g \quad (6)$$

$$g_D = 0 \quad \text{for } V_{ds} \geq V_g \quad (7)$$

and S_{Vs} is the 1/f voltage noise of R_{Dd} and R_{sS} together for a current I_D. According to Ref. 7, the 1/f noise in a correctly processed MOSFET is always dominated by the channel contribution but never by the series resistance, i.e., $g_D^2 S_{Vs} << S_{Ich}$. Therefore, S_{Ich} can be approximated

$$S_{I_{ch}} = (1 + R_{sS} g_m + R_{sS} g_D + R_{Dd} g_D)^2 S_I \quad (8)$$

where $(1 + R_{sS} g_m + R_{sS} g_D + R_{Dd} g_D)^2$ is a correction factor due to the series resistance contribution which is bias dependent.

In Fig. 4, S_{Ich} is calculated with eq. (8) from experimental results presented in Fig. 1. We see that S_{Ich} is proportional to $(V_{ds}/V_G^*)^2$ which is the well known trend for long-channel devices without an LDD structure. In a short-channel LDD MOSFET, raising V_{DS}, R_{Dd} has a more pronounced influence on S_I when the device is biased at low V_G^* than when it is biased at high V_G^*. The carrier concentration in LDD part increases as V_G^* increases which results in a decrease of R_{Dd}. By increasing V_{DS}, the value of R_{Dd} increases[6]. In a long-channel LDD device, the dependences $S_{Ich} \sim V_{ds}/V_G^*$ and $S_I \sim V_{DS}/V_G^*$ are almost the same (Fig. 2). The reason is that the series resistance is much smaller than R_{ch} which is proportional to the channel length. The contribution of R_{Dd} to V_{ds} and S_I is negligible. To compare the measured and recalculated (by eq. (8)) 1/f drain current noise in a long-channel LDD device, we find that there is no difference between them at low V_{DS}, the difference becomes considerable at high V_{DS}, when V_G^* remains constant. The surprising result is that uncorrected noise results suggest a saturation in current noise while from the recalculation of S_{Ich}, the channel noise is not yet in saturation. This shows again that R_{Dd} plays a more important role than R_{ss} when V_{DS} increases.

CONCLUSIONS

By taking into account the series resistance in the drain and source end with different dependence on V_{DS}, the 1/f drain current noise in an LDD MOSFET shows the same trend as that in a conventional long-channel MOSFET. The correction due to R_{Dd} becomes more significant at low V_G^* and high V_{DS}, and it is essential that $R_{Dd} > R_{ss}$ above the ohmic region. This shines a new light on the analysis of the degradation due to hot carriers at the drain side and resulting in noise increase[8, 9].

ACKNOWLEDGMENT

This work is supported by an ADEQUAT programme. The devices under test were supplied by Philips Research Laboratories, Eindhoven, The Netherlands.

REFERENCES

1. L.K.J. Vandamme and H.M.M. de Werd, Solid-State Electronics, **23**, pp. 325-329, 1980.
2. H.G. de Graaff and F.M. Klaassen, Compact transistor modelling for circuit design, Springer, Berlin (1990).
3. S. Kohyama, Very high speed MOS devices, Clarendon press, Oxford (1990).
4. M.I. Elmasry, Digital MOS integrated circuits II: with applications to processors and memory design, IEEE press (1992).
5. F.J. Lai and J.Y. Sun, IEEE Trans. Electron Devices, ED-32, pp. 2803-2811, 1985.
6. J.A.M. Otten and F.M. Klaassen, Proceedings ESSDERC 1992, pp. 703-706.
7. X. Li and L.K.J. Vandamme, Solid-State Electronics, **35**, pp. 1471-1475 (1992).
8. M. Stegherr, Solid-State Electronics, **27**, p. 1055 (1984).
9. Z.H. Fang, S. Cristoloveanu, and A. Chovet, IEEE Electron Device Letters, **EDL-7**, p.371 (1986).

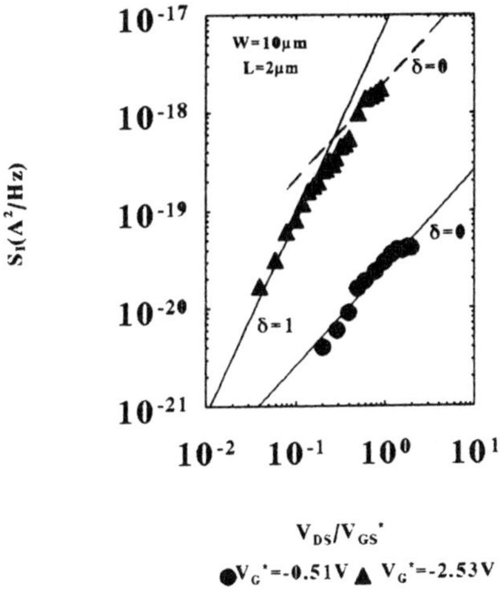

Fig.1 The measured S_I vs. V_{DS}/V_G^* in an LDD MOSFE

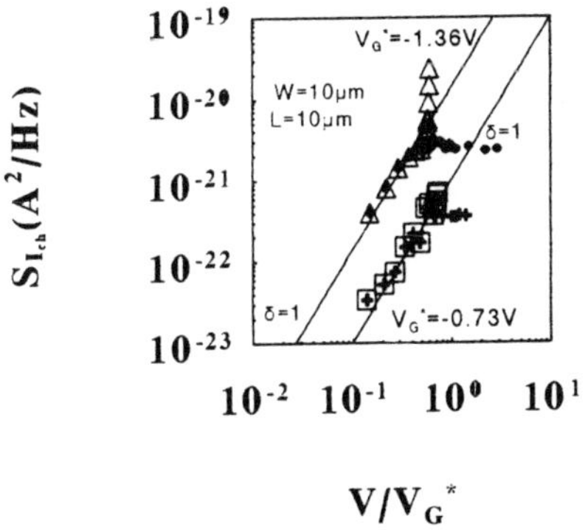

Fig. 2 The measured (+,•) and revised (□,△) 1/f current noise in a long channel LDD MOSFET.

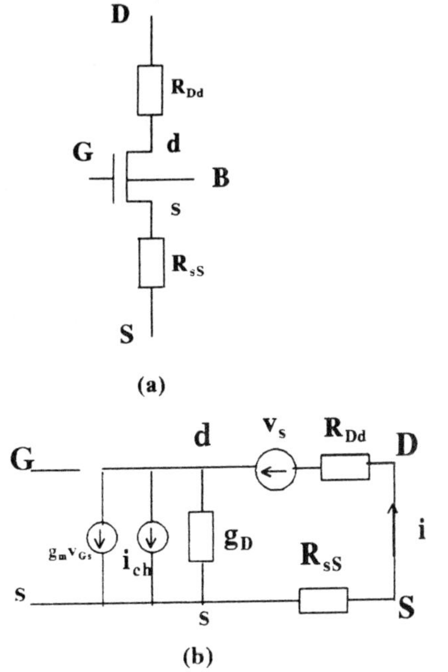

Fig. 3 (a) An LDD MOSFET consists of a MOST and series resistors.

(b) The noise equivalent circuit (low frequency).

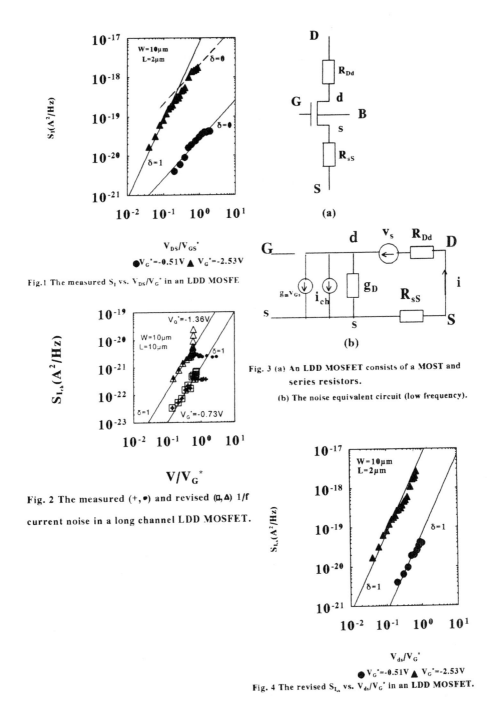

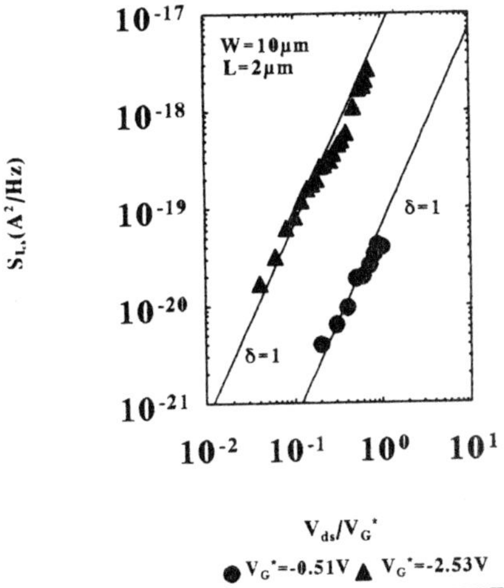

Fig. 4 The revised $S_{I_{ds}}$ vs. V_{ds}/V_G^* in an LDD MOSFET.

NOISE PROPERTIES OF MOSFETs PREPARED BY ZMR

N.B.Lukyanchikova, N.P.Garbar and M.V.Petrichuk
Institute of Semiconductor Physics
Ukrainian Academy of Sciences
Prospect Nauki 45 252650 Kiev-28 Ukraine

ABSTRACT

Noise spectra have been measured in p^+np^+, p^+pp^+ and n^+nn^+ MOSFETs prepared by ZMR. By analysing them, the conclusions have been made about the parameters and some peculiar features of the silicon layer and Si-SiO_2 interfaces with the gate and buried oxides.

Noise spectra investigations have been applied to examine silicon-on-insulator MOSFETs prepared with the help of zone melting recrystallization (ZMR). The structures studied were: p^+np^+, p^+pp^+ and n^+nn^+ with different doping concentration in the silicon layer. The thickness of the gate (front) oxide, buried (back) oxide and silicon layer were d_f = 120 nm, d_b = 1000 nm and t= 0.3 mcm, respectively. The length of the channel and the width of the gate were 15 mcm and 78 mcm for p^+np^+ structures and 550 mcm and 20 mcm for p^+pp^+ and n^+nn^+ structures. The noise has been also measured on the blocking contacts p^+-n or n^+-p to silicon layers investigated.

The current noise spectra S_i(f) of p^+np^+ MOSFETs have both generation-recombination and 1/f components either when a channel is under the gate oxide or when it is above the buried oxide (Fig.1a).

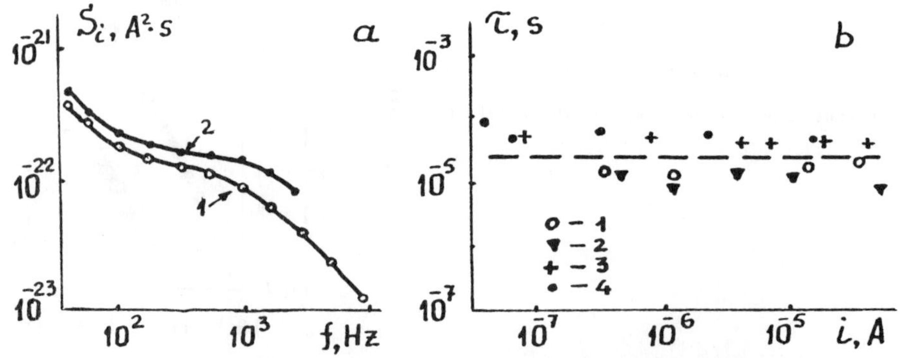

Fig.1. a - Current noise spectra of the p^+np^+ MOSFET measured at i = 10^{-5}A: 1 - U_{gf} = -3 V; 2 - U_{gb} = -25 V.

b - Dependences of the value of τ determined from the GR portion of the noise spectra on the channel current i changed by U_{gf} (1,2) or by U_{gb} (3,4).

It has been found that the value of the relaxation time τ for GR fluctuations observed is practically independent of the voltage on the front (U_{gf}) or back (U_{gb}) oxides (Fig.1b). By analysing the characteristics of the GR noise observed, the concentrations of traps N_t responsible for this noise have been determined. They are: $N_t = (1\div3)\cdot10^{14}$ cm^{-3} and $N_t = (0.6\div2)\cdot10^{15}$ cm^{-3} for the upper and lower portion of the silicon layer, respectively. As is seen from Fig.1b, those traps are characterised by $\tau \approx 2.5\cdot10^{-5}$ s.

As to the properties of 1/f noise, its peculiar features are shown in Fig.2 where S_U is the equivalent spectral density of the gate voltage fluctuations of 1/f type. It is seen that S_U does not depend on the voltage either on the front or on the back oxide. Besides, the ratio between the values of S_U measured under conditions where the channel is located near the front oxide and near the back oxide, respectively, is equal to the value $(d_f/d_b)^2$.

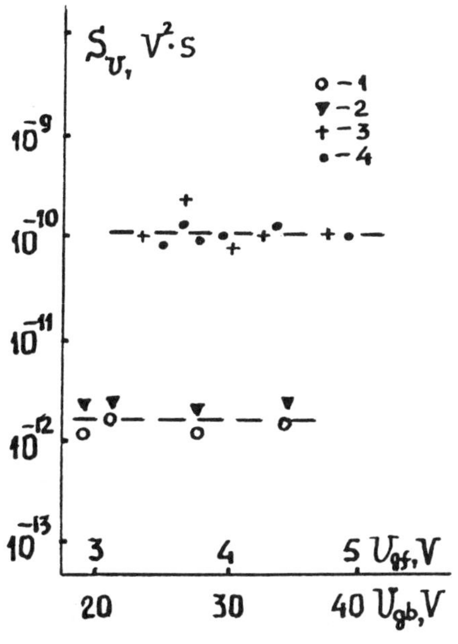

Fig.2. Dependences of S_U on U_{gf} (1,2) and on U_{gb} (3,4).

Such a behaviour is easily explained in the framework of the surface model of 1/f noise. Then, applying this model to the experimental results we come to the conclusion that the density of the surface states D_{ss} responsible for the 1/f noise observed does not depend on the energy depth of the surface levels and is the same for the upper and lower interfaces Si-SiO_2. The value $D_{ss} = 1.4\cdot10^{10}$ cm^{-2}eV^{-1} has been obtained.

The noise of a burst type has been observed in n^+nn^+ MOSFETs. The level of this noise appears to be changed when the source is used as a drain and the drain is used as a source while the drain current remains constant. This unipolarity of the noise suggests that the noise is a contact noise generated near the ohmic contacts to the layer. The noise considered remains unipolar and does not change significantly as the gate voltage U_{gf} is varied from $U_{gf} = 0$ to $U_{gf} = -12$ V where the strong inversion occurs at the gate interface. But the unipolarity of the noise disappears when the depleting voltage U_{gb} of the order of −30 V is applied to the buried oxide. As U_{gb} increased, the noise decreases significantly. The behaviour described is illustrated by Fig.3. Only 1/f noise is detected under strong inversion conditions near both interfaces.

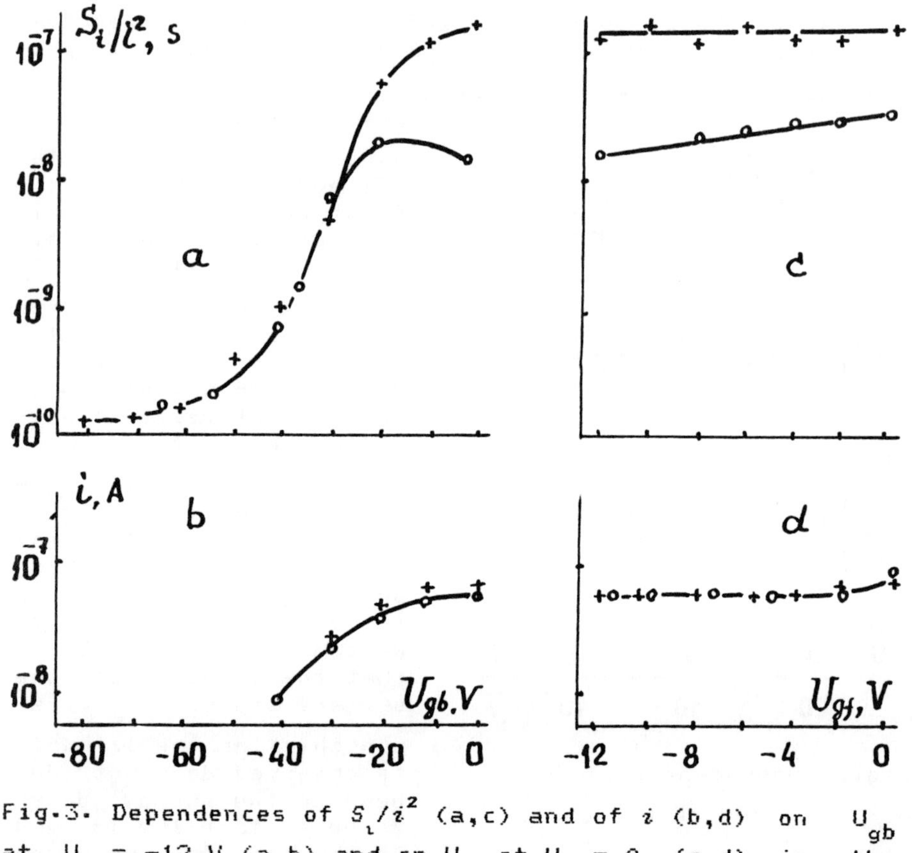

Fig.3. Dependences of S_i/i^2 (a,c) and of i (b,d) on U_{gb} at $U_{gf} = -12$ V (a,b) and on U_{gf} at $U_{gb} = 0$ (c,d) in the n^+nn^+ MOSFET; crosses and circles correspond to the opposite directions of the channel current; f = 30 Hz.

The dependence of the noise level on the polarity of the voltage applied between the source and drain contacts has been also revealed in p^+pp^+ MOSFETs. But it should be noted that this unipolarity manifests itself at more high currents in this case and the noise observed is of

1/f type. It has been found that the value of S_i/i^2 increases with increasing U_{gf} and the unipolarity of the noise is also increased untill the inversion layer is formed at the front surface. Under inversion conditions the values of S_i/i^2 for both current directions become independent of U_{gf}. At the same time, the unipolarity of the noise disappears when the positive voltage $U_{gf} \geq 20$ V is applied to the back surface (Fig.4).

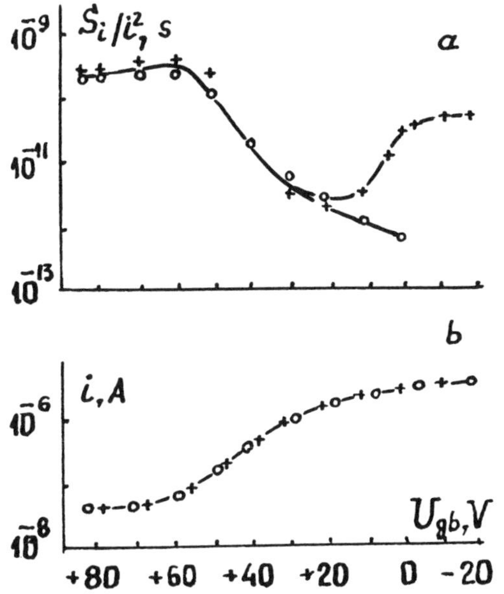

Fig.4. Dependences of S_i/i^2 (a) and of i (b) on U_{gb} in the p^+pp^+ structure; U_{gf} = +12 V; f=30 Hz; crosses and circles correspond to the opposite directions of the channel current.

The conclusion is that the source of the contact noise observed in structures of both types is situated near the buried oxide.

It is interesting that the noise in p^+pp^+ structures appears to be increased not only as U_{gf} increases but also as U_{gf} rises and becomes constant at strong inversion (Fig.4). This noise seems to be also connected with the fluctuations near the contacts.

It follows from above mentioned that some specific processes occur at the ohmic contacts to the silicon layer prepared by ZMR method and the noise can be effectively used for their studying.

Noise investigation of blocking contacts has shown that the breakdown which has been observed in some of them at too low reverse voltage is accompanied by the excess white noise $S_i \gg 2ei$ that is typical of impact ionization processes. This suggests that just such processes may be responsible for the low voltage breakdown observed.

Therefore, the noise measurements have been shown to be the powerfull tool for studying the properties of SOI MOSFETs prepared by ZMR and for controlling their parameters and quality.

MODELING OF HIGH-FREQUENCY NOISE OF MOS TRANSISTORS OPERATING IN WEAK INVERSION

D.H. Song, J.B. Lee, H.S. Min, and Y.J. Park
Department of Electronics Engineering and
Inter-Univ. Semiconductor Research Center, Seoul National University
San 56-1 Shinlim-Dong, Kwanak-Ku, Seoul 151-742, Korea

ABSTRACT

In this paper the high frequency noise mechanism of MOS transistors operating in weak inversion is re-investigated. We propose a theory that the noise source of the white noise in weak inversion is identified as thermal noise in the linear region, given by the steady-state Nyquist theorem rather than by the existing channel thermal noise theory, and shot noise in the saturation region. Furthermore, we give conclusive experimental evidence which supports the theory. Also, a simple formula for the high frequency noise in weak inversion for both long- and short-channel devices over the whole range of drain bias is introduced.

The high frequency noise mechanism associated with the drain current in weak inversion has been discussed by several authors but has not been fully understood[1–4]. There are two existing theories. Fellrath [1] has reported that the high frequency noise in weak inversion is shot noise at least in the saturation region, but using the existing channel thermal noise theory[5], Reimbold and Gentil [3] have shown that the high frequency noise should be thermal noise both in the linear and saturation regions even though the devices have the shot noise behaviors in the saturation region. In this paper, we will show from the experimental results of MOSFETs with nonuniform channel that the high frequency noise in the linear region is thermal noise, which should be given by the steady-state Nyquist theorem, and also from the experimental results of both long- and short-channel devices in weak inversion we will show that the high frequency noise in the saturation region is shot noise.

We will give a brief review of the existing theory [5]. According to the existing channel thermal noise theory used in Ref. 5, the spectral density of the drain thermal noise current $S_{i_d}(\omega)$ is given by

$$S_{id}(\omega) = 4KT\frac{W}{L^2}\int_0^L \mu_n(x)Q_n(x)dx, \tag{1}$$

where K is the Boltzman constant, T the absolute temperature of the device, W the channel width, L the channel length, $\mu_n(x)$ the carrier mobility at position x, and $Q_n(x)$ the inversion layer charge density per unit area at position x. In weak inversion for a device with low surface state density Eq. (1) can be rewritten as[3]

$$S_{id}(\omega) = 2qI_{Dsat}(1 + exp(-\frac{qV_D}{KT})), \tag{2}$$

where I_{Dsat} is the drain saturation current and V_D is the dc drain voltage. Reimbold and Gentil showed[3] that Eq. (2) was in good agreement with experimental results from low drain bias to high drain bias in a long channel MOS transistor. But for the case of a short-channel MOS transistor in saturation, the experimental noise was about 20 percent higher than the theoretical one. This discrepancy was attributed to the poor saturation of the drain current at high drain voltages caused by drain induced barrier lowering (DIBL). According to the generalized Nyquist formula[6], the spectral density of the drain noise current is expressed as[4]

$$S_{id}(\omega) = 4KTg_d(1 - \frac{I_D}{2g_d^2}\frac{\partial g_d}{\partial V_D}), \tag{3}$$

where g_d is the output conductance and I_D is the dc drain current. Eq. (3) is expected to explain the high frequency noise in both long- and short-channel devices because for long-channel devices Eq. (3) reduces to Eq. (2)[4], and for short-channel devices the influence of DIBL on the drain noise current can be incorporated in Eq. (3) by taking into account DIBL effects when calculating g_d. According to Eq. (3), the drain noise current is channel length dependent, and should have a more suppressed high drain voltage plateau $(S_{id}(\omega)/S_{id}(\omega)|_{V_D\to 0} < 0.5)$ as the channel length gets shorter. But contrary to this, the experimental results of Ref. 3 and ours (See Fig. 4) show that $S_{id}(\omega)/S_{id}(\omega)|_{V_D\to 0} > 0.5$ over the whole range of drain voltages in short-channel MOSFETs. Thus neither Eq. (2) nor Eq. (3) can explain the experimental results of high frequency behavior of the drain noise current for short-channel devices in weak inversion. Neither can the pure shot noise theory[1] explain the experimental results for low drain bias[3].

Recently, it has been shown that the spectral density of the drain thermal noise current in MOSFETs operating in the linear region should be given by the steady-state Nyquist theorem rather than by Eq. (1) [7,8,9]. The steady-state Nyquist theorem has been extended to explain the thermal noise in long-channel devices operating in the saturation region. According to the extended steady-state Nyquist theorem, we have [9]

$$S_{id}(\omega) = 4K\overline{T}_n\frac{I_D}{V_D} \tag{4}$$

with

$$\overline{T}_n = \int_0^L \frac{T_n(x)}{g_o(x)}dx / \int_0^L \frac{dx}{g_o(x)}, \tag{5}$$

where $T_n(x)$ is the electron temperature at x, and $g_o(x)$ is the channel conductance per unit length at x. We have already explained that Eq. (1) is physically unacceptable [9]. But since both Eq. (4) and Eq. (1) can explain the high frequency noise behaviors of MOSFETs with uniform channel operating in the linear region, we still need some conclusive experimental evidence to prove the correctness of Eq. (4) in the linear region. Since Eq. (4) and Eq. (1) give different values of thermal noise for a MOSFET with nonuniform channel at a given bias, we measured the high frequency noise in MOSFETs with nonuniform channel (devices ♯1 and ♯2), whose geometry and device parameters are given in Fig. 1 and Table I, respectively. Also we evaluated Eq. (4) and Eq. (1) using the device simulator SNU-2DE[10]. As expected, Figs. 2 and 3 show that Eq. (4) is in good agreement with the experimental results in the linear region for both weak and strong inversion. Thus we have conclusive experimental evidence that Eq. (4) is correct for MOSFETs in the linear region. Fig. 4 shows that thermal noise alone, even with hot electron effects taken into account, can not explain the high frequency noise of MOSFETs in weak inversion, especially in a short-channel device at high drain bias.

The similarity in conduction mechanism between the collector current in BJTs and the drain current in MOSFETs in weak inversion[11] suggests that MOSFETs operating in the saturation region of weak inversion should give rise to shot noise. Thus, we expect that shot noise and thermal noise are competing in MOSFETs operating in weak inversion, so that $S_{id}(\omega)$ can be expressed as

$$S_{id}(\omega) = 4K\overline{T}_n\frac{I_D}{V_D} + 2qI_D\Gamma^2, \qquad (6)$$

where Γ^2 is a shot noise suppression factor given by

$$\Gamma^2 = \frac{(1+e^{-\frac{qV_D}{KT}}) - 2\frac{KT}{qV_D}(1-e^{-\frac{qV_D}{KT}})}{1-e^{-\frac{qV_D}{KT}}}. \qquad (7)$$

Eq. (7) is obtained by equating Eq. (2) with Eq. (4) with $\bar{T}_n = T$, because Eq. (2) can explain the experimental results for long-channel devices[3]. This Γ^2 for long-channel devices is assumed to be vaild in short-channel devices. Our experiments (Figs. 4 and 5) show that Eq. (6) is in good agreement with the experimental results of both short- and long-channel MOSFETs (device ♯3 and ♯4) in weak inversion over the whole range of drain bias. As we see in Figs. 4 and 5, thermal noise is dominant in the linear region, and shot noise is dominant in the saturation region with $\Gamma^2 = 1$.

REFERENCES

1. J. Fellrath, Rev. Phys. Appl., **13**, 719 (1978).
2. S.T. Liu and A. van der Ziel, Appl. Phys. Lett., **37**, 950 (1980).
3. G. Reimbold and P. Gentil, IEEE Trans. Electron Devices, **ED-29**, 1722 (1982).
4. G. Ghibaudo, phys. stat. sol. (a), **113**, 223 (1989).
5. F.M. Klaassen and J. Prins, Philips Res. Rep., **22**, 505 (1967).
6. M.S. Gupta, Phys. Rev. A **18**, 2725 (1978).
7. H.S. Min, J. Appl. Phys., **64**, 6339 (1988).
8. H.S. Min, Solid-State Electron., **32**, 295 (1989).
9. D.H. Song, J.B. Lee, H.S. Min and Y.J. Park, in *Proc. 11th Int. Conf. on Noise in Physical Systems and 1/f Fluctuations*, Kyoto, 1991, p.269.
10. W.S. Choi, J.G. Ahn, Y.J. Park and H.S. Min, in *Proc. 4th Workshop on Numerical Modeling of Process and Devices for Integrated Circuits*, Seattle, 1992, p.103.
11. R.J. Overstraeten, G.J. Declerck, and P.A. Muls, IEEE Trans. Electron Devices, **ED-22**, 282 (1975).

TABLE I. The device parameters for the test devices.

	Device ♯1	Device ♯2	Device ♯3	Device ♯4
L_1/L_2	$20\mu m/20\mu m$	$5\mu m/5\mu m$	$0\mu m/2\mu m$	$0\mu m/10\mu m$
W	$40\mu m$	$1500\mu m$	$30\mu m$	$1500\mu m$
T_{ox}	$1300Å/250Å$	$1300Å/250Å$	$1300Å/250Å$	$1300Å/250Å$
V_T	$0.135V$	$0.135V$	$0.334V$	$-0.097V$
N_{sub}	$5\times10^{14}cm^{-3}$	$5\times10^{14}cm^{-3}$	$1\times10^{15}cm^{-3}$	$5\times10^{14}cm^{-3}$

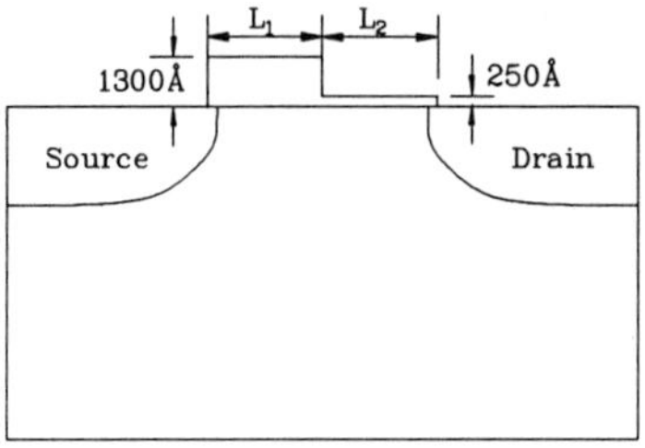

Fig. 1. Geometry of the devices under test.

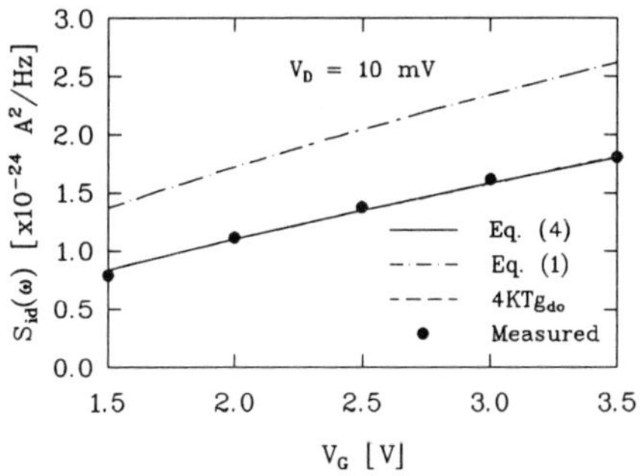

Fig. 2. Noise measurement and simulation of the device ♯1 in strong inversion near thermal equilibrium.

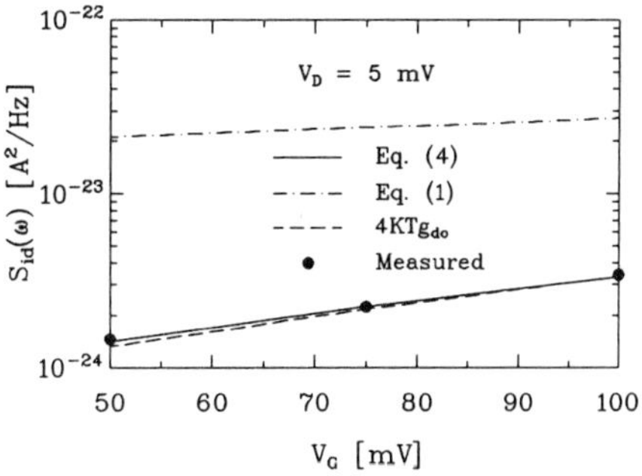

Fig. 3. Noise measurement and simulation of the device ♯2 in weak inversion near thermal equilibrium.

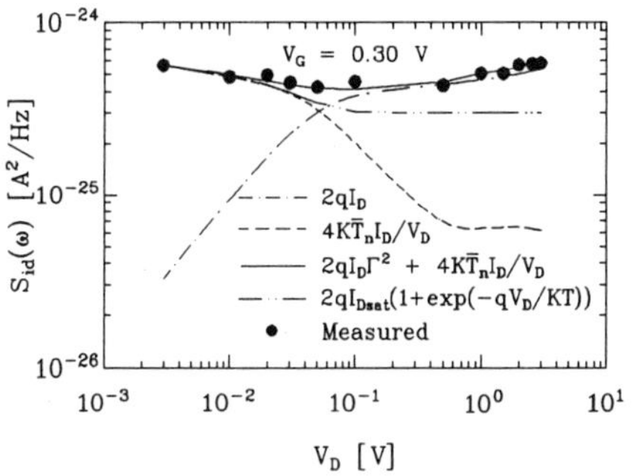

Fig. 4. White noise of the short-channel MOS device (device ♯3) in weak inversion.

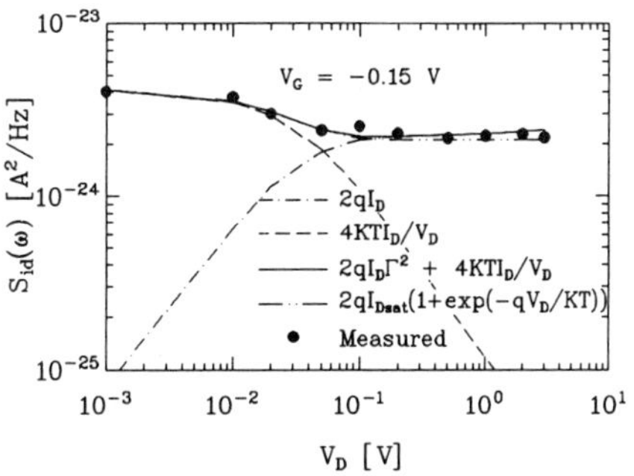

Fig. 5. White noise of the long-channel MOS device (device ♯4) in weak inversion.

LOW-FREQUENCY NOISE AND RANDOM TELEGRAPH SIGNALS IN 0.35μm SILICON CMOS DEVICES

O. Roux-dit-Buisson, G. Ghibaudo and J. Brini
LPCS/ENSERG, 23 rue des martyrs, BP 257, 38016 Grenoble, France.

G. Guégan
DMEL/LETI, CENG, 17 rue des martyrs, BP 85X, 38031 Grenoble, France.

ABSTRACT

A detailed investigation of the low frequency noise in Si MOS devices issued from a 0.35 μm CMOS technology is conducted. The normalized drain current noise WLS_{Id}/I_d^2 has been systematically measured at a fixed frequency (10Hz) and constant normalized drain current. It is found that, for large area devices (> 7-10μm^2), the sample-to-sample noise level variation lies around a factor of 2-3, while, for the smallest devices (0.1-0.3μm^2), the sample-to-sample noise level variation can exceed 3 decades. In large area devices, the 1/f noise can easily be described by a classical carrier number fluctuation model using the concept of dynamic flat band voltage. In the case of intermediate or small areas, a multi-RTS component scheme has to be employed.

INTRODUCTION

The impact of miniaturization on the performance of silicon MOS devices is a key issue for ULSI integrated circuits. Among other limitations, the low frequency noise and fluctuations may constitute a serious constraint with respect to the signal-to-noise ratio or the noise margin for the use of MOS devices in analog and digital circuits. Despite much efforts, very few attention has been paid to point out the influence of scaling-down on the low frequency noise level in small area MOS devices, and, in particular, in newly evolved deep submicron Si technologies.

Therefore, in this this work is presented a detailed investigation of the low frequency noise of Si MOS devices issued from a 0.35 μm CMOS technology. This gives us the opportunity to study the noise and fluctuations in MOS transistors with areas monotonically distributed between 50μm^2 and 0.1μm^2.

EXPERIMENTAL DETAILS

The devices used throughout this work have been fabricated at LETI (Grenoble) according to a LDD CMOS process with electron beam or deep-UV lithography and a SILO isolation technique. The gate oxide thickness is about 70Å, the gate width W and length L vary from 0.4-25μm and 0.3-20μm, respectively. The combination of all test pattern geometries provides 17 areas from 0.1 to 50μm^2.

The low frequency noise (10mHz-10kHz) and Random Telegraph Signal (RTS) measurements have been conducted using an ONO-SOKKI spectrum analyzer loaded by a low noise voltage amplifier and an EG&G model 181 current-voltage converter.

RESULTS AND DISCUSSION

The normalized drain current noise WLS_{Id}/I_d^2 has been systematically measured at a fixed frequency (10Hz) and constant normalized drain current ($LI_d/W=0.4\mu A$) on several (6-7) chips with the same test pattern, giving access to 17 different gate areas. It is found that, for large area devices (> 7-10μm^2), the sample-to-sample noise level variation lies around a factor 2-3, while, for the smallest devices (0.1-0.3μm^2), the sample-to-sample noise level variation can exceed 3 decades (see Fig. 1). It should be mentionned that, for such a 0.35μm CMOS technology, the threshold voltage variation in linear region does not exceed 60-90mV for all the geometries, emphasizing the very good optimization of the scaling down process with good control of the narrow and short channel effects (see Fig. 2).

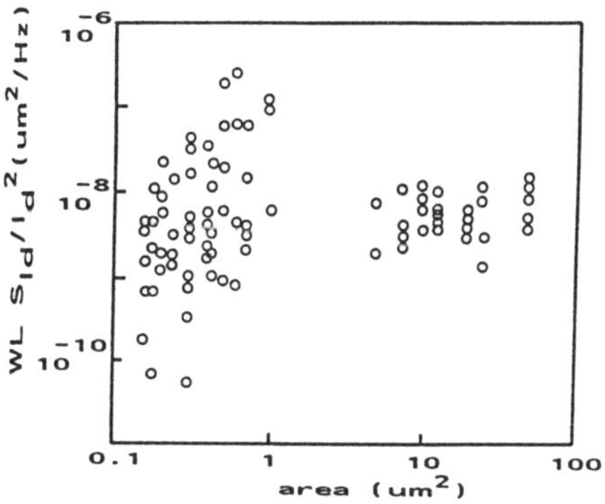

Fig. 1 : Experimental plot of the normalized drain current noise level versus device area for 17 different sample geometries

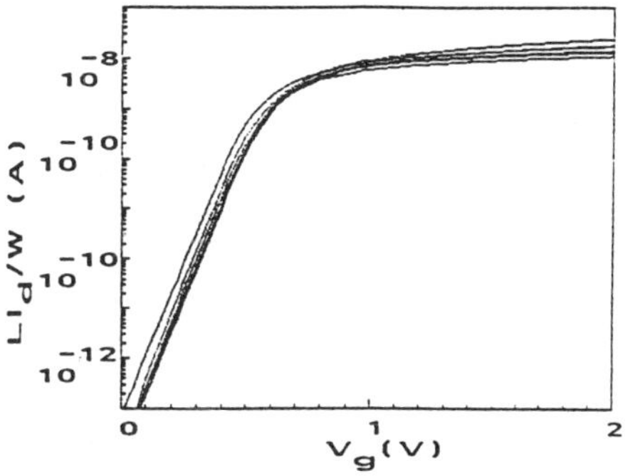

Fig. 2 : Typical normalized transfer characteristics as obtained for various geometries (W/L=25/.3, 25/1, 25/5, .4/.3, 1/.3, 2/.3).

A detailed study of the spectrum and time domain data carried out on typical devices clearly demonstrates that the strong sample-to-sample variation in the noise level is mainly originated from the drastic change of the noise type with scaling-down. In effect, in sufficiently large area devices (>5-7μm^2), the noise is typically 1/f like, whereas, for small area devices ($\lesssim 1\mu m^2$), the noise results from the contribution of one or several RTS components (see Fig. 3).

In large area devices, the 1/f noise can easily be described by a classical carrier number fluctuation model using the concept of dynamic flat band voltage [3]. This results in the fact that the dependence with gate and drain voltages of the normalized drain current noise is well correlated to those of the corresponding transconductance to drain current ratio squared [1].

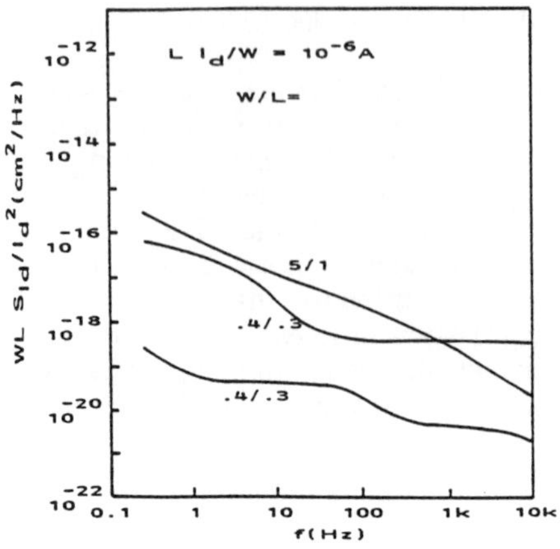

Fig. 3 : Typical example of drain current spectra as obtained for various devices.

For very small area devices where only one RTS component is active, the lorentzian noise spectrum can be calculated provided the emission and capture time constants as well as the RTS amplitude dependence with biases can be modeled. Most of the time, the modified SHR statistics including tunnel assisted process is well appropriate to describe properly the variations with gate and drain voltages of the time constants. Moreover, as for large area devices, the concept of dynamic flat band voltage can also successfully be used to model the RTS amplitude variation both with gate and drain voltages [2]. These time constant and RTS amplitude approaches enables in turn a reasonable modeling of the noise spectrum to be obtained for various biases.

In the case of intermediate areas, a multi-RTS component scheme has to be employed [3,4]. In this situation, the strong sample-to-sample noise level dispersion can be satisfactorily evaluated by a low frequency noise model in which the low frequency fluctuations in a Si MOS transistor results from the superposition of several RTS fluctuators with randomly distributed time constants. In such a case, the normalized drain current noise is obtained as [4] :

$$S_{Id}/I_d^2 = \frac{q^2}{(W\ L\ C_{ox})^2} \cdot \frac{g_m^2}{I_d^2} \sum_k 4\,A_k \frac{\tau_k}{1+\omega^2\tau_k^2} \qquad (1),$$

where q is the electron charge, C_{ox} is the gate oxide capacitance, g_m is the gate transconductance, τ_k is the effective time constant of the kith RTS, $\omega=2\pi f$ is the angular frequency and $A_k=f_k(1-f_k)$ with f_k being the occupancy factor of the kith RTS.

A good representation of the noise level dispersion diagram can therefore be obtained assuming that the oxide traps are uniformly distributed in space and in energy with a given volume trap state density $N_t \simeq 10^{17}/eVcm^3$ (see Fig. 4).

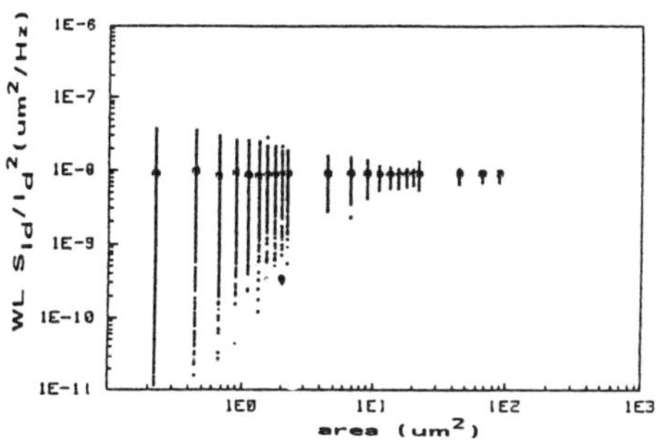

Fig. 4 : Theoretical plot of the normalized drain current noise dispersion versus device area.

CONCLUSION

The drastic change in the type of low frequency noise due to the emergence of RTS components in scaled down devices, renders the modeling of noise more complicated than in large area devices. The complexity of the modeling results essentially from the fact that, in small area transistors, the spectrum is composed by the superposition of a small numbers of lorentzian spectra. The characteristic parameters (time constant and amplitude) of each lorentzians are in addition extremely difficult to predict accurately because of the sample-to-sample dispersion and of the bias dependencies.

REFERENCES

1. G. Ghibaudo, Sol. State Electron., 32, 563 (1989).
2. O. Roux, G. Ghibaudo, J. Brini, Sol. State Electron., 35, 1273 (1992).
3. M. Kirton and M. Uren, Adv. Phys., 38, 468 (1989).
4. G. Ghibaudo, O. Roux-dit-Buisson, J. Brini, Phys. Stat. Sol. (a), 132, 501 (1992).

RANDOM TELEGRAPH SIGNALS IN SMALL GATE-AREA P-MOS TRANSISTORS*

John H. Scofield and Nick Borland
Department of Physics, Oberlin College
Oberlin, OH 44074 USA

Daniel M. Fleetwood
Sandia National Laboratories, Dept. 1332
Albuquerque, NM 87185-5800 USA

ABSTRACT

We report the observation of random telegraph signals (RTS) in the channel resistances of nominally 1.25 µm x 1.25 µm, enhancement-mode pMOS transistors fabricated using the AT&T 1-µm radiation hardened technology. Devices were operated in strong inversion in the linear regime. Measurements, performed for temperatures ranging from 77 to 300 K and various gate voltages, show that capture and emission times are both thermally activated and that the capture time depends strongly on the gate voltage. Results suggest that the unfilled trap is charged and that, after capturing a hole, the trap relaxes to a lower energy. Basic features of a model are discussed.

EXPERIMENTAL RESULTS AND DISCUSSION

Metal-oxide-semiconductor (MOS) transistors are known to exhibit relatively large levels of low-frequency 1/f noise.[1] Much evidence now suggests that this noise is related to the capture and emission of charge carriers by localized defects at or near the Si/SiO_2 interface.[2,3,4] The drain voltage of very small gate-area devices, especially at low temperatures, shows random switching between two discrete levels, apparently arising from the capture and emission of a single charge carrier.[2,5,6,7,8] Such random telegraph signals (RTS), observed in small gate-area devices, have been shown to superpose to give 1/f noise in larger area devices.[2] Thus, information gained from the study of RTS's in MOSFETs should be helpful in understanding the origins of 1/f noise in these devices.

We have investigated six RTS's in two relatively small gate-area (≈ 1.25 µm x 1.25 µm) p-channel, enhancement mode MOS transistors operated in strong inversion. Devices have an oxide thickness of 18 nm and were fabricated using the AT&T 1-µm radiation hardened technology.[9] To our knowledge these devices have the lowest defect-density and are the most radiation-tolerant of any devices used for such studies. Temporal fluctuations ($\delta V_d(t)$) in the drain voltage (V_d) were observed when devices were operated in their linear regimes with fixed gate voltage (V_g) and drain current (I_d); the source lead was grounded during all measurements.[10] The measurement conditions were similar to those we have used previously for noise measurements on large area devices.[3] Measurements were performed for sample temperatures (T) between 77 and 300 K and effective gate-voltages (V_g-V_{th}) ranging from -200 mV to -2 V, where V_{th} is the threshold voltage. The measurement bandwidth was from 0.03 Hz to 30 kHz and typically -100 mV < V_d < 0. For these devices, RTS's were very reproducible even after many days and multiple temperature cycles.

For each device it was possible to find a range in T and V_g for which V_d was observed to randomly switch between two discrete levels, similar to behavior reported by others.[2,5-8] The drain-voltage switching scaled with the I_d indicating switching in the channel resistance, $\delta R_{ch} = \delta V_d/I_d$. No attempt was made to use the drain-voltage dependence to locate the trap along the channel.[8] The RTS's were characterized by their resistance changes ΔR_{ch} and their mean times in the "high-" and "low-resistance" states. We found the dependence on V_g to be consistent with the idea that the high resistance states were associated with the trapping of a single charge carrier. We thus identified the mean time in the high resistance state as the trap emission time (τ_e). The mean time in the opposite state was identified as the trap capture time (τ_c).

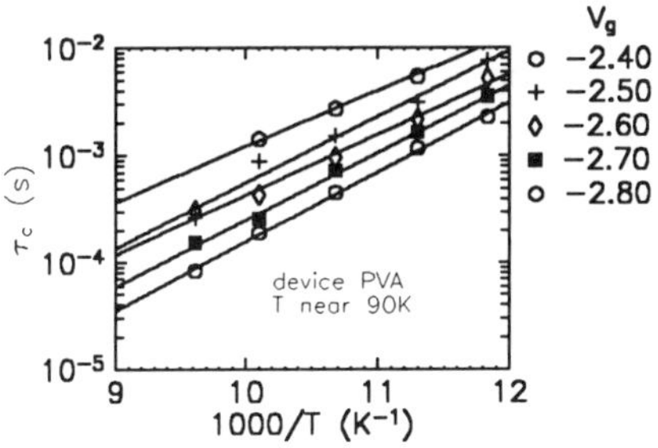

Figure 1. Semilog plot of capture time versus inverse temperature at fixed gate voltage.

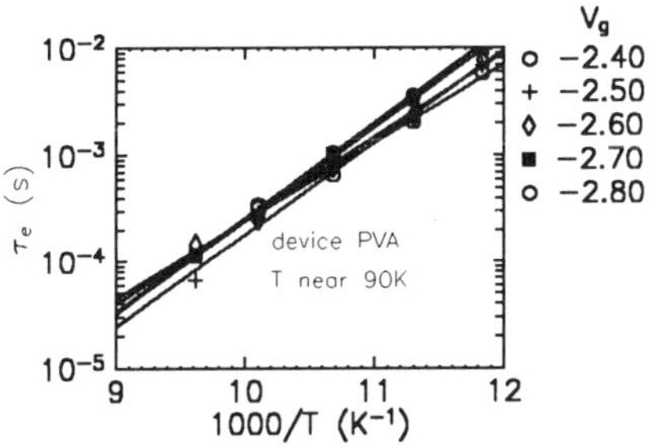

Figure 2. Semilog plot of emission time versus inverse temperature.

The duty cycles of the RTS's were found to depend primarily on V_g while the switching rates depended primarily on T, similar to the findings of others.[2,5] The corner frequency increased with T, leaving the measurement bandwidth with a change of 20-30 K.

Here we display data from one trap. Data from other traps were similar. We observed that, for fixed V_g, both τ_c and τ_e varied with T in a manner consistent with thermal activation, i. e.

$$\tau_j = \tau_{oj} \exp(E_j/kT), \qquad (1)$$

where j stands for capture or emission, E_j is the activation energy, and τ_{oj} is the attempt time. Typical capture and emission time data are illustrated in Figures 1 and 2.

Within experimental error, the activation energies for both capture and emission were independent of gate voltage. This is shown in Figure 3. For this particular RTS the activation energies were found to be $E_c \approx (115\pm10)$ meV and $E_e \approx (150\pm10)$ meV, respectively. The data of Figures 1 and 2 may be re-plotted to show how the RTS varies with gate voltage at fixed temperature. This is shown in Figure 4 for T = 90 K, confirming the strong dependence of τ_c and

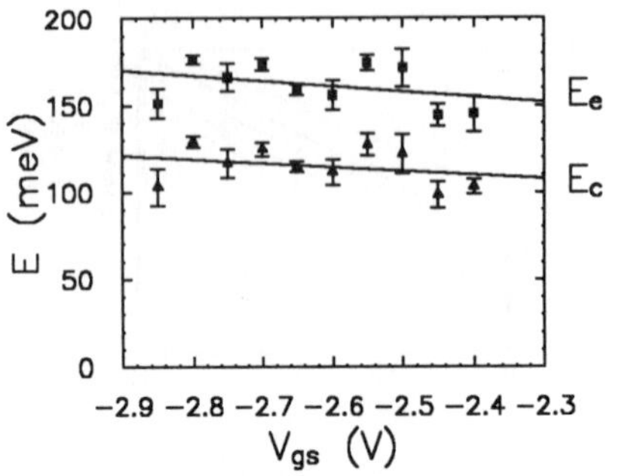

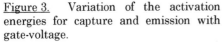

Figure 3. Variation of the activation energies for capture and emission with gate-voltage.

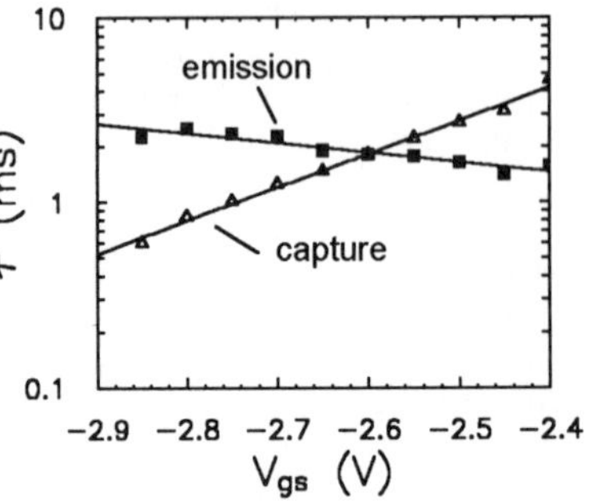

Figure 4. Gate voltage dependencies of the capture and emission times for fixed T = 90K.

weak dependence of τ_e on V_g. Referring to the above equation, this means that both prefactors may be written as

$$\tau_{j0} = \zeta_j \exp(V_g/\phi_j), \tag{2}$$

where ζ_j and ϕ_j are fit parameters independent of both T and V_g. We note, however, that there are large uncertainties in extrapolated intercepts for graphs like those in Figures 1 and 2.

The data suggest the following model. We assume that the RTS arises when a majority carrier is captured and emitted by a single trap, located 0-3 nm from the Si/SiO_2 interface. The empty trap level, E_t, is located below the silicon valence band edge at the interface. To be captured, a hole must first be excited to an energy E_t in the silicon valence band, then tunnel to the localized trap state in the oxide. The thermally activated behavior comes from the T-dependence of the Fermi-Dirac hole distribution. Thus, we identify the energy difference $E_c \approx E_t - E_f$ as the activation energy for capture.

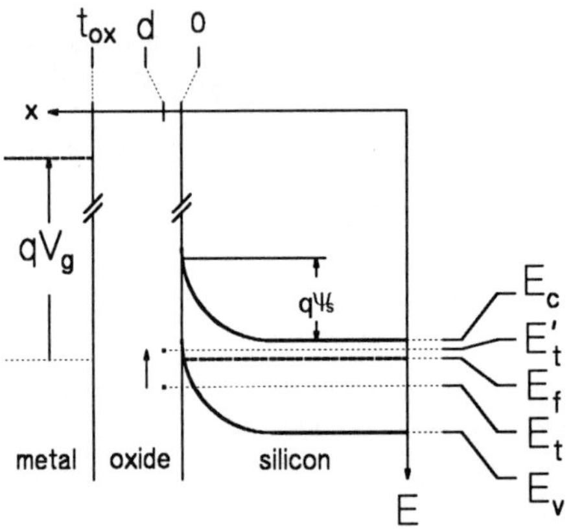

Figure 5. Band diagram showing hole trap energy levels.

Since the hole is not immediately emitted, the filled trap is assumed to undergo a lattice relaxation resulting in the lowering of the localized hole state to a new energy, E_t', with $E_t' < E_f$.[11,12,6,7] For emission to proceed, the lattice atoms must rearrange themselves, raising the trap level above E_f. This process would typically depend strongly on lattice temperature, with an activation energy $E_e = E_f - E_t'$.[11] These ideas are illustrated in Figure 5.

Several unresolved issues remain. For instance, one would expect both E_t and E_t' to vary with oxide field (i.e., gate voltage) whereas the data do not support this. Since only the capture time varies significantly with V_g we speculate that the empty trap is negatively charged while the filled trap is neutral. The V_g-dependence of τ_c might enter both through E_t and also though a V_g-dependent tunneling rate.

Very recent data for one trap shows that both capture and emission times become independent of lattice temperature below 15 K. This suggests that lattice motion other than thermal, perhaps zero-point motion or configurational tunneling, is involved in the lattice transition.

In conclusion, we have observed highly reproducible random telegraph signals in small gate-area pMOS transistors at temperatures down to 77 K. We find both capture and emission times to depend strongly on temperature (i.e., thermally activated) while only the capture time varies strongly with gate voltage. We conclude that the unoccupied trap is charged, and suggest a model involving lattice relaxation of the filled trap.

The authors would like to thank B. Mukhergee for help with some measurements and one of us (JHS) expresses appreciation to C. Rogers and K. Farmer for useful conversations.

REFERENCES

(*) This work was supported by the U. S. Department of Energy through Contract No. DE-AC04-76DP00789.

1. See, for instance, A. van der Ziel, Adv. Electron. Electron Phys. 49, 225 (1979).
2. M. J. Kirton and M. J. Uren, *Advances in Physics* **38**, 367-468 (1989).
3. John H. Scofield, T. P. Doerr, and D. M. Fleetwood, *IEEE Trans. Nucl. Sci.* **36**, 1946 (1989).
4. D.M. Fleetwood and J.H. Scofield, *Phys. Rev. Lett.* **64**, 579 (1990).
5. K. S. Ralls, W. J. Skocpol, L. D. Jackel, R. E. Howard, L. A. Fetter, R. W. Epworth, and D. M. Tennant, *Phys. Rev. Lett.* **52**, 228 (1984).
6. M. Schulz and A. Papas, in *Noise in Physical Systems and 1/f Fluctuations*, ed. T. Musha, S. Sato, and M. Yamamoto (Ohmsha, Ltd, Tokyo, 1991) p. 265.
7. K. R. Farmer, in *Insulating Films on Semiconductors*, ed. W. Eccleston and M. Uren (Adam Hilger, Briston, 1991) pp. 1-18.
8. Philip Restle, *Appl. Phys. Lett.* **53**, 1862 (1988).
9. John H. Scofield and D. M. Fleetwood, *IEEE Trans. Nucl. Sci.* **38**, 1567 (1991).
10. Note that most reports of RTS's in MOSFETs have usually been associated with switching in the drain current with constant voltage bias [2].
11. L. D. Jackel, W. J. Skocpol, R. E. Howard, L. A. Fetter, R. W. Epworth, and D. M. Tennant, in *Proceedings of the 17th International Conference on the Physics of Semiconductors*, ed. by J. D. Chadi and W. A. Harrison (Springer-Verlag, New York, 1985), pp.221-224.
12. M. J. Kirton and M. J. Uren, *Appl. Phys. Lett.* **48**, 1270, (1986).

CRITICAL EXAMINATION OF THE RELATIONSHIP BETWEEN RANDOM TELEGRAPH SIGNALS AND LOW-FREQUENCY NOISE IN SMALL-AREA Si MOST's

E. Simoen and B. Dierickx
IMEC, Kapeldreef 75, B-3001 Leuven, Belgium

ABSTRACT

This paper investigates the relationship between Random Telegraph Signals (RTS's) often observed in small-area Si MOST's and the corresponding low-frequency (LF) noise spectrum. From the study of hot-carrier (HC) stress induced changes follows that RTS can be used as a sensitive tool for HC degradation studies. Secondly, it is demonstrated that the noise in a small-area MOST is largely reduced, when cycled from accumulation in inversion, indicating that RTS is the dominant LF noise source.

INTRODUCTION

According to the McWorther theory[1] the fundamental origin of low-frequency 1/f noise in Si MOST's is trapping-detrapping through interface-near oxide-traps. The spectrum of such a single Generation-Recombination (GR) center is Lorentzian and takes the general form[2,3]:

$$S_{ID} = (\Delta I_D)^2 \frac{\tau}{1 + (2\pi f \tau)^2} \qquad (1)$$

with:

$$\frac{1}{\tau} = \frac{1}{\tau_e} + \frac{1}{\tau_c} \qquad (2)$$

S_{ID} is the drain current (I_D) noise spectral density, ΔI_D the drain current step induced by the oxide trap, f the measurement frequency and τ the effective GR time constant, which is determined both by the emission τ_e and by the capture time constant τ_c. Averaging over a large number of uncorrelated oxide traps, which are unavoidably present in a "large-area" Si MOST, yields the well-known 1/f noise spectrum, as convincingly demonstrated by Uren et al. [3,4]. Scaling down the device feature size, the number of active oxide traps reduces accordingly and the spectrum becomes predominantly Lorentzian, while at the same time Random Telegraph Signals (RTS's) are observed in the drain current. Therefore, in this paper, the relationship between the LF noise spectrum and the RTS characteristics (amplitude ΔI_D; capture and emission time) is thoroughly investigated, aiming at a better understanding of the RTS phenomenon.

RESULTS AND DISCUSSION

One attractive way to validate *in situ*, i.e. in a single device, the McWorther theory is to gradually increase the number of oxide traps in a small-area Si MOST by hot-carrier (HC) degradation and to monitor the corresponding LF noise spectrum. It is well-known that HC degradation increases the number of interface/oxide traps in a damaged region near the drain [5]. From this, one would expect the LF noise spectrum to change from Lorentzian-like, i.e. dominated by one or a few oxide traps, to a more 1/f-like behaviour. Unfortunately, it is hard to create new RTS's by the HC stress procedure used[6,7]. On the other hand, from a detailed study of the RTS characteristics prior to and after stress, a systematic change can be deduced, for instance in the amplitude ΔI_D. This change is a measure of the local damage of the Si-SiO_2 interface.

Both n-channel and p-channel devices have been stressed and characterized, as descibed in Refs. 6,7. RTS has been measured in linear operation (constant drain voltage V_{DS}) and as a function of V_{DS}, both for "positive" or forward (F) operation and for reverse (R) operation, resulting in the RTS amplitude asymmetry.

In the linear region, a systematic reduction of the fractional RTS amplitude $\Delta I_D/I_D$ is observed (Fig. 1a). The opposite trend is found for the pMOST's (Fig. 1b). To a first approximation, the same trend is observed in the dc characteristics: the normalized transconductance g_m/I_D reduces for n- and increases for pMOST's after HC stress[6,7]. The change of the amplitude asymmetry is more complex[6,7], although to a first approximation the same tendency dominates, i.e. an increase for pMOST's (Fig. 2) and a reduction for n-channel devices.

These HC stress induced changes can be understood as follows. As shown previously, the RTS amplitude is in its most generalized form given by [8,9]:

$$\frac{\Delta I_D}{I_D} = \frac{\Delta\sigma \, \Delta L \, \Delta W}{\sigma \, LW} \tag{3}$$

with LW the device area (length times width); $\Delta L \Delta W$ the area which is cored out by the RTS and σ, $\Delta\sigma$ respectively the channel conductivity and the change in the channel conductivity due to the presence of a single interface trap. The degradation, resulting from the application of HC stress, of the normalized RTS amplitude then corresponds to:

$$\delta(\Delta I_D/I_D) \approx \frac{\Delta I_D}{I_D} \frac{\delta L}{L} \frac{\sigma_2 - \sigma}{\sigma_2} \tag{4}$$

with δL the extent of the damaged region ($\approx$0.1 µm) and σ_2 the conductivity in the damaged region. In other words, the change in the RTS amplitude is according to eq. (4) proportional to the variation in the channel conductance due to the presence of the HC damaged region.

Alternatively, it can be demonstrated that [6,7]:

$$\delta\left(\frac{\Delta I_D}{I_D}\right) \approx -\frac{\Delta I_D}{I_D} \frac{\delta(C_{ox}+C_{it})}{C_{ox}+C_{it}} \tag{5}$$

with C_{ox} and C_{it} respectively the oxide and the interface-trap capacitance per unit of area. Based on eq. (5), it is concluded that for nMOST's, the creation of interface traps is dominant[5], resulting in an increase of C_{it}, or a reduction of ΔI_D. In the case of pMOST's, electron trapping is dominant, yielding a reduction of C_{ox}, or an increase of ΔI_D in linear operation.

The normalized drain current noise $\langle i_d \rangle / I_D$ at a constant frequency f and corresponding with an RTS is characterized by the occurrence of peaks (Fig. 1b and 3), which are related to the Lorentzian nature of the noise spectrum. By filling in the measured capture and emission constants and amplitude in eqs. (1)-(2), this peak-shaped behaviour is reconstructed satisfactorily. In other words, these features can be used to extract the parameters of the corresponding RTS - it is for instance demonstrated that the peak maximum occurs for the condition $\tau_e = \tau_c$. After stress, the peaks are shifted and show a higher amplitude for pMOST's (Fig. 1.b and 3), which is related to the reported change in the RTS amplitude and to the change in the time constants. The latter can be estimated from the LF noise peaks at different f.

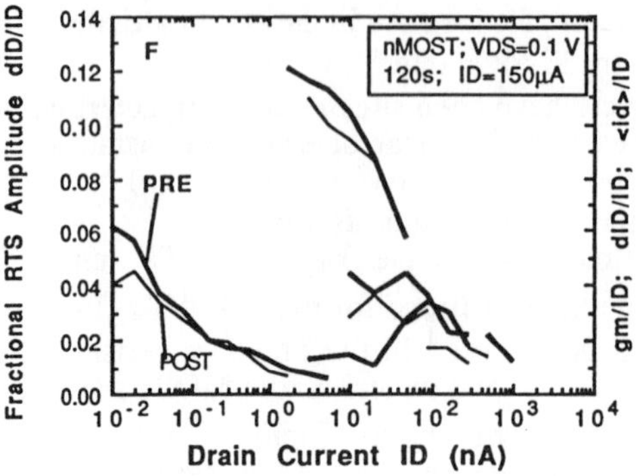

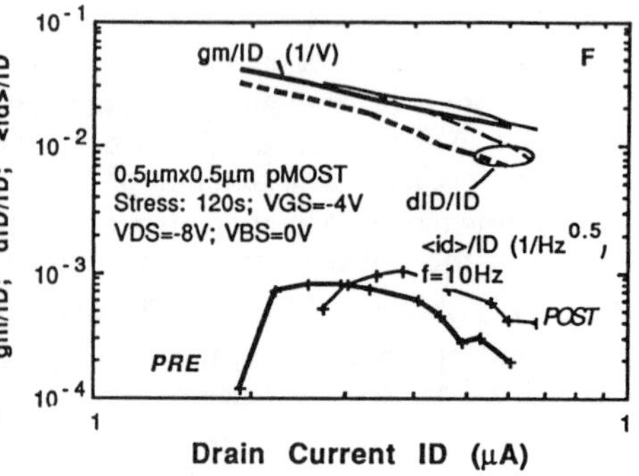

Fig. 1a. HC stress induced reduction of the fractional RTS amplitude in linear operation (V_{DS}=0.1 V) for a set of RTS's in the same nMOST. Stress was for 2 min., at a gate voltage V_{GS}=4 V and V_{DS}=8 V. Forward stress.

Fig. 1b. HC stress induced increase of the fractional RTS amplitude and of g_m/I_D for a pMOST in linear operation (V_{DS}=-50 mV). Stress was for 2 min. at V_{GS}=-4 V; V_{DS}=-8 V and zero substrate bias V_{BS}. F stress.

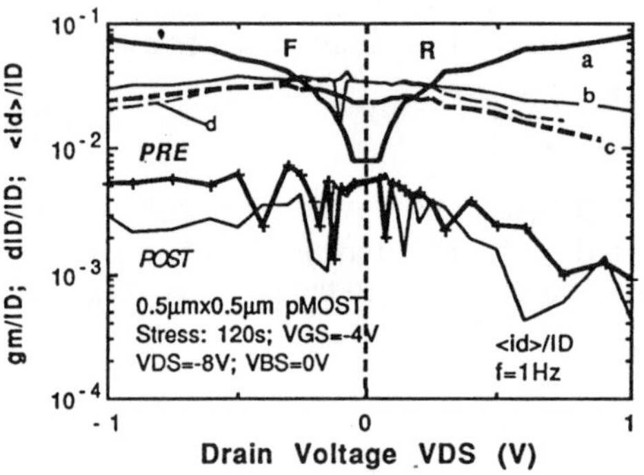

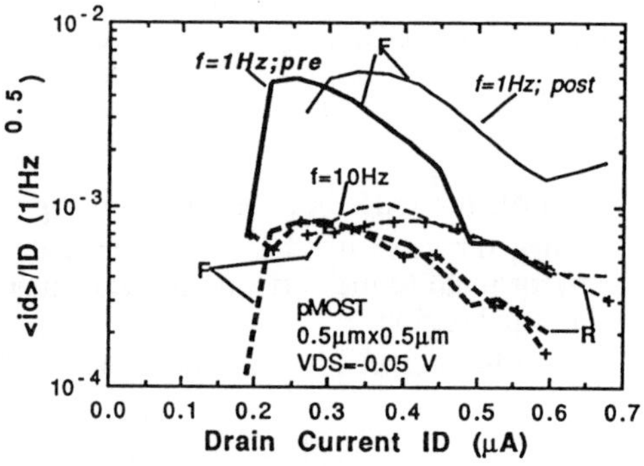

Fig.2. Influence of forward stress on the "noise" asymmetry for a pMOST in saturation. V_{GS}= -1.8V. A forward stress was applied.
Curve a : g_m/I_D pre and b: post.
Curve c: $\Delta I_D/I_D$ pre and d: post.

Fig. 3. Pre- and post HC stress LF noise characteristics at constant f, for a small-area pMOST in linear operation (V_{DS}=-50 mV).
A forward stress was applied.

In a second approach, the LF noise behaviour of submicron MOST's is examined by cycling the device from inversion into accumulation[10,11]. This technique was originally proposed by Bloom and Nemirowski[10] and applied to large-area Si MOST's. There it is shown that the LF noise is considerably reduced by cycling the MOST between inversion and accumulation, i.e. between V_{GS1} (on) and V_{GS2} (off; in accumulation), with some frequency f_{cycle}. A sample and hold circuit keeps track of the MOST output voltage only during the time windows of "normal" operation, corresponding with V_{GS1}. The output of the sample and hold circuit is fed to a HP3562A spectrum analyzer.

From Fig. 4 clearly follows that a strong reduction of the LF noise is observed in the small-area nMOST, similar as for large-area devices. However, in this case, the Lorentzian-like spectrum, corresponding with an RTS, is reduced by cycling operation to a white LF noise spectrum. Monitoring the RTS time constants, one concludes that for sufficiently large f_{cycle}, the trap is no longer able to follow the cycling. This means that the centre remains occupied by a carrier, which has been trapped during the off period, so that it no longer contributes to the LF noise (Fig. 5). From this, it is inferred that RTS is the dominant LF noise source, at least in the small-area devices studied. At the same time, this technique offers a simple way to eliminate the RTS contribution and to investigate the residual LF noise.

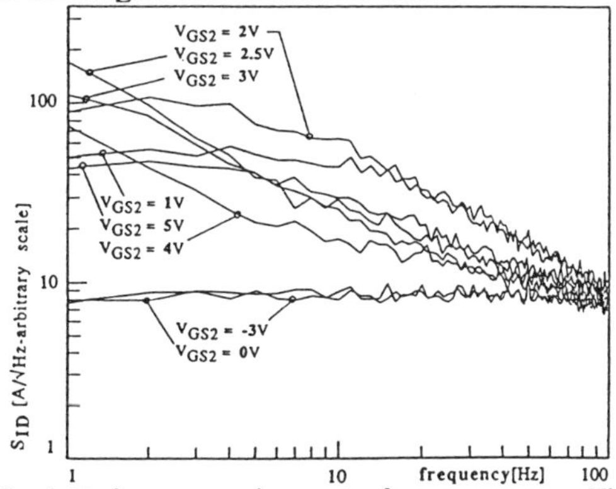

Fig. 4. Drain current noise spectra for a 0.4x0.8 μm nMOST in linear operation (V_{DS}=0.1 V). f_{cycle}=1 kHz, I_D=40 nA and V_{GS1}=2.13 V.

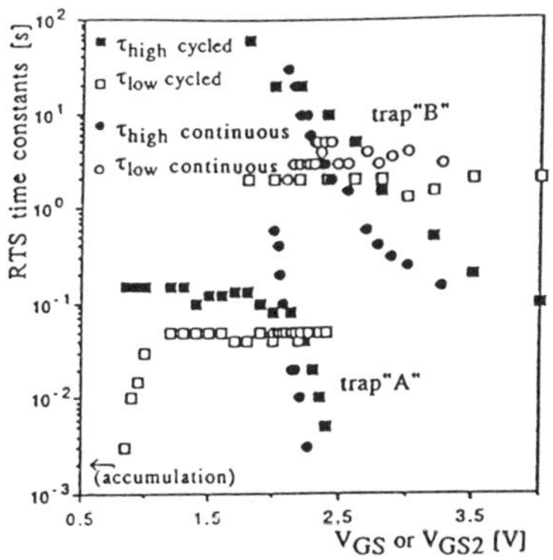

Fig. 5. Time constants of the two RTS's observed in the nMOST of Fig. 4. τ_{high} and τ_{low} under "continuous" operation (V_{GS1}= V_{GS2}) and under "cycling" operation (V_{GS1} =2.13 V and V_{GS2} on the x axis).

CONCLUSIONS

From the present study clearly follows the close relationship between RTS and LF noise in Si MOST's. However, the observed HC-stress induced changes are still a challenge for present-day theory. Furthermore, the RTS-related LF noise peaks provide an alternative, spectroscopic means for studying the relevant trap parameters.

REFERENCES

1. A.L. McWorther, "Semiconductor Surface Physics", Ed. by R.H. Kingston (Philadelphia: University of Philadelphia) 207 (1957).
2. S. Machlup, J. Appl. Phys. 25, 341 (1954).
3. M.J. Kirton and M.J. Uren, Adv. Physics 38, 367 (1989).
4. M.J. Uren, D.J. Day and M.J. Kirton, Appl. Phys. Lett. 47, 1195 (1985).
5. J.Y. Choi, P.K. Ko, C. Hu and W.F. Scott, J. Appl. Phys. 65, 354 (1989).
6. E. Simoen, B. Dierickx and C. Claeys, Paper accepted by Appl. Phys. A.
7. E. Simoen, B. Dierickx and C. Claeys, Microelectronic Engineering 19, 605 (1992).
8. E. Simoen, B. Dierickx, C. Claeys and G. Declerck, IEEE Trans. Electron Devices ED-39, (1992).
9. O. Roux, B. Dierickx, E. Simoen, C. Claeys, G. Ghibaudo and J. Brini, Microelectronic Engineering 15, 547 (1991).
10. I. Bloom and Y. Nemirowski, Appl. Phys. Lett. 58, 1664 (1991).
11. B. Dierickx and E. Simoen, J. Appl. Phys. 71, 2028 (1992).

THE KINK-RELATED LOW-FREQUENCY GENERATION-RECOMBINATION NOISE IN SILICON-ON-INSULATOR MOST's

E. Simoen, U. Magnusson*, A.L.P. Rotondaro and C. Claeys
IMEC, Kapeldreef 75, B-3001 Leuven, Belgium
*presently at the Institute of Microelectronics, Kista, Sweden

ABSTRACT

This paper reports the results of a detailed investigation of the kink-related excess low-frequency noise in partially-depleted Silicon-on-Insulator MOST's. As is shown, the noise overshoot amplitude is proportional to the density of active generation-recombination centres in the depletion region of the transistor. Furthermore, the amplitude is inversely proportional to the measurement frequency f and to the device effective length. From the study of a large number of devices fabricated in different SOI technologies, at different temperatures, a general model is derived, which takes into account the close relationship with the multiplication current.

INTRODUCTION

One of the problem issues, related to the implementation of partially-depleted (PD) Silicon-on-Insulator (SOI) MOS transistors for analog applications is the occurrence of the kink and the corresponding excess low-frequency (LF) noise[1]. As shown in Fig. 1, a drastic increase of the LF noise is observed in the kink region, which is particularly important at lower frequencies. At the same time, a Lorentzian spectrum is observed, indicating the Generation-Recombination (GR) nature of the excess noise.

In the present paper, this feature is examined systematically for a large number of SOI MOST's, fabricated in different CMOS technologies, on different SOI substrate types. The influence of various physical (frequency f, device length L, the operation temperature T, the back-gate bias V_{BG}, etc.) and technological parameters (gate oxidation temperature, gate-oxide thickness t_{ox}, etc.) is investigated. A model will be presented, relating the excess noise amplitude to the effective density N_T of GR centres in the depletion region of the transistor, while the cut-off frequency is proportional to the inverse effective GR lifetime τ.

EXPERIMENTAL

The devices studied have been processed in different SOI technologies: 3 and 1 μm SOI CMOS on ZMR and laser-recrystallized substrates; 1 μm SOI CMOS on SIMOX substrates (the gate oxide thickness t_{ox}=20 nm; the film thickness t_f=180 nm) and on 0.5 μm CMOS SOI on SIMOX material (t_{ox}=15 nm; t_f=100 nm). In the latter technology also fully depleted (FD) devices are available for comparison. In the 1 μm technology on SIMOX, various splits with different gate-oxidation temperature and edge isolation technology (LOCOS or MESA-LOCOS) have been processed[2,3].

RESULTS

It has been previously demonstrated that the increase in the noise of a PD SOI nMOST at the kink position is closely related to the injection of holes in the floating film[1]. This follows first of all from the fact that when the film of the device is grounded, no such effect is observed[3,4]. By grounding the film, most of the injected holes are drained to the film contact and hence the major cause of the GR noise is removed. By applying a positive back-gate (or substrate) bias this feature is reduced considerably, as indicated in Fig. 2, while for a negative V_{BG}, the LF excess noise increases.

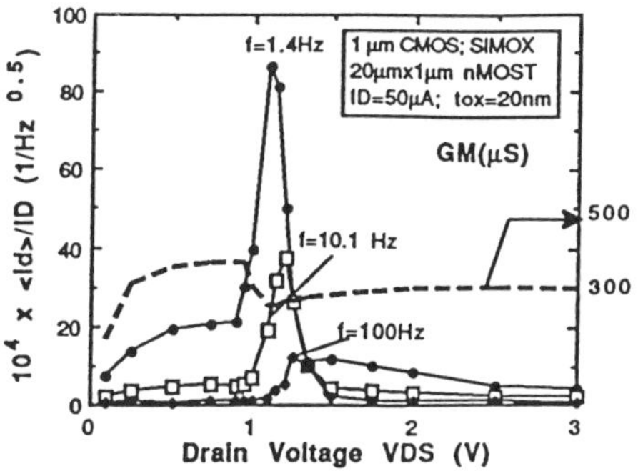

Fig.1. Normalized drain current noise for a 20 µm x1 µm PD SOI nMOST, for three different f. Also shown is the transconductance g_m correspon-ding with a fixed I_D=50 µA.

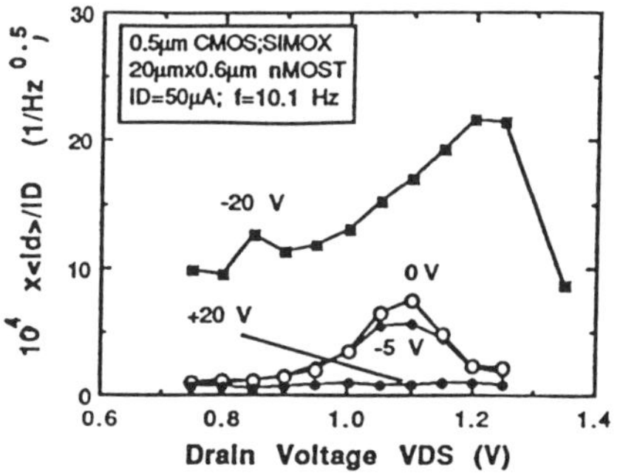

Fig.2. Influence of the back-gate bias on the LF excess noise in a 20 µmx0.6 µm PD SOI nMOST at f=10 Hz.

As shown in previous work, the noise overshoot generally increases upon cooling, showing a maximum at 77 K, compared with room temperature or 4.2 K[4,5]. By reducing the effective length L_{eff}, the noise-overshoot maximum shifts to a lower drain voltage (V_{DS}) position[4,6] and is tightly connected to the saturation voltage V_{DSSAT} (Fig. 3). The noise overshoot amplitude is proportional to 1/f (see e.g. Fig. 1)

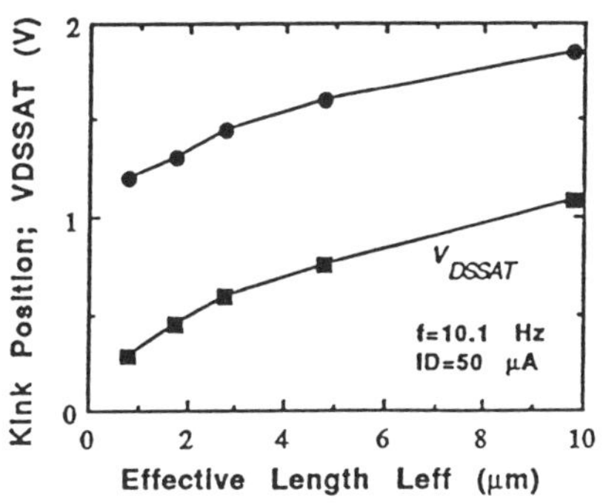

Fig.3. The noise overshoot position as a function of the effective device length, for f=10.1 Hz. Also shown is the calculated saturation voltage V_{DSSAT}. The I_D=50 µA; the device width is 20 µm.

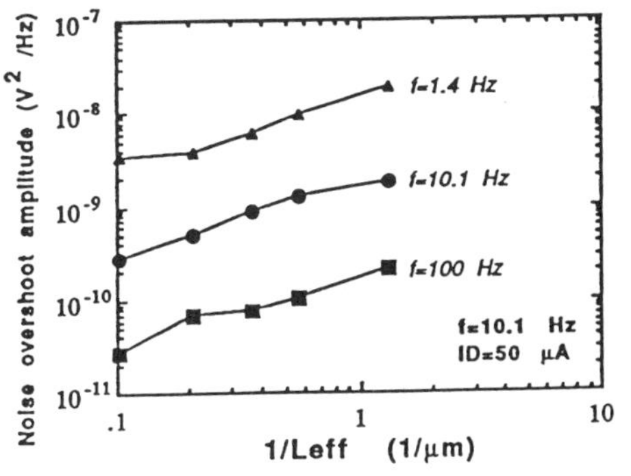

Fig.4. Dependence of the noise overshoot amplitude on L_{eff} and f, for a set of W=20 µm PD SOI nMOST's. I_D=50 µA.

and to $1/L_{eff}$ (Fig. 4). Furthermore, it has been shown recently that the noise overshoot amplitude is little affected by ionizing radiation, which seriously degrades the interface-related LF noise behaviour[2,7].

Note finally that in fully depleted devices both the drain-current kink and the noise overshoot disappear, except when a sufficiently negative back-gate bias is applied, which sets the back-interface in accumulation. In that case - and particularly at 77 K - a noise overshoot may be noted again. From all these arguments, it is inferred that the noise overshoot in SOI MOST's is related to the injection of majority carriers in the film, where they interact with defect centres.

As demonstrated before[3-5], a clear impact of the local film defectiveness exists on the excess-noise amplitude. In general, the noise peak at 300 K is larger for SIMOX substrates, than for ZMR or laser-recrystallized SOI substrates. On the other hand, there is little influence of the gate-oxidation temperature, or of the isolation technology (LOCOS, MESA-LOCOS) on the excess-noise behaviour[3]. Finally, the LF noise-overshoot amplitude scales proportionally with the oxide thickness.

MODEL

A model for the LF noise overshoot in PD SOI MOST's should at least include the close relationship which exists with the multiplication current and should in some way or another relate the excess-noise amplitude to the density of GR centres in part of the depletion region. For that purpose, a model earlier proposed for the kink-related noise in bulk and SOI transistors at liquid helium temperatures[5,8] is adapted. The basic features are schematically represented in Fig. 5, whereby the approach, proposed in[9,10] is followed. Only these GR centers will contribute to the excess noise which correspond with a value of the GR time constant τ which is close to the cut-off time constant $1/\pi f$. This yields an expression of the form:

$$\frac{S_{ID}}{I_D^2} = \frac{g_m^2}{I_D^2} \frac{q^2 N_T w_d}{C_{ox}^2 WL} \frac{4}{(\tau_c + \tau_e)\left[\left(\frac{1}{\tau_e} + \frac{1}{\tau_c}\right)^2 + (2\pi f)^2\right]} \qquad (1)$$

Hereby is S_{ID} the drain-current noise spectral density; I_D the drain current; g_m the transconductance; q the electronic charge; N_T the trap density; WL the device area; C_{ox} the front oxide capacitance per unit area and w_d the effective width where the trap centres contribute significantly to the GR noise. This is a region delineated schematically by the Lorentzian function in Fig. 5, around the crossing point.

The next step is to relate the capture (τ_c) and the emission time constant (τ_e) to the multiplication current I_M. As for the noise-overshoot model at cryogenic temperatures[5], it is again assumed that the effective GR time constant is inversely proportional to I_M:

$$\tau = \left(\frac{1}{\tau_e} + \frac{1}{\tau_c}\right)^{-1} = \frac{\alpha}{I_M} \qquad (2)$$

As shown previously [9,10], the peak-maximum is found for $\tau_e = \tau_c = 1/\pi f$. Filling in this value in eqs. (1)-(2) yields for the noise overshoot amplitude the following expression:

$$\Delta S_{IDmax} = g_m^2 \frac{q^2 N_T w_d}{C_{ox}^2 WL} \frac{1}{4\pi f} \qquad (3)$$

Note first of all that eq. (3) correctly reproduces the experimental length dependence reported in Fig. 4. Furthermore, filling in reasonable estimates for w_d, acceptable values for N_T can be derived, as shown elsewhere[3,4], so that the feature lends itself for material characterization purposes. Filling in the standard expression for I_M, enables to accurately fit the data[4], as in Fig. 6.

CONCLUSIONS

From the foregoing, it is concluded that the noise overshoot can be used to evaluate the quality (or the defectiveness) of the SOI film. Given the frequency dependence of the noise overshoot, there is a spectroscopic potential, i.e. it is in principle possible to extract the trap parameters, i.e. the energy level in the band-gap and the hole capture cross section.

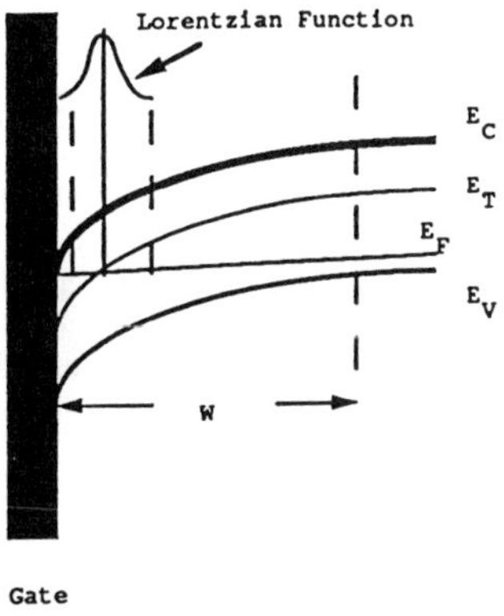

Fig.5. Schematical representation of the kink-related GR noise mechanism. Only traps in the neighbourhood of the intercept between the Fermi level E_F and the trap level E_T contribute to the noise.

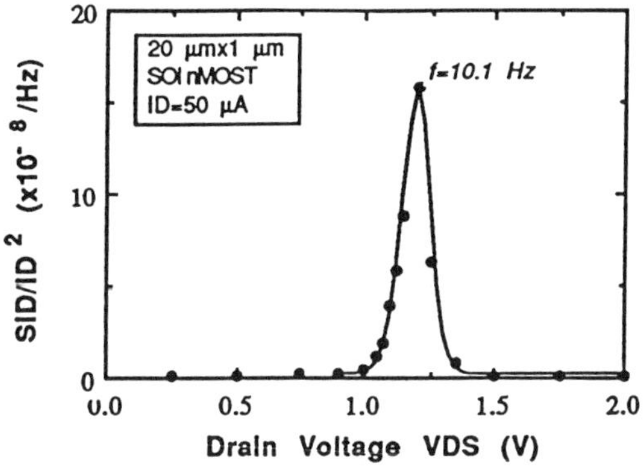

Fig.6. Fit of the model to the experimental noise overshoot, for f=10.1Hz and I_D=50 µA. WxL = 20 µmx1 µm.

REFERENCES

1. J. Chen, P. Fang, P.K. Ko, C. Hu, R. Solomon, T.-Y. Chan and C.G. Sodini, Proc. of the 1990 IEEE SOS/SOI Technol. Conf., 40 (1990).
2. E. Simoen, U. Magnusson, G. Van den bosch, P. Smeys, J.P. Colinge and C. Claeys, Paper to be published in the Proceed. of the 2nd ESA Electronic Component Conference, ESTEC, Noordwijk (The Netherlands), 24-28 May, 1993.
3. E. Simoen, U. Magnusson and C. Claeys, Appl. Surf. Sci. 63, 285 (1993).
4. E. Simoen, U. Magnusson, A.L.P. Rotondaro and C. Claeys, Paper to be published in IEEE Trans. Electron Devices.
5. E. Simoen, B. Dierickx and C. Claeys, J. Appl. Phys. 72, 1416 (1992).
6. E. Simoen and C. Claeys, Paper submitted for publication in Solid-State Commun.
7. E. Simoen and C. Claeys, Paper submitted for publication in Appl. Phys. Lett.
8. E. Simoen and B. Dierickx, Solid-State Electron. 35, 1455 (1992).
9. F. Scholz, J.M. Hwang and D.K. Schroder, Solid-State Electron. 31, 205 (1988).
10. D.C. Murray, A.G.R. Evans and J.C. Carter, IEEE Trans. Electron Devices 38, 407 (1991).

CHANNEL NOISE AND NOISE FIGURE OF M.O.S INTEGRATED TETRODES IN LOW-FREQUENCY RANGE

M. VALENZA, A. HOFFMANN, D. RIGAUD
Centre d'Electronique de Montpellier (CNRS URA 391)
Université Montpellier II
34095 Montpellier Cedex 5,France

ABSTRACT

Channel noise of MOS integrated tetrodes is investigated versus frequency and bias. The behaviour is analysed in the four operation modes taking into account the variations of the small signal parameters. Voltage-gain and noise figure are simultaneously measured. The two inputs are investigated. Experimental results and their interpretations lead to the bias range giving the best compromise between high voltage-gain and low noise figure.

INTRODUCTION

Because of their integrated cascode configuration MOS tetrodes are particulary attractive for R.F amplification, automatic gain control, mixer applications ...

The noise takes a predominant part in these linear and non linear applications and previous papers[1,2] have studied its behaviour in the high frequency range. As MOS transistors exhibit high noise levels at low frequency (1/f noise) we have carried out a noise study of MOS integrated tetrodes up to 100 KHz. From small signal properties of the device, noise equivalent circuits have been developped to analyze the channel noise and the noise figure behaviour.

TETRODE MODEL

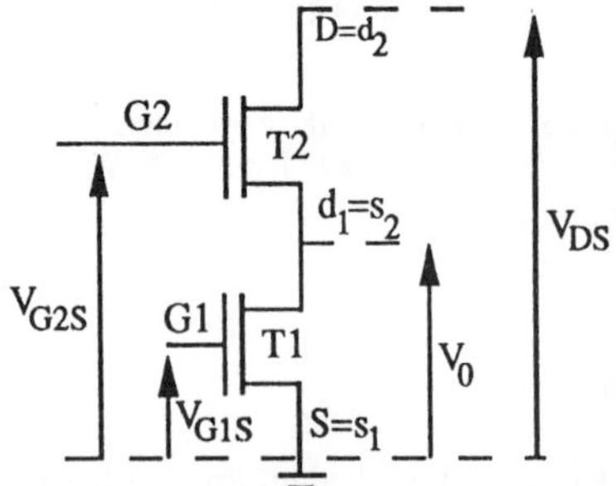

figure 1

The MOS tetrode can be described by two MOS transistors (T1 and T2) in a series configuration (see fig 1). Since the two gates G_1 and G_2 can be biased independently, four conditions of operation can arise where each transistor can be saturated (Pentode mode : P) or non saturated (Triode mode : T). The small signal behaviour of the tetrode can be obtained with the help of the low frequency equivalent circuit of each transistor. The conductance G_D and the transconductances G_{M1} and G_{M2} can be expressed as :

$$G_D = \frac{g_{d1}\, g_{d2}}{g_{d1}+g_{d2}+g_{m2}} \qquad G_{M1} = \frac{g_{m1}(g_{m2}+g_{d2})}{g_{d1}+g_{d2}+g_{m2}} \qquad G_{M2} = \frac{g_{m2}\, g_{d1}}{g_{d1}+g_{d2}+g_{m2}} \qquad (1)$$

where the small symbols are related to elementary transistors (T1 or T2) by the subscripts (1 or 2). The subscripts of capital symbols indicate the active input of the tetrode.

Then the voltage gain of the device can be calculated taking into account an external load resistance R_L. We obtain :

$$G_{Vj} = -\frac{G_{Mj} R_L}{1 + G_D R_L} \quad \text{where } j = 1 \text{ or } 2 \qquad (2)$$

The noise equivalent circuit of the tetrode is obtained by adding a noise current generator between drain and source of each elementary transistor. The spectral densities

$S_{ic1}(f)$ and $S_{ic2}(f)$ of these generators take into account thermal noise and 1/f noise which is the main noise source at low frequencies.

Using equations (1) classical noise developpements lead to the channel noise spectral density $S_{iT}(f)$ of the tetrode :

$$S_{iT}(f) = \left(\frac{G_{M1}}{g_{m1}}\right)^2 S_{ic1}(f) + \left(\frac{G_{M2}}{g_{m2}}\right)^2 S_{ic2}(f) \qquad (3)$$

When T1 is saturated (g_{d1} and $G_{M2} \rightarrow 0$) the tetrode noise is only due to this transistor. Equation (3) shows that the noise of the tetrode can be represented by two noise voltage generators located at the inputs of the device. Their spectral densities are respectively $S_{VG1}(f)=S_{ic1}(f)/g_{m1}^2$ and $S_{VG2}(f)=S_{ic2}(f)/g_{m2}^2$.

The noise figure of the tetrode used as an amplifier can be expressed as :

$$F = 1 + \frac{1}{4kTR_g}\frac{S_{iT}(f)}{G_{Mj}^2} \qquad (4)$$

j indicates the active input . Rg is the input load resistance.

EXPERIMENTAL RESULTS

Measurements have been performed on N channel BF982 MOS tetrodes produced by RTC. The active dimensions of each transistor are $2000 * 4\ \mu m^2$. The threshold voltage is $V_T = -0.78$ V. I-V characteristics and small signal parameters are obtained with the HP 4142B measurement system . Noise measurements are performed with the help of low noise Brookdeal 5004 amplifiers and an HP 3562A FFT analyser. The HP 3562A has been also used for voltage-gain measurements.

The transconductances behaviour versus the tetrode polarization has been studied because of their influences on the voltage-gain and the noise figure. Typical variations are shown in figures 2a and 2b for V_{DS} = 8 V (saturation of the I-V characteristics). Similar plots are obtained in the ohmic range. Maxima values are obtained when the transition between the P-P and T-P modes or between the P-T and T-T modes occur. G_{M1} values are higher than G_{M2} ones because in this configuration T1 acts as a load located in the source of T2 and induces a feedback.

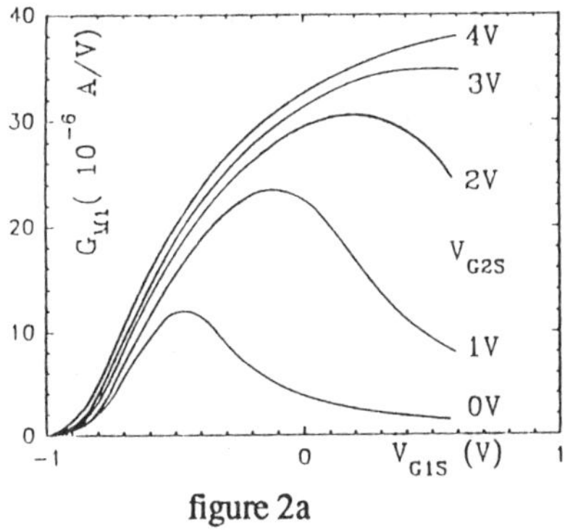

figure 2a

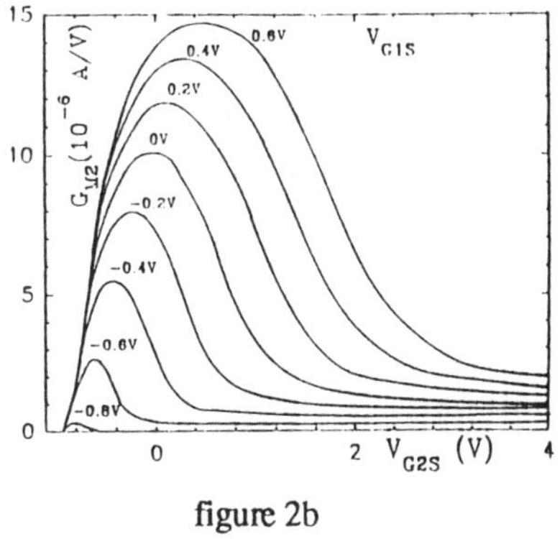

figure 2b

Channel noise measurements have been performed versus frequency at various bias points of the tetrode. 1/f noise is dominant. The variations of current spectral density versus I_D are reported figures 3a and b respectively in the ohmic and saturated ranges. When I_D (or V_{G1S}) increases the channel noise reaches a maximum value, decreases and increases again.

Voltage-gain and noise figure have been measured versus frequency for various quiescent points. The noise figure versus frequency exhibits 1/f law and F-1 varies as 1/Rg, showing that only the channel noise is involved in the noise figure measurements at low frequencies. Figures 4 and 5 show typical results obtained with G_1 or G_2 as active input.

INTERPRETATIONS

When T1 is saturated (P-T mode at low drain bias or P-P mode at high drain bias) $I_D = K\, V_{G1}^{*2}/2$ (with V_G^* in place of $V_{GS}-V_T$). From eq(1) $G_{M1} = g_{m1}$, $G_{M2} = 0$ and from equation (3) $S_{iT}(f) = S_{ic1}(f)$. As $S_{iT}(f)$ exhibits 1/f noise we can write[3,4] $S_{ic1}(f) = \alpha_0\ I_D V_0 / f$ where α_0 is a characteristic parameter of 1/f noise and $V_0 = V_{D1S}$.

Since T1 is saturated $V_0 = V_{G1}^*$ and $S_{iT}(f) = \frac{\alpha_0}{f}\frac{K}{2} V_{G1}^{*3} = \frac{\alpha_0}{f}\sqrt{\frac{2}{K}}\, I_D^{3/2}$ (5)

From Mc Whorter's model[5] α_0 varies as $1/V_{G1}^*$ and $S_{iT}(f)$ as V_{G1}^{*2} or as I_D.

When T1 is non saturated the tetrode channel noise can be expressed in the T-T mode (low drain bias) as :

$$S_{iT}(f) = \frac{\alpha_0}{f}\, k\, \{V_{G1}^* V_0 - \frac{V_0^2}{2}\}\, \{V_0 \frac{1 - a^2}{(1+a)^2} + V_D \frac{a^2}{(1+a)^2}\} \qquad (6)$$

and in the T-P mode (high drain bias) as :

$$S_{iT}(f) = \frac{\alpha_0}{f}\frac{k}{2}(V_{G2}^* - V_0)^2\, \{V_0 \frac{1 - a^2}{(1+a)^2} + V_{G2}^* \frac{a^2}{(1+a)^2}\} \qquad (7)$$

where $a = (V_{G1}^* - V_0)/(V_{G2}^* - V_0)$ $\quad G_{M1}/g_{m1} = 1/(1+a)$ $\quad G_{M2}/g_{m2} = a/(1+a)$

Equations (6) and (7) allow us to analyse the shape of the curves obtained in figures 3a and 3b. The maximum value takes place at the transition of the two involved modes and the minimum value is due to 1/f noise of T1 and T2 balanced by $(1 - a^2)/(1+a)^2$ and $a^2/(1+a)^2$ terms which vary also with the bias[6].

In saturation range, when G1 is the active input the gain voltage reaches its maximum value during the P-P to T-P mode transition as G_{M1} (see fig. 4a). F follows the variations of G_{M1} and $S_{iT}(f)$. It is an increasing function of V_{G1S} but is roughly constant when G_{V1} is maximum. At this point, G_{V1} and F remain constant for large values of V_{G2S} (see fig. 4b) but are damaged for low G2 biases.

When G2 is the active input G_{V2} is lower because the feedback due to T1. In the T-P mode this feedback decreases since T1 is no longer saturated and G_{V2} increases with V_{G1S}. The noise figure is a decreasing function of V_{G1S} but keeps a high value when T1 is saturated (see fig. 5a). For a given V_{G1S} bias it is possible to obtain simultaneously high voltage-gain and low noise figure by convenient values of V_{G2S} (see fig. 5b).

In the ohmic range where the P-T and T-T modes are involved similar behaviours have been obtained but the input G2 exhibit poor noise figure and voltage-gain.

CONCLUSION

The low frequency noise behaviour of the MOS tetrodes has been investigated. The contribution of each transistor to the channel noise of the tetrode has been pointed out when the biases vary. For high drain currents the cascode configuration leads to non expected variations.

The best compromise between the voltage gain and the noise figure can be obtained when G1 or G2 are the active inputs at high drain bias.

REFERENCES

1 F.M. Klaassen and J. Prins, Philips Reserch Reports, 23 , 478 (1968)
2 H. Chenming and R.S. Muller IEEE Trans. Electron Devices, 18 , 418 (1971)
3 L.K.J. Vandamme, Solid-State Electronics, 23 , 317 (1980)
4 A. Van Der Ziel, Noise in solid state devices and circuits, Wiley-Interscience Publication (1986)
5 L.K.J. Vandamme et al, Solid-State Electronics, 28 , 1049 (1985)
6 M.Sadiki, Thesis, University of Montpellier, June 1992

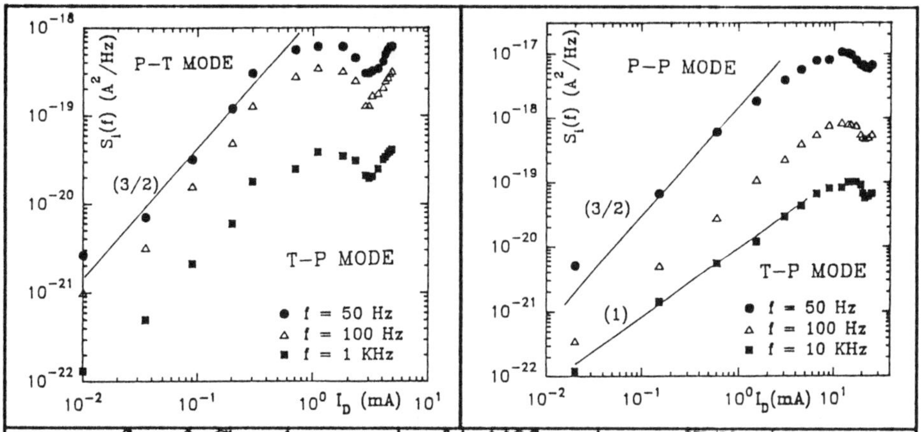

figure 3 : Channel current noise of the MOS tetrode versus ID(V_{G1S})
a)V_{DS} = 0.3 V ; V_{G2S} = 0 V b) V_{DS} = 8 V ; V_{G2S} = 1 V

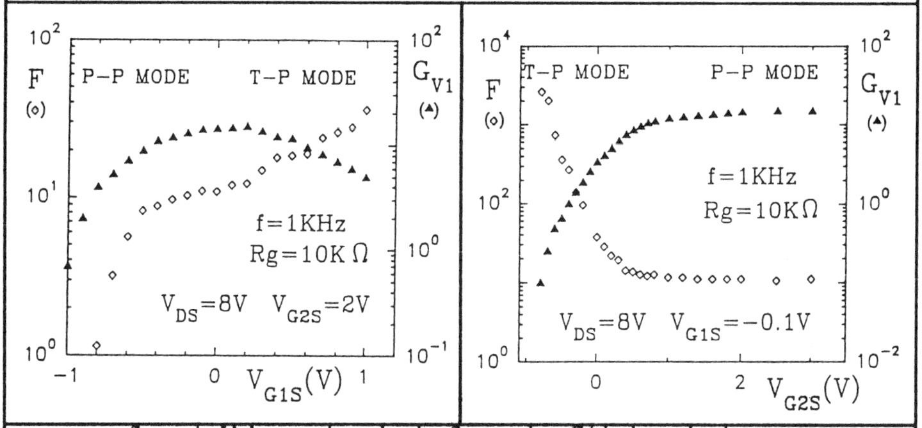

figure 4 : Voltage-gain and noise figure when G1 is the active input
a) Versus V_{G1S} b) Versus V_{G2S}

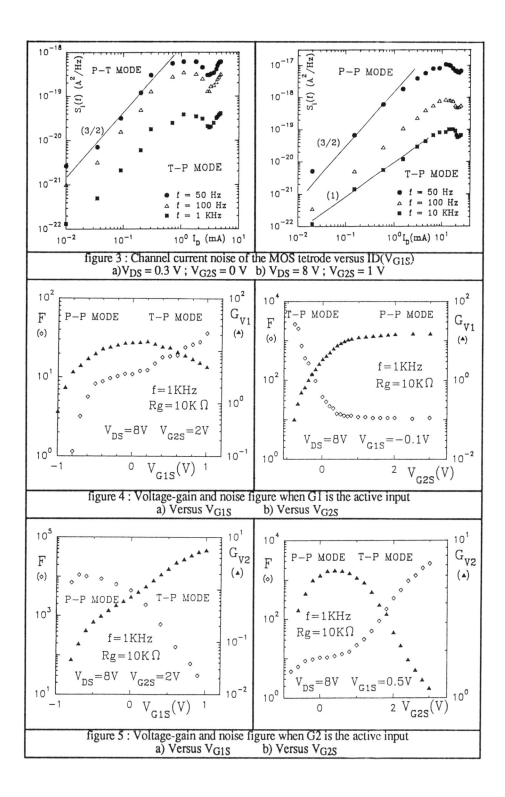

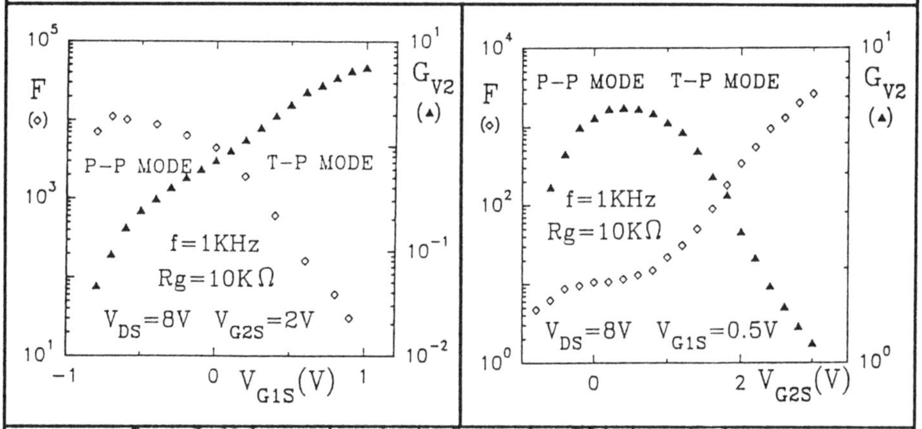

figure 5 : Voltage-gain and noise figure when G2 is the active input
a) Versus V_{G1S} b) Versus V_{G2S}

CORRELATION BETWEEN $1/f$ NOISE (SLOW STATES) AND CHARGE PUMPING (FAST STATES) IN DEGRADED MOSFET'S

Jen-Tai Hsu, Xiaoyu Li, and C.R. Viswanathan
Department of Electrical Engineering
University of California
Los Angeles, California 90024, U.S.A.

ABSTRACT

In this paper, low frequency noise measurements made on MOS transistors, which have been stressed, is reported. The devices degraded by Fowler-Nordheim (F-N) injection as well as by Hot-Carrier-Injection (HCI) were studied. We found that the Hooge's parameter, which is directly proportional to the slow states, as determined from noise measurements, correlates with fast interface states determined from charge pumping and subthreshold swing measurements.

INTRODUCTION

Device degradation study uses different techniques and one of them is the low frequency noise characterization in MOS transistors [1,2,3]. Charge pumping current measurement is also used sometimes to study the damage in these devices [5]. It is known that $1/f$ noise is sensitive to the slow states at the interface, while the charge pumping current measures the interface states that can respond at the gate pulse frequency (fast states). In this paper, we report a correlation between slow and fast interface states on samples degraded by Fowler-Nordheim (F-N) injection and hot-carrier (HC) stressing.

EXPERIMENTAL

MOS transistors fabricated in a VLSI processing facility were used in this study. The devices were conventional LDD nMOSFET's with W/L equal to $10/0.5\mu$m, and a gate oxide thickness of 12nm. The devices were subjected to four different stressing conditions: positive and negative Fowler-Nordheim(F-N) tunneling injection, Drain Avalanche Hot Carrier injection (DAHC), and Channel Hot Electron injection (CHE). In F-N tunneling injection, a constant current density of $1\text{m}A/cm^2$ was used. In positive F-N injection the current is injected from the gate into the oxide while in negative F-N injection the current is of opposite polarity. The DAHC injection was performed with V_d equal to $8V$, and V_g equal to $V_d/2$; while in the CHE injection, the stress condition corresponded to $V_d = V_g = 7.5V$. In order to make a meaningful comparison among different degradations, the observed decrease in the peak transconductance g_m was kept the same by adjusting the stress time in each type of stress. In each stress condition, the low frequency noise was measured in the linear region with a drain voltage of 0.5V. The noise measurements were performed using a custom made low noise amplifier and HP3561A dynamic signal analyzer [4]. The charge pumping technique that we used in this study is the constant pulse height with varying gate base level as in [5]. The frequency of the pulse was kept at 100KHz and pulse height was 4V with a duty-cycle of 50 %.

RESULTS AND DISCUSSION

The input referred noise spectra measured with various gate voltages is shown in Fig.1 for the virgin device and in Fig.2 for the DAHC degraded devices. The

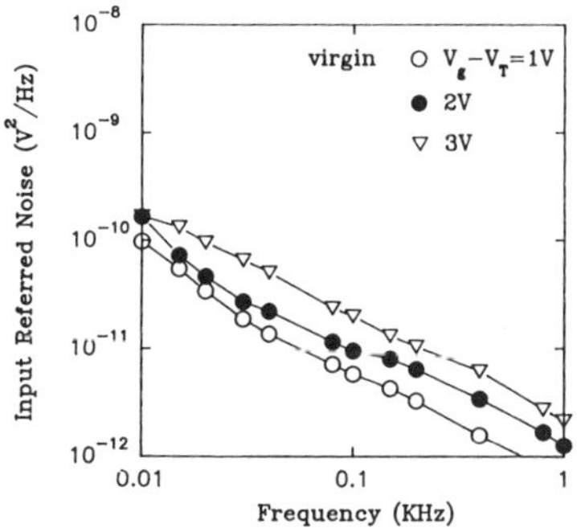

Figure 1: Plot of the input referred noise spectra with different gate voltages for virgin devices.

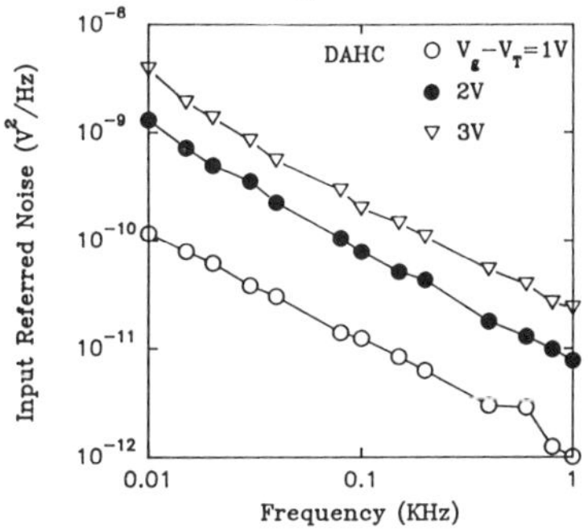

Figure 2: Plot of the input referred noise spectra with different gate voltages for the DAHC degraded device.

noise in the virgin sample increases with $V_g - V_T$. A gate voltage dependence is observed suggesting that the mobility fluctuation is the dominant noise mechanism. In the DAHC degraded sample, the noise increases significantly with $V_g - V_T$. The noise increases by nearly a factor of 30 from its value at $V_g - V_T = 1V$. Such a pronounced increase in the noise with $V_g - V_T$ was not observed in samples degraded by F-N injection. In Fig.3, the input referred noise for the virgin and for the four different degraded devices is shown under same bias conditions. It is seen that DAHC produces the largest increase in low frequency noise in comparison with the other three stress conditions. The spot noise measured at 1 KHz for a gate voltage of 2 volts above the threshold is given in Table 1 for the virgin and stressed samples. The input-referred noise of MOSFET's operating in the linear region can be expressed as [6]:

$$S_{V_g}(f) = \frac{q}{C_{OX}} \frac{\alpha_H}{WLf}(V_g - V_T) \qquad (1)$$

where S_{V_g} is the input referred noise, and α_H is the Hooge's parameter. Hooge's parameter (α_H) is largest for DAHC degraded samples as shown in Table 1. The results of charge pumping current measurements are shown in Fig.4 for the stressed and virgin samples. We can extract the fast interface state density from charge pumping using the expression

$$I_{cp} = fq^2WL\bar{D_{it}}\Delta\psi_s \tag{2}$$

where $q\Delta\psi_s$ is the region of the bandgap over which the interface states are measured and $\bar{D_{it}}$ is the average interface state density over this energy range. Subthreshold swing was also measured for these devices and the interface state density is obtained by the change in the subthreshold swing from the virgin sample.

$$\Delta S = \frac{kT}{q}\frac{q\Delta D_{it}}{C_{OX}}ln10 \tag{3}$$

Parameters	$Virgin$	CHE	$DAHC$	$FN(-)$	$FN(+)$
$S_{V_g}(V^2/Hz)$ $(1KHz, V_g - V_t = 2V)$	1.3×10^{-12}	1.9×10^{-12}	7.8×10^{-12}	2.9×10^{-12}	2.6×10^{-12}
$\alpha_{\rm H}$	6.3×10^{-5}	1.3×10^{-4}	1.1×10^{-3}	1.5×10^{-4}	8.5×10^{-5}
$D_{it}(1/cm^2eV)$ (from charge pumping)	3.7×10^{10}	2.6×10^{11}	7.8×10^{11}	2.3×10^{11}	4.6×10^{11}
$\Delta D_{it}(1/cm^2eV)$ (from ΔS)	0	1.1×10^{11}	2.3×10^{11}	7.7×10^{10}	1.5×10^{11}

Table 1 : List of input referred noise spectra, Hoodge's parameter (α_H) and fast interface trap density calculated from I_{CP} and ΔS for virgin and four different degraded devices.

Table 1 lists the results obtained from the noise measurements in comparison with those obtained in charge pumping and subthreshold swing measurements. The fast interface trap density extracted from both charge pumping and subthreshold swing techniques shows good correlation with the noise measurements among all samples, except for the negative F-N injection. The devices degraded by negative F-N injection have larger $1/f$ noise but lower charge pumping current and lower subthreshold swing change than the devices degraded by positive F-N injection. This result suggests that negative gate F-N injection, when compared with positive gate F-N injection, creates more slow states than fast interface states. During the negative F-N injection, hole traps are created by the electrons injected from the gate. The trapped holes give rise to a strong local electric field which

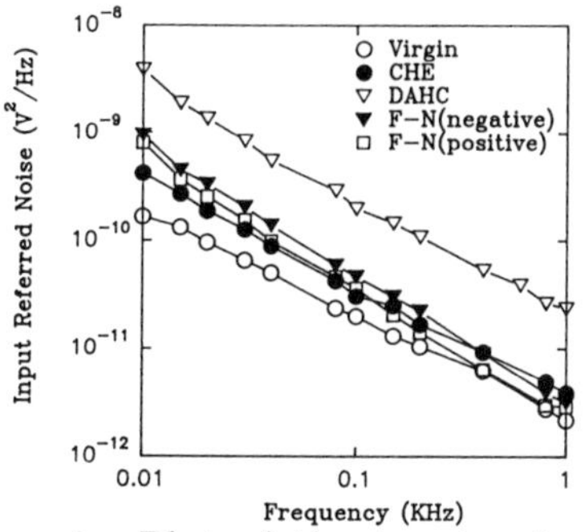

Figure 3: Plot of the input referred noise spectra for virgin and four different degraded devices under the same bias condition ($V_g - V_t = 3V$).

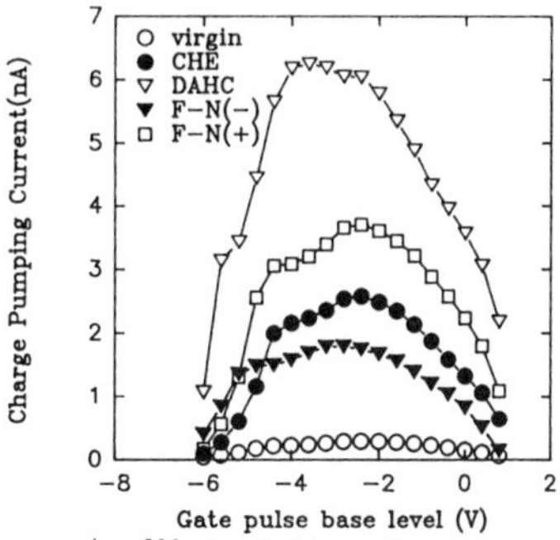

Figure 4: Plot of the charge pumping current for virgin and four different degraded devices.

may modify the energy or the capture cross section of nearby defects [7] causing slow states that interact with the channel current. This can also be explained by the larger hysteresis observed immediately after negative gate F-N than positive gate F-N injection. Hence the negative F-N injection results in a higher low frequency noise. However, the DAHC degradation, in which most traps generated are interface type, is different from the F-N injection where both oxide and interface traps are generated. This explains why in DAHC degradation, both slow interface traps measured from $1/f$ noise, and fast interface traps measured from charge pumping are the largest among the four degraded devices.

CONCLUSION

In conclusion, we performed both $1/f$ noise and charge pumping techniques to report a correlation between slow and fast interface states on the MOSFET's degraded by F-N and HC injection. More slow states appear to be generated in negative gate F-N than in positive gate F-N injection . DAHC degradation causes the largest $1/f$ noise and charge pumping current.

REFERENCES

[1] Z.H.Fang et. al., IEEE Electron Device Letters, vol. EDL-7, p.371,1986.
[2] J.M.Pimbley et. al., IEEE Electron Device Letters, vol. EDL-5, p.345, 1984.
[3] J. Chang and C.R. Viswanathan, Proc. of the 10th International Conf. on $1/f$ noise in Physical Systems, Budapest, Hungary, 1989.
[4] J. Chang, M.S. Thesis, UCLA, 1988.
[5] P. Heremans et. al., IEEE Trans. on Electron Devices, vol. 36, p.1318, 1989.
[6] F.N. Hooge, Physica, 83B, p.14, 1976.
[7] T.L.Meisenheimer et. al., IEEE Trans. on Nuclear Science, vol.37, p.1696, 1990.

IX. OPTICAL DEVICES AND OTHERS

PHOTO DETECTORS APPROACHING IDEAL AMPLIFICATION

R. P. Jindal
AT&T Bell Laboratories, Whippany, New Jersey 07981

ABSTRACT

Progress in the understanding of how to design high performance photo detectors will be described. The interplay between the evolution of new concepts, material limitations and performance requirements will be elucidated. Directions for future growth in this field will also be highlighted.

INTRODUCTION

Recovery of information, whether it be in the form of receiving data at the destination of a communication link or the measured output of an experiment, is intimately tied to the fundamental problem of extracting a weak signals embedded in fluctuations. The fundamental component of any such detection/measurement process involves the amplification and detection of the quantity of interest. Amplification is necessary to preserve the signal to noise ratio at the detection point. However, every amplification process results in some degradation of the signal to noise ratio. Although as a fundamental problem this issue has been examined over several decades, now due to its technological importance, it has enjoyed considerable attention over the past 10 years. The purpose of this work is to discuss the concepts that have been conceived over the last fifty some years for the design and development of photo detectors to achieve near ideal amplification.

PHOTO DETECTOR CLASSIFICATION

Based on their mode of operation, independent of the technology used for fabrication, photo detectors can be classified into two broad categories. In a photo conductive detector the effect of incident photons is the modulation of device impedance. The photo voltaic type of detectors respond to the incident light by the generation of a voltage across their terminals. Both types are extensively used in a variety of applications depending upon, among other factors, their speed and noise performance. Detectors with built in amplification such as photo transistors and avalanche detectors use one of the above principles for detection followed by amplification. This amplification is achieved by conventional device action observed in field effect and bipolar devices or through carrier multiplication. The overall performance of amplifying detectors is effected by the noise generated both during the detection and the amplification process.

GENERAL MULTIPLICATION NOISE THEORY[1]

The response of a device to individual photons of light is highly dependent upon the physical properties of the medium including its band structure, physical size and the carrier species taking part in the process. As will be explained later, the gain and the fluctuations in the gain generated by this response are further dependent upon the mode in which the device is operated. In general, the device response can be either highly localized in time with respect to the input or be distributed over a relatively large period. According to the classical school of thought, it was usually suspected that the later case was not very useful since the signal characteristics will be smeared out in time. It will be shown that this is really not the case.

Fig. 1 Model of a photo detector

Let us analyze the following situation. Consider a photo detector as shown on the left represented by a black box. Photons constituting the light signal are incident upon this detector at random instants of time generating a response given by x(t). Therefore, let x(t) be a random variable composed of a sum of P primary events f(t) occurring in time T.

Then mathematically speaking,

$$x(t)=\sum_{k=1}^{P} f(t-t_k) \text{ and } \overline{x(t)}=\alpha\int_{-\infty}^{+\infty} f(t)dt \tag{1}$$

where α is the average rate of occurrence of the events f(t) and is given by $\lim_{T\to\infty}\frac{P}{T}$. Now let each event f(t) be composed of M subevents g(t) where M is a random variable. Physically this would correspond to the gain exhibited by the device. Then it can be shown[1] that the spectral density of the variable x(t) is given by

$$S_x(\omega)=2\alpha\left|G(j\omega)\right|^2\left\{\overline{M}+\overline{\sum_{r\neq s=1}^{M} Cos(\omega\Delta t_{rs})}-\left|\overline{\sum_{r=1}^{M} e^{-j\omega\Delta tr}}\right|^2\right\}+\left|\overline{\sum_{r=1}^{M} e^{-j\omega\Delta tr}}\right|^2 S_p(\omega) \tag{2}$$

Here Δt_{rs} is the temporal separation of subevents r and s and Δt_r is the delay between the primary event and the rth subevent. Note that so far we have made no assumptions about the statistics of the input to the detector. This expression has two interesting limits which we shall refer to as the classical (deterministic) limit and the new (random) limit.

Deterministic Limit: Assume that the subevents are spaced in time such that $\omega M \Delta t_{rs} << 1$. Under these conditions the above expression reduces to

$$S_x(\omega)=2\alpha\left|G(j\omega)\right|^2 \text{var}(M)+\overline{M}^2 S_p(\omega) \tag{3}$$

This expression has a simple interpretation that the total fluctuation in the output is a sum of two terms. The first term is proportional to the fluctuations associated with the gain process var(M) while the second term is nothing but the amplified input noise. This expression was first derived by Shockley et al[2] in 1938. To completely predict the behavior for a specific device one now has to evaluate $\overline{M}$ and var(M).

Random Limit: Assume that Δt_{rs} and Δt_r are random variables distributed uniformly in time over a time interval τ_p. Then for a measurement time $T >> \tau_p$ the randomness does not manifest itself since the subevents will appear instantaneous and hence we recover (3). However, in the reverse situation where $\tau_p >> T$ the randomness will exhibit itself. In this case the three averages in (2) vanish giving

$$S_x(\omega)=2\alpha\left|G(j\omega)\right|^2\overline{M} \tag{4}$$

In this limit the fluctuations are independent of var(M) and are equivalent to shot noise generated by a process occurring at an average rate $\alpha\overline{M}$. Under the condition $\tau_p \alpha >> 1$ the physical situation corresponds precisely to this description. This is analogous to the idea of cutting apart clustered events for short measurement times and recovering a Poisson process which, for a shot-noise-driven doubly stochastic process, has been proved by Saleh and Teich[3]. The implications of these results will be discussed later in detail.

We shall next discuss photo detector research under the two broad categories summarized above.

DETECTOR RESEARCH IN THE DETERMINISTIC LIMIT

Van Vliet et al[4,5] have presented a general theory of carrier multiplication noise, in what turns out to be the deterministic regime, encompassing and extending the previous work of Tager[6], McIntyre[7] and Personick[8] and Lukaszek et al[9]. This theory deals with both situations when either a single carrier species (either electrons or holes) or both carrier types take part in the multiplication process. Let N be the number of possible ionizing collisions per primary carrier transiting through the device. Then for the case when N approaches infinity this theory recovers McIntyre's[6] results. In this regime one effectively ignores the threshold of energy needed for a pair production and hence the probability of ionization is a continuous function of the distance traveled by the ionizing carrier. This turns out to be a very good approximation when device dimensions are very large compared to the mean free path between two ionizing collisions of the ionizing carrier. We shall refer to this as the Infinite Medium Avalanching. For electron initiated avalanche process, the above theory gives

$$\mathrm{var}(M) = \overline{M}\left(\overline{M}-1\right) + k\left(\overline{M}-1\right)^2 \overline{M} \tag{5}$$

where $k=\beta/\alpha$ β and α are the ionization probabilities per unit length for the holes and electrons respectively. This formula has been used very successfully to explain the noise behavior of avalanche photo detectors where the ionizing region is large compared to the carrier mean free path. The above formula also gives rise to the conventional wisdom that to achieve low noise amplification one must use materials which have highly asymmetric electron to hole ionization ratios. Also, the avalanche process must be initiated by the more strongly ionizing species. The best performance that one can obtain in this case is given for the case when $k=0$ when one is left with only the first term in (5). Physically this implies that the ionization should be initiated by the most ionizing species and that the relative ionizing power of the other carrier species should approach zero.

For the case when N approaches unity the theory recovers the results obtained by Lukaszek and van der Ziel and Chenette[9]. We shall refer to this as Finite Medium Avalanching[16]. In this case we get

$$\mathrm{var}(M) = \overline{M}\left(\overline{M}-1\right)\left\{1+\frac{2(\mu-\lambda)}{(1+\lambda)(1+\mu)}\right\} \tag{6}$$

Here μ and λ are the a priori probabilities of ionization by a hole and by an electron respectively for one traversal of the ionizing region. This expression is again for an electron initiated process. This regime of device design has three interesting cases.

Case I: Consider the case when the probability of ionization by the primary (avalanche initiating) species i.e. electron in this case approaches unity. ($\lambda \rightarrow 1$). Then (6) reduces to

$$\mathrm{var}(M) = \overline{M}^2\left(1-\frac{2}{\overline{M}}\right) \tag{7}$$

The variance is therefore better than the best result obtainable for $N \to \infty$ case.

Case II: Next we consider the case when electrons and holes ionize equally. ($\mu \to \lambda$). Then (6) reduces to

$$\mathrm{var}(M) = \overline{M}\left(\overline{M} - 1\right) \tag{8}$$

The variance is now identical to that for the best result obtained for the infinite medium case.

Case III: Finally consider the case when the probability of ionization by the secondary species i.e. holes in this case approaches unity. ($\mu \to 1$). Then (6) reduces to

$$\mathrm{var}(M) = \overline{M}^2\left(1 - \frac{1}{\overline{M}^2}\right) \tag{9}$$

The variance is slightly larger than the best $N \to \infty$ case.

It is therefore observed that the performance for the finite medium avalanching are close to the best results obtainable from the infinite medium avalanching.

Apart from the work of van Vliet[4,5] several others authors[10,11]have later presented unifying formulations for calculating the gain and its variance in multiple device structures with their respective strong points and limitations and hence preferable for specific situations.

DETECTOR DESIGN BASED ON INFINITE MEDIUM AVALANCHING

Almost all of the detector device research has been solely targeted to follow the road chartered by the infinite medium theory. In pursuing this design philosophy two constraints have to be satisfied. The first constraint is that the carrier multiplication process has be initiated by the carrier species with higher ionizing power. This is easily arranged. The second constraint to be met is that the ionizing power of the two carrier species i.e. holes and electrons should be vastly different. This is naturally satisfied in the case of silicon. Consistent with this, high performance silicon photo detectors have been realized albeit operating at short wavelengths. For longer wave lengths (1.3 - 1.5 μm) where the coupling efficiency of silicon is poor one must work with compound semiconductor material systems which have a higher coupling efficiency due to their narrower band gap. However, these material systems have electron to hole ionization ratios close to unity which is detrimental to the noise performance.

The key to getting around this problem is to realize that even though the ionization coefficients of electrons and holes may be equal for a uniform band structure device the situation may be entirely different for a device involving layered structures and band discontinuities. The role of using band discontinuities to provide enhanced ionization was first pointed out by Chin et al.[12] This same concept was further developed to provide enhanced orderly ionization in the proposal for the staircase photo diode[13]. However clear successful demonstrations of such devices have yet to be made. Part of the problem lies in the complex processing that is required to make these device structures which require extreme uniformities and control during the fabrication process.

DETECTOR DESIGN BASED ON FINITE MEDIUM AVALANCHING

Although the first experimental evidence of this phenomena was observed by Jindal[14,15] as early as 1983, the suggestion to use this effect for high performance detector design was made later by Jindal[16] and Hollenhorst[17]. However, even since then, actual device design and fabrication in this area remains virtually unexplored and holds great potential.

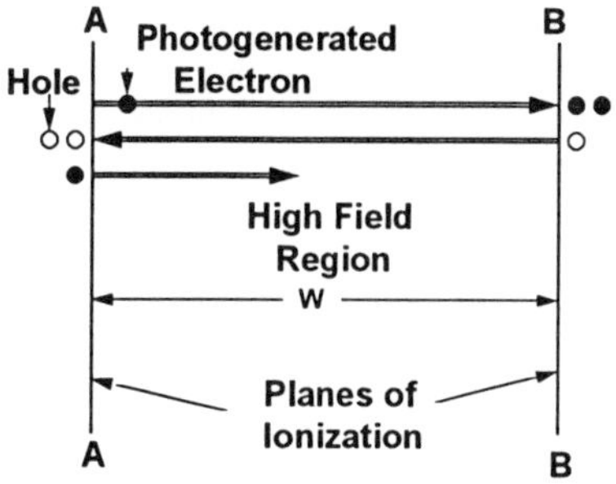

Fig. 2 Idealized section of an FMA detector

Although no specific structures have been proposed or been built yet, the conceptual basis[16] of such a device structure is described here. It essentially consists of a high electric field region defined by two planes of ionization "A" and "B". The device operates in the following manner. The process is initiated by the generation of an electron hole pair by a photon. The photo generated electron travels through the high field region and generates an electron hole pair at the ionization plane "B". Here, the two electrons are collected and placed out of circulation and it is the hole that enters back the high field region. This hole now travels back towards plane "A" where it produces another electron hole pair. Now the two holes are collected and it is the electron that enters the high field region. This process continues till one of the carriers fails to produce ionization. The total device gain is the sum of all the pairs generated in this process. Not only do such devices provide superior noise performance but also operate at lower supply voltages resulting in improved reliability and systems compatibility. This is a highly desirable feature from the systems perspective. Since these devices require nearly equal ionizing power from both carrier species this scheme is ideally suited for long wavelength detectors which, due to band gap restrictions are suited for compound semiconductor material systems.

DETECTOR RESEARCH IN THE RANDOM LIMIT

Next we shall examine the consequences of the random limit discussed earlier. This regime is radically different from the deterministic limit in that the device response is *not* assumed to be instantaneous in relation to the (photon) input. In practice this is always the case. The assumption of instantaneous response is valid if the measurement time is large compared to the device response time. However, constant migration towards higher operating speeds prompts us to not avoiding this dilemma by proposing even higher speed devices but rather taking the distributed nature of the response into account when assessing the device performance. When evaluating the detection process in this regime two situations stand apart.

(i) In the first situation the goal is to accurately measure the magnitude of a constant signal incident on the device. This is referred to as level detection. Due to space limitations we will only summarize the results the details appearing elsewhere[18]. In such a measurement if the accuracy is limited by the finite integrating time τ the detection system and not by the available measurement time, one can use a random detector to attain an arbitrary improvement in the signal-to-noise ratio over conventional detection techniques.

(ii) In the second situation the challenge is to detect individual pulses of light making up the bits of information being transmitted in a fiber optic communication system. This is referred to as pulse detection. Due to its complex nature, this problem is not easily amenable to analytical treatment. An essential component of this mode of operation is that the response of the photo detector is spread over time τ_p which is large compared to the measurement time. Using this scheme, we have shown both semi-analytically and numerically[19] that the detector can be made practically immune to fluctuations in the conventional gain of the detector.

DETECTOR DESIGNS BASED ON THE RANDOM LIMIT

Although at first it seemed almost impossible that a physical phenomena could exist which would implement random (non instantaneous) multiplication, the first theoretical implementation of it proposed by Jindal[20] turned out to be extremely simple. In fact the device structure illustrated in

Fig. 2 can be easily operated in the non instantaneous (random) mode. To achieve this mode of operation one would have to operate under the physical situation where the probability of ionization by both the electron and the hole approached unity. Then this multi step process of ionization would continue generating the non instantaneous response distributed in time. When operating this detector in the non instantaneous mode one would terminate this response after a fixed measurement time such as the time slot assigned to a bit in a fiber optic data stream. Assuming full shot noise associated with the incident optical signal, when operated in the classical (deterministic mode), this finite medium avalanching device provides noise performance 3 dB below the ideal limit. Since the electrons and holes ionize equally the situation is naturally compatible with materials used for implementing long wavelength detectors where the electron to hole ionization ratio is close to unity. Operated in the non instantaneous (random) mode, this device can bridge the last 3dB gap to approach the ideal limit of noise free amplification. Excellent noise performance coupled with this dual mode of operation makes this device specially attractive for light wave system applications.

It was later realized that this non instantaneous mode of operation had very general applicability and could be applied to conventional devices[21] and hence improve their performance as well.

CONCLUSIONS

Error free conversion of signals from the optical to the electronic domain will continue to be an important focus for future device research. This will be true in spite of the emphasis on all optical processing because of electronic information handling at the destination. Fluctuations will therefore continue to be among one of the most important aspects in future device, technology and systems research. The pull from current and future applications involving high performance systems will continue to keep this field active for the foreseeable future.

REFERENCES

1. R.P. Jindal, IEEE EDL, 8 315 (1987).
2. W. Shockley and J. R. Pierce, Proc. IRE, 26 321 (1938).
3. B. E. A. Saleh and M. Teich, Proc. IEEE, 70 229 (1982).
4. K. M. van Vliet, L. M. Rucker, IEEE TED, 26 46 (1979).
5. K. M. van Vliet, A. Friedmann and L. M. Rucker, IEEE TED, 26 752 (1979).
6. A. S. Tager, Sov. Phys. Solid State, 6 1919 (1965).
7. R. J. McIntyre, IEEE TED, 13 164 (1966).
8. S D. Personick, BSTJ, 50 167 (1971).
9. W. Lukaszek, A. van der Ziel and E. R. Chenette, SSE 19 57 (1976).
10. M. Teich, K. Matsuo and B. E. A. Saleh, IEEE JQE 22 (1986).
11. J. N. Hollenhorst, IEEE TED 37 3333 (1990)
12. R. Chin, N. Holonyak, G. E. Stillman, J. Y. Tang and K. Hess, Electron. Lett., 16 467 (1980).
13. G. F. Williams, F. Capasso and W. T. Tsang, IEEE-EDL 3 71 (1982).
14. R. P. Jindal, 42nd Device Research Conference, Santa Barbara (1984).
15. R. P. Jindal , IEEE TED 32 1047 (1985).
16. R. P. Jindal, IEEE-TED 34 301 (1987).
17. J. N. Hollenhorst, APL, 49 516 (1986).
18. R. P. Jindal, JAP 63 2824 (1988).
19. R. P. Jindal, JAP 64, 6845 (1988).
20. R. P. Jindal, , IEEE EDL 10 49 (1989).
21. R. P. Jindal, JAP 68 324 (1990).

Noise Characterization of Novel Quantum Well Infrared Photodetectors

by

D. Wang, Y. H. Wang, G. Bosman, and S. S. Li
Department of Electrical Engineering
University of Florida
Gainesville, FL 32611

ABSTRACT

Dark current noise measurements between 10 and 10^5 Hz were carried out on four different types of III-V quantum well infrared photodetectors designed for 8 - 12 μm IR detection at T = 77 K. At frequencies between 10^2 and 10^4 Hz noise plateau levels stemming from the trapping and detrapping of electrons in the quantum wells were observed in devices with superlattice barriers. From this data the bias dependent noise gain and electron trapping probability were calculated. Devices with bulk barriers only showed frequency dependent excess noise, most likely associated with interface state trapping, in our spectral window.

INTRODUCTION

Methods to detect infrared (IR) radiation have been studied ever since its discovery by William Herschel in 1800.[1] Among the various kinds of infrared detectors, those responding to radiation in the wavelength regions of 1-3 μm, 3-5 μm and 8-14 μm (the so-called atmospheric windows) receive most attention. Detectors operating in the wavelength region of 8-14 μm are especially important for imaging since the temperature of the human body and environments suitable for human life are around 300 K resulting in a peak wavelength of 10 μm in their radiation spectrum. Various device structures have been designed and fabricated, such as photoconductive detectors, metal-insulator-semiconductor (MIS) detectors and silicide Schottky barrier detectors. A poor quantum efficiency for silicide Schottky barrier detectors is the main restriction for its usage. MIS detectors are often limited by surface states which can exist in the thin insulating layer or at the semiconductor insulator interface.[2] In terms of semiconductor material, selectively doped silicon or germanium may be used to fabricate detectors. However, because the photon capture process involves an impurity center, even for heavily doped devices, the absorption coefficient is relatively low.The ternary compound

$Hg_{1-x}Cd_xTe$ (MCT) is another common material choice. Although MCT detectors have the highest performance among IR detectors in terms of detectivity(D^*), some shortcomings, such as mechanical softness and sensitivity to elevated temperatures, are fundamental.
After West and Eglash observed infrared absorption resulting from an intersubband transition in a GaAs/AlGaAs quantum well structure[3], quantum well infrared detectors (QWIPs) have become increasingly popular[(4-8)]. Despite its large dark current in comparison with a MCT detector, a QWIP has higher production yield, lower fabrication cost, fewer material defects, energy band selectivity and predictable spectral response. All these advantages may make QWIPs good candidates for long wavelength infrared detection.

DEVICE DESCRIPTION AND EXPERIMENTAL SETUP

The energy band diagram of a step-bound-to-miniband (SBTM) QWIP, sample C, is presented in figure 1.

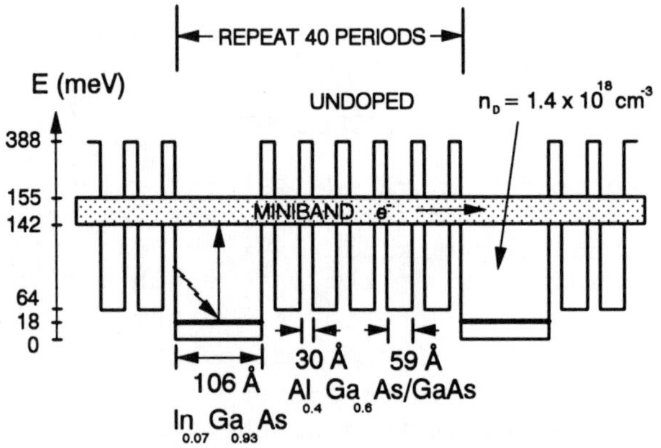

Fig. 1. Energy band diagram of a step-bound-to-miniband QWIP.

Note that the InGaAs well is doped, the superlattice barrier is undoped and that due to the specific material choice the bottom of the superlattice is a step up from the bottom of the InGaAs well. The superlattice barrier introduces a miniband at the energy levels indicated. The other device structures which we investigated are bound-to-miniband (BTM) transition QWIPs without a build-in step (samples A and B). Device structure D employs a bulk AlGaAs barrier and therefore does not have a miniband. The photon induced electron transition is described as bound-to-continuum (BTC). Calculations of the energy states were carried out by using a multilayer transfer matrix method.[9] In order to determine accurately the intersubband transition energies, the effects of nonparabolicity[10], electron-electron interaction (exchange energy) E_{xch} and depolarization effects E_{de} were taken into account.(11,12) The results of these calculations and the material parameters are listed in table I.

Table I QWIP Device parameters

Sample	A	B	C	D
SL barrier	$In_xAl_{1-x}As$	$Al_xGa_{1-x}As$	$Al_xGa_{1-x}As$	-
x	0.52	0.32	0.40	
Width(A)	35	78	30	
SL well	$In_xGa_{1-x}As$	GaAs	GaAs	-
x	0.53			
Width(A)	50	26	59	
L_B (A)	460	598	478	875*
QW	$In_xGa_{1-x}As$	GaAs	$In_xGa_{1-x}As$	GaAs
x	0.53		0.07	
Width(A)	110	85	106	110
N_D ($10^{18}/cm^3$)	0.5	0.4	1.4	5.0
Periods	20	30	40	40
Substrate	InP	GaAs	GaAs	GaAs
ΔE_C (meV)	500	260	388	190
Step (meV)	0	0	64	-
E_{BS} (meV)	37	35	18	17,85
E_{MB} (meV)	167-201	166-174	142-155	-
Transition	BTM	BTM	SBTM	BTC
Mesa (μm^2)	1.92 10^5	4 10^4	4.9 10^4	4 10^4

The symbol L_B represents the total superlattice or bulk barrier width between two adjacent photonic active quantum wells (QW), and ΔE_C, E_{BS}, and E_{MB} stand for the conduction band offset in the quantum wells, the bound state energy level, and the miniband energy position ,respectively. Other symbols have their usual meaning.
A standard low noise receiver[13] consisting of a Brookdeal 5004 low noise amplifier and a HP 3561A dynamic signal analyzer was used to measure the dark current noise between 10 and 10^5 Hz of the QWIPs. The QWIPs were submerged into liquid nitrogen during the measurements. Current-voltage characteristics were measured using a HP 4145B semiconductor parameter analyzer.

EXPERIMENTAL RESULTS AND DISCUSSION

The dark current and differential resistance of our samples were determined as a function of bias voltage at T=77 K. The differential resistance was obtained by differentiating the measured current-voltage characteristic. Due to the relatively high differential resistance, especially at low bias, low bias noise

measurements may fall below the detection limit of our setup for some devices.
Typical current noise density spectra of sample C are plotted in figure 2.

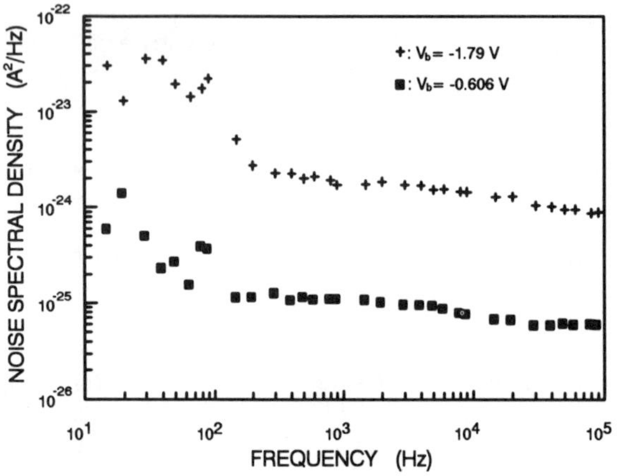

Fig. 2. Current noise spectral density of sample C at T = 77 K

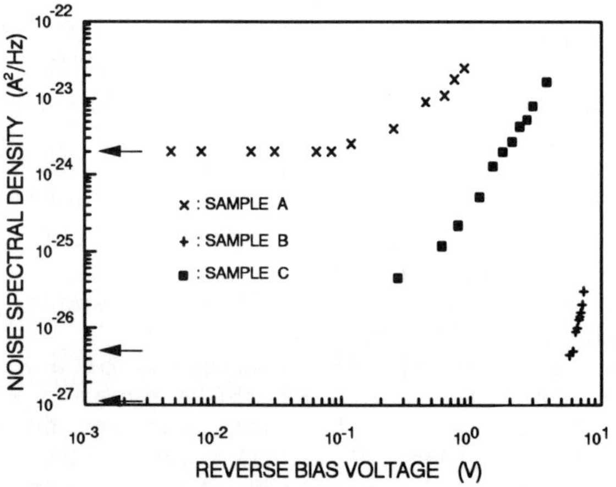

Fig. 3. Current noise spectral density as a function of reverse bias voltage measured at T = 77 K.

The strong signal at low frequencies is attributed to pick up. At midrange the noise spectral density is affected by RC parasitic effects, but after corrections were made, frequency independent noise levels result. These noise plateaus are attributed to electron trapping and detrapping in the quantum wells. Samples A and B show similar behavior, whereas sample D shows frequency dependent excess noise over the entire frequency span. The latter is supposedly due to electron trapping in interface or bulk barrier trapping states. In figure 3 the noise plateau levels are plotted as a function of reverse bias voltage. The arrows indicate current noise levels calculated using the Nyquist expression $S_i = 4\ k\ T\ dI/dV$. Note that sample A displays thermal noise up to 10^{-1} V after which g-r noise becomes dominant. As explained above, the low bias noise data of samples B and C could not be measured due to their large dynamic resistance, but seem to approach Nyquist values asymptotically. For the bias dependent current noise component it holds that[15] $S_i = 4\ q\ I\ g$, where g is called the noise gain and is about equal to the ratio of electron lifetime over electron transit time, i.e.,τ_0/τ_d. The noise gain of samples A, B, and C is calculated and plotted in figure 4.

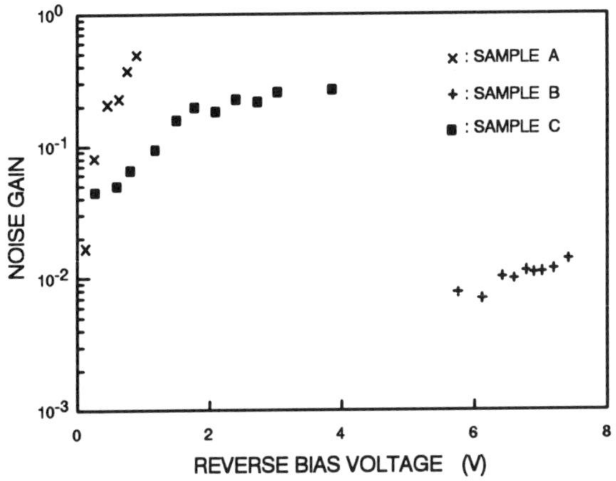

Fig. 4. Noise gain versus reverse bias voltage.

Note that for samples A and B the gain increases with increasing bias voltage whereas in sample C saturation sets in. Liu [14] expressed the noise gain in terms of the electron trapping probability p and the number of quantum wells N. We use from his paper $g = (1-p)/(N\ p)$ which Liu refers to as optical gain. His noise gain expression however, results in non-physical values of p larger than 1. With the help of this equation and the data presented in table I and figure 4 we are able to calculate p as a function of bias voltage. The results are presented in figure 5. The quantum well barrier lowering, due to the applied field, in the direction of electron emission increases (1-p), which is the probability for excited carriers to leave the well region. As a result p will have to decrease in accordance with the observations presented in this figure.

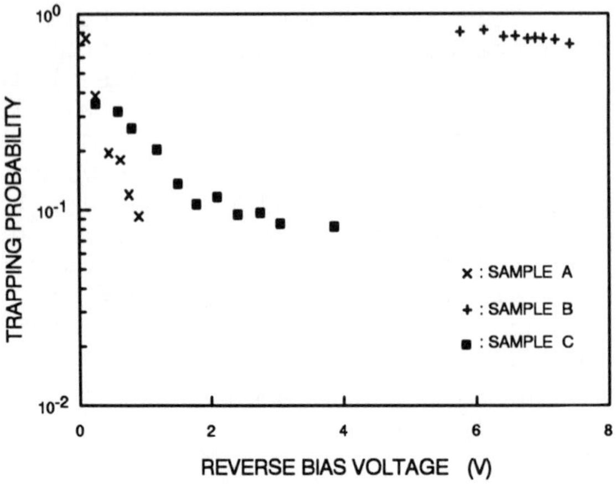

Fig. 5. Electron trapping probability versus reverse bias voltage.

REFERENCES

1. W. Herschel, Phil. Tran. R. Soc. 90, 284, 477 (1800).
2. M. A. Kinch, R. A. Chapman, A. Simmons, D. D. Buss, and S. R. Borrello, Infrared Physics. 20, 1 (1980)
3. L. C. West and S. J. Eglash, Appl. Phys. Lett. 46, 1156 (1985).
4. B. F. Levine, C. G. Bethea, G. Hasnain, V. O. Shen, E. Pelve, R. R. Abbott, and S. J. Hsieh, Appl. Phys. Lett. 56, 851 (1990).
5. B. F. Levine, C. G. Bethea, G. Hasnain, J. Walker, and R. J. Malik, Appl. Phys. Lett. 53 , 296 (1988).
6. Larry S. Yu, Y. H. Wang, Sheng S. Li, and Pin Ho, Appl. Phys. Lett. 60, 992 (1992).
7. Larry S. Yu, Sheng S. Li, and Pin Ho, Appl. Phys. Lett. 59, 2712 (1991).
8. Y. H. Wang, Sheng S. Li, Appl. Phys, Lett. 62, 621 (1993).
9. A. K. Ghatak, K. Thyagarajan, and M. R. Shenoy, IEEE J. Quantum Electron. QE-24, 1524 (1988)
10. A. Raymond, J. L. Robert, and C. Bernard, J. Phys. C 12, 2289 (1979).
11. K. M. S. V. Bandara, D. D. Coon, and O. Byungsung, Appl. Phys. Lett. 53, 1931 (1988).
12. M. Ramsteiner, J. D. Ralston, P. Koidl, B. Dischler, H. Ennen, J. Appl. Phys. 67, 3900 (1990).
13. J. S. Parker and G. Bosman, IEEE Tran. Electron Devices, 39, 1282 (1992).
14. H. C. Liu, Appl. Phys. Lett. 61 2703 (1992).
15. Aldert van der Ziel, Noise in Solid State Devices and Circuits (Wiley-Interscience, New York, 1986), p. 18-23.

DIFFUSION CURRENT INDUCED 1/f NOISE IN HgCdTe PHOTODIODES

INVITED PAPER

G. M. Williams, E. R. Blazejewski, and R. E. DeWames
Rockwell International Science Center
1049 Camino Dos Rios, Thousand Oaks, CA 91360, USA

ABSTRACT

Experiments indicate that excess low frequency noise (1/f noise) is related to current components in semiconductor diodes. Our measurements on HgCdTe photodiodes support this interpretation. We find that for the diffusion current component, the 1/f noise current power spectrum is very nearly proportional to the current squared. This disagrees with the predictions of models for 1/f noise in diodes which start from Hooge's relation for 1/f noise in conductors. Although Hooge's relation also employs a current squared dependence, association of the minority carrier density with N (the number of current carriers in the sample) in Hooge's relation leads to a predicted linear relation between noise power and current in diodes. This predicted linear relationship is not obtained in our experiments.

INTRODUCTION

Excess low frequency noise is a poorly understood but widely observed phenomenon.[1] Excess low frequency noise is usually called 1/f noise, since the noise power spectrum often varies approximately as inverse frequency. This paper is concerned with 1/f noise in diodes, although the phenomenon has been observed in a wide range of physical systems. 1/f noise in semiconductor photodiodes is a topic of practical interest, since it is detrimental to device applications.

Frequency independent (white) noise sources in diodes are better understood than 1/f noise sources. White noise sources can often be related by theory to the mean value of current components, based on noise theory and the physics responsible for the respective diode current components.[2] White noise current sources in diodes are often closely related to shot noise (i. e., one-half to full shot noise) of the independent current components responsible for total diode current, so that the dominant current component is usually the one responsible for measured white noise. This result greatly facilitates the reconciliation of theory and experiment. However, this is not the situation which pertains to 1/f noise in HgCdTe diodes; for example, the 1/f noise generated by diffusion current was not observed by early workers in the field, even though diffusion current could be made to be the dominant current component.[3-6]

This paper reports experimental results obtained on diffusion current induced 1/f noise in infrared sensitive photovoltaic HgCdTe detectors, and discusses them in terms of 1/f noise theories as applied to p/n junctions. Diffusion current generated 1/f noise was first reported in HgCdTe diodes in 1985,[7] and has also been reported in InSb[8] and InGaAs/InP[9] diodes. To avoid the possible complications of interference from other current components, we have investigated photocurrent induced 1/f noise. Photocurrent results from a carrier diffusion process and can be observed at zero diode bias, allowing separation from other current components which are induced by the application of bias voltage.

Although there is no general physical model which accounts for 1/f noise in electronic devices, Hooge[10] proposed an empirical relation to account for 1/f noise in resistors

$$S_I = \frac{\alpha_H I^2}{fN} \tag{1}$$

where S_I is the noise power, a_H is a constant parameter, I is the current, f is the frequency, and N is the number of carriers in the sample. Hooge's relation was deduced from measurements on homogenous samples; however, it has been proposed that extension of this relation to samples in which the carrier density is spatially varying may be useful in accounting for 1/f noise in diodes.[11-14] We will see, however, that as currently formulated, these extensions result in predictions which are not in accord with our experiments.

EXPERIMENT

We have investigated 1/f noise in HgCdTe diodes produced by a variety of material growth and device fabrication techniques.[15] Measurements are performed with the diodes held at low temperatures in cryogenic dewars. An aperture in the detector cold shield is provided to allow illumination. For photocurrent dependent measurements, sufficient rangeof current is obtained from illumination by room temperature objects and variable field of view in the case of LWIR (10 - 12 μm cutoff wavelength) diodes; however, MWIR (3 - 5 μm cutoff wavelength) diodes may need to be illuminated by a heated blackbody, for instance. 1/f noise in the source of illumination is a concern which may be addressed by noting correlation in noise generated in simultaneously illuminated detectors.

The detectors are connected directly to a room temperature transimpedance amplifier (TIA) for noise measurements. The output of the TIA is coupled to a buffer voltage amplifier which provides appropriate signal levels to a spectrum analyzer. The spectrum analyzer produces the resulting noise power spectral density. This system allows direct measurement of current fluctuations, as opposed to measuring voltage fluctuations across a resistor in series with the diode. The usual frequency range measured is 0.25 to 100 Hz. At the highest useful TIA gain of 10^{-10} amps/volt, the input referred measurement system noise is about 4×10^{-30} amps^2/Hz within this range of frequencies, if we neglect peaks due to 60 Hz line or dewar microphonic resonance frequency interference. The cryogenic dewar is placed on a vibration isolation table to minimize microphonic interference.

RESULTS

1/f noise in HgCdTe diodes appears to be related to current components, and is not directly related to total terminal current or applied bias voltage. This is inferred based on the results of the following experiment, which was also performed by Tobin.[3] We show in Fig. 1 the current-voltage (I-V) and differential resistance-voltage characteristics of an LWIR HgCdTe photodiode at 78.5K in the dark (full cold shield) and under illumination. Measurements of noise were made at the four points labeled A-D in Fig. 1; the results are tabulated as follows:

Point in Fig. 1	Bias Voltage (mV)	Current (amps)	Noise Power at 1 Hz (amps^2/Hz)
A	0	0	1.0×10^{-26}
B	0	3.28×10^{-7}	7.6×10^{-22}
C	-17	2.20×10^{-8}	1.1×10^{-22}
D	-17	3.56×10^{-7}	8.6×10^{-22}

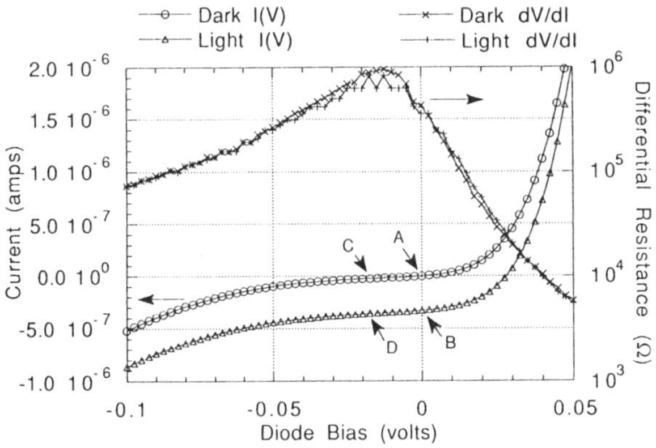

Fig. 1. Current and differential resistance vs. applied bias for an LWIR HgCdTe photodiode in the dark and under illumination. Diode temperature is 78.5K; junction area is 2.5×10^{-5} cm^2.

As is expected for an unilluminated diode at zero bias, the noise at point A agrees with Johnson noise calculations based on the junction resistance. For this case only, the noise spectrum is frequency independent at 1 Hz; all other noise power table entries refer to 1/f noise magnitudes. The increase in 1/f noise at zero bias with photocurrent observed at point B in comparison to point A immediately suggests an association between diode 1/f noise and current, while discouraging attempts to account for noise in terms of applied voltage. Comparison of the data at point C, reverse bias case without photocurrent, to the data at point B shows that the dark current induced 1/f noise is disproportionately large compared to the photocurrent induced 1/f noise, if we consider the magnitudes of the respective currents involved. Since we will later show that photocurrent induced 1/f noise is proportional to the square of the photocurrent, we deduce that 1.25×10^{-7} amps of photocurrent would give the same 1/f noise as the 2.2×10^{-8} amps of bias induced current, so the dark current is much more effective in producing 1/f noise. This comparison of the data at point B and C then further supports the idea that 1/f noise in diodes is related to current components, and also suggests that different current components generate 1/f noise with unequal efficiency. The last entry in the table at point D, corresponding to combined bias and photocurrent, shows that the noise power is the sum of the individual noise powers from points B and C, demonstrating that current sources act independently (no correlation) in producing diode 1/f noise. This result is completely consistent with the concept that diode 1/f noise is driven by independent current sources, each of which has its own particular 1/f noise generation mechanism. We note that the relationship between 1/f noise and diode current components is then analogous to the description of white noise in diodes, which is also associated with independent current components,[2] as mentioned in the introduction. Of course, the physics relating 1/f noise and current components is quite different from the physics relating white noise to its current components.

Contrary to some predictions,[16] 1/f noise is present when a photodiode is operated at the open circuit voltage. In Fig. 2 we show a plot of 1/f noise at 1 Hz vs. applied bias for a diode under illumination and in the dark. Under illumination, the data shows no tendency for 1/f noise to disappear as the forward bias approaches the open circuit voltage of +17.5 mV, where the net total current is zero. This result is consistent with the association of diode 1/f noise with current components, and not the total junction current. At small bias while under illumination, the 1/f noise is dominated by photoinduced 1/f noise and is independent of applied voltage. At large bias magnitude, dark current induced 1/f noise dominates, increasing faster with forward bias voltage because of the relatively large bias induced dark current.

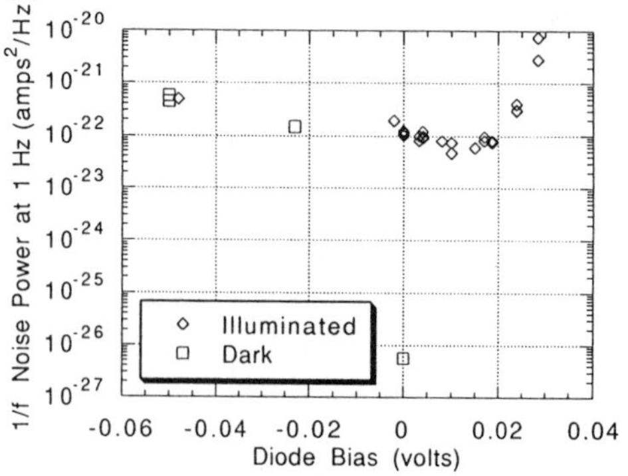

Fig. 2. Noise power at 1 Hz vs. applied bias for an LWIR HgCdTe photodiode in the dark and under illumination. Diode temperature is 78K; junction area is 2.5×10^{-5} cm^2. The open circuit voltage is 17.5 mV, and the photocurrent is 5.5×10^{-8} amps.

If we accept that diode 1/f noise is related to current components, the functional relationship between 1/f noise and these current components is of primary interest. In Fig. 3 we show a plot of 1/f noise power vs. photocurrent at zero bias for an MWIR and an LWIR HgCdTe diode. The data clearly suggests that the noise power is proportional to the square of the photocurrent, as the trend line in the figure indicates.

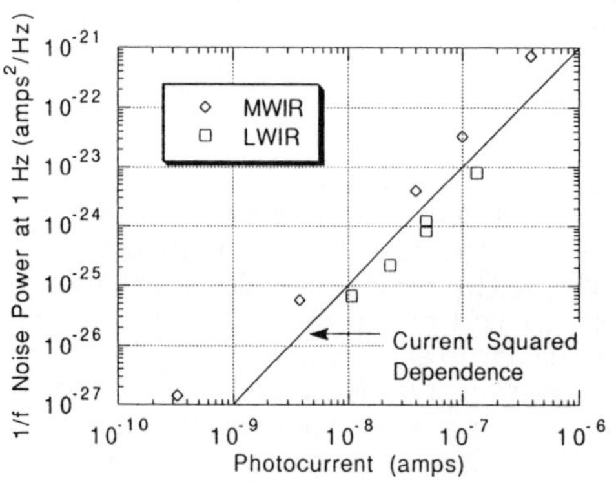

Fig. 3. 1/f noise power at 1 Hz vs. photocurrent for an LWIR and MWIR HgCdTe photodiode. Both diodes at 78K. LWIR junction area 2.5×10^{-5} cm^2, MWIR junction area 1.6×10^{-4} cm^2.

DISCUSSION

The inferred dependence of 1/f noise in diodes on current components can complicate attempts to relate measured diode 1/f noise to measured diode current if, as the data suggest, current components generate 1/f noise with substantially unequal efficiency. The problem is that applied junction bias voltage excites all current components, so it is reasonable to conclude that situations may arise where the dominant current component is not the dominant 1/f noise producing component. Consider that analysis of the I-V characteristic in Fig. 1 shows that the dominant dark current components are diffusion current and tunneling current. Some generation-recombination current must also be present. All current components have surface and bulk contributions, while tunneling currents also come in band-to-band and trap-assisted varieties. Because the application of bias voltage excites all current components through different functional relationships, analysis of the bias dependence of 1/f noise is a complex issue when the efficiency of 1/f noise generation by these current components is without theoretical or experimental basis.

This difficulty can be circumvented in one particular case - the study of photocurrent induced 1/f noise. Zero applied diode bias can be maintained so that no current component other than photoinduced diffusion current is present. A further advantage is that diffusion current sources in diodes have a sound theoretical basis. The absence of 1/f noise at zero bias and without photocurrent (at least at the magnitudes discussed in this paper as measured at 1 Hz) make this technique viable.

We now consider the significance of the observed photocurrent squared dependence of 1/f noise power spectral density. This empirical result is reminiscent of Hooge's empirical result, Eq. 1, for 1/f noise in conductors, which also shows a current squared dependence. However, predictions of the 1/f noise behavior of diodes which are based on Hooge's relationship[11-14] currently assume that the minority carrier density is to be associated with N (the number of carriers in Hooge's relationship). Since the minority carrier density is proportional to diffusion current in diodes, these theories predict a net linear relationship between 1/f noise power and current, since the direct proportionality of N to I cancels an I in the numerator of Eq. 1. So we must conclude that either these theories do not apply to our experiments, or they require modification.

The above theories may not apply in at least two ways. First, the measured 1/f noise may simply be of the Hooge type in the series resistance of the diode. However, this explanation is unlikely, since the currents are relatively small and the sample is large. Based on the measured noise power and current, and assuming nominal doping densities of 2×10^{15} cm^{-3} and relevant device volumes derived from a 10 μm layer thickness, 5 mm distance to contact, a 1 mm effective device width, Hooge parameters of 40 to 400 would be required to account for the data. These numbers are 5 to 6 orders of magnitude larger than those measured in HgCdTe.[17] In addition, we have measured photocurrent induced 1/f noise on the same diode using different ground contacts to vary the series resistance (by a factor of 1.23), and demonstrated that 1/f noise is unaffected. From Eq. 1, a change in 1/f noise power by this same factor should have been observed if the series resistance was responsible for the noise. The second case in which the theories may not apply is if the noise is of surface origin, which also has a current squared dependence.[18] We have shown, however, from the investigation of variable area detectors that the photocurrent induced 1/f noise is a bulk effect,[15] at least for HgCdTe active layers grown on non-lattice matched substrates. Therefore, we conclude that these two objections cannot be used to rule out the possibility that the proposed models apply to our experiment. The dependence of 1/f noise power on the square of diffusion current therefore suggests that if Hooge's relation is to be of use in describing 1/f noise in diodes, the factor N must be chosen such that it is independent of current. This is not the way in which present attempts to account for 1/f noise in diodes treat the factor N.

CONCLUSIONS

Our results support evidence which suggests that 1/f noise in HgCdTe diodes is driven by current components, and so is not necessarily directly related to the total diode current. We present a method of studying the diffusion current component without interference from other leakage current mechanisms, and find that diffusion current induced 1/f noise is proportional to the diffusion current squared.

These results demonstrate that theories for diodes which are based on Hooge's relationship for 1/f noise in homogenous conductors do not correctly account for the diffusion current dependence of induced 1/f noise, because they equate N (the number of current carriers in the sample in Hooge's relation) to diode minority carrier density. Although even in conductors clear identification of the physical significance of the factor N has remained problematic,[17] a factor like N must be included in the relationship between noise and current when noise power is proportional to the square of the current to insure that independent noise sources will add in quadrature.[15] Identification of the physical phenomena which relate to the factor N for diffusion current is key to developing a better understanding of this 1/f noise process.

ACKNOWLEDGMENT

We thank R. Zucca for a careful review of the manuscript, and W. E. Tennant for his support and encouragement.

REFERENCES

1. P. Dutta and P. M. Horn, Reviews of Modern Physics, **53**(3), 1981, p. 497.
2. M. J. Buckingham, *Noise in Electronic Devices and Systems*, Halsted Press, New York, 1983.
3. S. P. Tobin, S. Iwasa, and T. J. Tredwell, IEEE Trans. ED, **27**(1), 1980, p. 43.
4. W. A. Radford and C. E. Jones, J. Vac. Sci. Technol. A **3**(1), 1985, p. 183.
5. H. K. Chung, M. A. Rosenberg, and P. H. Zimmerman, J. Vac. Sci. Technol. A **3**(1), 1985, p. 189.
6. J. Bajaj, G. M. Williams, N. H. Sheng, M. Hinnrichs, D. T. Cheung, J. P. Rode, and W. E. Tennant, J. Vac. Sci. Technol. A **3**(1), 1985, p. 192.
7. R. E. DeWames, M. Hinnrichs, J. Bajaj, and G. M. Williams, Bulletin of the American Physical Society, **30**(6), June 1985, p. 1155.
8. N. B. Lukyanchikova, B. D. Solganik, and O. V. Kosogov, Solid State Electronics, **16**, 1973, p.1473.
9. L. He, Y. Lin, A. D. van Rheenen, A. van der Ziel, A. Young, and J. P. van der Ziel, J. Appl. Phys. **68**(10), 1990, p. 5200.
10. F. N. Hooge, Phys. Lett. A, **29**, 1969, p. 139.
11. T. G. M. Kleinpenning, Physica 98B, 1980, p. 289.
12. T. G. M. Kleinpenning, J. Vac. Sci. Technol. A, **3**(1), 1985, p. 176.
13. A. van der Ziel, P. H. Handel, X. L. Wu and J. B. Anderson, J. Vac. Sci. Technol. A, **4**(4), 1986, p. 2205.
14. X. L. Wu, J. B. Anderson, and A. van der Ziel, IEEE Trans. ED, **ED-34**(9), 1987, p. 1971.
15. G. M. Williams, R. E. DeWames, J. Bajaj and E. R. Blazejewski, accepted for publication, Journal of Electronic Materials, June, 1993.
16. A. van der Ziel, *Noise in Physical Systems and 1/f Noise - 1985*, A. D'Amico and P. Mazzetti (eds.), Elsevier Science Publishers B. V., 1986, p. 11.
17. F. N. Hooge, Physica B, **162**, 1990, p. 344.
18. W. W. Anderson and H. J. Hoffman, J. Vac. Sci. Technol. A, **1**(3), 1983, p. 1730.

1/f TRAPPING NOISE THEORY WITH UNIFORM DISTRIBUTION OF ENERGY-ACTIVATED TRAPS AND EXPERIMENTS IN GaAs/InP MESFETs BIASED FROM OHMIC TO SATURATION REGION

M. CHERTOUK, A. CHOVET*, A. CLEI

France Telecom/ CNET / Groupement Microelectronique
196 Avenue Henri Raverra BP 107 F-92225 Bagneux, France
* Lab. PCS (U.R.A CNRS 840), ENSERG, BP 257-38016 Grenoble Cedex France

ABSTRACT- The theory of the number-fluctuation 1/f noise due to a uniform distribution of energy-activated traps is presented for the first time with a clear experimental support. The expressions of the drain current noise are derived as a function of an effective density of traps at the channel-buffer interface, taking into account electrical and technological parameters of the GaAs MESFETs biased in the ohmic region as well as in the nonohmic region. In the ohmic region, the drain current noise is proportional to V_{DS}^2 and independent of gate bias, and correlation between the calculated effective density of traps and structural quality of GaAs layers has been observed. In the non-ohmic region, the drain current noise increases with V_{DS} and becomes constant in the saturation region, which is mainly related to the variations of the channel resistance with drain bias.

INTRODUCTION

The most generally accepted theory relates the 1/f noise in MOS transistors to fluctuations ΔN of the number of carriers N in the channel. These fluctuations are caused by tunneling of free-charge carriers into oxide traps close to the $Si\text{-}SiO_2$ interface [1]. Another model relates the 1/f noise to fluctuations of the mobility due to lattice scattering; this model, based on Hooge's empirical relation, corresponds to a bulk effect. The ΔN model is more often related to surface trapping effect. However, some difficulties exist to discriminate between number-fluctuation and mobility-fluctuation 1/f noise in GaAs MESFETs [2] because there is a lack of theoretical models describing the 1/f noise in GaAs MESFETs; moreover the observed noise strongly depends on substrate and active layer qualities.

In this paper, we present a theory for a number-fluctuation 1/f noise using a uniform distribution of energy-actived traps. Applications of this theoretical model to GaAs MESFETs grown on InP substrates are also reported. Notice that the aim of GaAs growth on InP substrates is to combine the GaAs electronics devices with the InP optoelectronics devices on the same InP wafer for OEICs, because InP-based FET technology is not as mature as GaAs technology. In view of integrated photoreceiver applications, low 1/f noise component is needed to get high sensivity receivers.

THEORY OF NUMBER-FLUCTUATION 1/f NOISE

A generation-recombination low-frequency noise for single level traps with one time constant is given by [3]:

$$S_N(\omega) = \frac{4\tau\overline{\Delta N^2}}{1+(\omega\tau)^2} \tag{1}$$

where $\overline{\Delta N^2}$ is the variance of the number N of free carriers (in the channel) and τ, the time constant of traps, may correspond to a thermally activated process. In order to calculate the spectral density of carrier number fluctuations $S_N(\omega)$ due to traps located at energy levels distributed between E_{T_1} and E_{T_2} (from the conduction band), a normalized energy distribution $g(E_T)$ is required verifying :

$$\int_{E_{T_1}}^{E_{T_2}} g(E_T)dE_T = 1 \tag{2}$$

The spectral density $S_N(\omega)$ is then given by [4]:

$$S_N(\omega) = \int_{E_{T_1}}^{E_{T_2}} g(E_T)4\overline{\Delta N^2}\frac{\tau}{1+(\omega\tau)^2}dE_T \tag{3}$$

Under the hypothesis that all these traps are equally participating to the noise, i.e $g(E_T)$ distribution is uniform, we have :

$$g(E_T)=\frac{1}{E_{T_2}-E_{T_1}} \qquad \textit{for} \quad E_{T_1}<E_T<E_{T_2}$$

$$g(E_T)=0 \qquad \textit{elsewhere}$$

For a thermally activated process, the trapping time constant is related to its activation energy E_T by [5]:

$$\tau=\tau_0\exp\left(\frac{E_T}{kT}\right) \tag{4}$$

Since $dE_T = kT\,(d\tau/\tau)$, it follows after integrating, for $\frac{1}{2\pi\tau_2}\ll f\ll\frac{1}{2\pi\tau_1}$:

$$S_N(f)=\frac{kT}{E_{T_2}-E_{T_1}}\frac{\overline{\Delta N^2}}{f} \tag{5}$$

For a homogeneous device the fluctuations of the total number of carriers N is related to resistance R and current I fluctuations by:

$$\frac{S_N(f)}{N^2}=\frac{S_R(f)}{R^2}=\frac{S_I(f)}{I^2}=\frac{\gamma}{fN^2} \tag{6}$$

where γ is a 1/f noise parameter related to the variance of the carrier number:

$$\gamma=\frac{kT}{E_{T_2}-E_{T_1}}\overline{\Delta N^2} \tag{7}$$

Assuming a binomial distribution for the trap occupancy, the maximum noise contribution corresponds to the maximum variance, i.e:

$$\overline{\Delta N^2}=\overline{\Delta(N-N_0)^2}=\frac{N_T\Delta V}{4}$$

where $N_T(m^{-3})$ is the effective density of traps which contribute to the noise, and ΔV the volume in which the traps are efficient. Defects in GaAs and their effect on noise performance have been extensively observed and reported in the literature. A variety of these defects are present at the interface between the substrate and channel[6] and experimental results have confirmed the fact that the buffer traps can generate 1/f low-frequency noise [7]. DLTS experiments have also shown that in GaAs material a few electron traps (EL) exist with activation energy in the 0.2 - 0.8 eV range [8] below the conduction band.

From a band diagram of the channel-buffer layer junction affected by traps, the volume where the traps are efficient can be written: $\Delta V = WL_G(\lambda_2-\lambda_1)$, where W and L_G are the width and length of the channel, λ_1 and λ_2 correspond to the distance over which $E_C(x)-E_F\le E_{T_1}$ or E_{T_2} respectively in the depletion region of the channel-buffer junction; we have:

$$E_{T_1}-E_C=\frac{q^2N_B}{2\varepsilon}\lambda_1^2 \qquad E_{T_2}-E_C=\frac{q^2N_B}{2\varepsilon}\lambda_2^2 \tag{8}$$

where N_B is the (low) impurity concentration in the buffer layer; from (7), (8) and expression of $\overline{\Delta N^2}$, we get:

$$\gamma=\frac{\sqrt{kT}}{\sqrt{E_{T_2}}+\sqrt{E_{T_1}}}\frac{N_TWL_G\lambda_D}{4} \tag{9}$$

with λ_D the Debye length in the buffer layer ($\lambda_D=\sqrt{2\varepsilon kT/q^2N_B}$).

In a MESFET biased in the ohmic region, the total number of carriers in a n-channel is given by: $N=L_G^2/q\mu_nR$, with R the channel resistance and μ_n the electron mobility. If the source and drain resistances are negligible with regard to the channel resistance (which is generally true in a recessed MESFET), $V_{DS}=R.I_D$ and it follows from (6) and (9):

$$S_{I_D}(f) = \frac{N_T W \lambda_D \sqrt{kT}}{4\left(\sqrt{E_{T1}}+\sqrt{E_{T2}}\right)} \frac{(q\mu_n V_{DS})^2}{L_G^3} \frac{1}{f} \tag{10}$$

Then, the spectral density of drain current fluctuations is proportional to μ_n^2; note that this result has already been experimentally observed for epitaxial and implanted GaAs MESFETs, where a bulk contribution to the 1/f noise is described [9].

For undoped GaAs buffer, the residual impurity concentration [10] is about 2 10^{14} cm^{-3}; the Debye length in the buffer is then λ_D= 0.43 μm. Therefore, the measurements of the drain current fluctuations (S_{I_D}) at a given frequency of the 1/f noise in the ohmic region of a MESFET allow to estimate the effective trap density N_T using eq(10).

It is worth noting that, when Hooge relation is used to describe the 1/f noise, eq(6) is written under the form:

$$\frac{S_{ID}}{I_D^2} = \frac{\alpha_H}{Nf} \qquad i.e \qquad \alpha_H = \frac{\gamma}{N}$$

Then the corresponding Hooge parameter for the 1/f noise analysed here, using Eq(9) and writing $R = R_0/(1-\sqrt{(V_{BI}-V_{GS})/V_P})$, with R_0 the open channel resistance, is given by:

$$\alpha_H = \frac{N_T \lambda_D}{4 N_D a} \frac{\sqrt{kT}}{\sqrt{E_{T1}}+\sqrt{E_{T2}}} \frac{1}{(1-\sqrt{(V_{BI}-V_{GS})/V_P})} \tag{11}$$

where a is the active layer thickness and N_D the carrier density in the channel. It appears that the Hooge parameter will dependent on the quality of the GaAs layers (trap density) and technological parameters of the device (doping level); it is independent of the length and width of the channel. It also appears that a low value of Hooge parameter corresponds to high layer thickness and large doping level in the channel. Several authors have presented measured values of the Hooge parameter for GaAs and have found α_H values in the 10^{-6}-10^{-3} range [11]. For classical MESFETs with 1μm gate length, the channel doping is generally $2\times10^{17}cm^{-3}$ so that for a layer thickness of 0.12 μm, V_P=2V.

Using eq(11) to calculate the Hooge parameter at V_{GS}=0V, we find that an effective density of traps of 10^{15} cm^{-3} gives $\alpha_H=2\times10^{-3}$ while $N_T=10^{12}cm^{-3}$ gives $\alpha_H=2\times10^{-6}$.

In the linear region, when the drain voltage is small ($V_{DS} \ll V_P+V_{GS}-V_{BI}$), the number of carriers is independent of V_{DS}. Increasing the drain bias produces a non-uniformity of the depletion layer along the channel, then a variation of the number of carriers in the sample. We divide the channel of the MESFET into elementary volumes containing dN carriers and through which a constant current I_D is passed. With h the depletion layer at a distance x from the source (fig.1), the total number of carriers in the channel (between source and drain) can be written:

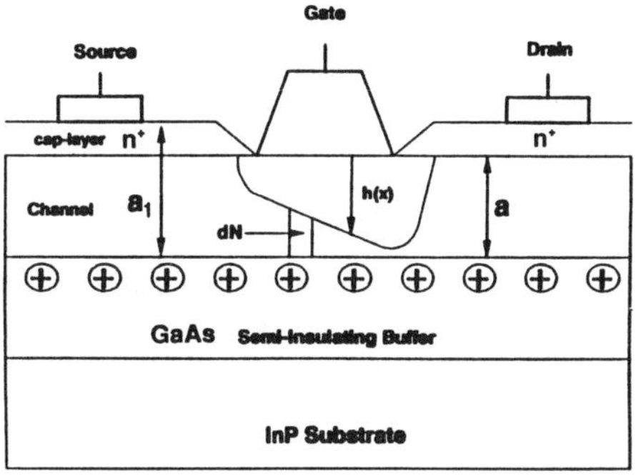

Fig.1: Cross Section of our GaAs MESFET on InP substrate

$$N(V_{DS}, V_{GS}) = N_D W \int_0^{L_G} (a-h)dx \tag{12}$$

Using $dV(x)= I_D\, dx / (q\mu_n N_D(a-h)W)$ and, from Poisson equation, $V_{BI} - V_{GS} + V(x) = q N_D h^2/2\varepsilon$, we have $dV(x)=qN_D h dh/\varepsilon$, where ε is GaAs permittivity. Putting:

$$Z = \frac{h}{a} = \sqrt{\frac{(V_{BI}-V_{GS}+V)}{V_P}} \qquad U = \frac{Z_D}{Z_S} = \sqrt{\frac{V_{DS}+V_{BI}-V_{GS}}{V_{BI}-V_{GS}}}$$

and integrating eq(12), we obtain the following eq(13):

$$N(V_{DS}, V_{GS}) = \frac{2L_G^2}{I_D R_0^2} \frac{(V_{BI}-V_{GS})}{q\mu_n} \left(\frac{1}{2}(U^2-1) - \frac{2}{3}Z_S(U^3-1) + \frac{1}{4}Z_S^2(U^4-1) \right) \tag{13}$$

where Z_S is the reduced depletion layer of the Schottky gate at the source. When calculating drain current for different drain and gate biases we use the expressions given by Sze [12].

Thus, using these last equations, we can calculate the total number of carriers in the channel and the drain current for different biases V_{DS} and V_{GS}, and get a first order approximation of the spectral density of current drain fluctuations from eq(6), knowing the parameter γ or the effective density of traps.

Experimental Results and Discussion

The low-frequency noise measurements (10 Hz to 100 kHz) performed on GaAs MESFETs grown on InP substrates show 1/f spectra over the whole frequency range. It has been demonstrated experimentally that the 1/f noise corresponds to fluctuations of the number of carriers in the channel, related to trapping phenomena due to traps located in the depletion region of the buffer-channel interface [13].

The drain current noise spectral density was measured using a low-noise amplifier and a digital spectrum analyser. Experiments were performed at room temperature on MESFETs with different thicknesses of undoped buffer layers. The technological parameters of MESFETs used in this work are given in table (1); we took E_{T1}=0.2eV and E_{T2}=0.8eV. Devices gate length and width were respectively 1μm and 50 μm. The doping level is extracted from C(V) measurements and the channel thickness was derived by fitting the resistance of the channel for different gate biases. The drift mobility were deduced from Hall mobility measurements in these layers ($\mu_n=0.85\ \mu_H$).

Buffer Thickness	$N_D(cm^{-3})$	a(μm)	$\mu_n(cm^2/v.s)$	$N_T(cm^{-3})$	FWHM (Arcsec)
1 μm	$1.4\ 10^{17}$	0.150	2800	$1.38\ 10^{14}$	370
2 μm	$1.5\ 10^{17}$	0.152	3100	$7.00\ 10^{13}$	230
3 μm	$1.5\ 10^{17}$	0.122	3600	$3.25\ 10^{13}$	195
5 μm	$1.8\ 10^{17}$	0.118	3700	$2.40\ 10^{13}$	185

Table 1. Parameters used in this model for the devices under investigation together with the corresponding N_T and FWHM (DDX) values deduced from measurements

In the ohmic region, the measurements of drain current noise at f=1kHz and V_{DS}=20mV for different gate biases and several buffer thicknesses is plotted in fig.2. It is observed that S_{I_D} is indenpedent of V_{GS}, and the increase of buffer thickness produces a decrease of 1/f noise magnitude; this has to be related to the improvements of the structural quality of GaAs channel layers with increasing buffer thickness. By fitting this results with eq(10) and using parameters of MESFETs (mobility, length and width of the gate), we derived the effective density of traps also given in table.1. As shown in fig.3 the increase of the buffer thickness is related to a decrease of trap density, which appears to be nearly unchanged for thickness above 3μm. The difference in lattice constants of 3.8% between GaAs and InP is accommodated by the formation of high linear density of misfit dislocations at the heterointerface [15]. It has already been demonstrated by the measurements of the full width at half maximum (FWHM) of the X-ray double diffraction that an increase of the buffer layer thickness produces a significant improvement of structural quality of GaAs layers and remains constant for thickness above 4μm [16] as shown in the insert of fig.3. We can conclude that, this results show a good correlation between the structural quality of GaAs layers and density of traps participating to the 1/f noise.

From the values of channel resistance and drain current noise as functions of gate bias, we can calculate the experimental variation of the Hooge parameter with gate bias in the ohmic region for 1, 2 and 3μm buffer thicknesses. The Hooge parameter decreases with gate bias (fig.4); this behaviour can be explained by the increasing number of carriers in the channel with increasing gate bias. The experimental results are very well interpreted by the above given theory, eq(11), which proves that for devices under investigation, we are actually dealing with ΔN fluctuations, so that the Hooge formulation is not physically well-adapted.

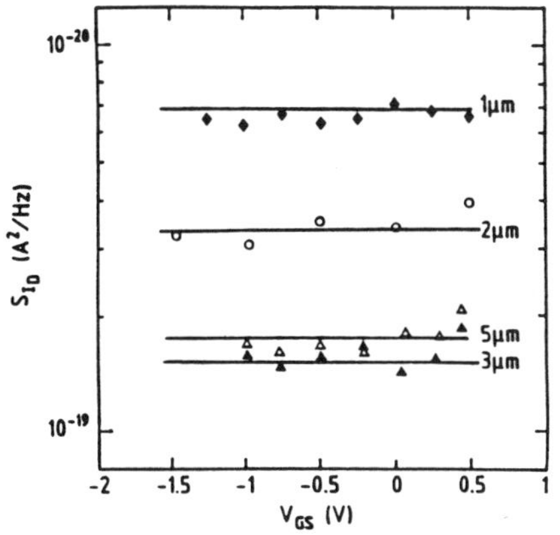

Fig.2: The drain current spectral density (at f=1kHz) vs the front gate voltage V_{GS}, at V_{DS}=20mV (ohmic region) for different GaAs undoped buffer layer thicknesses, grown on InP substrates.

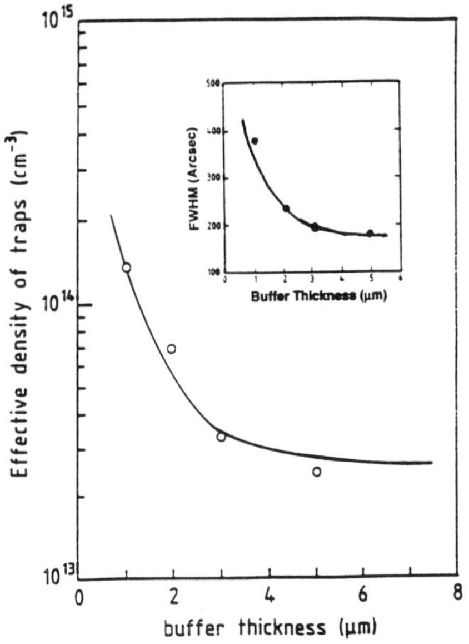

Fig.3: Effective trap density at the channel-buffer layer interface vs buffer thickness, as deduced from 1/f noise measurements (to be compared with the FWHM signal of X-ray Double Diffraction, in insert).

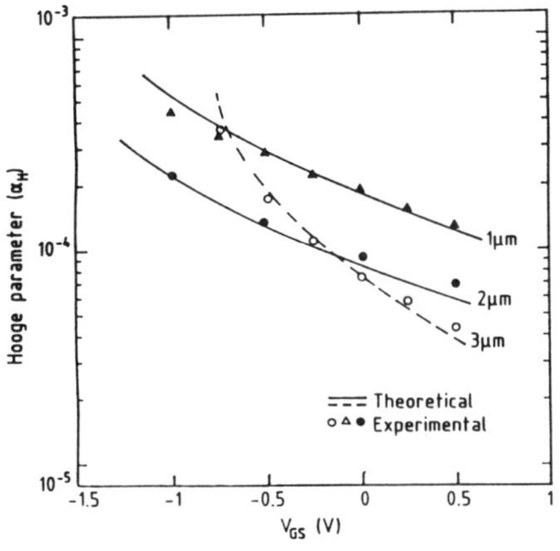

Fig.4: The Hooge parameter (considered as a noise index) as deduced from the experimental measurements and the presented theory eq(11), in the ohmic region, for different buffer thicknesses.

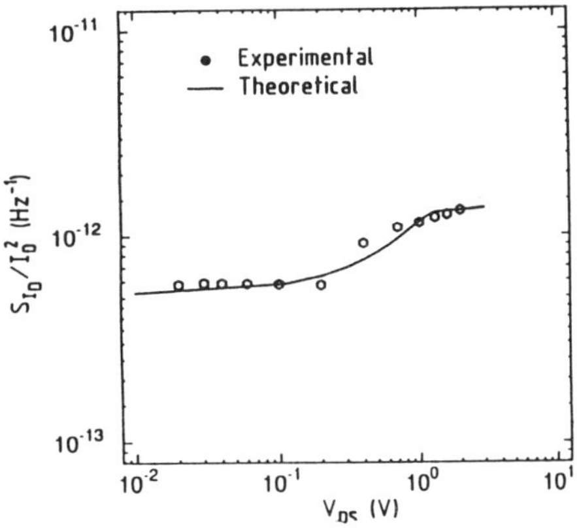

Fig.5: Experimental and theoretical variations of the relative drain current noise (at f=1kHz and V_{GS}=0V) vs drain bias; buffer layer thickness 1µm.

The dependence of drain current noise on drain bias has been also investigated experimentally. In sub-saturation and saturation regions a 1/f noise spectrum is observed. Figure 5 gives the experimental curves of the relative noise S_{I_D}/I_D^2 vs V_{DS} at f= 1kHz and V_{GS}= 0V for device with 1μm thick buffer layer. S_{I_D}/I_D^2, which is independent of V_{DS} in the ohmic region, then increases with V_{DS} in the sub-saturation region and becomes constant in the saturation region. The experimental results are rather well interpreted by the above given model (eq(13) and (6)) as shown in fig.4. In addition, this general model also fits very well the ohmic region (eq(10)).

In the ohmic region, the theoretical model is sufficient to determine the density of traps giving rise to the 1/f noise. In addition, the general model can also be used to describe the 1/f noise behaviour from the ohmic region to the saturation region. It appears that the measured values are fully consistent with results obtained from our model (fig.5). This confirms the validity of the 1/f trapping noise model in GaAs MESFETs.

CONCLUSION

A theory of 1/f trapping noise using uniform distribution of energy-activated traps has been developed for the first time. It gives proportionality between the 1/f noise magnitude and an effective density of traps at the channel-buffer interface. Application of this theoretical model to GaAs MESFETs has been successfully made for noise study taking into account thickness and doping level of the channel. This theory is consistent with the experimental influences of the main parameters: gate and drain biases, density of traps (through buffer thicknesses). The experimental and calculated results are in good agreement for the whole range of drain bias. This confirms the validity of the 1/f trapping noise theory in GaAs MESFETs.

ACKNOWLEDGMENT

The authors wish to thanks Dr. Scavennec for contunuous encouragement.

REFERENCES

1. G. Reimbold, IEEE Trans. Elec. Devices, vol. ED. 31, pp. 1190-1198, 1984.
2. C. H. Suh ,A. Van Der Ziel and R. P. Jindal, Solid-State Electronics, Vol. 24, pp. 717-718, 1981
3. A. Van Der Ziel, Noise in Solid State Devices and Circuits, J. Wiley, New York, p. 121, 1986
4. A. Van Der Ziel, Noise in Solid State Devices and Circuits,J. Wiley, New York,p. 126, 1986
5. Davis C. Look, Electrical Characterization of GaAs Materials and Devices, John Wiley & Sons, New York, Chap.4, p. 200. 1989.
6. T. W. Hickmott, IEEE Transactions Electron Devices, vol. ED.31, pp. 54-62. 1984.
7. P. Canfield, L. Forbes, IEEE Tranactions Electron Devices, vol. ED-33, pp.925-928, 1986.
8. G. M. Mitonneau, A. Mircea, Electronics Letters, vol. 13, pp. 191-194, 1977
9. P. A. Folkes, Appl. Phys. Lett. vol. 55, pp. 2217-2219, 1989.
10. John F. Wager, Angus J. Mc Camant, IEEE Transactions on Electron Devices, Vol. ED-34, pp. 1001-1007, 1987
11. Kuang Han Duh, A. Van Der Ziel, IEEE Transactions on Electron Devices, Vol. ED-32, pp. 662-666, 1985
12. S. M. Sze, Physics of Semiconductor Devices, 2nd Ed, J. Wiley an Sons, 1981, p.136.
13. M. Chertouk, A. Chovet, A. Clei, ESSDERC 92 Conf, Leuven (Belgium) 14-17 Sept. 1992, Proc. Microelectronic Engineering, vol. 19, pp. 755-758, 1992
14. Christopher Kocot and Charles A. Stolte, IEEE Transactions on Microwave Theory and Techniques, Vol. MTT-30, pp. 963-968, 1982
15. R. Azoulay, A. Clei, L. Dugrand,N. Draira, G. Leroux and S. Biblemont, Journal Crystal Growth, Vol.27, pp. 1003-1013. 1991.
16. M. Chertouk, A. Clei, R. Azoulay, G. Leroux, WOCSDICE San Rafael 92 (Spain) 24-27 may, 1992

DEPENDENCE OF PHOTOCONDUCTION NOISE ON TEMPERATURE: A CHECK OF BARRIER-TYPE THEORY OF PHOTOCONDUCTIVITY IN CdS BASED DEVICES

A. Carbone and P. Mazzetti

Dipartimento di Fisica del Politecnico di Torino
Corso Duca degli Abruzzi 24 - 10129 Torino (Italy)

ABSTRACT

A theory of current noise in photoconducting devices developed in a previous paper has been checked against measurements of photoconductance noise in a CdS based device as a function of temperature. It is shown that the theory reproduces the esperimental results without the introduction of free parameters in the low frequency range where the photo-induced component dominates the whole power spectrum.

INTRODUCTION

In previous papers[1,2] a theory of current noise in photoconducting CdS based devices has been developed on the basis of a barrier model of the photoconduction mechanism. According to this model the electrical conductance of the device is determined by foto-sensitive potential barriers whose height depends on the positive trapped chargc created by light[3,4]. In the present case it is assumed that the light sensitive barrier is localized in proximity of the metal contacts. The main result of this theory concerns the presence of a photoinduced noise component related to the spontaneous fluctuation of the height of this potential barrier, due to the fluctuation of the trapped charge produced by light. An interesting aspect of the theory is represented by the fact that it allows the evaluation of both the amplitude and shape of the power spectrum of the photoinduced noise component from experimental data concerning conductance and relaxation time measurements vs. light intensity and wavelength, without the introduction of adjustable parameters. Since at light intensity above 10^{11} photons s^{-1} cm^{-2} this component dominates the whole photocurrent noise spectrum in the low frequency range (below 1kHz typically), a comparison with the experimental results can be easily made.

In papers[1,2] theoretical and experimental noise power spectra were reported for various light intensities and wavelengths at room temperature. A fair to good agreement was found both for the shape and the intensity of the noise spectra. In the present paper a new check of the theory is made by comparing its results with the experimental ones taken at a lower temperature (-45°C), where a rather drastic change of the photocurrent relaxation time τ_d takes place. As in the case of the room temperature, τ_d is nearly independent of the light wavelength λ except when $\lambda = \lambda_c$, where an abrupt change of τ_d of almost one order of magnitude is observed. λ_c is the critical light wavelength, corresponding to a photon energy equal to the photoconductor energy gap. For CdS λ_c is equal to500nm , and thus experimental and theoretical results are reported for two values of λ slightly above and slightly below λ_c , i.e. 510nm and 490 nm.

PHOTOCONDUCTION MODEL AND NOISE POWER SPECTRUM

In this section we very briefly summarize the results of the theory developed in [1]. Reference is made to the relative photoconductance fluctuation power spectrum under steady illumination :

$$\psi_G(\omega) = \frac{1}{G^2}\left[g \cdot \Delta g \cdot \tau_g \frac{<|S(\omega)|^2>}{\tau_g^2} \cdot n_d + 2 \cdot (\Delta g)^2 \frac{|<S(\omega)>|^2}{\tau_g^2} n_d \cdot \Sigma_j \frac{a_j \tau_d^{(j)}}{1+\omega^2 \tau_d^{(j)2}} \right]. \quad (1)$$

The meaning of the symbols is the following:

- g is the contribution to the conductance G of the device of a single electron in the conduction band,
- Δg is the average increment of conductance related to the change of barrier height due to the excess ionization of a single deep donor center (or due to a trapped hole) during its lifetime τ_d,
- τ_g is the average lifetime of an electron in the conduction band of the photoconductor, related to trapping processes in shallow centers,
- n_d is the average number of the ionized deep donor centers or trapped holes in the illumination condition determining the conductance G,
- $a^{(j)}$ is the relative weight of the ionized centers of type j, whose lifetime is $\tau_d^{(j)}$, in the same illumination conditions.

Finally the quantities $<|S(\omega)|^2>$ and $|<S(\omega)>|^2$ represent respectively the average of the square modulus and the square modulus of the average of the Fourier transform of a square conductance pulse of unitary amplitude and duration $\tau_g^{(i)}$. The distribution of the $\tau_g^{(i)}$'s, which are the individual electron lifetimes in the conduction band, is discussed in [1] and corresponds to the one given in the literature to describe g-r and $1/f$ noise components. The second term within square brackets in eq. (1) represents the photoinduced noise component produced by the barrier fluctuation while the first term is an intrinsic noise generated by trapping-detrapping processes in shallow centers of free electrons within the photoconducting material. As stated in the previous section, the photoinduced component dominates the whole noise spectrum in the low frequency range, i.e. approximately below 2KHz, and contains only quantities obtainable from experiments.

Actually $|<S(\omega)>|^2/\tau_g^2$ is very nearly a constant, whose value is $1/2\pi$, Δg is, by definition, the derivative of the conductance G with respect to n_d, and n_d can be expressed in terms of the photon flux n_f [1]:

$$n_d = n_f \cdot \eta_\lambda \cdot \tau_d \quad (2)$$

In this equation η_λ is the quantum efficiency coefficient for the considered wavelength and τ_d the average photocurrent relaxation time, defined as the area to height ratio of the relaxation pulse following a brief pulse of light superimposed to the bias illumination. The photoinduced noise spectrum $\psi_G^{ph}(\omega)$ thus becomes

$$\psi_G^{ph}(\omega) = \frac{1}{\pi} \cdot \left(\frac{\Delta g}{G}\right)^2 \cdot n_d \cdot \Sigma_j \frac{a_j \tau_d^{(j)}}{1 + \omega^2 \tau_d^{(j)2}} \qquad (3)$$

with $\Sigma_j a_j = 1$.

Let us assume, as a first approximation, that the photocurrent relaxation is characterized by a single exponential decay with time constant given by τ_d. Eq. (3) becomes:

$$\psi_G^{ph}(\omega) = \frac{1}{\pi} \cdot \left(\frac{\Delta g}{G}\right)^2 \cdot n_d \cdot \frac{\tau_d}{1 + \omega^2 \tau_d^2} \qquad (4)$$

If the temperature T is changed and noise spectra are taken at a constant value of the device conductance G by adjusting the light intensity, then in eq. (4) the quantity which would be mostly affected by the temperature change is τ_d.

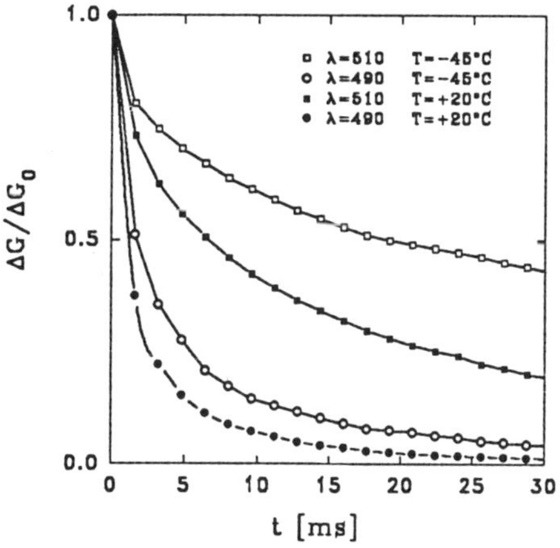

Fig.1 Normalized photocurrent relaxation pulses at $T = 20°C$ and at $T = -45°C$ for two λ values slightly above and slightly below the critical wavelength λ_c.

Actually, at medium to high illumination values ($n_f > 10^{11}$ photons $s^{-1} cm^{-2}$) when the relation between G and n_d becomes linear owing to the feedback effect of the injected electron charge on the barrier height[1], if electron mobility is little affected by the change of the temperature, both Δg and n_d can be considered independent of T. The strong variation of τ_d with temperature is shown in fig. 1, where the photocurrent relaxation pulse are reported for two λ values slightly below and slightly above λ_c at $T = 20°C$ and at $T = -45°C$. In both cases the value of $G \gg \Delta G_o$ was $5.5 \cdot 10^{-6} S$. According to eq. (4), in the range of frequencies where $\omega^2 \tau_d^2 \gg 1$ (frequencies larger than a few Hz typically), when temperature is changed, the spectral power density of the photoinduced component behaves as $1/\tau_d(T)$.

Since τ_d generally increases when T decreases, a reduction of the photocurrent noise is expected when temperature is lowered, such a behaviour is in full agreement with the experiments. Quantitative results are reported in figs. 2 and 3, which show the spectra calculated from eq. (3) by evaluating Δg and n_d from photoconductance vs. light intensity and wavelength measurements and by taking into account the distribution of the time constants $\tau_d^{(j)}$. Numerical values refer to $\Psi_G(f) = 4\pi \Psi_G(\omega)$

These quantities, together with their relative weights a_j were obtained by fitting the non exponential relaxation pulses with a sum of exponential functions. Even if in principle this fitting is unique, in practice good fittings can be obtained with different sets of values of $\tau_d^{(j)}$ and a_j. Since these different sets give only slight differences in the theoretical spectra, it is possible to choose a set of values of $\tau_d^{(j)}$ such that their inverses are uniformly distributed within the range of ω values delimiting the power spectrum and to evaluate the a_j by fitting the relaxation pulse. All these data for $T = 20°C$ and $T = -45°C$ are given in the figure captions. The conductance G was in both cases G=5.5 10^{-6} S.

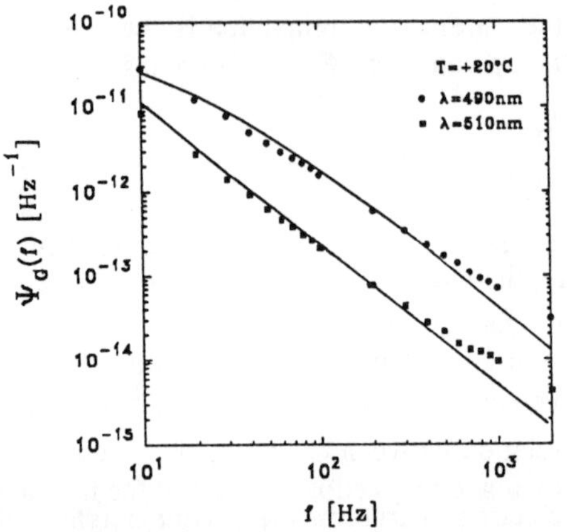

Fig.2 Experimental and calculated power spectra of the relative conductance fluctuation at $T = 20°C$ for $\lambda = 510nm$ and for $\lambda = 490nm$. Experimental values of the parameters for $\lambda = 490nm$ are: $n_f = 2\ 10^{12}\ s^{-1}\ cm^{-2}$, Δg = 2.5 10^{-15} S, n_d= 7 10^9; $a^{(1)}$=0.339, $a^{(2)}$=0.468, $a^{(3)}$=0.193; $\tau_d^{(1)}$=71ms, $\tau_d^{(2)}$=7.7ms, $\tau_d^{(3)}$=1.6ms and for $\lambda = 510nm$ are: n_f=5.5 $10^{11}\ s^{-1}\ cm^{-2}$, Δg = 2.5 10^{-15} S, n_d=7 10^9; $a^{(1)}$=0.777, $a^{(2)}$=0.216, $a^{(3)}$=0.007; $\tau_d^{(1)}$=400ms, $\tau_d^{(2)}$=40ms, $\tau_d^{(3)}$=4ms.
The calculated spectra refer to the photoinduced noise component only.

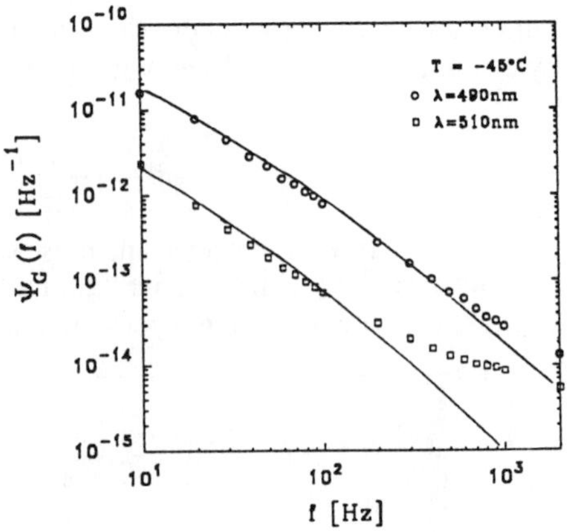

Fig.3 Experimental and calculated power spectra of the relative conductance fluctuation at $T = -45°C$ for $\lambda = 510nm$ and for $\lambda = 490nm$. Experimental values of the parameters for $\lambda = 490nm$ are: n_f= 1.1 $10^{12}\ s^{-1}\ cm^{-2}$, Δg = 2.5 10^{-15} S, n_d= 7 10^9; $a^{(1)}$=0.657, $a^{(2)}$=0.243, $a^{(3)}$=0.1; $\tau_d^{(1)}$=71ms, $\tau_d^{(2)}$=7.7ms, $\tau_d^{(3)}$=1.6ms and for $\lambda = 510nm$ are: n_f=5.5 $10^{12}\ s^{-1}\ cm^{-2}$, Δg = 2.5 10^{-15} S, n_d=7 10^9; $a^{(1)}$=0.979, $a^{(2)}$=0.020, $a^{(3)}$=0.001; $\tau_d^{(1)}$=400ms, $\tau_d^{(2)}$=40ms, $\tau_d^{(3)}$=4ms.

CONCLUSIONS

In this paper we have presented a set of preliminary experiments on the photoconductance noise in CdS based photoconducting devices concerning the behaviour of the noise power spectrum vs. temperature. The results have been compared with the theoretical ones obtained by a theory developed in a previous paper and a good quantitative agreement has been found without the introduction of free parameters. A more complete check over a more extended range of values of light intensity and temperature is going to be developed.

REFERENCES

1. A. Carbone and P.Mazzetti (submitted to Phys. Rev B)
2. A. Carbone and P.Mazzetti, Proc. of 11th Int. Conf. of Noise in Phys. System and 1/f Noise, p.317, (1991)
3. R.L. Petritz, Phys.Rev. 104, 1508, (1956)
4. R.H. Bube, Photoconductivity of Solids, John Wiley & Sons, Inc., New York, London, (1964), p.357

SOURCES OF 1/f NOISE IN HgCdTe METAL-INSULATOR-SEMICONDUCTOR CHARGE INTEGRATING STRUCTURES

Wenmu He and Zeynep Çelik-Butler
Southern Methodist University, Department of Electrical Engineering
Dallas, Texas 75275 USA

ABSTRACT

We investigated the possible origin of 1/f fluctuations in $HgCd_{0.71}Te_{0.29}$ (E_g=0.254 eV at 90 K) Metal-Insulator-Semiconductor (MIS) charge integrating structures. These structures are used for infrared detection, usually in conjunction with a correlated double sampling (CDS) circuit. By varying the reset voltage, the inject pulse voltage in the CDS circuit and the charge integration time, we were able to change the charge well size, the magnitude of the dark current, and the position of the electron Quasi-Fermi levels at the semiconductor - insulator interface, which enabled us to investigate the role of each in generating 1/f fluctuations in these structures. Bulk and surface mechanisms were considered. In the light of the experimental data, dark current originated bulk mechanisms were ruled out as the origin of 1/f fluctuations. Interface - trap originated charge fluctuations theory, on the other hand yielded values for HgCdTe - ZnS interface states which were comparable to that obtained by other independent techniques.

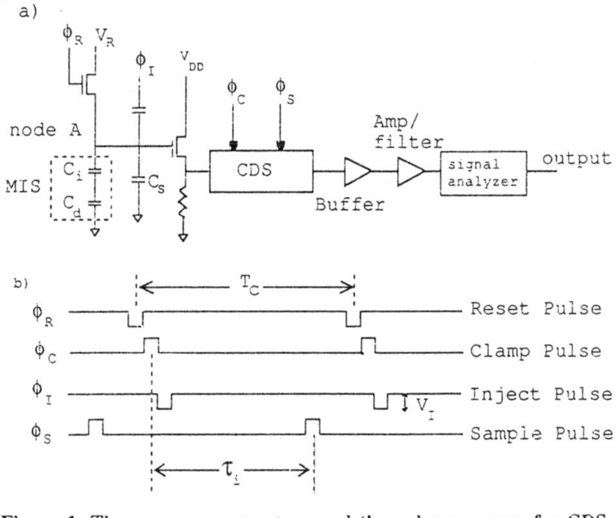

Figure 1. The measurement set-up and the pulse sequence for CDS.

INTRODUCTION

The MIS structure becomes a detector when a voltage is applied to the metal gate such that a depletion is formed under the dielectric in the semiconductor. This layer can be also viewed as a potential well for minority carriers where charge is collected due to generation from the infrared radiation through the transparent metal gate. Carriers are also generated in the inversion layer through what is called a dark current. Charge accumulated due to radiation should be far in excess of the charge due to the dark current for the IR detector to operate properly.

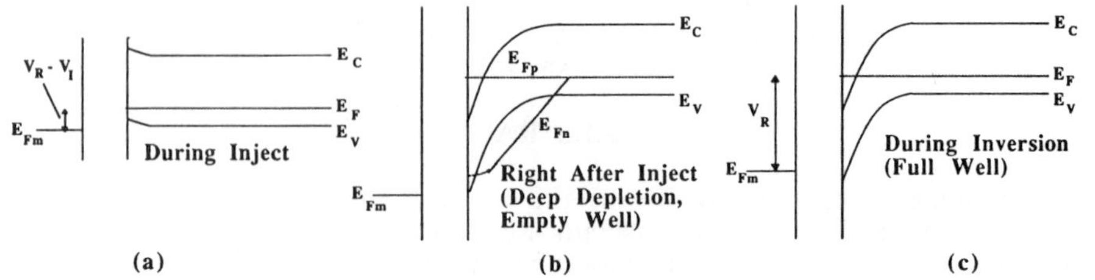

Figure 2. The energy band diagram for a p- substrate MIS structure during inject pulse (a), right after inject pulse (b), and after integration of charge (c).

The most common signal processing technique used in conjunction with these devices is the Correlated - Double-Sampling (CDS) Technique[1] which samples the charge in the potential well before and after the integration of charge due to infrared radiation (IR), thus providing an output voltage that is proportional to the difference of voltages between these times. Figure 1 shows the operation and diagram of the CDS circuit. The energy band diagram of the HgCdTe - ZnS interface is shown in Figure 2 during different stages of charge integration in dark. Inject pulse is used to repel the inversion charge away from the semiconductor - insulator interface such that a deep depletion is present (empty well) right after the inject operation. This prepares the well for charge integration due to IR generated carriers or due to the carriers generated by dark current if the detector is not subjected to any IR radiation. As charge accumulates in the well, the energy band diagram relaxes to the inversion state.

In our experiment, we measured the output voltage noise power spectral density as a function of reset voltage V_R and inject pulse voltage V_I at different charge integration times, τ_i, while the device is under 90 K (=device temperature) thermal background radiation (dark) or subjected to 300 K IR radiation. The system and photon noise contributions were subtracted from the measurements to obtain the MIS noise. Then the transfer function of the CDS circuit and the equivalent capacitance of the circuitry were used to convert the voltage noise to charge noise. The results and analyses are presented in the next section.

RESULTS AND ANALYSIS

The MIS structures were fabricated in Texas Instruments on LPE grown $HgCd_{0.71}Te_{0.29}$ (E_g=0.254 eV at 90 K) which was intrinsically Hg-vacancy doped p-type. The surface was passivated with anodic sulfide followed by 1200 Å of ZnS deposition as a gate dielectric. The gate metal was 35 Å aluminum. The transparent gate dimensions were 10 x 20 mils (25.4 x 50.8 μm). All reported measurements were done at 90 K. The clocking frequency f_c was 23 KHz. The pulse durations were less than 1 μsec. Two different clamp to sample times (τ_i) were used: 4μsec and 30 μsec. 30 μsec corresponded approximately to full well under 300 K background radiation. The system noise was subtracted from the measured spectral density. At least 8 averages were taken for each noise spectrum.

Figure 3 shows charge noise power spectral density at 2 Hz versus inject voltage V_I for 4 μs integration time when the device is under 300 K background radiation.

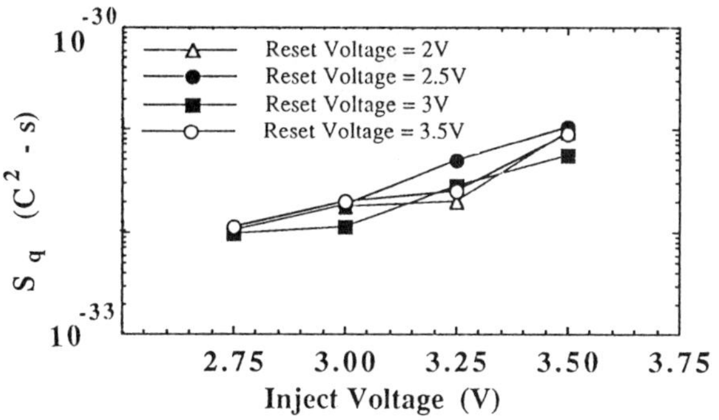

Figure 3. Charge noise power spectral density at 2 Hz for 4 μs integration time under 300 K radiation.

The noise magnitude does not change significantly with reset voltage whereas it shows a noticeable increase with inject voltage. Although not shown here, similar trend was observed for the frequency exponent γ of the $1/f^{\gamma}$ spectral form. γ was observed to increase from 0.8 to 1.6 as the inject voltage was varied from 2.75 to 3.5 V, while changes in the reset voltage was found to cause no effect on γ. Similar observations were made for measurements in dark.

The role of dark current was also investigated for 1/f fluctuations in these structures. In our devices, the dominant source of dark current was depletion region generated minority carriers. Since this current source is a function of the depletion region width and therefore the reset voltage, we saw an increase in the measured dark current with increasing reset voltage. (Figure 4) There was no corresponding increase, however, in the measured 1/f noise magnitude. When the inject voltage dependence was investigated, the dark current was found to be independent of V_I, whereas charge noise power showed a strong functional dependence on V_I. These observations support the fact that 1/f fluctuations in these MIS structures do not originate from depletion region generated minority dark current.

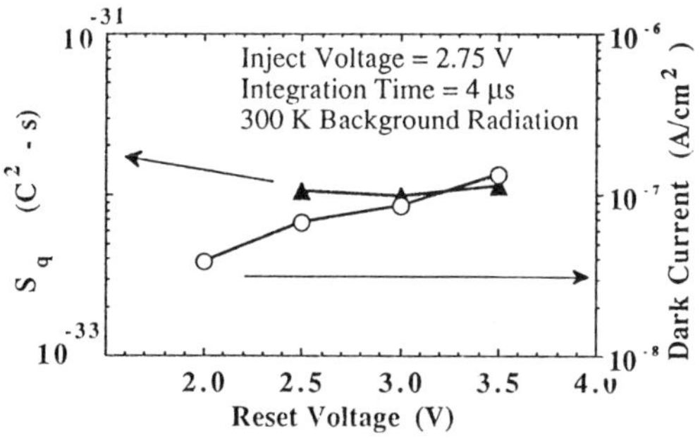

Figure 4. Dark current density and charge noise power spectral density at 2 Hz vs reset voltage.

We also investigated the possible surface origin of the observed charge fluctuations. Using McWhorter Theory[2], the charge noise power spectral density was calculated based on trapping and detrapping of charge from HgCdTe - ZnS interface states [3,4]. The effective interface trap density N_{teff} was extracted (Figure 5) using the expression:[3]

$$S_q = \frac{q^2 AkT}{\alpha f} N_{teff} \tag{1}$$

Where, q is the elementary charge, A is the device area, k is the Boltzmann constant, T is the temperature, and α is a parameter that depends on dielectric barrier height. α was taken[4] as 4.6 x 10^7 cm^{-1}. The computed N_{teff} from the equation above is an average value for that energy range that the Quasi-Fermi level E_{Fn} for electrons sweeps during charge integration. Since in dark, the quasi-Fermi level sweep is small (less than 0.1 meV for 30 µs integration time), S_q values measured in dark were used to obtain more accurate interface state values. The trap distribution in Figure 5 is about an order of magnitude higher in value than that measured by Schiebel[5] on MISFETs of similar band-gap HgCdTe. It is, however, in the same range with the interface state values measured on much smaller band-gap HgCdTe by conventional capacitance and conductance methods[6] if the traps are assumed to be distributed in the dielectric up to a few angstroms.

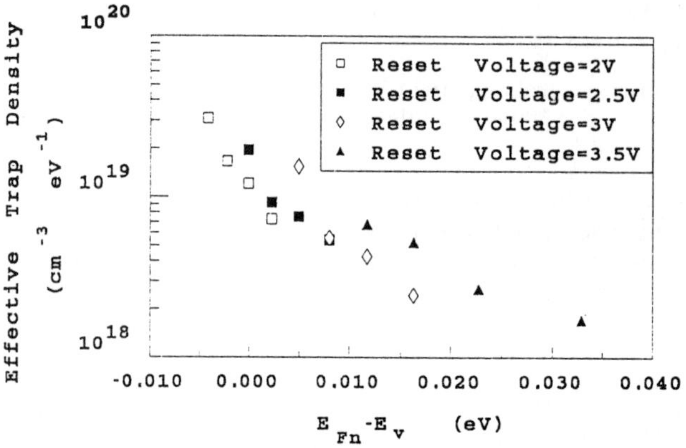

Figure 5. Effective trap density extracted from dark noise measurements with 4us integration time.

N_{teff} changes with surface treatment. Therefore, there might not be much scientific merit in making direct comparisons between different structures.

In conclusion, we investigated the origin of 1/f noise in the integrated charge in HgCdTe MIS infrared detectors. Our experimental findings agree withe predictions of the McWhorter theory.

ACKNOWLEDGEMENT

This material is based in part upon work supported by The Texas Advanced Technology Program under Grant No. THECB ATP003613-004 and the National Science Foundation under Grant No. ECS-9116209. We would like thank David Boyd, Mark Wadsworth and Sebastian Borrello of Texas Instruments Inc. for their contributions to this investigation.

REFERENCES

1. W. H. White, D. R. Lampe, F. C. Blaha, and I. A. Mack, IEEE J. Solid-State Circuits, **SC-9**, 1 (1974).
2. A. L. McWhorter, *Semiconductor Surface Physics*, ed. R. H.. Kinston (University of Pennsylvania Press, 1957), pp. 207-228.
3. J. L. Melendez, "Extraction of Insulator Trap Densities from 1/f Noise Measurements in Integrating Metal-Insulator-Semiconductor Devices," M.S. Thesis, Massachusetts Institute of Technology, 1991.
4. W. He, and Z. Çelik-Butler, in press.
5. R. A. Schiebel, Solid-St. Electronics, **32**, 1003 (1989).
6. M. J. Yang, C. H. Yang, M. A. Kinch, J. D. Beck, Appl. Phys. Lett., **54**, 265 (1989).

ULTRASHORT COMBINED INTERFEROMETER IN MULTIMODE RESONANT BAR GRAVITATIONAL WAVE DETECTORS

V. V. Kulagin
Sternberg Astronomical Institute, Moscow State University
Universitetsky prospect 13, 119899, Moscow, Russia

ABSTRACT

High sensitive optical transducer for resonant bar gravitational wave detectors with three mechanical modes is proposed. Sensitivity of such gravitational antenna to metric perturbation could be about the potential limit of 10^{-20} for helium temperature bar.

INTRODUCTION

Potential sensitivity of resonant bar detectors (minimal detectable metric perturbation h_p due to gravitational wave) is about

$$h_p = 10^{-20} \tag{1}$$

for helium temperature bar and usual parameters of antenna. However, such sensitivity did not obtained yet because of noises of modern transducers. The present paper is dedicated to optical transducer which make it possible to achieve the sensitivity (1) in Weber type detector with three mechanical modes.

BAR DETECTOR WITH SEVERAL MECHANICAL MODES

One of the way to enhance the sensitivity is to use several small masses binded elastically to the bar[1]. Then for optical readout system one could obtain the following sensitivity

$$h_{min} = h_p a^{-1/(4n)} \tag{2}$$

where n - total number of mechanical modes, $a=G_B/G_L$, G_B and G_L - spectral densities of the bar Brownian noise and photon noise. For modern parameters of antenna factor a is about 10^{-12}-10^{-16} (for laser power 10 mwt and number of reflections N=10). Then one could obtain that n must be larger than 6 for potential sensitivity could be achieved. However, sensitivity (2) is obtained for optimal mass ratio, and then for the smallest mass one could get (M -

bar mass)

$$m = Ma^{(n-1)/n} \tag{3}$$

and for n=6 one could obtain $m=10^{-10}M$ that is hardly achievable in experiment.

OPTICAL READOUT SYSTEM WITH VERY HIGH SENSITIVITY

Another way to achieve the potential sensitivity is to use two or three mechanical modes detector with optimal mass ratio and very sensitive optical readout system, such as ultrashort (0.01 mm) Fabry-Perot interferometer with very large number of reflections N and operation point on the slope of transmission curve[2] (one mirror of optical resonator is attached to the smallest mass and another one to the bar end). Then theoretically sensitivity could be large enough (up to potential), however one must take into consideration frequency noise of the laser. For relative frequency stability q about 10^{-12} and minimal length of interferometer one could obtain (λ - wavelength of the laser, L - bar length)

$$h_{min} = 2q\lambda/L = 10^{-18} \tag{4}$$

that is two orders of magnitude larger than h_p.

ULTRASHORT FABRY-PEROT-MICHELSON INTERFEROMETER

Limitation (4) could be removed in Michelson interferometer with two ultrashort Fabry-Perot resonators in its arms. Then for sensitivity one could obtain

$$h_{min} = 2q\lambda/(LN) \tag{5}$$

In this case one must adjust Fabri-Perot resonators to the top of the transmission curve so that phase modulation of transmitted laser light would take place. Transmitted beams of Fabry-Perot resonators from two arms must be mixed at the photodetector. It is worth mentioning that reference Fabry-Perot resonator could have another physical length than measuring one. The only condition is that optical lengths of two resonators are equal. Therefore for reference resonator one could use stable monolithic resonator with several number of reflections and appropriate length. For example if the length of ultrashort resonator is about 0.01 mm, then for N=10 000 the length of reference resonator could be 2.5 cm for 4 reflections.

CONCLUSION

In conclusion the ultimate sensitivity of resonant bar antenna with three mechanical modes and foregoing optical readout system could be about 10^{-20} for helium temperature bar. Useful properties of such optical readout system are easyness of adjustment, low sensitivity to seismic and acoustic noises and possibility of integral realization.

REFERENCES

1. J. P. Richard, J. Appl. Phys., 64, 2202 (1988).
2. K. Tsubono, Conf. on Neitrino Astrophysics, Takayma, 17-22 October (1992).

BURST NOISE IN FORWARD CURRENT OF LATTICE-MISMATCHED InP/InGaAs/InP PHOTODETECTORS

D. Pogany, S. Ababou and G. Guillot
Laboratoire de Physique de la Matière (URA CNRS 358), INSA de Lyon, Bât. 502, 20, Av. A. Einstein, 69621 Villeurbanne cedex, France

ABSTRACT

Burst noise (BN) data of forward biased lattice-mismatched InP/InGaAs/InP photodetectors are presented. Voltage and temperature BN dependencies in both time and frequency domains suggest that the BN is due to an excess current which flows through a localized leakage site and is modulated by an action of a metastable defect.

INTRODUCTION

Burst noise (BN) has previously been observed in forward [1,2] and reverse [2] biased p-n junctions. We have recently reported on low frequency noise including the burst noise in reverse biased lattice-mismatched InP/InGaAs/InP photodetector arrays.[3,4] In this paper, the investigation of BN in forward direction is presented.

RESULTS

A typical diode has been choosen for detailed BN analysis in both time and frequency domains. The diode structure is described in Ref.5. In figure 1 we present current (I_F) and burst noise amplitude (I_{BN}) dependences on forward bias. Both I_F and I_{BN} vary exponentially with the bias up to 0.3V and then, I_F is limited by the diode series resistance and I_{BN} tends to saturate. In the exponential region, we approximate I_F and I_{BN} by [1]

$$I_{F(BN)} = K \exp\left[-\frac{\left(\Delta E_{F(BN)} - qU_F\right)}{m_{F(BN)} k_B T}\right] \qquad (1)$$

where K is a weakly temperature dependent coefficient, k_B the Boltzmann constant, T the temperature, q the electron charge, U_F the forward bias, $\Delta E_{F(BN)}$ and $m_{F(BN)}$ (m_F =1.04, m_{BN} =1.22 at 297K) are respectively the thermal activation energies and ideality factors of the current (F) and BN amplitude (BN). ΔE_F determined from the saturation current temperature dependence is 0.74±0.02eV. The Arrhenius plot of I_{BN} *vs* 1/T at two biases and the apparent activation energies ΔE_A (I_{BN} =Kexp[$-\Delta E_A/k_BT$]) are shown in figure 2. Using (1), ΔE_{BN} has been estimated from ΔE_A at 0.3V (Fig.2) to be 0.58±0.03eV. In figure 3, a typical BN waveform is shown. It can be seen that the BN is superposed on a chaotic (1/f) noise background and that the BN pulses are clustered into groups of pulses giving rise to intermittent longer pulses of the "on" BN state (t_+). We have measured the pulse width distributions by using a computer controlled system and a software elimination of the effect of the chaotic background

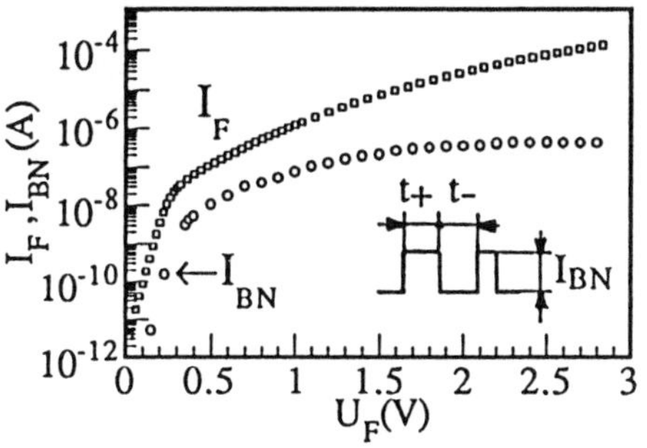

Fig. 1. Current I_F and BN amplitude I_{BN} as function of forward bias, T=297K.

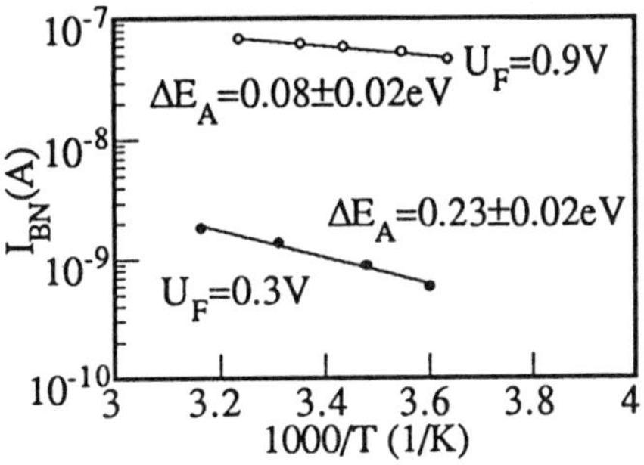

Fig. 2. Temperature dependences of BN amplitude at two different biases.

noise. The pulse width distributions in both BN states exhibit nearly exponential statistics (fig.4). From the slope of the distribution we have determined the mean pulse widths $<t_+>$ and $<t_->$. In figure 5, voltage dependences of $<t_{+,-}>$ show that $<t_+>$ decreases exponentially with the applied bias, while $<t_->$ is bias independent. Besides, both $<t_+>$ and $<t_->$ increase with decreasing temperature as seen in the Arrhenius plot in figure 6. Activation energies of $<t_+>$ and $<t_->$ are 0.86±0.03 and 0.91±0.03 eV

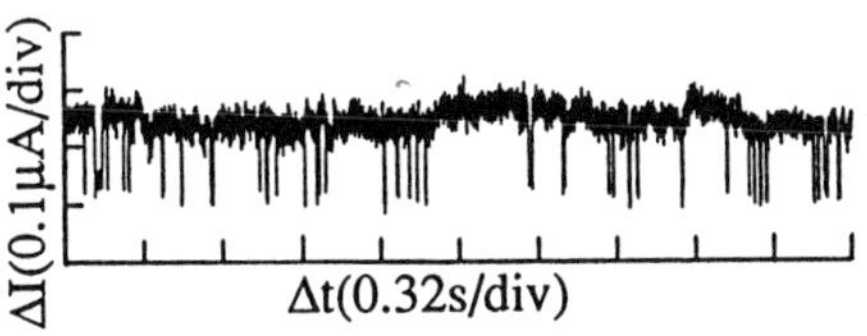

Fig. 3. Typical current fluctuation ΔI *vs* time Δt at U_F=1.3V, T=297K.

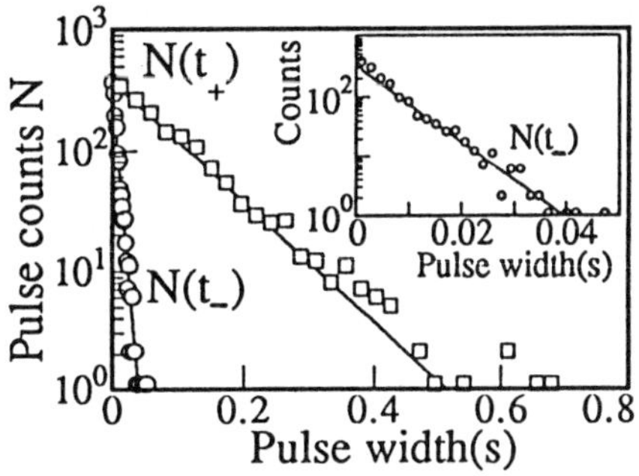

Fig. 4. Pulse width distributions $N(t_{+,-})$ at 1.3V with exponential fits. The inset is the time expansion of $N(t_-)$, T=297K .

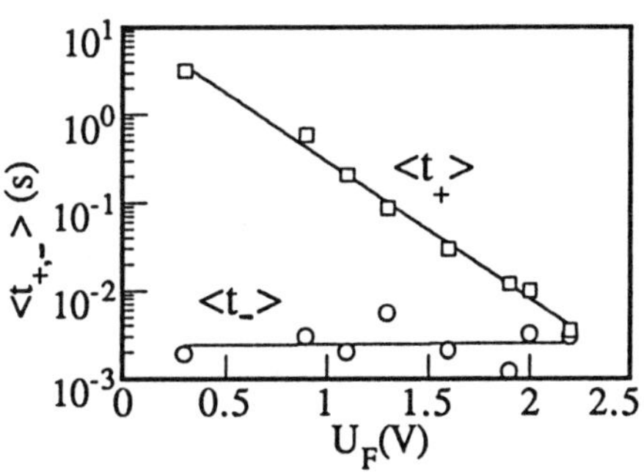

Fig. 5. Voltage dependences of mean pulse widths $<t_{+,-}>$ at T=297K.

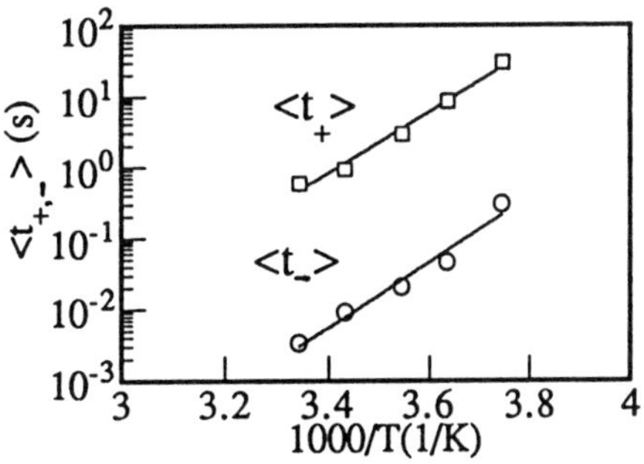

Fig. 6. Temperature dependences of mean pulse widths $<t_{+,-}>$ at U_F=0.9V.

respectively. Noise power spectral density (S_I) measurements have been performed at room temperature for different bias polarizations (fig.7). At each bias, the spectrum shows a distinct Lorentzian-like component superposed on a $1/f^b$ noise background (b varies between 1.15 and 1.34 in the frequency range f=0.1-10Hz). The values of both the thermal and shot noises are negligible. The theoretical Lorentzian shape depends on BN parameters as follows [6]

$$S_I(f) = \frac{4(I_{BN})^2}{(<t_-> + <t_+>)} \frac{<t>^2}{1+[2\pi <t> f]^2} \quad (2)$$

where $<t>^{-1} = <t_+>^{-1} + <t_->^{-1}$. Note, that the theoretical turnover frequency f_t ($f_t = 1/(2\pi <t>)$ is limited by the shortest time constant ($<t_->$ in our case). So, the fact that f_t does not move significantly with the applied bias (fig.7) is consistent with bias independence of $<t_->$ (fig.5). However, two differences from the theoretical prediction are apparent. First, the slope of the Lorentzian of the measured spectra after the turnover frequency has $1/f^c$ character with c between 1.1-1.3 in 1k-10kHz range, which is far from the predicted value of c=2.[6] Second, the measured f_t (600-800 Hz) is shifted to higher frequencies compared to the assumed value of $f_t = 1/(2\pi <t_->) = 40\text{-}160$Hz for $<t_-> = 1\text{-}4$ ms taken from figure 5. These differences could be explained by the theory of the clustering Poisson process because the pulses grouped into clusters (fig.3) can result in $1/f^c$ (c=0-2) spectrum behavior and can also shift the turnover frequency.[7,8,9]

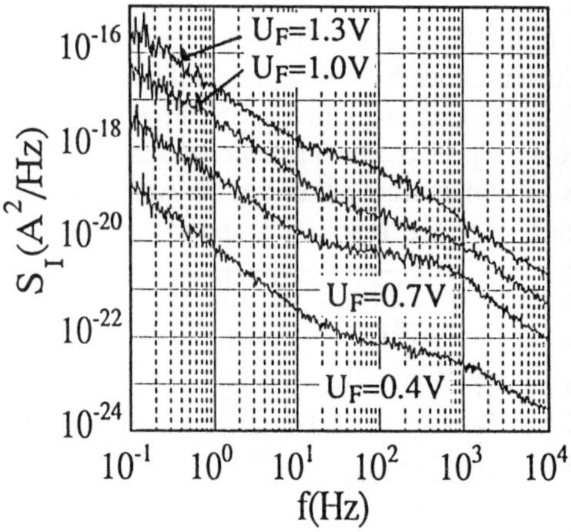

Fig. 7. Noise power spectral density S_I *vs* frequency f at four different biases.

DISCUSSION

We suppose that the BN is caused by a modulation of an excess current (EC) which is superposed on the dominant diode diffusion current ($m_F \approx 1$). We consider that I_{BN} is directly related to this EC. Moreover, the value of ΔE_{BN} (0.58eV) lower than ΔE_F (0.74eV which is related to InGaAs gap) suggests that the EC flows through a leakage path with a reduced barrier.[1,2] This can be due to a band-gap narrowing at a localized site caused by a dislocation crossing the p-n junction.[2] Indeed, dislocations are present in the studied structures.[5] On the other hand, the voltage and temperature dependences of $<t_{+,-}>$ suggest that the modulation of the EC can be caused by a charge fluctuation of a defect located at or near the leakage site.[1] The decrease of $<t_+>$ with U_F could be related to voltage dependent capture, while $<t_->$ can be a bias independent emission time constant.[1] However, the large activation energies of $<t_{+,-}>$

are difficult to be explained by this model because a very large defect lattice relaxation have to be considered.[6] Therefore, we may take into account a BN model where the charge change is controlled by a structural reconfiguration of the defect and not by the carrier capture and/or emission.[10] In this case, the mean pulse width activation energies correspond to barrier energies for the defect reconfiguration.[10] Here, the decrease of $<t_+>$ with U_F can be explained by a defect reconfiguration induced by an energy transfer from excess current carriers to the defect.[10] On the other hand, an alternative BN model can be considered where BN is due to a defect assisted tunneling through a bistable defect (located in the high field region due to a dislocation) with different tunneling rates in each defect state.[11] However, we have not a clear evidence about the tunneling nature of the excess current although an exponential behavior of tunneling current in forward direction is reported.[12] Independently on the used BN model, we suppose that the clustering of BN pulses can be related to a complex metastable character of the defect controlling the BN.[6,7]

In conclusion, the present study shows that useful informations are obtained from the BN behavior in forward direction. Investigations are in progress to establish the physical nature of the forward excess current and its relation to the previously revealed EC in the reverse direction which clearly exhibits tunneling components.[3,4]

ACKNOWLEDGMENTS

The authors thank Dr. M. A. Py and Dr. Z. M. Shi from the Institute for Micro- and Optoelectronics, Swiss Federal Institute of Technology in Lausanne, Switzerland for making possible to use facilities for the noise measurements in frequency domain.

REFERENCES

1. S. T. Hsu, R. J. Whittier and C. A. Mead, Solid-St. Electron. 13, 1055 (1970)
2. G. Doblinger, in "Noise in Physical Systems" (ed. by D. Wolf), Proc. 5th Int. Conf. on Noise, Bad Nauheim, Germany, Springer Verlag, Berlin, 64 (1978)
3. D. Pogany, S. Ababou and G. Guillot, Proc. 5th Conf. on InP and Rel. Mat., IEEE Catalog #93CH3276-3, 611 (1993), Paris, France
4. D. Pogany, S. Ababou and G. Guillot, to be published in Proc. 23rd ESSDERC' 93 Conference, September 13-16, Grenoble, France
5. D. Pogany, F. Ducroquet, S. Ababou and G. Brémond, J. Electrochem. Soc. 140, 560 (1993)
6. M. J. Kirton and M. J. Uren, Adv. Phys. 38, 367 (1989)
7. M. Athiba Azhar and K. Gopala, Jpn. J. Appl. Phys. 31, 391 (1992)
8. F. Grüneis and T. Musha, Jpn. J. Appl. Phys. 25, 1504 (1986)
9. V. B. Orlov, Solid-St. Electron. 35, 1827 (1992)
10. K. R. Farmer and R. A. Buhrman, Semicond. Sci. Technol. 4, 1084 (1989)
11. G. I. Andersson, M. O. Andersson and O. Engström, J. Appl. Phys. 72, 2680 (1992)
12. J. A. Del Alamo and R. M. Swanson, IEEE Electron Device Lett. 7, 629 (1986)

The contribution of electrode to 1/f noise in SPRITE LWIR detectors

Zheng Wei-Jian and Zhu Xi-Chen
Kunming institute of physics, Yunnan, 650223, P.R.China

Ⅰ. INTRODUCTION

SPRITE detector with special configuration is a good object for studying 1/f noise phenomena. Our laboratory have obtained the Hooge parameter around $\alpha_H=3.4\times10^{-5}$ for n-MCT(x=0.2) from the experiments of SPRITE and photoconductor[1,2], which are different from $\alpha_H=5\times10^{-3}$ given by Dr. H.I. Hanafi[3] in varieties MCT devices. About electrode noise, there are a lot of works [4,5,6,7] in both theories and experiments.

This paper represents the dc characteritics, low frequency noise of SPRITE detector, and their relationship, and provides a proof of influence of electrode to the 1/f noise levels.

Ⅱ. EXPERIMENTS

Several samples of LongWave InfraRed SPRITE with horned contacts are made from x=0.2 n-MCT. The C-contact is grounded; The O-contact is a potential probe without current for signal read-out; The B-contact hold positive potential to maintain a electric field (typically:30V/cm) in the filament. SPRITE performance depends on the bias field strongly, the electrode has to afford to a large current density. The contact noise often dominates the device noise, and makes degeneration of SPRITE performances.

The I-V characteritics both for B-C contacts and for O-C contacts are tested. Fig.2 shows four groups of typical data. For 92S65-2BC it's linear, but for 92S30-7BC nonliear, particularly at low bias field. The resistance R is given by:

$$R_i = \frac{\Delta V_i}{\Delta I_i} = \frac{V_{i+1} - V_i}{I_{i+1} - I_i} \quad (i=-n+1,\ldots,-1,0,1,\ldots,n-1) \qquad ----(1)$$

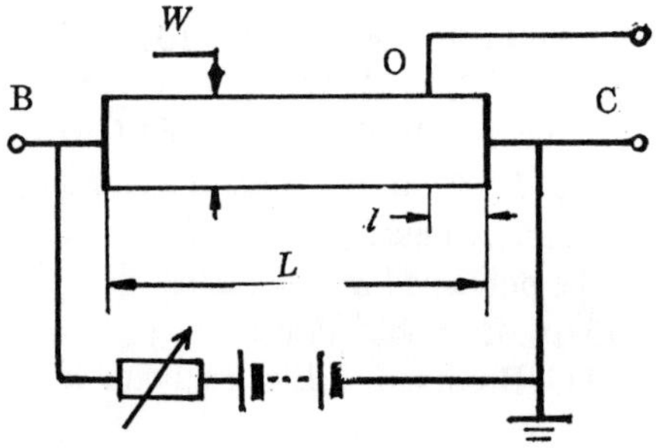

Fig.1 diagram of SPRITE configuration

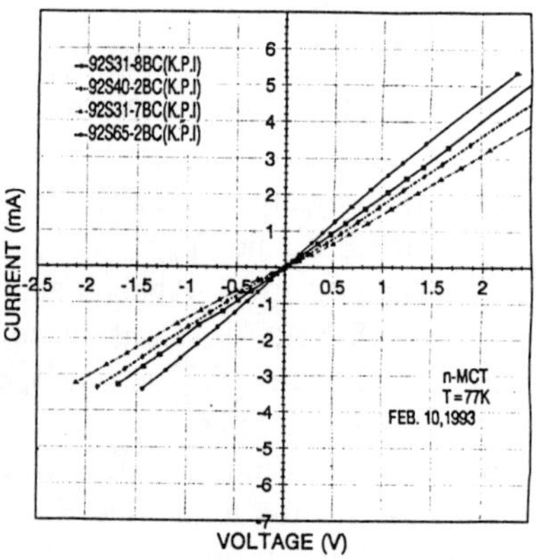

Fig.2 I-V charactertics of SPRITE (B-C)

Fig.3(a) indicates that I–V characteritics is straight line then leads to smooth resistance curves nearby zero–bias. At high field, R~I is upward a little due to sweepout of carriers. On the contrary, the nonlinear I–V characteritics for poor contact brought R~I curves with sharp peak in near–zero bias field.(Fig.3(b),(c)).

A linearity parameter is defined as:

$$\sigma = \frac{R_n - R_o}{R_o} \qquad ------(2)$$

Generally, the low freequence noise spectrum intensity can be expressed for homogenous samples :

$$S_v(f) = k \frac{V^{\gamma}}{f^{\beta}} \qquad ------(3)$$

where, $\gamma=2$ and $\beta=1$ is common for regular device. k is a constant associated with device characteritics.

The spectrum of low frequency noise was measured at a constant dc bias voltage (e.g: E=30V/cm) (Fig.4(a),(b)). The poor contact induces additional 1/f noise, and β arises to about 1.5~2.5. But the spectum of ohmic contact should hold 1/f law (β ≈0.85~1.15).

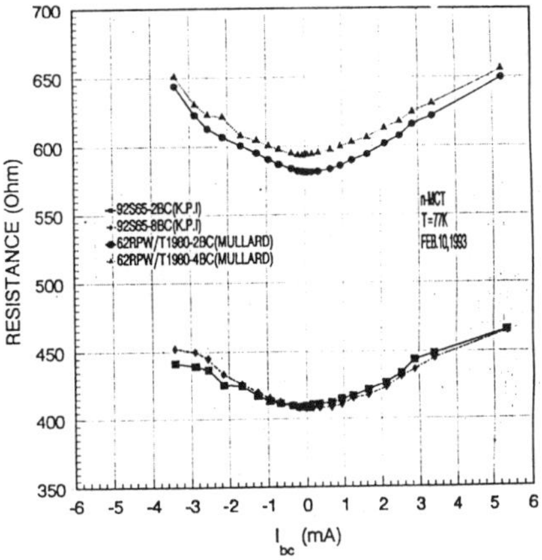

Fig.3(a) Resistertics of B–C (Normal)

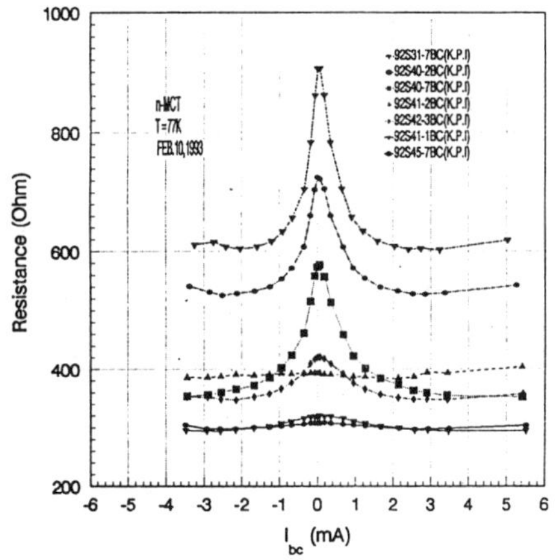

Fig.3(b) Resistance of B–C (Abnormal)

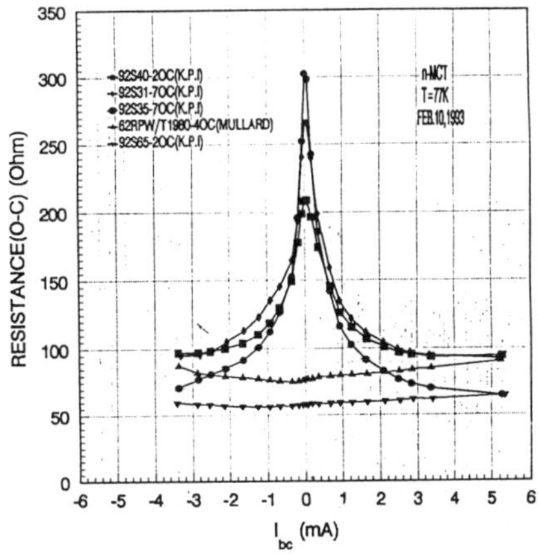

Fig. 3(c) Resistance of O–C contacts

The curves of S_υ versus bias voltage at 30.8Hz and 77K are plotted in Fig.5(a),(b) . For poor contact it's observable that the voltage noise drops to a certain value with increasing bias field, then increases again. The noise spectrum intensity of good contact rises steadly as general.

The S_υ/V^2 results are displayed in Fig.6(a),(b). For good contacts S_υ/V^2 tested almost is a constant between B–C contacts, but for poor contacts S_υ/V^2 reduces dramaticly with increasing bias field. Neverthless, for O–C contacts, S_υ/V^2 is a function of electric field strength for both good and poor contact. (Fig.6(b))

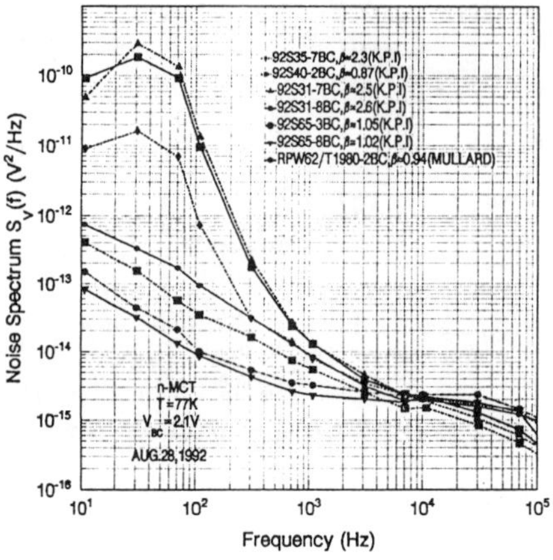

Fig.4(a) Noise spectrum of B–C contacts

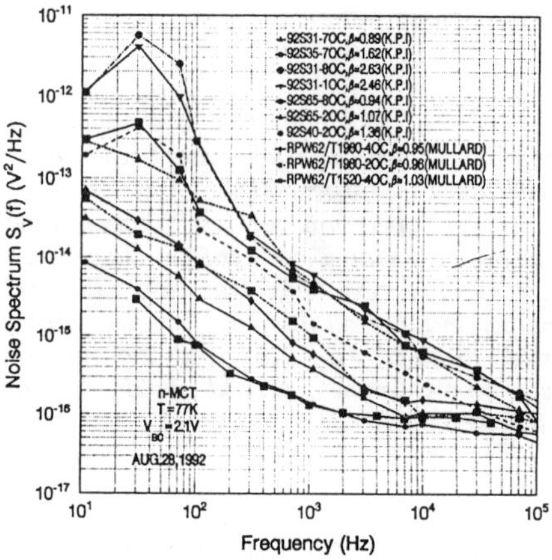

Fig.4(b) Noise spectrum of O–C contacts

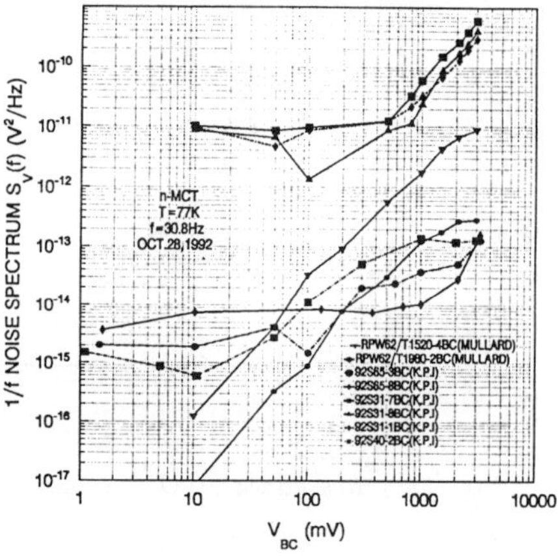

Fig.5(a) S_υ vs. bias of B–C contacts

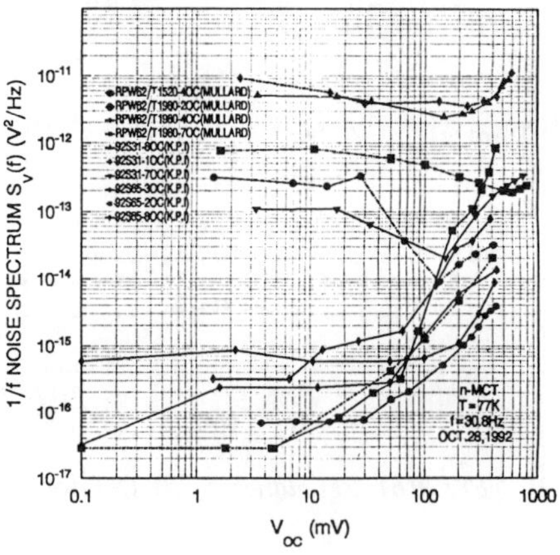

Fig.5(b) S_υ vs. bias of O–C contacts

Ⅲ.THE RESULTS AND DISCUSSION

We discuss the Hooge empirical formula[8] *as:*

$$\frac{S_R(f)}{R^2} = \frac{\alpha_H}{N\,f} \qquad \text{- - - - - (4)}$$

If R independs of electric field, then $\delta V = I\,\delta R$ *for homogenious sample and ohmic contacts, we have:*

$$\frac{S_R(f)}{R^2} = \frac{S_\upsilon(f)}{V^2} = \frac{\alpha_H}{N\,f} \qquad \text{- - - - (5)}$$

Thus, in equation (3) the parameter γ *should be 2.*

1. *The resistance R in equation (4) is the function of V (refers to Fig.3(b)), which causes that* S_υ/V_2 *curves in Fig. 6(a) has marked rising for B–C contacts. At high field (* $L/\mu_a E << \tau$ *), carrier injection at poor contacts*[6] *can be neglected, the adding 1/f noise is minimized due to carrier sweepout.*
2. *In most samples* S_υ/V^2 *of O–C contacts depends on the applied field for both good and poor contact due to carrier sweepout. The carrier in the read-out region will be more than other place of the filament. The total number N increase with increasing bias field. According to equation(4)* S_υ/V^2 *will be decreased.*

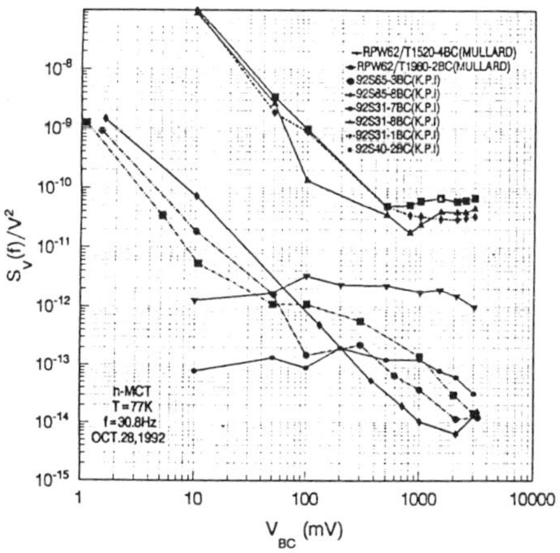

Fig.6(a) S_υ/V^2 *of B–C contacts*

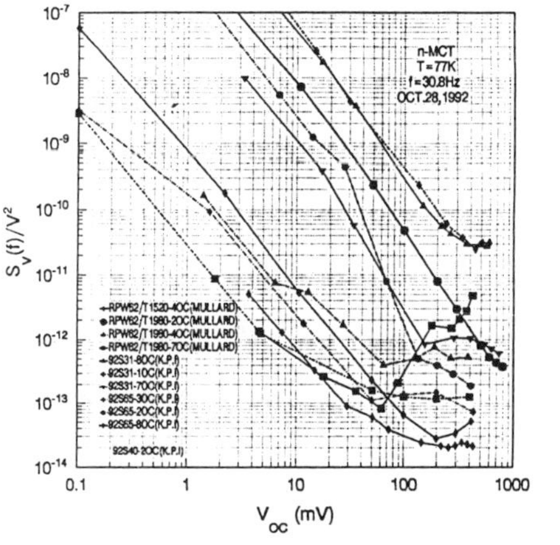

Fig.6(b) S_υ/V^2 *of O–C contacts*

3.The performance and noise charactertics of testing samples are listed in Tab.I, thermal carrier concentration n_o=5×10^{14} cm^{-3} and excess carrier n_b=5×10^{14} cm^{-3} generated by 300K background and 30° FOV. D is blackbody detectivity.The bias field is 30V/cm.The subscript 1 refers to B-C contacts, 2 refers to O-C contacts. In Tab.Ⅰ, α_{H1} is calculated for B-C contacts and N=n_o LWd, α_{H2} for O-C contacts and N=(n_o+n_b) lWd. It's obvious that ohmic contact devotes to high performance and low 1/f noise, the Hooge parameters are α_H~10^{-3} to 10^{-5}. The poor contact yields the larger Hooge parameter than Hooge constant (2×10^{-3}).*

TAB.I PERFORMANCE and NOISE CHARACTERTICS

Element No.	D*	β_1	β_2	α_{H1}	α_{H2}	σ_1	σ_2
RPW62/T1520-4	1.3×10^{11}	---	1.03	--------	5.6×10^{-5}	<10%	<10%
92S65-8	8.8×10^{10}	1.02	0.94	1.2×10^{-4}	8.2×10^{-5}	20%	18%
92S65-2	8.0×10^{10}	0.98	1.07	6.0×10^{-4}	2.4×10^{-4}	18%	17%
92S31-7	7.2×10^{10}	1.03	0.89	6.0×10^{-3}	2.6×10^{-4}	-130%	-53%
RPW62/T1980-4	7.0×10^{10}	0.94	0.95	3.9×10^{-3}	9.8×10^{-5}	11%	10%
92S40-2	5.6×10^{10}	0.87	1.36	1.4×10^{-3}	1.8×10^{-4}	-30%	-8%
92S31-8	2.3×10^{10}	2.60	2.63	4.2×10^{-2}	1.2×10^{-2}	-193%	-178%
92S31-1	1.9×10^{10}	2.50	2.46	3.2×10^{-1}	1.1×10^{-3}	-218%	-255%

Ⅳ.CONCLUSION.

The low frequence noise of n-MCT SPRITE LWIR detecter shows some complicate. For the poor contact, the electrode noise dominates 1/f noise. The noise spectrum does not hold 1/f law, the performance of device is degraded, and S_v/V^2 depends on applied electric field.

The SPRITE detector with good contact have a 1/f spectrum as expected. The Hooge parameter α_H ≈10^{-3}~5×10^{-5}, less than the constant given by Dr. Hooge in 1969.

But for O-C contacts of SPRITE, the S_v/V^2 dependend on bias for both good contact and poor contact, because of sweepout and integration of carriers. The further studies are expected, for existing noise theory couldn't give a perfect explanation.

REFERENCES:

1. *Zhu Xi-Chen, Proc. 10th ICNF(1989)*
2. *Zheng Wei-Jian & Zhu Xi-Chen, Infrared Phy. Vol.33, No.1, pp27-31(1992)*
3. *H.I.Hanafi et al., Solid St.Electron, Vol.21,(1978)*
4. *T.G.M.Kleinpenning, Physica,V ol.77B,(1974)*
5. *l.K.J.Vandamme,Proc.7th ICNF, pp183-192,(1983)*
6. *H.I.Hanafi & A.Van Der Ziel, IEEE Trans., Vol.ED-25, No.9, p1141,(1978)*
7. *A.Van Der Ziel Proc. IEEE, Vol.76, No.3,(1988)*
8. *F.N.Hooge, Phys. Lett., 29A No.11, pp642-643,(1969)*
9. *C.T.Elliott et al.,Infrared Phys. Vol.22, pp31-42,(1982)*
10. *A. Van Der Ziel, Noise in Measurements,(John Wiley & Sons Inc.),(1976)*

X. LASERS

ELECTRICAL AND OPTICAL NOISES IN OPTOELECTRONIC TRANSMITTERS AND RECEIVERS

R. ALABEDRA, B. ORSAL

Centre d'Electronique de Montpellier, Université Montpellier II, 34095 Montpellier cedex 2 , FRANCE.

ABSTRACT

In the first part of this paper the electrical noise of the photodiodes are summarized. In the second part the electrical and optical noise of the laser diodes are presented taking into account the correlation between the both noise sources. In the last part the prospective aspects for the new researches in noise of optical amplifiers are briefly reviewed.

INTRODUCTION

During the last decade the improvements in semiconductor lasers, receivers, detectors arrays, amplifiers and other optoelectronic devices are very significative. Semiconductor optoelectronic devices have now imported all major fields of information technology. The applications of optoelectronics are storage, processing,transmission, memory, computer, etc... Signal and data transmission need a major resolution with the advent of optical fiber communications. The optoelectronic devices are Light Emiting Diode (L.E.D.), Avalanche Photo Diodes (A.P.D.), Charge Coupled Devices (C.C.D.) etc...The optical components are optical fibers, optical amplifiers, filters, directional couplers and the set of linear and non-linear optical passive components.

So the basic studies are very extensive with the advent quantum-Well optoelectronic Devices, for example the Multi-Quantum-Well Avalanche Photodiodes (M.Q.W.A.P.D.) or Quantum Well Lasers (Q.W.L.D.). The investigations on III-V and II-VI materials are very important in order to obtain the optical spectrum from X rays to infra-red wavelengths.

The main purpose of this paper is to present some current ideas on the electrical and optical noises of photodetectors and laser diodes in their applications in optical fiber telecommunications, in consequence an introduction to noise of Erbium doped fiber optical amplifiers.

I NOISE IN PHOTODIODES

I-1 Basic consideration

The photodiodes are light current transductors and so can detect optical signals. A general behaviour of photodiodes biased in reverse condition has basically three processes :

i) carrier generation by absorbed photons.

ii) carrier transport without or with multiplication by impact ionization.

iii) interaction of photocurrent with the external circuit to provide the output signal.

Several parameters are required in order to design P.I.N. photodiodes as photodetectors in optical communication systems. There parameters are for example :

- a strong electric field to serve to separate the carriers,
- a narrow space charge to reduce the transit time,
- nevertheless a wide enough space charge in order to have a good quantum efficiency,
- a small active surface to reduce the capacitance of junction favorising the gain-bandwidth product,
- a hight absorption coefficient α_{ab}(expressed in cm^{-1}) at the working wavelength λ,
- a smaller dark current as possible,
- no-tunneling effect,
- no 1/f noise in obscurity current and photocurrent,
- a good responsitivity $\sigma_0(\lambda)$ in A/W.

I-2 Noise in photodiodes

In this section we shall separate the noise study in two parts : the noise of the device under obscurity condition and the noise under illuminated condition because we will distinguish between the dark current and the photocurrent. We also shall consider at this present time only excellent devices which one can find currently in the modern systems.

I-2-1) Noise in P.I.N. photodiode

I-2-1-1) Noise in obscurity condition

For an optimized device without leakage current, as well known in the classical description[1], the device exhibits a shot noise given by :

$$S_i(f)_{dark} = 2\,qI_{dark} \qquad (A2/Hz) \qquad (1)$$

Of course the shot noise sets a lower limit to the noise in dark current.

I-2-1-2) Noise under illumination condition

I-2-1-2-a) P.I.N. photodiodes illuminated by typical light sources (filament Lamps or Lambertian sources)

In this case the light sources do not bring with their noise sources and the total number of photo electrons produced during any time interval is much smaller than N Atomes, and the individual ionization processes are statistically independent. In other words the number of photons obeys Poisson statistics. The total noise in this situation will be the shot noise as following :

$$S_i(f)_{total} = S_i(f)_{dark} + S_i(f)_{phot} \qquad (A2/Hz) \qquad (2)$$

Let be

$$S_i(f)_{total} = 2q\,[\,I_{dark} + I_{phot}] \qquad (A2/Hz) \qquad (3)$$

Where it is assumed no correlation between the photogenerated and thermal generated carriers. The ideal situation is obtained when the dark current is much smaller than photocurrent. For P.I.N. Silicium photodiode at $\lambda = 0{,}9\ \mu m$ the dark current is about 10^{-11} A and the working photocurrent about 10^{-8} or 10^{-7} A. This is not always the case for infra-red photodiodes designed in small gap materials. Of course the shot noise is the lower limit to the total noise

I-2-1-2-b) P.I.N. photodiode illuminated by non-typical light sources

When P.I.N. photodiodes are illuminated by laser diodes the beam light can bring a noise source which is caracterized by Relative Intensity Noise (R.I.N.).

This term describes the fluctuations of laser diodes optical power by following relation [2] :

$$R.I.N. = \frac{S_{iph} - 2\,qI_{ph}}{I_{ph}^2} \qquad (Hz^{-1}) \qquad (4)$$

So if the P.I.N. photodiodes are illuminated with a thermal lamp, S_{iph} is equal to shot noise and R.I.N. becomes equal to 0. This optical noise source will be discussed in the second part of this paper.

At last the recent theoretical analysis [3,4,5] of the squeezing phenomenon and especially the balanced homodyne detection scheme used for its observation, have led a renewed interest in the shot noise present in the photocurrent and photoelectric detectors [3]. Quantum mechanical description is used as a general expression for noise in the photocurrent even in the case of non-classical states of the radiation. In this case the total noise may actually fall below the shot noise (classical limit) that indicating the presence of non-classical features of the radiation field such as antihuching or squeezing [6,7].

I-3 Noise in Avalanche P.I.N. photodiodes

I-3-1-) Basic consideration

The Avalanche photodiodes are P.I.N. photodiodes which exhibit an internal multiplication of carriers by impact ionization. There devices are strongly reverse biased in order to obtain the high Electric field region between 5.10^4 and 6.10^5 $V.cm^{-1}$ according to the type of materials. Several theoretical and experimental studies have been reported for various structures[8,9]. In low noise photodetection application only multiplication process at low electric field is very useful in order to obtain a great difference between ionization coefficients $\alpha(E)$ and $\beta(E)$ (in cm^{-1}) of electrons and holes respectively.

Currently there are two groups of A.P.D. structures.

i) classical A.P.D. homojunction or heterojunction such as S.A.M.A.P.D. that is Separated Absorption Multiplication Avalanche Photodiode where the ionization by impact is non localized [8,9].

ii) Multi Quantum Well Avalanche Photodiode or Superlatice Avalanche detector where the ionization by impact are localized in stages [10,11].

Briefly the principal theories of ionization process by impact have been developped by Schokley [12], Wolf [13], Baraff [14], Ridley [15] and so on... for various values of electric field in order to obtain the expressions of $\alpha(E)$ and $\beta(E)$.

I-3-2) Noise in non localized multiplication process

It is well known that the A.P.D noise conventional expression is refered to the PhotoMultiplier Tube (PMT) noise given by :

$$S_i(f) = 2qI_{inj}\,(M)^2 \qquad (A^2/Hz) \tag{5}$$

Where I_{inj} is the injected current and M the average multiplication factor

Therefore in the A.P.D. , where the multiplication process concerns the both types of carrier, the noise is given by :

$$S_i\,(f) = 2\,qI_{inj}\,(M)^2 \,.\, F(M) \tag{6}$$

Where F(M) is called the exess noise factor. The excess noise factor is a useful quantity because it compactly represents the statistical properties of the gain fluctuations that introduce multiplication noise; in all cases F(M) strongly depends from values $\alpha(E)$, $\beta(E)$ or $k = \beta(E) / \alpha(E)$. In the literature several authors have proposed[8,9] various theories relative to the statistical treatment of the avalanche process and the noise in A.P.D. in order to determine F(M). For example the Mc Intyre theory[8] permits, statistically speaking, all the possibilities of the impact ionization compatible with thickness W, the α and β ionization coefficients without any limit of the random impact ionization number given by the probability laws. In the case of Van Vliet theory[9] based on discrete device physics, the M and variance [M] are obtained by method of recurrent generating functions. In this approach the noise is always lower than the Mc Intyre limit,the latter being approached to within 5% for gains of the order of 100 or higher and the noise versus (M) shows break points when the regime changes from N to N+1 possible ionizations per carrier transit.

I-3-3) Noise in multilayer avalanche photodiodes

An essential requirement for low noise avalanche photodiodes is to have a large difference between the ionization coefficients α and β. There are two ways to obtain this large difference. First : the main mean to obtain high **k or 1/k** materials is to use the III-V or II-VI ternary or quaternary alloys which exhibit a high k for a given value of stochiometric composition such as $Ga_{1-x}Al_xSb$ APD[16] or $Hg_{1-x}Cd_xTe$[17]. In this case the ionization coefficients α and β material parameters depend on the semiconductor band structure. Second : one can obtain high k values by designing new class of APD using the Band Structure Engineering such as Multi quantum- Well Avalanche Photodiode (MQWAPD)[10] or Staircase Avalanche Photodiodes[18].

The staircase A.P.D. is a sophistical structure with a periodical distribution of graded gap wells. We have a ballistic only by electrons ionization process repeated at each stage. The noise approach proposed by Capasso[10] is the same like for the P.M.T but the variance of the random gain at each stage is $\delta(\delta-1)$ if δ is the fraction of electrons which do not impact-ionize. Thus the excess noise factor after Capasso is:

$$F\,(N,\delta) = 1 + \frac{\delta\,[1-(2-\delta)^{-N}]}{(2-\delta)} \tag{7}$$

Where N is the number of stages.

Recently a Generalized Excess Noise Factor for Avalanche Photodiodes of Arbitrary Structure is proposed by Hakim Saleh and Teich[11]. The authors consider a generic multilayer avalanche photodiode model that admits arbitrary variation of the band gap, dark generation rate and ionization coefficients within each stage of the device. Their formalism follows the usual assumption that the ability of a carrier to ionize other carriers is independent of the carrier's history. Their calculations make use of Mc Intyre's general approach and they relax the restriction that the multiplication assume a Bernouilli form with infinitesimal small success probability in the limit of an infinitesimal distance. Their generalized multilayer structure used for their calculation is shown in figure 1[11].

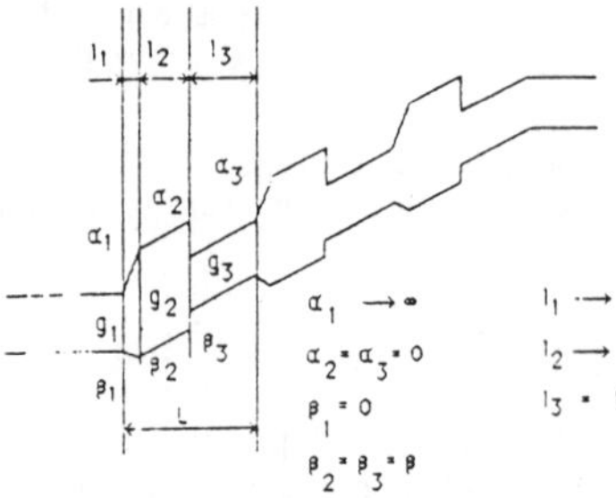

Figure 1 : Generic model used for generalized Excess noise. Factor for Avalanche Photodiode after Hakim et al[11] with number M of identical stages of width L and number N of substages of width l[11].

This theory is applied to :

i) Conventional Avalanche Photodiode APD (see table 1)
ii) MultiQuantum-Well Avalanche Photodiode MQWAPD
iii) Staircase Avalanche Photodiode.

by assuming that the number of photons at the input of the detector is a Poisson distribution. For example with this theory in the continuous limit, the formula (22-d)of[11] is a version of the expression obtained by Mc Intyre[8]. That incorporates both injected and dark generated carriers. The carrier's history is taked into account, $\alpha(x)$ and $\beta(x)$ become $\alpha(x,x')$ and $\beta(x,x')$ respectively to reflect the ionization probabilities of a carrier at the point x when it was generated at the point x'. In this case the authors introduce a "dead space" which prohibits the carrier from multipliyng within a certain distance of its birthplace[19].

In this paper we give some results obtained on three MBE Grown MultiQuantum Wells A.P.D. with thickness of 0.5, 0.2 and 2 µm and consisting of 25, 10 and 20 periods of GaAs Wells and $Al_{0.4}Ga_{0.6}As$ barriers with thickness of 100 - 100 Å and 500 - 500 Å respectively in order to compare Mc Intyre's[8] and Hakim's[11] theories. These results are reported on table 1 in the case of an electron injection[20]

Table 1[20]

		$k = \frac{\alpha}{\beta}$: Mc Intyre[8]	$k_s = \frac{Q}{P}$: Hakim[11]
PIN MQW de 100Å/100Å	ratio k, k_s	$0,15 \leq k \leq 0,25$	$0,1 \leq k_s \leq 0,25$
	Multiplicationf factor	7< M< 14 3.10^5 V/cm < F < 3,2 10^5 V/cm	5 < M < 18 2,7 10^5V/cm< F <3,3 10^5V/cm
Schottky MQW de 100Å/100Å	ratio k, k_s	$0,1 \leq k \leq 0,15$	$0,1 \leq k_s \leq 0,25$
	Multiplication factor	$M \approx 4$ $F \approx 3,7\ 10^5$ V/cm	$M \approx 4$ $F_m = 4\ 10^5$ V/cm
PIN MQW de 500Å/500Å	ratio k,k_s	$0,15 < k < 0,25$	$0,1 \leq k_s \leq 0,2$
	Multiplication factor	4 < M < 10 1,8 10^5V/cm< F < 2,2 10^5V/cm	2 < M < 10 1,5 10^5V/cm <F< 2 10^5 V/cm

Where P and Q in Hakim's theory represent the electron and hole ionization probabilities per stage respectively.

In conclusion we notice that the results obtained by both theories are of the same order of magnitude. It is certainely due to the accuracy of the design of the devices. The second remark is that one must keep in the mind that the quality of the interfaces between each stage provide a non negligible 1/f noise [21].
In spite of everything it seems that Hakim's theory[11] is well suited to study of multilayer A.P.D. (Multiplication, current and noise...).

II ELECTRICAL AND OPTICAL NOISE IN LASER DIODE AND THEIR CORRELATION

Several kinds of noise are often generated in semi conductor lasers because of the wideband response characteristics of carrier density fluctuations. 1/f noise, mode hopping noise or mode partition noise are troublesome : these kinds of noise impede attempts to improve optical coherence[22].
Some studies have been made in order to introduce the low and medium frequency noise as a characterisation parameter with the fluctuations of the carrier density of lasers.In addition to the static characterization, noise measurements are performed on these devices. We show in the low and medium frequency range (1 Hz < f < 10 MHz) the Terminal Electrical Noise (TEN) of several laser structures:V-Groove , D.F.B , RIDGE , S.Q.W. lasers, etc. So additional data can be obtained in order to specify and characterize semiconductor lasers, for instance using electrical noise to qualify the noise behaviour of laser diodes[23].

On one hand, the Terminal Electrical Noise (**TEN**) is due to the fluctuations of the laser voltage $V_d(t)$ and is given by the voltage noise spectral density $\mathbf{S_{Vd}}$ ($\mathbf{V^2/Hz}$). In our case, $\mathbf{S_{Vd}}$ can be written as[23] :

$$S_{V_d}(f) = 4kTR_S + \left(M\frac{kT}{q}\right)^2 < \frac{|\Delta n(f)|^2}{n_0^2\,\Delta f} > \tag{8}$$

$$M = 2 + \frac{n_0}{2\sqrt{2}\mathrm{Vol}}\left(\frac{1}{N_V} + \frac{1}{N_c}\right) \tag{9}$$

where N_C , N_V are the effective conduction and valence band densities, k is the Bolzman's constant, T is the absolute temperature, n_0 is the steady state carrier density, $\Delta n(f)$ is the fluctuation of carrier density, Vol is the volume of the active layer and R_S is the series resistance. The first term gives the thermal noise due to the series resistance R_S.

On the other hand, the optical noise is due to the fluctuations of the optical power P_{opt} of the laser, related to the fluctuations of the detected photocurrent $I_{ph}(t)$, and given by the photocurrent spectral density $\mathbf{S_{Iph}(A^2/Hz)}$[24] :

$$S_{Iph}\,(A^2/Hz) = \sigma^2 \,.\, S_{Popt}\,(W^2/Hz) \tag{10}$$

In order to analyse the noise behavior of the laser, we used the experimental set up shown in another work.
In the first channel we measure $\mathbf{S_{Vd}}$ thanks to a voltage amplifier, connected in parallel with the laser diode. With the second channel we measure $\mathbf{I_{ph}}$ and $\mathbf{S_{Iph}}$, through a standard InGaAs PIN photodiode, respectively via the DC and AC outputs of a current amplifier. Possible optical feedback due to the second channel is suppressed by an optical isolator when we measure simultaneously the electrical noise spectral density $\mathbf{S_{Vd}}$ and the photocurrent spectral density $\mathbf{S_{Iph}}$[24].

II- 1 An important result : 1/f Optical noise

The 1/f noise in the light output of laser diodes imposes a maximum achievable signal to-noise ratio (S/N) in high frequency narrow-band applications. The 1/f noise in the light output has been explained by Fronen and Vandamme in terms of uncorrelated fluctuations in gain and spontaneous emission[25]. The dependence of the spectral noise density S_P on the average output power <P> is expressed by

$$S_p \; \alpha \; \langle P \rangle^m \qquad (11)$$

Two possible noise sources were put forward : fluctuations in the optical absorption coefficient or in the number of carriers[25]. Either source can explain the observed values of m. Experiments on a multimode laser show, for the value of the exponent, m =2/3 in the LED region, m = 5/2 in a narrow transition region, m = 4 in the superradiative region and $0 \leq m \leq 1$ in the laser region. A steeper increase in the noise with m = 6 to 7 has been found in the superradiative region of monomode lasers. This deviates very considerably from the calculated and experimentally observed dependence with m = 4 in gain-guided lasers.

They showed that m = 6 to 7 is caused by the onset of external cavity modes. This situation can be prevented by using an optical isolator. To explain the 1/f noise in the laser region, Fronen and Vandamme have made the assumption that the coherent emission makes no significant contribution to the 1/f noise in the total emission. Subsequently, the noise above threshold depends only on the noncoherent emission and the noise obeys the relation which is just below threshold[26].

II-2 Experimental results: electrical noise study (1 Hz < f < 10 MHz)

Spectral densities of the electrical voltage noise as a function of the frequency are shown in *figure 2-a* at 20°C and for five laser currents I_L equal to 9, 11, 26, 30 and 65 mA. For 9 and 65 mA, the noise spectra are classical : 1/f electrical noise ($f < 10^3$ Hz) and white noise. For I_L=30 mA we observe the *mode hopping noise*, which is attributed to the suppression of one mode because it is spent lasing by the other mode. The insert shows time dependence of voltage and optical noises, given by $V_d(t)$ and $I_{ph}(t)$ respectively.

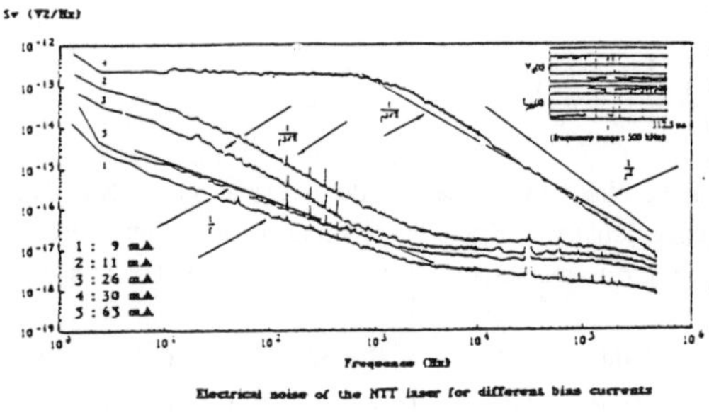

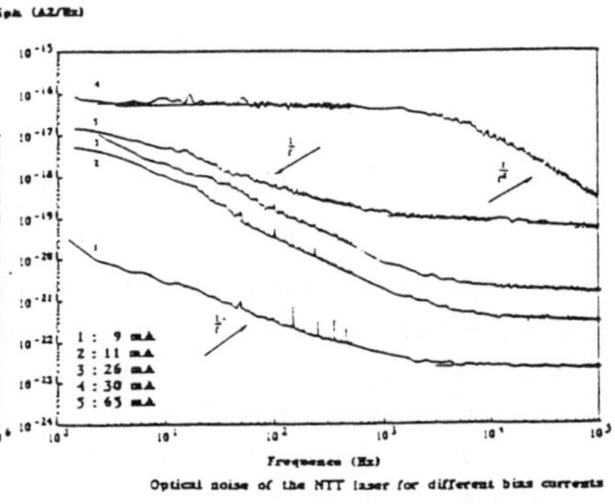

Figure 2-a

Figure 2-b

II-3 Experimental results: optical noise study (1 Hz < f < 10 MHz)

Spectral densities of the optical noise as a function of the frequency are shown in *figure 2-b*. Comparing *figures 2-a* and *2-b*, we notice, for each polarization current, a tight correlation between the two noise sources.

The *figures 3-a* and *3-b* show respectively the electrical noise S_{Vd} and the optical noise S_{Iph} versus laser current I_L. These noises are strongly correlated. The fine structure observed at high polarizations in the behavior of both noises can be explained by the *mode hopping noise* phenomenon (see *figures 2*) described by Ohtsu and al[22].

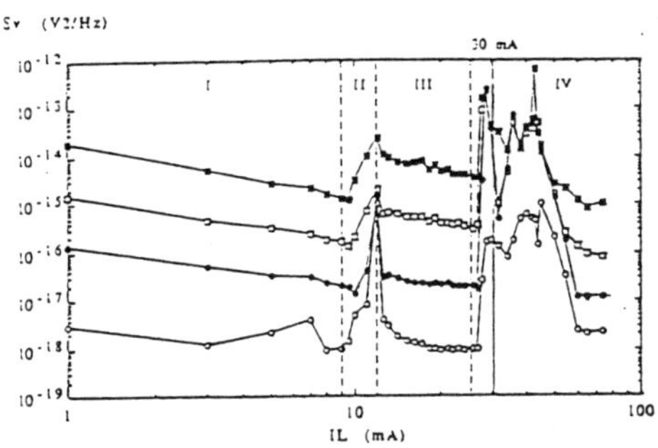

Figure 3-a

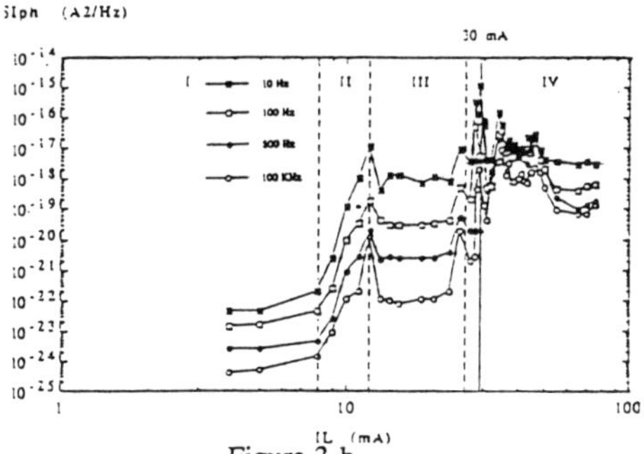

Figure 3-b

II-4 Rate equations

All these results can be explained by using the noise equivalent circuit of a semiconductor multimode laser derived by the rate equation taking into account Mc Cumber's, Harder's, Yamada's and Andrekson's theories[23,27,28]. The noise equivalent circuit of a semiconductor multimode laser diode is derived from the rate equations including Langevin noise sources fn(t), fs(t) and 1/f noise sources $f_{n1/f}$, $f_{s1/f}$ taking into account Rob Fronen and Vandamme's theory[25] as :

$$\frac{dn}{dt} = \frac{i}{q} - \frac{n}{\tau_s} - \sum_{k=1}^{p} g_k s_k + f_n(t) + f_{n1/f}(t) \tag{12}$$

$$\frac{ds_k}{dt} = g_k s_k + \gamma_k \frac{n}{\tau_s} - \frac{s_k}{\tau_{ph}} + f_{sk}(t) + f_{sk1/f}(t) \qquad 1 \le k \le p \tag{13}$$

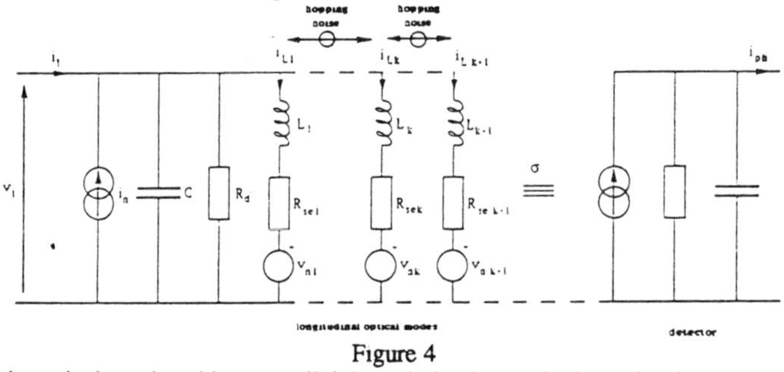

Figure 4

Each optical mode adds a parallel branch in the equivalent circuit where the current i_{Lk} corresponds to the photons of the signal in the kth mode .The gain of the kth mode in the linear approximation is given by [5] :

$$g_k = g_{ok} + A_k n 1_k \tag{14}$$

By computation we obtain the noise sources $\mathbf{i_n}(t)$ and $\mathbf{v_n}(t)$ in relation to the fluctuations of carrier number n and photon number **s** :

$$i_n(t) = C\frac{MV_T}{n_o}[f_n(t) + f_{n1/f}(t)] \tag{15}$$

$$v_{nk}(t) = q\, g_{ok}\, L\, [f_{sk}(t) + f_{sk1/f}(t)] \tag{16}$$

V_T is KT/q, C is the usual diffusion capacitance of the junction and g_{ok} (s^{-1}) is the steady state optical gain of the kth mode in the equivalent circuit .

$\mathbf{Z_{eq}}$ is the low frequency equivalent impedance, **R** the differential resistance, $\mathbf{R_{se}}$ the resistance related to the spontaneous emission ratio γ_k : $\gamma_k = 10^{-4}$. $\mathbf{I_{th}}$ is the threshold current.

We have computed the spectral density S_{v_d} (f) of the electrical noise :

II-5 Electrical and Optical Noise Spectral Densities

From the intrinsic equivalent circuit and the knowledge of intrinsic parameters, the noise modulation characteristic can be computed.

$$S_{v_d}(f) = \frac{\overline{V_d^2}}{\Delta f} \quad \text{is the voltage noise spectral density} \tag{17}$$

$$S_{\Delta s}(f) = \frac{\overline{S_n^2}}{\Delta f} \quad \text{is the photon number noise spectral density,} \tag{18}$$

where $S_n(t)$ is the instantaneous photon number. All these parameters have been computed in a previous work. At low and medium frequency: 10Hz<f<10 MHz, we can consider that $L\omega \ll R_{se}$ and $\frac{1}{C\omega} \gg R$.

a) Computed spectral density S_{v_d}(f) of the electrical noise:

$$Svd_{1/f} = 4kTR_s + \frac{\alpha}{Nf} R_s^2 I_L^2 + \frac{Zeq^2}{\Delta f}\left(\overline{i_n^2} + \sum_{k=1}^{p} \frac{\overline{v_{nk}^2}}{Rse_k^2} + 2\sum_{k=1}^{p} \mathrm{Re}\left\{ \frac{\overline{i_n v_{nk}^*}}{Rse_k} \right\} \right) \tag{19}$$

$$\frac{1}{Zeq} = \frac{1}{R_d} + \sum_{k=1}^{p} \frac{1}{Rse_k} \quad \text{with } R_d = \frac{MV_T}{qn_0\left(A_k s_0 + \frac{1}{\tau_s}\right)} \quad \text{and } Rse_k = \frac{g_k M V_T}{q\tau_s s_0 g_{0k}\left(A_k s_0 + \frac{g_k}{\tau_s}\right)} \tag{20}$$

$$\overline{i_n^2} = \left(C\frac{MV_T}{n_0}\right)^2 [\, \overline{f_n^2} + \overline{f_{n1/f}^2} \,] \tag{21}$$

$$\overline{v_{nk}^2} = (q\, g_{ok}\, L)^2 [\, \overline{f_{sk}^2} + \overline{f_{sk1/f}^2} \,] \tag{22}$$

$$\overline{i_n v_{nk}^*} = C\frac{MV_T}{n_0} q\, g_{ok}\, L\, [\, \overline{f_n f_{sk}^*} + \overline{f_{n1/f} f_{sk1/f}^*} \,] \tag{23}$$

b) Computed spectral density $S_{i_{Ph}}$(f) of the optical noise:

$$S_{I_{ph_{1/f}}}(f) = R.I.N.*I_{ph}^2 = I_{ph}^2 \left(\frac{\frac{\overline{i_n^2}}{\Delta f} + \frac{1}{R_d^2}\frac{\overline{v_n^2}}{\Delta f} - \frac{2}{R_d}\frac{\overline{v_n * i_n}}{\Delta f}}{(qg_0s_0)^2\left(1+\frac{R_{se}}{R_d}\right)^2} \right) \tag{24}$$

We observe that the equations S_{vd} and S_{iph} depend on the same noise sources $\frac{\overline{i_n^2}}{\Delta f}$, $\frac{\overline{v_n^2}}{\Delta f}$ and $\frac{\overline{i_n v_{nk}^*}}{\Delta f}$. Here, we consider only the 1/f and white noise sources and the cross spectrum $\overline{v_{nm} \cdot v_{nl}} = 0$, but when the hopping noise is present, $\overline{v_{nm} \cdot v_{nl}}$ is not equal to zero because the population of each mode is not independent. Our results clearly show the expected correlation between the electrical noise and the optical noise, particularly at the onset of stimulated emission and with an excess noise source which can be due to hopping noise or 1/f noise.

c)Mode Hopping Noise: The intensity fluctuations of different modes are negatively correlated because the different modes are competing for gain from the electron "tank"[22]. This so called competition noise significantly reduce the signal to noise ratio in communication systems. Longitudinal mode hopping is associated with output power fluctuations and gives exccss noise both in the optical intensity noise and in the electrical noise (see § III). It is attributed to the suppression of one mode by the decrease in carrier density because it is spent lasing by the other mode. This phenomenon is driven by the randomly generated spontaneous emission that works as the triggering force for lasing the other mode. Intensity fluctuations follow the statistics of a Poisson process. The profile of the power spectral density of the noise can be approximated as a Lorentzian one[22]. This fact introduces excess noise sources which can be detected by electrical and optical noise measurements (*figures 3-a* and *3-b*) on the low and medium frequency range.

III NOISE of OPTICAL AMPLIFIERS

The amplifier noise model is based on the work by Simon[29] and Olsson[30] .The spontaneous emission power at the output from an optical amplifier is given by:

$$P_{sp}=N_{sp}\,(G-1)\,h\nu\,B_o \tag{25}$$

B_o is the optical bandwidth,P_{sp} the spontaneous emission power,N_{sp} the spontaneous emission factor,$h\nu$ the photon energy, G the optical gain. For an ideal amplifier $N_{sp}=1$,for an Erbium doped fiber amplifier $N_{sp}=2.2$ and for a semiconductor laser amplifier N_{sp} ranges from 1.4 to 4 depending both on the pumping rate and the operating wavelength.This model is based on a talk given by P.S.Henry[31].We assume an optical amplifier and a detector with unity quantum efficiency.After square law detection in the receiver,the signal power S is given by:

$$S=(GI_sC_1C_2L)^2 \tag{26}$$

I_s is the photocurrent equivalent of amplifier input power,C_1 amplifier input coupling efficiency,C_2 amplifier output coupling efficiency, L Optical loss between amplifier and receiver. The total noise N_{tot} is:

$$N_{tot}=N_{shot}+N_{sp\text{-}sp}+N_{s\text{-}sp}+N_{th} \tag{27}$$

The different terms are:

$$N_{shot}=2B_e qC_2L(GI_sC_1+Isp) \quad (28)$$

$$N_{sp\text{-}sp}=4B_e qC_1C_2{}^2L^2GI_{sp}C_1/B_o \quad (29)$$

$$N_{s\text{-}sp}=I_{sp}{}^2C_2{}^2L^2B_e(2B_o-B_e)/B_o{}^2 \quad (30)$$

N_{th} is the receiver noise, q is the electronic charge,B_e the electrical bandwidth,I_{sp} the photocurrent equivalent of the spontaneous emission power, N_{shot} the shot noise, $N_{sp\text{-}sp}$ the spontaneous-spontaneous beat noise, $N_{s\text{-}sp}$ the signal-spontaneous beat noise. A schematic of amplifier model is shown in figure 5.

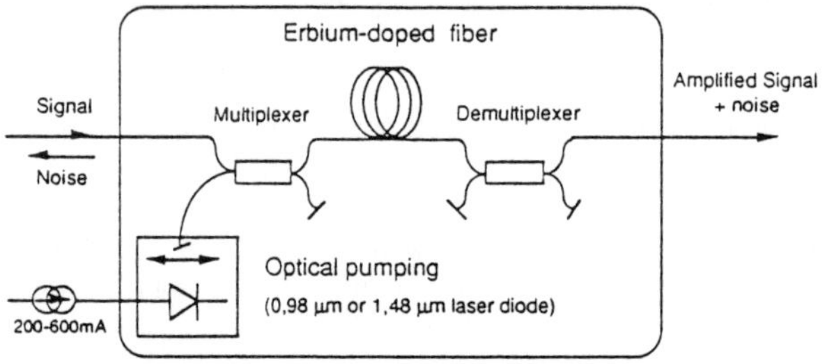

Figure 5 : Schematic of amplifier model.

An application : optical preamplifier

Of particular interest for preamplifier applications is the receiver sensitivity dependence of the amplifier gain, noise figure, and optical bandwidth. The application of optical preamplifiers is at very high data rates where avalanche photo detectors are limited by their gain-bandwidth product. To make the calculations realistic we have used measured values for the receiver and amplifier parameters. The value chosen for the thermal noise current corresponds to a base receiver sensitivity of 25 dBm. At low amplifier gains the receiver is limited by the thermal noise and consequently the receiver sensitivity improves 1 dB for every decibel of gain. For higher gains, the signal spontaneous and spontaneous-spontaneous beat noise becomes dominant and the best achievable receiver sensitivity depends of the optical bandwidth when $B_o = 2B_e$ and is in this case equal to -39.4 dBm.

CONCLUSION

Researches in photonics and photon-electron interactions, as for example the correlation between electrical noise and optical noise, will become more and more important. Indeed optic will take place in the very great number of modern systems. For instance, in the case of optical fibre telecommunication networks, the noise limits the number of photons per bit which is about 400 photons for a given Bit Error Rate of 10^{-9}, while the theoretical limit is about 40 photons.

REFERENCES

1 Van der Ziel A. : Proceeding of the I.E.E.E, pp1178, Vol 58,N°8 August 1970.
2 Joindot : Phd. Montpellier, September 1990.
3 Paul H.: Journal of Modern Optics, Vol 35, n°7, pp1225-1235, 1988.
4 Glauber R.J : Quantum Optics and Electronics, p65, Edited by C. de Witt, A. Blandin and C. Cohen-Tannoudji (New York : Gordon and Breach) 1965.
5 Yurke B. : Phys. Rev., A.35, pp 300, 1985.
6 Slusher R.E., Hollberg L.W., Yurke B, Mertz J.C. and Valley J.F : Phy. Rev. lett. 55, 2409, 1985.
7 Wu L.A., Kimble H.J., Hall J.L. and Wu H.F : Phys. Rev., Lett. 57, 2520, 1986.
8 Mc Intyre R.J. : IEEE Trans Electronic Device, Vol ED, pp164-168, 1966.
9 Van Vliet K.M., Rucher L.M. : IEEE Trans Electron Devices, Vol ED-6, n°5, pp 746-764, 1979.
10 Capasso F., Tsang W.T. and Williams G.F. : IEEE Trans Electron. Device, Vol ED-30, pp 381-390, 1993.

11 Hakim N.Z., Saleh B.E.A., Teich M.C. : IEEE Trans electron devices Vol 37, n°3, pp 599-610, March 1990.
12 Schokley N. : Solid State Electronics, Vol 2, n°1, pp 35-67, 1961.
13 Wolff P.A. : Phytsical Rewiew, Vol 95, n°6, pp1415, Septembre 1954.
14 Baraff G.A. : Phy. Rev., Vol 1284, n°6, pp 2507-2517, 1962.
15 Ridley B.K. : J. Phys (16, n°17, pp3373-88) June 1983.
16 Hildebrad O., Kuchart W., Benz K.W., Pilkum M.H. : IEEE, Journal of Quantum Electron. Vol QE17-n°2, February 1981.
17 Orsal B., Alabedra R., Maatougui A., Flachet J.C. : IEEE Trans Electron Devices, Vol 38, n°8, pp1748-1756, August 1991.
18 Capasso F., Williams G.F., Tsana W.T. : Tech. Dig. IEEE Specialist Conf. on Light Emitting Diodes and photodetecgtors (Ottawa - Hull Canada) pp 166-167, 1982.
19 Hakim N.Z., Saleh B.E.A., Teich M.C. : Journal of Lightwave Technology, Vol 10 n°4, pp 458-468, April 1992.
20 Kibeya S. : Phd. thesis, Montpellier, April 1992.
21 Kibeya S., Orsal B., Alabedra R., Vandamme L.K.J. : Proceeding Inter-Conf. Noise in Physical Systems and 1/f Fluctuations-Edited by T. Musha, S. Sato, M. Yamamoto, pp 325-329, Kyoto, Japan, 1991.
22 Ohtsu M. : Highly Coherent Semiconductor Lasers, Edited by Culshaw, A; Rogers, H. Taylor. Artech House Inc. Norwood M.A. 02062 U.S.A., 1992.
23 Andrekson P.A., Anderson P., Alping A., Eng S.T. : Journal of Lightwave Technology, Vol LT 4, n°7, pp 804-812, July 1986.
24 Orsal B., Alabedra R., Signoret P., Letellier C. : Proceedings of S.P.I.E, E.C.O.4 "Infrared Materials and Devices" The Haye, the Netherlands, Vol 1512 A, March 1991.
25 Fronen R.J., Vandamme L.K.J. : I.E.E.E. J.Q.E. Vol 24, n°5, May 1988.
26 Orsal B., Peransin J.M., Daulasim K., Signoret P., Joindot J. : INCF' 93, August 16-20, 1993 Saint Louis, Missouri, U.S.A.
27 Yamada M. : I.E.E.E., J. Quantum Electron, Vol QE-22, pp1052-1059, 1986.
28 Harder C., Katz J., Yariv A. : I.E.E.E., J. Quantum Electron, Vol QE 18, pp 333-337, 1982.
29 Simon J.C. : J. Lightwave Technology, Vol LT 5, n°5,pp1286-1295, 1987.
30 Olsson N.A. : I.E.E.E., J. L.T., Vol 7, pp 1071-1082, July 1989.
31 Henry P.S. : I.E.E.E., J. Quantum Electron, Vol QE 21, n°12, pp 1862-1869, 1985.

Electrical and Optical noise of high power Strained Quantum Well lasers used in Erbium doped fiber amplifiers.

B. ORSAL, K. DAULASIM, P. SIGNORET, R. ALABEDRA, J-M. PERANSIN
Centre d' Electronique de Montpellier - CEM (CNRS URA 391) - Université Montpellier II -
34 095 Montpellier Cedex - France

ABSTRACT

The optical amplifiers are formed by an Erbium-doped fiber, optically pumped medium by a high power semiconductor laser emitting in the 980 nm region. The pump power is efficiently injected into the Erbium fiber using a precision wavelength division multiplexing fiber optic coupler . The aim of this paper is to propose a noise study of InGaAs / GaAs ridge single quantum well (SQW) lasers as pump lasers .

INTRODUCTION

Several kinds of noise are often generated in the semiconductors lasers because of the wideband response characteristics of carrier density fluctuations: 1/f ,hopping mode or partition noises are very troublesome in the various applications of the Erbium doped Fiber Amplifier [1,2]. These noise levels impede from improving the optical coherence and the signal to noise ratio . Several studies have been made in order to introduce the low and medium frequency noise for transmitters (for example DFB lasers), but yet nothing concerning the strained SQW lasers. See figure1.

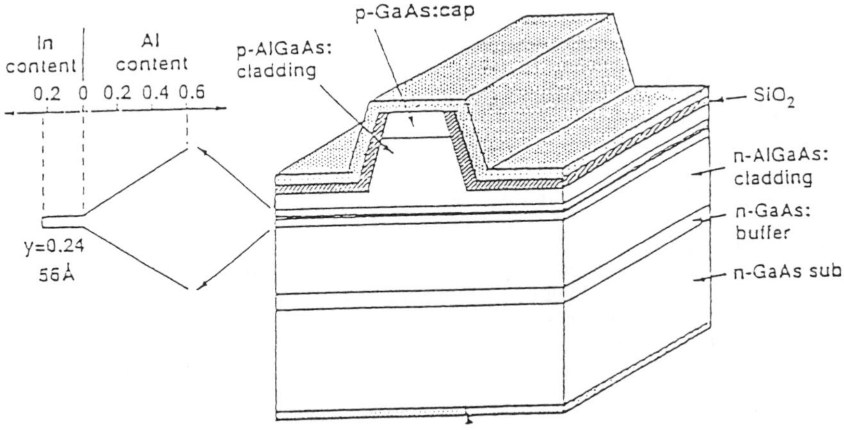

Figure 1 : Schematic diagram of the 980 nm InGaAs/GaAs Strained Single Quantum Well Laser.

EXPERIMENTAL RESULTS:

On one hand , the Terminal Electrical Noise (**TEN**) is due to the fluctuations of the laser voltage $V_d(t)$ and is given by the voltage noise spectral density $\mathbf{S_{Vd}}$ **(V^2/Hz)**, as [3,4] :

$$S_{V_d}(f) = 4kTR_S + \left(m\,\frac{kT}{q}\right)^2 < \frac{|\Delta n(f)|^2}{n_0^2\,\Delta f} > \qquad (1)$$

$$\text{with} \qquad m = 2 + \frac{n_0}{2\sqrt{2}\mathrm{Vol}}\left(\frac{1}{N_V} + \frac{1}{N_c}\right) \qquad (2)$$

where Nc ,Nv are effective conduction and valence band densities, k is the Bolzman's constant, T is the temperature, n_0 is the steady state carrier density, $\Delta n(f)$ is the low and medium frequency fluctuation of carrier density and R_S is the series resistance. The first term gives the thermal noise due to R_S.

On the other hand, the optical noise is due to the fluctuations of the optical power $\mathbf{P_{opt}}$ of the laser, related to the fluctuations of the detected photocurrent $\mathbf{I_{ph}(t)}$, and given by the photocurrent spectral density $\mathbf{S_{Iph}}$:

$$S_{Iph}\ (A^2/Hz) = \sigma^2 \,.\, S_{Popt} \qquad (W^2/Hz) \qquad (3)$$

σ is the apparent photodetector sensitivity. The measurements at low and medium frequency of the electrical and optical noises are shown figures 2 and 3. We oberve the presence of the "mode-hopping" phenomenon in the strained SQW lasers at 30 mA and T=20°C. It is found the same frequency dependence both on the electrical and the optical noise spectra for different currents. The electrical mode-hopping noise has a well marked Lorentzian dependence. The optical mode-hopping noise has been observed firstly and described by Ohtsu et al [2].

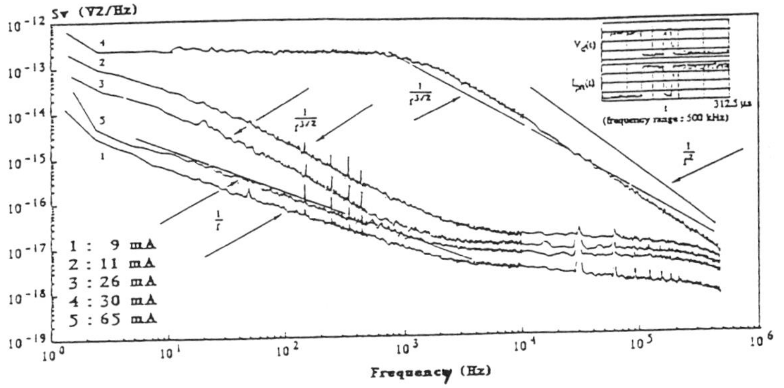

Figure 2 : Electrical noise of SQW laser for different bias currents.

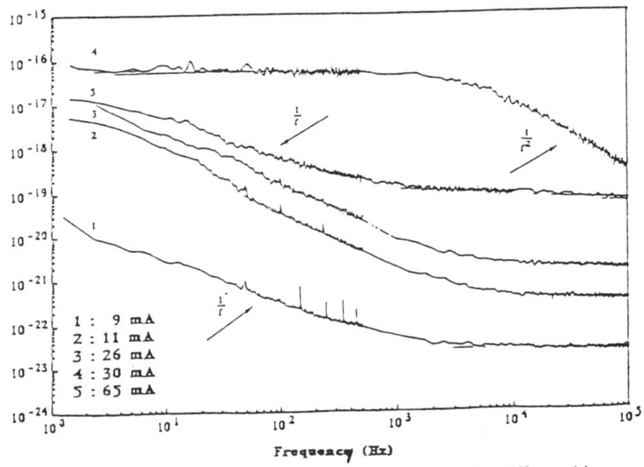

Figure 3 : Optical noise of SQW laser for different bias currents.

DISCUSSION

a)-White noise: Physical Interpretation

To explain the experimental results, we can use a theoretical relation based on Haug's model given by [5,6]:

$$S_i = 4qI_L\left(\frac{L_n}{d}\right)^2 + 2qI_L + 4q^2E_{vc}S_o + S_{i\Delta S} \quad (4)$$

Where I_L is the current of the laser diode, L_n is the minority carrier diffusion length, d is the active layer thickness, E_{vc} is the absortion rate [5,6]. S_0 is the steady state photon number.

Involving the following noise sources :

- the thermal noise due to the thermal fluctuations of injected minority carrier : $4qI_L\left(\frac{Ln}{d}\right)^2$
- the noise source due to the effect of the absorption : $4q^2E_{vc}S_o$
- the pure shot noise effect : $2qI_L$
- the light-field fluctuations : $S_{i\Delta S}$

If we transform the relation (4) in voltage spectral noise density by using the value of the differential resistance R_d we obtain for the white noise:

$$S_v = \left(\frac{mkT}{qI_L} + R_s\right)^2 \left[4qI_L\left(\frac{L_n}{d}\right)^2 + 2qI_L\right] + \left(\frac{mkT}{qI_L} + R_s\right)^2 \left[4q^2E_{vc}S_o + S_{i\Delta S}\right] + 4KTR_s \quad (5)$$

We note the strong correlation between the electrical and optical noises in the white region.

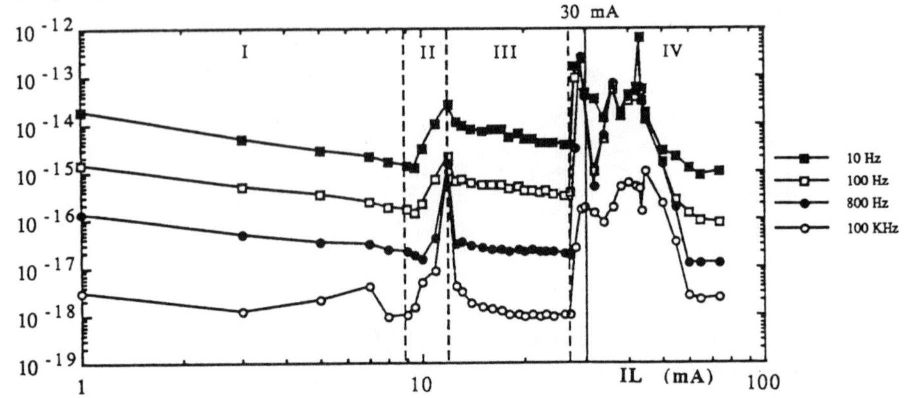

Figure 4 :Electrical -Voltage noise of the strained Single Quantum Well (S.Q.W.) laser versus laser current I_L ar T=20°C and I_{th}=11,5mA

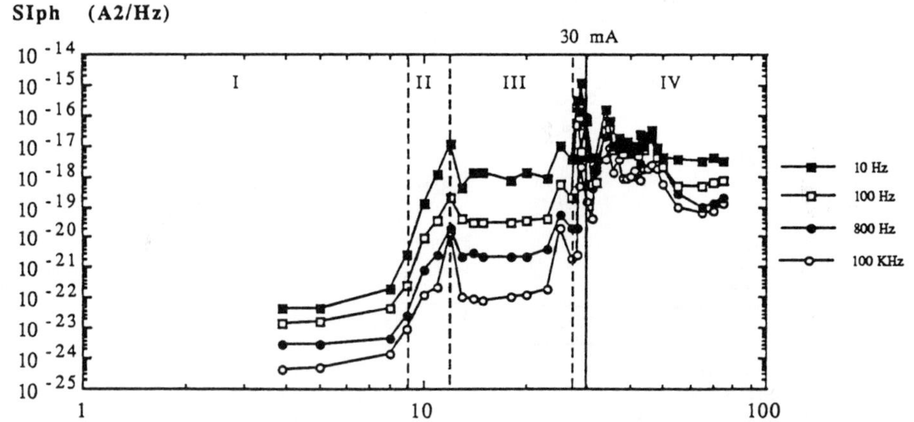

Figure 5 :Optical noise of the strained Single Quantum Well (S.Q.W.) laser versus laser current I_L ar T=20°C and I_{th}=11,5mA

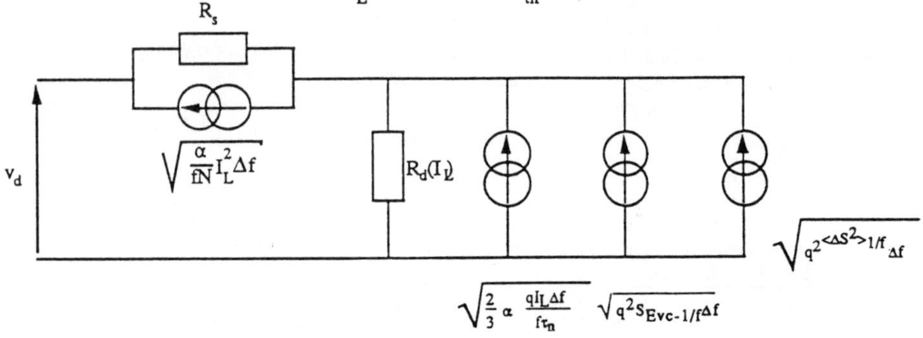

Figure 6 :1/f noise equivalent circuit

b)-1/f noise:

For the 1/f noise measured at 10,100,800Hz,we propose a model based on the same idea where 1/f noise sources are independent of white or hopping noise sources.In this case , the equivalent circuit is similar to the one proposed by Harder et al in which 1/f sources take place of white noise sources [5].The term ($\alpha/Nf*I^2*R_s^2$) can be neglected because $\alpha/N=10^{-12}$ and Rs=3.5Ω.See Figure 6.

$$S_{v1/f}=\left(\frac{mkT}{qI_L}+R_S\right)^2\left[\left(\frac{2}{3}\alpha\frac{qI_L}{f\tau_n}\right)+\left(q^2 S_{Evc-1/f}+q^2 <\Delta S^2>_{1/f}\right)\right] \qquad (6)$$

- $\left(\frac{2}{3}\alpha\frac{qI_L\,\Delta f}{f\tau_n}\right)^{1/2}$: noise current due to carrier diffusion given by the Kleinpenning's model [8] ,
- $\left(q^2 S_{Evc-1/f}\,\Delta f\right)^{1/2}$: excess noise current due to fluctuations of the absorption coefficient E_{vc},
- $\left(q^2<\Delta S^2>_{1/f}\Delta f\right)^{1/2}$: excess noise current due to 1/f fluctuations of the photon number S.

Figure 6 :1/f Noise Equivalent Circuit

The first product varies as I_L^{-1}as predicted by LKJ.Vandamme when the laser current is lower than the treshold current I_{th} : **Region N°I**[7,8]. It is associated to the fact that current noise spectral density $S_{I1/f}$ varies as I_L. The second term gives the excess noise due to the fluctuations of absorption coefficient E_{vc}, around the threshold (see figures 4 and 5) :**region N°II** . The high level observed at 12mA is due to mode hopping noise which appears around the threshold. The third term is caused by the excess noise of the light field because fluctuations of the photon number:**region N°III**.(12mA<I_L<26mA):The very high levels mentionned in the **regionIV** are due to the Hopping Mode effect, particularly around 30,36,45mA.

The same behaviour of the optical noise $S_{Iph1/f}$ is observed in the four regions :

region N°I : increasing of $S_{Iph1/f}$ as $I_L^{3/2}$: spontaneous emission.

region N°II : increasing of $S_{Iph1/f}$ as I_L^{15} around the threshold I_{th}=11.5mA, just before the stimulated emission. This behaviour is due the hopping effect which appears around 12 mA and disappears after.

region N°III.(12mA<I_L<26mA):We observe a good saturation of 1/f optical noise as predicted by Fronen and Vandamme[10]. It corresponds to a saturation of the electrical noise $S_{v1/f}$. It is the LASER regime which corresponds to a good saturation of 1/f optical noise. In this case, the white noise level tends to the SHOT NOISE of the photocurrent.

region N°IV:Hopping noise is shown in figure 5 :picture n°4 , I_L=30 mA,see also figure 4.

The correlation between the optical and electrical noises is computed for low and medium frequencies using the **coherence function** [4]:

$$\gamma^2_{Iph-Vd}(f)=\frac{|S_{Iph-Vd}(f)|^2}{S_{Iph}(f)\,.\,S_{Vd}(f)} \qquad (7)$$

S_{Iph-Vd} is the noise cross spectrum between S_{Vd} and S_{Iph} .

CONCLUSION

In summary,the noise measurements as a function of current I_L confirm that it is possible through only the electrical noise to detect the presence of hopping noise as seen on the figure 4.We have shown that the Terminal electrical noise (T.E.N.) is highly correlated to the optical noise and the electrical noise measurement is a good image of the behaviour (or the defects) of the light-field fluctuations (or the optical power).It could be used for "in situ" noise characterisation of laser diodes without any optics and accompagnying alignment which might introduce undesired optical feedback[4].

REFERENCES :

1. WALKER.R.G.,STEELE.R.G.andWALKER N.G,IEEEJof Lightwave Technology vol.LT8, N°9,pp 1409-1413, 1990
2. OHTSU M. and TERAMACHI Y., IEEE J. of Quantum Electronics, vol.25, pp 31-38, January 1989.
3. ANDREKSON P.A., ANDERSSON P., ALPING A. and ENG S.T., IEEE J. of Lightwave Technology, vol.LT4, N°7, July 1986
4. ORSAL B, ALABEDRA R, SIGNORET P.and LETHELIER C,Proc of S.P.I.E,The Hague The Netherlands, vol.1512 A,March,1991
5. HAUG H., Zeitschrift Für Physik 206, 163-176, 1967.
6. HARDER C.,KATZ J.,MARGALIT S.,SCHACHAM J and YARIV A., IEEE J. of Quantum Electron.,vol. 18, N°3, pp 333-337, 1982.
7. ORSAL B.,PERANSIN J.M,DAULASIM K., SIGNORET P.,JOINDOT I. I.C.N.F.'93, August 16-20, 1993 St Louis,Missouri,U.S.A..
8. VANDAMME L.K.J. and RUYVEN L.J.V., Noise in physical Systems and 1/f noise, pp 245-247, Elserier science publishers BV, 1983.
9. KLEINPENNING T.G.M.., J.Vac.Sci.Technol. A3 (1),Jan / Feb 1985
10. R.J.FRONEN and L.K.J.VANDAMME,I.E.E.E.,J.Q.E..,Vol.24,N°5,May 1988.

Correlation between 1/f optical and electrical noises in InGaAsP substrate Buried Crescent (PBC) laser diodes

B. ORSAL*, J-M. PERANSIN *, K. DAULASIM*, P. SIGNORET*, I. JOINDOT**
* Centre d' Electronique de Montpellier - CEM (CNRS URA 391) - Université Montpellier II - 34 095 Montpellier Cedex - France
** Centre National d'Etudes des Télécommunications - France Télécom - Route de Trégastel - BP 40 - 22 301 Lannion Cedex - France

In this work, the influence of the PBC electrical parameters on the 1/f electrical and optical fluctuations is discussed. The frequency range is:(1Hz-200KHz).The correlation between these two types of noise is studied at 10Hz. The device used in our measurements is a P-Inp substrate -1.3 µm InGaAsP / InP buried crescent (figure 1).

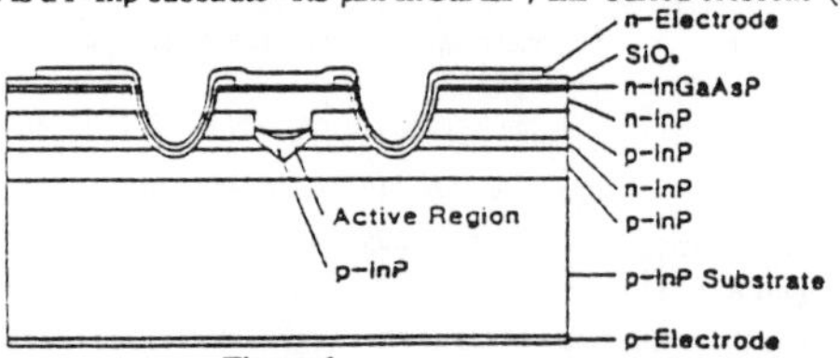

Figure 1

It has a high characteristic temperature ($T_0 \approx 60K$) which is mainly attributed to the following reason : the leakage currents flowing through the blocking layer adjacent to the active region are much weaker than in a simple Buried Crescent (BC) laser [1]. This point is very important for noise behaviour, because in our measurements the intrinsic junction noise can be separately investigated. The laser is an index guided HITACHI HL 1343 MF. We show on figure 2 the principle of measurement [2].

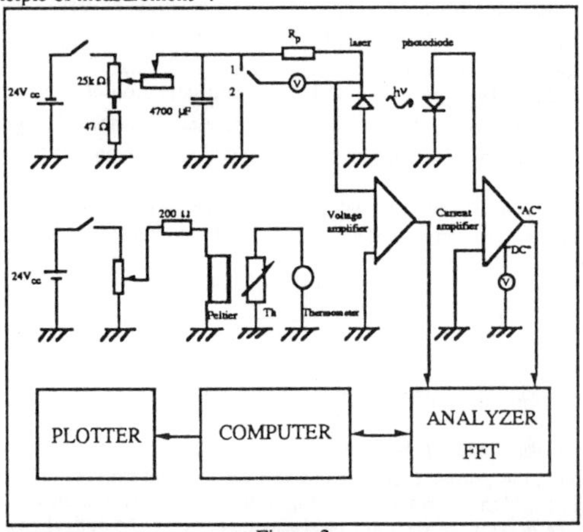

Figure 2

On one hand , the Terminal Electrical Noise (**TEN**) is due to the fluctuations of the laser voltage $\mathbf{V_d(t)}$ and is given by the voltage noise spectral density $\mathbf{S_{Vd}}$ (V^2/Hz), as :

$$S_{V_d}(f) = 4kTR_S + \frac{\alpha}{Nf} R_S^2 I_L^2 + S_{V_1}(f) \quad \text{where} \quad S_{V_1}(f) = \frac{\overline{V_1^2(f)}}{\Delta f} \qquad (1)$$

V_1 is the RMS noise voltage of the laser junction. We must consider that the 1/f noise source related to series resistance $\mathbf{R_s}$ is given by the Hooge's relation [3]. The laser current $\mathbf{I_L}$ is the mean value of the instantaneous current i_1: $I_L = \langle i_1 \rangle$, α is the Hooge parameter, N is the number of carriers and f the frequency On the other hand, the optical noise is due to the fluctuations of the optical power of the laser $\mathbf{S_{Popt}}$, related to the fluctuations of the detected photocurrent $I_{ph}(t)$, and given by the photocurrent spectral density $\mathbf{S_{Iph}}$. σ is the apparent sensitivity of the detector (A/W).

$$S_{Iph}\ (A^2/Hz) = \sigma^2 \cdot S_{Popt}\ (W^2/Hz) \qquad (2)$$

The noise equivalent circuit of a semiconductor multimode laser diode (figure 3) is derived from the rate equations including Langevin noise sources $f_n(t)$, $f_s(t)$ and 1/f noise sources $f_{n1/f}$, $f_{s1/f}$ taking into account Rob Fronen's theory [4] as :

$$\frac{dn}{dt} = \frac{i}{q} - \frac{n}{\tau_s} - \sum_{k=1}^{p} g_k s_k + f_n(t) + f_{n1/f}(t) \qquad (3)$$

$$\frac{ds_k}{dt} = g_k s_k + \gamma_k \frac{n}{\tau_s} - \frac{s_k}{\tau_{ph}} + f_{sk}(t) + f_{sk1/f}(t) \qquad 1 \le k \le p \qquad (4)$$

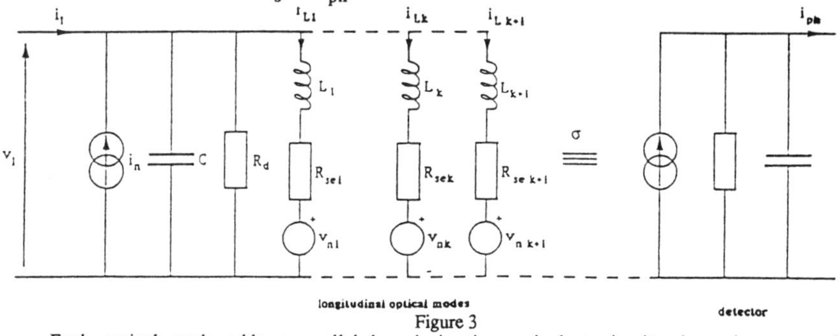

Figure 3

Each optical mode adds a parallel branch in the equivalent circuit where the current $i_{L\,k}$ corresponds to the photons of the signal in the kth mode .The gain of the kth mode in the linear approximation is given by [5] :

$$g_k = g_{ok} + A_k n \qquad (5)$$

By computation we obtain the noise sources $i_n(t)$ and $v_n(t)$ in relation to the fluctuations of carrier number n and photon number s :

$$i_n(t) = C \frac{mV_T}{n_o} [\, f_n(t) + f_{n1/f}(t) \,] \qquad (6)$$

$$v_{nk}(t) = q\, g_{ok}\, L\, [\, f_{sk}(t) + f_{sk1/f}(t) \,] \qquad (7)$$

$$M = 2 + \frac{n_o}{2\sqrt{2}\mathrm{Vol}} \left(\frac{1}{N_V} + \frac{1}{N_c}\right) \qquad (8)$$

where Nc and Nv are the effective conduction and valence band densities, n_o is the steady state carrier density, $\Delta n(f)$ is the fluctuation of carrier density, V_T is KT/q, C is the usual diffusion capacitance of the junction and g_{ok} (s^{-1}) is the steady state optical gain of the kth mode in the equivalent circuit .

Z_{eq} is the low frequency equivalent impedance, R the differential resistance, R_{se} the resistance related to the spontaneous emission ratio γ_k : $\gamma_k = 10^{-4}$. I_{th} is the threshold current.

We have computed the spectral density $S_{v\,1/f}$ of the 1/f electrical noise :

$$Svd_{1/f} = \frac{\alpha}{Nf} R_s^2 I_L^2 + \frac{Zeq^2}{\Delta f} \left[\overline{i_n^2} + \sum_{k=1}^{p} \frac{\overline{v_{nk}^2}}{Rse_k^2} + 2 \sum_{k=1}^{p} \mathrm{Re} \left\{ \frac{\overline{i_n v_{nk}^*}}{Rse_k} \right\} \right] \qquad (9)$$

At low frequency (f=10 Hz), we can consider that $L\omega \ll R_{se}$ and $\frac{1}{C\omega} \gg$ R.

$$\frac{1}{Zeq} = \frac{1}{R_d} + \sum_{k=1}^{p} \frac{1}{Rse_k} \quad \text{with } R_d = \frac{MV_T}{qn_o\left(A_k s_0 + \frac{1}{\tau_s}\right)} \quad \text{and } Rse_k = \frac{\gamma_k M V_T}{q\tau_s s_0 g_{0k}\left(A_k s_0 + \frac{\gamma_k}{\tau_s}\right)} \qquad (10)$$

$$\overline{i_n^2} = \left(C \frac{MV_T}{n_o}\right)^2 [\, \overline{f_n^2} + \overline{f_{n1/f}^2} \,] \qquad (11)$$

$$\overline{v_{nk}^2} = (q\, g_{ok}\, L)^2 [\, \overline{f_{sk}^2} + \overline{f_{sk1/f}^2} \,] \qquad (12)$$

$$\overline{i_n v_{nk}^*} = C \frac{MV_T}{n_o} q\, g_{ok} L \left[\overline{f_n f_{sk}^*} + \overline{f_{n1/f} f_{sk1/f}^*} \right] \tag{13}$$

In our case: $\overline{f_{n1/f}^{\,2}} \gg \overline{f_n^2}$ and $\overline{f_{s1/f}^{\,2}} \gg \overline{f_s^2}$

DISCUSSION:

a) $I_L < I_{th}$: The experimental results are plotted on figure 4 at 10 Hz: 1/f electrical noise spectral density $S_{vd1/f}$ varies as I_L^{-1} when the spontaneous emission is dominant, as predicted by Vandamme [6]. The current noise spectral density S_i is proportional to the laser current I_L. It can be explained by the fluctuations of diffusion coefficient in the diffusion region [7]: ($S_i \propto I_L$; $S_v \propto I_L^{-1}$). We have made measurement of optical noise (figure 5) given by S_{Iph} : see relation [2]. Experiments on this multimode laser give the value of the exponent m= 3/2 in the L.E.D. region as predicted by R.Fronen et al [3]. We have simultaneously measured the correlation by using the coherence function of the spectrum analyser given by :

$$\gamma^2_{Iph\text{-}Vd}(f) = \frac{\left| S_{Iph\text{-}Vd}(f) \right|^2}{S_{Iph}(f).S_{vd}(f)} \tag{14}$$

where $S_{Iph\text{-}Vd}$ (f) is the cross spectral density between the optical noise S_{Iph} and the electrical noise S_{vd}. We have plotted the result of γ^2 versus I_L/I_{th} in figure 6. We observe a high correlation between these two noises when the spontaneous emission is dominant. This behaviour is due to the spontaneous carrier recombination which is associated to 1/f noise due to fluctuations in free carrier concentrations[3]. See relations (3),(4),(12),(11). With respect of relation (1), we obtain [2] (with $\frac{\alpha}{N}$ =4.9 10^{-12} and Rs=4,4 Ω):

$$\gamma^2_{Iph\text{-}V_d}(f) = \frac{1}{1 + \dfrac{4kTR_s + \frac{\alpha}{fN} R_s^2 I_L^2}{S_{v_1}(f)}} \tag{15}$$

The theoretical results computed by using the electrical noise values are in concordance with the experimental results. This result shows that 1/f electrical and 1/f optical noises are strongly correlated beyond the threshold .We observed this behaviour with another type of laser (RIDGE LASER) [8].

b) $I_L > I_{th}$: We note a decreasing of γ^2 when I_L/I_{th} is higher than unity due to decreasing of 1/f noise spectral density $S_{v1}(f)$ around the threshold current. This behaviour is due to the fast decreasing of $\mathbf{R_{se}}$, which tends to a low computed value (R_{se} = 0.1 Ω) when I_L tends to the threshold current I_{th} [4]. When I_L/I_{th} is higher than 1.2, we observe a saturation of the electrical noise $S_{v_d}(f)$. It does not increase as I_L^2 because the term $\frac{Zeq^2}{\Delta f} \sum_{k=1}^{p} \frac{\overline{v_{nk}^2}}{Rse_k^2}$ is dominant. This fact is related to a saturation of the coherence function which corresponds to a saturation of the optical noise spectral density $S_{Iph}(f)$. The fluctuations of the photon number and those of spontaneous emission saturate when I_L/I_{th} is higher than unity [5]. In relation (9) we note that S_{vd} is directly related to the fluctuations of photon number, if we consider all the optical mode fluctuations given by $\frac{\overline{v_{nk}^2}}{\Delta f}$, $1 \le k \le p$. This relation shows that the electrical noise is correlated to the fluctuations of optical power by means of different optical mode noise sources.

CONCLUSION:

1/f electrical noise spectral density depends on the intrinsic noise sources v_{nk}. We have shown the 1/f electrical noise $S_{v_d}(f)$ is correlated to the 1/f optical noise $S_{Iph}(f)$ in L.E.D. region. Above threshold, we note a decreasing and the saturation of the intrinsic noise $S_{v1}(f)$ which corresponds to the decreasing and saturation of R_{sek} when I_L is higher than the threshold. This phenomenon is related to the saturation of the optical noise $S_{Iph}(f)$ which is connected to the fluctuations of the photon number given by: $\frac{\overline{v_{nk}^2}}{\Delta f}$, $1 \le k \le p$.

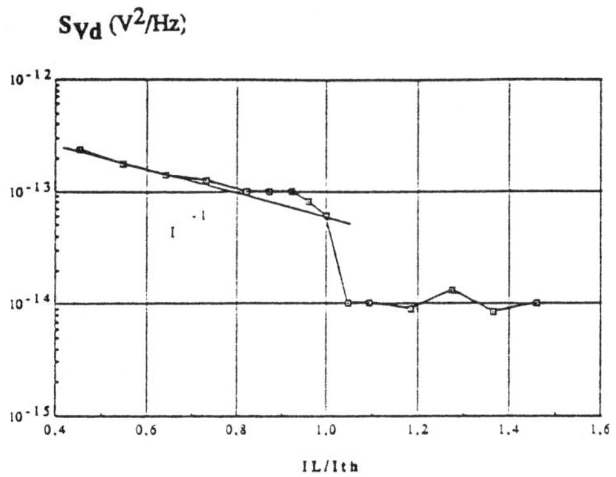

Figure 4

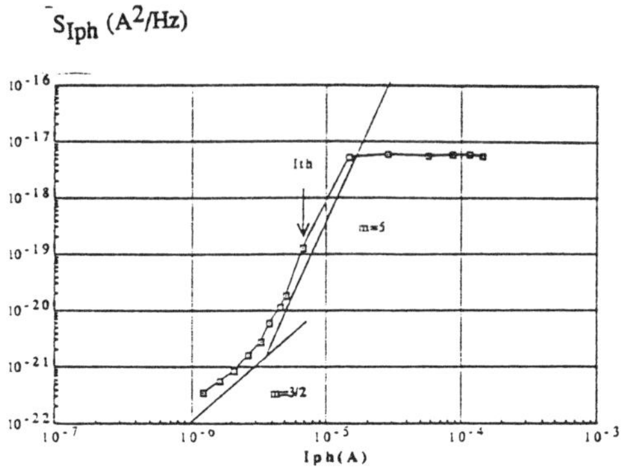

Figure 5

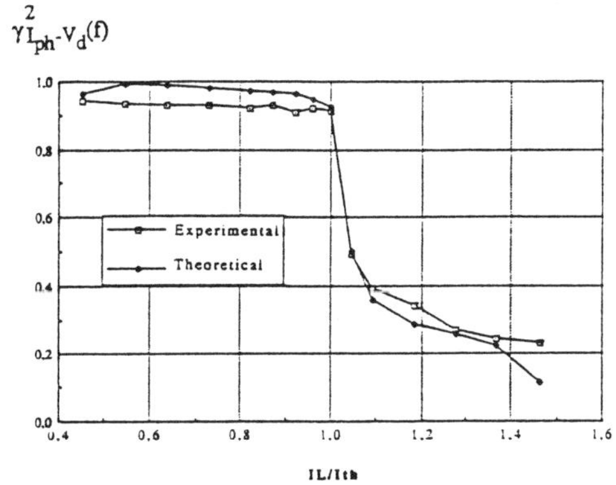

Figure 6

REFERENCES :

1. SAKAKIBARA Y., HIGUCHI H., OOMURA E., *High power 1.3 µm InGaAsP p Substrate Buried Crescent Lasers*, IEEE J. of Lightwave Technology, vol LT3, N°5, October 1985.
2. ORSAL B., ALABEDRA R., SIGNORET P. and LETHELIER C., *InGaAsP/InP Distributed Feedback Lasers for long wavelength optical communication systems, λ = 1.55 µm : Electrical and Optical noises study*, Proceedings of S.P.I.E., E.C.O.4, "Infrared Material and Devices", The Hague The Netherlands, vol. 1512 A, March 1991.
3. HOOGE F.N., KLEINPENNING T.G.M., VANDAMME L.K.J., Rep. Prog. Phys. 44, pp 479-532, 1981.
4. FRONEN R.J., *Wavelength dependence of 1/f noise in the light output of laser diodes : an experimental study*, IEEE J. of Quantum Electron., vol. 26, pp 1742-1746, October 1990.
5. HARDER C. , KATZ J., MARGALIT S., SCHACHAM J. and YARIV A., *Noise equivalent circuit of a semiconductor laser diode* , IEEE J. of Quantum Electron., vol. 18, N°3, pp 333-337, 1982.
6. VANDAMME L.K.J. and RUYVEN L.J.V., *1/f noise used as a reliability test for laser diodes*, Noise in physical systems and 1/f noise, pp 245-247, Elsevier Science Publishers BV 1983.
7. KLEINPENNING T.G.M.., *1/f noise in p-n junction diodes* ,J. Vac.Sci.Technol. A3 (1),Jan / Feb 1985
8. ORSAL B.,DAULASIM K., SIGNORET P.,ALABEDRA R.,PERANSIN J.M, *Electrical and Optical noise of high power S.Q.W.l asers in Erbium doped fiber amplifiers*, I.C.N.F.'93, 12th International Conference on Noise in Physical Systems, August 16-20, 1993, St Louis Missouri,U.S.A..

NOISE ANALYSIS OF QUANTUM WELL SEMICONDUCTOR LASERS AT LOW FREQUENCY

H. Dong, Y. Lin, A. D. van Rheenen, and A. Gopinath
Department of Electrical Engineering, University of Minnesota
Minneapolis, MN 55455.

ABSTRACT

The noise characteristics of quantum well single mode semiconductor lasers are analyzed by developing a noise equivalent circuit from the rate equations. The spectral intensity of the noise in the pump current and the light intensity of a GaAs/AlGaAs quantum well single mode ridge waveguide laser are measured. 1/f noise and generation-recombination noise are observed. These noise sources are introduced into the noise equivalent circuit and empirically modeled. The theoretical results fit the experiment well.

INTRODUCTION

Noise in semiconductor lasers is one of the most important problems encountered in laser applications. The analysis of noise in semiconductor lasers has been based on the quantum mechanical Langevin equation method[1] and the density matrix method[2]. We derived a noise equivalent circuit to analyze the noise. This model is based on the rate equations which describe the operation of quantum well(QW) single mode semiconductor lasers. We also measured the amplitude noise of the pump current and of the light intensity of a GaAs/AlGaAs quantum well single mode ridge waveguide laser. We empirically model the observed noise spectra by inserting appropriate noise source in the equivalent circuit model.

NOISE EQUIVALENT CIRCUIT

We deal with the rate equations for the carriers and photons of a quantum well single mode semiconductor laser analogous to the approach in reference[3]. Assuming that the time derivatives in the rate equations are zero, we obtain the steady state quantities which determine the operation of the laser diode. Next, fluctuations of all the steady state variables are introduced. Substituting all of the perturbed variables into the rate equations, we obtain a set of differential equations for the noise characteristics of the laser diode, which is equivalent to an electrical circuit (see Figure 1). The elements of this circuit are determined by the device parameters and steady state variables. From the noise equivalent circuit, we obtain the transfer function between voltage noise and current noise.

MEASUREMENT AND COMPARISON

We measured the pump current noise and optical intensity noise of a high-speed GaAs/AlGaAs quantum well single mode ridge waveguide laser.

The device structure is shown in Figure 2. There are three quantum wells in this device, and single mode operation is obtained by properly designing the width and the ridge height of the device. The 45% mole fraction of the Al in the cladding layer creates a strongly guided mode in the transverse direction. The weakly guided mode is in the lateral direction since the optical confinement in the lateral direction is only obtained by the ridge. The threshold current is about 6mA.

The noise is measured from 1Hz to 100kHz. Figures 3 and 4 show the pump current noise and optical noise, separately, for pump currents around the threshold current. The 1/f noise and g-r noise components are observed in both the current noise spectra and the optical noise spectra. The current noise is rapidly decaying as 1/f before it meets the bump of the g-r noise center. The knee frequency of g-r center is near 70kHz. Figure 5 shows the pump current noise intensity at 1Hz as a function of the pump current. Before the current reaches the threshold current, the 1/f noise has a linear dependence on current. After it reaches the threshold current, the noise approximately depends quadratically on the current. The optical noise (Fig.4) is also decaying as 1/f until it meets the first Lorentzian shaped bump. This bump comes from the RC time constant of the detector. After the first bump, the noise still decays as 1/f before meeting the second bump which is associated with the g-r center in the pump current noise by the coupling of the rate equations. This noise after the last bump is from the preamplifier since the device noise is so small that the preamplifier noise becomes significant.

We modeled the noise contributions as A/f, $B/[1 + (f/f_B)^2]$, and C, where f is frequency, A, B, f_B, and C are constants whose values are determined by comparison with experimental data. The solid lines in Figures 3 and 4 show the fit. The fit in Fig. 4 were obtained by multiplying the spectrum of the current noise by the transfer function we derived from the equivalent circuit.

CONCLUSION

We have developed a noise equivalent circuit that shows the connection between the observed noise in the light output and pump current of a GaAs/AlGaAs quantum well single mode ridge waveguide laser diode. The equivalent circuit is of significance for the understanding and analysis of the noise features of the diode. At low frequency, the equivalent circuit shows low pass behavior for 1/f noise and g-r noise.

We measured the pump current and optical AM noise of a quantum well single mode ridge waveguide laser. Generation-recombination noise and 1/f noise are observed. We also observed the effect of the parasitic capacitance from the detector.

We modeled the 1/f noise and g-r noise by fitting with experimental data. This helped us to understand and analyze the dominant noise sources in a laser diode at low frequencies. The comparison between the theoretical and experimental results shows the agreement.

All of our measurements are carried out at room temperature. By varying the temperature, we should be able to determine the trap parameters, density, and activation energy, of the g-r center. This is a useful method to characterize the material defect. Also, our equivalent circuit parameters are temperature dependent. Future work will focus on the temperature dependence of the noise of quantum well semiconductor lasers, providing a more solid foundation for the proposed model.

REFERENCES

1. H. Haug, Z. Phys., 200, 57 (1967).
2. M. O. Scully and W. E. Lamb,Jr., Phys. Rev., 159, 208 (1967).
3. Ch. S. Harder, B. J. Van Zeghbroeck, M. P. Kesler, H. P. Meier, P. Vettiger, D. J. Webb, and P. Wolf, IBM J. Res. Develop. 34, 568 (1990).

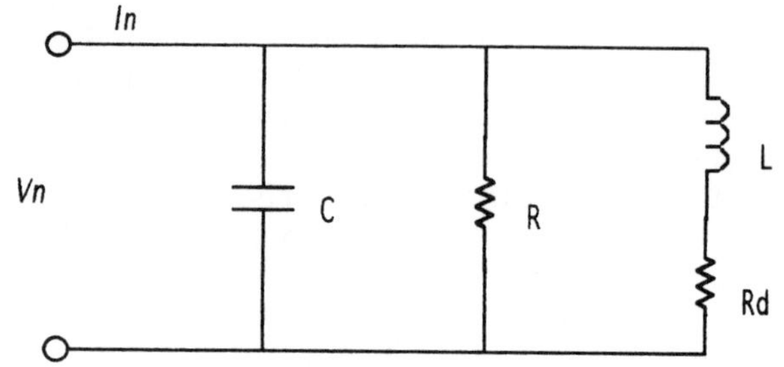

FIGURE 1. Noise equivalent circuit for a quantum well single mode semiconductor laser diode.

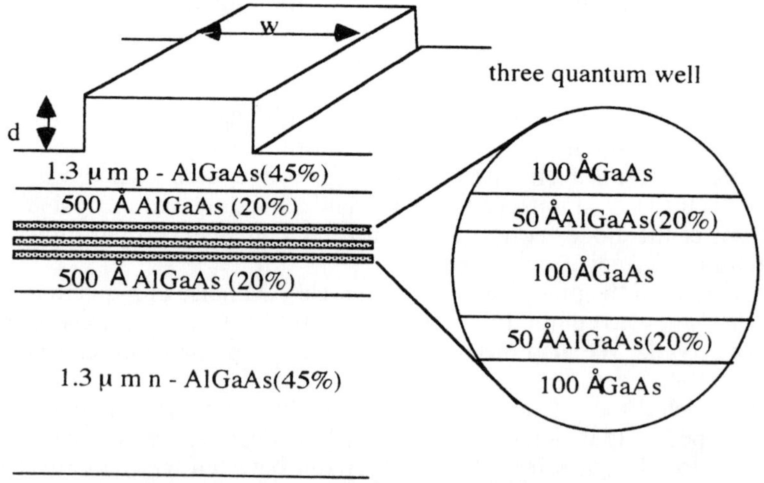

Figure 2. The structure of a high-speed GaAs/AlGaAs quantum well single mode ridge waveguide laser.

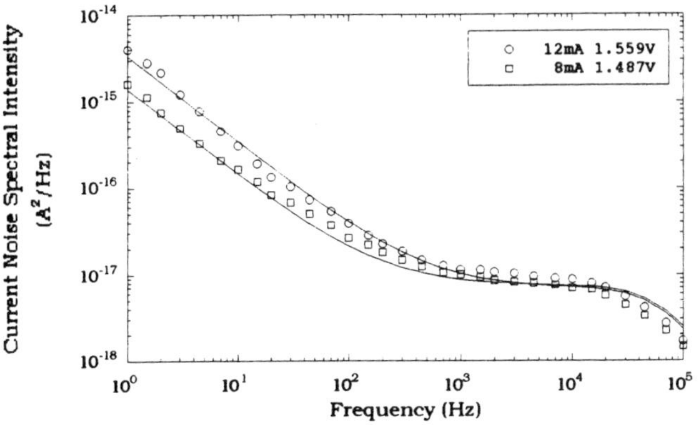

FIGURE 3. The pump current noise spectra vs. frequency. The symbols are from the experiment and solid lines are from the theoretical approaches.

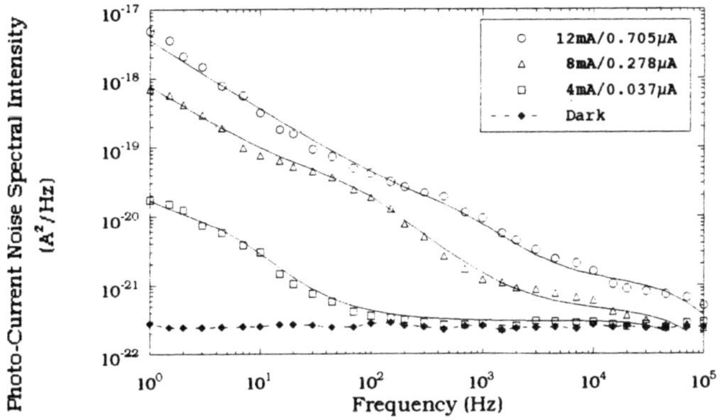

FIGURE 4. The optical noise spectra vs. frequency. The symbols are from the experiment and solid lines are from the theoretical approaches.

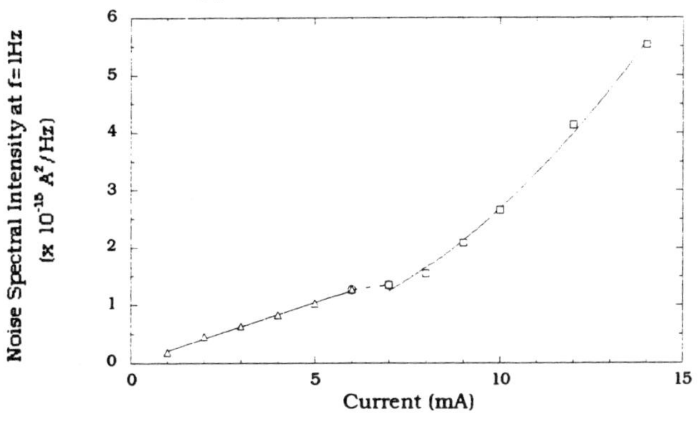

FIGURE 5. The pump current noise vs. the pump current at 1Hz.

XI. QUANTUM SYSTEMS

DISSIPATIVE QUANTUM NOISE IN A PARAMETRIC OSCILLATOR

Peter Hänggi and Christine Zerbe
University of Augsburg, Institute of Physics
Memmingerstr.6, D-86159 Augsburg, Germany

ABSTRACT

In this paper we investigate exact solutions for a parametric quantum oscillator. Without dissipation we focus on the propagator and the variances of momentum and position. In the presence of dissipation (Ohmic heat bath) we apply the influence-functional method due to Feynman and Vernon to obtain exact expressions for the time evolution of the reduced density matrix. Knowing this density matrix we calculate and discuss the variances in presence of friction.

1 INTRODUCTION

The potential of the system has the form

$$V(x,t) = \frac{1}{2}m(a - b\cos\Omega t)x^2. \tag{1}$$

This potential has several possible applications. One major application is the study of the Paul trap in the quantum regime[1]. Another suggested application is the generation of squeezed states of light[2], and particulary interesting is the application to the topic of tunneling through a barrier with a time dependent barrier-width.

Although the system under consideration can be used as an amplifier it should be distinguished from the so-called 'parametric amplifiers' studied by Louisell, Yariv, and Siegman[3] and Mollow and Glauber[4]. Their model consists of two harmonic oscillators coupled bilinearly via a time dependent parameter which oscillates at the combination frequency of the two individual oscillators. The Hamiltonian for this system is time dependent, but elimination of one harmonic oscillator in the Heisenberg equations of motion leads to a differential equation with constant coefficients.

2 THE QUANTUM PARAMETRIC OSCILLATOR WITHOUT DISSIPATION

Introducing the scaled variables $\bar{x} = \sqrt{m\Omega/2\hbar}x$ and $\bar{t} = \Omega t/2$ the dimensionless Schrödinger equation for eq. (1) reads

$$i\dot{\Psi}(\bar{x},\bar{t}) = \left[-\frac{1}{2}\partial_{\bar{x}}^2 + \frac{1}{2}\omega^2(\bar{t})\bar{x}^2\right]\Psi, \tag{2}$$

with $\omega^2(\bar{t}) = \bar{a} - 2\bar{b}\cos 2\bar{t}$, $\bar{a} = 4a/m\Omega^2$, $\bar{b} = 2b/m\Omega^2$. In this section we will use only scaled variables and *henceforth omit the overbars.*

The periodicity of the Hamiltonian leads to Floquet form solutions of the Schrödinger equation. A solution $\Psi_n(x,t)$ of eq.(2) can be factorized as

$$\Psi_n(x,t) = \exp(-i\epsilon_n t)\phi_n(x,t), \qquad \phi_n(x,t) = \phi_n(x,t+\pi). \tag{3}$$

ϕ_n is called Floquet function, ϵ_n a Floquet- or quasienergy. Because of the linearity of the system ϕ_n and ϵ_n are fully determined through the solutions of the corresponding classical problem, i.e.

$$\ddot{x} + \omega^2(t)x = 0. \tag{4}$$

It is *not possible* to obtain the solution in explicit form, but with $\omega^2(t)$ defined like in (2) this is the well-studied Mathieu-equation. Depending on the value of the parameters a and b the solution of (4) can be bounded or increasing with time. Whenever we need explicit solutions of (4) we calculate them numerically.

There are different approaches to the quantum mechanical problem. The Floquet functions and Floquet energies for the three regions were given first by Perelomov and Popov[5]. In the stable region a discrete spectrum of qusienergies exist. In the unstable regions and at the boundaries between these regions the spectrum becomes continous.

The propagator for this system, obtained first by Husimi[6] , can be derived in a variety of ways[7]. One possibility is based on Feynman's path integral method. Given the solutions of eq. (2) it is also possible to construct the propagator directly in terms of a *spectral representation*, i.e.

$$K(x_f,t_f|x_i,t_i) = \sum_{n=0}^{\infty} \phi_n(x_f,t_f)\phi_n^*(x_i,t_i)\exp[-\frac{i}{\hbar}\epsilon_n(t_f-t_i)]. \tag{5}$$

For a continous spectrum the sum becomes an integral and we have to take into consideration possible degeneracies of the quasienergy spectrum. Doing so we obtain for the propagator

$$K(x_f,t_f;x_i,t_i) = \sqrt{\frac{1}{2\pi i X(t_f)}} exp\left[\frac{i}{2X(t_f)}(x_f^2\dot{X}(t_f) - 2x_f x_i + x_i^2 Y(t_f))\right] =$$

$$= e^{-i\frac{\pi}{2}[m(t_f)-m(t_i)]}\sqrt{\frac{1}{2\pi i|X(t_f|)}} exp\left[\frac{i}{2X(t_f)}(x_f^2\dot{X}(t_f) - 2x_f x_i + x_i^2 Y(t_f))\right]. \tag{6}$$

X and Y are special solutions of (4) with the initial conditions

$$X(t_i) = 0, \qquad \dot{X}(t_i) = 1, \qquad Y(t_i) = 1, \qquad \dot{Y}(t_i) = 0. \tag{7}$$

$m(t)$ is the number of zeros of X in the interval $[0,t]$, $m(0) = 0$ and we used the definition of the root

$$X^{\frac{1}{2}}(t) = |X|^{\frac{1}{2}} e^{i\frac{\pi}{2}m(t)}.$$

But this propagator (6) is valid only for times $t_f \neq t_n$. With t_n we denote the time when the m-th zero of X occur. To calculate the propagator at these so called *caustics* we use the semigroup property of the propagator

$$K(x_n, t_n; x_0, 0) = \int dx_c\, K(x_n, t_n; x_c, \pi) K(x_c, \pi; x_0, 0). \tag{8}$$

This relation holds for any time order of $0, \pi, t_n$. It is not necessary that $\pi < t_n$. We have chosen a special time $t_c = \pi$ because then we can employ the following relations for the solutions of the Mathieu equation

$$Y(\pi - t) = Y(\pi)Y(t) - \dot{Y}(\pi)X(t), \qquad X(\pi - t) = X(\pi)Y(t) - Y(\pi)X(t). \tag{9}$$

For the propagator at a caustic we find explicitly the result

$$K(x_n, t_n; x_0, 0) = e^{-i\frac{\pi}{2}m(t_n)} \frac{1}{\sqrt{|Y(t_n)|}} exp\left[\frac{i}{2}\frac{\dot{Y}(t_n)}{Y(t_n)} x_n^2\right] \delta(x_n - Y(t_n)x_0). \tag{10}$$

It was shown before with various methods that a time dependent harmonic oscillator generates squeezed states[2]. To study its squeezing properties explicitly we compute the variances of the operators x and p with the Heisenberg equation of motion. The mean values for this linear system follow the solutions of the classical equation (4).
The variances $U(t) \equiv \langle x^2\rangle - \langle x\rangle^2$, $V(t) \equiv \frac{1}{2}\langle xp + px\rangle - \langle x\rangle\langle p\rangle$ and $W(t) \equiv \langle p^2\rangle - \langle p\rangle^2$ satisfy the coupled set of equations

$$\dot{U} = 2V, \qquad \dot{V} = W - \omega^2(t)U, \qquad \dot{W} = -\omega^2(t)V. \tag{11}$$

By eliminating $V(t)$ and $W(t)$ from eqs. (11) we find an equivalent third-order equation for $U(t)$

$$\dddot{U} + 4\omega^2(t)\dot{U} + 2\left\{\frac{d}{dt}[\omega^2(t)]\right\} U = 0 \tag{12}$$

This equation is solved with $U(t) = W(0)X^2 + U(0)Y^2 + 2V(0)XY$ where X and Y are defined as before in (6). Because of (11)

$$V(t) = W(0)X\dot{X} + U(0)Y\dot{Y} + V(0)(X\dot{Y} + \dot{X}Y) \text{ and}$$

$$W(t) = W(0)\dot{X}^2 + U(0)\dot{Y}^2 + 2V(0)\dot{X}\dot{Y}.$$

We see that the variances are bounded or increasing with time like the solutions of the Mathieu-equation in the corresponding region. Depending on the chosen parameters the form of the results varies strongly. In figs. 1 we plot the time variation of U for a fixed value of a. We start at $t = 0$ with a wavepacket with minimum uncertainty $U(0) = 1/2r\sqrt{a - 2b}$, $V(0) = 0$, $W(0) = r\sqrt{a - 2b}/2$. r is a parameter which characterizes the amount of squeezing of the state; $r = 1$ refers to the unsqueezed state. Fig. 1a shows U for a squeezed state and different

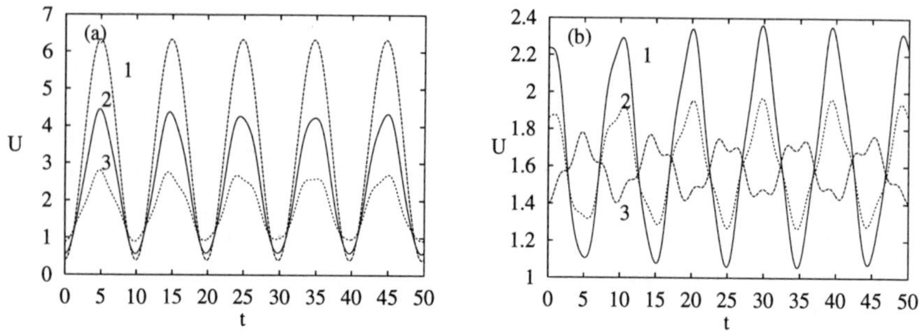

Figure 1: Variance U for various amplitudes of the parametric modulation and squeezing parameters for $a = 0.1$, (a): $b = 0$ (1), 0.025 (2), 0.04 (3) and $r = 4$ (b): $r = 1$ (1), 1.2 (2), 1.6 (3) and $b = 0.025$

amplitudes of the parametric modulation b. It can be seen how the variation of b changes the form and also the amplitude of the oscillations. In fig. 1b variances are plotted for different squeezing parameters r.

3 THE DAMPED QUANTUM PARAMETRIC OSCILLATOR

To describe damping we couple our system linearly to an environment[8]. This environment is modeled as a linear system consisting of a set of noninteracting harmonic oscillators . The Hamiltonian of the coupled system assumes then the following form

$$H = H_A + H_I + H_B$$

$$\begin{aligned} \text{with } H_A &= \frac{p^2}{2m} + \frac{1}{2}(a - b\cos\Omega t)mx^2, \qquad H_B = \sum_{n=1}^{N} \frac{p_n^2}{2m_n} + \frac{1}{2}\omega_n^2 m_n x_n^2, \\ H_I &= x\sum_{n=1}^{N} c_n x_n + \frac{c_n^2}{2m_n\omega_n^2}x^2. \end{aligned} \tag{13}$$

H_A and H_B are the Hamiltonians of the parametric quantum oscillator and the bath oscillators, respectively. The first term in H_I couples the system to the bath. This coupling leads to a frequency shift of our system, that is removed with the second term in H_I. To gain explicit results we consider from now on an Ohmic heat bath. As we are not interested in the dynamics of the environment we elimimate the bath and calculate the *exact reduced density operator* of the system at time t, i.e.

$$\rho_R(x_f, y_f, t) = \int dx_i dy_i J(x_f, y_f, t|x_i, y_i, 0)\rho_R(x_i, y_i, 0), \tag{14}$$

where J is calculated by the influence-functional method. Introducing the sum and difference variables $q = x - y$, $r = \frac{1}{2}(x + y)$ yields for J

$$J(q_f, r_f, t|q_i, r_i, 0) = \frac{\dot{u}_2(t,0)}{2\pi\hbar} \exp\left[\frac{-1}{\hbar}\{a_{11}(t)r_i^2 + [a_{12}(t) + a_{21}(t)]r_i r_f + a_{22}(t)r_f^2\}\right]$$

$$\times \exp\left[\frac{i}{\hbar}\{[\dot{u}_1(t,0)r_i + \dot{u}_2(t,0)r_f]q_i - [\dot{u}_1(t,t)r_i + \dot{u}_2(t,t)r_f]q_f\}\right]. \qquad (15)$$

a_{ij} is given through

$$a_{ij}(s) = \frac{1}{2}\int_0^s \int_0^s ds_1 ds_2 v_i(t,s_1) v_j(t,s_2) K(s_1 - s_2)$$

with the noise kernel

$$K(s) = \int_0^\infty \frac{d\nu}{\pi} m\gamma\nu \coth(\frac{\nu\hbar}{2k_BT}) \cos(\nu s). \qquad (16)$$

The set $\{u_1, u_2\}$ determines the solution of the equation of motion

$$\ddot{u} - \gamma\dot{u} + (a - b\cos\Omega s)u = 0 \qquad (17)$$

with the conditions $u_1(t,0) = 1$, $u_1(t,t) = 0$, $u_2(t,0) = 0$, $u_2(t,t) = 1$ and $v_{1,2}(t,s) = u_{1,2}(t,s)\exp(\gamma s)$. To arrive at this form we assumed that the system and environment were initially ($t_0 = 0$) *uncoupled* and the bath was in equilibrium at temperature T. Knowing the density matrix we are able to calculate expectation values of the variables. The mean values of space and coordinate follow the trajectories of a damped classical parametric oscillator. Next we study the time development of the variances $U(t)$, $V(t)$ and $W(t)$. The Ohmic damping leads to a divergence in W, just as with a damped quantum oscillator[9]. We introduce an abrupt high frequency cutoff ν_c of the bath frequencies ν in the frequency integral in (16) to remove this divergence. This is correct as long as we consider only times that are large compared to ν_c^{-1}.The results are plotted in figs 2. Fig.2a shows $m\Omega U/2\hbar$ for increasing modulation amplitude $|b|$. The initial values of the variances are like in section 2. For curves labeled (1), (2) and (3) b has values that lead to decaying solutions of the damped Mathieu equation in eq.(17), i.e. $\langle x(t)\rangle \to 0$ as $t \to \infty$. After a short time, U becomes a constant for curve labeled (1), whereas for curves (2) and (3) U becomes a *periodic function* which oscillates with the frequency Ω. This frequency is *not affected* by the strength of the friction γ. The amplitude of the oscillations is increasing with increasing modulation strength $|b|$. For b in the unstable region the variances become unbounded also, as can be seen in curve labeled (4). In fig.2b we start with a squeezed state and compare it with the unsqueezed state. As an interesting result we find that the effect of initial squeezing relaxes on a fast time scale. This relaxation time depends only weakly on temperature, but depends on the strength of Ohmic friction γ.

We also like to point out here that the system dynamics $x(t)$ of the coupled system in eq. (13) obeys an exact equation of motion. The Heisenberg operator

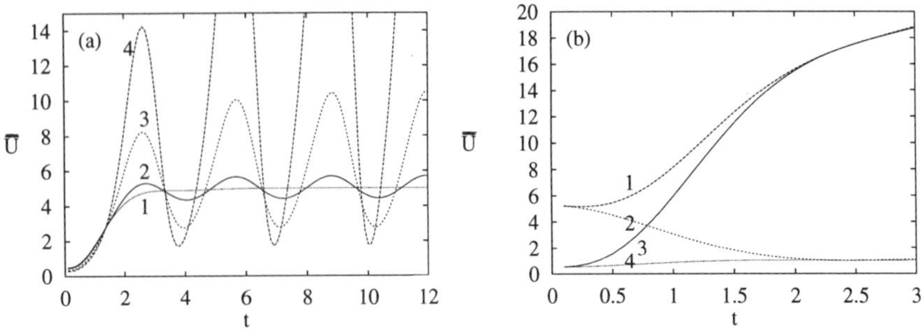

Figure 2: Variance $\bar{U} = m\Omega U/2\hbar$ for various amplitudes of the parametric modulation and different squeezing parameters, $\nu_c = 50$, $a = 1$, $\Omega = 2$, $\gamma = 1$, $\omega_0^2 = a$,
(a): $|b| = 0$ (1), 0.1 (2), 1 (3), 2 (4) and $r = 1$, $k_BT/\hbar\omega_0 = 5$,
(b): $r = 1$ (3,4), 0.1 (1,2), $k_BT/\hbar\omega_0 = 1$ (2,4), 20 (1,3) and $|b| = 0.1$

$x(t)$ obeys - after elimination of the bath degrees of motion - an *exact Quantum Langevin equation*, which in terms of the initial time of preparation reads explicitly

$$m\ddot{x} + m\int_{t_0}^{t} \gamma(t-s)\dot{x}(s)ds + m(a - b\cos\omega t)x + m\gamma(t-t_0)x(t_0) = \xi(t) \qquad (18)$$

where $m\gamma(t) = \sum_n c_n^2 \cos\omega_n t/m_n\omega_n^2$; $\frac{1}{2}\langle\xi(t)\xi(0) + \xi(0)\xi(t)\rangle = \hbar K(t)$, with $K(t)$ given in eq. (16). The limit of Ohmic friction - without cutoff - is obtained from (18) by setting $\gamma(t-t_0) = 2\gamma\delta(t-t_0)$. We note that (18) presents a suitable starting point to investigate initial and long-time (aged) correlation function properties of the damped parametric quantum oscillator[9].

REFERENCES

1. M. Combescure, Ann. Inst. Henri Poincare **A44**, 293 (1986)
   L.S. Brown, Phys. Rev. Lett. **66**, 527 (1991)
2. C.F. Lo, J. Phys. **A23**, 1155 (1990)
   L.S. Brown, Phys. Rev. **A36**, 2463 (1987)
3. W.H. Louisell, A. Yariv and A.E. Siegman, Phys. Rev. **124**, 1646 (1961)
4. B.R. Mollow and R.J. Glauber, Phys. Rev. **160**, 1076 (1967)
5. A.M. Perelomov and V.S. Popov, Teor. Mat. Fiz. **1**, 360 (1969)
6. K. Husimi, Progr. Theor. Phys. **9**, 381 (1953)
7. J. Rezende, J. Math. Phys. **25**, 3264 (1984)
8. R.P. Feynman and F.L. Vernon, Ann. Phys. (USA) **24**, 118 (1963)
   R. Zwanzig, J. Stat. Phys.**9**, 215 (1973)
   H. Grabert, P. Schramm and G.Ingold, Phys. Rep.**168**, 115 (1988)
9. P. Riseborough, P. Hänggi, and U. Weiss, Phys. Rev. **A31**, 471 (1985)

LOW-FREQUENCY NOISE IN SCANNING TUNNELING MICROSCOPY MEASUREMENTS

F. Bordoni,
University of L'Aquilla, Italy

S.V. Savinov, A.V. Stepanov, V.I. Panov, I.V. Yaminsky
Advanced Technologies Center, Moscow State University,
119899 Moscow, Russia

ABSTRACT

A stable scanning tunneling microscope is constructed for low frequency noise measurements. Its electronic noise is much lower than noise of tunnel junction. The experimental data are obtained with PtIr tip and gold sample. $1/f^{\alpha}$ type noise is predominant at low frequencies. The estimated value of α is 1.1 ± 0.2.

INTRODUCTION

Scanning tunneling microscopy is well established method of local surface investigation with atomic or molecular resolution. It can provide the profound knowledge about the local electronic density, local electronic conductivity, local barrier height and other surface properties necessary for the solution of fundamental and applied problems. $1/f^{\alpha}$ type low frequency noise is the main fundamental restriction on the sensitivity of scanning tunneling microscope and other surface and force devices with tunnel sensor.[1,2]

EXPERIMENTAL

We have constructed a high stable scanning tunneling microscope with a rigid tripode piezoactuator (X, Y and Z bars are 30mm in length and 3.5mmx3.5mm cross section). The frequency of the first mechanical resonance of the actuator is about 4kHz. Under ordinary laboratory conditions no resonant peaks due to seismic vibrations of actuator or other mechanical parts can be seen in the noise spectrum of the tunnel current (fig. 1). Thermal drift during the measurements was always less than a few angstroms per minute. Usually the output of high voltage amplifiers is applied to the electrodes of actuator in order to sustain the appropriate distance between the tip and the sample. The electrical noise of high

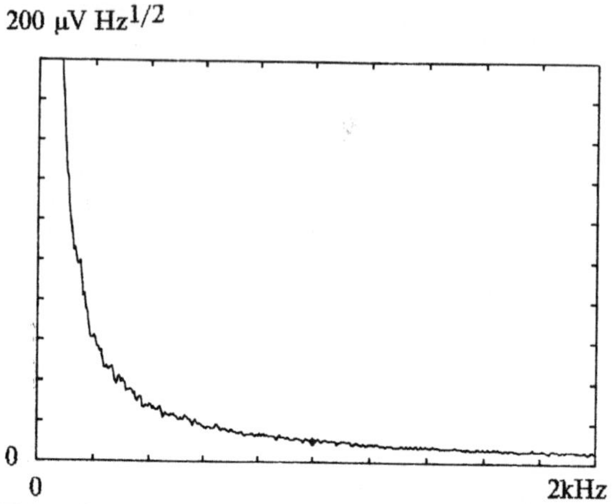

Fig.1. Noise spectum of the tunnel current

voltage signal may modulate the tip-sample separation and consequently the tunnel current. In present measurements the fine mechanical adjustment up to the hundreds of angstroms was implemented to avoid the use of high voltage signals. The intensity of electric field in the piezoceramics was less than 30 V/cm, the hysteresis of piezoceramics was less than 1%. The creep of ceramics is difficult to control, the main precaution is to avoid the appearance of any short or abrupt pulses on the electrodes of piezoceramics.

We use a precise current-voltage transducer (preamplifier) with the transimpedance gain K_{IU}=80 mV/nA. The currrent and voltage noise of the precise operational amplifier was: $I_{e.n.}$= 10^{-6} nA $Hz^{1/2}$ and $U_{e.n.}$=10 nV $Hz^{1/2}$ in the frequency band 1Hz-100kHz. For the typical values of the input resistance R= 10-100 MΩ the resulting value of preamplifier noise was of the order of 10^{-6} nA $Hz^{1/2}$. The excess low frequency noise of the preamplifier was about 10 times smaller than the intensity of low frequency noise of the tunnel junction. The measurements were conducted at normal air pressure with sample of pure gold (layer thickness 1 μm) evaporated on silicon. The tip was mechanically cutted $Pt_{0.8}Ir_{0.2}$ wire. The gold surface is quite chemically inert in ambient conditions, so there was no oxide layer formation on its surface.

NOISE MEASUREMENTS

The tunnel junction noise is composed of "shot" noise of tunnel current, Johnson noise of the junction and excess low frequency noise with $1/f^{\alpha}$ power spectrum. Preamplifier noise described by current noise generator $I_{e.n.}$ and voltage noise generator $U_{e.n.}$ is added to the total noise of the system. The resulting spectral density of the current noise is

$$S_I = 2eI + 4k_bT/R_{eff} + I_{e.n.} + U_{e.n}/R_{eff} + S_{exc}$$

where I - tunnel current, k_b - Bolttzmann's constant, $R_{eff} = R_tR_e/(R_t+R_e)$ is the total resistance of the dynamic resistance of tunnel junction R_t and the feedback resistance of the preamplifier R_e connected in parallel. Effective resistance of tunnel junction is $R=R_oe^{-2kz}$, z - tip-sample separation, the decay inverse length k is defined by the average barrier height between the sample and the tip. The excess noise S_{exc} is the low frequency noise with the power spectrum of $1/f^{\alpha}$ type. The estimated value of α in our experiments was 1.1±0.2 (fig.2.). For the frequencies below 1kHz the "shot" noise and Johnson noise of tunnel junction are approximately equal and usually smaller than 10^{-5} nA $Hz^{1/2}$.

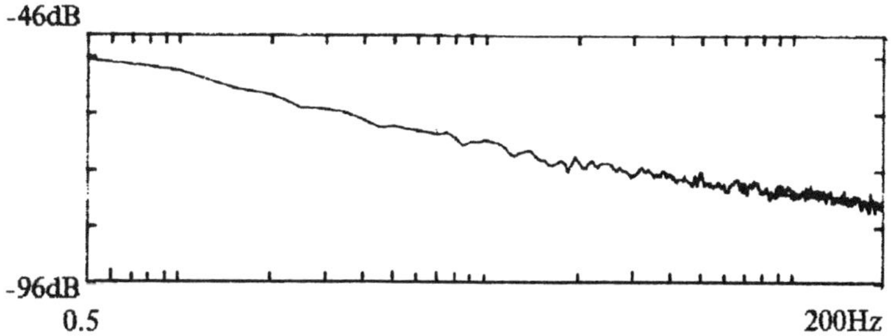

Fig.2. The dependence of 10 lg (I_n/I_{no}) vs frequency

During STM measurements in the constant current mode feedback is used to sustain constant tip-sample separation. In this case the tunnel current noise I_n at the feedback output signal U_z is given by:

$$U_z(j\omega)=K(j\omega)\ K_{IU}\ I_n\ /(1+K(j\omega)\ K_{UI}K_{IZ}K_{ZU}),$$

$K(j\omega)$ - the gain of the feedback.

Due to the presence of integrator in a feedback loop $K(j\omega)$~10^5...10^6 at low frequencies. The voltage-displacement coefficiency of actuator K_{ZU}=2nm/V. K_{IZ} is defined from tunnel current vs tip-sample separation dependence: K_{IZ} ~10 nA/nm. The proper adjustment of the feedback alows to eliminate the distortions of the tunnel noise spectrum at the feedback output:

$$U_z=I_n\ /(K_{IZ}K_{ZU})\ \sim I_n\,.$$

DISCUSSION

The nature of $1/f^{\alpha}$ noise of the tunnel current is not clear. We suppose the presence of two different sources of low noise fluctuations. The first one is originates from the random character of electron tunneling between the tip and the sample. The tunneling probability may vary in time due to surface degradation, oxidation process or appearance of contamination on the tip or the sample. The variation of probability lead to random modulation of tunnel current. The second cause of $1/f^{\alpha}$ noise may be an electron scattering in the region of current spreading. Because of the small dimensions of spreading area (of the same order as the tip-sample separation) the fluctuations of its resistance may be quite noticeable. The random changes of resistance lead to additional tunnel current fluctuations.

The $1/f^{\alpha}$ type noise is often seen as abrupt jumps of the tunnel current resulting in image distortions. The tunnel current jumps (burst $1/f^{\alpha}$ noise) are typical for STM measurements. Abrupt changes can be seen in images directed along the scan line, most probably produced by the changes of the tip geometry and its electronic structure.

The tunnel sensor is often regarded to be unpromising for different force and surface applications. But it is worth noting that its sensitivity up to 10^{-5} nm $Hz^{1/2}$ is sufficient enough for many practical applications. It is easy to obtain perfect large scale images using atomic force microscope with tunnel sensors of large scale and we managed to obtain molecular resolution images as well[3]. The sensitivity of presented tunnel sensor of small displacements is better than 10^{-5} nm $Hz^{1/2}$.

We now plan tunnel noise measurements with different clean and contaminated surfaces to clear up the mechanism of low frequency fluctuations. Measurements with and without the use of feedback will be done.

REFERENCES

1. M.F. Bocko and K.A. Stephenson, J.Vac. Sci.Technol. **B9(2)**, 1363 (1991)
2. E.Stoll and O. Marti, Surf. Sci. **181**, 222 (1987).
3. Yu. Moiseev, V. Panov, S. Savinov, I. Yaminsky, P. Todua and D. Znamensky , Ultramicroscopy, **44**, 304 (1992).

1/f NOISE OF STM TUNNEL PROBE AS A FUNCTION OF TEMPERATURE

F. Bordoni * , G. De Gasperis † , G. Ferri *

* Dipartimento Ingegneria Elettrica
† Dottorato di Ingegneria in Ingegneria Elettronica
Universitá degli Studi di L'Aquila
67040 Monteluco di Roio - L'AQUILA - **ITALY**

Abstract

The tunneling probe used in Scanning Tunneling Microscopy (STM) has been recently proposed as displacement sensor for resonant gravitational-wave antennas [1].
At present the performances of the tunnel probe as displacement sensor are limited by the $1/f$ noise in the tunnel current which, at room temperature, lowers of three orders of magnitude the expected sensitivity in the KHz range: preliminary results indicate that at liquid helium temperature the tunnel probe is a shot noise limited device. To understand the origin of the $1/f$ noise we are carrying on systematic measurements of the tunnel current noise as a function of the temperature and for different materials; the experimental results will be presented during the conference. Due to the quantum mechanical nature of the tunnel probe we believe that noise measurements as a function of the temperature can give also some useful information on the fundamental origin of the $1/f$ noise.

Introduction

The sensitivity ($3x10^{-17}\frac{m}{\sqrt{Hz}}$) of displacement sensor used in gravitational-wave resonant antennas, permits the detection of near events which are quite occasional (estimate rate in our galaxy one event every 30 years). One way to increase the sensitivity is that of coupling the antenna (which can be modelled as a harmonic oscillator) to a second resonator with the same resonant frequency but a much lower mass. Energy conservation yields a gain in the amplitude vibration of the second resonator which is equal to $\sqrt{\frac{m_a}{m_b}}$ where m_a and m_d are respectively the mass of the antenna and of the coupled resonator. To obtain large amplitude gain the coupled transducer should have reduced geometrical dimensions making very promising the use of a tunnel probe which has proved to work with extremely small masses down to atomic dimensions. Moreover the tunnel probe avoids the back-action of the following electronic amplifier [1] making possible to achieve, even with antennas working at room temperature, high sensitivity. The dependence of the tunnel current, as a function of the distance s, is

$$i(s) = i(s_0)e^{-2ks} \tag{1}$$

where s_0 is the average distance and k a constant which depends on the materials used but it is always of the order of $10^{10}m$. The responsivity of the tunnel probe as position detector is then

$$Resp \cong -2ki_0 A/m \tag{2}$$

where i_0 is the d.c current corresponding to the distance s_0. For a reasonable d.c. current of $10\mu A$ and assuming the device shot-noise limited, a sensitivity for $S/N = 1$ of $10^{-17}\frac{m}{\sqrt{Hz}}$ is

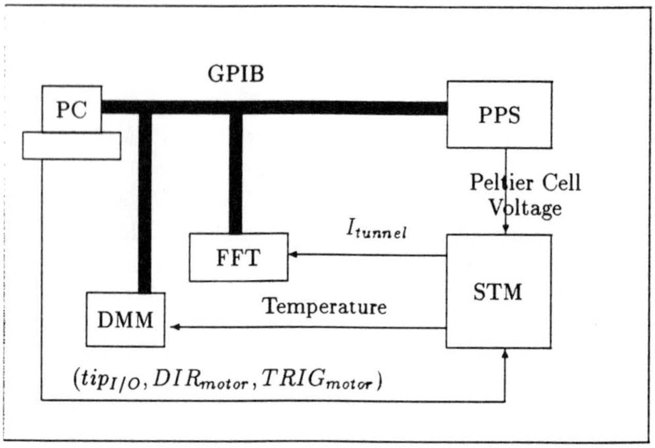

Figure 1: *Experimental apparatus used to make noise spectra in function of the temperature, at different tunnel current magnitudes. PC = Personal Computer, DMM = Digital Multi Meter, FFT = FFT Analyzer, PPS = Programmable Power Supply, STM = Scanning Tunneling Microscope*

achievable. Unfortunately the performances of the tunnel probe as a displacement sensor are limited by the $1/f$ noise which, at the room temperature and in the KHz range, lowers of three orders of magnitude the predicted sensitivity.
To understand the physical origin of the noise and in view of a possible application of the detector to criogenic antennas, we modified a home made STM, to measure the noise of the tunnel current as a function of the temperature.

Measuring Apparatus

A block diagram of the experimental apparatus is shown in fig.(1).
The STM, based on a Binnig-Smith tube scanner [2], is inside a vacuum chamber, isolated from the sismic noise of the building by means of a vibration isolation system. Two micrometric screws, with different range, control the tip-sample approach; the fine motion screw is driven by a D.C. motor.
All the measurements can be programmed by a PC through two control buses: a GPIB bus and a digital bus. On the GPIB bus the PC is the master that controls the following instruments: a FFT Analyzer which records the noise spectra in the frequency range 0-100 KHz and sends them to the PC for further elaborations, a digital multimeter to measure, by means of a transistor sensor, the working temperature and a programmable power supply which drives a Peltier cell which permits to vary the temperature of the sample in the range 0-100 C.
The digital control bus from the PC is implemented with a commercial digital I/O board and actuates the following STM signals: $tip_{I/O}$ which forces the tip towards the sample to a proper distance to obtain the tunnel current, DIR_{motor} which imposes the direction of the D.C. motor to control the coarse mechanical approach and the $TRIG_{motor}$ which permits to feed the motor with a single step.

Experimental Results

Preliminary measurements have been carried out using gold sputtered on silicon sample and

electrochemical etched tungsten tips or commercial platinum-iridium tips. We have taken data for different d.c. tunnel currents at three different working temperatures: low (0 C), medium (20 C) and high (100 C). All the measurements, to reduce the surface contamination of the sample, has been obtained operating in a nitrogen atmosphere.
Each set of measurements is carried on in four steps: tip/sample approach, temperature adjusting, frequency spectrum acquisition, data recording.
Step 1: the coarse approach is manually done using the coarse micrometric screw. When the distance between the tip and the sample is a few hundred microns the PC feeds, with a train of single pulses the DC motor, until the analogical control loop senses and locks the tunnel current.
Step 2: the PC sends a GPIB command to the power supply of the Peltier cell to increase or decrease the temperature of the sample. During this phase, the variation of distance between the sample and the tip, due to thermal expansion, is corrected by the PC. A safe sequence of operations is that of making the measurements starting from the higher value of temperature in order to avoid contacts between the tip and the sample due to the finite range of the control loop.
Step 3: when the programmed temperature is reached and its maximum variation is below 1 C the PC sends a GPIB command to the FFT analyzer to start the acquisition of the noise spectrum in the frequency range 0-100 KHz. The spectra are obtained averaging 128 samples in 400 channels each with a band width of 250 Hz.
Step 4: at the end of the acquisition, the noise spectrum is sent to the PC by the GPIB bus and then stored into a floppy disk.
Typical behaviours of the noise spectra obtained with tungsten tip and gold sample are reported in fig.(2A), (2B) and (2C).
In fig.(2A) the noise spectra of the tunnel current as a function of the d.c. value at room temperature are shown. The noise increases with the value of the current, but the dipendence is faster than its square root. A plateau at 100KHz only for the current value of 1nA can be individuate; its value is greater than the corresponding shot noise. The spectra show two slopes for currents of 20nA and 80nA, while for the 1nA spectra the slope is unique.
As shown in fig.(2B), the shape of the noise spectra does not depend on the temperature. The minimum of the noise is obtained at room temperature while the maximum occurs for 100 C. This behaviour suggests that noise measurements at fixed frequencies and as a function of the temperature should be carried on.
As shown in fig.(2C), the tip bias affects only the noise value. The noise is lower for a positive bias of the tip i.e. when the tunnel current is from the gold sample.

Acknowledgements

We are grateful to Fabrizio Mancini, Stefano Ricci and Ferdinando Feliciangeli for their invaluable help in setting up and fixing the experimental apparatus. We also would like to acknowledge the invaluable contribution at the earlier stage of the work of Loredana D'Auria.

Bibliography

1. F.Bordoni, M.Karim, M.F.Bocko, Tang Mengxi, Phys. Rev. D, 42, 2952, (1990)
2. G.Binnig, D.P.E.Smith, Rev. Sci. Instrum. 57, (1688), (1986)

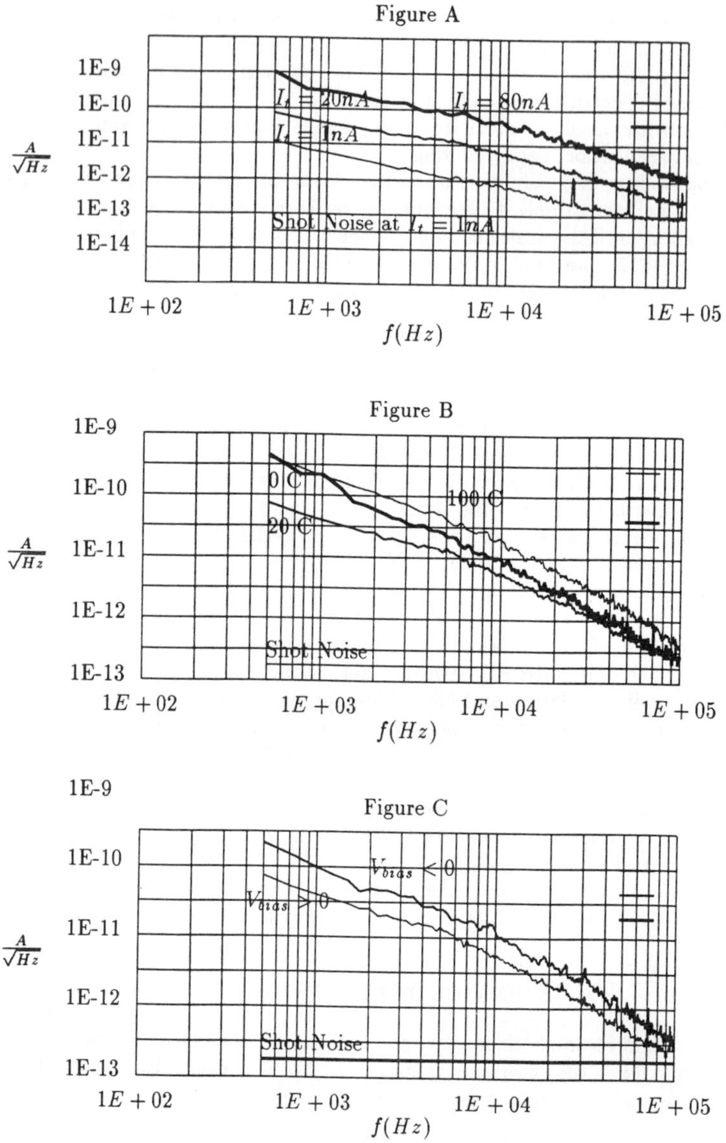

Figure 2: **A**: Noise spectra at room temperature for different values of d.c. current, at room temperature, $V_{bias} > 0$. **B**: Noise spectra obtained with a d.c. current of $20nA$ and at the temperatures of 0 C, 20 C, 100 C, $V_{bias} > 0$. **C**: Noise spectra obtained for the same tunnel current of $20nA$, at room temperature, but with opposite tip-to-sample polarization V_{bias}.

XII. RANDOM PROCESSES AND STOCHASTIC SYSTEMS

DIFFUSION NOISE AT THE ELECTROKINETIC CONVERSION.

Antohin A. Y., Kozlov V.A.
Moscow Institute of Physics and Technology
Mailing address: Nagornaya 8-31, Dolgoprudniy, Moscow Region, 141700, Russia.

At the equilibrium state the noise of a capillary filled with low conductive liquid is determined by Johnson noise and by hydrodynamics fluctuations that are presented at any liquid. This noise has flatten form at low frequencies and presented in equilibrium ensemble where the average current is absent. In such state, current fluctuations do not contain a component due to fluctuations in concentration of charge carriers. Another situation arises in the presence of electrical current through a capillary. In that case the concentration fluctuations give contribution to the power spectrum density of noise and what's more independently on Johnson noise[1]. A new essentially arises at the hydrodynamics level of description. This is spatial dependence of elementary process.

Let us consider the process of electrokinetic conversion in a single capillary of radius R, length L. The liquid has constant dynamic viscosity ξ and dielectric constant ε. It is presumed that the capillary surface is equally charged and has potential ψ_o. The density of solution is constant so that

$$\mathrm{div}\, V = 0 \tag{1}$$

A pressure drop ΔP across the capillary will give rise to a Poiseuille flow inside the capillary with a velocity profile

$$V(r) = -\,0.25\Delta P\;(r^2 - R^2)/\xi \tag{2}$$

The concentration fluctuations are calculated in the framework of Langevin method. The deviation $\delta n(r,r)$ from the average concentration n_o can be found from non-equilibrium thermodynamics. This can be done by introducing a random driving term f to the linearized convective diffusion equation

$$dn/dt = -\mathrm{div} j/e \tag{3}$$

$$j = neV + \sigma E \tag{4}$$

By taking into consideration that L>>R we neglect by end effects and consider only fluctuations in r-direction. As a consequence of it, a term which is proportional to the velocity of flow vanishes. This takes place because the velocity vector is perpendicular to gradient of concentration and they scalar production equals to zero. This result in the following equation

$$d\delta n(r,t)/dt = -D\Delta\delta n(r,t) - \sigma\Delta\psi + f \qquad (5)$$

Where σ is the conductivity, ψ potential across the capillary, D diffusion coefficient. The average value of f is zero and its second moment is given by the following expression

$$\langle ff\rangle = -4D\nabla_r\nabla_{r'}\left[n(r)\delta(r-r')\right]\delta(t-t') \qquad (6)$$

The concentration fluctuations will give rise in fluctuations of potential across the capillary. So, the additional fluctuating flow in r-direction will take place. The correlation between δn and $\delta\psi$ is given by Poisson-Boltsman equation

$$-e\delta n - e(n^+ - n^-) = \varepsilon\varepsilon_o\Delta(\psi + \delta\psi) \qquad (7)$$

By using Einstein correlation

$$\mu = kTD \qquad (8)$$

we can find

$$\delta n = 2n_o e\delta\psi/kT \qquad (9)$$

By solving the set of equations (5), (7)-(9) we can find the needed correlation between δn and f

$$\left[1.5æ^2D + i\omega\right]\delta n = f \qquad (10)$$

Where æ is the reverse Debai radius. Thus

$$\langle\delta n\delta n\rangle = \frac{\langle ff\rangle}{\omega^2 + (1.5æ^2D)^2)} \qquad (11)$$

Therefore, the p.s.d. of current fluctuations has the following form

$$S_D = \frac{(2\pi e)^2\int_0^R\int_0^R\langle ff\rangle rVr'V(r')drdr'}{\omega^2 + (1.5æ^2D)^2} \qquad (12)$$

The solution of Poisson-Boltsman equation can be written through Bessel's functions

$$n_o(r) = \frac{-\varepsilon\varepsilon_o \text{æ}^2 \psi_o I_o(\text{æ}r)}{e I_o(\text{æ}R)} \tag{13}$$

Where I_o is a Bessel function of zero order. The average current with the concentration distribution given by (13) is written as follows

$$I = \pi\varepsilon\varepsilon_o\psi_o R^2 \left[1 - \frac{2}{\text{æ}R} \frac{I_1(\text{æ}R)}{I_o(\text{æ}R)} \right] \tag{14}$$

For the ratio of excess noise to the electrical current we now find fomula

$$\frac{S_D}{I^2} = \frac{eDG(\text{æ}R)}{2\pi^2\varepsilon\varepsilon_o\psi_o R^2 L} \frac{1}{\omega^2 + (1.5\text{æ}^2 D)^2} \tag{15}$$

where

$$G(\text{æ}R) = \left. \frac{\int_0^R I_o(x)(x^2 - (\text{æ}R)^2)dx}{x^4 I_o(x)\left[1 - 2I_1(x)/xI_o(x)\right]} \right|_{x=\text{æ}R} \tag{16}$$

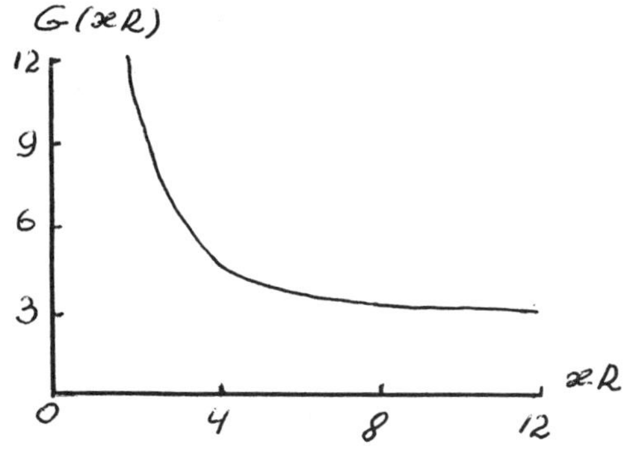

Function G(æR) has significant dependence on electrokinetic radius of a capillary (figure 1). Let us evaluate the contribution of obtained noise to the whole noise in comparison with the Johnson noise. The typical parameters for EKC are $R=1\mu m$, $\text{æ}R=2$, $D=10^{-9} m^2/s$, $L=2mm$, $\psi_o=25\mu V$, $\varepsilon=10$. Thus for single capillary we have

$$S_I = 2\ 10^{-39} A^2/Hz + 4\ 10^{-3} I^2 A^2/Hz \tag{17}$$

Therefore, at the current of 10^{-18}A through a capillary, the excess noise results in the same contribution to p.s.d. as Johnson noise does.

REFERENCES.

1. J. Keizer. Statistical Thermodynamics of Nonequilibrium Processes (Springer-Verlag, 1987).

COMPUTER-AIDED ANALYSIS OF PRECISION OSCILLATING SYSTEMS WITH WIDE-BAND FLUCTUATIONS IN CIRCUIT PARAMETERS

N.V.Demin, Post grad.stud.
State University, N.Novgorod 603600, Russia

ABSTRACT

The method for the analysis of linear inertial circuits affected by wide-band (up to carrier frequency) fluctuations in parameters is worked out. AM-FM noise caused by capacity fluctuations in *LC*-oscillator and in the capacitive oscillatory circuit is determined.

INTRODUCTION

Different methods for the analysis of fluctuations in oscillators are known. Here the method by *Yakimov*[1] for computer-aided analysis is developed. This method uses the analytical signal by Gabor. Fluctuations are presented by a column vector with two elements: $m(t)$ - AM noise (relative to the carrier), and $\varphi(t)$ - PM noise:

$$\delta\dot{X}(t) = [m(t),\ \varphi(t)]^T .$$

Here "T" means the transposition of the matrix.

The circuit is divided on nonlinear inertialess, and linear inertial elements, described by matrices [2×2] determining the transformation of AM-PM noise in the signal with frequency ω_o. Each noise source

$$\dot{e}(t)=(e_m(t)+je_\varphi(t))\cdot exp(j\omega_o t)$$

in the circuit is described by the column vector:

$$\dot{E}(t) = [e_m(t),\ e_\varphi(t)]^T.$$

Fluctuations in parameters of elements are accounted in similar way. Thus, output signal of each element is determined by matrix algebra. Data on matrices and matrix relations are put in a computer. The method of *Haus and Adler*[2] is used giving the resulting matrix of AM-PM noise spectra. The peak and pedestal of the signal spectral line may also be evaluated [3].

The method has no restrictions on noise modulation frequencies. The problem only was with the account of wide-band fluctuations in parameters of linear inertial circuits (e.g., capacitors and inductances). Just this problem is solved in the present paper.

THE MAIN IDEA OF THE METHOD

Let us consider, as an example, the capacitor having non-perturbed capacitance C_o and relative fluctuations $\delta C(t)$. The input current $i(t)$ is represented as an analytic signal:

$$\dot{i}(t) = (1+\delta\dot{I})\dot{I}_o \cdot exp(j\omega_o t) .$$

Here $\dot{I}_o$ is the complex amplitude of the signal; $\delta\dot{I}(t) = m_i(t)+j\varphi_i(t)$ -relative fluctuations in the amplitude.

Simplifying the problem, we suppose the spectrum of capacity fluctuations to be nonzero only in finite frequency band, and these fluctuations are small (if necessary, the considered time interval is to be restricted):

$$< \delta C^2 >_F = 0 \quad \text{for} \quad 2F > f_o = \omega_o/2\pi; \quad \overline{\delta C^2} \ll 1 . \quad (1)$$

AM-PM noise is to be satisfied to similar conditions; thus, we can reject small effects of the second order.

The capacitor is considered to be inertial element. But the relation between the applied voltage and the accumulated charge is inertialess:

$$v(t) = \frac{1}{C(t)} \int i(t)dt. \quad (2)$$

The following relation for relative fluctuations in complex amplitudes of the current and voltage may be carried out from Fourier-transform of eq.(2):

$$\dot{V}_o\delta\dot{V}(\Omega) = \dot{Z}_C(\omega_o+\Omega)\cdot \dot{I}_o\delta\dot{I}(\Omega) - \dot{V}_o\delta C(\Omega) . \quad (3)$$

Here $\Omega=\omega-\omega_o$ is the analyzing frequency, $\dot{Z}_C=(j\omega C_o)^{-1}$ -the impedance, $\dot{V}_o$ -the complex amplitude of the voltage $v(t)$. One can see from eq.(3) that the smallness of fluctuations allows to describe perturbations in reactance by the additive voltage noise source having amplitude presented by the last term in eq.(3): $\dot{V}_n= -\dot{V}_o\delta C(\Omega)$. The supposition about quasi-static character of capacitance fluctuations was not used in eq.(3). Restrictions on the spectrum of fluctuations are linked only with simplification (1).

Relation (3) may be written in other way:

$$\dot{I}_o\delta\dot{I}(\Omega) = \dot{G}_C(\omega_o+\Omega)\cdot \dot{V}_o\delta\dot{V}(\Omega) + \dot{I}_n(\Omega) ,$$

where $\dot{G}_C=j\omega C_o$; that means perturbations in the capacitance are accounted by the current noise source having the amplitude

$$\dot{I}_n(\Omega) = \dot{G}_C(\omega_0+\Omega)\cdot\dot{V}_n(\Omega) . \qquad (4)$$

Carrying out analogous to (2,3) relations for perturbations in the inductance $L(t)=(1+\delta L(t))\cdot L_0$ and resistance $R=(1+\delta R(t))\cdot R_0$, one can formulate the universal algorithm for the account of wide-band fluctuations in linear circuits parameters. Let us note that conditions (1) are used in the common analysis as well.

AN ACCOUNT OF FLUCTUATIONS IN PARAMETERS

The account of fluctuations in linear circuit parameters under AC excitation is consisted in the following. An element subjected to fluctuations in its parameters is replaced by the equivalent circuit containing the non-perturbed element and the voltage noise source

$$\dot{v}_n(t) = \dot{V}_n^{(\lambda)}(t)\cdot exp(j\omega_0 t),\ \lambda=C,L,R ,$$

in series with this element; or non-perturbed element and the current noise source

$$\dot{i}_n(t) = \dot{I}_n^{(\lambda)}(t)\cdot exp(j\omega_0 t) .$$

in parallel with it. Fourier-transforms for amplitudes of voltage noise sources are:

$$\dot{V}_n^{(C)}(\Omega)= -\dot{V}_0\delta C(\Omega),\ \dot{V}_n^{(L)}(\Omega)= \dot{V}_0\delta L(\Omega),\ \dot{V}_n^{(R)}(\Omega)= \dot{V}_0\delta R(\Omega)$$

for capacitance, inductance, and resistance; $\dot{V}_0$ is non-perturbed amplitude of the voltage on the element.

The current noise source amplitude is evaluated by relation analogous to eq.(4):

$$\dot{I}_n^{(\lambda)}(\Omega) = \dot{G}_\lambda(\omega_0+\Omega)\cdot\dot{V}_n^{(\lambda)}(\Omega) ,$$

where $\dot{G}_\lambda(\omega)$ is conductance of non-perturbed element. The further analysis is made up by the method of linear circuit analysis based on Kirchhoff equations.

FLUCTUATIONS IN *LC*-OSCILLATOR

As the simplest example, let us consider the feedback circuit of *LC*-oscillator (see fig.1). Here the current noise source accounts perturbations in the resonator capacity $C(t)=(1+\delta C(t))C_o$. Fourier-transform of the source amplitude is $\dot{I}_{nC}(\Omega)= -j(\omega_o+\Omega)\dot{U}_o C_o \delta C(\Omega)$, where $\dot{U}_o$ means non-perturbed complex amplitude of the voltage on the capacitor. Using Kirchhoff equations one can get the following equation for the feedback circuit:

$$\dot{I}_o \delta\dot{I}(\Omega) = \dot{G}_T(\omega_o+\Omega)\cdot\dot{V}_o\delta\dot{V}(\Omega) + \dot{I}_{nC}(\Omega) \ , \qquad (5)$$

where $\delta\dot{I}(\Omega)$ and $\delta\dot{V}(\Omega)$, $\dot{I}_o$ and $\dot{V}_o$ are relative fluctuations and non-perturbed values of complex amplitudes of the amplifier output current $i(t)$ and the feedback voltage $v(t)$ fundamental harmonics;

$$\dot{G}_T(\omega) = (1 + jQ(\omega/\omega_1 - \omega_1/\omega))/(R_{oe}n_T)$$

is the complex conductivity of the oscillator feedback circuit; ω_1, R_{oe} and Q -resonance frequency and resistance, and the quality factor of the resonator; $n_T = M/L$ is the transformation factor.

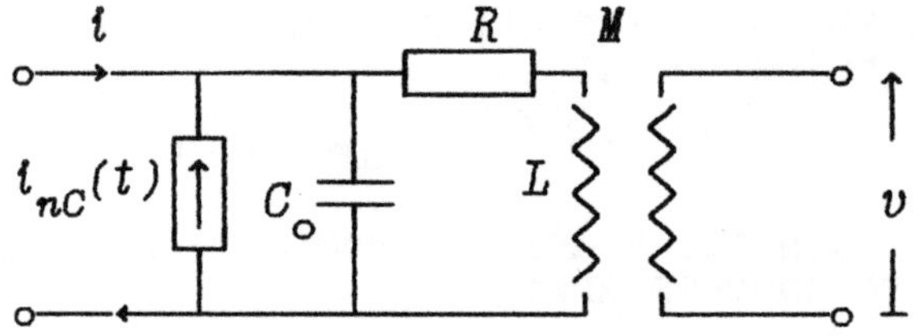

Fig.1. The feedback circuit of *LC*-oscillator.

The transformation of the feedback voltage into the current by inertialess amplifier does not affect the phase noise, but AM-noise is weighted by factor γ_1 been the normalized local slope for the fundamental harmonic:

$$\varphi_i = \varphi_v \quad ; \quad m_i = \gamma_1 m_v \quad . \qquad (6)$$

AM-frequency noise caused by perturbations in the resonator capacitance may be evaluated from eq.(5) by the usage of the method by *Yakimov* [1] and accounting eq.(6):

$$m_v(\Omega)=\dot{T}_{mC}(\Omega)\cdot\delta C(\Omega); \quad \nu_v(\Omega)=j\Omega\cdot\varphi_v(\Omega)=\dot{T}_{\nu C}(\Omega)\cdot\delta C(\Omega).$$

Transformation factors $\dot{T}_{mC}$ and $\dot{T}_{\nu C}$ may be determined up to analyzing frequencies $|x| \leqslant 1/2$, where $x = \Omega/\omega_o$ is the normalized analyzing frequency. But these factors practically do not differ from values found in quasi-static approximation ($|x|<<1$) giving the known result:

$$m_v^{st}(\Omega) = -(1-\gamma_1)^{-1}\delta C(\Omega) \; ; \quad \nu_v^{st}(\Omega) = -(\omega_o/2)\cdot\delta C(\Omega) \; .$$

It should be noted that capacity perturbations affect the frequency much stronger than the amplitude of oscillations. Factors $\dot{T}_{mC}$ and $\dot{T}_{\nu C}$, in contrast with those quasi-static values, account cross-transformation effects caused by imaginary parts of the factors. As far as in *LC*-oscillator cross-transformations are small, the difference between the factors at high analyzing frequency and near the zero frequency is small as well. But some oscillator circuits exist having cross-transformation stronger than in *LC*-circuit. The usage of such circuits may lead to the increase in transformation factors at high analyzing frequencies.

CAPACITIVE OSCILLATORY CIRCUIT ANALYSIS

Let us analyze the capacitive oscillatory circuit having the feedback with perturbations in capacity $C_2(t)$ shown in fig.2. Carrying the analysis similar to made for *LC*-oscillator, one can find transformation factors determining the output voltage AM-FM noise.

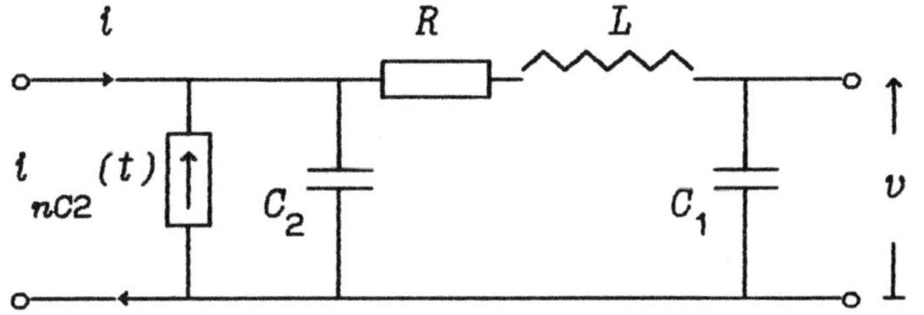

Fig.2. The feedback of capacitive oscillatory circuit.

Absolute values of these factors have the sense of sensitivities of AM-FM noise to capacity fluctuations. Dependence of these sensitivities on the normalized frequency $x = \Omega / \omega_o$ is shown in fig.3; full lines are drawn for AM-noise, and broken lines - for frequency noise normalized on value $\omega_o/2$. Following values of parameters were used: Q = 50, γ_1 = 0.7; $n = C_1/(C_1+C_2)$ = 0.8 (curves 1) and n = 0.9 (curves 2). Note that the increase in the difference between the capacitance C_2 and the serial capacitance of C_1 and C_2 determining the oscillation frequency leads to the increase of the difference between these factors at zero and high frequencies.

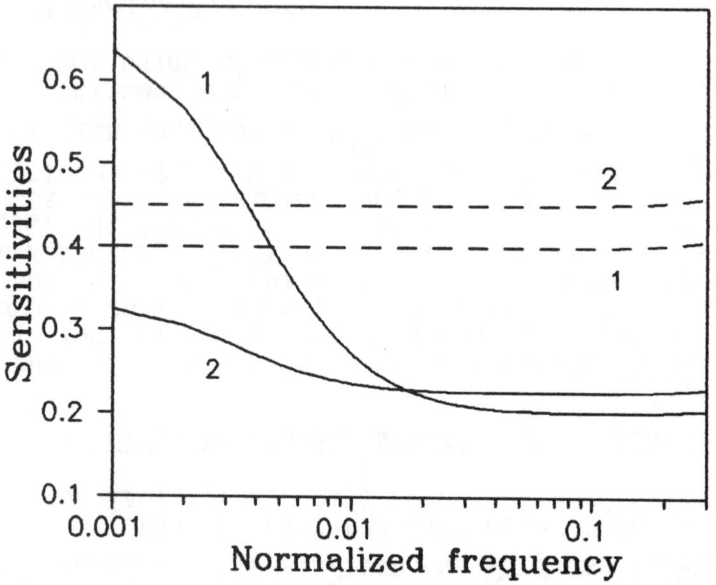

Fig.3. Modulation sensitivities for AM-FM noise.

CONCLUSION

The method suggested here for the account of fluctuations in parameters of linear inertial circuits gives the possibility to investigate amplitude and phase perturbations of signals in wide analyzing frequency band.

Author is thankful to The Netherlands Organization for Scientific Research (NWO) and to Prof. F.N.Hooge for the support of investigations on the considered problem.

REFERENCES

1. A.V.Yakimov, Radio Eng. Electron.Phys. 30 (1985) No.12 (Russian original, p.2361).
2. H.A.Haus and R.B.Adler, Circuit Theory of Linear Noisy Networks (The Technol.Press,Massach.Inst.of Technol. and J.Willey&Sons, Inc., New York Chapman and Hall, Ltd, London, 1959).
3. A.N.Malakhov, Fluctuations in Autooscillatory Systems (Nauka-Press, Moscow, 1968, in Russian).

NOISE-ENHANCED HETERODYNING

M I Dykman
Department of Physics, Stanford University, Stanford, CA 94305, USA.

D G Luchinsky[1], P V E McClintock, N D Stein and N G Stocks[2]
School of Physics and Materials, Lancaster University, Lancaster, LA1 4YB, UK.

ABSTRACT

A new form of heterodyning is reported, related to stochastic resonance, in which a heterodyne signal can be *enhanced* by adding noise.

One of the important physical problems of information processing and transfer is how to control the signal-to-noise ratio (SNR). Usually, this ratio decreases with increasing intensity of noise. However, under certain circumstances it behaves in the opposite way. The phenomenon of the noise-induced increase of SNR was called stochastic resonance (SR) [1]. It has attracted much attention recently (see [2]). Most of the data on SR has been obtained for bistable systems driven by noise and by a low-frequency periodic force. The onset of SR in these systems is related to the fact that the probabilities W_{nm} of transitions between co-existing stable states $(n, m = 1, 2)$ increase exponentially, in the case of Gaussian noise, with increasing noise intensity D. A low-frequency external periodic force $A_0 \cos \Omega t$ modulates the activation energies and as a result the transition probabilities W_{nm} are modulated too. In turn, the modulation of W_{nm} gives rise to a modulation of the populations of the stable states. For a symmetrical double-well potential, the force periodically makes one of the wells deeper than the other, and the system occupies it with a larger probability. As a result, the amplitude of the oscillations is proportional to the relatively large difference $x_1 - x_2$ in the values of the coordinate in the stable states x_n.

When the above mechanism comes into play through the onset of fluctuational transitions, the amplitude of the periodic signal increases with increasing noise intensity, in a certain range of D, and the SNR increases with D, too. It works, provided (i) the stationary populations of the states in the absence of the periodic force are nearly equal to each other [3], and (ii) the frequency of the force is much smaller than the reciprocal relaxation time t_r^{-1} of the system, so that the transitions are likely to occur within the period $2\pi/\Omega$.

The frequency-selective response of bistable systems, and also the fact that the SNR increases with increasing noise intensity, makes it interesting to apply the idea of SR to heterodyning so as to obtain a form of the phenomenon that is *enhanced* rather than suppressed by noise. In heterodyning, two high-frequency fields, one of them a signal and the other a reference field, are mixed nonlinearly to generate a signal at a difference frequency. In this paper we report and discuss

[1]Permanent address: VNIIMS, Andreevskaya nab. 2, Moscow, 117965, Russia.
[2]Present address: Dept. of Engineering, University of Warwick, Coventry, CV4 7AL, UK.

a new form of the phenomenon, *noise-enhanced heterodyning* (NEH), that occurs in bistable systems and is highly frequency selective. We have investigated it theoretically and by analogue electronic simulation.

We shall illustrate the effect on a model of an overdamped bistable system driven by three time-dependent forces representing respectively the reference and input signals, and the noise. The motion of the system is described by the equation

$$\frac{dx}{dt} = -U'(x) + A_{\text{ref}} x \cos \omega_0 t + A_{\text{in}}(t) \cos[\omega_0 t + \phi(t)] + f(t) \tag{1}$$

Here, the term $\propto A_{\text{ref}}$ is the reference signal of a given frequency ω_0 (the corresponding force is applied multiplicatively), and the term $\propto A_{\text{in}}(t)$ is the modulated high-frequency input signal (applied additively). The functions $A_{\text{in}}(t)$ and $\phi(t)$ are slowly varying as compared with $\cos \omega_0 t$, and it is their variation in time that has to be revealed via heterodyning. The heterodyning can be characterized by the low-frequency signal at the output, $x(t)$, for $A_{\text{in}} =$ const and $\phi = \Omega t +$ const, with $\Omega \ll \omega_0$, i.e., for a monochromatic input signal with the frequency $\omega_0 + \Omega$ slightly different from the frequency ω_0.

We shall assume that the double-well potential of the system $U(x)$ has equally-deep wells, corresponding to standard SR, and is of the form

$$U(x) = -\frac{1}{2}x^2 + \frac{1}{4}x^4 \tag{2}$$

The minima of the potential (2) (the stable states of the system) lie at $x_n = (-1)^n$, $n = 1, 2$, and the characteristic (dimensionless) relaxation time of the system $t_r \equiv 1/U''(x_n) = 1/2$. The analysis of heterodyning in bistable systems is not limited to the particular form of Eqs.(1),(2). However, the explicit expressions take on a simple form for this model. They are further simplified in the case where the frequencies of the input and reference signals are high compared with the reciprocal relaxation time of the system, $\omega_0 \gg t_r^{-1}$.

The term $f(t)$ in (1) is a zero-mean Gaussian noise. In view of the possible applications we will allow for noise that consists of two independent components, of low and high frequency respectively, with the latter being randomly modulated vibrations at frequency ω_0 (which might result from the scattering of the signal at frequency ω_0):

$$f(t) = f_{lf}(t) + f_{hf}(t), \quad f_{hf}(t) = \text{Re}\left(\tilde{f}_{hf}(t)\exp(-i\omega_0 t)\right), \tag{3}$$

The power spectrum of the low-frequency noise $f_{lf}(t)$ is assumed to be flat up to $\omega \sim \omega_c \gg t_r^{-1}$ (ω_c may be small compared to ω_0).

For $\omega_0 \gg t_r^{-1}$ the motion of the system consists of fast oscillations at frequency ω_0 (and its overtones) superimposed on a slower motion. To first order in ω_0^{-1} the equation for the smooth part of the coordinate, $x^{(sm)}$, is of the form

$$\dot{x}^{(sm)} = -U'(x^{(sm)}) + A(t)\sin\phi(t) + f^{(0)}(t), \quad A(t) = \frac{A_{\text{ref}}}{2\omega_0} A_{\text{in}}(t) \tag{4}$$

where $f^{(0)}(t) = f_{lf}(t) - (A_{\rm ref}/2\omega_0){\rm Im}\tilde{f}_{hf}(t)$ is the noise which, in view of the above comments, will be assumed white (its correlation time $\sim \omega_c^{-1} \ll t_r$). The noise intensity D is composed of a weighted sum of low and high frequency contributions.

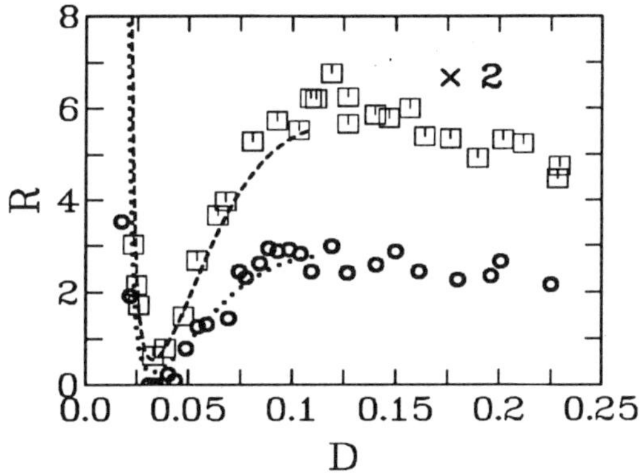

Fig.1 NEH for white (circles) and high-frequency (boxes) noise. The dashed and dotted lines are theory.

The dynamics of the system described by Eqs.(2), (4), for $A(t), \phi(t)$ varying slowly over the time t_r, has been investigated in detail in the context of stochastic resonance. For sufficiently small A the general analysis of the response can be done in terms of linear-response theory [3]. In the particular case of a monochromatic force of frequency Ω, i.e., $A(t)\sin\phi(t) = A\sin\Omega t$, a δ-shaped spike arises in the power spectrum of the coordinate $x^{(sm)}$ at $\omega = \Omega$ on top of the broad spectrum.

As in standard stochastic resonance, NEH can be characterized by the ratio R of the intensity (area) of this spike (which is just proportional to the squared amplitude of forced vibrations at frequency Ω) to the value of the power spectrum at the same frequency for $A = 0$

$$R = \pi \frac{A^2(x_2 - x_1)^2}{16D^2} \frac{W^2 + \Omega^2 t_r^2 \tilde{D}^2}{W + \Omega^2 t_r^2 \tilde{D}}, \quad \tilde{D} = 4D/(x_2 - x_1)^{-2}. \tag{5}$$

Eq.(5) was shown in [4] to hold for arbitrary Ω/W ($W \propto \exp(-\Delta U/D)$ is the probability of the interwell transitions) to lowest order in $\Omega t_r, \tilde{D}t_r, W/\tilde{D} \ll 1$. It follows from (5) (and also from the more general expression for the SNR) that, in the range of the noise intensities where $\Omega\tilde{D}t_r \lesssim W \lesssim t_r^{-1}$, the SNR *increases* with increasing noise intensity. The fact that this increase is quite sharp - nearly exponential - means that noise-enhanced heterodyning would be expected to arise in a bistable system, whether driven by a low- or a high-frequency noise (or both).

The onset of NEH has been investigated experimentally by means of analog electronic simulation. The design will be discussed elsewhere. In Fig.1 the experimentally measured values of the heterodyne SNR are compared with the theoretical predictions for low- and high-frequency noises. It is clearly seen that, in both cases, there is an interval of noise intensities where the SNR sharply increases with D. Below and above this interval the SNR decreases with increasing noise intensity. The experimental data are in satisfactory agreement with the linear-response theory; we note that the latter does not contain any adjustable parameters (we have used an expression [4] that allows for the corrections $\sim D/\Delta U$

omitted in (5)). It has been shown experimentally that R is proportional to the squared amplitudes of the reference and input signals, and also that $R \propto \omega_0^{-2}$ over a broad range of ω_0 up to $\omega_0 \approx 2t_r^{-1}$ as shown in Fig.2.

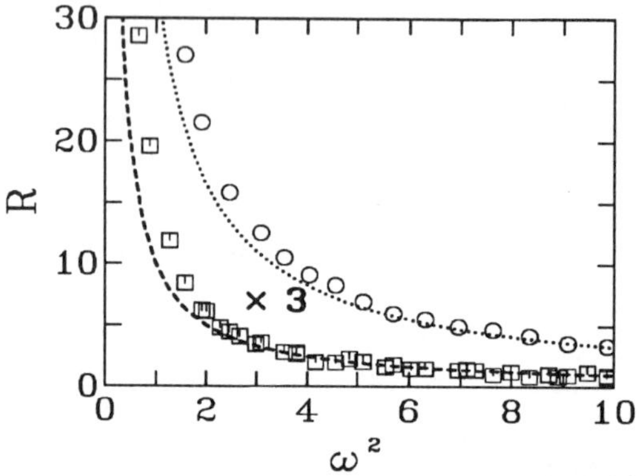

Fig.2 NEH dependence on the squared frequency of the reference signal for $\Omega = 0.0031$ with noise intensities D=0.015 (circles) and D=0.14 (boxes). The dashed and dotted lines $\propto 1/\omega_0^2$.

Of particular interest is the dependence of the SNR on the modulation frequency Ω. It follows from (5) that in the interesting range where R sharply increases with noise intensity, i.e. in the range $D \ll \Delta U$, R also increases quite sharply with Ω, from the value that corresponds to the SNR in the neglect of the *intrawell* motion, $R = R_0 \approx \pi A^2 W/4D\tilde{D}$ for very small Ω, up to the value that corresponds to the SNR in the neglect of the *interwell* transitions, $R \approx \pi A^2/4D$. As in the case of SR [2], a simple way to avoid frequency dispersion is to apply two-state filtering where the quantities of interest are the values of the coordinate coarse-grained over the vicinities of the stable states. For $\Omega \ll t_r^{-1}$ the value of the SNR at the output of a two-state filter is given approximately by R_0.

In conclusion, we have demonstrated, theoretically and experimentally, that bistable systems can be used to obtain heterodyning in which not only the amplitude of the signal at the output, but also the signal-to-noise ratio increases with increasing intensity of the noise.

The work was supported by the Science and Engineering Research Council (UK), by the EC, by the Royal Society of London, and by the Gosstandart of Russia.

REFERENCES

1. R. Benzi, A. Sutera and A. Vulpiani, *J. Phys. A* **14**, L453 (1981); C. Nicolis, *Tellus* **34**, 1 (1982); R. Benzi, G. Parisi, A. Sutera and A. Vulpiani, *Tellus* **34**, 10 (1982).

2. Special issue of *J. Stat. Phys.* **70**, no. 1/2 (1993).

3. M.I. Dykman, R. Mannella, P.V.E. McClintock, and N.G. Stocks, *JETP Lett.* **52**, 141 (1990).

4. M.I. Dykman, R. Mannella, P.V.E. McClintock, and N.G. Stocks, *Phys. Rev. Lett.* **68**, 2985 (1992).

FLUCTUATIONAL TRANSITIONS AND CRITICAL PHENOMENA IN A PERIODICALLY DRIVEN NONLINEAR OSCILLATOR SUBJECT TO WEAK NOISE

M I Dykman
Department of Physics, Stanford University, Stanford, CA 94305, USA.

R Mannella
Dipartimento di Fisica, Università di Pisa, Piazza Torricelli 2, 56100 Pisa, Italy.

D G Luchinsky[1], P V E McClintock, N D Stein and N G Stocks[2]
School of Physics and Materials, Lancaster University, Lancaster, LA1 4YB, UK.

ABSTRACT

Fluctuation-induced transitions between coexisting periodic attractors in a periodically driven nonlinear oscillator have been investigated theoretically and by analogue electronic experiment. Calculations and measurements of the corresponding activation energies are in good agreement, and have enabled the position of the kinetic phase transition (KPT) line to be established over its full range.

The kinetics of an oscillator is a long-standing and important problem of classical and quantum statistical physics, for two reasons. On the one hand, many real physical systems can be well modelled by such oscillators and, on the other, comparatively simple solutions can be obtained for models of this kind, in particular for an underdamped nonlinear oscillator (see [1]). Interesting new phenomena arise if an underdamped nonlinear oscillator is driven by a nearly resonant force. Since the frequency of the eigenvibrations of the oscillator depends on their amplitude, there is a certain range of the forcing amplitude in which there are co-existing stable states of vibration with comparatively small and large amplitudes respectively. The corresponding eigenfrequencies are self-consistently in comparatively bad or good resonance with the field frequency ω_F. The bistability of a periodically driven oscillator has been observed for cyclotron motion of an electron in a quadrupole trap [2]. It has been discussed also in the context of optical bistability [3], and in acoustics and engineering.

An oscillator bistable in a periodic field provides an example of a bistable system far from thermal equilibrium, for which the co-existing attractors are limit cycles. The quantities of particular interest and importance for nonequilibrium bistable systems are the probabilities W_{nm} of fluctuational transitions between the stable states ($n, m = 1, 2$). For weak intensity of the fluctuations (induced by external noise, or resulting from coupling to a thermal bath) these probabilities are very much smaller than the characteristic reciprocal relaxation time(s) of the system t_r^{-1}. In the general case, the probabilities W_{12} and W_{21} of the transitions $1 \rightarrow 2$ and $2 \rightarrow 1$ are strongly different (exponentially different, for Gaussian

[1]Permanent address: VNIIMS, Andreevskaya nab. 2, Moscow, 117965, Russia.
[2]Present address: Dept. of Engineering, University of Warwick, Coventry, CV4 7AL, UK.

noise), as are the stationary populations $w_{1,2}$ of the stable states ,

$$w_1 = W_{21}/W, \quad w_2 = W_{12}/W, \quad W = W_{12} + W_{21} \tag{1}$$

Only within a narrow range of the parameters of the system are w_1 and w_2 of the same order of magnitude. In this range a kinetic phase transition occurs and although the fluctuations about each stable state are small, large fluctuations then come into play, related to transitions between the states. A feature of these fluctuations is that they are infrequent, the characteristic time scale being $\sim W^{-1}$, and they would therefore be expected to give rise to high supernarrow peaks (of width $\sim W \ll t_r^{-1}$)in the spectral density of the fluctuations (SDF) and in the susceptibility at the frequency of the co-existing limit cycles and its overtones (we note that a nonequilibrium system can amplify a weak signal, i.e., the imaginary part of the susceptibility can be negative) [1]. The intensity of the peaks should (and for the SDF does [4]) display an extremely sharp dependence on the distance to the transition point. Since the supernarrow peaks in the susceptibility are due to fluctuations (caused by the noise driving the oscillator), then, in a certain range of the noise intensity B, the amplitude of a signal due to an additional weak force should *increase* with increasing B. This is a prerequisite for stochastic resonance (SR) - an interesting phenomenon that has attracted much attention recently in view of various applications [5]. In contrast to the "conventional" SR observed at low frequencies, in the case of a periodically driven oscillator there arises [6] a high-frequency form of stochastic resonance (HFSR).

We present below some results on fluctuations in a nonlinear oscillator obtained recently by means of analog simulation and theoretically. The model investigated is described by the equation

$$\ddot{q} + 2\Gamma\dot{q} + \omega_0^2 q + \gamma q^3 = F\cos\omega_F t + f(t) \tag{2}$$

$$\Gamma, |\delta\omega| \ll \omega_F, \quad \delta\omega = \omega_F - \omega_0, \quad \langle f(t)f(t')\rangle = 2\Gamma B\delta(t-t')$$

Here, $f(t)$ is Gaussian white noise. For small friction coefficient Γ and small frequency detuning of the driving force with respect to the eigenfrequency of the oscillator ω_0, and also for a comparatively small field amplitude F, the motion of the oscillator is primarily vibrations at a frequency ω_F with slowly varying amplitude and phase. The dynamics of the latter depends on the values of two dimensionless parameters, β and η, and also on the dimensionless noise intensity α,

$$\beta = 3|\gamma|F^2/32\omega_F^3|\delta\omega|^3, \quad \eta = \Gamma/|\delta\omega|, \quad \alpha = 3|\gamma|B/16\omega_F^3\Gamma \tag{3}$$

(the kinetics of slow variables is basically the same whether the model (2) is used or fluctuations are assumed to be due to the coupling to a thermal bath; the main point is that the power spectrum of the noise $f(t)$ should be flat in a range $\gtrsim |\delta\omega|$ about ω_F). The range of β, η where the two attractors coexist as calculated and measured in the experiment is enclosed by the approximately triangular region shown in Fig.1. On the upper branch of $\beta(\eta^2)$ the small-amplitude limit cycle (the stable state 1) merges with an unstable one and disappears, whereas on the lower branch this occurs to the large-amplitude limit cycle (the stable state 2). A fluctuating periodically driven nonlinear oscillator was one of the first physical systems without detailed balance for which the probabilities of transitions between coexisting stable states were analyzed [7]. It follows from the theory [7] that, to logarithmic accuracy, the dependence of W_{nm} on the noise intensity is of activation type,

$$W_{nm} = \text{const} \times \exp(-R_n/\alpha) \quad (4)$$

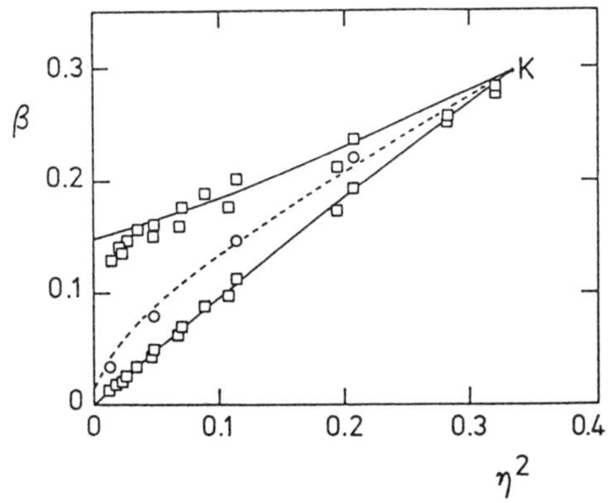

Fig.1 The phase diagram as calculated (full lines) and measured (squares [4]). The calculated KPT line (dashed) is compared with experimental measurements (circles).

The values of the activation energies $R_{1,2}$ for the transitions from the states 1,2 depend on the parameters β, η and are given by the solution of a variational problem. Such an activation dependence on the noise intensity for a system without detailed balance has indeed been observed in our experiments. The experimental mean first passage times (MFPT) and activation energies $R_{1,2}$ are shown in Figs.2 and 3. As expected, the activation energy for escape from the small-amplitude limit cycle decreases with increasing β (i.e., with increasing field amplitude) until R_1 becomes equal to 0 at the bifurcation point where the stable state disappears. Conversely, R_2 increases with increasing β. The data are in reasonably good agreement with the results of the numerical solution of the variational problem for $R_{1,2}$: there are no adjustable parameters in either the theory or the experiment. It follows from the expressions

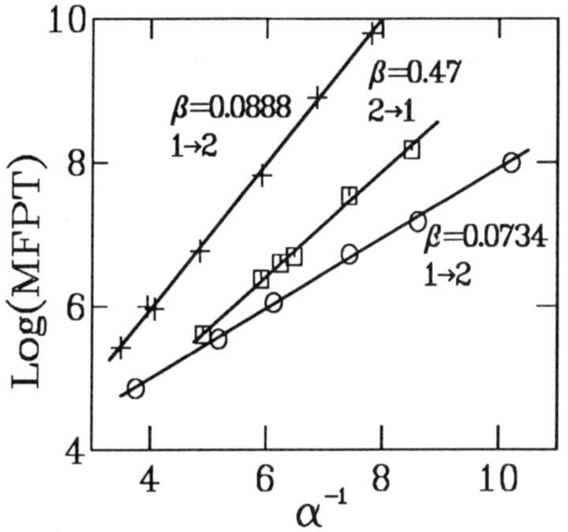

Fig.2 Dependence of the experimental MFPT on reciprocal noise intensity for $\eta^2 = 0.033$.

(1),(4) that for the most of the values of β, η the ratio of the stationary populations of the states $w_1/w_2 \propto \exp\left(-(R_2 - R_1)/\alpha\right)$ is either exponentially large or small, and only for $R_1 \approx R_2$ are the populations of the same order of magnitude.

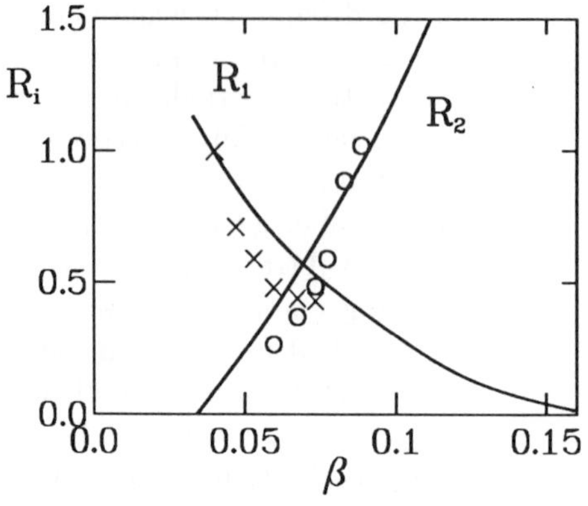

Fig.3 The experimental and theoretical activation energies for $\eta^2 = 0.033$.

The KPT phase-transition line $\beta_c(\eta^2)$ as given by the condition $R_1 = R_2$ is shown dashed in Fig.1. It was found analytically for small values of η^2 (we note that β_c is a nonanalytic function of η^2 for small η^2, $\beta_c(\eta^2) - \beta_c(0) \propto \eta$) and in the vicinity of the spinode point K, and it was evaluated numerically in between using the calculated values of $R_{1,2}$. The experimental points (circles), which were obtained from the condition $W_{12} = W_{21}$, are seen to be in good agreement with the theory. The supernarrow peaks in the SDF, corresponding to those anticipated [7] in the susceptibility of the system, were observed previously [4].

In conclusion, we note that the calculated and measured activation energies are in satisfactory agreement and that they have enabled us to establish the position of the KPT line (Fig.1) over the full range of η^2 from zero up to the spinode point K.

The work was supported by the Science and Engineering Research Council (UK), by the EC, by the Royal Society of London, and by the Gosstandart of Russia.

REFERENCES

1. M.I. Dykman and M.A. Krivoglaz, in *Soviet Physics Reviews*, edited by I.M. Khalatnikov (Harwood Academic, N.Y., 1984), vol.5, p. 265.
2. G. Gabrielse, H. Dehmelt and W. Kells, *Phys. Rev. Lett.* **54**, 537 (1985).
3. H.M. Gibbs, *Optical Bistability: Controlling Light with Light*, (Academic Press, New York, 1985); P.D. Drummond and D.F. Walls, *J. Phys. A* **13**, 725 (1980); Chr. Flytzanis and C.L. Tang, *Phys. Rev. Lett.* **45**, 441 (1980); J.A. Goldstone and E. Garmire, *Phys. Rev. Lett.* **53**, 910 (1984).
4. M.I. Dykman, R. Mannella, P.V.E. McClintock, and N.G. Stocks, *Phys. Rev. Lett.* **65**, 48 (1990).
5. See, e.g. special issue of *J. Stat. Phys.* **70**, nos. 1/2 (1993).
6. M.I. Dykman, D.G. Luchinsky, R. Mannella, P.V.E. McClintock, N.D. Stein, and N.G. Stocks, see p 492 of [5].
7. M.I. Dykman and M.A. Krivoglaz, *Sov. Phys.- JETP* **50**, 30 (1979).

RESONANT CROSSING PROCESSES IN BISTABLE SYSTEMS

L. Gammaitoni
Istituto Nazionale di Fisica Nucleare, Sez. di Perugia and Dipartimento di Fisica dell' Universita', I-06100 Perugia (Italy)

F. Marchesoni
Istituto Nazionale di Fisica Nucleare, Sez. di Perugia, I-06100 Perugia and Dipartimento di Matematica e Fisica, Universita' di Camerino, I-62032 Camerino (Italy)

E. Menichella-Saetta and S. Santucci
Dipartimento di Fisica, Universita' di Perugia, I-06100 Perugia (Italy)

ABSTRACT

The crossing processes in an overdamped double-well potential driven by an external colored noise and a weak sinusoidal time-dependent modulation are investigated by means of analogue simulation. A new resonance phenomenon (*resonant crossing*) is revealed.

The crossing-time distribution in an overdamped bistable potential driven by an external white noise and a weak sinusoidal time-dependent modulation was successfully employed in the past to characterize the switch dynamics under Stochastic Resonance[1-5] (SR) condition. At SR the amplitude of the periodic component of the system response reaches a maximum when the forcing period is about twice the average crossing-time induced by the noise, T_k, in the absence of the modulation. When looking at the crossing-time distribution, the SR condition causes the peak at half the forcing period to increase up to a maximum[2,3]. Such a condition is usually attained by keeping both the frequency and the amplitude of the periodic forcing term fixed and tuning the noise intensity.

Recently it has been shown[6] that a new, peculiar resonant behavior of the crossing processes takes place when the correlation time of the noise term is finite (colored noise). For the sake of simplicity let us consider an overdamped bistable quartic oscillator

$$\dot{x} = -V'(x) + \varepsilon(t) + A\cos(\omega t), \quad \text{where} \quad V(x) = -\frac{a}{2}x^2 + \frac{b}{4}x^4$$

The random noise $\varepsilon(t)$ is taken gaussian, zero-mean valued and exponentially time correlated with correlation time τ. The switch dynamics of x(t) between its stable values is controlled by the interplay of random fluctuations and periodic drive.

On sampling the signal x(t) with an appropriate time-base, one can determine a random point process t_n as follows: data acquisition is triggered at time $t_0=0$ when x(t) crosses, for instance, x_- with negative derivative; t_1 is the subsequent time when x(t) crosses first x_+ with positive derivative; t_2 is the time when x(t) switches back to negative values by re-crossing x_- with negative derivative, and so on. Sequences of 50000 switch times $T_n = t_{n+1} - t_n$ (with n>0) have been distributed for different values of the crossing level x_b, bias parameters A and ω and noise correlation time τ, to obtain the normalized switch-time distributions.

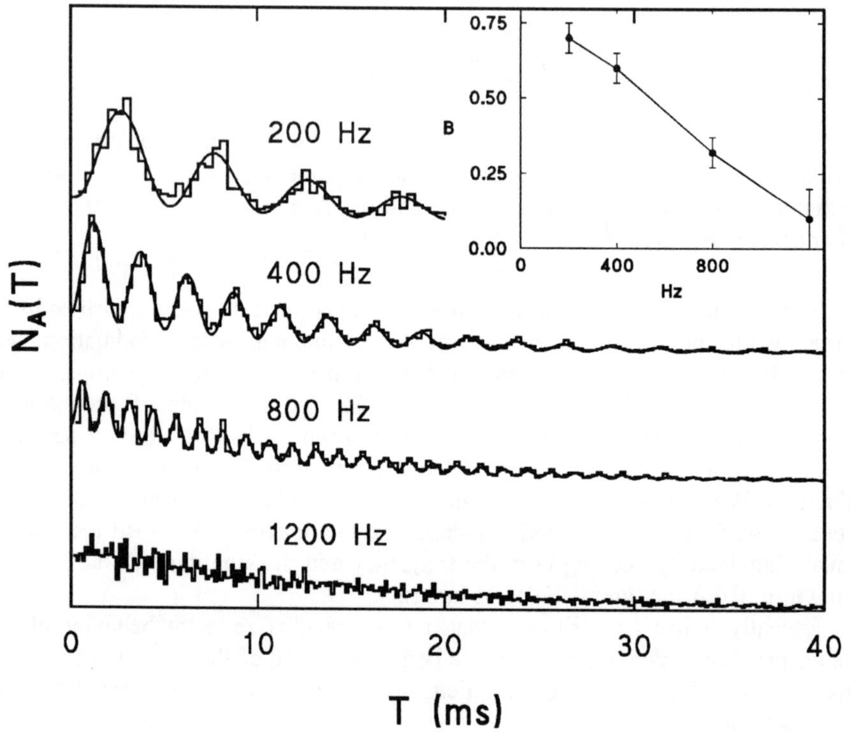

Fig. 1: $N_A(T)$, aτ=0.05, x_b=0, Ax_0/D=3 for four values of ν. Inset: B *vs* ν.

The results of our simulation work can be summarized as follows:

(i) the switch-time distribution $N_0(T)$ in the absence of modulation (A=0) has been obtained for small ($a\tau << 1$) to large ($a\tau >> 1$) noise-correlation times and crossing level x_b varying between 0 and the most probable x value $x_m(\tau)$. $N_0(T)$ falls off exponentially on a characteristic time scale (of the order of T_K), which increases with x_b, markedly in the weak color limit. In the presence of a periodic forcing the switch-time distributions $N_A(T)$ exhibit a peak structure (Fig. 1 and 2) superimposed on the relevant unperturbed distribution $N_0(T)$. The peaks are located at around the odd multiples of π/ω, $T_m = (2m+1)\ \pi/\omega$. At high frequency the

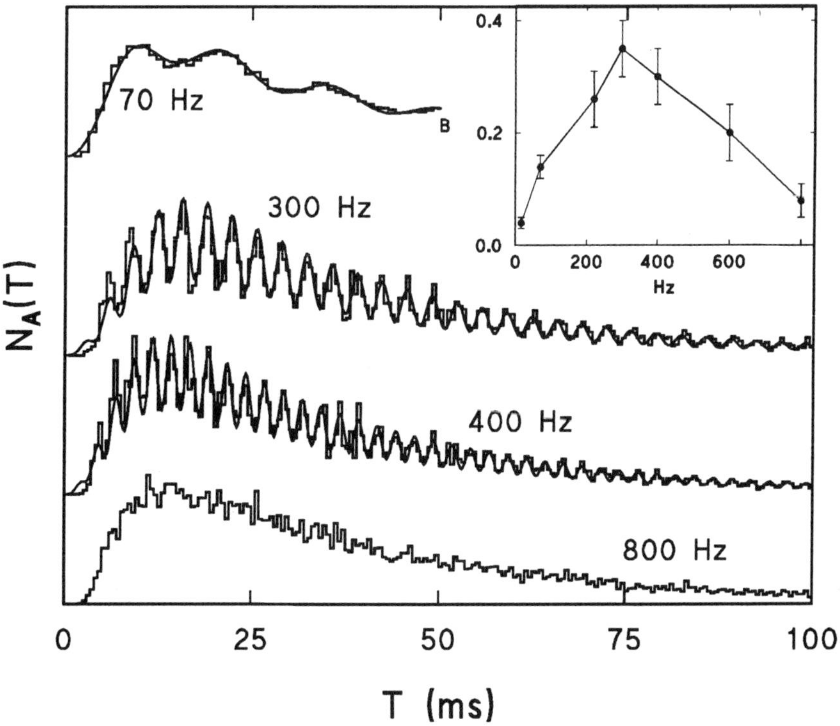

Fig. 2: $N_A(T)$, $a\tau=100$, $x_b=x_m$, $Ax_0/D=0.006$ for four values of ν. Inset: B *vs* ν.

distribution peaks increase with m and, then, decrease exponentially. This property can be detected only for values of ω such that the first peak of $N_A(T)$ occurs on the rising branch of $N_0(T)$.

(ii) on increasing ω well above the SR value, a qualitative difference in the peak structure of $N_A(T)$ for $a\tau << 1$ and $a\tau >> 1$ becomes noticeable. In the white noise limit (Fig. 1) the peak height, with respect to the background, is maximum at SR and, then, decreases with increasing ω, i.e. with the number of detectable peaks. For very large values of ω, $N_A(T)$ approaches the corresponding exponential distribution $N_0(T)$.

In the strongly colored limit, the peak height is much smaller than for $\tau=0$; moreover, it grows through a maximum at $\omega \approx \sqrt{2a/\tau}$ (Fig. 2), thus, indicating a resonance effect in the colored *intra-well* dynamics. The appearance of an enhanced peak structure in $N_A(T)$ for a certain value of ω can be viewed, indeed, as the signature of a higher order in the random point process t_n.

The resonant crossing process can be analyzed quantitatively by introducing a fitting law[6] $N_A = \mathcal{N}_A\ [1-B(A,\omega)\cos(\omega T)]\ N_0(T)$. The fitted values of $B(A,\omega)$ lie on characteristic resonance curves as shown in Fig. 2 (inset). A theoretical interpretation of these results is reported in Ref. 6.

In conclusion, we have shown that the crossing process in an overdamped bistable stochastic system driven by a periodic modulation is characterized by three different time-scales, in coincidence with the inter-well escape time, the noise correlation-time and the intra-well time-constant; the spacing between adjacent peaks is controlled by the forcing frequency, whereas the position of the highest peaks is determined by the noise correlation-time and their sharpness is an apparent signature of the colored intra-well dynamics. These properties define one new resonance effect induced by colored noise: we propose to term such a phenomenon *resonant crossing*.

Acknowledgment: Work supported in part by the Istituto Nazionale di Fisica della Materia.

REFERENCES

1. R. Benzi, G. Parisi, A. Sutera and A. Vulpiani, Tellus 34 10 (1982)
2. L. Gammaitoni, F. Marchesoni, E. Menichella-Saetta and S. Santucci, Phys. Rev. Lett. 62 349 (1989) and C. Presilla, F. Marchesoni and L. Gammaitoni, Phys. Rev A40 2105 (1989)
3. T. Zhou, F. Moss and P. Jung, Phys. Rev A42 3161 (1990)
4. A. Simon and A. Libchaber, Phys. Rev. Lett. 62 349 (1989)
5. G. Nicolis, C. Nicolis and D. McKernan, J. Stat. Phys. 70 125 (1993)
6. L. Gammaitoni, F. Marchesoni, E. Menichella-Saetta and S. Santucci, submitted (1993)

Self-Consistent Calculation of Shot Noise in a Double-Barrier Resonant Tunneling Structure in the Presence of Magnetic Field

Mirza M. Jahan and A. F. M. Anwar
Electrical and Systems Engineering Department
The University of Connecticut
Storrs, CT 06269-3157

ABSTRACT

Shot noise is calculated in a Double Barrier Resonant Tunneling Structure (DBRTS) by taking the space charge accumulated inside the quantum well into account. The calculation is self-consistent in nature and obtained by simultaneously solving Schrödinger and Poisson's equations. The calculation manifests the suppression of the shot noise in the the positive differential resistance region and an enhancement in the negative differential resistance region of the DBRTS. The behavior is explained in terms of the fluctuation of the eigen energy of the structure due to the stored charge in the quantum well.

INTRODUCTION

Shot noise is observed in almost all electronic devices and is due to the fluctuations in device current. The fluctuations are attributed to the randomness of the carrier flowing through the device. In resonant tunneling device, the current is a very strong function of eigen energies and thus current gets modulated by any parameter which changes eigen energies. One such parameter is space charge developed inside the quantum well. In devices where eigen energies are dependent on current, significant deviations from full shot noise may occur. In DBRTS, as the current determines the amount of space charge, it is intuitively expected that it may show the suppression of shot noise.

To calculate shot noise, the auto- and cross-correlations among allowed eigen energies inside the quantum well have to be computed. In this paper, these correlation terms are computed by calculating space charge for every allowed energy level separately and then evaluating the dependence among different eigen energies that arises due to the contribution through the space charge.

THEORY

The self-consistent calculations are performed by using the logarithmic derivative. Once the logarithmic derivative $\Xi(x, E_x)$ is obtained as a function of distance x along the structure and the incident electron energy E_x, the group velocity $v_g(x, E_x)$[1] is calculated as

$$v_g(x, E_x) = \frac{1}{2} Re[\Xi(x, E_x)] \tag{1}$$

The determination of the current density at every energy E_x , $J(E_x)$, enables the evaluation of the space charge $n(x, E_x)$:

$$n(x, E_x) = \frac{J(E_x)}{v_g(x, E_x)} \tag{2}$$

The space charge is calculated for each allowed eigen energy and the contribution of the space charge to the potential profile is determined by solving the Poisson's equation. Accordingly, the potential profile is modified and the effect of the modified potential profile on the other energy levels and the level itself is determined. The process is repeated to evaluate:

$$(\Delta J)^2 = \sum_{i=1}^{N} \sum_{j=1}^{N} [\Delta J(E_x^i, E_x^j)]^2 \tag{3}$$

where, $\Delta J(E_x^i, E_x^J)$ is the difference of current densities corresponding to energy level E_x^i due to the space charge accumulated for energy level E_x^j, and N is the total number of allowed eigen energies. ΔJ is used to calculate shot noise in the structure[2].

Shot noise power spectrum S is obtained as

$$S = \frac{\Delta J^2}{\delta f} \tag{4}$$

where, δf is the measurement bandwidth. Shot noise factor γ is calculated as

$$\gamma = \frac{S}{2qJ} \tag{5}$$

RESULTS AND DISCUSSIONS

Results for a symmetric DBRTS are presented in this section. The barrier and the quantum well are each 50Åwide. The barriers height is assumed to be 0.275 eV. The effective mass of the electron is $0.067m_0$ and $0.096m_0$ in the quantum well and in the barrier, respectively and m_0 is the electron rest mass.

In Figs 1 and 2, theoretical current voltage characteristics and shot noise factor as a function of bias voltage are plotted. As expected, current starts increasing with bias voltage in the positive differential resistance region (PDR) followed by the current peak at 0.13 Volt. Any further increase of bias voltage causes a reduction of current (NDR) until it reaches the valley current at 0.17 Volt. The rise in current beyond the valley is caused by the onset of current flow through the second quasibound level in the quantum well. Fig.2 shows a suppression of shot noise factor in the PDR and an enhancement in the NDR. However, for a given voltage, shot noise becomes less as magnetic field is increased. This is because of the fact that effect of magnetic field and electric field on the eigen energy is opposing in nature. The suppression and enhancement can be explained in terms of the behavior of quasibound level of the quantum well. As space charge gets accumulated in the quantum well, it will reduce the effective voltage across the well. As a result, the quasibound level in the quantum well moves up along the energy axis. However, it leads to a decrease in the total current in the PDR as the quasibound level is moving away from the Fermi level. On the other hand, in the NDR, as the quasibound state moves up along the energy axis due to space charge, it moves towards the Fermi level and hence total current will be increased. More current implies more shot noise and greater shot noise factor and vice versa. This explains the suppression and the enhancement of shot noise factor in the PDR and NDR, respectively.

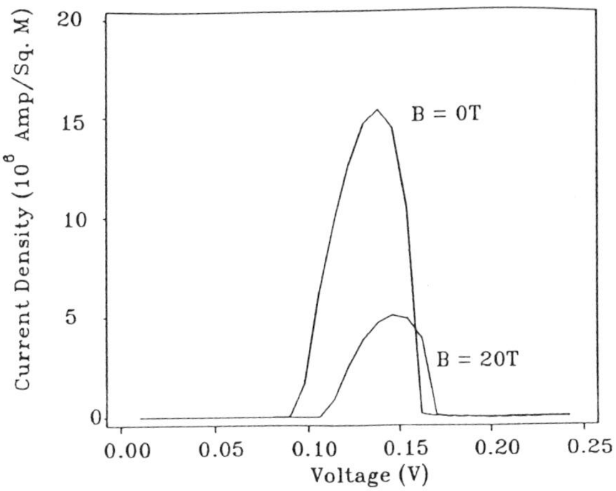

Figure 1: Current-voltage characteristics of a DBRTS with magnetic field as a parameter.

Fig.3 shows the space charge that is stored in the quantum well as a function of bias voltage. It is to be noted that the peak in the space charge occurs at the voltage corresponding to the peak current. Thus, the greatest contribution to the shift of quasibound level due to space charge occurs when the space charge becomes a maximum (0.13 Volt). Since, this voltage is situated just at the onset of the negative resistance region, eigen energies will be shifted up by a considerable amount and again there will be relatively few electrons available to tunnel through causing a reduction of shot noise factor. From Fig.2, it is also noted that the enhancement of shot noise gets peaked not at the voltage corresponding to peak current rather at a higher voltage. It can be explained in the following manner. The current becomes maximum when the first eigen energy is aligned with the Fermi level of the emitter. Moreover, the space charge moves the eigen energy up along the energy axis. Thus, effectively, eigen energy will be aligned when device is operated in the negative resistance region . This is why, shot noise becomes the maximum in the negative resistance region not at the onset of negative resistance region.

CONCLUSION

The computation of shot noise shows suppression in the PDR and enhancement in the NDR of a DBRTS. The two opposite behaviors are explained in terms of space charge stored in the quantum well.

REFERENCES

1. A. F. M. Anwar, A. N. Khondker and M. R. Khan, J. Appl. Phys. 65(7), p. 2761 (1989).

2. E.R. Brown, C.D. Parker, A.R. Calawa and M.J. Manfra, Dev. Research Conf., Colorado, 1991.

3. Mirza M. Jahan and A.F.M. Anwar, submitted to Phys. Rev. B.

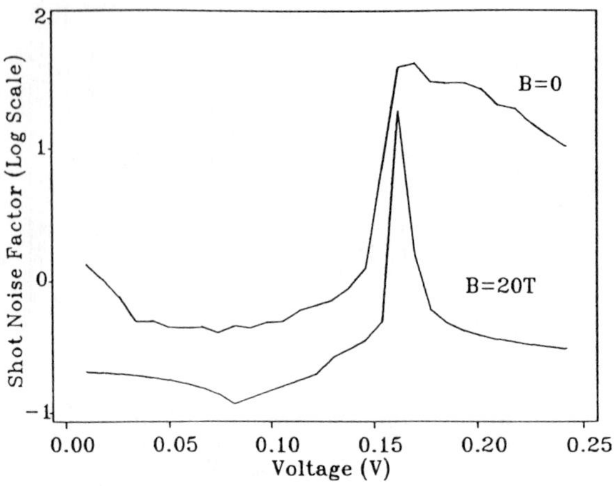

Figure 2: Shot noise factor is plotted as a function of bias voltage with magnetic field as a parameter.

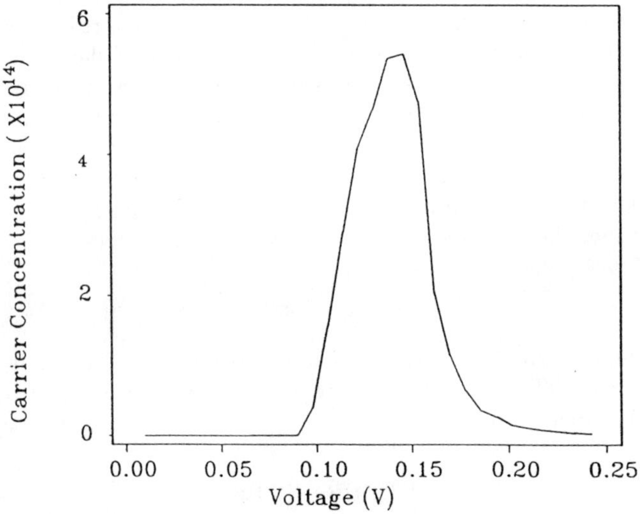

Figure 3: Space charge is plotted as a function of bias voltage.

PHONON NUMBER FLUCTUATION OF A CHAIN OF PARTICLES

M. Koch, R. Tetzlaff, and D. Wolf
Institiut für Angewandte Physik der Universität Frankfurt
Robert-Mayer-Strasse 2-4
D-6000 Frankfurt a.M.

INTRODUCTION

As reported previously the phonon number $N_q(t)$ of a chain of nonlinear coupled particles shows fluctuations with power spectra of the $\frac{1}{f^\alpha}-$ type for a certain vibration mode q [2,3]. In order to clarify the basic physical mechanisms of these fluctuations we examined the influence of a cubic term of the atomic potential in a chain of $N+1$ equidistant atoms assuming the mutual displacements to be small with respect to the atom distance.

Taking into account only nearest neighbour interactions we obtain for the longitudinal motion of the atoms the set of equations

$$m\ddot{y}_i = c(y_{i+1} + y_{i-1} - 2y_i) + k\left[(y_{i+1} - y_i)^2 - (y_i - y_{i-1})^2\right], \quad i = 1, \ldots, N-1 \quad (1)$$

with $y_0 = y_N = 0$ for $N+1$ particles with equal masses m of a chain with fixed ends [1]. According to the physical conditions the quadratic terms in (1) have to be considered as small perturbations.This statement is equivalent to the assumption of a small value of the coupling constant k, if the distance between adjacent particles is set to unity.

The phonon number belonging to a certain mode q of vibration is proportional to the portion $E_q(t)$ of the total vibration energy and can be approximated in the above case by

$$E_q(t) = \frac{1}{2}\dot{a}_q^2(t) + 2a_q^2(t)\sin^2\frac{\pi q}{2N}, \quad (2)$$

where a_q is the spatial Fourier coefficient of the particles' displacements at the time t.

In order to enable a closed treatment of the problem, the set of differential equations (1) of the chain is transformed into one partial differential equation (pde) for a continuous string. Assuming the total number N of the particles to tend to infinity and simultaneously the space h to tend to zero while the length of the chain $L = Nh$ remains constant one gets for $m = c = 1$ in (1) the pde

$$\frac{\partial^2 y(x,t)}{\partial t^2} = (\frac{\partial^2 y(x,t)}{\partial x^2})\left(1 + \epsilon\frac{\partial y(x,t)}{\partial x}\right) \quad (3)$$

with $\epsilon = \frac{2k}{N}$ and for the normalized particle positions $x = \frac{\xi}{L}$ along the string.

As outlined in [2], the velocity $v(x,t)$ can be determined for the initial conditions $y(x,0)$ and $v(x,0) = 0$; in particular for $y(x,0) = \sin\frac{\pi x}{L}$ and $v(x,0) = \frac{\partial y(x,t=0)}{\partial t} \equiv 0$ we get

$$v(x,t) = \frac{1}{3\epsilon}\left[(1 + \epsilon a\pi\cos\pi x_B)^{\frac{3}{2}} - (1 + \epsilon a\pi\cos\pi x_A)^{\frac{3}{2}}\right] \quad (4)$$

with

$$x_A = x - t + \frac{\epsilon a}{8}(\sin\pi x_A - \pi(x_B - x_A)\cos\pi x_A - \sin\pi x_B) + \mathcal{O}(\epsilon^2) \quad (5)$$

and

$$x_B = x + t + \frac{\epsilon a}{8}(\sin \pi x_B + \pi(x_B - x_A)\cos \pi x_B - \sin \pi a_A) + \mathcal{O}(\epsilon^2).$$

RESULTS

For various small values of k with $\epsilon << 1$ the displacements $y_i(T)$ of the i -th particle in a chain of $N+1 = 65$ particles were calculated by integration of (4). For selected values of T the sequences of displacement values $y_i(T), i = 0, \ldots, 64$ were Fourier transformed yielding the Fourier coefficients $a_q(T)$ for $q = 1, 2, 3, \ldots$ in order to determine the modal energy $E_q(T)$, which is proportional to the number $N_q(T)$ of phonons with wavelength $\lambda = \frac{2N}{q}$.

Some representative results of $N_q(T)$ vs. the discrete variable T are shown in Figures 1, 2 and 3 for the first modes, where $T = n\frac{T_L}{128}$ is given in units of the period $T_L = 2N\sqrt{\frac{\sigma}{S}}$ of the linear vibrating string with the tension S and the mass density σ.

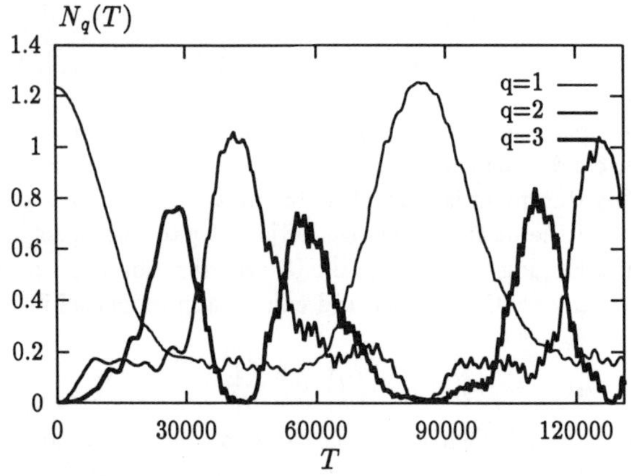

Figure 1: Number $N_q(T)$ of phonons vs. T for $k = 0.1$. The Energy is concentrated on the first mode at $T = 0$.

The Figures 1, 2 and 3 demonstrate the typical behavior of the phonon number $N_q(T)$ for three different initial conditions with $k = 0.1$. An important result is the exchange of vibration energy between the different modes in dependence of time. In the first case the energy is concentrated completely on the first mode at $T = 0$. For increasing time the energy flows to the adjacent modes in a periodic manner, first mainly to mode $q = 3$ and then predominately to mode 2 and so on.
The periodicity is determined by the mode $q = 1$.

In the second case, which offers the same periodic behavior, the energy is distributed on the first three modes as shown in Figure 2. The structure of the energy peaks is more complicated; again the period is determined by the mode $q = 1$. Between the corresponding maxima a group of four distinct maxima of N_q for $q = 2$ can be observed.

In the third case, the energy in the initial state is equally distributed on the first two modes.

The energy transfer from one mode to another generally takes a large number of time periods T_L, which is the larger the smaller the coupling constant k is.

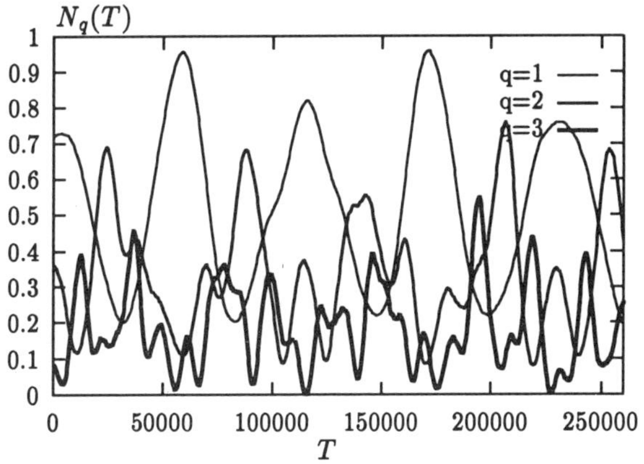

Figure 2: Number $N_q(T)$ of phonons vs. T for $k = 0.1$ with a initial distribution of energy on the first three modes at $T = 0$.

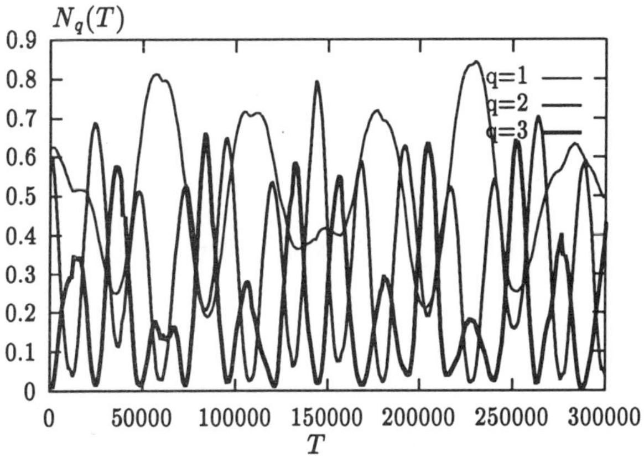

Figure 3: Number $N_q(T)$ of phonons vs. T for $k = 0.1$ with an equipartition of energy between the first two modes at $T = 0$.

Finally, the power spectral density of the phonon number fluctuations has been calculated for a variety of different coupling constants between $k = 0.01$ and $k = 0.25$. In Fig. 5 a typical power spectrum $S(f)$ is shown for $k = 0.1$. For all treated cases we found $\frac{1}{f^2}-$ spectra.

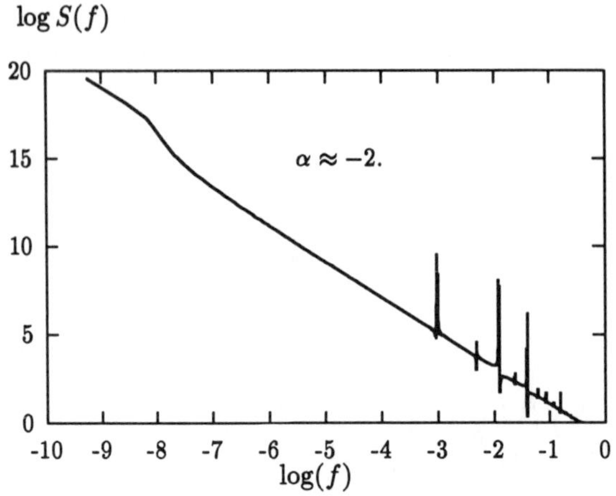

Figure 5: Power spectrum $S(f)$ vs. f for $k = 0.1$

CONCLUSION

The examination of the one dimensional vibrating system of particles, coupled by a superposition of linear and nonlinear forces such that the nonlinear influences remain small compared with the linear ones requires long times of observation to show the nonlinear effects.

Up to now, for reasons of numerical accuracy and stability a simulation of the vibration for such long times runs into difficulties [4] even in the case, when massive parallel computing is used. The approach presented here offers a semi analytic treatment which also leads to stable solutions for long observation times too.

The emphasis of the present activities is put of the further development of powerful simulation techniques for the solution of more general partial differential equations. This new approach based on Cellular Neural Networks offers new possibilities [5].

REFERENCES

1. Fermi, E., Pasta, J., and Ulam, S.: Studies of Non Linear Problems in Collected Papers of Enrico Fermi, pp. 978– 988 (1965)

2. Koch,M.,Tetzlaff,R.,and Wolf,D.: Dynamics of a Nonlinear Chain for Large Observation Times; Proceedings of the 5th van der Ziel Symposium , in print

3. Musha,T., Gabor, B., and Shoj, M.: $\frac{1}{f}$ Phonon Number Fluctuation in Quartz Observed by Laser Light Scattering; Phys. rev. letters 64(1990), p.2394

4. Musha,T., and Kobayashi,R.: Computer Experiment on Dynamics of Nonlinear Lattice; Proc 11th Noise in Physical Systems and $\frac{1}{f}$ Fluctuations Kyoto 1991, pp. 559– 562

5. Szolgay,P., Vöröss,G., and Eröss,G.: On the Applications of the Cellular Neural Network (CNN) Paradigm in Mechanical Vibrating Systems; Research Report: Computer and Automation Institute of the Hungarian Academy of Sciences, DNS–8–1992, MTA–SZTAKI

RELATIONS BETWEEN THE MICROSCOPIC LIFE TIME, RELAXATION LIFE TIME AND NOISE TIME CONSTANT

B. KOKTAVY

TECHNICAL UNIVERSITY OF BRNO
Zizkova 17, 602 00 BRNO, CZECH REPUBLIC

ABSTRACT

In this paper we deal with the probability densities of random time intervals between neighbouring electron transitions and their pertinent mean values in a system that is described by an n-dimensional g-r process.

On the basis of this theory the microscopic carrier life time, relaxation life time and noise time constant and their mutual relations have been expressed. Whereas there is no direct relation between the microscopic and relaxation life times, the relaxation time is - under certain conditions - equal to the noise time constant.

INTRODUCTION

We deal with the relation between the following 3 characteristics:

a) the microscopic life time τ_m,
b) the relaxation life time τ_r,
c) the noise time constant τ_n.

These characteristics are in substance statistical characteristics of a generation-recombination (g-r) process and can be, therefore, derived from the general theory of the g-r process[1, 3].

Let us consider first a uni-dimensional g-r process, in which random electron transitions between the levels 1 and 2 take place. The electron concentrations on these levels will be denoted $\xi_1(t)$ and $\xi_2(t)$. If it holds $\xi_1(t) + \xi_2(t) = \text{const.}$, then the system can be described by one random quantity $\xi(t)$.

Let the transition probability of an electron from level 2 to level 1 in the time interval $(t, t + \Delta t)$ be denoted $g(x, t)\Delta t + 0(\Delta t)$, where $0(\Delta t)$ tends to zero for $\Delta t \to 0$. Similarly, the probability of transition from level 1 to level 2 will be denoted $r(x,t)\Delta t + 0(\Delta t)$. Furthermore, let us denote $q(x,t) = g(x,t) + r(x,t)$. The expression $q(x,t)\Delta t + 0(\Delta t)$ represents the probability of an event consisting in a change in the level occupation within the time interval $(t, t + \Delta t)$. Let the time interval between arbitrary neighbouring transitions be τ, the time interval between an arbitrary transition and the following transition $1 \to 2$ τ_{12}, or $2 \to 1$ τ_{21}. The statistical characteristics of these random quantities depend, for stationary processes, on the state of the system. For non-stationary processes, they depend on the time, too.

STATISTICAL CHARACTERISTICS OF QUANTITIES τ, τ_{12}, τ_{21}

We look for the probability density of random quantities τ, τ_{12} a τ_{21}. The probability of the event $\tau \geq$ s-t, supposing that the system occurs in the state x, is

$$P\{\tau \geq s-t \mid \xi(t)=x\} = \lim_{n\to\infty} P\{A_1 . A_2 ... A_n\} = \lim_{n\to\infty} \prod_{i=1}^{n} [1-q(x,t_i)\Delta t] , \tag{1}$$

where A_i (i = 1, 2, ..., n) is the event consisting in no change of the system state in the time intervals $(t_i - \frac{\Delta t}{2}, t_i + \frac{\Delta t}{2})$ and $\Delta t = \frac{s-t}{n}$. We expand the logarithm of P and neglect quantities containing the second and higher powers. Then we get

$$P\{\tau \geq s-t \mid \xi(t)=x\} = \exp[-\int_t^s q(x,\theta)d\theta] . \tag{2}$$

The probability density of the random quantity τ is

$$w(x,t,s) = q(x,s) \exp[-\int_t^s q(x,\theta)d\theta] . \tag{3}$$

For a stationary process the mean value of the quantity τ is

$$\tau_0 = \int_0^\infty s\, w(x,s)ds = \frac{1}{q(x)} = \frac{1}{g(x)+r(x)} . \tag{4}$$

Similarly, for random quantities τ_{12} and τ_{21} it is

$$\tau_{120}(x) = \frac{1}{r(x)} , \quad \tau_{210}(x) = \frac{1}{g(x)} , \quad \frac{1}{\tau_{120}(x)} + \frac{1}{\tau_{210}(x)} = \frac{1}{\tau_0(x)} . \tag{5}$$

MICROSCOPIC LIFE TIME

Now we look for the probability of an event, where the occupation time τ_1 on the first level by the r - th electron is equal or greater than s - t

$$P\{\tau_1 \geq s-t \mid \xi_{ir}(t) = 1\} = \lim_{n\to\infty} \prod_{i=1}^{n} [1 - \frac{r(x,t_i)}{x_1} \Delta t] = \exp[-\int_t^s \frac{r(x,\theta)}{x_1} d\theta] . \tag{6}$$

The probability density

$$w_1(x,t,s) = \frac{r(x,s)}{x_1} \exp[-\int_t^s \frac{r(x,\theta)}{x_1} d\theta] . \tag{7}$$

We call the mean value $m_1\{\tau_1\} = \tau_{m1}(x)$ the microscopic life time of a carrier on the first level, on condition that the system is in the state x. For a stationary process it holds

$$\tau_{m1}(x) = \int_0^s s\, w_1(x,s)\,ds = \frac{x_1}{r(x)} \quad . \tag{8}$$

Similarly, for the second level we get

$$w_2(x,t,s) = \frac{g(x,s)}{x_2} \exp\left[-\int_t^s \frac{g(x,\theta)}{x_1}\,d\theta\right] , \qquad \tau_{m2}(x) = \int_0^s s\, w_2(x,s)\,ds = \frac{x_2}{g(x)} \quad . \tag{9}$$

Comparing the results of (5), (8), (9) we can see that

$$\tau_{m1} = x_1\, \tau_{120} \quad , \qquad \tau_{m2} = x_2\, \tau_{210} \quad . \tag{10}$$

For a system having n levels ($n \geq 2$), among which there are random electron transitions

$$\tau_{mi} = \frac{x_i}{r_i(x,t)} , \qquad r_i(x,t) = \sum_{k=1,k\neq i}^{n} p_{ik}\,(x,t) \quad , \tag{11}$$

where $p_{ik}(x,t)$ is the probability of the transition of an electron from the i-th to the k-th level per unit time.

THE RELAXATION LIFE TIME

We denote $g^{(v)}$ and $r^{(v)}$ the generation and recombination rates, respectively, which result from natural processes. Furthermore, let $g_E = g^{(f)} - r^{(f)}$ the resulting transition rate induced by an external radiation. If $x_i = x_{i0} + \Delta x_i$, where x_{i0} is the equilibrium electron concentration on the i-th level, then the relaxation life time of a carrier on the i-th level is[2]

$$\tau_{ri} = \frac{\Delta x_i}{r_i^{(v)} - g_i^{(v)}} \quad . \tag{12}$$

As the microscopic life time is

$$\tau_{mi} = \frac{x_i}{r_i} = \frac{x_{i0} + \Delta x_i}{r_i^{(v)} + r_i^{(f)}} \quad , \tag{13}$$

it is evident that there exists no direct relation between the relaxation and microscopic life times. They would equal only in the case, where

$$\Delta x_i = x_{i0} \frac{r_i^{(v)} - g_i^{(v)}}{r_i^{(f)} + g_i^{(v)}} \quad . \tag{14}$$

THE NOISE TIME CONSTANT

In the analysis of an n-dimensional process the quantities $\lambda_1, \lambda_2, ..., \lambda_n$, play an important role[3,4]. We get them as roots of the equation $|a_{rk} - \lambda\ \delta_{rk}| = 0,\ \ r, k = 1, ..., n$, where

$$a_{rk} = [\frac{\partial a_r}{\partial x_k}]_{x=x_0}\ ,\qquad a_r(x,t) = \sum_{s=1}^{n} [p_{sr}\ (x,t) - p_{rs}\ (x,t)]\ . \tag{15}$$

We define the noise time constant $\tau_{ni} = -\frac{1}{\lambda_i}$. We show that there exists a relation between the noise time constant τ_{ni} and the relaxation life time τ_{ri} .

For a uni-dimensional process it is $x_1 + x_2 =$ const., $a_1 = g - r\ ,\ \ a_2 = r - g$.

$$\tau_{n1} = \tau_{n2} = \frac{1}{r'(x_{10}) - g'(x_{10})}\ , \tag{16}$$

where x_{10} is the mean value of the electron concentration on the level 1. We can write for $\Delta x_1 << x_{10}$

$$\tau_{n1} = \tau_{n2} = \frac{\Delta x_1}{r(x_1) - g(x_1)} = \tau_{r1} = \tau_{r2}\ . \tag{17}$$

It follows that for a two-level semiconductor in thermodynamic equilibrium and in the low-injection mode the noise time constant equals the relaxation life time.

For an n-dimensional process it holds for the i-th level and small Δx_i

$$\tau_{ri} = \frac{\Delta x_i}{r_i^{(v)}(x) - g_i^{(v)}(x)} = \frac{\Delta x_i}{-\sum_k a_{ik}\ \Delta x_k} = \frac{1}{-\sum_k a_{ik}\ \frac{\Delta x_k}{\Delta x_i}}\ . \tag{18}$$

In the case where $a_{ik} << a_{ii}$, $(k \neq i)$ is, then $\tau_{ri} = -\ (a_{ii})^{-1}$. In this case, however, it is $\lambda_i = a_{ii}$. Then the relaxation life time τ_{ri} equals the noise time constant τ_{ni}.

REFERENCES

1. A.T. Bharucha-Reid, Elements of the Theory of Markov Processes and their Applications (MC Graw-Hill, N.Y., 1960), p.106.
2. J.S. Blakemore, Semiconductor Statics (Pergamon Press, Berlin, 1962), p. 200.
3. B. Koktavy, The General Process of Generation-Recombination in Semiconductors, Sborník Vysokého učení technického v Brně (Scientific Papers of the Technical University of Brno), 1978, No 1-4, p. 9.
4. B. Koktavy: A contribution to the study of stochastic processes in semiconductors. PhD Thesis, Technical University Brno, 1973.

LEVY PROCESSES IN NEW YORK STOCK EXCHANGE

R. N. Mantegna
Dipartimento di Energetica ed Applicazioni di Fisica, Universita' degli Studi, Viale delle Scienze, I-90128, Palermo, Italy.

ABSTRACT

We perform a statistical analysis of the New York stock exchange composite index. We show that the daily variations of the price index are distributed on a Lévy stable probability distribution. The spectral density of the price index is close to the one expected for a Brownian motion and the autocorrelation function of the daily variations is a fast decaying function. By studying the time evolution of the standard deviation of successive price variations we detect enhanced diffusion.

INTRODUCTION

The academic analysis and modeling of the Stock Exchange prices started at the beginning of this century thanks to the work of L.Bachelier[1] . In the sixties, new detailed studies clarified and substantiated the so-called random walk hypothesis[2]. The interest toward complex systems, the emergence of new paradigms (chaos and fractals) and the availability of extensive data files have recently motivated several people to re-examine the time evolution of price series of financial markets[3,5].

A STATISTICAL ANALYSIS

In this communication, we report a statistical study of a very important price index of the most important Stock Exchange in the world: the New York Stock Exchange (NYSE). We perform statistical analyses of the NYSE Composite Index. The analyses are performed directly on the price index $\{X_n\}$ deflated by the Consumer Price Index (CPI) (fig. 1). The investigated period is June 64-December 90 . The file consists of 6688 daily records. The time evolution of the daily variations of the price index

$$I_n = X_n - X_{n-1} \tag{1}$$

shows intermittence (fig. 2). The spectral density of the price index $\{X_n\}$ as a function of the frequency is power-law decreasing

$$S(f) \propto \frac{1}{f^{\beta}} \tag{2}$$

with $\beta = 1.90 \pm 0.03$ (fig. 3), this result is quite close to the one expected for a random walk ($\beta = 2$). The autocorrelation function of the daily variations

$< I_n I_{n+t} >$ is a fast decaying function with a correlation time shorter than one trading day.

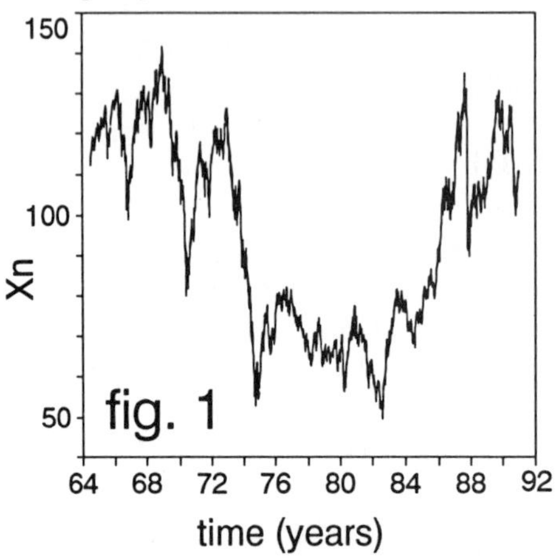

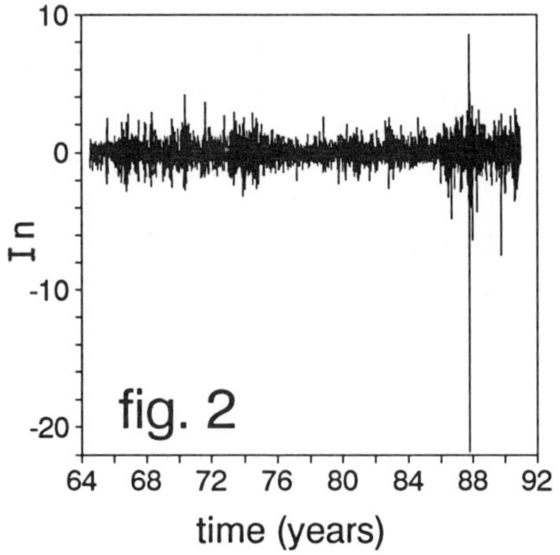

Fig. 1. Daily records of the New York Stock Exchange composite index deflated by the U.S. consumer price index.

Fig. 2. Time evolution of the daily variations of the NYSE composite index. An intermittent behavior on a time scale of years is present.

The probability distribution of the 1-day variations $P(I,1)$ is well fitted by a symmetrical Lévy stable distribution [6]

$$L_{\alpha,\gamma}(I) = \pi^{-1} \int_0^\infty \exp(-\gamma q^\alpha) \cos(qI) dq \tag{3}$$

characterized by $\alpha = 1.50$ and $\gamma = 0.28$ in the investigated I_n range (-10,10) (fig. 4). The dynamics of the stochastic process is investigated by studying the probability distributions for several different t-days variations:

$$I_{n,t} = X_n - X_{n-t} \tag{4}$$

we determine $P(I,t)$ with t ranging from 2 to 31 trading days (for t>3 a smoothing is necessary to reduce the noise). By analyzing the probability of returning $P(0,t)$ we can obtain an independent estimation of the α parameter. It is known that for Lévy stable processes we have

$$L_{\alpha,\gamma}(0,t) = \frac{\Gamma(1/\alpha)}{\pi\alpha(\gamma t)^{1/\alpha}} \tag{5}$$

so that the slope of the logarithm of the probability of returning as a function of the logarithm of the time is equals to $-1/\alpha$. By fitting the first ten points of our data ($t \leq 10$ trading days) we obtain $\alpha = 1.49$ (fig. 5) a value almost coincident with the one obtained by fitting the $P(I,1)$ distribution (fig. 4). This analysis allows us to state that a Lévy process can model the time evolution of the NYSE composite index for $t \leq 10$ trading days. The limited number of the records does not allow to conclude about the behavior of the central region of the probability distribution for t>10.

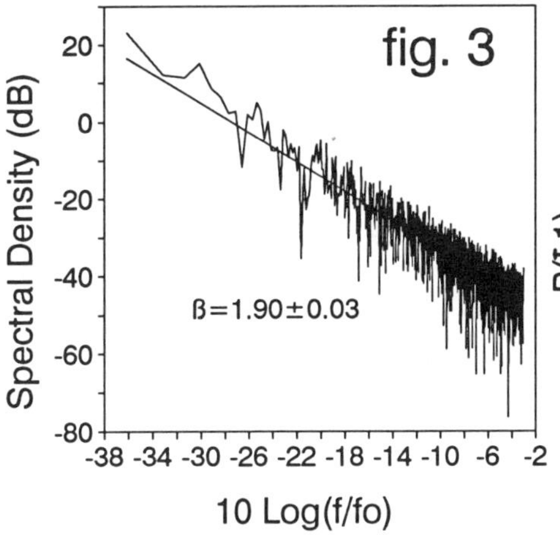

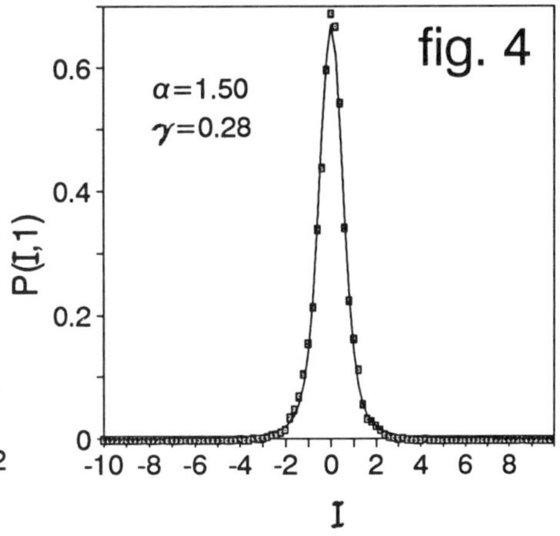

Fig. 3. Spectral density of the NYSE composite index. The power-law (eq. (2)) is evident from the log-log plot. The parameter β is obtained by performing the best linear fitting of the displayed values.

Fig. 4. Probability distribution of the NYSE composite index daily variations (black square). The curve shown in the figure is the Lévy stable probability distribution of order $\alpha = 1.50$ and scale factor $\gamma = 0.28$ which best fits the data.

Finally, we study the diffusive behavior of the process by investigating the time evolution of the standard deviation of the t-days variations. In agreement with the behavior already observed in the Milan stock exchange[3], we detect enhanced diffusion[7]

$$\sigma(t) \propto t^{\mu} \tag{6}$$

i.e. a more than Gaussian diffusion ($\mu > 0.5$) of the NYSE composite index time evolution. To avoid bias, we perform an analysis which compares the time evolution of the standard deviation of the t-days variations of the file (time-ordered data) with the time evolution of the standard deviation of a file obtained by mixing randomly in time the same records of the previous file (scrambled data)[3]. By performing best fittings of the two different time evolutions of standard deviations, we obtain $\mu = 0.519 \pm 0.005$ for time-ordered data and $\mu = 0.493 \pm 0.005$ for scrambled data (fig.6). The difference between the two results shows that the value $\mu = 0.519$ is statistically reliable so that we can conclude that enhanced diffusion is present in the time evolution of the NYSE Composite index.

CONCLUSIONS

The time evolution of a price index of a financial market can be investigated with the tools used to analyze stochastic processes. By studying the statistical properties of the NYSE composite index we observe that the time evolution of the price index is close to a Brownian motion, but important differences are observed: a) The probability distribution is well fitted by a Lévy stable distribution in the region close to the origin, so that the probability of returning is expected different from the one characterizing a Gaussian process. b) Enhanced diffusion is detected by studying the time evolution of the standard deviation of successive variations of the NYSE composite index. This last result supports the hypothesis of the existence of a long range memory in the time series of the price indices[3,7] . Further work is in progress.

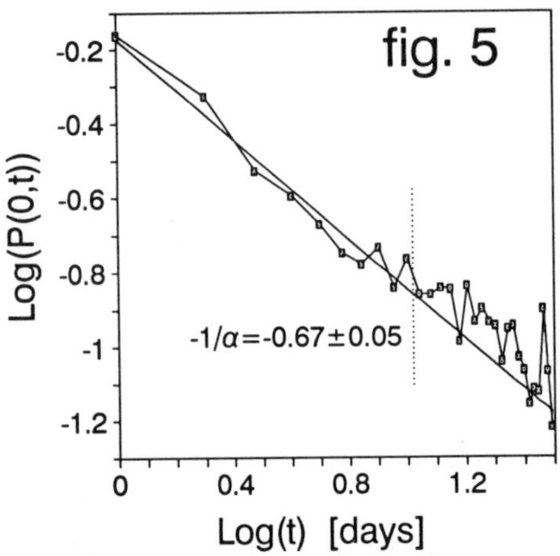

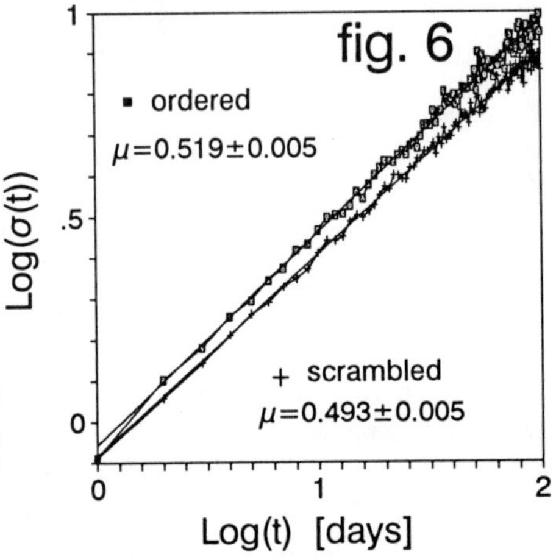

Fig. 5. Logarithm of the probability of returning as a function of the number of trading days. The straight line is the best fitting of the first ten points. From the slope of the best fitting line we obtain $\alpha = 1.49$.

Fig. 6. Logarithm of the time evolution of the standard deviation of the t-days variations of the NYSE composite index for the time-ordered data (■) and for scrambled data (+) (see text for details).

REFERENCES

1. L.Bachelier, Theorie de la Speculation (Gauthier-Villars, Paris , 1900).
2. P.H.Cootner, Ed., The Random Character of Stock Market Prices (MIT Press, Cambridge, Ma, 1964).
3. R.N.Mantegna, Physica A179, 232 (1991).
4. W.Li, Int. Journal of Bifurcation and Chaos (Sept,1991).
5. H.Takayasu, H.Miura, T.Hirabayashi, and K.Hamada, Physica A184, 127 (1992).
6. P.Levy, Theorie de l'Addition des Variables Aleatoires (Gauthier-Villars, Paris, 1937).
7. M.F.Shlesinger, J.Klafter, and Y.M.Wong, J. Stat. Phys. 27, 499 (1982).

AN EFFICIENT ALGORITHM FOR LEVY STABLE STOCHASTIC PROCESSES

R. N. Mantegna
Dipartimento di Energetica ed Applicazioni di Fisica, Universita' degli Studi,
Viale delle Scienze, I-90128, Palermo, Italy.

ABSTRACT

We present a new algorithm generating stochastic processes with a probability distribution very close to a Lévy stable probability distribution characterized by the parameter α. The parameter α can be selected within the range $0.3 < \alpha < 2$. The algorithm is very efficient for $0.75 \leq \alpha \leq 1.95$.

INTRODUCTION

In recent years, a growing interest has been devoted to Lévy stable stochastic processes [1,2]. The stable distributions are characterized by the following property: linear combinations of stable distributions of the same kind are themselves stable distributions. This property, which is well known in the case of Gaussian processes is valid for a wider family of stochastic variables if one rcleases the hypothesis of a finite second moment [3]. The distribution function of a symmetric Lévy stable process with zero mean is given by:

$$L_{\alpha,\gamma}(x) = \pi^{-1} \int_0^\infty \exp(-\gamma q^\alpha) \cos(qx) dq \tag{1}$$

The parameter α ranges between 0 (excluded) and 2 (included) while γ has to be a positive real number. $L_{\alpha,\gamma}(x)$ has not analytical form except for $\alpha = 2$ (Gaussian distribution), $\alpha = 1$ (Cauchy distribution) and $\alpha = 3/2$. Lévy stable distributions are, for large values of the stochastic variable x , power-law distributions:

$$L_{\alpha,\gamma}(x) \propto x^{-(\alpha+1)} \tag{2}$$

Distributions with power-law tails are quite frequent in physics and in complex systems. They have been observed in disordered systems, fractals [1,4], dynamical systems [2,5], polimerlike [6] and economic systems [7]. In spite of this widespread interest there has not been an algorithm generating Lévy stable processes defined all over the x range until now.

THE ALGORITHM

In this communication we present an algorithm that allows to obtain in a fast and accurate way Lévy stable processes characterized by arbitrary values of α ranging between 0.75 and 1.95 . Lévy stable processes characterized by values of α defined within the intervals $0.3 < \alpha < 0.75$ and $1.95 < \alpha < 1.99$ are also obtainable but with a longer calculation time. For the sake of simplicity, we set the parameter γ equals to one, however with a simple linear transformation[8] a stochastic process characterized by an arbitrary positive value of γ can be obtained. The numerical procedure is the following: first of all, we generate two independent stochastic Gaussian variables z_1 and z_2 with standard deviation σ_1 and σ_2 respectively. By performing the transformation:

$$y = \frac{z_2}{\mid z_1 \mid^{1/\alpha}} \tag{3}$$

we obtain a stochastic process $\{y\}$ with a probability distribution $P_\alpha(y)$ characterized by the same asymptotic behavior for large values of y of a Lévy stable process of order α and scale factor $\gamma = 1$ if [8] :

$$\sigma_1(\alpha) = 1 \tag{4}$$

$$\sigma_2(\alpha) = \left[\frac{\Gamma(1+\alpha)\sin(\pi\alpha/2)}{\Gamma((\alpha+1)/2)\,\alpha\,2^{(\alpha-1)/2}}\right]^{1/\alpha} \tag{5}$$

with this choice a full agreement is observed between the probability distributions $P_\alpha(y)$ and $L_{\alpha,1}(y)$ when $|y| > 10$ for any value of α. On the other hand, in the region close to the origin ($|y| < 10$) $P_\alpha(y)$ deviates from $L_{\alpha,1}(y)$. This mismatch is zero for $\alpha = 1$ (i.e. the transformation $y = z_2/|z_1|$ generates a stochastic process with a Cauchy distribution [9]) and increases monotonically both for $\alpha > 1$ and for $\alpha < 1$. We minimize this difference by performing the nonlinear transformation

$$x = y(k(\alpha) - 1)\exp\left(-\frac{|y|}{c(\alpha)}\right) + y \tag{6}$$

with

$$k(\alpha) = \frac{\alpha\,\Gamma((\alpha+1)/2\alpha)}{\Gamma(1/\alpha)}\left[\frac{\alpha\Gamma((\alpha+1)/2)}{\Gamma(\alpha+1)\sin(\pi\alpha/2)}\right]^{1/\alpha} \tag{7}$$

As shown elsewhere [8] we determine $k(\alpha)$ by equalizing

$$P_\alpha(x=0) = \frac{P_\alpha(y=0)}{k(\alpha)} = L_{\alpha,1}(x=0) \tag{8}$$

We are not able to write down an analytical form of the parameter $c(\alpha)$ as a function of α. However we are able to determine the integral equation which solution is the parameter $c(\alpha)$. This parameter speeds up the convergence of the stochastic process $\{x\}$ to a Lévy stable stochastic process [8]

$$\frac{1}{\sigma_2(\alpha)}\int_0^\infty q^{1/\alpha}\exp\left[-\frac{q^2}{2} - \frac{q^{2/\alpha}c(\alpha)^2}{2\sigma_2(\alpha)^2}\right]dq =$$

$$\int_0^\infty \cos\left[\left(\frac{k(\alpha)-1}{e}+1\right)c(\alpha)q\right]\exp(-q^\alpha)dq \tag{9}$$

In fig. 1 we show the probability distribution $P_{1.4}(x)$ of the stochastic process $\{x\}$ obtained by using equations (3) and (6). The probability distribution is determined by collecting 10^5 random variables x characterized by the parameters $\alpha = 1.4$, $\sigma_2(1.4) = 0.759679$, $k(1.4) = 1.44647$ and $c(1.4) = 2.8315$. The agreement with the Lévy stable distribution $L_{1.4,1}(x)$ (the continuous curve in the plot) is quite remarkable all over the entire range of the x variable. We also show the logarithm of the probability distribution to point out the behavior on the wings (fig. 2). From the picture, it is evident that the agreement is excellent.

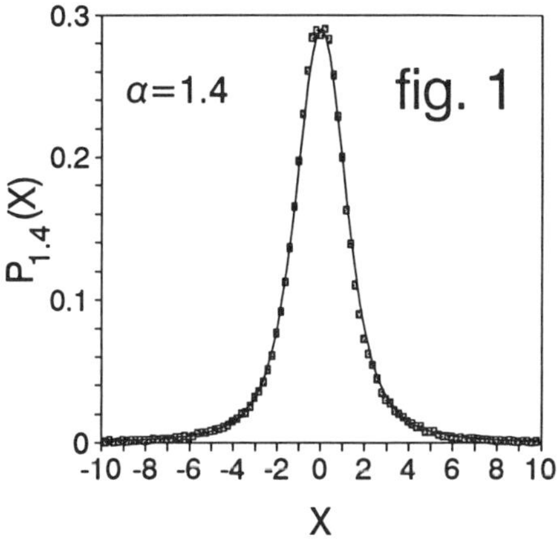

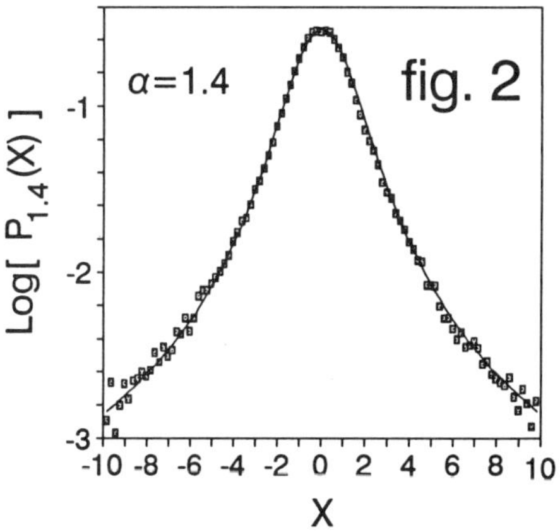

Fig. 1. $P_{1.4}(x)$ probability distribution (▪) of the stochastic process generated by the algorithm of eqs. (3) and (6) with $\alpha = 1.4$ (see text for the other parameters). The curve is the $L_{1.4,1}(x)$ Lévy stable distribution.

Fig. 2. The logarithm of the same probability distributions of fig. 1 .

In fig. 3 we show a set of the logarithm of probability distributions obtained with the above reported algorithm. In the figure the control parameter α is ranging from $\alpha = 0.8$ to $\alpha = 1.9$ with a step $\Delta\alpha = 0.1$. Each distribution is obtained by accumulating 10^5 independent realizations of the process $\{x\}$ for a selected value of α. The values of $\sigma_2(\alpha)$, $k(\alpha)$ and $c(\alpha)$ are determined by using eqs. (5), (7) and (9) respectively. In fig. 4 we compare the results of our numerical simulations with the Lévy stable distributions. The comparison is performed by superimposing the contour lines of the surface shown in fig. 3 with the corresponding contour lines obtained from the Lévy stable probability distributions calculated from eq. (1) by numerical integration. In the figure the smooth curves are the contour lines of the Lévy distributions whereas the noisy curves are the contour lines of our numerical simulations. The agreement is very good on the whole investigated range ($-10 \le x \le 10$ and $0.8 \le \alpha \le 1.9$).

CONCLUSIONS

The algorithm is very efficient for values of α ranging from 0.75 to 1.95 . In this interval a solution of the integral equation exists and the nonlinear transformation of eq. (6) has a single-value inverse function. The algorithm is also working for $0.3 \leq \alpha < 0.75$ and $1.95 < \alpha \leq 1.99$ but in these intervals the mismatch between $P_\alpha(x)$ and $L_{\alpha,1}(x)$ can be avoided only by performing a relatively time consuming convergence of the stochastic process $\{x\}$ towards the Lévy stable process. This can be achieved by calculating a process which is a linear combination of n independent stochastic variables $\{x\}$.

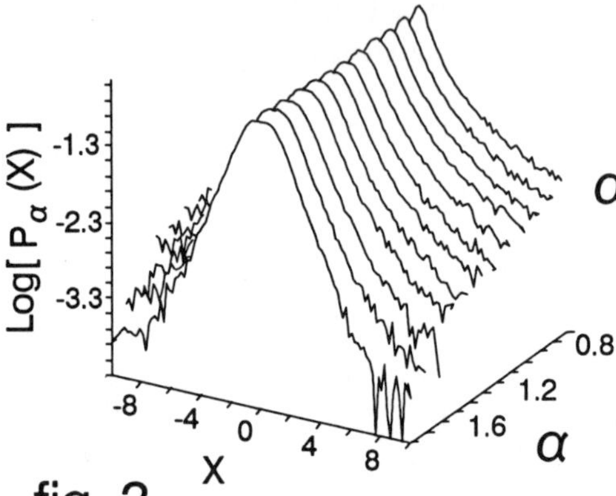

fig. 3

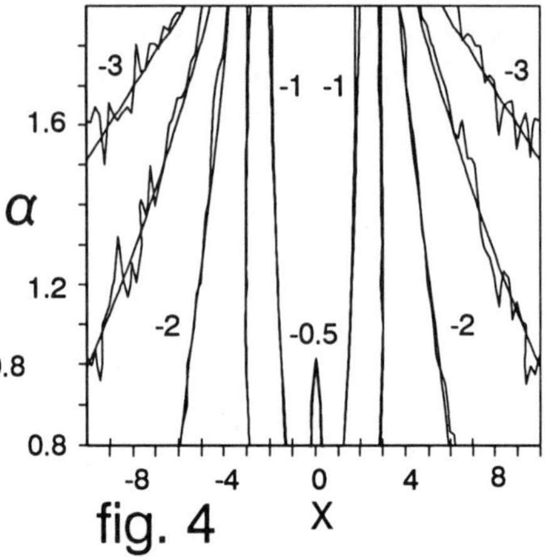

fig. 4

Fig. 3. Set of logarithm of probability distributions obtained with the algorithm of eqs. (3) and (6). α (y-axis) is ranging from 0.8 to 1.9 with a 0.1 step. We accumulate 10^5 realizations for each curves (the control parameters are determined by using eqs. (5), (7) and (9)).

Fig. 4. Contour lines of the 3D surface shown in fig. 3 (noisy curves) superimposed to the contour lines of Lévy stable distributions (smooth curves). The numbers in the picture are the z-value of the contour lines.

REFERENCES

1. B. Mandelbrot, The Fractal Geometry of Nature (Freeman, San Francisco 1982).
2. M.F. Shlesinger and J. Klafter, in On Growth and Form, Edited by H.E. Stanley and N. Ostrowsky (Nijhoff, Amsterdam, 1985) pag.279 ; J.P. Bouchaud and A. Georges, Physics Reports 195, 127 (1990).
3. P.Levy, Theorie de l'Addition des Variables Aleatoires, (Gauthier-Villars, Paris 1937).
4. S. Havlin and D. Ben-Avraham, Adv. Phys. 36, 695 (1987).
5. R.N. Mantegna, J. Stat. Phys. 70, 721 (1993).
6. A. Ott, J.P. Bouchaud, D. Langevin and W. Urbach, Phys. Rev. Lett 65, 2201 (1990).
7. R.N. Mantegna, Physica A179, 232 (1991).
8. R.N. Mantegna, (to be published).
9. A. Papoulis, Probability, Random Variables, and Stochastic Processes (McGraw-Hill, N.Y., 1984).

The effects of internal fluctuations on a class of nonequilibrium statistical field theories

Mark M. Millonas
Complex Systems Group, Theoretical Division and Center for Nonlinear Studies,
MS B258 Los Alamos National Laboratory, Los Alamos, NM 87545
& Santa Fe Institute, Santa Fe, NM

Abstract

A class of models with applications to swarm behavior as well as many other types of spatially extended complex biological and physical systems is studied. Internal fluctuations can play an active role in the organization of the phase structure of such systems. In particular, for the class of models studied here the effect of internal fluctuations due to finite size is a renormalized *decrease* in the temperature near the point of spontaneous symmetry breaking.

In this paper I introduce a class of models which is in line with the basic processes acting in a variety of systems in nature, particularly biological ones. Here we study what will be called *stigmergic processes* as a generalization of the concept of stigmergy introduced by Grassé[2] in the context of collective nest building in social insects. The more generalized idea of a stigmergic process is realized here in systems composed of three basic ingredients. The first ingredient is a *particle dynamics* which obeys a Markov process on some finite state space $\mathcal{X}$. The particle density $\rho(\mathbf{x},\tau)$ obeys the Master equation

$$\frac{\partial\rho(\mathbf{x},\tau)}{\partial\tau} = \int_{\mathcal{X}} \{W_\tau(\mathbf{x}|\mathbf{y})\rho(\mathbf{y},\tau) - W_\tau(\mathbf{y}|\mathbf{x})\rho(\mathbf{x},\tau)\}\, d^D\mathbf{y}, \tag{1}$$

where $W_\tau(\mathbf{x}|\mathbf{y})$ is the probability density to go from state $\mathbf{y}$ to $\mathbf{x}$ at time τ. The second element is a *morphogenetic field* $\sigma(\mathbf{x},\tau)$, representing the environment which the particles both respond to, and act on. We will study one of the simplest situations, a fixed one-component pheromonal field which evolves according to

$$\frac{\partial\sigma(\mathbf{x},\tau)}{\partial\tau} = -\kappa\,\sigma + \eta\,\rho, \tag{2}$$

where κ measures the rate of evaporation, breakdown or removal of the substance, and η the rate of emission of the pheromone by the organisms. Lastly, some form of *coupling* is made between the particles and the field. This coupling takes the form of a behavioral function which describes how the particles move in response to the morphogenetic field, and in turn, how the particles act back on this field.

In the region of a nonequilibrium phase transition the morphogenetic field, and hence the transition matrix, changes very slowly on scales typical of the particle field relaxation time since in this region the unstable modes will exhibit critical slowing down and will relax on a time scale much longer than the time scale of the stable modes. The particle modes can then be adiabatically eliminated from the picture, and we obtain the stochastic order parameter equation

$$\frac{\partial \sigma(\mathbf{x},\tau)}{\partial \tau} = \kappa\, \sigma + \eta \rho_s[\boldsymbol{\sigma}] + \eta\; g[\boldsymbol{\sigma}]\xi(\mathbf{x},\tau), \tag{3}$$

where $\rho_s[\boldsymbol{\sigma}]$ is the quasi-stationary particle density, $g[\boldsymbol{\sigma}]$ is a function describing the fluctuations of the quasi-stationary particle density about its mean value, and $E\{\xi(\mathbf{x},\tau)\} = 0$, $E\{\xi(\mathbf{x},\tau)\xi(\mathbf{x}',\tau')\} = \delta(\mathbf{x}-\mathbf{x}')\delta(\tau-\tau')$.

For the purposes of this paper we will consider the case where the transition matrix takes the form $W(\mathbf{x}|\mathbf{y}) \propto f(\sigma(\mathbf{x}))\, p(|\mathbf{x}-\mathbf{y}|)$, where f is some weighting function describing the effect of the field σ on the motion of the particles, and $p(|\mathbf{x}-\mathbf{y}|)$ is a probability distribution of jumps of length $r = |\mathbf{x}-\mathbf{y}|$. Transition matrices of this type obey detailed balance, $W(\mathbf{x}|\mathbf{y})f(\sigma(\mathbf{y})) = W(\mathbf{y}|\mathbf{x})f(\sigma(\mathbf{x}))$. In this case we can define a partition function

$$Z = \left\{\frac{1}{V}\int d^D\mathbf{x}\; f(\sigma(\mathbf{x}))\right\}^N, \tag{4}$$

where V is the volume of the state space $\mathcal{X}$, and N is the total number of particles. A one-to-one analogy with a thermodynamic system with energy $U(\sigma(\mathbf{x}))$ and temperature $T = \beta^{-1}$ can be made if we set $f(\sigma(\mathbf{x})) = \exp(-\beta U(\sigma(\mathbf{x})))$, where any parameter T can be regarded as a temperature parameter if $f(\sigma(\mathbf{x});\alpha\, T) = f^{-\alpha}(\sigma(\mathbf{x});T)$. Statistical quantities of interest can be calculated from the partition function according to the usual prescriptions. In a closed system the mean particle density and dispersion in the energy state ϵ are given by

$$E\{\rho_\epsilon\} = \frac{N}{VZ}\exp(-\beta\epsilon), \quad E\left\{(\Delta\rho_\epsilon)^2\right\} = \frac{E\{\rho_\epsilon\}}{\mu_\epsilon}\left(1 - \frac{\mu_\epsilon}{N}E\{\rho_\epsilon\}\right), \tag{5}$$

where μ_ϵ is the volume of the system in energy state ϵ. The slaved particle field in energy state ϵ can then be represented, to lowest order in the fluctuations, by $\rho_\epsilon[\boldsymbol{\sigma}] = E\{\rho_\epsilon[\boldsymbol{\sigma}]\} + \sqrt{E\{(\Delta\rho_\epsilon)^2[\boldsymbol{\sigma}]\}}\ \xi(\mathbf{x}, t)$.

We introduce the dimensionless parameter $\bar{\rho} = N/V$, the mean density of particles, and $\upsilon = \mu^-/\mu^+$, the ratio of the volume of the field $\sigma(\mathbf{x})$ in the σ^- state, to the volume in the σ^+ state. We also define the function $R(\sigma^+, \sigma^-) = f(\sigma^+)/f(\sigma^-)$. In the mean field approximation a Langevin equation

$$\frac{dm}{dt} = -m + F(m) + \frac{1}{\sqrt{N}} Q(m)\ \xi(t) \tag{6}$$

for the order parameter m can be derived, where

$$F = \bar{\rho}(1+\upsilon)\frac{R^\beta - 1}{R^\beta + \upsilon}, \quad Q^2 = \frac{\bar{\rho}^2(1+\upsilon)^4}{\upsilon}\frac{R^\beta}{(R^\beta+\upsilon)^2}, \tag{7}$$

and where the F and Q are determined as functions of m by

$$R(m) = R\left(\bar{\rho} + \frac{\upsilon\, m}{1+\upsilon}, \bar{\rho} - \frac{m}{1+\upsilon}\right). \tag{8}$$

The order parameter m is analogous to a gas-liquid order parameter, and represents the difference in the values of the field in the σ^+ and σ^- states after spontaneous symmetry breaking. The behavior of this system is described by the potential function

$$\Phi(m) = \int^m \frac{m - F(m)}{Q^2(m)}\, dx + \frac{1}{N} \ln Q, \tag{9}$$

where the phases m_i of the system are determined by the conditions $\Phi'(m_i) = 0$, $\Phi''(m_i) > 0$.[1]

In the continuum limit ($N \to \infty$) it can be shown that the critical value of the mean density $\bar{\rho}_c$ at which spontaneous symmetry breaking occurs is given by the condition $-\bar{\rho}_c\, U'(\bar{\rho}_c) = T$. Generally $\bar{\rho}_c$ is will increase with increasing temperature. The relative stability of two phases m_1 and m_2 is determined by the relative potentials $\Phi(m_1)$ and $\Phi(m_2)$ for each phase. Even in the continuum limit the details of the fluctuations cannot be neglected due to the presence of the factor $Q^2(m)$ under the integral in 9, and the relative stability of the phases will depend on the precise details of the internal fluctuations. Similar observations have been made elsewhere by Landauer and others.[3]

When N is finite, the situation is still more complicated. It is clear that the possible values of the order parameter and the phase structure do not remain unchanged under the influence of internal fluctuations. The criterion for spontaneous symmetry breaking in this case is $-\bar{\rho}_c\, U'(\bar{\rho}_c) = \hat{T}$, where $\hat{T}$ is the renormalized temperature $\hat{T} = \gamma(N)T$ where $\gamma(N) = \sqrt{N + (N/2)^2} - N/2$. This is precisely the continuum condition except that the finite size fluctuations have the effect of renormalizing the temperature by the factor $\gamma(N) < 1$. The effect of increasing the internal fluctuations through decreasing the total number of particles has the effect of *decreasing the temperature.* We thus arrive at the seeming paradox that increased internal fluctuations may produce increased order.

More details may be found in previously published papers[4] where the properties of an ant swarm are analyzed in depth, and it is also shown how the collective behavior of real ants[5] can be understood in terms of such models.

References

[1] W. Horsthemke, & R. Lefever, *Noise Induced Transitions*, Springer-Verlag (1984).

[2] P. P. Grassé, *Experientia* **15**, 356 (1959); E. O. Wilson, *The Insect Societies*, Belknap (1971).

[3] R. Landauer, *J. Stat. Phys.* **53**, 233 (1988); N.G. Van Kampen, *IBM J. Res. Dev.* **32**, 107 (1988).

[4] M. M. Millonas, *J. Theor. Biol.* **159**, 529 (1992); In: *Cellular Automata and Cooperative Systems*, (N. Bocara, E. Goles, S. Martinez & P. Pico, eds.), Kluwer (1993); In: *ALIFE III* (C.G. Langton, ed.), Santa Fe Institute: Addison-Wesley (1993); Phys. Rev. E (to appear, 1993).

[5] J.-L. Deneubourg, S. Aron, S. Goss and J. M. Pasteels, *J. Insect Behavior* **3**, 159 (1990).

ABOUT THE STATE MODEL FOR THE ANALYSIS OF LEVEL-CROSSING INTERVALS OF RANDOM PROCESSES

T.Munakata
Faculty of Engineering, Tamagawa University, Tokyo, Japan

D.Wolf
Institut für Angewandte Physik der Universität Frankfurt a.M.,FRG

ABSTRACT

As well known the calculation of probability density $P_n(\tau)$ of the sum of n+1 level crossing intervals of random processes is difficult. Previously known methods based on the quasi-independent assumptions, multi-state model, or others are not sufficient for the approximations of $P_0(\tau)$ in case of narrow band processes. In this paper 6-state model is modified by introducing newly defined relaxation states, and the model represents good approximations of $P_0(\tau)$ and $P_1(\tau)$ also in case of Gaussian processes having middle- and narrow-bandpass power spectrum density.

INTRODUCTION AND THEORY

Up to now the calculation method for the probability density $P_0(\tau)$ of the level crossing interval τ and the probability density $P_1(\tau)$ of the sum τ of two adjacent level crossing intervals of random processes have not been found. Many approximation methods based on a statistical independence of the intervals were known. Recently a 6-state model based on a multi-state model have been proposed by the authours [1,2].

For Gaussian processes a 6-state model represents very good approximations of P_0 and P_1 for level crossing intervals with zero and positive level I. For negative level I, some approximation errors are observed, if the process has middle- or narrow-band-pass power spectrum. In this paper we propose a model, which can describe the properties of narrow- and middle- bandpass processes, by introducing some relaxation states to the 6-state model.

As shown in Fig. 1a, 4 states S_1, S_2, S_3, and S_4 are defined in a following manner. Given an upward crossing of I in (t,t+dt), then the process in

- S_1 remains above the level I at least until the time $t+\tau$;
- S_2 is found below the level I at the time $t+\tau$ after just one subsequent downward crossing ;
- S_3 is found above the level I at time $t+\tau$ after at least one subsequent upward crossings during the time interval τ ;
- S_4 is found below the level I at time $t+\tau$ after at least two subsequent downward crossings during the time interval τ .

Inside the states S_1 and S_3 two relaxation states S_{1b} and S_{3b}, resp., are assumed, and $S_1=S_{1b}+S_{1a}$ and $S_3=S_{3b}+S_{3a}$ are hold. As illustrated in Fig. 1a the process having arrived at a state S_{1b} or S_{3b} remains in this state for certain constant time T_b or T_{d1}, resp., before the process proceeds to S_{1a} or S_{3a}. Differs from a 6-state model, the transition into S_{3b} occurs not only by upward crossing from S_2 and S_4, but also from S_{3a} by some inner transition of state S_3. Same is occurs from S_{1a} to S_{1b} in the state S_1. Inner transition from S_{1a} to S_{1b} is proportional to the state probability $S_1(\tau)$ and $W(\tau-T_{a1}, I=0)$ of Rice function for upward crossing at zero-level and at delayed time $\tau-T_{a1}$. In the other hand the inner transition from S_{3a} to S_{3b} is proportional to $S_1(\tau)$ and Rice function $W(\tau-T_{c1}, I)$ for level I and at delayed time $\tau-T_{c1}$. Constant value c_1 controls the strength of inner transitions, and for zero and positive level I, by setting

$c_1=0$, the model represents the same results to those of 6-state model.
Then the state probabilities and P_0 are described as follows;

$$S_{1a}(\tau) + S_{1b}(\tau) = S_1(\tau) \;, \quad S_{3a}(\tau) + S_{3b}(\tau) = S_3(\tau) \,, \tag{1}$$

$$S_1(\tau) + S_3(\tau) = P_+(\tau) \;, \quad S_2(\tau) + S_4(\tau) = 1 - P_+(\tau) \,, \tag{2}$$

$$P_0(\tau) = S_{1a}(\tau)\, Q(\tau, I)/[\, S_{1a}(\tau) + S_{3a}(\tau)\,] = \lambda\, [S_1(\tau) - S_{1b}(\tau)] \,, \tag{3}$$

$$\lambda = Q(\tau, I)/[\, P_+(\tau) - S_{1b}(\tau) - S_{3b}(\tau)\,] \,, \tag{4}$$

$$S_1(\tau) = 1 - \int_0^{\tau} P_0(t)\, dt \,, \tag{5}$$

$$S_{1b}(\tau) = \int_{\tau - T_b}^{\tau} C_1\, S_1(t)\, W(t - T_{a1}, I=0)\, dt \,, \tag{6}$$

$$S_{3b}(\tau) = \int_{\tau - T_{d1}}^{\tau} [\, C_1\, S_1(t)\, W(t - T_{c1}, I) + W(t, I)\,]\, dt \,, \tag{7}$$

$$P_+(\tau) = 1 + \int_0^{\tau} [\, W(t, I) - Q(t, I)\,]\, dt \,. \tag{8}$$

For the calculation of P_1 the state S_1 and S_2 also S_3 and S_4 are added together, (see Fig. 1b), and they are described as

$$\underline{S}_{1a}(\tau) + \underline{S}_{1b}(\tau) = \underline{S}_1(\tau) \;, \quad \underline{S}_{3a}(\tau) + \underline{S}_{3b}(\tau) = \underline{S}_3(\tau) \,, \tag{9}$$

$$\underline{S}_1(\tau) + \underline{S}_3(\tau) = 1, \tag{10}$$

$$P_1(\tau) = \underline{S}_{1a}(\tau) W(\tau, I)/[\, \underline{S}_{1a}(\tau) + \underline{S}_{3a}(\tau)\,] = \lambda\, [\underline{S}_1(\tau) - \underline{S}_{1b}(\tau)] \,, \tag{11}$$

$$\lambda = W(\tau, I)/[\, 1 - \underline{S}_{1b}(\tau) - \underline{S}_{3b}(\tau)\,] \,, \tag{12}$$

$$\underline{S}_1(\tau) = 1 - \int_0^{\tau} P_1(t)\, dt \,, \tag{13}$$

$$\underline{S}_{1b}(\tau) = \int_{\tau - T_b}^{\tau} C_1\, \underline{S}_1(t)\, W(t - T_{a2}, I=0)\, dt \,, \tag{14}$$

$$\underline{S}_{3b}(\tau) = \int_{\tau - T_{d2}}^{\tau} [\, C_1\, \underline{S}_1(t)\, W(t - T_{c2}, I) + W(t, I)\,]\, dt \,. \tag{15}$$

RESULTS AND DISCUSSION

Based on this model probability densities P_0 and P_1 of level crossing intervals are numerically calculated for the Gaussian processes having 8th order Butterworth type middle- and narrow-bandpass power spectrum densities. Bandwidth is denoted by a parameter $k = f_L/f_H$ as a ratio of low cutoff frequency f_L and high cutoff frequency f_H. Fig. 2 show the approximation P_{0x} calculated by this model, and $P_0(6)$ of 6-state model together with measured P_0 and Rice function Q. The relaxation times are shown in Table 1. Coinsidence between P_{0x} and measured P_0 is very good for both middle bandpass (k=0.5) and narrow bandpass (k=0.8) cases. Especially for the 2nd peak and 2nd valley position of Q function, the model represents very good approximations that cannot be satisfied by 6-state model.

The values listed in Table 1 are deternined in the following way,

$$T_{a2} = L_2 + L_1 - T_b, \qquad T_{a1} = L_2 - T_b,$$
$$T_{c2} = L_1 + L_2 + l_a + l_b, \qquad T_{c1} = L_1 + L_2 + l_b, \tag{16}$$

where the quantities L_1 and L_2 are calculated from the i th peak position t_{wxi} and t_{qxi} of Rice functions $W(\tau,I)$ and $Q(\tau,I)$, resp., as follows;

$$L_1 = t_{wx1} - t_{qx1} ,$$
$$L_2 = t_{wx2} - t_{wx1} \qquad \text{for } k= 0.8$$
$$= t_{qx3} - t_{wx1} - t_{qx3}|_{I=0} + t_{wx2}|_{I=0} \qquad \text{for } k= 0.5 . \tag{17}$$

The values l_a and l_b are the constant for all level I, and they are

$$l_a = 1.7 \text{ and } l_b = -1.3 \quad \text{for } k = 0.8,$$
$$l_a = 1.5 \text{ and } l_b = 2.7 \quad \text{for } k = 0.5. \tag{18}$$

Other three quantities T_{d1}, T_{d2} and T_b are determined by compareing the first moment of the solution P_{0x} with the theoretical moment μ_+. T_{d2} is constant for all level I, and this value can be determined at I=0 . By using first moment of P_1 the quantity T_b can be determined, then next T_{d1} can be determined for P_0.

The method of parameter determination used here was found empirically, and more theoretical methods are required.

REFERENCES

1. T. Munakata, and D. Wolf, On the Distribution of the Level Crossing Time Intervals of Random Processes, Noise in Physical Systems and 1/f Noise, Proc. 7th Int. Conf. on Noise in Physical Systems, Montpellier, (Editor M. Savelli, North-Holland Publ. Co., Amsterdam, 1983), p. 49-52.
2. T. Munakata, Mehr-Zustände-Modelle zur Beschreibung des Pegelkreuzungs-verhaltens stationärer stochastischer Prozesse, (Dissertation, Universität Frankfurt am Main ,FRG, 1986.)

TABLE 1

	k = 0.8					k = 0.5				
I	0.0	-0.5	-1.0	-1.5	-2.0	0.0	-0.5	-1.0	-1.5	-2.0
T_{a1}		2.5	3.0	2.9	2.5		3.7	4.3	4.8	5.0
T_b		4.3	3.8	4.0	4.5		3.2	3.3	3.3	3.6
T_{c1}		8.3	7.6	7.3	7.1		12.5	12.6	12.7	12.8
T_{d1}	2.3	3.0	3.4	3.6	4.2	2.6	3.2	3.7	4.0	4.4
T_{a2}		5.3	5.1	4.6	3.9		6.6	6.6	6.7	6.5
T_{c2}		10.0	9.3	9.0	8.8		14.0	14.1	14.2	14.3
T_{d2}	5.0	5.0	5.0	5.0	5.0	5.6	5.6	5.6	5.6	5.6
L_1		2.8	2.1	1.7	1.4		2.9	2.3	1.9	1.5
L_2		6.8	6.8	6.9	7.0		6.9	7.6	8.1	8.6
c_1	0.0	0.58	0.9	1.0	1.0	0.0	0.2	0.67	1.0	1.0

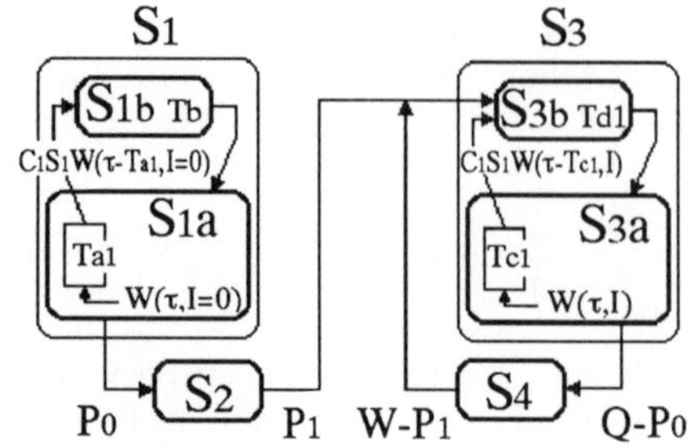

Fig.1a Diagram of model for P0.

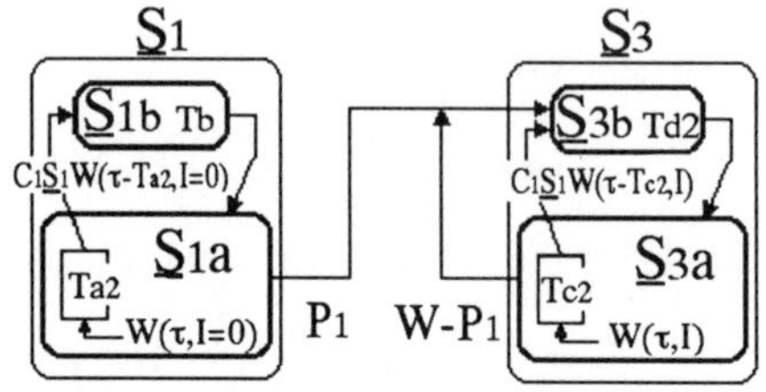

Fig.1b Diagram of model for P1.

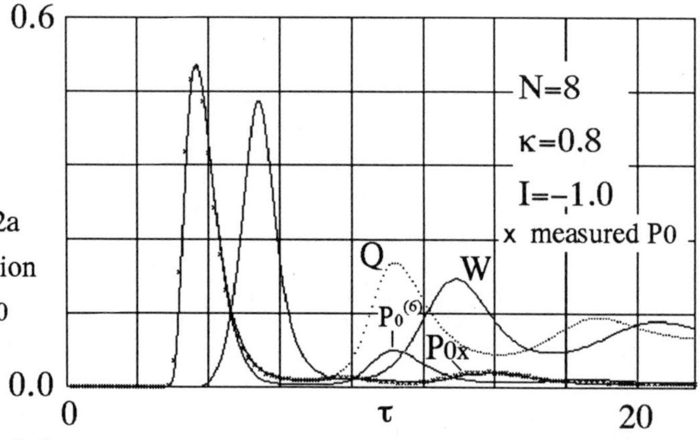

Fig. 2a Solution for P0

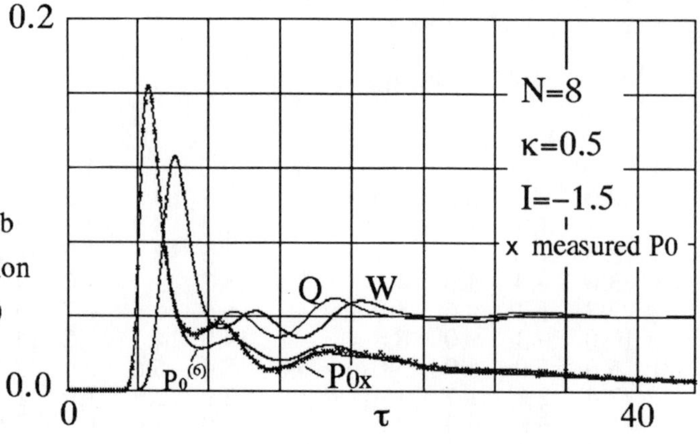

Fig. 2b Solution for P0

Noise Sampled Signal Transmission in an Array of Schmitt Triggers.

Eleni Pantazelou and Frank Moss
Department of Physics
University of Missouri at St. Louis
St. Louis, MO 63121 USA

Dante Chialvo
Department of Neurosurgery
SUNY Health Science Center
Syracuse, NY 13210 USA

ABSTRACT

In this paper we discuss the noise driven dynamics of an array of Schmitt Triggers (ST's) subject to a weak signal. The signal is subthreshold, that is, when applied without noise, it cannot cause state changes in the ST's. Each ST is subject to a Gaussian noise which is uncorrelated with that of its neighbors. In this realization, the ST's are not coupled but rather take their signal input from a common bus. Their outputs are summed to reproduce the input signal. Possible VLSI applications, and the motivation for this experiment are discussed.

INTRODUCTION

The Schmitt trigger (ST) is the simplest possible electronic realization of a bistable system. It is therefore an ideal device for studies on stochastic bistable dynamics and particularly *Stochastic Resonance*[1-4] (SR), a nonlinear statistical phenomenon whereby noise can enhance the transmission of information in bistable systems. It has recently been shown that the residence time probability densities[5] in simple bistable systems, driven by noise and a weak periodic signal, and particularly the ST, bear a close analogy to the inter spike interval histograms (ISIH's) long measured on stimulated sensory neurons in experimental neurophysiology[6]. Moreover, SR has recently been demonstrated in globally coupled arrays of bistable elements[7]. A rich dynamics with neurophysiological implications has been observed on a two dimensional array of coupled excitable cells[8]. It therefore seemed interesting to construct an array of ST's driven by noise and a weak periodic signal. A further motivation was that as the individual elements on VLSI chips are reduced to the limits in size, statistical noise, both classical and quantum, becomes a problem. Each element is thus subject to its own more-or-less independent noise. SR may offer a way to "design around" this limiting noise, and indeed, to make use of the noise for signal enhancement rather than simply living with its inevitability.

EXPERIMENTAL REALIZATION AND TYPICAL RESULTS

We have built an amplifier consisting of ten ST's which transmit a weak signal in parallel. Each ST is subject to a Gaussian, wide band noise which is incoherent with the noises of all other ST's. The correlation function of the noise is given by $\langle \xi(t)\xi(s)\rangle = (D/\tau)\exp[-|t-s|/\tau]$, where D is the noise intensity and τ is the noise correlation time. The ST's switch more-or-less randomly depending on the relative strengths of the signal and noise: a dynamics similar to that of SR. The outputs of the ten ST's are summed in a final amplifier where the *analog* input signal is recreated more-or-less faithfully depending on the noise and number of ST's. The individual ST's behave like digital signal samplers, where the samples are taken at random times, as dictated by the noise. We present measurements of the signal amplification and phase shift[9] through the system and the residence time probability density of the individual ST's. Figure 1 shows a

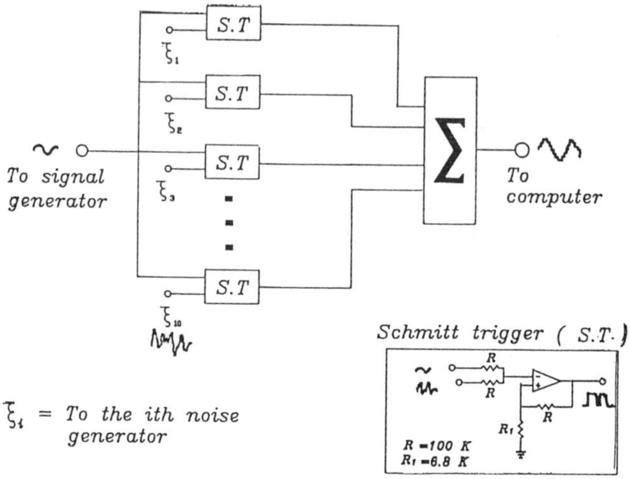

Fig. 1. Schematic diagram of the array of Schmitt Triggers. The inset shows a single ST. The ξ_i are the incoherent noises.

diagram of the system. The incoherent noises are obtained from the delayed outputs of a single noise generator connected to ten delay lines in series. The noise generator was a Quan-Tech Model 420 which produces Gaussian noise with a 200 kHz bandwidth. This noise was passed initially through a linear filter to fix the correlation time at $\tau = 42\ \mu s$. The delay lines were EG&G RD5107A charge transfer, "bucket brigade" devices each operating at a delay of 1.0 ms. The delay time of each delay line was, therefore, much longer than the noise correlation time so that each delayed output was incoherent with the others. A typical output signal from the summer and its power spectrum are shown in Fig. 2. The summed output is a discretized reproduction of the input sine wave but with noise and "sampled" at irregular (noisy) times. The output of any individual ST is a two-state switching waveform with stochastic switch times. The fidelity of the output would improve significantly with more individual ST's in the array, since the discretization would be on a finer scale. The signal feature is represented by the sharp spike and its 3rd harmonic (a signature of SR systems) as shown on the power spectrum of the output.

RESULTS

We have measured the stochastic equivalent of the complex transfer function for the amplifier, that is its gain and phase as functions of the signal frequency. Figure 3 shows the system gain versus signal frequency for three noise levels. The noise level, which establishes a Kramers rate for each ST and thus a mean sampling rate for the signal,

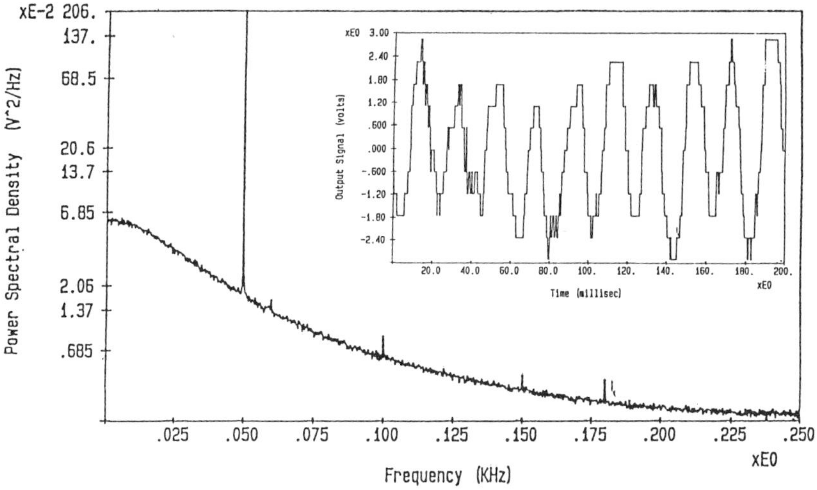

Fig. 2. The output at the summer showing the "noisy discretization" in tenths (upper) and its power spectrum (lower). The amplitude of the signal feature A_0 can be compared to the amplitude of the input signal A_i.

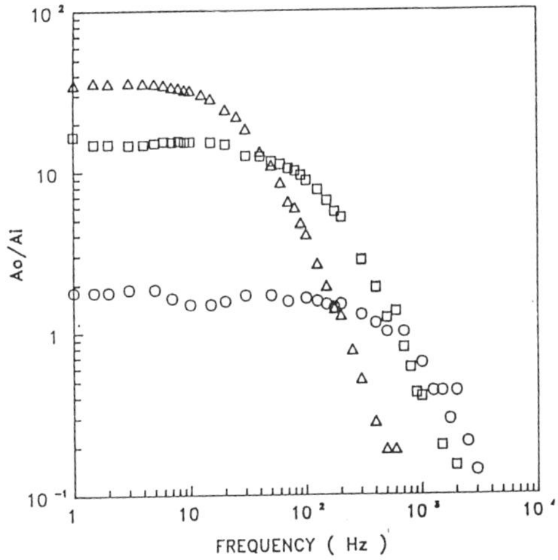

Fig. 3. The gain function at three noise levels: 0.5 (triangles), 0.7 (squares) and 1.5 (circles) V-rms for an input A_i = 1.4 V-pk which was smaller than the switch threshold of 2.0 V-pk.

determines a Nyquist (upper cutoff) frequency. Larger noise levels lead to larger switching rates for the individual ST's and hence to larger Nyquist frequencies. Larger noise levels result in lower low frequency gains due to the increased randomization of the sampling process. The gain-bandwidth product is not a constant.

We have also measured the phase shifts as functions of the signal frequency as shown

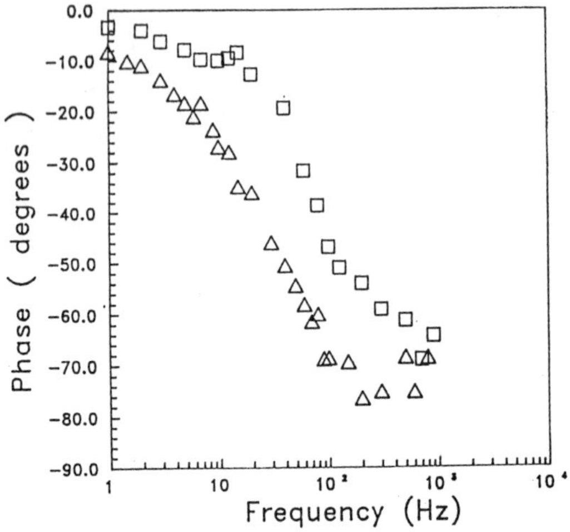

Fig. 4. The phase lags versus the frequency for two noise levels: 0.5 (triangles) and 0.7 (squares) V-rms. These results were obtained by signal averaging the summed output and comparing it directly to the input signal.

in Fig. 4. As shown in the Figure, the phases only lag, and they vary from 0^0 to -90^0 as would be expected from a linear amplifier. It was not practical to measure the phase lags for the largest noise level, because of the inordinately long averaging time required.

In conclusion, we have shown how a set of bistable switches driven by noise plus a subthreshold signal can operate as an amplifier based on digital sampling of the input signal at random times governed by the noise. A clearly defined Nyquist frequency is established by the Kramers rate of the noise. The amplifier has a transfer function which is similar to that of a linear amplifier.

Work supported by ONR grant N00014-90-J-1327 and by NATO grant 0770/85

REFERENCES

1. F. Moss, "Stochastic Resonance: From the Ice ages to the Monkey's Ear"; in *Some problems in Statistical Physics*, edited by George H. Weiss (SIAM, Philadelphia, in press)
2. P. Jung and P. Hanggi, Europhys. Lett **8**, 505 (1989)
3. P. Jung, Z. Phys. B **16**, 521 (1989)
4. P. Jung and P. Hanggi, Phys. Rev. A **41**, 2977 (1990)
5. T. Zhou, F. Moss and P. Jung, Phys. Rev. A **42**, 3161 (1990)
6. A. Longtin, A. Bulsara and F. Moss, Phys. Rev. Lett. **67**, 656 (1991)
7. P. Jung, U. Behn, E. Pantazelou and F. Moss, Phys. Rev. A **46**, R1709 (1992)
8. D. R. Chialvo, Santa Fe Institute preprint (1993)
9. P. Jung and P. Hänggi, Z. Phys. B. **90**, 255 (1993)

TWO-MODE MODEL OF THE CHAOTIC STELLAR PULSATION

Dmitry M. Vavriv and Yury A. Tsarin
Institute of Radio Astronomy, 4 Krasnoznamennaya street, Kharkov, 310002, Ukraine

ABSTRACT

The interacting modes model describing the W Vir variables behavior is considered. It is demonstrated, that nonisochronism plays an essential role in appearance of multifrequency regimes and stochastization processes. The model considered supposedly is the simplest one which describes such an interesting phenomenon, as multistability. Thus, it seems reasonable to say, that nonisochronism has to be taken into account in investigation of the stellar eigenmodes interaction.

INTRODUCTION

According to present-day status of research a number of laws of the W Vir variables behavior is explained by arising of the stellar radial oscillations which is caused by exciting of one or more hydrodynamical eigenmodes.However, to our knowledge, one of the most enigmatic phenomena, the irregular chaotic pulsations of some stars of considering type has defied explanation in terms of finite-mode models. Up to now the mathematical modeling of such pulsations occurrence could be explained only by direct solving of the stellar radiative hydrodynamic equations[1, 2]; and one of the most principal questions was in fact unresolved: (i) is the occurrence of such pulsations conditioned by small number of interacting modes or (ii) is finite-mode consideration principally unacceptable in this case. In this paper we prove that the chaotic pulsations occurrence may be caused by the interactions of only two radial stellar oscillation modes, and such regimes can be realized even at small nonlinearities of the radiant heat exchange processes. The interacting modes equations describing the chaotic stellar pulsations regimes follow immediately from the radiative hydrodynamic equations using the standard simplifying. We have studied into the chaotic stellar pulsations occurrence scenarios and its interplay with regular stellar behavior.

MODEL

A usual method of solving the radiative hydrodynamic problem is as follows[3, 5]. The hydrodynamic equations on the assumption of spherical symmetry take the form:

$$\frac{d^2R}{dt^2} = -4\pi R^2 \frac{\partial P}{\partial m} - \frac{Gm}{R^2} \equiv -g(R,s)$$

$$\frac{ds}{dt} = -\frac{1}{T}\frac{\partial L}{\partial m} \equiv h(R,s) \qquad (1)$$

$$\frac{\partial R^3}{\partial m} = \frac{3}{4\pi\rho}$$

with boundary conditions at inner and outer boundaries of the pulsating envelope, respecting to our assumptions of radius, luminosity and temperature of burning core and their observed values. These equations relate the Lagrange radius R, the pressure P, the density ρ, the specific entropy s of a spherical shell and are the third order ones with respect to time and the fourth order ones to m and must be supplemented by the equtions of state. The time independent solution (R_0, s_0) to the above set of equations can be obtained by assuming $\frac{d}{dt}=0$ and integrating the derived boundary problem for ordinary differential equations system. Further, the system (1) must be linearized in the neighborhood of (R_0, s_0) and only variations $\delta R \equiv R - R_0$, $\delta s \equiv s - s_0$ have to be considered. For these variations the system (1) takes the form:

$$\frac{dz}{dt} = \mathbf{A}z + \mathbf{N}_2 zz + \mathbf{N}_3 zzz\ldots \qquad (2)$$

where $z \equiv (\delta R, \delta v, \delta s)$, $\delta v \equiv \frac{d\delta R}{dt}$, **A** is linear nonadiabatic (LNA) operator, $\mathbf{N}_2$ and $\mathbf{N}_3$ are quadratic and cubic nonlinearities operators of the system (1) respectively. In this paper we consider the resonance between the fundamental tone and the second overtone of operator **A** with the frequency relationship $\Omega_2 : \Omega_1 = 2:1$. As may be inferred from the existing data, this resonant case is of prime interest for the stellar pulsation in generall and the irregular ones specifically[1,2]. For this case the model (2) can be rearranged to the cubic approximation form[4]:

$$\frac{da}{d\tau} = (\delta_1 - \delta_3 a^2)a - kab\sin\varphi$$

$$\frac{db}{d\tau} = \delta_2 b + ka^2 \sin\varphi \qquad (3)$$

$$\frac{d\varphi}{d\tau} = -\Delta + \beta b^2 + k(a^2/b - 2b)\cos\varphi,$$

Here a, b and φ are the oscillations amplitudes of the active and passive modes and the phase difference respectively; τ is the "slow" time, δ_1 and δ_2 are the real parts of the eigenvalues, Δ is the detuning between eigenfrequencies Ω_1 and Ω_2; δ_3, k and β are the coefficients describing nonlinear dissipation, resonant interaction and nonisochronism respectively represented complicate integrals along stellar radius which depend upon interacting modes eigenfunctions. Our model is more general in comparison with previously considered[1,2,4] because the nonisochronism was taken into account. We used the same order parameters in our model as in the ones, regarded previously and characteristic of W Vir stars and now we shall consider the main properties of our model. The non-zero values of stationary amplitudes A and B for this case are described by the resonance curve equations obtained by solving the equations (3) at zero values of the time derivatives:

$$\Delta_{1,2} = -\frac{\beta(\delta_1 - \delta_3 A^2)^2}{\delta_2} \pm (\delta_2 + 2\delta_1 - 2\delta_3 A^2)\sqrt{\frac{k^2 A^2}{\delta_2(\delta_3 A^2 - \delta_1)} - 1},$$

$$B^2 = -\frac{(\delta_1 - \delta_3 A^2)^2}{\delta_2} \qquad (4)$$

Its typical view is shown in the Fig. 1.The sign "±" indicate the two independent

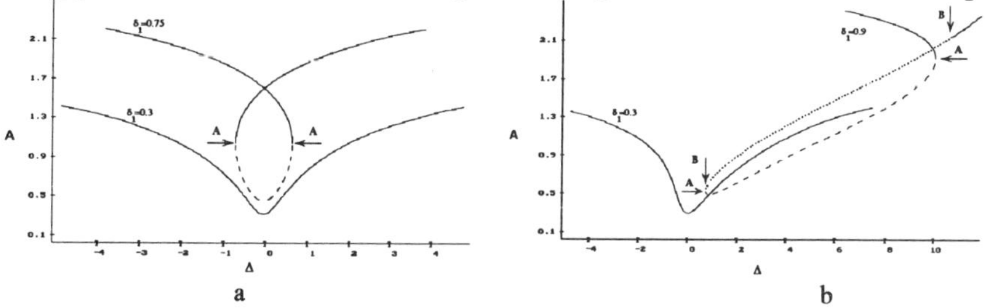

Fig.1 The resonance curve for following parameters: δ_2=-1, δ_3=0.1, k=2; a) β=0 b) β=5

branches of it which correspond to two different possible signs of the phase difference and switch one to one at Φ=0. The fig 1a demonstrates its typical form for isochronous case (β=0). From this figure it is clear that in this case only the stable node-focuses (solid lines) and, for sufficiently large δ_1, saddle (dashed lines) stationary states may exist in the system (3). The latter appears when the resonance curve becomes multivalued and the behavior of system (3) and also the governing hydrodynamical system (1) essentially depends upon its evolution i.e. the system is multistable. Any other regimes, excepting stationary equilibrium states that correspond to the stationary amplitude periodical regimes of the hydrodynamical system, have not been detected. The resonance curve typical view for nonisichronous case is shown in the Fig. 1b. One can see that for sufficuenly large δ_1 the stationary states in the system (3) become unstable, that is marked by dotted line between the points B. Generally speaking, the Hopf bifurcation appears in the point B and the stable limit cycle comes into being, that correspond to biperiodic pulsations in the initial system (1). For detail study of the system (3) dynamics refer to the bifurcation diagram (Fig. 2). This diagram constructed on the basis of the numerical experiments enables us to trace stellar dynamics regimes in a wide range of the parameters change. The multistability region is bounded by bold solid lines marked A-A according to the points A on the resonance curve (Fig. 1). The line marked B-B denotes the Hopf bifurcation and corresponds to the points B on the Fig. 1b. Inside the region enclosed by it the stable limit cycle exists. The dashed line labelled D designate its period doubling bifurcation. The dashed area marked by C is the one of chaotic motion. Above the dashed area an attractor (or cycle) becomes unstable and oscillations jump to the stable branch of resonance curve (solid line on the Fig 1b).

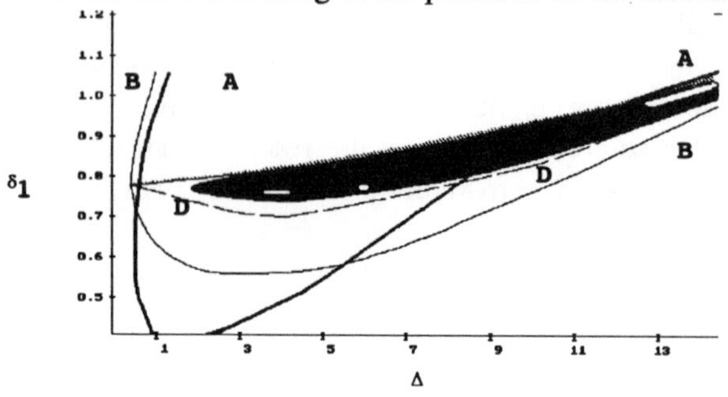

Fig.2 Bifurcation diagram δ_2=-1, δ_3=0.1, Δ=8, k=2

CONCLUSION

In our work the mechanism of the transition to chaos through a series of period doubling bifurcations in a case of resonance between the fundamental tone and the second overtone for W Vir models has been considered. An investigation of obtained interacting modes equations indicates that the nonisochronism plays a crucial role in the stochastization processes. In general, a large body of nonlinear oscillations theory research[6,7] supports this fact not only in the 2:1 resonant case but in other interacting modes cases. However, we believe that the model considered is the simplest one by using which it is possible, at least qualitatively, to investigate insufficiently explored phenomena, such as appearance of pulsations, their transformation into irregular regimes, multistability and other nonlinear stellar dynamic effects.

REFERENCES

1. J.R. Buchler and G. Kovacz Astrophys. Journ.,303, 749, 1986
2. J.R. Buchler and G. Kovacz Astrophys. Journ. (Letters), 320, L57, 1987
3. J.P. Cox Theory of Stellar Pulsations (Princeton University Press, Princeton, New Jersey, 1980)
4. G. Kovacz and J.R. Buchler Astrophys. Journ., 346, 898-905, 1989
5. R.F. Stellingwerf Current numerical tecniques for pulsation. (in The Numerical Modelling of Nonlinear Stellar Pulsations, ed. by J.R. Buchler), 1990, Kluwer, Dordrecht
6. S. Tousi and A.K. Bajaj Trans. ASME: J. Appl. Mech., 52, 446, 1985
7. D.M. Vavriv, Yu. A. Tsarin and I.Yu. Chernesov Radiotekhnika i electronika, 36, 2015-2023, 1991

ACCURACY OF 1/F-LIKE SPECTRUM DECOMPOSITION ON THE SUM OF LORENTZIANS

A.L.Mladentzev, Ir., and *A.V.Yakimov*, Prof.
State University, N.Novgorod 603600, Russia

ABSTRACT

1/f noise is considered as a superposition of processes having Lorentz type spectrum with different corner frequencies and variances. Processes are associated with the motion of defects within a sample. Thus, the decomposition of the 1/f-like spectrum on the sum of Lorentzians allows to determine electro-physical parameters of defects existing in the considered sample. Here the problem of the accuracy evaluation for parameters of each Lorentzian extracted from analyzed spectrum is solving.

INTRODUCTION

The spectrum of the 1/f noise is considered as a superposition of Lorentz spectra differing by corner frequencies and variances (see, e.g.[1]). The decomposition of spectrum on the sum of Lorentzians opens the new way for the identification of defects existing in the sample.

The first problem is the test of the 1/f-like spectrum. This spectrum must have "wavy" shape due to the presence of Lorentz components. The accuracy ε_{th} of the spectrum measurement is determined by known methods. This value is compared with the estimation ε_{ex} reached after fitting of the "real" spectrum by the power law. If ε_{th} is smaller than ε_{ex}, the spectrum may be decomposed as the sum of Lorentzians. Variances and characteristic (corner) frequencies of Lorentzians are to be determined.

The second problem is the variance and characteristic frequency accuracy evaluation for every Lorentzian selected from the spectrum. The similar problem for the spectrum approximation by the power law was treated by authors in paper [2]. The formulated problem we have solved analytically for a single Lorentzian. In the case of a number of Lorentzians the numerical treatment is necessary. The corresponding procedure is worked out. It takes into account the possibility of edge effects. That is, in the low-frequency part of the measured spectrum the tail may exist from Lorentzians having characteristic frequencies small comparing with the lowest analyzing frequency; at the high-frequency part the plateau takes place due to Lorentzians with very high characteristic frequencies and due to the presence of an additive noise of the measuring setup.

ANALYSIS OF SINGLE LORENTZIAN

Consider the relaxation type spectrum (Lorentzian):

$$S(f) = A\cdot\nu/(\nu^2+f^2) \ , \qquad (1)$$

Here f is the analyzing frequency, A -dimensionless amplitude, and ν -characteristic (corner) frequency. While measuring such spectrum, the problem arises to determine values A and ν, and to estimate their accuracies.

To find A and ν, one can use the known least square method minimizing the mean square error of the approximation:

$$\varepsilon^2 = \frac{1}{N}\sum_{i=1}^{N}(\Delta S_i/S_i)^2 \ ,$$

where N is the number of spectrum samples S_i at frequencies f_i, $i=\overline{1,N}$; $\Delta S_i=S_i-A\cdot\nu/(\nu^2+f_i^2)$ -the discrepancy.

The minimization of ε^2 leads to the following system of equations:

$$F_1 \equiv \sum_{i=1}^{N}\frac{\Delta S_i}{S_i^2}\,\frac{1}{\nu^2+f_i^2} = 0 \ ,$$

$$F_2 \equiv \sum_{i=1}^{N}\frac{\Delta S_i}{S_i^2}\,\frac{f_i^2-\nu^2}{(\nu^2+f_i^2)^2} = 0 \ . \qquad (2)$$

Solving this system, one can find values A and ν. Let us go to the accuracy evaluation for these values.

Let δs_i be the relative error of the spectrum measurement. Then the spectrum is as follows:

$$S_i = \langle S_i\rangle\cdot(1+\delta s_i) \ , \ i=\overline{1,N} \ ,$$

where $\langle S_i\rangle$ is the spectrum determined by eq.(1) at frequency f_i. Error δs_i leads to the error in A and ν:

$$A = \langle A\rangle\cdot(1+\delta a) \ , \quad \nu = \langle\nu\rangle\cdot(1+\delta\nu).$$

Here $\langle A\rangle$ and $\langle\nu\rangle$ are exact values of these parameters, δa and $\delta\nu$ -relative errors. Evaluating total differentials of functions (2), one can find dependences $\delta a=\delta a(\delta s_1,\delta s_2,\ldots,\delta s_N)$, $\delta\nu=\delta\nu(\delta s_1,\delta s_2,\ldots,\delta s_N)$ for small values δs_1, $\delta s_2,\ldots,\delta s_N$. Taking $\langle S_i\rangle=\langle A\rangle\cdot\langle\nu\rangle/(\langle\nu\rangle^2+f_i^2)$ and denoting $b_i=(f_i^2-\langle\nu\rangle^2)/(f_i^2+\langle\nu\rangle^2)$, we get:

$$\delta a = \frac{1}{N\cdot\sigma_b^{\ 2}} \sum_{i=1}^{N} (\overline{b^2} - \bar{b}\cdot b_i)\cdot\delta s_i \ ,$$
$$\delta\nu = \frac{1}{N\cdot\sigma_b^{\ 2}} \sum_{i=1}^{N} (b_i - \bar{b})\cdot\delta s_i \ , \qquad (3)$$

where $\sigma_b^{\ 2} = \overline{b^2} - \bar{b}^2$; the bar denotes an arithmetic averaging over existing samples. Eqs.(3) provide the final solution of the problem: knowing the statistic of δs_i, one may find σ_a^2 and σ_ν^2.

ANALYSIS RESULTS

Let us suppose that relative measurement errors at different frequencies are mutually non-correlated and have fixed variance: $\langle\delta s_i \cdot \delta s_j\rangle = \sigma_S^2 \cdot \delta_{ij}$. Here δ_{ij} is Kronecker delta. Variances of errors (3) in this case are:

$$\sigma_a^2 = (\sigma_S^2 /N)\cdot(\overline{b^2}/\sigma_b^2) \ , \quad \sigma_\nu^2 = (\sigma_S^2 /N\sigma_b^2) \ . \qquad (4)$$

An example. The quality $N=31$ of frequencies are chosen equidistantly from $f_l=1Hz$ up to $f_h=3112Hz$. Normalized on σ_S^2 functions $\sigma_a^2 = \sigma_a^2(\langle\nu\rangle)$ and $\sigma_\nu^2 = \sigma_\nu^2(\langle\nu\rangle)$ are shown in fig.1 by full and broken lines. One can see that the region of frequencies $\langle\nu\rangle$ exists where accuracies are nearly fixed and have minimum values. This circumstance, after crude spectrum estimation, allows to choose analyzing frequencies on the manner giving the minimal error in fitting parameters.

Following relations may be used for estimations (4) if the equidistant in logarithmic scale choice of frequencies is made:

$$\overline{b^2}=1+[(\nu^2\cdot 20\lg e)/\Delta y]\cdot[(f_h^2+\nu^2)^{-1}-(f_l^2+\nu^2)^{-1}] \ ,$$
$$\sigma_b^2 = \overline{b^2}-1+2Z-Z^2 \ ; \quad Z=(10/\Delta y)\cdot\lg[(f_h^2+\nu^2)/(f_l^2+\nu^2)] \ . \qquad (5)$$

Here $\Delta y=10\cdot\lg(f_h/f_l)$ is the band width (in dB).

Variances (4) are minimal in the centre of logarithmic scale band (at $\nu = \surd(f_l\cdot f_h)$):

$$\sigma_{a\ min}^{\ 2} = \sigma_S^{\ 2}/N \ , \quad \sigma_{\nu\ min}^{\ 2} = \sigma_{a\ min}^{\ 2} /(1-8{,}69/\Delta y) \ . \qquad (6)$$

These variances are considerably higher at band edges ($\nu = f_l$, $\nu = f_h$).

Eqs.(6) are valid if $f_h >> f_l$. For additive condition $f_h >> \nu$ the following simplification of eqs.(5) may be used:

$$\overline{b^2}=1-20\cdot \lg e/\Delta y, \quad \sigma_b^2=\overline{b^2}-1+4Y-4Y^2 \;;\; Y=(y_h-y_\nu)/\Delta y \;,$$

where $y_h=10\cdot \lg f_h$, $y_\nu=10\cdot \lg \nu$. Due to the usage of the approximation $\delta s_i<<1$, $i=\overline{1,N}$, in eq.(4), standards σ_a and σ_ν are linear functions of σ_s.

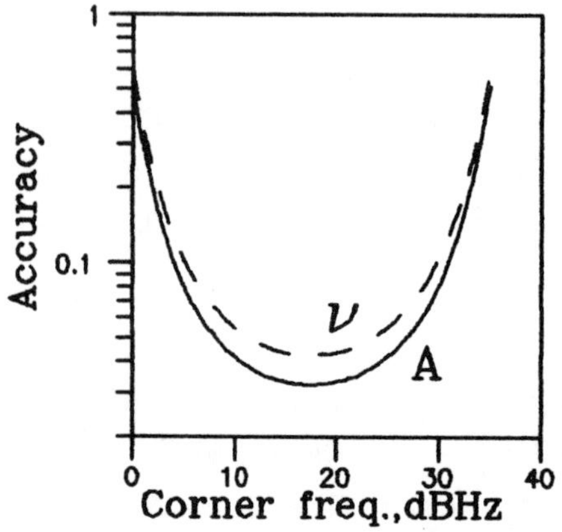

Fig.1. Variances σ_a^2 and σ_ν^2, normalized on σ_s^2 .

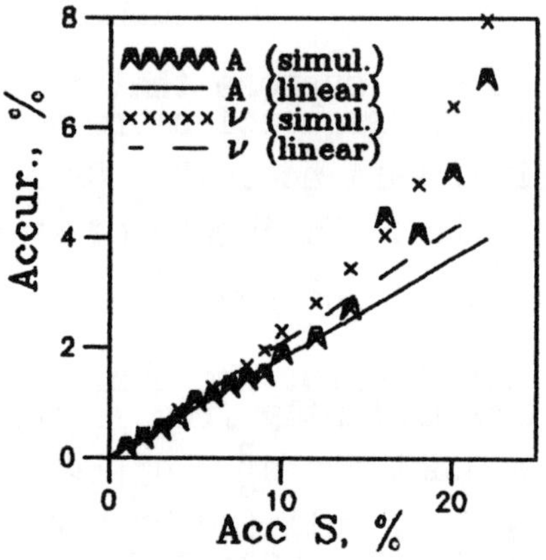

Fig.2. Spectrum simulation results.

The approximation validity limits are illustrated by fig.2 for the case $N=31$, $f_l=1Hz$, $f_h=3112Hz$, $\nu=100Hz$, $A=3$. Direct lines correspond to (4); points - Monte Carlo simulation data. Random values δs_i follow to Gauss law. Each point is the averaged result over *100* found values of A and ν. One can see that for experimental error within *10%* the linear approximation works well. If the spectrum measurement accuracy is *10%* then errors of A and ν determination do not exceed *2%*.

As an example, fig.3 presents solutions of eqs.(2) for simulated spectra containing experimental errors *10*, *20* and *30* percents. Curves are results of approximation; points show values of simulated spectra.

One can check the approximation correctness (in other words, found values A and ν validity) mentioning that for taken measuring accuracies the approximation accuracies are *10*, *25* and *45* percents, respectively.

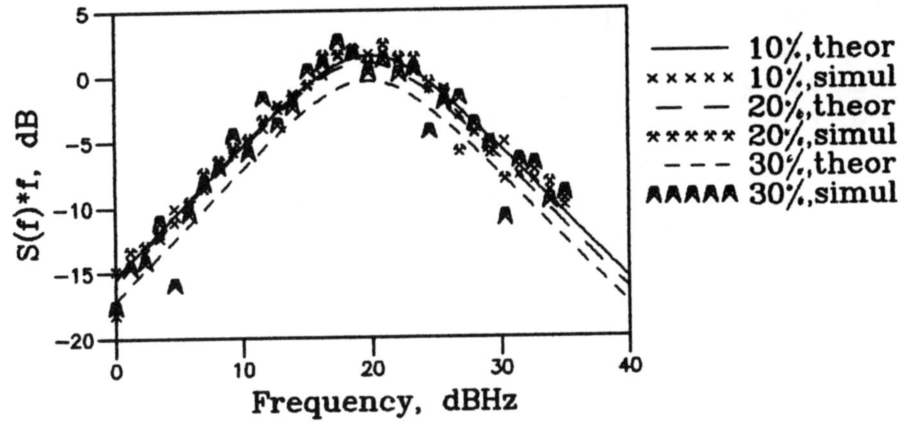

Fig.3. Approximation of simulated spectra.

1/f-LIKE SPECTRUM DECOMPOSING

The spectrum of 1/f-like noise may be presented as the following superposition of Lorentzians:

$$S(f) = D/f^2 + S_o + \sum_{m=1}^{M} A_m \nu_m / (\nu_m^2 + f^2). \qquad (7)$$

The first term here respects to the high-frequency "tail" of Lorentzians with small corner frequencies comparing with f_l; the second term in eq.(7) is plateau caused by Lorentzians having high comparing with f_h corner frequencies; the last term accounts Lorentzians forming the 1/f-like spectrum in the region of analyzed frequencies; M is the number of these Lorentzians determined by parameters A_m and $\nu_m \in [f_l, f_h]$.

Testing experiments [3] on gas-sensing SnO_2 films were carried out to check the spectrum decomposing procedure in the frequency band $[1.25, 5\cdot 10^3]Hz$. The spectrum measurement accuracy estimation was $\varepsilon_{th} \approx 0.2\ dB$. Approximation of spectrum data by the power law $S(f) \sim f^{-\gamma}$ gives the discrepancy up to $\varepsilon_{ex} = 0.6\ dB$. This result was considered as the base for the spectrum decomposing in the accordance with eq.(7). Up to three Lorentzians on the background of f^{-2}-type low-frequency tail and high-frequency plateau S_o were determined. The error in parameters for eq.(7) were found to be rather high. Nevertheless, the discrepancy between spectrum (7) and the measured spectrum did not exceed the measurement accuracy εth.

Reached results show that the decomposing procedure of 1/f-like spectrum on to sum of Lorentzians is quite real. It is desirable to analyze the noise in wide temperature band taking in the mind the goal to increase the decomposing procedure accuracy and, as a result, to determine electro-physical parameters of defects existing in the sample. That means, found dependences of parameters in spectrum (7) on the temperature allows to get additive estimations for these parameters. On other turn, the analysis of the spectrum temperature evolution may lead to relations necessary for the determining of types of defects existing in the sample.

CONCLUSION

The way to check the accuracy of 1/f-like spectrum decomposition on the superposition of Lorentzians is shown. Knowing the measurement accuracy, one can find accuracies of values A and ν. The spectrum treatment is to be considered as successful if the found discrepancy between measured and evaluated spectra does not exceed the measurement accuracy. In other case the spectrum is to be approximated by another function.

Authors are thankful to The Netherlands Organization for Scientific Research (NWO) and to Prof. F.N.Hooge for the support of investigations on the considered problem.

REFERENCES

1. P.Dutta and P.M.Horn, Rev.Modern Phys.53, 497 (1981).
2. A.L.Mladentzev and A.V.Yakimov, Radiophys. Quant. Electron, to be published.
3. N.V.Demin, A.V.Ikonnikov, A.L.Mladentzev, V.B.Orlov, G.P.Pashev and A.V.Yakimov, to be published.

METHOD OF NOISE FIGURE MEASUREMENT WITH ELIMINATION OF LOSSES OF INPUT MATCHING CIRCUIT

Jacek Cichosz , Lech Hasse , Alicja Konczakowska , Ludwik Spiralski
Technical University of Gdańsk,
Faculty of Electronics, Department of Measuring Equipment
ul. G. Narutowicza 11/12, 80-952 Gdańsk, Poland

ABSTRACT

The main sources of instrumentation errors (inaccuracy of source admittance setting, inherent noise of measuring system, losses of input matching circuit) occuring at the noise figure measurement have been analysed. They have great influence on the accuracy of estimation of noise parameter set of linear twoports. Selected results obtained by means of the elaborated system for the noise figure measurement are presented. The system was applied for the measurement of noise properties of two-gates MOSFETs at the frequency f=200 MHz.

INTRODUCTION

Noise properties of a linear twoport in a high frequency range are precisely determined by the basic set of four noise parameters [F_o, G_o, B_o, R_n] defined according to the equation describing the noise figure F[1]:

$$F(f) = F_o(f) + \frac{R_n(f)}{G_s(f)}\left\{\left[G_s(f) - G_o(f)\right]^2 + \left[B_s(f) - B_o(f)\right]^2\right\} \qquad (1)$$

where: F_o - minimal value of noise figure (for noise matching),
$Y_o = G_o + jB_o$ - optimal source admittance which enables to obtain the minimal noise figure F_o,
$Y_s = G_s + j\,B_s$ - source admittance,
R_n - equivalent noise resistance characterizing deterioration of noise properties when Y_s is not equal to the optimal admittance Y_o.

According to theoretical considerations for the complete determination of noise behaviour of a twoport it is sufficient to carry out the four measurement of noise figure at different measurement conditions [G_o, B_o]. Basing on results of measurements the values of four noise parameters are evaluated by a chosen estimation procedure[2]. However, usually the greater (redundant) number of measurements than four is performed to minimize the statistical error of estimation method.

Generally, two kinds of error can be distinguished: instrumentation and statistical (related to the estimation procedure). In the paper only the former errors, especially essential in the case of low available power gain and low noise figure of a twoport under test, have been considered. They are uncorrelated and additive, so the total instrumentation error ΔF of noise figu re measurement can be written as follows:

$$\Delta F = \Delta F_a + \Delta F_b + \Delta F_c \qquad (2)$$

where the component ΔF_a is due to inaccuracy of source admittance setting, ΔF_b is caused by the inherent noise of measuring system, and ΔF_c is due to losses of matching circuit. Each of them has been analysed in relation to the measuring system.

INACCURACY OF SOURCE ADMITTANCE SETTING

Usually values G_{sx} and B_{sx} as measurement conditions (taken into account to calculations of a set of four noise parameters) differs from the actual value of source admittance $Y_s = G_s + jB_s$. In that case an error ΔF_a is given by the relation:

$$\Delta F_a = \pm\sqrt{(\Delta F')^2 + (\Delta F'')^2} = \pm\left\{\left[\frac{\partial F}{\partial B_s}\Delta B_s\right]^2 + \left[\frac{\partial F}{\partial G_s}\Delta G_s\right]^2\right\}^{1/2} \tag{3}$$

where:

$$\frac{\partial F}{\partial B_s} = 2\frac{R_n}{G_s}(B_s - B_o) \tag{4}$$

$$\frac{\partial F}{\partial G_s} = \frac{R_n}{G_s^2}\left[\left(G_s^2 - G_o^2\right) - (B_s - B_o)^2\right] \tag{5}$$

$$\Delta B_s = B_s - B_{sx} \ , \qquad \Delta G_s = G_s - G_{sx} \tag{6}$$

The components $\Delta F'$ and $\Delta F''$ calculated for the bipolar transistor BF180 at different relative values of ΔB_s and ΔG_s (10, 25, 50%) are shown in Fig.1. For $G_s < G_o$ values of $\partial F / \partial B_s$ depend on G_s, therefore measurements at $G_s > G_o$ are recommended. It is easy to recognize that the increase of ΔB_s (Fig.1.a) gives in a result the increase of $\Delta F'$ especially at $G_s < G_o$. The increase of ΔG_s (Fig.1 b) gives as a consequence the increase of $\Delta F''$ especially at $G_s < G_O$ and $B_s \neq B_o$, but this component of error is not equal to zero for $B_s = B_o$ and for different values of G_s ($\Delta F''$ is equal to zero only at an optimal source admittance Y_o . In Fig.2 the errors ΔF_a at different relative values of $\Delta B_s = \Delta G_s$ are presented. The error ΔF_a increases at $G_s < G_o$. Minimal value of ΔF_a at $G_s > G_o$ can be obtained at $B_s \approx B_o$. From equations (1) and (3) for $G_s = 0.05S$ and $B_s = -0.03S$ we have, for example:

$\Delta F_a = 5.24 \qquad F = 5.1 \qquad \Delta F_a / F = 1.027 \qquad \Delta B_s = \Delta G_s = 10\%$

$\Delta F_a = 13.12 \qquad F = 5.1 \qquad \Delta F_a / F = 2.572 \qquad \Delta B_s = \Delta G_s = 25\%$

ΔF_a is nearly constant for $G_s \to G_o$ and for different B_s. For values $G_s < G_o$ the error ΔF_a rapidly increases; for example, at $G_s = 0.01S$,$B_s = -0.03S$,$\Delta B_s = \Delta G_s = 10\%$, the value of ΔF_a is equal to 40.31 (F=6.3) and we have $\Delta F_a / F = 6.398$. It is important factor because relative errors ΔB_s and ΔG_s equal 10% can be found in practice.

For the same conditions we obtain $\Delta F_a / F = 10\%$ at $\Delta B_s = \Delta G_s = 1\%$ or $\Delta F_a / F = 1\%$ at $\Delta B_s = \Delta G_s = 0.1\%$.

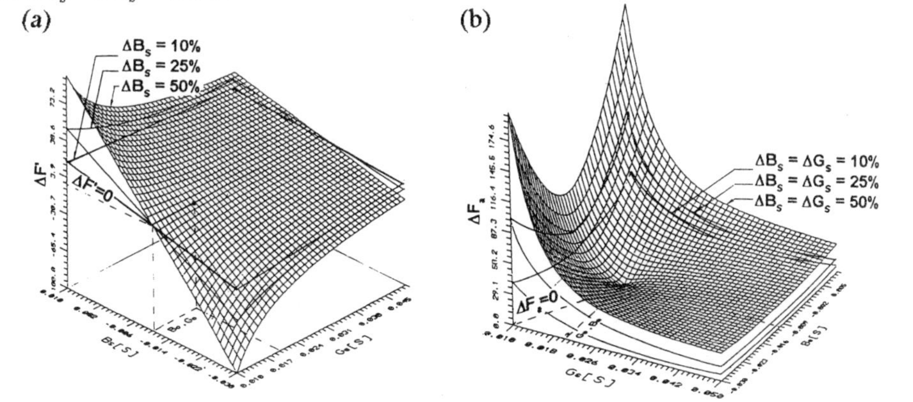

Fig.1. Graphical representation of components $\Delta F'$ (a) and $\Delta F''$ (b) versus Y_s.

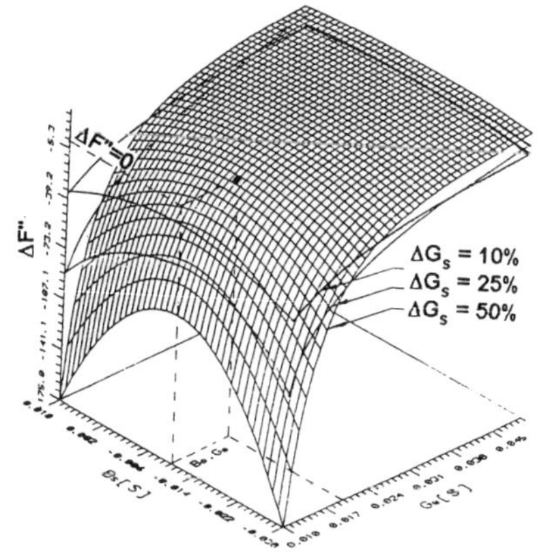

Fig.2. Error ΔF_a versus Y_s

One can conclude that the minimum of ΔF_a can be obtained at $G_s > G_o$ and at $B_s \approx B_o$. Values of G_{sx}, B_{sx} should be possibly equal to G_s and B_s. Basing on those considerations one can conclude that an accurate determination of actual values of G_s and B_s proved to be very important.

INHERENT NOISE OF MEASURING SYSTEM

Knowing avallable power gain of measured twoport and noise figure of measuring system one can evaluate ΔF_b by means of the Friis' equation . Those magnitudes can be measured using methods elaborated by the authors[34] and rely on the intrinsic increase of noise power from the noise gnerator to cause the noise figure increase by $1/2\ kT_O$ or $1\ kT_0$.

LOSSES OF INPUT TUNED MATCHING CIRCUIT

Standard noise generatots with typical output impedance (e.g. 50Ω) are commonly used for noise figure measurement in the high frequency range (Fig.1). Input tuned circuits setting a source admittance Y_s connected to them have inherent losses T. Mea-

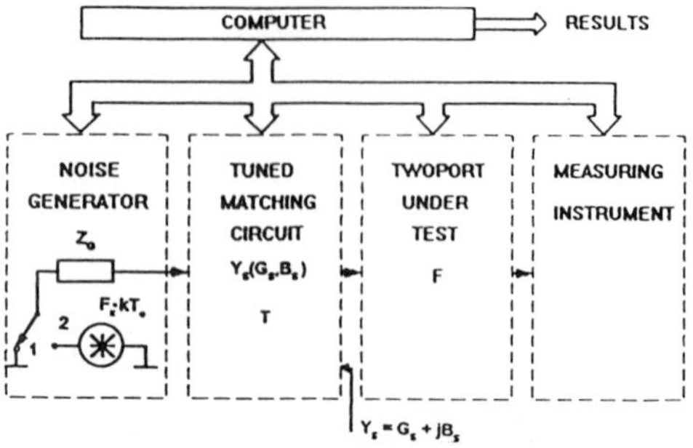

Fig.3. Block diagram of noise figure measurement system in the high frequency range

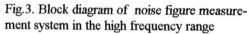

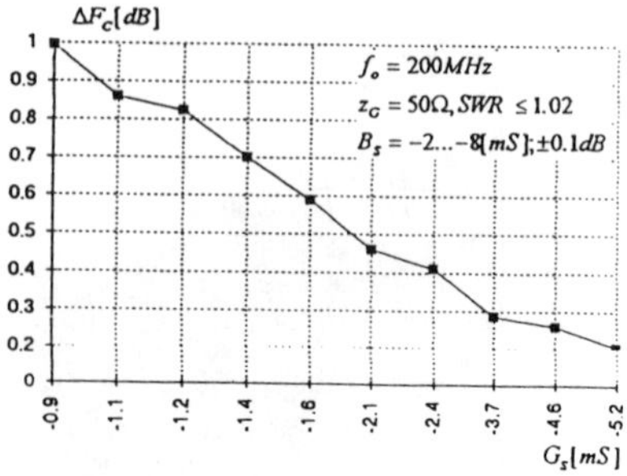

Fig.4. Dependence ΔF_c versus G_s for the input matching circuit

sured noise figure has a value F_x different from the actual F. A correction component $\Delta F_c[\text{dB}] = F_x - F$ depends only on losses of tuned matching circuit. The value of ΔF_c can be a function of Z_s (Fig.4). In the frequency range to hundreds of MHz the losses of the matching circuit can be determined with an accuracy of ± 0.1dB.

CONCLUSIONS

In the computer-aided system for the noise figure measurement it is possible by means of the programmable matching circuit not only to set (to adjust) the measuring conditions of a twoport under test but also to take into account the influence of different factors on an accuracy of measurement. The application of presented techniques gives the possibility to improve an accuracy of noise figure measurement for different values of noise source impedance and finally to determine the four noise parameter set of linear twoport with an error smaller than 0.5 dB.

REFERENCES

1.H.Haus et al., Proc.IRE, 48, 69-74 (1960).

2.L.Spiralski, J.Cichosz, M.Zieliński, Nachrichtentechnik-Elektronik, 29, 298-301 (1969).

3.L.Spiralski, Noise In Physical Systems and 1/f Noise-1985 (North-Holland, Amsterdam, 1986), p.313-315.

4.L.Spiralski, L..Hasse, Z.Zdybel, PolishPatent Nr 95035, Method od elimination of inherent noise of measuring system at noise figure measurement.

XIII. CHAOS AND FRACTAL

FEIGENBAUM UNIVERSALITY UNDERLIES INTERMITTENCY AND ANOMALOUS EVENTS IN HADRON COLLISIONS

A.V.Batunin
Institute for High Energy Physics, Protvino,142284, Russia

ABSTRACT

Unusually-high-particle-density events find natural explanation within bifurcation mechanism of hadroproduction proposed earlier to account for intermittency in particle collisions. The comparison between theory and experiment is performed for an anomalous event with dN/dy > 100 of NA22 Collaboration, for Si-AgBr interaction in cosmic rays of JACEE Collaboration and for the pseudorapidity distribution of the charged particle group centers with a peak at $|\eta| = 0.3$ of NA23 and NA27 Collaborations.

INTRODUCTION

One of nowadays hadron physics puzzles is the dynamical origin of intermittency in multiplicity distributions of charged particles [1]. In this paper we propose a new method of investigating intermittency which follows from the bifurcation mechanism (BM) of hadroproduction [2].

BM relates intermittency to nonlinearity of a differential equation (still hypothetical) governing the interaction between quarks and gluons at the hadronization stage when low momentum transfer dominates and QCD perturbation theory seems to be inapplicable. Hadron generation in the framework of BM is analogous to appearance and development of turbulence in fluid through a series of period-doubling bifurcations (PDB) and the mentioned above governing equation (GE) is expected to be a somewhat modified Navier-Stokes equation.

We apply our method to the description of three well-known experiments chosen due to their most statistical significance.

BM AND NEW METHOD OF STUDYING INTERMITTENCY

The main BM propositions [2] are briefly listed below.
There exists nonlinear GE whose solutions depend on one parameter – the energy of collision in a given event $\sqrt{\hat{s}}$ ($< \sqrt{s}$). The number of splittings of a phase space trajectory (PST) corresponding to a given solution determines the number of intermediate particles (clans) which, in turn, decay into observable particles. The number of particles k per decay may vary, however, the averaged over all events $< k >$ at any fixed energy keeps constant for all energies.

On PST, a one-dimensional Poincare map is defined

$$z_{n+1} = f_s(z_n), \qquad (1)$$

where z_n means the point of intersection of PST and some straight line z in the phase space while the subscript s indicates the dependence of map (1) on energy.

The basic assertion is that map (1) has a unique quadratic maximum if one takes the rapidity axis as the direction z. In other words, after some variable transformation one can write map (1) in the following form:

$$x_{n+1} = \lambda x_n(1 - x_n), \quad x_n \in [0,1], \tag{2}$$

where λ is the energy-dependent governing parameter and the correspondence between the variable x_n and the nth clan rapidity y_n is defined by the relation

$$2|x_n - 0.5| \equiv \xi_n \Leftrightarrow \tilde{y}_n \equiv |y_n|/y_{max}. \tag{3}$$

Here $y_{max} = ln(\sqrt{s}/2m_{cl})$, m_{cl} is the kinematical cut parameter, the factor 2 indicates pair clan production. Note the center of rapidity interval $y = 0$ corresponds to the point $x = 0.5$ of map (2).

Map (2) is well studied [3]. In particular, at fixed λ from the interval $\Lambda = [3, \lambda_\infty = 3.5699..]$ the sequence $[x_n]$ sets to one of the limit stable (LS) 2^m-cycles $(m = 0, 1, 2, ..)$ as $n \to \infty$. The cycle size doubles (i.e., PDB $2^m \to 2^{m+1}$ occurs) when λ passes through a given value $\lambda = \lambda_m \in \Lambda$. The rate of the λ_m convergence towards λ_∞ is determined by the first Feigenbaum constant δ:

$$(\lambda_\infty - \lambda_m) \sim \delta^{-m}, \quad m \gg 1, \tag{4}$$

so that the number of LS 2^m-cycle elements satisfies the following equation

$$2^m \sim (\lambda_\infty - \lambda_m)^{-\Delta_F}, \quad \Delta_F = ln2/ln\delta = 0.449.. . \tag{5}$$

In BM, the number of LS 2^m-cycle elements is equal to the number of clans n_{cl}, which determines the mean multiplicity of charged particles $< n_{ch} >=< k > n_{cl}$. On the other hand, fit over experimental $< n_{ch} >$ data in pp and $\bar{p}p$ collisions in the $\sqrt{s}$-range 5 – 900 GeV gives the following dependence [4]:

$$< n_{ch} >\sim \sqrt{s^\Delta}, \quad \Delta = 0.449 \pm 0.018 . \tag{6}$$

Then, it is easy to obtain from (5)–(6) the correspondence

$$(\lambda_\infty - \lambda) = a/\sqrt{s}, \quad a \simeq 3m_\pi, \quad \lambda \in \Lambda, \tag{7}$$

where m_π is π-meson mass. Thus, we know which of LS 2^m-cycles corresponds to a given energy of collision and we can use our knowledge to compare the positions ξ_n of the cycle elements on the interval $[0,1]$ and the positions of clans on the normalized rapidity axis $\tilde{y}_n$. For every partition of the unit interval into bins $\delta\xi$ we find the dependence of the number of the cycle elements N in a bin $\delta\xi$ on the bin position, which corresponds to the observed dependence dn_{ch} / $d|y|$. The only difficulty is that there are the clan positions on the normalized rapidity axis which theoretically predicted but not the positions of observed particles themselves.

That is why one should either reconstruct clan positions dividing observed particles into groups and then compare group center positions with BM-predictions, or select events with maximal particle density $dn_{ch}/d|y|$ interpreting a high density group as a clan beginning to decay.

COMPARISON WITH EXPERIMENT

The first way is suitable if there are a lot of events at the same energy. The experiment of NA23 and NA27 collaborations [5] where 33228 events in pp collisions at $\sqrt{s} = 26\ GeV$ were studied, satisfies this condition. In each event of this experiment, 3-particle groups were picked out and group center positions on the pseudorapidity axis η were determined. Then the $|\eta|$-axis was divided into bins $d|\eta| = 0.1$ and the dependence of the number of group centers within a given bin on the bin position was plotted, see fig. 1a.

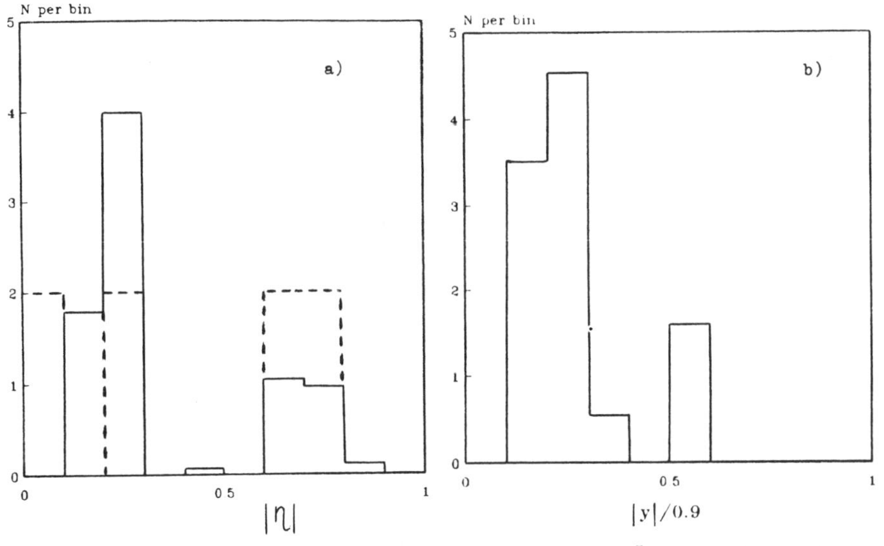

Fig.1. Solid line - 3-particle group center distribution [5] on pseudorapidity axis after subtraction of the background (8) and normalization to 8 events, dashed line - theoretical 8-cycle element density distribution at the bin size $\delta\xi = 0.1$ – a); experimental charged particle distribution at $\delta|y| = 0.1$ in AE [6] after subtraction of the background (9) – b).

The group center distribution was found to have a pronounced narrow peak at $|\eta| = 0.3$ which is more than four standard deviations from smooth background. In the framework of BM, this peak finds a natural explanation as well as a less pronounced second peak at $|\eta| \in [0.6, 0.8]$. In fact, to the energy $\sqrt{s} = 26\ GeV$ the 8-cycle corresponds, see form.(7). Plotting for it the distribution $N(\xi)$ and comparing it with the experimental histogram where a smooth fit ("background")

$$f_0(|\eta|) = 310(1 - exp[|\eta| - 3]), \tag{8}$$

is subtracted, we have found that the peak positions coincide (see fig.1a) if one takes $m_{cl} = \sqrt{s}/5.5$ ($y_{max} = 1$). The height excess of the first peak and suppression of the second one in comparison with the theoretical calculations we attribute to the phase space influence still not taking into account in BM.

The smooth background in BM arises, first, as a result of contributions from 2- and 4-cycles, since $\sqrt{\hat{s}}$ varies in different events even at fixed $\sqrt{s}$.. Second, the clans can decay not only into 3 particles but also into 2, 4, 5, ... ones. Therefore, the partition of all particles only into 3-particle groups can distort the true clan positions spreading their distribution along the pseudorapidity axis.

Next, we apply the second way to the prominent anomalous event (AE) in π^+p interaction at $\sqrt{s} = 22\ GeV$ of NA22 collaboration [6]. In this AE, 10 out of 26 charged particles are concentrated within very narrow rapidity interval $\delta y = 0.098$ that is 10^3 times greater than the probability of such event expected from the extrapolation of the rest of the data. Subtracting background $f_0(|y|)$,

$$f_0(|y|) = 1.64(1 - exp(|y| - 2.5)). \tag{9}$$

and taking $m_{cl} = \sqrt{s}/4.9$ ($y_{max} = 0.9$) we see that again the main peaks in theoretical and experimental distributions coincide, compare figs. 1a,b. Thus, the separate AE is consistent both with the high statistics results on group center positions [5] and BM predictions.

Finally, we apply our method to the description of AE in cosmic rays: $Si - AgBr$ interaction at the energy $4.1 \pm 0.7\ TeV/nucleon$, observed by JACEE collaboration [5]. In this AE, 1010 ± 30 charged particles were detected, with the particle density distribution on the pseudorapidity axis being extremely inhomogeneous, see fig.2a. Statistical probability of such event is turn out to be less than 9 per cent [8] which compels us to find a nonstatistical origin of this AE. Subtracting background $f_0(\eta)$,

$$f_0(\eta) = 184[(1 - exp(-5.5 - \eta))(1 - exp(-5.5 + \eta))]^{8.1}, \tag{10}$$

and taking $m_{cl} = \sqrt{s}/30$ ($y_{max} = 2.7$), we see that the theoretical and experimental main peak positions coincide at $\xi \in [0.30, 0.33]$, see figs.2a,b. Note that in this case (nucleus-nucleus collision), the background is generated by uncorrelated nucleon-nucleon interactions.

CONCLUSION

We have shown that the so-called "anomalous" events are quite natural from the point of view of bifurcation mechanism of hadronization via intermediate particles – clans. Anomalous events are nothing else but a manifestation of fractal structure of LS 2^m-cycles formed by the clans on the (pseudo)rapidity axis. In such events we see clans immediately after they begin to decay into observable particles.

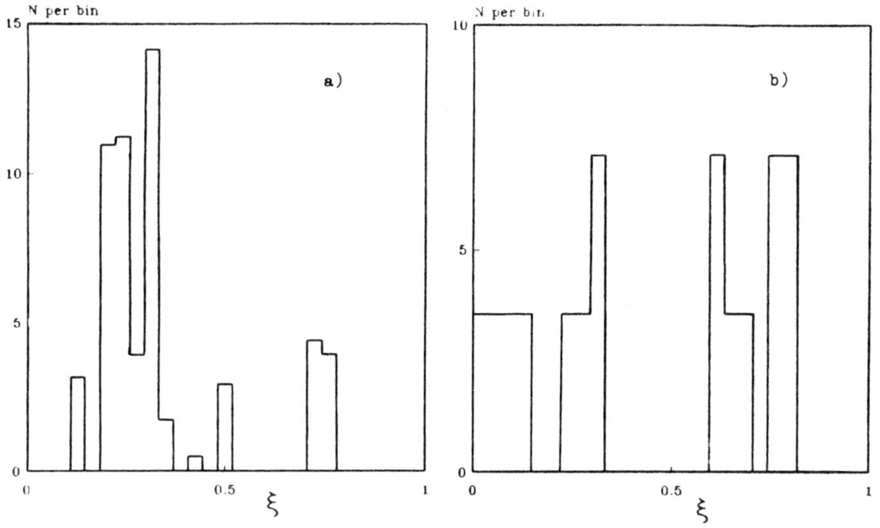

Fig.2 Experimental charged particle distribution [7] at the bin size $\delta\eta = 0.1$ after subtraction of the background (10), $\xi = |\eta|/2.7$ – a); theoretical 64-cycle element density distribution (corresponding to $\sqrt{s} = 4.1$ TeV) at the bin size $\delta\xi = 1/27$ normalized to 57 elements to compare with the experiment [7] – b).

REFERENCES

1. A.Bialas and R.Peschanski, Nucl. Phys. B273, 703 (1986).
2. A.V.Batunin, Sov. J. Nucl. Phys. 55, 2525 (1992).
3. M.J.Feigenbaum, J. Stat. Phys. 19, 25 (1978); ; 21, 669 (1979).
4. A.V.Batunin and O.P.Yushchenko, Mod. Phys. Lett. A5, 2377 (1990).
5. I.M.Dremin et al., Mod. Phys. Lett. A5, 1743 (1990).
6. NA22 Collab. M.Adamus et al., Phys. Lett. B185, 200 (1987).
7. JACEE Collab. T.H.Burnett et al., Phys. Rev. Lett. 50, 2062 (1983).
8. F.Takagi, Phys. Rev. Lett. 53, 427 (1984).

SOFT TURBULENCE IN THE ATMOSPHERIC BOUNDARY LAYER

Imre M. Jánosi
Department of Atomic Physics, Eötvös University
Budapest, Puskin u. 5-7, H-1088, Hungary
Fax: + 36-1-266-0206
E-mail: JANOSI@LUDENS.ELTE.HU

Gábor Vattay
Department of Solid State Physics, Eötvös University
Budapest, Múzeum krt. 6-8, H-1088, Hungary
Fax: + 36-1-266-7509
E-mail: VATTAY@LUDENS.ELTE.HU

In this work we compare the spectral properties of the daily medium temperature fluctuations with the experimental results of the Chicago Group, in which the local temperature fluctuations were measured in a helium cell. The results suggest that the dynamics of the daily temperature fluctuations is determined by the soft turbulent state of the atmospheric boundary layer, which state is significantly different from low dimensional chaos.

There is a widespread interest in forecasting and modeling of weather. Although all the basic mechanisms that govern the dynamics of the atmosphere have been well known since quite a long time, the detailed understanding and adequate characterization of the fluctuations of various statistical quantities of the lower atmospheric boundary layer is still not complete. For example, there were several attempts to describe the dynamics of daily medium temperature fluctuations. A typical viewpoint is that the underlying mechanism is fully stochastic in nature and can be considered as an autoregressive process. A completely different viewpont suggests that the apparent irregularities may be attributed to a deterministic chaotic behavior, although serious doubts have arisen on the existence of low dimensional chaos in the long time behavior of the atmosphere (climatic attractor), as well as in the processes over very short time scales.

The present investigation is based on temperature measurements by the Hungarian Meteorological Service performed at twenty different meteorological stations covering the area of Hungary for the period 1.1.1951 – 31.12.1989. The detailed analysis has been published in Ref. [1]. We produced the fluctuation time series by substracting the deterministic part from the daily medium temperature data. The histogram of the fluctuation amplitudes has a pronounced Gaussian distribution[1]. The power density spectrum can be obtained by well established methods, Figure 1 shows a typical result.

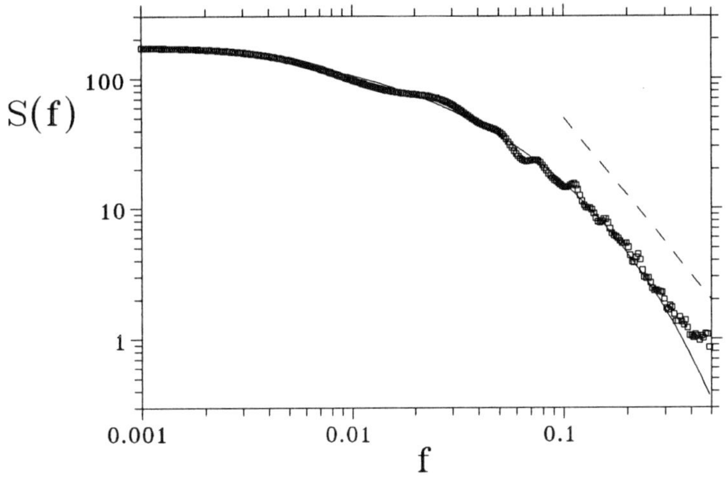

FIG. 1. *Unnormalized power density spectrum of the temperature fluctuations measured in Szombathely. The frequency unit is 1/day. The solid line is the fit given by Eq. (1), the dashed line illustrates the* $1/f^2$ *behavior on a restricted frequency range.*

All of the spectra of the time series measured at different meteorological stations can be fitted by the function

$$P(f) = P_s \exp\left[-\left(\frac{f}{f_s}\right)^{\beta}\right] \quad , \tag{1}$$

where $P_s = 222 \pm 16$, $f_s = 0.017 \pm 0.005$, $\beta = 0.54 \pm 0.03$, and the deviations indicate slight meteorological station dependence. Surprisingly, this form of the fitting function (1) with the same exponent β, and the Gaussian fluctuation-histogram are exactly the same as found in the helium experiment[2] in the soft turbulent region (see later). We note that the high frequency part of the power spectrum may be fitted by a power law on a very restricted frequency range (approximately half decade) with an exponent ~ -2, which inspired the Markovian stochastic models.

The shape of the power spectrum and the Gaussian fluctuation distribution do not rule out theoretically the existence of a low dimensional meteorological attractor. However, the embedding process[3] applied to our data did not show any saturation of the correlation dimensions. This observation completely agrees with the measurement of Talkner, Weber, and Roser[4], in which they could not find a weather attractor with a dimension less than 10 from a longer temperature time series .

Our conclusions are based on the detailed investigations of the Chicago Group, in which they have measured the local temperature fluctuations in gaseous helium at very wide parameter ranges[2,5]. The control parameter in these experiments is the dimensionless Rayleigh number[5] R. Four different domains in R were observed.

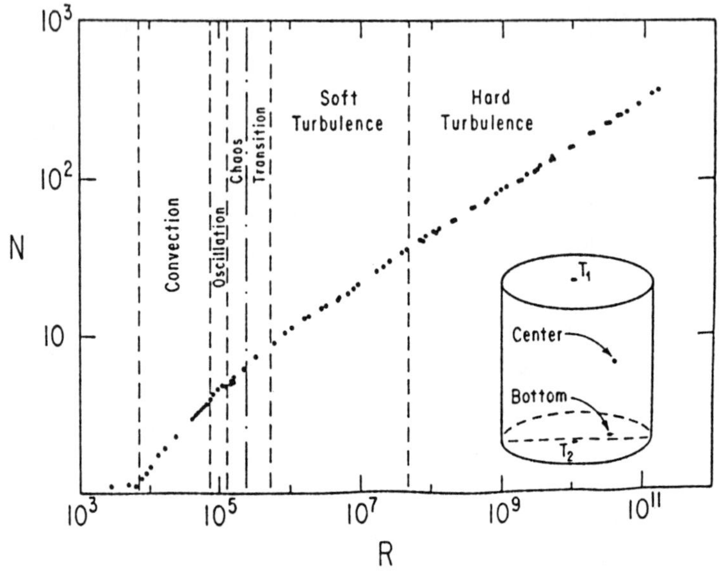

FIG. 2. *Nusselt number versus Rayleigh number measured in a helium cell, the domains for the various transitions are defined. [F. Heslot, B. Castaing, and A. Libchaber, Phys. Rev.* **A 36**, *5870 (1987).]*

In the first domain, the onset of convection, the onset of an oscillatory instability, and the onset of a chaotic state could be easily identified[5]. The correlation dimension D of the chaotic state[3] was determined[5], which was $D \approx 2$ at the onset of chaos, and increased rapidly reaching a value of $D = 4$ at $R = 2 \times 10^5$. The second domain from $R = 2.5 \times 10^5$ to $R = 5 \times 10^5$ was a transition region, where the coherence function[5] between the bolometers decreased rapidly and finally disappeared. The third domain is known as the soft turbulent regime, and the transition from soft to the so called hard turbulence occurs at $R \approx 10^8$, which seems to be universal value. The main differences between hard and soft turbulence are the following: The probability distribution function of the local temperature fluctuations in hard turbulence is exponential while for soft turbulence is Gaussian. Moreover, the power spectrum of the local temperature fluctuations is streched-exponential in soft turbulence [Eq. (1)], while the low-frequency range of the power spectra clearly exhibits a power law behavior with an exponent −7/5 in hard turbulence[2].

We have performed a detailed comparison of the two measurement in Ref. [1]. The conclusions are the following:

1. One can estimate the characteristic height L of the convecting atmospheric layer using the fitted cutoff frequency f_s of Eq. (1) and known parameters, such as the thermal diffusivity and the Prandtl number of the air. The result ($L \approx 30 - 500$ m) is in agreement with the accepted values of the thickness of the air layer influenced by the daily cycle of temperature change.

2. The atmospheric boundary layer is usually considered as a layer of infinite aspect ratio. We think that this problem might be resolved by observations, which suggest that the vertical and horizontal sizes of the medium scale ($\sim 100 - 500$ m) convective eddies are approximately equal.

3. The spectral properties and the probability distribution of the daily medium temperature fluctuations suggest that the atmospheric boundary layer exhibits a soft turbulent thermal convection.

4. As the soft turbulent state occurs after several transitions from chaotic state, connected with the increase of the number of effective degrees of freedom, it is unlikely that the typical atmospheric dynamics exhibits low dimensional chaos.

This work has been supported by the Hungarian Scientific Research Foundation (OTKA) under Grant No. 521. and 2091.

---

[1]I. M. Janosi, G. Vattay, Phys. Rev. **A 46**, 6386 (1992).

[2]X. Z. Wu, L. Kadanoff, A. Libchaber, M. Sano, Phys. Rev. Lett. **64**, 2140 (1990).

[3]P. Grassberger, I. Proccacia, Physica D **9**, 189 (1983).

[4]P. Talkner, R. Weber, W. Roser, in *Annual Report of Paul Scherrer Institut*, (Villigen, 1990), p. 78.

[5]F. Heslot, B. Castaing, A. Libchaber, Phys. Rev. **A 36**, 5870 (1987).

Stochastic chaos: An analog of quantum chaos

Mark M. Millonas
Complex Systems Group, Theoretical Division and Center for Nonlinear Studies,
MS B258 Los Alamos National Laboratory, Los Alamos, NM 87545
& Santa Fe Institute, Santa Fe, NM

Abstract

Some intriging connections between the properies of nonlinear noise driven systems and the nonlinear dynamics of a particular set of Hamilton's equation are discussed. A large class of Fokker-Planck Equations, like the Schrödinger equation, can exhibit a transition in their spectral statistics as a coupling parameter is varied. This transition is connected to the transition to non-integrability in the Hamilton's equations.

In this paper we will be concerned with diffusion processes on $\Re^n$ described by the set of coupled stochastic differential equations

$$dq^i(t) = -\partial^i\Phi(\mathbf{q})dt + \sqrt{g}dW^i(t), \qquad i = 1, ..., n, \tag{1}$$

where $\Phi(\mathbf{q})$ is a potential bounded from below, the $W^i(t)$ are uncorrelated Wiener processes, and g is a diffusion coefficient. In this case the evolution of the probability density $\rho(\mathbf{q}, t)$ on $\Re^n$, is described by the Fokker-Planck equation

$$\partial_t\rho = \frac{g}{2}\Delta\rho + \nabla\cdot(\rho\nabla\Phi). \tag{2}$$

Using the time separation ansatz $\rho(\mathbf{q}, t) = \rho(\mathbf{q})e^{-\lambda t/g}$, we can write Eq. (2) as an eigenvalue equation $\mathcal{L}\rho_\lambda(\mathbf{q}) = -\lambda\rho_\lambda(\mathbf{q})$, where $\mathcal{L} = \frac{g^2}{2}\Delta + g\nabla^2\Phi + g\nabla\Phi\cdot\nabla$. After the change of basis $\rho(\mathbf{q}) = e^{-\Phi/g}\Psi(\mathbf{q})$ we obtain

$$\mathcal{H}\Psi_\lambda(\mathbf{q}) = \lambda\Psi_\lambda(\mathbf{q}), \tag{3}$$

where $\mathcal{H} = -e^{\Phi/g}\,\mathcal{L}\,e^{-\Phi/g} = -\frac{g^2}{2}\Delta + \hat{\Phi}(\mathbf{q})$, is a Hermitian Schrödinger type operator with the transformed potential $\hat{\Phi} = \frac{1}{2}(\nabla\Phi)^2 - \frac{g}{2}\nabla^2\Phi$. The problem of solving Eq. (2) has been reduced to the problem defined by equation (3).

For small g the WKB solutions of equation (3) are given by

$$\Psi_\lambda(\mathbf{q}) = \sum_\alpha c_\alpha |\nabla S_\alpha|^{-\frac{1}{2}} \exp\left(\frac{i}{g} S_\alpha(\mathbf{q}, \lambda)\right), \tag{4}$$

where the $S_\alpha(\mathbf{q}, \lambda)$ are the solutions of the Hamilton-Jacobi equation $\frac{1}{2}(\nabla S_\alpha)^2 + \hat{\Phi} = \lambda$. The solutions of this equation are given by the integrals $S_\alpha(\mathbf{q}, \lambda) = \int^{\mathbf{q}} \mathbf{p}_\alpha \cdot d\mathbf{q}$ where the integration in along the classical trajectories of Hamilton's equations of motion

$$\dot{\mathbf{p}} = -\frac{\partial H}{\partial \mathbf{q}}, \quad \dot{\mathbf{q}} = \frac{\partial H}{\partial \mathbf{p}}, \tag{5}$$

with

$$H(\mathbf{p}, \mathbf{q}) = \frac{1}{2}\mathbf{p}^2 + \hat{\Phi}(\mathbf{q}). \tag{6}$$

The time reversal symmetry of equations (5) with Hamiltonian (6) insures that the eigenfunction (4) are real since the solutions of the Hamilton-Jacobi equation will come in pairs $\pm S_\alpha$. The dynamics of (5) determine the solution of (3) through (4), and the solutions of (2) are given by

$$\rho_\lambda(\mathbf{q}, t) = \exp\left(-\frac{\Phi + \lambda t}{g}\right) \Psi_\lambda(\mathbf{q}). \tag{7}$$

Thus the properties of the Fokker-Planck equation (2) are connected to the dynamics of the system with Hamiltonian (6) in a manner somewhat analogous to the relation of a quantum mechanical system to its classical counterpart.

One question we might ask is how the behavior of (2) is affected by the degree of chaos in the equations of motion (6). Such effects, in the quantum mechanical case ((5) affecting (3)), are often referred to as *quantum chaos*, which is usually defined as the characteristics of quantum systems whose classical analogues exhibit chaos. The statistical properties of the eigenvalues of such systems are such characteristics, and the level spacing distribution $P(S)$, giving the probability of level separation S (measured in units of the local mean spacing), provides one such statistical property. Berry & Tabor[1] have shown that nearly all quantum systems whose classical analogues are integrable will have a Poisson level spacing distribution $P(S) = \exp(-S)$, indicating the statistical independence of neighboring energy levels. On the other hand, it is now understood that the eigenvalues of systems whose classical analogues are chaotic exhibit level repulsion. That is, $P(S) \to 0$ as $S \to 0$.[2] It is expected that systems with time-reversal symmetry whose classical analogues are globally chaotic will have a Wigner level spacing distribution, $P(S) = \pi S/2 \exp(-\pi S^2/4)$, indicating a linear level repulsion as $S \to 0$.[3] Since the eigenvalues of the Fokker-Planck operator, $\mathcal{L}$, with potential Φ are the negative of the eigenvalues of a Hamiltonian, $\mathcal{H}$, with potential $\hat{\Phi}$, the spectral statistics of the Fokker-Planck equation (2) would then be expected to provide a signature of the dynamics of the equations of motion (5). To explore these ideas Millonas and Reichl[4] studied a family of two-dimensional Fokker-Planck equations with potentials

$$\Phi_\epsilon(x,y) = 2x^4 + \frac{3}{5}y^4 + \epsilon xy(x-y)^2. \tag{8}$$

The system needed to be at least two-dimensional in order to observe chaos in equations (5). When $\epsilon = 0$ the system is completely integrable, since it decouples into two one-dimensional systems. They observed the transition (as ϵ is varied) in the level spacing statistics of the Fokker-Planck operator as the dynamics of equations (5) changes from completely integrable ($\epsilon = 0$) to almost globally chaotic ($\epsilon = 0.14$). *Stochastic chaos* can then be defined, at least for the case of diffusion in a time-independent potential, as *the properties of stochastic systems described by Eq. (2) when the equations of motion (5) exhibit chaos.* In particular, given a family of potentials $\hat{\Phi}_\epsilon$ where the dynamics of (5) varies from globally integrable to globally chaotic as ϵ is increased, we would expect the spectral spacing distribution of the λ's to exhibit a corresponding transition from Poisson to Wigner level spacing statistics.

An entirely separate problem is the question of the direct physical relevance of the dynamics of (5) to the underlying microscopic dynamics as described by (1). One thing is clear: chaos in (5) is emphatically *not* related to chaos in the dynamics generated by (1) with $g = 0$, that is $\dot{\mathbf{q}} = -\boldsymbol{\nabla}\Phi(\mathbf{q})$. When there is no noise the individual trajectories just follow the gradient of the potential along the route of steepest descent stopping at any local minimum in Φ, so what would normally be considered the underlying microscopic dynamics is trivial, and never chaotic. Thus, there is no simple physical relationship between the dynamics of (5) and the dynamics of (1). A deeper analysis shows that eqs. (5) are the imaginary-time equations of motion for the most probable, or optimal trajectories. These ideas can be extended to the case where there is no detailed balance, but in that case no meaningful analytic continuation of the most probable trajectories is possible. There are still however the optimal trajectories which obey a set of

Hamilton's equations with Hamiltonian $\mathcal{H} = 1/2(\mathbf{p} - \mathbf{A})^2 + 1/2\nabla \cdot \mathbf{A}$, where $\mathbf{A}$ is the nongradient force field which replaces $-\partial^i\Phi(\mathbf{q})$ in Eq. 1 in the case where there is no detailed balance. The long-time behavior of systems with or without detailed balance can be calculated in the low-noise asymptotic limit from a knowledge of the optimal trajectories with energy $\lambda = 0$. These are the instanton trajectories which line on the unstable manifold of the Hamiltonian system. This manifold is smooth as a consequence of the center manifold theorem, so chaos will not play a role. However this manifold may have singular projections onto the configuration space in the case where detailed balance is broken, resulting in a rich nonlinear behavior of nonequilibrium stationary states.[5] The most surprising result presented here is than even in the case where there is detailed balance, and the underlying dynamics is completly integrable, chaos will play a role in determining the time-dependent properties of such systems. It appears that there is a deep analogy between quantum dynamics and stochastic dynamics through their relationship to the properties of these conservative dynamical systems. This connection, once made, opens up the study of stochastic processes to a whole range of new tools and concepts.

References

[1] M.V. Berry and M. Tabor. Proc. R. Soc. Lond. A. **356**, 375 (1977).

[2] M.V. Berry, in *The Wave-Particle Dualism*, S. Diner et. al. Eds. (Reidel, 1984); T. H. Seligman and H. Nishioka (Eds.), *Proceedings of the International Conference of Cuernavaca* (1986), Lecture Notes in Physics **263** (Springer-Verlag, 1986); M. C. Gutzwiller, *Chaos in Classical and Quantum Mechanics* (Springer-Verlag, 1990); L. E. Reichl, *The Transition to Chaos in Conservative Classical Systems: Quantum Manifestations* (Springer-Verlag, 1992).

[3] E. P. Wigner, SIAM Rev. **9**, 1 (1967).

[4] M. M. Millonas and L.E. Reichl, *Phys. Rev. Lett.* **68**, 3125 (1992).

[5] V. Smelianskiy and M. M. Millonas, & M. Dykman, V. Smelianskiy and M. M. Millonas (to appear); also R. Maier and D. Stein (to appear).

CHAOS DEATH DUE TO NOISE-INDUCED SWITCHING

V. B. Ryabov
Institute of Radio Astronomy
4 Krasnoznamennaya St., 310002 Kharkov, Ukraine

ABSTRACT

The general results pertaining to chaotic and regular motions in a class of systems termed quasilinear oscillators that are described by weakly nonlinear Duffing-type equations with multifrequency external forces are presented. The case of primary resonance is studied in detail, and the external signal was chosen to have the form of two-frequency oscillation. Various bifurcation phenomena are detected and traced, including homoclinic bifurcations, period doubling sequences, intermittency, attractor crises, hysteretic effects, etc.

INTRODUCTION

A wealth of complicated features have been recently discovered in simple, but rather universal, mathematical models, like nonlinear oscillators with harmonic and quasiperiodic forcing. The possibility of coexisting of several attractors, both regular and chaotic, in the phase space of these systems makes especially attractive their investigation, because of feasibility to provide the new type of switching devices exploiting the multistability properties of nonlinear elements.

MULTISTABILITY AND MANIFOLDS

The following equation has been proven to be the adequate mathematical model of many physical systems

$$\frac{d^2x}{dt^2} + \omega_0^2 x = \varepsilon\left(F_1\cos(\omega_1 t) + F_2\cos(\omega_2 t) - 2\delta\frac{dx}{dt} - \gamma x^3\right) \tag{1}$$

The behavior of its solutions under single frequency excitation can be chaotic at rather large values of external force magnitude. However, it has been shown recently[1-3] that two-frequency excitation of such system may result in considerable lowering of chaos onset. Following the well-known scheme of averaging procedure we put $x = U\cos(\omega_1 t) + V\sin(\omega_1 t)$ and obtain the system of averaged equations that permits to investigate in detail the solutions in the vicinity of different resonances. The case of principal resonance is investigated in the present report. Here we have in the limit $\varepsilon << 1$

$$\frac{dU}{d\tau} = -\delta U + \left[\Delta + \beta\left(U^2 + V^2\right)\right] + P_2\sin(\Omega\tau) \tag{2a}$$

$$\frac{dV}{d\tau} = -\delta V - \left[\Delta + \beta\left(U^2 + V^2\right)\right] + P_1 + P_2 \cos(\Omega\tau) \tag{2b}$$

where

$$\Omega = (\omega_2 - \omega_1)/\varepsilon;\ \ \Delta = (\omega_0 - \omega_1)/\varepsilon;\ \ \tau = \varepsilon t;\ \ \beta = 3\gamma/(8\omega_1);\ \ P_1 = F_1/(2\omega_1);\ \ P_2 = F_2/(2\omega_1)$$

At $P_2 = 0$ (one frequency excitation) in the certain region of parameters Δ and P_1 in the phase space exist three equilibrium positions - two foci and a saddle (herefrom we shall call them primary). When P_2 becomes non vanishing the situation depicted in Fig.1 takes place. Here, in the area restricted by Curve 1 two attractors appear in the vicinity of the focus corresponding to the lower branch of response curve[1]. Alongside with them the attractor corresponding to the upper branch of the response curve exists. Chaotic attractors arise independently in the vicinity of the foci through period doubling bifurcations (Curves 4,3 and 2,5) and may coexist. Such situation occurs, for example, at $\delta=1$, $\beta=-1.54$, $\Delta=15.05$, $\mu=-2.8$, $P_1=3.75$, $\Omega=8.4$, and the cross-section of corresponding attractors (two chaotic - A2, A3, and periodic - A1) is shown in Fig.2.

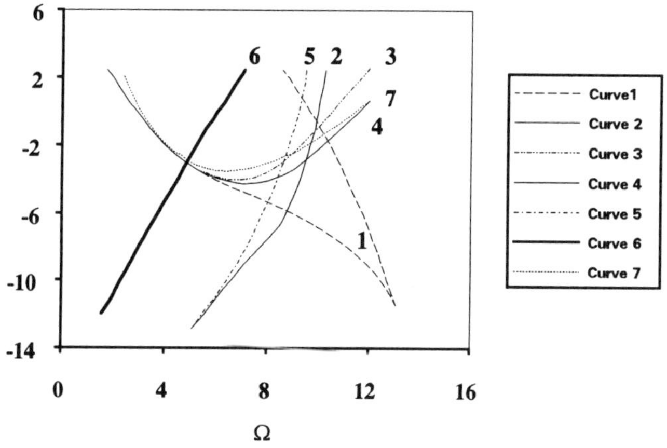

Fig.1. Phase diagram for system (2). Curve 1-boundary of multistability area; 2,4-first period-doubling; 5,3-second period-doubling; 6-boundary of bistability area; 7-homoclinic bifurcation (Melnikov criterion, eq.(3)). $\mu=20\lg(P_2 / P_1)$.

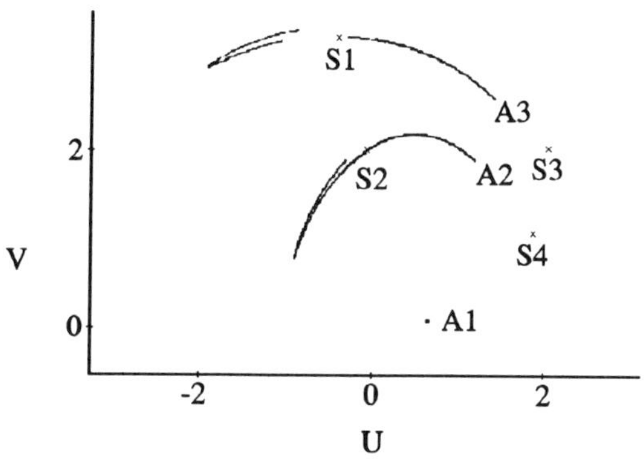

Fig.2. Poincare cross-section for system (2). A1-A3 - coexisting attractors, S1-S4 - period-one saddle points.

It was recently shown[2,3] that chaos onset in this system can be derived analytically by means of Melnikov's method through detecting the parameter values at which the manifolds of the perturbed primary saddle point intersect transversely (curve 7 in Fig.1).

However, it should be noted that chaos occurs at much less values of P_2 than

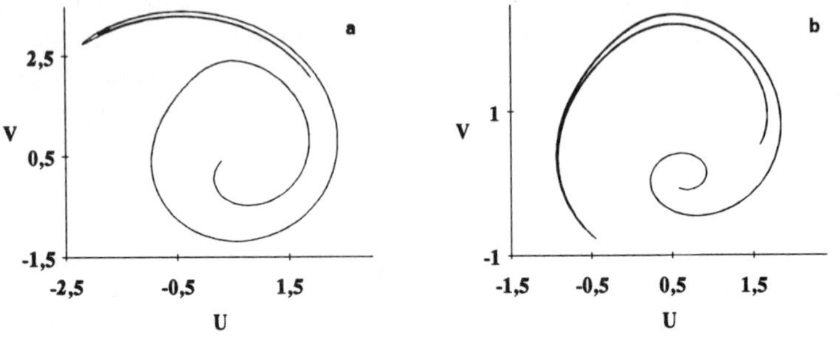

Fig.3. Unstable manifolds of periodic unstable orbits S1(a) and S2(b).

can be expected from the Melnikov criterion. The corresponding saddle orbit intersects the plane of Fig.2 at the point S3, and it is clear that the homoclinic structure associated with it is not attracting. The saddle points S1 and S2 correspond to the period one orbits that lost their stability at Curves 2 and 4 of Fig.1, i.e., at period doublings. So, these new period-one saddle points play the principal role in dynamics of the system.

Moreover, the strange attractors A2 and A3 exist in the very narrow band in Ω-μ plane, adjacent to period-doublings. Both chaotic attractors undergo the crisis, caused by the extremely complicated structure of the phase space. The numerical experiments indicate that the unstable manifolds of periodic orbits S1 and S2 coinciding with the two chaotic attractors shown in Fig.2 demonstrate the sudden

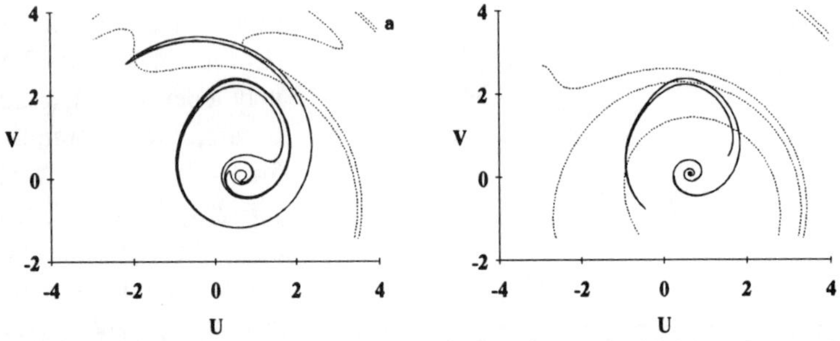

Fig.4. Stable (dotted line) and unstable (solid line) manifolds of unstable orbits S3(a) and S4(b)

transition and at Ω=8.25 and look like those depicted in Fig.3. In order to illustrate the structure of the phase space, the stable and unstable manifolds of saddle orbits S3 and S4 are also given in Fig.4. The fine fractal structure of basin

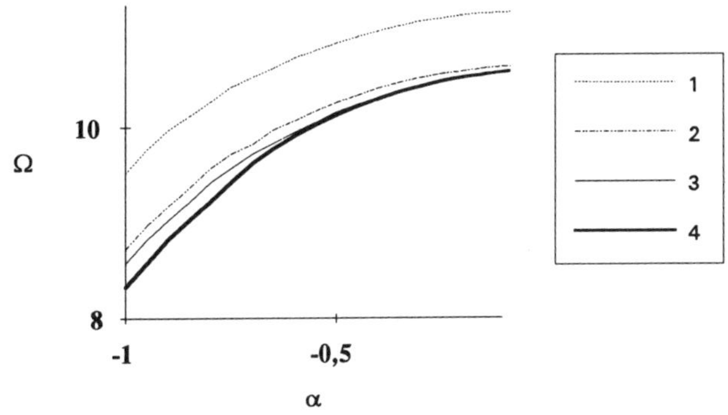

Fig.5. Phase plane of system (2) at the same parameter values as Fig.1. Curve 1-first period doubling, 2-second period doubling, 3-strange attractor appearance, 4-chaos collapse.

boundaries of different attractors makes the system in hand extremely sensitive to the influence of noise and can completely deteriorate the stability of strange attractors. It becomes evident by decreasing the dissipation parameter δ, when both strange attractors are being destroyed. This situation is depicted in Fig.5 where the solid heavy line denotes the collapse of chaotic oscillations. Below this line only periodic attractor exists.

CONCLUSION

It has been demonstrated that the new saddle orbits, arising at non zero values of external force, can play a crucial role in the dynamics of a weakly nonlinear oscillator. The intersection of their manifolds may lead to the formation of strange attractors at considerably lower levels of excitation than can be expected on the basis of Melnikov criterion. It has been shown that the decrease in dissipation may lead to collapse of strange attractors, making the system extremely sensitive to the influense of noise, when the very small fluctuation may cause the switching from chaotic ro regular oscillations.

ACKNOWLEDGMENT

The author is very much indebted to Prof. D.M.Vavriv and Dr. A.B.Belogortsev for useful discussions and to Dr. S.A.Sharapov for help in numerical experiments.

REFERENCES

1. A.B.Belogortsev, D.M.Vavriv and O.A.Tretyakov, Sov.Phys. JETP, 65, 737, 1987
2. K.Yagasaki, Physica D44, 445, 1990.
3. V.B.Ryabov, and D.M.Vavriv, Phys.Lett. A153, 431, 1991.

CHAOS AND STABILITY OF MICROWAVE CIRCUITS

D. M. Vavriv
Institute of Radio Astronomy, 4 Krasnoznamennaya St.,
310002 Kharkov, UKRAINE

ABSTRACT

The results of theoretical and experimental studies are summarized of the dynamical chaos effect on the microwave circuits field behavior. We discovered the possible existence of numerous chaotic states in weakly nonlinear single- and coupled-mode systems under quasilinear excitation conditions being dominant for many practical cases. It is shown that chaos may produce strong influence on the circuit dynamics, their stability and output noise level. These results are illustrated by a stability analysis of the parametric amplifiers and superconducting quantum interferometers (SQUIDs). The conditions and mechanisms of chaos arising, the properties of excited oscillations are discussed.

INTRODUCTION

The study of dynamical chaos was carried out for a long period without separate consideration of the weakly nonlinear systems. It was apparently so due to an opinion that the chaotic behavior was principally nonlinear phenomenon demanding the strong nonlinearity to be arised. For the most practical microwave devices and circuits this condition fails to occur and, therefore, the chaos onset is supposed to be not typical for this case. However, the recent results [1-3] have shown that, in fact, chaotic oscillations can arise under quasilinear excitation condition as well and, thus, can define the stability, output noise level, sensitivity of the devices to a great extent.

We have found that the weakly nonlinear physical systems are characterized by specific conditions and scenario of the chaotic motion onset. As it turned out, the chaotic instabilities can appear if, firstly, a n-dimensional torus with $n \geq 2$ exists in phase space of a physical system and, secondly, if nonisochronizm of oscillations goes over some critical value. Only under these conditions chaotic oscillations can arise in the weakly nonlinear limit. The transition to chaos is always conditioned by destruction of tori. Various analytical approaches such as the averaging method, the Melnikov's method, and the current Lyapunov exponents technique [4,5] were applied in order to find the analytical conditions of tori bifurcations and chaotic states arising.

The developed theory was applied to the problem of stability of microwave circuits and devices. We have revised the main classical results concerning field behavior in single- and two-mode oscillators [5], amplifiers [6], parametric oscillators and amplifiers [1,7]. We have found that most of these devices exhibit chaotic oscillations in some regions of control parameters. The theoretical studies were carried out simultaneously with the experiments on investigation of the above-mentioned devices in different wavelength bands (from meter to millimeter). The qualitative and quantitative relationship between theoretical and experimental results have been revealed. The present paper describes the results of the investigation of two types of devices: parametric amplifiers and SQUIDs.

PARAMETRIC AMPLIFIERS

A variety of microwave and optical devices fall into a class of parametric amplifiers wherein a signal amplification is due to a time modulation of some nonlinear reactance. These amplifiers, as it follows from the conventional theory of parametric devices, provide low noise amplification because they consist of an entirely reactive circuit and they utilize an ac power supply. At the same time, there are a lot of experimental evidences that parametric amplifiers possess low stability and their output noise level is in some cases much more higher compared with the theoretical expectations. We have found that the chaotic instabilities may give rise to such behavior. It is of great interest that exactly the reactive nature of the device and the application of the pumping oscillation are the main factors responsible for the unstable oparation of the amplifiers.

The dynamics of oscillations of parametric amplifiers is adeduately described within a scope of some universal mathematical models. For single-mode amplifiers the equations used can be reduced to the following system of averaged equations for the slowly varying amplitude a and phase φ

$$\begin{aligned} da/dt &= -\alpha_0 + m\ sin(2\varphi) + P\ sin(\varphi - \Omega t) \\ a(d\varphi/dt) &= \Delta + \beta a^2 + m\ cos(2\varphi) - (P/a)\ cos(\varphi - \Omega t) \end{aligned} \tag{1}$$

where α_0 is the damping coefficient, β is the nonlinearity parameter (nonisochronizm of oscillation), $\Delta = \omega - \omega_0$, $\Omega = \omega_s - \omega$, ω_0 is the natural frequency of the resonant circuit, P and ω_s are the amplitude and frequency of the signal wave, m and 2ω are the amplitude and frequency of pumping. This system permits to study the bifurcations of the two-dimensional tori of the the original physical system via studying bifurcations of the periodic orbit of the averaged equations. It is to be noted that when the external signal vanishes ($P = 0$) the system (1) exhibits only regular behavior.

In order to find the analytical conditions of chaos arising the Melnikov method was used. To apply this method to the system (2) we made use of the fact that at $\alpha_0 = P = 0$ the closed separatrix loop exists in the phase space of the system [7]. The results consist in the following condition defining the region in control parameters space where chaos may arise

$$\frac{P\sqrt{\beta}}{\alpha_0} \geq \frac{2\sqrt{m}}{\pi\Omega} \left| \frac{[\sqrt{m^2-\Delta^2} - \Delta arccos(\Delta/m)] ch(\pi\Omega/2\sqrt{m^2-\Delta^2})}{exp(\Omega arccos(\Delta/m)/2\sqrt{m^2-\Delta^2})} \right| \tag{2}$$

Proceeding from this condition and from the results of numerical and experimental studies we may conclude:
1.The chaotic oscilation excitation in the weakly nonlinear mode of amplifier operation results solely from the pumping oscillation interaction with the signal to be amplified.
2.The instability threshold with respect to the signal amplitude P_{th} is much more lower compared with other types of amplifiers (say, transistor ones).
3.The increase in both the nonisochronizm of oscillations and the quality factor of the resonant circuit leads to a lowering of the stochastic instability threshold.
4.The regions in control parameters space with chaotic behavior occur intermittently with the regions of the regular behavior.
5.The characteristic feature consists of the fact that several (more than two) different attractors can exist simultaneously in the phase space of the such system, and hence, this system is multistable one. We observed up to five attractors (regular and chaotic) existed simultaneously.

SUPERCONDUCTING QUANTUM INTERFEROMETERS

We examined the one-contact SQUID which consists of the high-quality circuit inductively coupled with a superconducting ring [8]. The circuit is excited by an external HF pumping oscillator. Such SQUIDs attract constant attention due to a possibility of achieving sensitivity approaching the Planck limit. However, the expected high-sensitivity level has not been reached in practice. We have shown that the stochastic instability arising in the weakly nonlinear limit can set a limit on the sensitivity of SQUIDs [10].

The adequate description of the SQUID dynamics appears to be possible within the framework of the following averaging equations for the amplitude a and phase γ in the resonant circuit [8]

$$
\begin{aligned}
da/d\tau &= -a - b\ sin\ \gamma \\
a(d\gamma/d\tau) &= -\Delta a + RJ_1(a)\ cos\ \phi_e - b\ cos\ \gamma
\end{aligned}
\tag{3}
$$

Here the following dimensionless parameters are used, b is pumping amplitude, $\Delta = (\omega - \omega_0)2Q/\omega$ is the pumping frequency shift, $\tau = t\omega/2Q$, $R = 2k^2QL \geq 1$, k^2 is the coupling coefficient of the tuned circuit with the superconducting ring, $Q \gg 1$, is the Q-factor of the resonant circuit, L is the ring inductance, $J_1(a)$ is the first-order Bessel function.

In a contrast to solutions of the system (3) considered earlier we have taken into account the slow time variation of the external magnetic flux ϕ_e, which looks like [9]

$$\phi_e = \phi_0 + \phi_1\ cos\ \Omega\tau \tag{4}$$

where ϕ_0 and ϕ_1 are the constant and alternating flux component, Ω is the dimensionless frequency.

The results of investigations indicate that the chaotic instability manifests itself even in the case of adiabatic, equilibrium and nonhysteristic processes in superconducting interferometer closed by a weak link. Such a mode of SQUID operation was being considered as the most promising one for the achievement of the magnetometer high-sensitivity levels. There are two main causes of the chaotic oscillations arising. The first one is the finite variation speed of magnetic flux (signal measured or bias flux). The chaotic instability appears if $\Omega \geq 1$. The second cause is that the superconducting interferometer induces the reactive nonlinearity in a SQUID circuit. The chaos arising has the following threshold by the nonlinearity parameter R and the amplitude of pumping b: $R_{th} = 4$; $b_{th} = 1.25$. The results obtained allow to assume that limited SQUID sensitivity observed in some experiments with $R \approx 10$ may be explained by the appearance of chaotic instability. The analysis has shown that avoiding of chaos and retaining the high sensitivity level requires the adequate value of R over the interval $1 \leq R \leq 4$ to be chosen.

CONCLUSION

The theoretical and experimental results support the viewpoint that dynamical chaos can significantly influence the stability and output noise level of microwave devices. The two main factors are responsible for the initiation of the chaotic instabilities: firstly, the reactive nonlinearity of the majority of real microwave devices and, secondly, the interaction of several independent frequencies excited in a circuit. The principal result is that the real spectrum of signals must be taken into account when analysing the circuits stability. This is due to the fact that systems exhibiting only regular behavior under the harmonic excitation can acquire a lot of chaotic states when the external force turns, for example, to a quasiperiodic one.

REFERENCES

1. I.Yu.Chernyshov, and D.M.Vavriv, Phys.Lett. A165, 117 (1992).
2. D.M.Vavriv, Proc. 11th Int. Conf. on Noise in Physical Systems and 1/f Fluctuations (Kioto, Japan, 1991). p.377.
3. V.B.Ryabov, and D.M.Vavriv, Phys.Lett. A153, 431 (1991).
4. D.M.Vavriv,and V.B.Ryabov, Zh. Vychislit. Mat. and Mat. Phys. 32, 1409 (1992).
5. A.B.Belogortsev, M.Poliashenko, O.A.Tretyakov, and D.M.Vavriv, Electronics Lett. 26, 1354 (1990).
6. D.M.Vavriv, and O.A.Tretyakov, Theory of Resonant Amplifiers with Distributed Interaction of O-type (Naukova Dumka Press. Kiev. 1989).
7. D.M.Vavriv, V.B.Ryabov, and I.Yu.Chernyshov, Zh.Tech.Fiz. 61, 1 (1991).
8. V.V.Danilov and K.K.Likharev, Zh. Tekh. Fiz. 32, 1110 (1975).
9. S.A.Bulgakov, V.B.Ryabov, V.I.Shnyrkov, and D.M.Vavriv, J.Low.Temp.Phys. 83, 241 (1991).

XIV. MISCELLANEOUS 1/*f* FLUCTUATIONS

NATURE OF THE BULK 1/f NOISE IN GaAs AND Si

N.V.D'yakonova, M.E.Levinshtein, S.L.Rumyantsev
A.F.Ioffe Institute, St.Petersburg
194021 Russia

ABSTRACT

It is demonstrated, that the bulk 1/f noise in semiconductors can appear as a result of fluctuations of the occupancy of levels formed density-of-states tail near the band edge. It is shown that in pure Si with value of Hooge parameter $\alpha=10^{-4}-10^{-6}$ and n-GaAs with $\alpha=10^{-4}-10^{-5}$ the nature of 1/f bulk noise is determined by this mechanism. Phenomenological model of the bulk 1/f noise in semiconductors is developed. This model explained well all main experimental data. The model predicts some new physical effects. All these effects have been observed experimentally.

INTRODUCTION

The 1/f noise (flicker noise) is observed for enormous number of a great variety of objects and physical systems. For the majority of the investigated objects the nature of the 1/f noise is not known even qualitatively. However the last decade has seen considerable progress in studies of the 1/f noise in semiconductors.The most physical processes used to account for the 1/f noise are related to the presence of various imperfections in the crystal lattice, such as impurities, structure defects, stresses, etc. Direct experimental evidence has recently been published to show that the level of the 1/f noise is related to the number of defects in metals and semiconductors[1-3].

In this report we present the results of 1/f noise investigations in Si and GaAs with Hooge parameter $\alpha\lesssim10^{-4}$ (this values are typical for good quality material used in modern semiconductor electronics). The model of bulk 1/f noise in semiconductors is proposed. This model predicts the existence of several new physical effects, connected with the same mechanism which provides the volume 1/f noise in semiconductors. These effects are: a nonmonotonic dependence of the 1/f noise on the intensity of illumination, a new mechanism of a persistent photoconductivity, an increase of the 1/f noise on deterioration of the structural quality of a material and retention of the mechanism of formation of the 1/f noise. All these effects have been observed experimentally[4].

INFLUENCE OF THE BAND-TO-BAND LIGHT ON 1/f NOISE IN GaAs[5]

A key experiment allowed to explain the nature of 1/f noise in GaAs looks rather simple. In Fig.1 the frequency dependencies of the relative spectral density of noise S_I/I^2 are shown . At room temperature (Fig.1,b) S_I/I^2 dependence (solid curve) has a form of 1/f noise. Hooge parameter $\alpha=10^{-4}$. Illumination with light of intensity reducing the resistivity by just 0.1-1% reduced considerably the noise at low frequencies and increased significantly the noise at relatively high frequencies. At low temperature (Fig.1,a) illumination increases the noise level in all frequency range. At high temperature (Fig.1,c) illumination does not affect the noise. This effect was observed for GaAs with donor concentration $N_d=10^{15}\ cm^{-3}$ [5] and $N_d=10^{17}\ cm^{-3}$ [6].

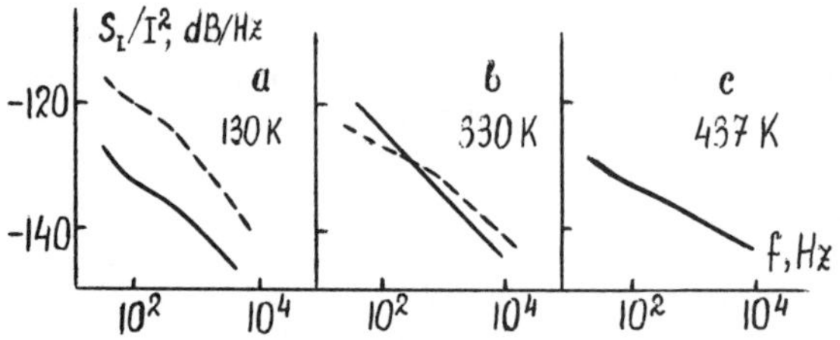

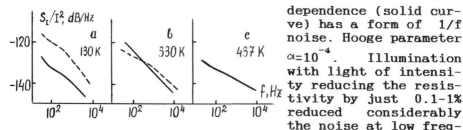

Fig.1. Frequency dependencies of S_I/I^2 at three temperatures in darkness (continuous curves) and during illumination (dashed curves).

Such kind of illumination effect on noise level one could easily explain for generation-recombination (GR) noise of local level E_t. It is well known that the noise value S_I for GR noise is determined by the electron occupancy F of the local level: $S_I \sim F^2(1-F)$. When the Fermi level lies well below E_t then this level is practically empty in darkness. Therefore, the appearance of holes as a result of illumination has no effect on the occupancy F of the level E_t. In this case, illumination cannot influence the noise level (Fig.1,c). If the Fermi level lies well above E_t this deep level is practically completely filled with electrons ($(1-F) \ll 1$) and the noise is weak. The holes created by illumination are captured by the level E_t. The F value is decreased. The noise should therefore grow. If the Fermi level lies somewhat higher E_t and the noise is close to its maximum (this occur if the occupancy is $F=2/3$), the appearance of holes alters the occupancy and can weaken the noise under illumination.

The feasibility of providing such a simple qualitative explanation allows to suggest that 1/f noise observed in

GaAs is a superposition of generation-recombination processes due to the presence of closely spaced levels or a region with a continuous spectrum of levels[7]. All experimental data can be explained if one supposes that this region is density-of-states tail in the band gap.

MODEL OF THE BULK 1/f NOISE IN SEMICONDUCTORS [8]

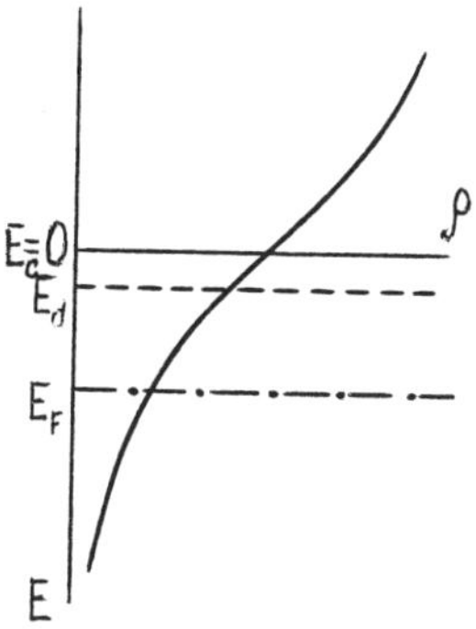

Fig.2.Density-of-states tail.

Figure 2 shows a qualitative dependence of the density of states $\rho(E)$ describing the situation under consideration. A shallow donor level E_d is ionized completely at the measurement temperature, so that the free-carrier density is $n_o=N_d$, the Fermi level E_F lies below E_d.

In accordance with the model, which postulates multiphonon capture, the time constant $\tau(E)$ representing relaxation of the occupancy of the levels is:

$$\tau(E)=\tau_{oo}\exp(E/E_1)F(E), \tag{1}$$

where E- energy, F- occupancy of a level with energy E, τ_{oo}- is the capture time constant of the levels at E=0.

The density-of-states tail falls exponentially with depth in the band gap:

$$\rho(E)=\rho(0)\exp(-E/E_o). \tag{2}$$

Then, the relative spectral density of fluctuations of the carrier density is:

$$\frac{S_n}{n_o^2}=\frac{4N_o}{VN_d^2}\int_0^\infty\frac{F^2(1-F)\tau_{oo}\exp(E/E_1-E/E_o)dE}{1+\omega^2(\tau_{oo}\exp(E/E_1))^2F^2\quad E_o}, \tag{3}$$

where N_o is the total density of levels in the tail.

If $E_1 \ll E_o$ and $E_1 \ll kT$ the frequency dependence of the spectral density of the noise is of the 1/f form. The Hooge parameter α is:

$$\alpha=N_F/N_d, \tag{4}$$

where N_F - is the total number of states in the tail below Fermi level.

NONMONOTONIC DEPENDENCE OF THE 1/f NOISE ON THE ILLUMINATION INTENSITY (GaAs)[9]

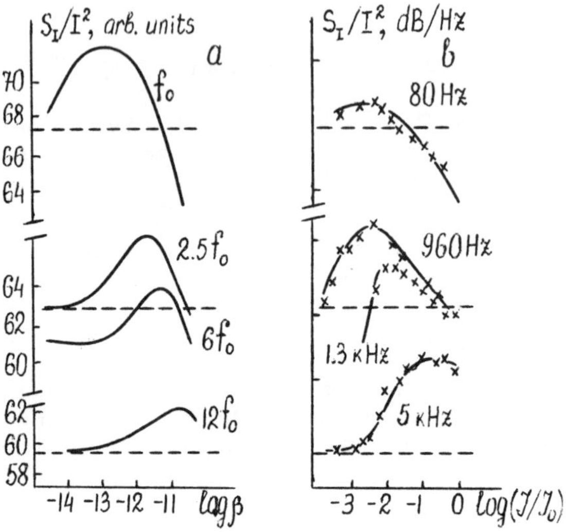

Fig.3. Theoretical (a) and experimental (b) dependencies of the S_I/I^2 on the intensity of illumination

In Fig.3a the calculated according the model dependencies of S_I/I^2 on intensity of illumination are shown (the value of β is proportional to intensity of illumination). One can see that for any frequency f these dependencies are nonmonotonic. The more is f the higher is illumination level providing a maximal noise level. In Fig.3b the experimental curves are shown. Good agreement between the results of calculations based on this model with the reported experimental data is in our opinion a serious evidence in support of the proposed model.

NEW MECHANISM OF LONG-TERM RELAXATION OF THE PHOTOCONDUCTIVITY[10]

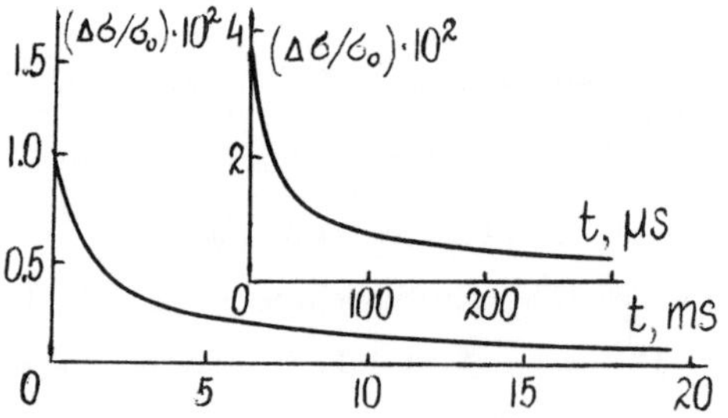

Fig.4. Kinetics of the photoconductivity decay excited with a GaAs laser (inset) and with light from an incandescent lamp.

It easy to see that from the proposed 1/f noise mechanism follows a new mechanism of long-term photoconductivity. This mechanism is determined by the capture of excess nonequilibrium electrons on the levels of the tail. Because there are an extremely wide range of the capture time τ, the model predicts the existence of conductivity relaxation with a very wide range of the time constants.

In Fig.4 the dependence of photoconductivity versus time is shown for GaAs at 300K. One can see a wide range of time constants from $\tau=10^{-7}$s up to $\tau=10^{-1}$s.

It was shown experimentally that this type of photoconductivity is practically insensitive to the electric field F and temperature T. It allows to distinguish this mechanism from two well known persistent conductivity mechanisms: barrier mechanism [11] and recombination mechanism via trapping levels[12].

THE CONNECTION BETWEEN 1/f NOISE VALUE AND STRUCTURAL QUALITY OF THE MATERIAL[13]

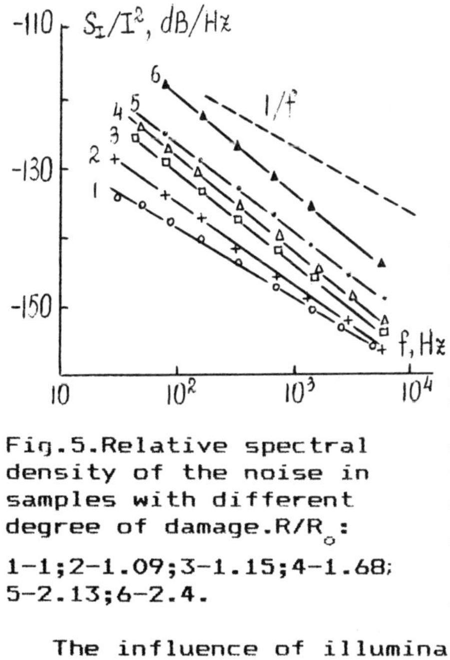

Fig.5.Relative spectral density of the noise in samples with different degree of damage.R/R_o: 1-1;2-1.09;3-1.15;4-1.68; 5-2.13;6-2.4.

In Fig.5 the frequency dependencies of S_I/I^2 are shown at different levels of GaAs damages. Structural defects were induced in GaAs by uniaxial stress. Irreversible changes of the sample resistance R/R_o were used as a measure of the crystal damage of the structure.

It should be pointed out that no changes in the mobility measured either at 77K or at 300K was observed even in samples with R/R_o=2.4 (curve 6, Fig.5). At the same time the 1/f level have been increased for R/R_o=2.4 in 100 times (at f=10^2Hz). So one can see that 1/f noise is very sensitive to the destructive damage.

The influence of illumination and temperature on the noise spectrum in samples with induced defects was similar to that in initial samples with high quality crystal structure [14]. The results obtained show that in GaAs with and without induced structural defects the origin of 1/f noise is the same.

1/f NOISE IN Si[15]

The 1/f noise has been investigated in n- type Si grown by the floating zone method and in similar, in respect of the electrical and noise parameters, Si samples prepared by the neutron-transmutation doping method (N_d=(1-2)$10^{13}cm^{-3}$). It was found that the surface 1/f noise predominated at room temperature over the bulk noise in the case of sample with $\alpha=2 \cdot 10^{-4}$. However, cooling changes this situation. In the temperature range 100K-150K the bulk 1/f noise predominated over the surface contribution. In Si samples at 100K-150K we have observed all effects which we revealed in GaAs: nonmonotonic dependence of the 1/f noise on the illumination intensity [16], slow relaxation of the photoconductivity [17], increase of the noise level as a result of deterioration of the structural quality, which does not alters the mechanism of the 1/f noise [16].

REFERENCES

1 J.Pelz, J.Clarke, W.E.King, Phys.Rev.,B38,10371(1988).
2 L.K.J.Vandamme, S.Oosterhoff,J.Appl.Phys.,59,3169(1986).
3.R.H.M.Clevers, J.Appl.Phys.,62,1877(1987).
4.N.V.D'yakonova, M.E.Levinshtein, S.L.Rumyantsev, Sov.Phys, Semicond., 25,1241(1991).
5.N.V.D'yakonova, M.E.Levinshtein, Sov.Tech.Phys.Lett.,14, 857(1988).
6.M.E.Levinshtein, S.L.Rumyantsev, Sov.Tech.Phys.Lett.,19, (1993), in print.
7.M.Skowronski, J.Lagowski, M.Milshtein, C.H.Kang, F.P.Dabkowski, A.Hennel, H.C.Gatos,J.Appl.Phys.,62, 3791(1987).
8.N.V.D'yakonova, M.E.Levinshtein, Sov.Phys.Semicond., 23,175(1989).
9.N.V.D'yakonova, Sov.Phys.Semicond.,25,219(1991).
10.N.V.D'yakonova, M.E.Levinshtein, S.L.Rumyantsev, Sov.Phys.Semicond.,23,1132(1989).
11.M.K.Sheinkman, A.Ya.Shik, Sov.Phys.Semicond., 10,128(1976).
12.S.M.Ryvkin (Photoelectric Effects in Semiconductors) Consultans Bureau, New York (1964).
13.M.E.Levinshtein, S.L.Rumyantsev, Sov.Phys.Semicond.,24, 1125(1990).
14.N.V.D'yakonova,M.E.Levinshtein, S.L.Rumyantsev, Sov.Phys.Semicond.,25,217(1991).
15.E.G.Guk,N.V.D'yakonova,M.E.Levinshtein, Sov.Phys. Semicond.,22,707(1988).
16.E.G.Guk, N.V.D'yakonova, M.E.Levinshtein, S.L.Rumyantsev,Sov.Phys.Semicond.,24,513,(1990).
17.N.V.D'yakonova, M.E.Levinshtein, S.L.Rumyantsev, Sov. Phys.Semicond.,24,956(1990).

ANALYSIS OF THE BURGERS EQUATION

K.Anton, R.Tetzlaff and D.Wolf

Institut für Angewandte Physik
der Johann Wolfgang Goethe-Universität
D-6000 Frankfurt am Main 11, Robert-Mayer-Straße 2–4,FRG

ABSTRACT

A nonlinear partial differential equation has been approximated by a set of difference equations and analysed in detail. Several solutions have been determined in simulation experiments for different excitations and parameter values. For special parameter configurations we found a power spectra close to $\frac{1}{f}$–noise.

INTRODUCTION

In a previous paper [4] we have studied the homogenous solution of the Burgers equation in detail. The inhomogenous solution showed $\frac{1}{f^m}$–spectra with $m > 1.4$ by taking white Gaussian noise (WGN) for the excitation force and different values of the parameter R.
In this paper the dynamical behaviour of the solution is analysed for various excitation forces. Therefore we have implemented the simulation system on a new computer system in order to obtain a higher frequency resolution.

SIMULATION AND RESULTS

As outlined in [1,2], the density $n(x,t)$ of moving objects is determined by the Burgers equation

$$\frac{\partial n(x,t)}{\partial t} - \frac{1}{R}\frac{\partial^2 n(x,t)}{\partial x^2} + n(x,t)\frac{\partial n(x,t)}{\partial x} = \mathcal{F}(x,t) \quad . \tag{1}$$

Since solutions of the inhomogenous Burgers equation can only be found numerically, we have approximated Eq.(1) by the set of difference equations

$$\frac{(n_{i,j+1} - n_{i,j})}{\Delta t} - \frac{(n_{i+1,j} - 2n_{i,j} + n_{i-1,j})}{\Delta x^2} + \frac{n_{i,j}(n_{i+1,j} - n_{i-1,j})}{2\Delta x} = \mathcal{F}_k(i,j) \tag{2}$$

$$i = 0,1,\ldots,k \quad ; \quad j = 0,1,\ldots \quad ,$$

and implemented on our computer system.
For three different excitations $\mathcal{F}_k(i,j)$, which are given in Tab. 1, the solution $n_{i,j}$ and its power spectra were determined numerically.

Table 1 : Excitation forces $\mathcal{F}_k(i,j)$

k	$\mathcal{F}_k(i,j)$	
1	A · sin(i ·Δ x + j ·f_0 · Δ t)	, j=1,2,3 ...
2	WGN	, j=1,2,3 ...
	with variance $\sigma = .1, .001, .0001$ and mean $\mu = 0$	
3	WGN	, j=1,250,500 ...
	with variance $\sigma = .001$ and mean $\mu = 0$	

In all these cases we measured $\frac{1}{f^m}$–spectra with $m \leq 2$. Figs. 1 and 2 show typical results for $\mathcal{F}_1(i,j)$. In this case the density $n_{i,j}$ is a nearly periodical function with small amplitude fluctuations. We obtain a power spectrum with slope $m = -2$ over a wide range of frequency for all taken parameter configurations. The measured power spectrum in Fig. 2 shows a distinct peak at the frequency $f_0 = 5.5$ kHz.

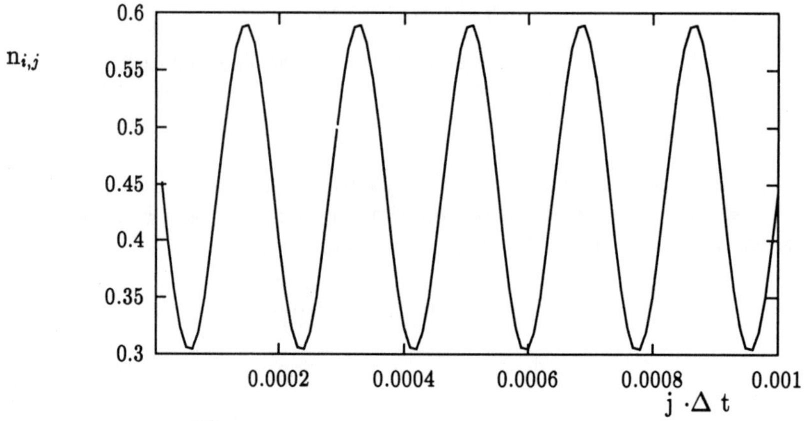

Figure 1 : Time signal for $\mathcal{F}_1(i,j)$ at $x = 0$

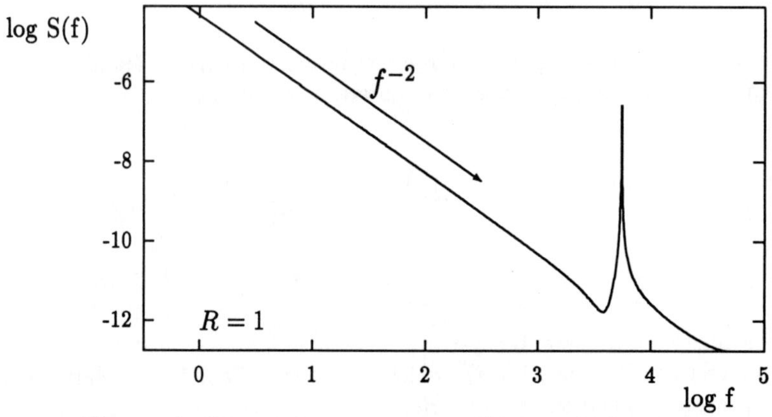

Figure 2 : Measured power spectrum S(f) for $\mathcal{F}_1(i,j)$ at $x = 0$

For the stochastic excitation $\mathcal{F}_2(i,j)$ as illustrated by Fig. 3 power spectra with slopes between $m = -1.23$ and $m = -1.41$ over 6 decades of frequency were found. From Fig. 4 can be seen that the slope m of the power spectrum $S(f)$ increases for decreasing variance σ of WGN.

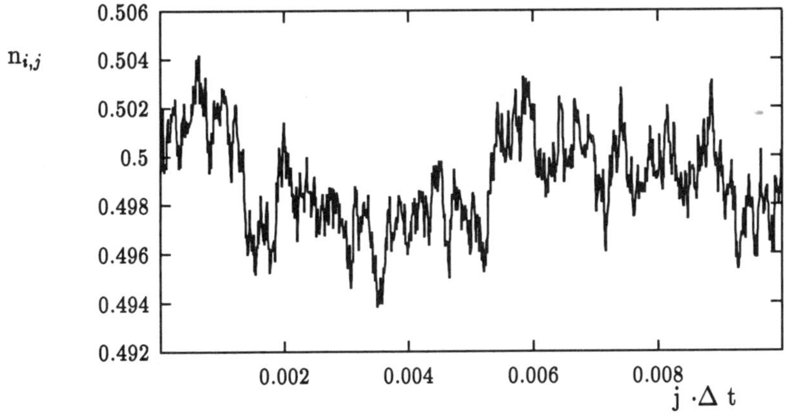

Figure 3 : Time signal for $\mathcal{F}_2(i,j)$ with $\sigma = .001$ at $x = 0$

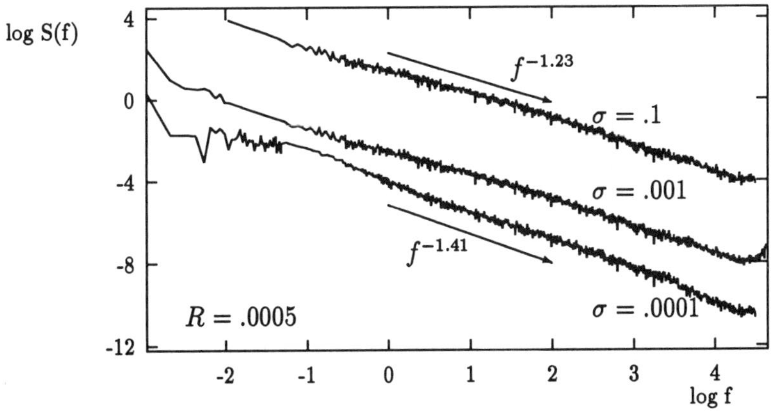

Figure 4 : Measured power spectra S(f) for $\mathcal{F}_2(i,j)$ at $x = 0$ for different variances σ of WGN

Finally with $\mathcal{F}_3(i,j)$, a time signal as shown in Fig.5, we obtained a power spectrum with a slope smaller then $m = -1.2$ over 4 decades of frequency as shown in Fig. 6

We can state that for certain excitation forces and decreasing parameter value R the inhomogenous solution of the Burgers equation represents fluctuations with power spectra close to $\frac{1}{f}$–type.

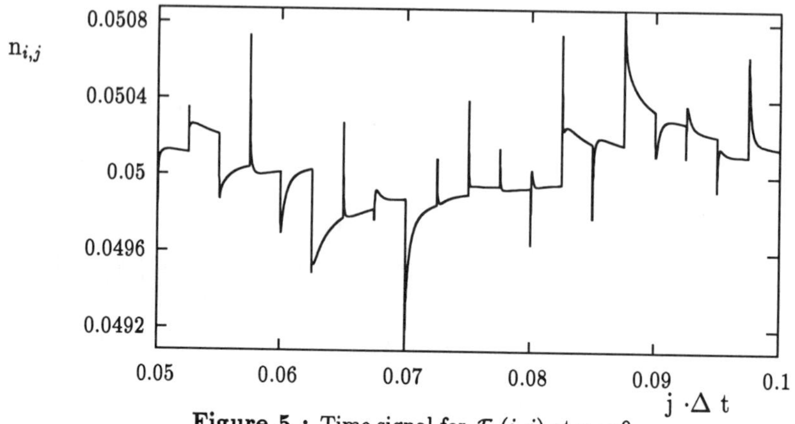

Figure 5 : Time signal for $\mathcal{F}_3(i,j)$ at $x = 0$

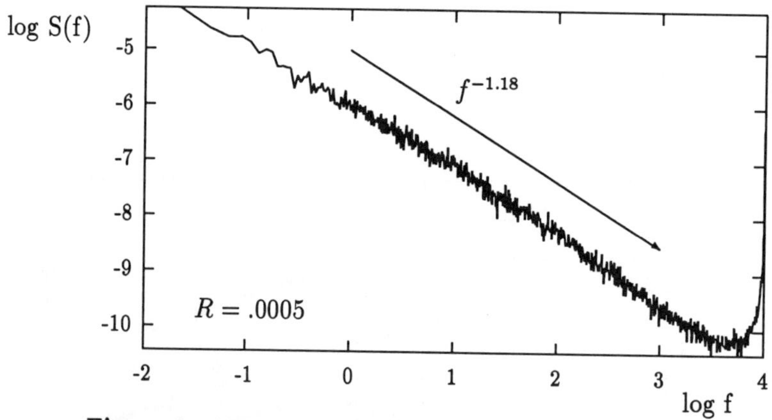

Figure 6 : Measured power spectrum S(f) for $\mathcal{F}_3(i,j)$ at $x = 0$

REFERENCES

1. T. Musha, H. Higuchi: The $\frac{1}{f}$ fluctuation of a traffic current on an expressway; Jap. J. Appl. Physics 15, pp. 1271–1275 (1976).

2. T. Musha, H. Higuchi: Traffic current fluctuation and the Burgers equation; Jap. J. Appl. Physics 17, pp. 811–816 (1978).

3. K. Anton,B. Weber, D. Wolf: Untersuchungen an nichtlinearen Differenzengleichungen als Mechanismen zur Erzeugung von $\frac{1}{f}$–Fluktuationen; 7. Aachener Symposium für Signaltheorie, Proceedings, pp. 142-147 (1990).

4. K. Anton,B. Weber,D. Wolf: Simulation of $\frac{1}{f}$–fluctuation; Proceedings of ICNF '91 Kyoto, pp. 579-582 (1991).

5. E. Hopf: The Partial Differential Equation $u_t + uu_x = \mu u_{xx}$; Comm. Pure Appl. Math. 3, pp.201-230 (1950).

$1/f$–NOISE IN THIN ALUMINIUM FILMS DAMAGED BY ELECTROMIGRATION

K. Dagge, J. Briggmann, A. Seeger, H. Stoll
Max–Planck–Institut für Metallforschung, Institut für Physik,
D-70569 Stuttgart, Germany

C. Reuter
Institut für Mikroelektronik Stuttgart, D-70569 Stuttgart, Germany

ABSTRACT

Electromigration damage in thin aluminium films is detected by the AC–measurement of $1/f$–noise. Since this technique avoids defect production by electromigration during the measurement it allows to study the effect of the electromigration pretreatment on the noise. A special phase–sensitive correlation method suppresses background noise and thermal noise. It is found that noise spectra remain nearly constant during electromigration damage until in a late stage of electromigration a sudden increase of noise magnitude takes place. Reversal of the damaging current leads to the partial recovery of this effect. The increase in noise is attributed to the creation of highly mobile defects.

INTRODUCTION

Electromigration damage is the main source of failure in metallic interconnections of electronic devices. It is caused by the interaction of charge carriers and the atoms of the metal film, leading to the creation of defects which eventually form cracks across the film[1]. The standard methods to study this process, e.g. electron microscopy and life–time tests, provide only limited information on the microscopic processes. Furthermore they are time consuming because they require statistical analysis, and lead to the destruction of the films. Therefore the development of new methods is highly desirable.

As low–frequency noise in metal films is correlated to the motion of defects in the crystal lattice[2,3,4], it can be used to examine the damage resulted by electromigration. So far noise measurements were performed while the electromigration process was going on. This was done by using a DC noise–measurement technique where the current testing the resistance fluctuations was at the same time used to damage the film[2,5,6]. Noise spectra observed during electromigration follow a $1/f^{\gamma}$–law with γ reaching from 1 to 2 in dependence of the applied stress (current, temperature). The $1/f^2$–component of the noise originates from the atom flow in the grain boundaries. The thermally activated motion of the defects caused by electromigration contributes to the $1/f$–component of the noise. Since both noise sources cannot be seperated experimentally in a DC noise–measurement set–up, the data is difficult to interpret especially in respect to microscopic processes. Time– and temperature–dependent experiments give rough estimates on activation energies of the electromigration process. The present work employs an AC technique that does not induce electromigration damage during noise measurements. Thus only the thermally activated motion of defects contributes to resistance fluctuations and information on the defects can be deduced from noise spectra.

EXPERIMENTAL METHOD

An alternating current applying a special phase–sensitive correlation technique[7] (Fig. 1) is used to detect the resistance fluctuations of metal films damaged by electromigration. The metal films ($2 \times 100 \times 0.8 \mu m$) consist of aluminium doped

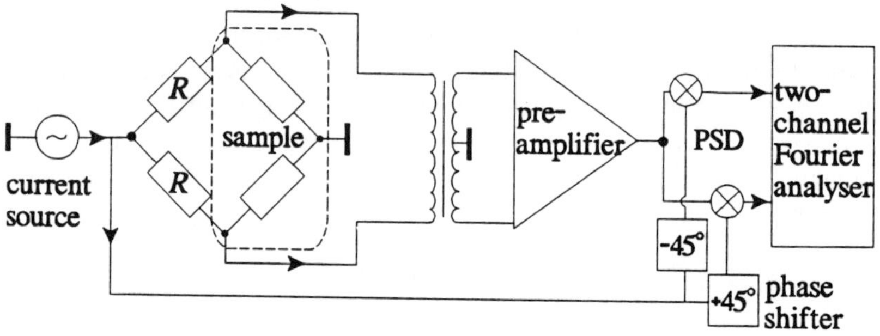

Figure 1: Set–up for noise measurement with a phase–sensitve AC correlation technique.

with 1% Si and 0.5% Cu on a silicon wafer and are structured in a five terminal geometry. Half of the measuring bridge is made up of the sample, the other half consists of two resistors (R) used to balance the bridge. The voltage across the sample is coupled to a transformer (Triad G4) and a low–noise preamplifier (Ithako 1201) and is then detected under $+45°$ and $-45°$ with respect to the phase of the alternating current with two phase sensitive detectors (PSD). By calculating the cross–correlation of these two channels in a FFT analyser, background noise and thermal noise are averaged out, so that only noise caused by resistance fluctuations is left over. The electromigration damage is induced by a direct current ($j = 2 \cdot 10^{10}$ A/m^2) and interrupted for the noise measurements. All experiments were performed at 30°C.

EXPERIMENTAL RESULTS

All measured noise spectra follow a $1/f^{\gamma}$–law with γ between 0.85 and 1.15. This suggests that the obseved spectra are caused by thermally activated resistance fluctuations and that they may be interpreted in terms of the model of

Dutta, Dimon, and Horn[8]. While resistance increases nearly continously during electromigration damage, noise remains almost constant in the first stages of electromigration damage (Fig. 2). At a late stage of electromigration (2% resistance increase), after typically about $1.2 \cdot 10^5$ s, a sudden increase in noise power by a factor of 2 to 5 occurs, while the resistance still increases continously (Fig. 2). After such an increase, noise again remains constant close up to the failure of the film.

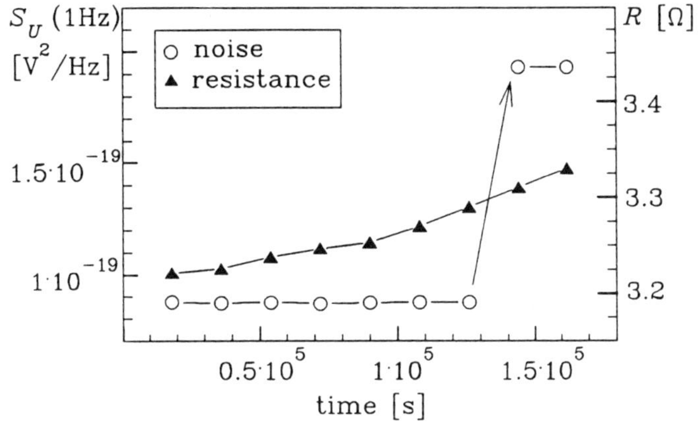

Figure 2: Typical development of the spectral density of noise power S_U at 1 Hz (open circles) and resistance (full triangles) under electromigration induced by a current density $j = 2 \cdot 10^{10}$ A/m^2 at 30°C.

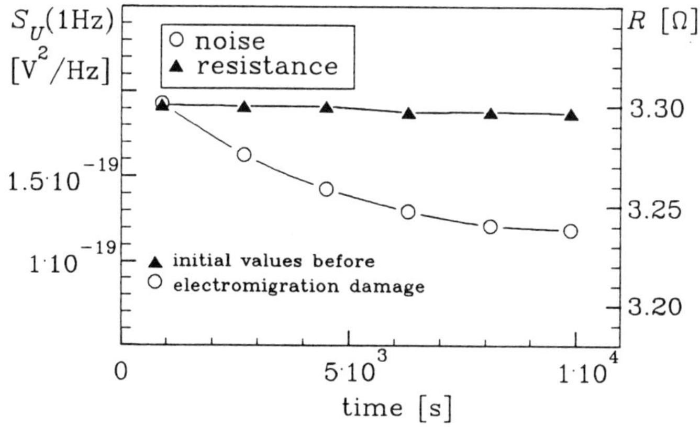

Figure 3: Recovery behaviour of the spectral density of noise power S_U at 1 Hz (open circles) and resistance (full triangles) upon reversal of current direction at a current density $j = -2 \cdot 10^{10}$ A/m^2 at 30°C, the initial values of resistance and noise of the undamaged sample are shown for comparison.

In an additional experiment the direction of the damaging current was reversed after the drastic increase in noise described above had taken place. In intervals of about $1.8 \cdot 10^3$ s the influence of the reversed current on noise and resistance was observed in order to get information on the recovery behaviour. As it is to be seen in Figure 3 the resistance remains nearly unchanged, whereas the increase in noise is partially recovered. Noise magnitude is reduced during $9 \cdot 10^3$ s of current reversal to approximately a factor of 1.2 of its original value before electromigration. The process is close to an exponential decay process with a rate constant of about 0.02 s^{-1}.

CONCLUSIONS

The presented experimental results show that $1/f$–noise is affected by electromigration damage. In all stages of electromigration noise follows a $1/f^\gamma$–law with γ close to 1, which may be explained by thermally activated resistance fluctuations within the framework of the Dutta, Dimon, and Horn theory[8]. The observed sudden increase in noise power is to be attributed to the creation of a highly mobile defect type which causes noise by thermally activated motion. Resistance seems to be insensitive against the creation of this defect type, which indicates that the defect observed in noise is a minority defect.

Since recovery of the noise increase upon reversal of current direction is observed, one may assume that the noisy defects are of a simple, recoverable type. The experimental results obtained so far seem to support their identification with dislocation loops. Further clues on the identification of the noisy defect may be found by the temperature dependence of the noise spectra and their analysis in terms of the distribution of activation energies according to Dutta, Dimon, and Horn[8].

It looks promising to attempt the statistical analysis of the onset of noise in order to test whether it is correlated with the lifetime of the metal film. If such a correlation can be found, it will provide us with a simple test for the reliability of metal films.

REFERENCES

1. I. A. Blech, J. Appl. Phys. 47, 1203 (1979).
2. R. H. Koch, J. R. Lloyd, and J. Cronin, Phys. Rev. Lett. 55, 2487, (1985).
3. J. Pelz and J. Clarke, Phys. Rev. Lett. 55, 738, (1985).
4. J. Briggmann, Dissertation Universität Stuttgart 1993.
5. T. M. Chen, P. Fang, J. G. Cottle, Noise in Physical Systems (ed. A. Ambrózy, Budapest, 1990), p. 515.
6. A. Touboul, F. Verdier, Y. Herve, Noise in Physical Systems (ed. T. Musha, Sato and M. Yamamoto, Kyoto, 1991), p. 73.
7. A. H. Verbruggen, H. Stoll, K. Heek, and R. H. Koch, Appl. Phys. A48, 233, (1989).
8. P. Dutta, P. Dimon, and P. M. Horn, Phys. Rev. Lett. 43, 646, (1979).

1/f NOISE IN LOW-TEMPERATURE-IRRADIATED ALUMINIUM FILMS

J. Briggmann, K. Dagge, W. Frank, A. Seeger, H. Stoll
Max-Planck-Institut für Metallforschung, Institut für Physik
Postfach 800 665, D-70506 Stuttgart, Germany

A. H. Verbruggen
Technische Universiteit Delft, Delfts Instituut voor Micro-Electronica en Submicron-technology (DIMES)
P. O. Box 5053, NL-2600 GB Delft, The Netherlands

ABSTRACT

1/f noise and electrical resistivity of thin polycrystalline aluminium films were measured before and after low-temperature electron irradiation. Due to irradiation with 3.7×10^{23} e^-/m^2 (1 MeV) the noise measured at 10 K increased by a factor of 6 whereas the electrical resistivity increased only by 25 %. Subsequent isochronal annealing at progressively higher temperatures caused the electrical resistivity and the *1/f* noise to recover. The observed recovery stages in the *1/f* noise measured at 10 K occurred at the same temperatures as the well-known recovery stages of the electrical resistivity of pure bulk aluminium irradiated by electrons at low temperatures. Noise measurements performed at 40 K, however, showed an additional recovery stage at an annealing temperature of about 70 K which has no equivalent in the electrical resistivity.

INTRODUCTION

Lattice defects may contribute to the *1/f* noise in metals [1-5]. Little, however, is known about the *nature* of the defects whose motion causes the noise. The fundamental descriptions of fluctuations in the conductivity of thin metal films due to the motion of lattice defects [6], the universal-conductance-fluctuation model and the local-interference model, are rather unspecific in this respect. The same may be said of the phenomenological model of Dutta, Dimon, and Horn [7], which explains the approximate *1/f* frequency dependence and the temperature variation of the noise in terms of thermally activated resistivity fluctuations with a broad distribution of activation energies. Low-temperature electron irradiation of thin metal films [5] is a powerful tool to study the contribution of specific defect types to the *1/f* noise, since only elementary intrinsic defects, viz. vacancies and self-interstitials, are introduced. The present paper reports on a study of Al films irradiated by 3.7×10^{23} e^-/m^2 (1 MeV) at temperatures of about 10 K.

EXPERIMENTAL SET-UP

The 100 nm thick samples were obtained by electron-beam evaporation of aluminium (99.99 %) on silicon substrates covered with Si_3N_4. The room-temperature evaporation under high-vacuum conditions yields grain sizes comparable with the sample thickness. The films were structured by a conventional lift-off procedure in order to obtain a five-terminal geometry with a total sample length of 300 μm and a sample width of 5 μm (see box 'sample pattern' in Fig. 1). The residual-resistivity ratio of the films is about 5 and is dominated by the limitation of the electron free path by the sample thickness and the grain size.

Noise measurements were performed by a special phase-sensitive ac correlation technique[8] (Fig. 1), which allows the measurement of the 1/*f*-noise spectra by suppressing the always present thermal-noise background. The sample is part of a Wheatstone bridge and is biased by an alternate current at a frequency of 2.56 kHz and a rms current density of 4×10^{10} A/m^2. After pre-amplification the bridge error signal is detected by two phase-sensitive detectors (PSD) properly set to a phase angle of ±45° in respect to the modulating current (see box 'phase setting' in Fig. 1). Signals which are correlated to the bridge input current (i.e. resistivity fluctuations of the sample) are added up due to their fixed phase shift and appear in the real part of the cross power spectrum. Uncorrelated signals with a random phase shift are averaged out.

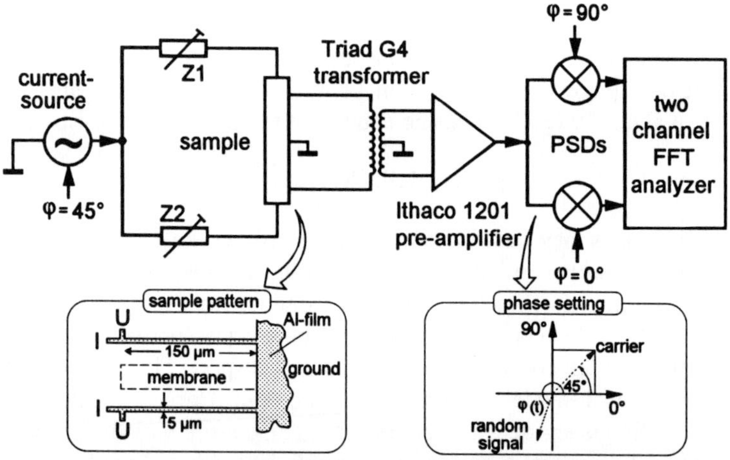

Fig. 1. Simplified schematic diagram of the noise measurement circuit. The two boxes show the used sample patterning and the correct phase setting of the two PSDs in respect to the bridge input circuit.

The samples were irradiated at 10 K with 1 MeV electrons in the high-voltage electron microscope (AEI-EM7) of the Max-Planck-Institut für Metallforschung. Small Si_3N_4 membranes, etched from the back into the silicon substrates (see box 'sample pattern' in Fig. 1), allowed us to exactly position the samples in the electron beam (250 μm diameter, 2.5×10^{19} e$^-$/m^{2}s). After irradiation the cold samples were transported by means of a special cryostatic system into a shielded cabin, where the annealing treatments and the noise measurements were carried out.

EXPERIMENTAL RESULTS

Due to irradiation with 3.7×10^{23} e^-/m^2 the $1/f$ noise of the samples measured at 10 K increased by about a factor of 6 whereas the electrical resistivity increased only by 25 %. Isochronal annealing (6×10^2 s) of the samples at progressively higher temperatures caused the electrical resistivity and the $1/f$ noise to recover. As a function of the annealing temperature the recovery of the $1/f$ noise measured at 10 K showed stages (Fig. 2) which occurred at the same temperatures as the well-known recovery stages of the residual resistivity of pure bulk aluminium electron-irradiated at low temperatures [9, 10]. After annealing above the Stage-III temperature (in Al at about 200 K) both the $1/f$ noise and the electrical resistivity returned to their values before irradiation.

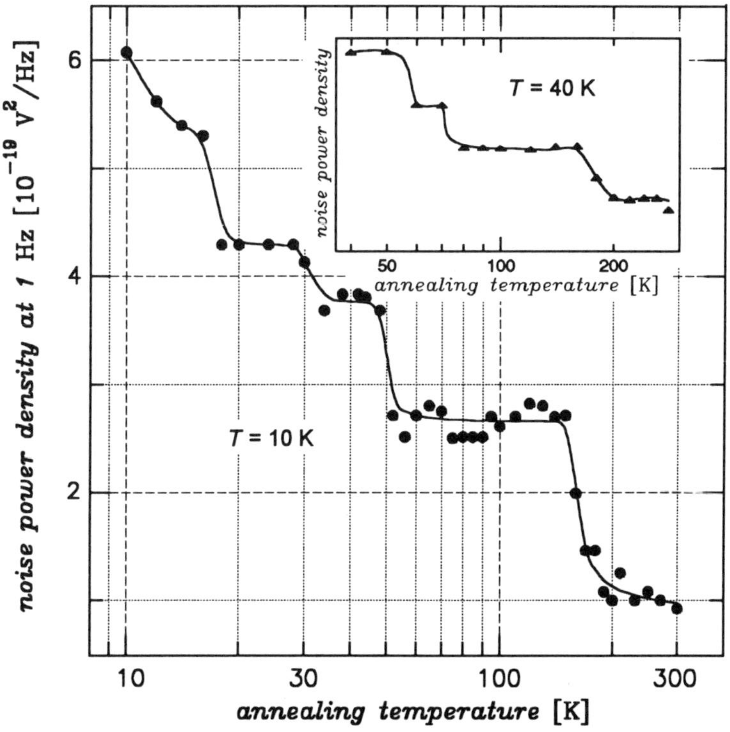

Fig. 2. Recovery of the irradiation-induced $1/f$ noise of thin aluminium films measured at 10 K (inset: measured at 40 K) due to isochronal annealing at progressively higher temperatures.

Noise measurements performed at 40 K showed an additional recovery stage in the $1/f$ noise at an annealing temperature of about 70 K (see inset in Fig. 2). This stage has no counterpart in the electrical resistivity and is presumably analogous to that found by Pelz and Clarke in 90 K noise measurements on copper films after electron irradiation at 90 K[5].

DISCUSSION AND CONCLUSIONS

Noise measurements on irradiated metal films can be used to study defects generated by particle irradiation. In the present case (low-temperature irradiation of Al films by 1 MeV electrons) the $1/f$ noise spectra may be explained, according to the Dutta–Dimon–Horn model[7], in terms of a broad distribution of activation enthalpies that control the motion of irradiation-induced defects interacting with other defects. As demonstrated by the existence of a recovery stage in the $1/f$ noise measured at 40 K that does not show up in the annealing curve of the residual electrical resistivity, noise measurements can give information on irradiation-induced minority defects that is hardly obtainable by other techniques.

Our results on electron-irradiated Al films strongly support the so-called two-interstitial model (TIM)[11, 12]. According to the TIM, the static crowdions (metastable self-interstitials extending along close-packed directions), which have been introduced by the 10 K electron irradiation, start to move via thermal activation at the end of recovery stage I (in Al at about 45 K). Owing to their migration in one dimension along close-packed directions, they may be confined by immobile defects on or close to their migration lines (for example, by dumbbell interstitials). In this way the crowdions may migrate to and fro between obstacles without annealing out, and therefore can contribute significantly to the $1/f$ noise subsequently measured above Stage-I temperatures. At higher temperatures the crowdions may be converted into the stable dumbbell configuration. Since conversion does not change the number of defects in the sample but reduces the number of mobile defects, this accounts for the observation of a recovery stage in the $1/f$ noise that is not present in the electrical resistivity.

REFERENCES

1. J. W. Eberhard and P. M. Horn, Phys. Rev. B18, 6681 (1978)
2. D. M. Fleetwood and N. Giordano, Phys. Rev. B28, 3625 (1983)
3. D. M. Fleetwood and N. Giordano, Phys. Rev. B31, 1157 (1985)
4. J. H. Scofield, J. V. Mantese, and W. W. Webb, Phys. Rev. B32, 736 (1985)
5. J. Pelz and J. Clarke, Phys. Rev. Lett. 55, 738 (1985)
6. N. Giordano, Rev. Solid State Sci. 3, 27 (1989)
7. P. Dutta, P. Dimon, and P. M. Horn, Phys. Rev. Lett. 43, 646 (1979)
8. A. H. Verbruggen, H. Stoll, K. Heek, and R. Koch, Appl. Phys. A48, 233 (1989)
9. K. R. Garr and A. Sosin, Phys. Rev. 162, 669 (1967)
10. R. L. Chaplin and H. M. Simpson, Phys. Rev. 163, 587 (1967)
11. W. Frank and A. Seeger, Crystal Lattice Defects 5, 141 (1974)
12. A. Seeger, H. Stoll, and W. Frank, Material Science Forum 15–18, 111 (1987)

RELATIONSHIP OF AM TO PM NOISE IN SELECTED MICROWAVE AMPLIFIERS, RF OSCILLATORS, AND MICROWAVE OSCILLATORS*

E.S. Ferre, L.M. Nelson , F.G. Ascarrunz, and F. L. Walls
Time and Frequency Division
National Institute of Standards and Technology
325 Broadway, Boulder, CO 80303

ABSTRACT

A study of AM and PM noise in microwave amplifiers, rf oscillators, and microwave oscillators was done to provide a basis for developing models for the origin of AM noise. We find that the 1/f and f^0 components of PM and AM noise in amplifiers and oscillators are often of equal magnitude, and that they probably have common sources, namely, a 1/f modulation of the gain element and thermal noise.

INTRODUCTION

To our knowledge there has been little quantitative work analyzing the relationship between amplitude modulation (AM) and phase modulation (PM) noise in amplifiers and oscillators. Most of the literature assumes that the AM noise is much smaller than the PM noise. We have therefore undertaken a detailed study of PM and AM noise in several microwave amplifiers, rf oscillators, and X-band oscillators to provide a basis for developing models for the origin of AM noise and its relationship, if any, to the PM noise.

To measure the noise in these devices we generally used two channel measurement systems with cross correlation spectrum analysis to reduce the noise contribution of the measurement process.[1] Based on this data we then suggest general noise models for these devices that include both AM and PM noise. The noise model for amplifiers is, to our knowledge, the first to show that the PM and AM noise added by an amplifier originates from two common sources. Specifically we show that for most of the amplifiers tested the 1/f (flicker) noise, which has always been assumed to be PM noise,[2] also produces AM noise of approximately equal amplitude. Thermal noise, the second source, also produces an equal amount of AM and PM noise. Our model for oscillators is also the first one to include both PM and AM noise. As in the amplifier case, we find that there is often 1/f and f^0 components of AM and PM noise of nearly equal amplitude. The AM noise and PM noise generally differ only at frequencies close to the carrier, where the PM noise varies as $1/f^3$ (flicker FM noise). In general the PM noise roughly follows the Leeson model.[3]

PM AND AM NOISE IN X-BAND AMPLIFIERS

A two channel measurement system with cross correlation spectrum analysis was used to measure the PM and AM noise of the amplifiers. The details of the measurement process have been discussed previously.[4] The results for PM noise in amplifier A are shown in Figure 1. This amplifier shows a 1/f behavior from approximately 1 Hz to 1 MHz from the carrier. Around 10 MHz from the carrier the PM noise is basically white at -173 dBc/Hz. The PM noise in this amplifier follows very closely Parker's PM noise model for linear amplifiers,[5] in which the spectral density of phase fluctuations, $S_\phi(f)$, is given by the expression

$$S_\phi(f) = \alpha_E \frac{1}{f} + \frac{2kTFG(f)}{P_o}, \qquad (1)$$

where α_E is the flicker noise coefficient, k is Boltzmann's constant, T is the temperature in kelvins, F is the noise figure of the amplifier, G(f) is the gain of the amplifier, and P_o is the output power of the carrier signal from the amplifier.

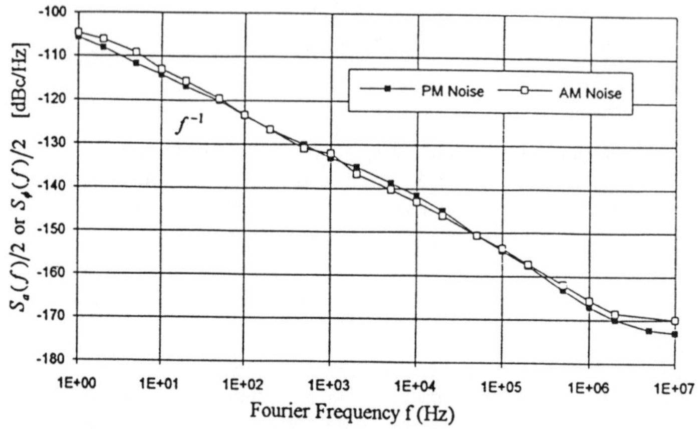

Figure 1. Single sideband AM noise $[S_a(f)/2]$ and PM noise $[S_\phi(f)/2]$ in amplifier A at 10.6 GHz.

The AM noise added by amplifier A, also shown in Figure 1, shows the same 1/f and white noise components as the PM noise. This suggests that the PM and AM noise of an amplifier originate from two common sources added to the signal. The 1/f component scales proportionally to the signal and the thermal component is independent of the signal. The AM noise for a linear amplifier is thus similar to the PM noise model given by Equation (1), that is,

$$S_a(f) = S_\phi(f) = \alpha_E \frac{1}{f} + \frac{2kTFG(f)}{P_o}. \tag{2}$$

The noise added by the rest of the amplifiers tested showed a similar behavior. Excess noise above the intrinsic flicker component, possibly caused by noise from the bias supplies in the amplifier, was observed in some samples.[4]

AM AND PM NOISE IN OSCILLATORS

PM and AM noise investigations in 5 MHz, 100 MHz and 10.6 GHz oscillators were made using two channel and three-corner-hat measurement techniques previously described.[4,6] Results from PM noise measurements in a 5 MHz oscillator, Figure 2, show three different power-law noise processes: flicker FM ($1/f^3$), flicker PM (1/f), and white PM (f^0). In general the PM noise follows the Leeson/Parker[3,5] PM noise model for oscillators given by

$$S_\phi(f) = \alpha_R \nu_o^4 (\frac{1}{f^3}) + \alpha_E f_\tau^2 (\frac{1}{f^3}) + \alpha_E (\frac{1}{f}) + \frac{2kTG(f)}{P_o}, \tag{3}$$

where the first term, characterized by α_R and the carrier frequency ν_o, is due to phase fluctuations in the resonator, the second and third terms are from the amplifier and the fourth term is the thermal noise of the amplifier.[5] The second term is usually negligible in quartz oscillators but may dominate in X-band oscillators. In some cases, depending on the noise and bandwidth of the resonator, there can be additional terms.[4,5]

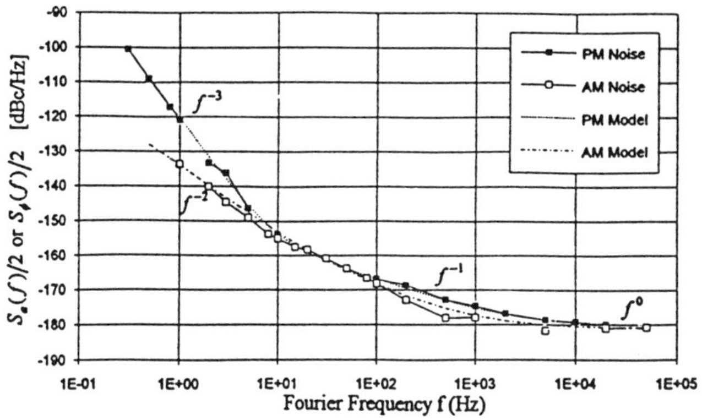

Figure 2. Single sideband AM noise $[S_a(f)/2]$ and PM noise $[S_\phi(f)/2]$ of a 5 MHz oscillator.

The AM noise in the 5 MHz oscillator, also shown in Figure 2, closely follows the PM noise from 10 Hz to 50 kHz. Thus, the flicker PM and the white noise processes seen in the PM noise are also part of the AM noise. At frequencies below 10 Hz, the AM noise exhibits a $1/f^2$ component not present in the PM noise. We believe this is caused by amplitude fluctuations in the amplitude control loop of the oscillator. As mentioned earlier the $1/f^3$ process observed in the PM noise is due to phase fluctuations in the resonator and thus is not expected in the AM noise. The AM noise of this oscillator can then be described by

$$S_a(f) = \beta(\frac{1}{f^2}) + \alpha_E(\frac{1}{f}) + \frac{2kTG(f)}{P_o}. \tag{4}$$

Figure 3 illustrates the PM and AM noise measured in a 100 MHz oscillator. The same noise processes observed in the PM noise of the 5 MHz oscillator (flicker FM, flicker PM and white PM) are present in this oscillator. In this specific case, the $1/f^3$ noise process was present in the 1 Hz - 500 Hz frequency range, while the 1/f was only observed from ≈ 1 kHz - 3 kHz. The AM noise is somewhat different since it only shows flicker PM and white PM noise. Although the AM noise is 1/f from 1 to 300 Hz, the level appears about 10 dB higher than we would predict based on the AM and PM noise at 3 kHz. The white FM noise observed in the 5 MHz oscillator is not present in this analysis range.

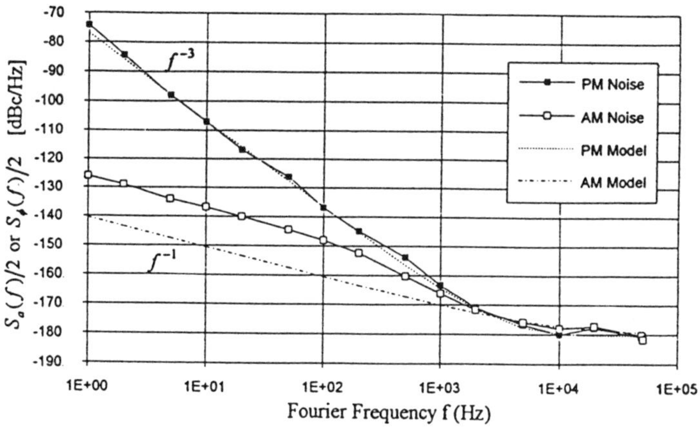

Figure 3. Single sideband AM noise $[S_a(f)/2]$ and PM noise $[S_\phi(f)/2]$ of a 100 MHz oscillator.

The PM noise in X-band oscillators showed similar characteristics to 5 MHz and 100 MHz oscillators. In general, all the X-band oscillators showed a $1/f^3$ component up to 1 MHz from the carrier. Figure 4 illustrates the PM noise in one of the 10.6 GHz oscillators tested. In this case the thermal noise is reached at approximately 100 MHz from the carrier. The AM noise, also shown in the figure, follows a 1/f behavior from 1 Hz to 500 Hz. Though excess noise is observed between 1 kHz and 1 MHz, it can be argued that the intrinsic AM noise behavior of the oscillator (excluding the excess noise) is 1/f until the thermal noise is reached. At high frequencies the two curves, AM and PM, converge to the thermal noise level.

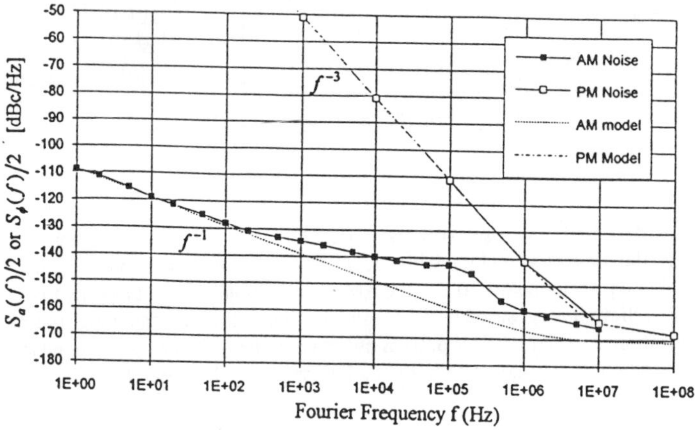

Figure 4. Single sideband AM noise $[S_a(f)/2]$ and PM noise $[S_\phi(f)/2]$ of a 10.6 GHz oscillator.

CONCLUSION

The noise measurements in the amplifiers and oscillators tested indicate a strong relationship between AM and PM noise. In both amplifiers and oscillators, the AM and PM noise showed a flicker noise component (1/f) and a white noise (f^0) component of similar magnitude. These findings suggest that the AM and PM noise in amplifiers and oscillators have two common sources: a 1/f noise modulation of the gain element and a thermal noise component. In oscillators, the PM noise had an additional $1/f^3$ component close to the carrier caused by frequency fluctuations in the resonator.[5] In some cases the AM noise had a $1/f^2$ component probably due to amplitude fluctuations in the oscillator's amplitude control loop.

ACKNOWLEDGMENT

The authors would like to thank Huascar D. Ascarrunz and Craig W. Nelson for their help in the measurements. They would also like to thank the Calibration Coordination group of JCTG/GMT-JLC and the Space and Terrestrial Communications Evaluation Division, Tactical Support Branch, Ft. Monmouth, New Jersey, for their financial support of this project.

REFERENCES

1. W.F. Walls, Proc. 46th Ann. Frequency Control Symp., pp. 257-261 (1992).
2. D.Halford, A.E.Wainwright & J.A.Borno, Proc. 22nd Ann. Frequency Control Symp., pp. 340-341 (1968).
3. D.B. Leeson, Proc. IEEE, Vol. #54, No. 1, pp. 329-330 (1966).
4. F.G. Ascarrunz, E.S. Ferre and F.L. Walls, Proc. 47th Ann. Frequency Control Symp. (1993), to be published.
5 T.E. Parker, Proc. 41st Ann. Frequency Control Symp., pp. 99-110 (1987).
6. L.M. Nelson, C.W. Nelson and F.L. Walls, Proc. 47th Ann. Frequency Control Symp. (1993), to be published.

SPECTRAL LINE SHAPE OF SIGNAL HAVING 1/F FLUCTUATIONS IN FREQUENCY,

A.G.Pashev, Dr.
State University, N.Novgorod 603600, Russia

ABSTRACT

The dependence both of the spectral line shape and the width for signals having the spectrum of frequency fluctuations of the type $1/\Omega^{a}$ $(1<a<3)$ on the value a is investigated. It is shown that the shape is non-Gaussian due to non-stationariness of fluctuations.

INTRODUCTION

The shape of a spectral line (SSL) is the classical description of signals generated in precision oscillatory systems. The problem of SSL is analyzed in many papers. Nevertheless, for the case of simultaneously existing both non-stationary (of 1/f type) and wide-band fluctuations in the frequency there are no algorithms for correct evaluation of this shape described in the literature. Even the shape of the spectral line for the signal with fluctuations in the frequency having diverged at low frequencies spectrum was not analyzed. These problems are considered in the present paper.

SPECTRAL LINE SHAPE EVALUATION METHOD

Let us consider the signal having the amplitude X_0, duration T, and fluctuations $\nu(t)$ in the frequency;

$$x(t)= X_0\cdot(1(t)-1(t-T))\cos[\omega_T t+\int_0^t(\nu(t)-\nu_T)dt] . \quad (1)$$

Here $1(\dots)$ is Heaviside function. The mean signal frequency ω_T differs from oscillation frequency ω_0 on the averaged over time T value ν_T of fluctuations in the frequency:

$$\omega_T = \omega_0 + \nu_T = \omega_0 + (1/T)\int_0^T \nu(t)dt .$$

The correlation function of signal (1) may be evaluated as follows:

$$K(\tau) = \int_{-\infty}^{+\infty} \langle x(t)x(t+\tau)\rangle d\tau = (X_0^2 T/2)K_0(\tau)\cos(\omega_T\tau).$$

Here the notation <...> means averaging over ensemble of realizations. The envelope $K_0(\tau)$ of the correlation function in the case of Gaussian fluctuations in the frequency is described by the following relation:

$$K_0(\tau)= (2/T) \int_0^{(T-\tau)/2} exp(-\tau^2\langle\Delta\nu^2(\tau,\theta)\rangle/2)d\theta, \quad \tau<T, \qquad (2)$$

Here

$$\Delta\nu(\tau,\theta) = (1/\tau) \int_{\theta-\tau/2}^{\theta+\tau/2} \nu(t)\cdot dt - \nu_T$$

is the deviation of averaged over interval $[\theta-\tau/2;\ \theta+\tau/2]$ frequency from the value ν_T ; $\theta = T/2-(t+\tau/2)$.

The variance of frequency increments $\langle\Delta\nu^2(\tau,\theta)\rangle$ may be evaluated if the spectrum $S_\nu(\Omega)$ of frequency fluctuations is known:

$$\langle\Delta\nu^2(\tau,\theta)\rangle = 2 \int_0^{\infty} G(\Omega)\cdot S_\nu(\Omega)\, d\Omega, \qquad (3)$$

here $G(\Omega) = sinc^2\beta + sinc^2\eta - 2sinc\beta\cdot sinc\eta\cdot cos(\Omega\cdot\theta)$, $\beta = \Omega T/\pi$, $\eta = \Omega\tau/\pi$, $sinc(z) = sin(\pi z/2)/(\pi z/2)$.

One can get SSL of signal $x(t)$ evaluating Fourier-transform from correlation function envelope (2) divided by $(X_0^2T/2)$ -the total energy of signal (1):

$$W(\Omega,T) = 1/(\pi T) \int_0^T K_0(\tau)cos(\Omega\tau)d\tau. \qquad (4)$$

This procedure allows to evaluate SSL presenting the signal energy distribution over analyzing frequencies centered at the mean frequency ω_T if spectrum $S_\nu(\Omega)$ is known. The following conditions are necessary: (a) fluctuations should be Gaussian; (b) the spectrum is increased slower than $1/\Omega^3$ if $\Omega \to 0$. In paper[1] the SSL energy width $\Pi = 1/W(0,T)$ versus duration T is evaluated.

NON-STATIONARY FLUCTUATIONS IN FREQUENCY

Let us consider two signals of duration T. Signal (1) has the following spectrum of frequency fluctuations:

$$S_\nu(\Omega) = \begin{cases} A\Omega^{-a}, & \Omega < \Omega_h,\ (1<a<3), \\ 0, & \Omega > \Omega_h, \end{cases} \qquad (5)$$

Spectrum $S_\nu^T(\Omega)$ of frequency fluctuations of other signal $x_T(t)$ is evaluated from the spectrum $S_\nu(\Omega)$ by the following relation eliminating the divergence in spectrum (5):

$$S_\nu^T(\Omega) = (1-sinc^2(\Omega T/\pi))\cdot S_\nu(\Omega), \tag{6}$$

In the analysis of SSL the traditional treatment may be used. Function $W(\Omega,T)$ is non-negative and fulfills to the normalization condition. Thus, it may be considered as the probability density function of some hypothetical random value ξ.

The description of the shape of the probability density function may be made by cumulant factors γ_n been the factors in Maclourin series expansion for characteristic function $\theta_\xi(\tau)$ [2]:

$$\gamma_{2n} = (-1/D_W)^n d^{2n}(ln\theta_\xi(\tau))/d\tau^{2n}\Big|_{\tau=0}, \tag{7}$$

here D_W is the second centered moment of function $W(\Omega,T)$.

In our case characteristic function $\theta_\xi(\tau)$ coincides with envelope (2) of the signal correlation function.

If signal $x(t)$ is considered, the following relations for the variance and the 4th cumulant factor may be achieved:

$$D_W = (2/T)\int_0^{T/2} \langle\Delta\nu^2(0,\theta)\rangle d\theta = D_W^T ,$$

$$\gamma_4 = \gamma_4^T + \gamma_4^\infty = \gamma_4^T + 3\left[\frac{2}{TD_W^2}\int_0^{T/2} \langle\Delta\nu^2(0,\theta)\rangle^2 d\theta - 1\right], \tag{8}$$

where D_W^T is the variance, and γ_4^T - the 4th cumulant factor for SSL $W_T(\Omega,T)$ of signal $x_T(t)$.

Note that the variance D_W is the limit of Allan variance[3] if number of samples on interval T tends to the infinity.

Analogous, but more complicated, relations may be achieved for higher order cumulant factors as well.

Factor $\gamma_4^T \to 0$ if $T\to\infty$ (e.g. $\gamma_4^T=(18\Omega_h)/(A\pi^2 T)$ if $a=2$). It may be shown that at the same time all the highest cumulant factors also tend to zero. Thus, the limiting SSL of signal $x_T(t)$ is Gaussian function.

The value γ_4^∞ turns to zero if the signal duration T is considerably greater than the correlation time of frequency fluctuations. In the case of diverging at low analyzing frequencies spectrum, frequency fluctuations have formally infinite correlation time, that means the non-zero value of the second term in eq.(8). For example, if $a = 2$ then $\gamma_4^\infty = 3/5$.

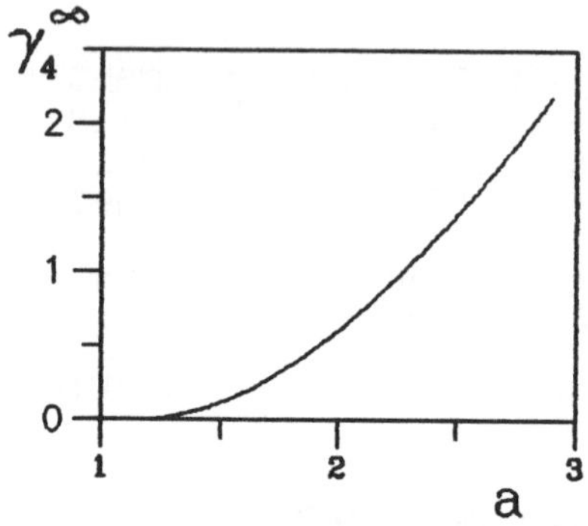

Fig.1. The limiting 4th cumulant factor versus spectrum exponent.

The dependence of the 4th cumulant factor γ_4^∞ for the limiting (at $T \to \infty$) SSL $W(\Omega)$ on the value of spectrum exponent a is presented in fig.1. One can see that γ_4^∞ is increasing function of value a characterizing the non-stationariness of frequency fluctuations. Analogous conclusion is valid for cumulant factors of higher orders as well. Thus, the difference between the limiting SSL and Gaussian function will increase if the divergence is increased in the spectrum of frequency fluctuations at $\Omega \to 0$.

SSL AND MEAN SOJOURN TIME FUNCTION OF FREQUENCY DEVIATIONS

The mean sojourn time function (STF) of frequency deviations $U_{\Delta\nu}(\Omega)$ gives mean time of sojourn of instant frequency deviation from value ω_T to be in small vicinity of argument Ω :

$$U_{\Delta\nu}(\Omega)=\lim_{\Delta\Omega \to 0}(2/T\Delta\Omega)\int_0^{T/2}\langle 1(\Delta\nu(0,\theta)-\Omega+\Delta\Omega/2)-1(\Delta\nu(0,\theta)-\Omega-\Delta\Omega/2)\rangle d\theta .$$

If signal duration T exceeds considerably the correlation time of frequency fluctuations, then STF coincides with probability density function of these fluctuations. In the case of Gaussian non-stationary frequency fluctuations the STF may be found by averaging of the probability density function over values θ:

$$U_{\Delta\nu}(\Omega)=(2/T)\int_0^{T/2}(2\pi<\Delta\nu(0,\theta)>)^{-1/2}\exp(-\Omega^2/(2<\Delta\nu^2(0,\theta)>))d\theta. \quad (10)$$

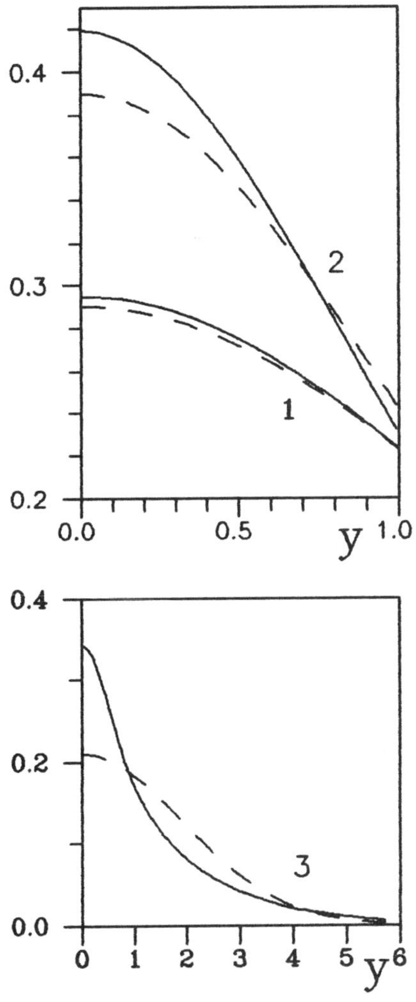

Fig.2. The limiting shape of spectral line and its Gaussian approximation.

For quasi-static stationary frequency fluctuations the SSL coincides with probability density function of these fluctuations [4] and, consequently, with STF. The computer analysis of signals with non-stationary frequency fluctuations shows that the limiting SSL coincides with STF inside the frequency region $|\Omega|<\Pi/2$, $\Pi=1/W(0)$. Thus, for the analysis of the limiting SSL (in the frequency region $|\Omega|<\Pi/2$) relation (10) may be used if the spectrum of frequency fluctuations follows to the power law ($S_\nu(\Omega) \sim |\Omega|^{-a}$, $1<a<3$). Accounting eq.(3), one can transform this relation to the following type:

$$W(\Omega)=1/(\sqrt{D_W})F_a(\Omega/\sqrt{D_W}). \quad (11)$$

The shape of function F_a is determined only by spectrum exponent a.

Examples of limiting SSL's (11) are given in fig.2 for signals with $a=1.2$ (curves 1), $a=2$ (2) and $a=2.9$ (3). The horizontal axis presents the normalized frequency $y=\Omega/\sqrt{D_W}$. Broken lines show Gaussian functions $g(\Omega)=(2\pi D_W)^{-1/2}\cdot\exp(-\Omega^2/2D_W)$.

These examples illustrate the conclusion made in the previous item (the last paragraph). If the divergence rate (value a) of the spectrum of fluctuations in the frequency is increased then the limiting spectral line shape becomes more and more non-Gaussian. That is, if $\alpha < 1.2$ then SSL practically coincides with Gaussian function; but for the case $\alpha=2.9$ the energy width of SSL evaluated in Gaussian approximation is 1.5 times greater than true value.

CONCLUSIONS

1. The algorithm for the spectral line shape (of signals having finite duration) evaluation at arbitrary analyzing frequencies is worked out (relations (2)-(4)). The spectrum of signal frequency fluctuations may follow to an arbitrary type.

2. Relation (11) for the limiting (at $T \to \infty$ and $|\Omega|<\Pi/2$) SSL is found. This relation is valid for signals having the spectrum of frequency fluctuations following to the power law diverged at low analyzing frequencies.

3. It is shown that the limiting SSL of signals having non-stationary frequency fluctuations is non-Gaussian. The non-Gaussian behavior leads to the decrease of the power width of SSL comparing with Gaussian approximation. This decrease is more pronounced if the exponent α characterizing the divergence in the spectrum of frequency fluctuations at low analyzing frequencies is increased.

Author is thankful to The Netherlands Organization for Scientific Research (NWO) and to Prof. F.N.Hooge for the support of investigations on the considered problem.

REFERENCES

1. A.G.Pashev, A.I.Saichev and A.V.Yakimov, Radio Eng. Electron Phys. 34 (Russian original p.2550, 1989).
2. A.N.Malakhov, Cumulant Analysis of Random Non-Gaussian Processes and Their Transformations (Soviet Radio Press, Moscow, 1978, in Russian).
3. J.A.Barnes et al., IEEE Trans. IM-20, 105 (1971).
4. A.N. Malakhov, Fluctuations in Self-Oscillating Systems (Nauka Press, Moscow, 1968, in Russian).

XV. MODELS AND SIMULATION OF $1/f$ FLUCTUATIONS

THE INTERPRETATION OF 1/f NOISE FROM THE POINT OF VIEW OF THE ELECTRON ENERGY PARADIGM

Gerhard Dorda
Siemens AG, Corporate Research and Development, Munich, Germany

ABSTRACT

A new model for the description of 1/f noise has been developed. It is deduced from the electron energy paradigm (EEP). A short description of the EEP enables us to show that 1/f noise reflects a basic energetic state of the electron. This has the same fundamental importance as the two well-known basic states, namely the wave and particle states.

THE ELECTRON ENERGY PARADIGM (EEP)

The EEP is an empirical relation [1] resulting from the experience gained in quantum transport phenomena [2-4], atomic light spectra, the umklapp energy of the electron spin and Hooge's empirical equation for the description of 1/f noise [5]. The EEP shown in Fig.1 links five fundamental reference energies $E_{o,n}$, n = 0,...4 , by means of the coupling constant $\alpha \approx 1/137$ raised to different powers, i.e.

$$E_{o,o} = \alpha E_{o,1} = \alpha^2 E_{o,2} = \alpha^3 E_{o,3} = \alpha^4 E_{o,4} \quad (1)$$

These $E_{o,n}$ values are reference energies related to quantum considerations and thus denoted by the index zero. They are evaluated on the basis of the relation for the square of the electron charge given by

$$e^2 = \alpha h \quad (2)$$

Based on the data of the quantum Hall effect [2] and of the quantum transport phenomena [3, 4] as well as on the fact that α cannot be quantized, equation (2) implies the quantum character of the magnetic flux density B, the voltage V and the current I. As has been already shown [1, 6, 7], the reference quantities B_o, V_o and I_o are deduced from the Rydberg energy Ry and the Rydberg constant

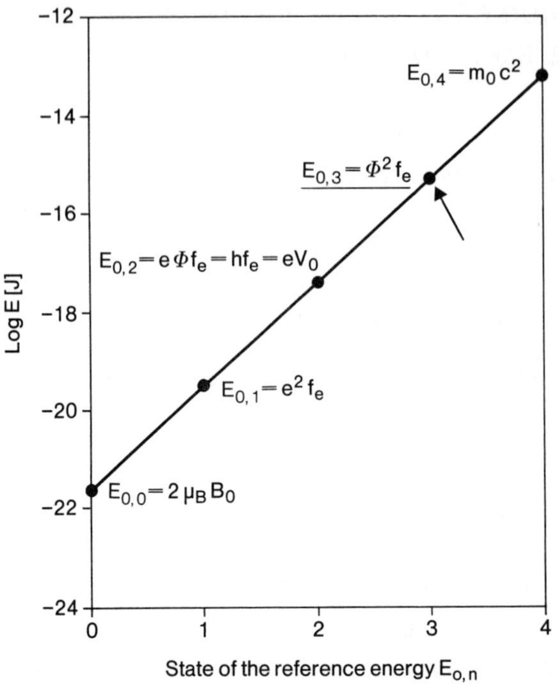

Fig. 1 Values for the five reference energies $E_{o,n}$ of the electron energy paradigm

$R = \frac{1}{2}(2a_e)^{-1}$, where a_e is the reference length. The generalization of these reference quantities is supported by experimental data [1, 6, 7]. Thus we can formulate:

1) the umklapp reference energy $E_{o,o} = 2\mu_B B_o = \alpha^2 eV_o$, where μ_B is the Bohr magneton, $B_o = \Phi/2\pi a_e^2$ the reference magnetic flux density, $\Phi = h/e = e/\alpha$ the flux quantum resulting from (2), $V_o = \Phi f_e$ the reference voltage, $a_e = 7.25$ nm the reference length, $f_e = (2\pi T_e)^{-1} = 6.58\text{x}10^{15}$ cps the reference frequency, $T_e = a_e/c$ the reference time and c the velocity of light [1, 6];

2) the reference energies $E_{o,1} = eI_o = e^2 f_e$ and $E_{o,2} = eV_o = e\Phi f_e = 2$ Ry $=$ $= 27.2$ *eV* obtained from Ohm's law given by $I_o = \alpha V_o$, where I_o and V_o are the reference current and voltage, respectively [1, 6, 7];

3) the rest mass energy of the electron $E_{o,4}$ related to $E_{o,2}$ by the known Rydberg relation given by $E_{o,4} = m_o c^2 = \alpha^2\, 2\text{Ry} = \alpha^2 eV_o$, where m_o is the rest mass of the electron;

4) the reference energy $E_{o,3}$. This will be deduced from Hooge's empirical equation for the description of 1/f noise and is given by $E_{o,3} = \Phi^2 f_e = \alpha^{-1} e V_o$.

As already stated, the evaluation of the EEP is based on $e^2 = \alpha h$. In another paper it was shown that formulation (2) has priority over the formulations $(e^2)_{SI} = 4\pi\varepsilon_o \alpha \hbar c$ in the SI system of units, where ε_o is the permittivity of vacuum, and $(e^2)_{CGS} = \alpha \hbar c$ in the CGS system of units [1, 6]. It is evident that $e^2 = \alpha h$ results in a new system of electromagnetic units based on the fundamental mechanical quantities expressed by the meter-joule-second system. For simplicity, the quantity of mass has been replaced by that of energy. We have called this system the MJS system of units [6].

The EEP suggests a highly interesting interpretation of electron energy, showing that the reference energies $E_{o,n}$ characterize five basic states of the electron. These differ by successive coupling of the basic yes/no or (+,-) information state ($E_{o,o}$) with time ($E_{o,1}$), with time and 1-dimensional space ($E_{o,2}$), with time and 2-dimensional space ($E_{o,3}$), and with time and 3-dimensional space ($E_{o,4}$) [1]. According to this interpretation, the 1/f noise-related state refers to the 2-dimensional state in space. This statement will be discussed in this paper.

THE REFERENCE ENERGY OF 1/f NOISE

Hooge was the first to formulate the following empirical relation in order to describe 1/f noise [5]

$$S_V = \alpha_{exp} V^2 (N_{el} f)^{-1} \qquad (3)$$

where $S_V = \Delta V^2/\Delta f$ represents the so-called spectral power density of the voltage fluctuations, V is the mean voltage at constant current, f is the frequency, N_{el} is a number assumed to represent the number of free charge carriers in the object under test and α_{exp} is a fitting parameter with no relation to the coupling constant. The noise quantity S_V, declared as the spectral power density, is in fact an energetical quantity, as indicated by measurements using thermoelements. This statement is confirmed by the application of the MJS unit system to equation (3), directly obtaining the dimension energy for V^2/f, as V^2 then has the dimension joule/second [6]. If we accept the priority of (2) as a fact of nature and use the SI system in the measurement, the correction factor $2\varepsilon_o c = 5\text{x}10^{-3}\ \Omega^{-1}$ must necessarily occur at V^2 due to the invariance of energy. The correction factor is obtained according to the transformulation equation

$$(V^2)_{MJS} = (2\varepsilon_0 c V^2)_{SI} \qquad (4)$$

In this context, it is highly remarkable that in most cases under study the fitting parameter α_{exp} reaches the order of magnitude of 10^{-3} [8]. This means that the empirical Hooge equation has to be represented in the MJS system by

$$S_V = (V^2)_{MJS} (N\ f)^{-1} \qquad (5)$$

A new interpretation is evidently required for the number N. Based on the general transport model, N_{el} of (3) in fact represents the number of phase-coherent states of the electron gas N_D [6, 7]. Thus N_D may be regarded as a quantum number assigned to the electron state. Using the reference frequency f_e for f and taking the ground state, i.e. $N_D = 1$, the relationship for the reference energy state $E_{0,3}$ of the 1/f noise is obtained from (5) in the following form

$$E_{0,3} = V_0^2\ f_e^{-1} = \Phi^2 f_e = \alpha^{-1} e V_0 = \alpha^{-1}\ E_{0,2} \qquad (6)$$

THE FITTING PARAMETER α_{exp}

As the experimental data show, the fitting parameter α_{exp} is not a constant, as it exhibits values in some cases even several orders of magnitude smaller than the theoretically expected value of $\alpha_{th} = 2\varepsilon_0 c = 5\times10^{-3}\ \Omega^{-1}$ [9]. To interpret these results, a description of the 1/f noise process will be given in detail.

In our model, 1/f noise represents a variation in energy related to the energy $E_{0,3}$. Based on the description of $E_{0,3}$ in the EEP, the (coupling) energy E_{1D} referring to a single domain with electrons in a phase-coherent state is given by

$$E_{1D} = I\ V\ (\alpha\ N_D\ f)^{-1} \qquad (7)$$

where V is the applied voltage, I the current and N_D the number of coherent domains in the object under test. It should be noted that here α means the coupling constant. Referring to the general transport model [7], the number N_D is approximately given by the number of 1-dimensional current filaments n_F, multiplied by the number of coherent domains n_D within one current filament. Following these ideas, 1/f noise is the result of a change in the coherent state of the electrons represented by an average change in the number N_D, i.e. for example by 1. Thus, using the MJS system of units, we can write

$$S_V = \left| \frac{IV}{\alpha f} \left(\frac{1}{N_D} - \frac{1}{N_D \pm 1} \right) \right| \approx \frac{IV}{\alpha f} \frac{1}{N_D^2} = \frac{V}{f} \frac{I}{\alpha (n_F / n_D)} \frac{1}{n_D^3 \, n_F} \qquad (8)$$

It can be simply shown that this equation exactly yields Hooge's empirical relation (3), when we introduce into (8)

1) Ohm's low formulated in the MJS unit system by [6, 7]

$$I = \alpha \sum_{n_F} \left(\frac{1}{n_D} \right) V \approx \alpha \frac{n_F}{n_D} V \qquad (9)$$

2) the relation

$$N_{el} = n_e \, n_D \, n_F \qquad (10)$$

and 3) when we write

$$n_D^2 / n_e = a_{th} / a_{exp} \qquad (11)$$

Here n_e expresses the average number of electrons in one domain. Note that relation (11) clearly explains the variability of a_{exp} by several orders of magnitude with respect to a_{th}. Moreover, it should be noted that the good agreement of (11) with the experimental data is given when we consider $a_{exp} / a_{latt} = (\mu / \mu_{latt})^2$, $a_{latt} < a_{th}$ and $n_D \sim 1/\mu$ [7, 9]. Here μ is the mobility and μ_{latt} the mobility with lattice scattering.

The fact that a_{exp} shows the same order of magnitude in the case of the current-related noise quantity $S_I = a_{exp} I^2 / Nf$, where I is the mean current at constant voltage [5, 8, 9], provides evidence that 1/f noise represents fluctuations in energy impressed by the external voltage and not by the current. This means that the current is only a consequence, and the voltage is the main releasing parameter. This can be simply shown when multiplying S_I and thus I^2 by the square of the sample resistance thus obtaining the same order of magnitude as would be given by measuring S_V on the studied system.

CONCLUSION

In conclusion we summarize that Hooge's empirical equation (3) for the description of 1/f noise can be interpreted by means of the reference energy $E_{o,3}$ of the EEP. We obtain the agreement of equation (3) with eqns. (6) - (8), when we apply $e^2 = \alpha h$ to the fitting parameter α_{exp} and interpret S_V as a change of energy E_{1D} caused by a variation in the number of coherent electron states.

Seen from the viewpoint of the different possible basic states of the electron expressed by the EEP, 1/f noise is a very important phenomenon. We have been familiar with electron dualism for a long time, i.e. the possibility of electrons existing in a 1-dimensional state in space, known as the wave state, and in a 3-dimensional state in space, known as the particle (i.e. mass) state. But our investigations show that 1/f noise discloses a third important state of the electron in space, namely its 2-dimensionality. Supported by the experimental evidence with 1/f noise and represented by the relation of $E_{o,3}$ to Φ^2, $E_{o,3}$ really expresses not only the behavior of the applied voltage but also includes its modification by an additional, inherent electric field. This results from imperfections in the material, such as ions or domain boundaries. These disturbances cause a modification of the applied electric flux, resulting inevitably in a kind of localization due to the crossing of two (i.e. 1- dimensional) electric fluxes (fields). Thus we can state that the most important feature of $E_{o,3}$ and thus of 1/f noise is its localization, which greatly hinders the transfer of this embedded energy from place to place.

REFERENCES

1. G. Dorda, Proc. of "Waves and Particles in Light and Matter", Int.Workshop on the Occasion of L. de Broglie's 100th Anniversary, Trani, Italy, Sept. 1992; F.Selleri ed., Hadronic Journal Press, Florida, to be publ.
2. K. von Klitzing, G. Dorda, and M. Pepper, Phys.Rev.Lett.45, 494 (1980)
3. B.J.van Wees, H.van Houten, C.W.J. Beenakker, J.G. Williamson, L.P. Kouwenhoven, D.van der Marel and C.T. Foxon, Phys.Rev.Lett.60, 848 (1988)
4. J. Hajto, A.E. Owen, S.M. Gage, A.J. Snell, P.G. Lecomber and M.J. Rose, Phys.Rev.Lett.66, 1918 (1991)
5. F.N. Hooge, Phys.Lett.29A, 139 (1969); Physica B83, 14 (1976)
6. G. Dorda, Physica Scripta T45, 297 (1992); Proc. of the 12th General Conf. of Cond. Matter Div. of the Europ.Phys.Soc., Prague 1992, session: New Develop.
7. G. Dorda, Supperlattices and Microstructures 7, 103 (1990)
8. L.K.J. Vandamme, Noise in Physical Systems and 1/f Noise, M. Savelli, G. Lecoy and J.P. Nougier eds. (North-Holland, Amsterdam, 1983), p. 183
9. F.N. Hooge, Physica B162, 344 (1990)

DISCRETE-TIME fGn AND fBm OBTAINED BY FRACTIONAL SUMMATION.

M.L Meade
Faculty of Technology, Open University, Milton Keynes, MK7 6AA, UK

ABSTRACT

This paper is concerned with the behaviour of random time series obtained by taking fractional sums of a delta-correlated random sequence. Models are developed for discrete-time processes having properties closely akin to the fractional Gaussian noises (fGn) and the fractional Brownian motions (fBm). It is shown that if the Mandelbrot parameter H is allowed to approach zero, the corresponding discrete-time fBm converges to a non-stationary 1/f-noise process, having a variance that increases indefinitely as the logarithm of the index n.

INTRODUCTION

Our approach closely follows that of Barton and Poor[1] who showed that continuous-time fBm can be defined in two stages. The first step involves a weighted integral of unit variance white noise $W(t)$, giving fGn, $W^H(t)$, with parameter $H \in (0,1)$:

$$W^H(t) = \frac{1}{\Gamma(H-1/2)} \int_{-\infty}^{t} (t-u)^{H-3/2}\, W(u)\, du \tag{1a}$$

Fractional Brownian motion $B^H(t)$ is then obtained from the finite integral:

$$B^H(t) = \int_0^t W^H(u)\, du \tag{1b}$$

DISCRETE-TIME fGn

Mandelbrot [2] has described a discrete-time approximation to the kernel in the fractional integral (1a). We suggest that a more natural choice is the weighting sequence $h^{H-1/2}$ that provides sums to fractional order $H-1/2$ of a bounded 'input' sequence [3]. Thus, in place of equation (1a), we write:

$$W_n^H = \sum_{k=-\infty}^{n} h_{n-k}^{H-1/2}\, W_k \tag{2}$$

where the right-sided weighting sequence $h^{H-1/2}$ is defined by:

$$h_n^{H-1/2} = \frac{\Gamma(n+H-1/2)}{\Gamma(H-1/2)\Gamma(n+1)} \tag{3}$$

and $\{W_n\}_{-\infty}^{\infty}$ represents white-noise samples with zero mean and unit variance.

The discrete-time and frequency-domain properties of $h^{H-1/2}$ have been described elsewhere[4]. We find that $h^{H-1/2}$ is square-summable for $H < 1$. W^H is then a stationary random process with finite variance given by

$$\mathcal{V}\left\{W_n^H\right\} = \lim_{n \to \infty} \sum_{k=0}^{n} \left(h_k^{H-1/2}\right)^2 = \frac{\Gamma(2-2H)}{\Gamma^2(3/2-H)}, \quad H \in (0,1) \tag{4}$$

We define $R_H = [2\Gamma(2H-1)\sin \pi H]^{-1} = H(2H-1)V_H$ where V_H is equal to Mandelbrot's constant of the same designation[1,2]. W^H is then characterized by the covariance sequence

$$c_m^H = c_{-m}^H = R_H \frac{\Gamma(m+H-1/2)}{\Gamma(m-H+3/2)}, \quad m = 0, \pm 1, \pm 2, \ldots \tag{5}$$

$$\approx R_H \,|m|^{2H-2} \text{ for } |m| >> H \tag{5a}$$

which corresponds to to the periodic spectral density:

$$S^H(\omega) = \left|2\sin(\omega/2)\right|^{1-2H} \tag{6}$$

$$\approx |\omega|^{1-2H} \text{ as } |\omega| \to 0 \tag{6a}$$

DISCRETE-TIME fBm

Discrete-time fBm, denoted by $\{B_n^H\}$, $H \in (0,1)$, is a non-stationary random process with $B_0^H = 0$ and

$$B_n^H = \sum_{k=1}^{n} W_k^H \tag{7}$$

The variance of the increments of B^H at index m can be calculated from either the discrete-time or frequency domains:

$$\mathcal{V}\left\{B_m^H\right\} = \sum_{n=1-m}^{m-1} (m - |n|)\, c_n^H \tag{8a}$$

$$= \frac{1}{2\pi} \int_0^{2\pi} \frac{4\sin^2(m\,\omega/2)}{|2\sin(\omega/2)|^{2H+1}}\, d\omega \tag{8b}$$

The result is an even sequence dependent only on m:

$$\mathcal{V}\left\{B_m^H\right\} = V_H\left[\frac{\Gamma(m + 1/2 + H)}{\Gamma(m + 1/2 - H)} - \frac{\Gamma(H + 1/2)}{\Gamma(1/2 - H)}\right] \tag{9a}$$

$$\approx V_H \mid m \mid^{2H} \text{ for } \mid m \mid >> H \tag{9b}$$

Since B^H is a process with stationary increments, the increments have variance $\mathcal{V}\left\{B_{n+m}^H - B_n^H\right\} = \mathcal{V}\left\{B_m^H\right\}$. As a result, the covariance of B^H has a distinctive structure. On setting $k = n + m$, we obtain:

$$\mathcal{V}\left\{B_n^H B_k^H\right\} = \frac{V_H}{2}\left[\mathcal{V}\left\{B_n^H\right\} + \mathcal{V}\left\{B_k^H\right\} - \mathcal{V}\left\{B_{k-n}^H\right\}\right] \tag{10a}$$

$$\approx \frac{V_H}{2}\left[\mid n \mid^{2H} + \mid k \mid^{2H} - \mid k-n \mid^{2H}\right], \mid n \mid, \mid k \mid >> H \tag{10b}$$

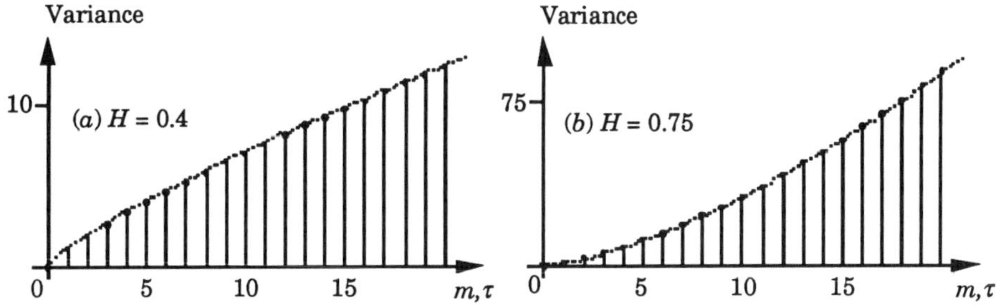

Fig. 1. The variance of $\left\{B_n^H\right\}$ compared with the graph of $V_H \mid \tau \mid^{2H}$ (dotted curve)

The asymptotic forms of equations (5), (6), (9) and (10) bear direct comparison with standard descriptions of continuous-time fGn and fBm. This is emphasized in Figure 1 which shows the variance from equation (9a) plotted against its continuous-time counterpart $V_H \mid \tau \mid^{2H}$. It is evident that the variance of $\left\{B_n^H\right\}$ conforms closely to the desired scaling property

$$\mathcal{V}\left\{B_M^H\right\} = \left(\frac{M}{m}\right)^{2H} \mathcal{V}\left\{B_m^H\right\}, \quad M, m = 1, 2, \ldots \tag{11}$$

Because there is no inherent upper limit on the value of the lag m, the increments can be regarded for practical purposes as a self-similar random process with parameter H. It should be noted, however, that it is only in the case of ordinary Brownian motion ($H = \frac{1}{2}$) that the discrete-time variance is strictly homogenous in m, thus ensuring that equation (11) holds exactly for all $M, n \geq 1$.

THE 1/f-NOISE BOUNDARY

Flandrin[5] has demonstrated that the non-stationary covariance function of continuous-time fBm is consistent with a time-averaged spectrum $|\omega|^{-1-2H}$. A straight-forward extension of Flandrin's analysis leads to the spectral model $|2\sin(\omega/2)|^{-1-2H}$, $H \in (0,1)$, which, we suppose, provides a useful characterization of discrete-time fBm. We note that a similar result is obtained by starting with the spectrum of stationary fGn in equation (6) and introducing a weighting factor $[2\sin(\omega/2)]^{-2}$, equivalent to the *steady-state* power frequency response of an elementary summation system[4].

The form of the time-averaged spectrum suggests that as $H \to 0$, non-stationary fBm takes on the character of 'true' 1/f-noise. It is perhaps surprising that this limiting form of behaviour seems always to be excluded from conventional treatments. In the present case we write:

$$\lim_{H\to 0} \mathcal{V}\left\{B_m^H\right\} = \lim_{H\to 0} V_H\left[\frac{\Gamma(m+1/2+H)}{\Gamma(m+1/2-H)} - \frac{\Gamma(H+1/2)}{\Gamma(1/2-H)}\right] \tag{12}$$

The limit is evaluated by applying l'Hôpital's rule, giving

$$\lim_{H\to 0} \mathcal{V}\left\{B_m^H\right\} = \frac{2}{\pi}\left[\psi(m+1/2) - \psi(1/2)\right], \quad m = 0, \pm 1, \pm 2, \ldots \tag{13}$$

where $\psi(x)$ is the logarithmic derivative of the gamma function.

It may be verified that the right-hand side of equation (13) is a strictly even function of m. Since the psi function $\psi(x)$ varies as $\ln x$ for large positive values of x, we see that the variance satisfies the asymptotic equality:

$$\lim_{H\to 0} \mathcal{V}\left\{B_m^H\right\} \approx \frac{2}{\pi}\ln|m|, \quad |m| >> 1 \tag{14}$$

At the boundary $H = 0$, therefore, the variance of discrete-time fBm and, by implication, the variance of its increments increases indefinitely and in logarithmic fashion for sufficiently large values of $|m|$. Such a conclusion is not unexpected, given the wealth of experimental evidence associating 1/f-type spectral properties with logarithmic growth in the time domain.

REFERENCES

1. R.J. Barton and V.H. Poor, *IEEE Trans. Inform. Theory*, vol. 34, pp 943 - 959, Sept.1988.
2. B.B. Mandelbrot and J.R. Wallis, *Water Resources Res*, vol. 5, pp 228 - 267, 1969.
3. M.L.Meade, *Proc. 10th Int. Conf. on Noise in Physical Systems*, pp 347-350, Budapest, 1989.
4. M.L. Meade, submitted to *Proc IEE*, 1993.
5. P.Flandrin, *IEEE Trans. Inform. Theory*, vol IT-35, pp 197-199, Jan. 1989.

1/F NOISE AND AGING DRIFT IN PARAMETERS CAUSED BY TWO LEVEL SYSTEMS IN SEMICONDUCTORS

A.V.Yakimov, Prof.
State University, N.Novgorod 603600, Russia

ABSTRACT

The 1/f noise is considered as the superposition of random telegraph processes. The nature of such processes is associated with TLS's (two level systems), i.e. point defects having two metastable states divided by the energy barrier with random height. The link between the noise and the aging of the device, displaying as the time drift in average values of its parameters, is found.

INTRODUCTION

Following to the idea suggested by *Malakhov*[1] fluctuations with the spectrum of 1/f type are considered as a consequence of processes leading to the aging of the sample. The further elaboration of this idea (see, e.g. [2,3]) resulted the uniting with the treatment by *Kogan* [4] interpreting the 1/f noise as a consequence of the presence of the so-called two-level systems (TLS's) in the sample. The nature of TLS's in semiconductors is not explicitly understood up to now. Nevertheless, it may be suggested that these are non-symmetric point defects described in monograph [5]. As an example, the account of rotating defects having the configuration of a dipole [6] gives the possibility [7] to explain experiments [8] with Corbino disk where the anisotropy of current 1/f noise was found.

On the other hand, the idea of an ensemble of TLS's gives a new incite on models by *Van der Ziel* [9] and *Du Pre*[10]. The 1/f spectrum synthesis by the superposition of simple Lorentzian spectra, suggested by these authors, acquired the explicit physical background. At the same time, it should be noted that outgoing from models [9,10] wide distribution in energy E, having a sense of an activation energy, was found to be necessary for the logarithmic drift in parameters of a device explanation [11]. Moreover, in paper [11] the model was used, based on systems completely analogous to the above mentioned TLS's. Thus, the model of TLS's having a random set of inner heights E opens the possibility to explain in the same frames both 1/f noise and logarithmic aging in radioelectron devices.

TWO LEVEL SYSTEM ANALYSIS

Let us suppose that the sample contains the defect randomly changing the configuration. The energy profile of the defect is shown in fig.1. The defect may be located in an arbitrary state: "*0*" or "*1*". These states are divided by the energy barrier of height E; the difference in depths of potential minimums is Δ_o. Following to the thermal vibrations of the lattice, defect is switched at random from one state to another one, and *vice versa*.

It is a reasonable assumption that electro-physical parameters of the defect are different in different states. That means, the random change in the defect configuration leads to change in the sample resistance. The last change, considered in the course of time, is of a random telegraph process (RTP) type having the magnitude $(\delta R)_1$ -the relative change in the resistance caused by switching of single defect (see fig.2).

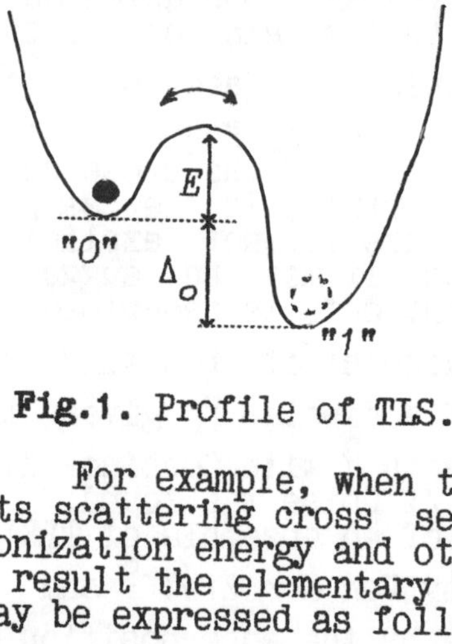

Fig.1. Profile of TLS.

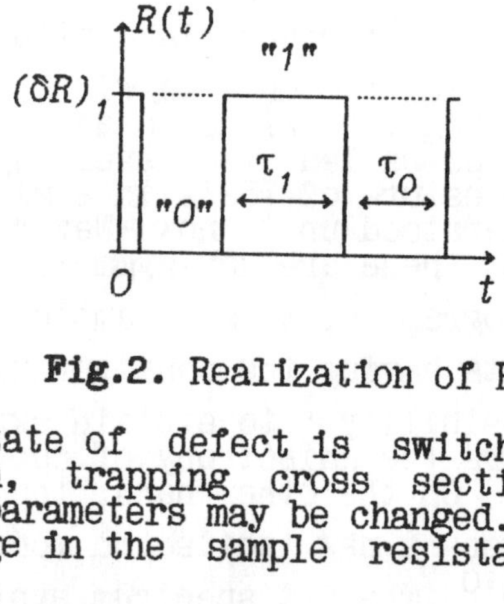

Fig.2. Realization of RTP.

For example, when the state of defect is switched, its scattering cross section, trapping cross section, ionization energy and other parameters may be changed. As a result the elementary change in the sample resistance may be expressed as follows:

$$(\delta R)_1 = -(\delta n)_1 - (\delta \mu)_1 , \quad (1)$$

that is, it is determined by changes both in the density and the mobility of carriers caused by the change of the state of single TLS. Relation (1) suggests an answer on the question (see, i.g.[12]): "*Is the 1/f noise to be a result of fluctuations in the density or in the mobility of carriers?*" It leads from the considered model that both values may be subjected to fluctuations. These fluctuations are mutually correlated if the primary noise source is associated with TLS's differed only by the energy profile, not by the change in electro-physical parameters.

Occupation times τ_0 и τ_1 of the defect in states "0" and "1" are random values. To switch from one state to another one the defect should cross the relatively high potential barrier (see fig.1). Due to this circumstance the distributions of these times may be considered as following to Boltzman law with different mean values:

$$\bar{\tau}_0=\tau_T\cdot\exp(E/k_BT)\ ,\quad \bar{\tau}_1=\bar{\tau}_0\cdot\exp(\Delta_0/k_BT),$$

here τ_T is the mean period of the lattice thermal vibrations; k_B -Boltzman constant; T -the temperature.

The spectrum of RTP corresponding to random switches in the sample resistance caused by the single TLS is of Lorentz kind:

$$S_{\delta R}(f|\nu) = \frac{A\nu}{\nu^2+f^2}\ ,$$

where

$$A = \frac{2\cdot\bar{\tau}_0\bar{\tau}_1}{\pi\cdot(\bar{\tau}_0+\bar{\tau}_1)^2}(\delta R)_1^2$$

is the parameter characterizing the intensity of the fluctuation process,

$$\nu = \frac{1}{2\pi}(\bar{\tau}_0+\bar{\tau}_1)/(\bar{\tau}_0\bar{\tau}_1)$$

-characteristic (or corner) frequency of the process.

An ensemble of defects, differing in the common case both by the intensity of the perturbation and corner frequency, may exist in the real sample. Every type of defect is described by two level system (see fig.1). In the assumption of fixed magnitudes $(\delta R)_1$ for all TLS's the resulted spectrum for total fluctuations in the sample resistance may be expressed as follows:

$$S(f) = N_D\cdot\int S_{\delta R}(f|\nu)\cdot W(\nu)d\nu\ ,$$

where $W(\nu)$ is the probability distribution for corner frequencies, N_D -the total number of defects in the sample. We put all intensities A to be fixed and take

$$W(\nu) = \frac{1}{\ln(f_h/f_l)}\cdot\frac{1}{\nu}\quad \text{for}\quad \nu\in[f_l,f_h],$$

where f_l is low and f_h -high corner frequencies. In this case the resulted spectrum is of the 1/f type in the frequency band from f_l up to f_h:

$$S(f) = G/f \quad , \qquad (2)$$

and $G = (\pi \cdot A \cdot N_D)/(2 \cdot \ln(f_h/f_l))$.

The distribution $W(\nu)$ is realized if all TLS's have fixed differences Δ_o and the distribution $W(E)$ for barrier heights is uniform in the energy range from value E_1 up to value E_2, corresponding to frequencies f_h and f_l.

Thus, the movement of defects in the solid state produces the stationary random process having the spectrum of 1/f type. Let us now consider nonstationary drift in the sample resistance.

LOGARITHMIC AGING

Let us suppose that after the sample manufacturing all defects turned out to be mainly in one selected state (for example, in state "0", see fig.1). In this case the distribution of defects over potential pits is nonstationary, and the non-zero (at average) stream of defects towards other state arises driving all the sample to the thermodynamic equilibrium.

At first, "fast" (having small relaxation time $\tau \sim 1/2\pi f_h$) defects begin to take part in the relaxation process. Then slow defects start to relax. As a result, the equilibrium will at the most be achieved during the time $t > 1/2\pi f_l$. The drift in the resistance mean value affected by the single defect with corner frequency ν follows to the exponential law:

$$\langle \delta R(t|\nu) \rangle = (\delta R)_1 \cdot (p - p_0) \cdot (1 - e^{-2\pi\nu t}) \ ,$$

here $p = (\ 1 + \exp(\Delta_0/k_B T)\ )^{-1}$ is the probability for the defect to be in the state "0" at the temperature T; p_0 -the same probability corresponding to the temperature of the previous treatment; the notation $\langle ... \rangle$ means the averaging over the ensemble of defects.

The dependence of p on the temperature is presented in fig.3. Here curves "1,2,3" correspond to differences between depths of pits in TLS Δ_o=10,20,30 meV. Horizontal and vertical lines show the TLS evolution during the sample cooling after the previous treatment.

The resistance drift caused by the going of all defects to the equilibrium is the superposition of exponents with different relaxation times. Under the assumption that the spectrum of the process is of type (2), the drift in the resistance at $2\pi t \in [1/f_h,\ 1/f_l]$ follows approximately to the logarithmic law:

$$\langle \delta R(t) \rangle = B \cdot \ln(2\pi f_h t) \ , \tag{3}$$

here $B = (p-p_0) \cdot N_D \cdot (\delta R)_1 / \ln(f_h/f_l)$.

The knowledge of value B allows to determine the time of the device stable operation:

$$t_S = t_1 \cdot (\exp(\langle \delta R_{max} \rangle / B) - 1) \ .$$

Here t_1 is the time passed after the sample manufacturing; $\langle \delta R_{max} \rangle$ -permitted maximum relative change in the device resistance.

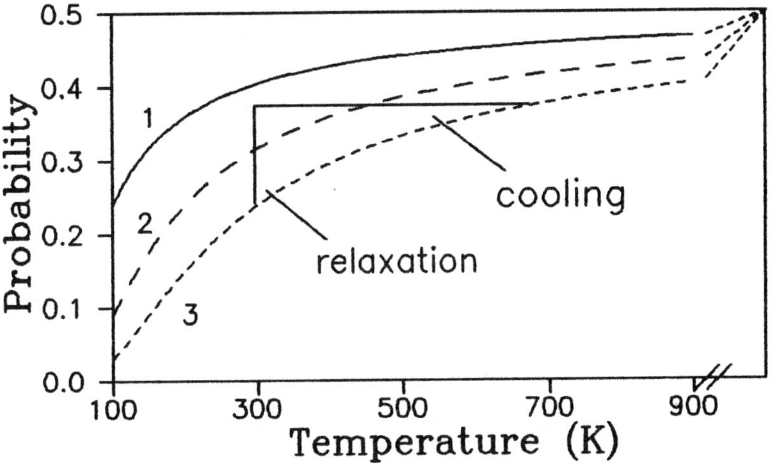

Fig.3. The evolution of TLS during the sample cooling.

Using relations (2) and (3) one can find the relation between the degradation (aging) rate and the value G determining the spectrum at 1 Hz:

$$B = G \cdot (p - p_0) / [p \cdot (1-p) \cdot (\delta R)_1] \ . \tag{4}$$

Thus, the increased noise value may indicate the increased aging rate, that is the possible non-reliability of the device. In other words, relation (4) shows the way for the selection of semiconductor devices on the basis of the information about the 1/f noise spectrum value, if the number N_D of defects in the sample is fixed.

In the course of the time the accumulation of defects, increasing the aging rate, may take place. This process, as one can see from relation (2), is immediately displayed through the noise intensity increase. Perhaps, just this mechanism may explain the known effect of sharp increase of the 1/f noise intensity before the burnout of the device (see, e.g. [13]). The test of the value N_D may be made by the test of the 1/f noise spectrum value.

Another interpretation of reached results is possible as well. Knowing both the spectrum at 1Hz and the value B (after the control of the drift in the total sample resistance), one can evaluate the factor for the relation between G and B in (4). This factor may be used for the type identification of defects existing in the solid state and having concrete values p_0, p and $(\delta R)_1$.

CONCLUSION

Thus, the model of moving defects allows to explain not only the nature the noise with spectrum of the 1/f type but also to get data about the reliability of semiconductor devices.

Author is thankful to The Netherlands Organization for Scientific Research and to Prof. F.N.Hooge for the support of investigations on the considered problem.

REFERENCES

1. A.N.Malakhov, Radiotekhnika i Elektronika 4, 54 (1959), in Russian.
2. A.V.Yakimov, Radiophys.Quant.Electron 23, 170 (1980).
3. V.B.Orlov and A.V.Yakimov, Physica B 162, 13 (1990).
4. Sh.M.Kogan, Sov.Phys.-Usp. 28, 170 (1985).
5. M.Lannoo and J.Burgouin, Point Defects in Semiconductors 1, Theoretical Aspects (Springer, Berlin, 1981).
6. A.G.Samoylovich and M.V.Nitzovich, Fiz. Tverd.Tela 5, 2981 (1963), in Russian.
7. V.B.Orlov and A.V.Yakimov, Solid State Electron. 33, 21 (1990).
8. M.H.Song and H.S.Min, J.Appl.Phys. 58, 4221 (1985).
9. A.van der Ziel, Physica 16, 359 (1950).
10. F.K.Du Pre, Phys.Rev 78, 615 (1950).
11. M.R.J.Gibbs and J.E.Evetts, Proc.4th Int.Conf. Rapidly Quenched Metals 1, 479 (1982).
12. F.N.Hooge, T.G.M.Kleinpenning and L.K.J.Vandamme, Rep.Progr.Phys. 44, 479 (1981).
13. A.van der Ziel and Hu Tong, Electronics 39,95 (1966).

BASIS OF UNIVERSAL EXISTENCE OF $1/F$ FLUCTUATION

—Mathematical proof and numerical demonstrations—

Toshio Kawai, Yoshiro Mihira, Makoto Sato, Maki Hayashi

Department of Physics, Faculty of Science and Technology, Keio University
3-14-1 Hiyoshi, Kohoku, Yokohama, 223 Japan

ABSTRACT

We define "$1/f$ fluctuation function". This function has a stability property that their sum and product are again $1/f$ functions. We prove this property mathematically and also demonstrate it numerically. It follows that nonlinear dynamical systems can have $1/f$ fluctuating solutions.

INTRODUCTION

The frequent appearance of fluctuation that has $1/f$ power density spectrum in the low frequency range is a riddle, and studies have been published in a series of proceedings of this conference. P.H. Handel, for example, is persistently developing theories for general understanding of $1/f$ noise by simple dimensional arguments or extensive analyses[1]. In search of a simpler answer we present another general viewpoint based on a property of chaotic function.

DEFINITION OF $1/F$ FLUCTUATION

A function of time

$$f(t) = \sum_{n=1}^{N} a_n \cos(\omega_n t + \delta_n) \tag{1}$$

will be called "chaotic" if the parameters a_n, ω_n and δ_n satisfy following criteria 1-3. If the function (1) further satisfies the fourth criterion, we define it to be the "$1/f$ fluctuation".

1. The phases, δ_n, are uniformly random in $[0, 2\pi)$.

2. The frequency range is limited by upper and lower cut off, $\Delta\omega < \omega_n < \Omega$, where $\Omega/\Delta\omega \equiv \mathcal{M} \gg 1$.

3. Spectral lines are so dense that many lines ω_n exist in any interval $(\omega, \omega + \Delta\omega)$. Here $\Delta\omega \equiv 2\pi/T$ is the resolution limit of the spectrum when the observation is performed during time T. We often meet such "chaotic" functions in dynamical systems $\ddot{x}_i = f_i(x_1 \cdots x_i \cdots x_I), (i = 1, \cdots, I)$, where nonlinearity of the interacting forces is sufficiently strong.

4. Power density spectrum defined by

$$p(\omega) = \sum_{\omega<\omega_n<\omega+M\Delta\omega} \frac{1}{2}a_n^2/(M\Delta\omega) \quad (2)$$

is proportional to $1/\omega$. Here $M \gg 1$, but $M\Delta\omega \ll \Omega$.

For convenience, we express the $f(t)$ in a complex form

$$f(t) = \sum_{n=-N}^{N} c_n e^{i\omega_n t}, \; where \; c_n = \frac{1}{2}a_n e^{i\delta n}, c_{-n} = \bar{c_n}. \quad (3)$$

We can group spectra within $\Delta\omega$ and express

$$f(t) = \sum_{m=\mathcal{M}}^{\mathcal{M}} C_m e^{i\omega_m t}, \quad where \;\; C_m = \sum_{m\Delta\omega<\omega_n<(m+1)\Delta\omega} c_n \quad and \quad \omega_m = m\Delta\omega. \quad (4)$$

The complex amplitude C_m is a result of many random walks in the complex plane, and its deviation is suppressed by summing over M successive $|C_m|^2$s

$$P_k = \sum_{\mu=1}^{M} |C_{kM+\mu}|^2. \quad (5)$$

Smooth power density spectrum $p(\omega)$ is obtained by connecting $\omega_k - P_k$ curve, where $\omega_k = kM\Delta\omega$.

SUM OF $1/F$ FLUCTUATION

Theorem 1. Sum of two $1/f$ fluctuations is again a $1/f$ fluctuation.

Suppose we have two $1/f$ functions

$$x_1(t) = \sum_{m=-\mathcal{M}}^{\mathcal{M}} C_m^{(1)} e^{i\omega_m t}, \; x_2(t) = \sum_{m=-\mathcal{M}}^{\mathcal{M}} C_m^{(2)} e^{i\omega_m t}. \qquad \square \quad (6)$$

The power spectrum of $x_1 + x_2$ is :

$$\begin{aligned} P_k^{(1)+(2)} &= \sum_{\mu=1}^{M} |C_{kM+\mu}^{(1)} + C_{kM+\mu}^{(2)}|^2 = \sum_{\mu=1}^{M} (|C_{kM+\mu}^{(1)}|^2 + |C_{kM+\mu}^{(2)}|^2) + crossterms \\ &= (P_k^{(1)} + P_k^{(2)})(1 + O(\frac{1}{\sqrt{M}})) \end{aligned} \quad (7)$$

as the phases of $C_m^{(1)}$ and $C_m^{(2)}$ are independent. Therefore $P^{(1)+(2)}(\omega) = P^{(1)}(\omega) + P^{(2)}(\omega)$.

POWER DENSITY OF PRODUCT

The product of $1/f$ functions (6) has power spectrum

$$P_k^{(1)(2)} = \sum_{m=kM+1}^{kM+M} |\sum_{\mu=-\mathcal{M}}^{\mathcal{M}} C_{m-\mu}^{(1)} C_\mu^{(2)}|^2. \quad (8)$$

Let us investigate the summand:

$$\begin{aligned} |\sum_{\mu} C^{(1)}_{m-\mu} C^{(2)}_{\mu}|^2 &= |\sum_{\mu}\sum_{\mu'} C^{(1)}_{m-\mu}\bar{C}^{(1)}_{m-\mu'} C^{(2)}_{\mu}\bar{C}^{(2)}_{\mu'}|^2 \\ &= \sum_{\mu}^{\mathcal{M}} |C^{(1)}_{m-\mu}|^2 |C^{(2)}_{\mu}|^2 + \sum_{\mu}^{\mathcal{M}}\sum_{\mu'\neq\mu}^{\mathcal{M}} C_{m-\mu}\bar{C}_{m-\mu'} C_{\mu}\bar{C}_{\mu'}. \end{aligned} \quad (9)$$

The first term on the right has $\mathcal{M}$ positive terms while the second term has $\frac{1}{2}\mathcal{M}(\mathcal{M}-1)$ terms having both signs (random walk). Absolute values of the two terms are of the same order. However, by summing M such terms, the first term dominates :

$$P^{(1)(2)}_k \underset{M\to\infty}{\longrightarrow} \sum_{m=kM+1}^{kM+M} \sum_{\mu}^{\mathcal{M}} |C^{(1)}_{m-\mu}|^2 |C^{(2)}_{\mu}|^2. \quad (10)$$

Thus we arrive at the lemma.

Lemma. Product of chaotic functions has the following power density spectrum

$$p_{xy}(\omega) = \int_{-\infty}^{\infty} p_x(\omega-\omega')p_y(\omega')d\omega'. \quad (11)$$

PRODUCT OF $1/F$ FLUCTUATION

Theorem 2. Product of $1/f$ fluctuation is again a $1/f$ fluctuation (approximately). We prove $p_{xy}(\omega) \fallingdotseq [c/\omega]$ when $p_x(\omega) = [1/\omega]$, $p_y(\omega) = [1/\omega]$.

Here we use notation [ ] to designate cut off :

$$[1/\omega] = 1/\omega \quad in\ \Delta\omega < |\omega| < \Omega\,; \quad = 0 \quad elsewhere. \quad \Box \quad (12)$$

This theorem can be proven in at least three different ways. The simplest one is the following. $[N/\omega]$ is a normalized pseudo δ−function, $\delta'(\omega)$, in the sense that most of its integral comes from the vicinity of the origin. If $x = \delta'(\omega)$ and $y = \delta'(\omega)$ then

$$p_{xy}(\omega) = \int \delta'(\omega-\omega')\delta'(\omega')d\omega' = 2\delta'(\omega). \quad (13)$$

If we normalize $f(t)$ so that $\int_{-\infty}^{\infty} p(\omega)d\omega = \frac{1}{2}$, and denote it by $\mathcal{F}(t)$, then

$$\prod_i a_i\mathcal{F}_i \overset{p}{=} (\prod a_i)\mathcal{F}, \qquad \sum_i a_i\mathcal{F}_i \overset{p}{=} \sqrt{\textstyle\sum_i a_i^2}\,\mathcal{F}, \quad (14)$$

where equality means that both sides have equal power-density spectrum.

NUMERICAL DEMONSTRATIONS

We prepare $1/f$ fluctuating functions

$$x(t),\ y(t) = \mathcal{N}\sum_{m=1}^{1024} \frac{1}{\sqrt{\omega_m}}\cos(\omega_m t + \delta_m) \quad (15)$$

where $\omega_m = m/1024$, and $\mathcal{N}$ is the normalization factor. Power density of x, $x+y$, xy, x^2 are obtained from Fourier transform of these functions of time. The results are shown in the figures, which demonstrate the "stability property" of the $1/f$ fluctuation.

CONCLUSION

If displacements x are $1/f$ fluctuations, then forces F (nonlinear functions of x) are also $1/f$ fluctuations owing to our theorems. These forces in turn cause $1/f$ fluctuation in displacements in the low frequency range where $x \propto F$. Thus $1/f$ fluctuation can generally be solutions of nonlinear dynamical systems.

REFERENCES

1) P.H.Handel, Noise in Physical Systems and $1/f$ Fluctuations (Ohmsha 1991) p.151

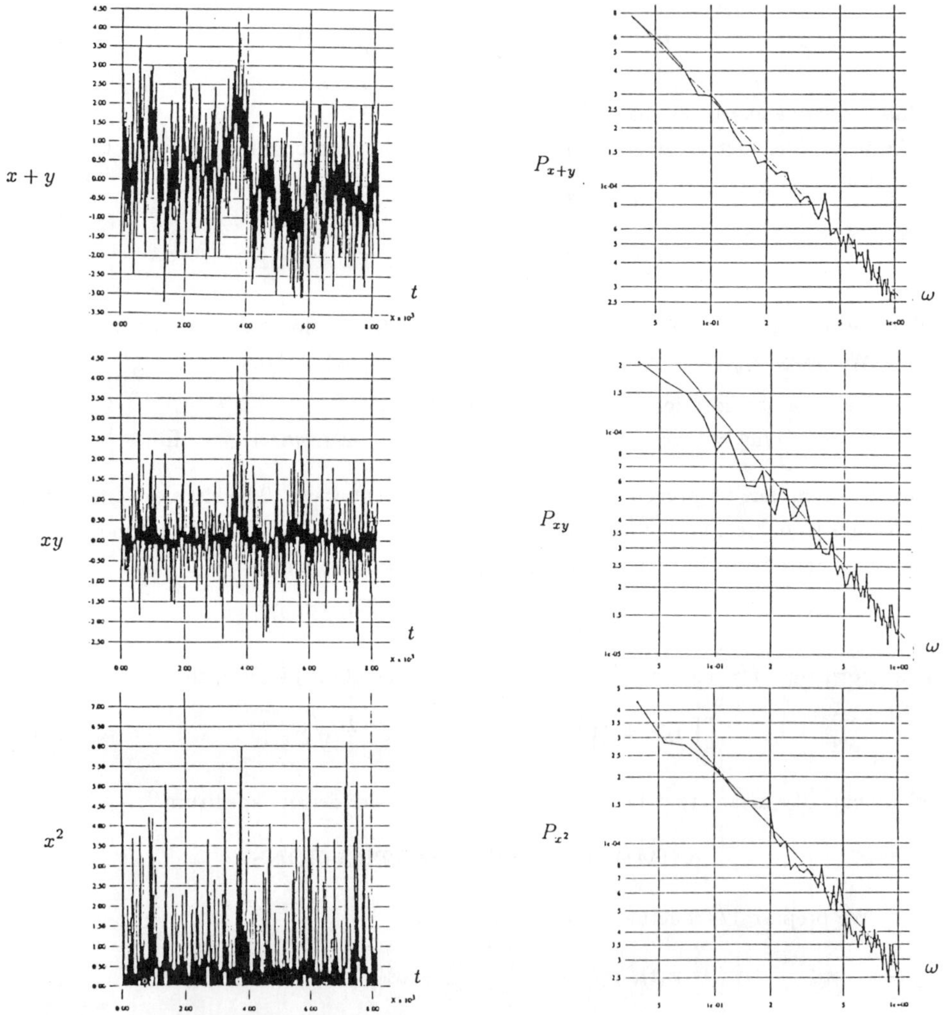

Figures. Sum and product of $1/f$ fluctuation $x(t)$ and $y(t)$.
Note that all power-density spectra are $1/f$, and agree with theory.

XVI. STATISTICAL PHYSICS OF NOISE IN SEMICONDUCTOR MATERIALS

TWO-POINT OCCUPATION - CORRELATION FUNCTIONS AS OBTAINED FROM THE STOCHASTIC BOLTZMANN TRANSPORT EQUATION (SEMI-CLASSICAL) AND THE MANY-BODY MICROSCOPIC MASTER EQUATION (QUANTAL)

Ayivi Huisso* and Carolyne M. Van Vliet

Department of Electrical and Computer Engineering
Florida International University, Miami, Florida 33199

ABSTRACT

Correlations in single particle state occupancies, such as $<n(\mathbf{k},\mathbf{r},t)n(\mathbf{k}',\mathbf{r}',0)>$ for a solid with extended states $|\mathbf{k}>$ can be obtained in two ways. First a stochastic Boltzmann equation provided with a Langevin source $\xi(\mathbf{k},\mathbf{r},t)$ can be employed using methods which essentially go back to Uhlenbeck and Ornstein[1]. Secondly one can use the quantum statistical many-body master equation for the reduced density operator $\rho^R(t)$ (after the Van Hove limit) and compute the correlations to any order from the associated equation for the generating functional.

1. STOCHASTIC BOLTZMANN EQUATION DESCRIPTION

The stochastic Boltzmann equation (or Boltzmann-Langevin equation) was introduced at about the same time (1970-1971) by Fox and Uhlenbeck[2] (Rockefeller U.) and by Van Vliet[3] (CRM, Univ. de Montreal). Correlations in state occupancies $n^p(y,t)$, where p is the species index and y the single particle attributes $\mathbf{k}$ or $(\mathbf{k},\mathbf{r})$ are obtained by a Green's operator solution of the linearized transport equations for the fluctuations,

$$[\frac{\partial}{\partial t} + \Lambda]\alpha(y,t) = \xi(y,t) \tag{1}$$

where $\alpha = \{\alpha^p\} \equiv \{\Delta n^p\}$, and $\xi = \{\xi^p\}$ is the Langevin source; both are vectors in a Hilbert space $\mathcal{H} = \mathcal{E}^s \otimes \mathcal{L}^2(\mathcal{D},R^l)$ where $y \in \mathcal{D}$. In the case of one species (s=1), Λ is the linearized Boltzmann operator $\Lambda = \mathcal{D} + B^l$, where $\mathcal{D}$ relates to the streaming part and B^l is the linearized collision operator. The Boltzmann-Langevin equation is a transformation in the Hilbert space $\mathcal{H}$. The formal solution of (1) is

* Permanent address: Centre de Recherches Mathématiques, Université de Montréal, Montréal, Qué H3C 3J7, Canada.

$$\alpha(y,t) = \int_0^{t+0} dt' g_y(t,t')\xi(y,t') + g(t,0)\alpha(y,0), \qquad (2)$$

where g_y is the Green's operator (propagator) $\theta(t-t')\exp[-\Lambda(t-t')]$, $\theta(t)$ being the Heaviside function. To obtain the correlations one follows the procedure of Ornstein and Uhlenbeck. For the sources we assume white noise, i. e.,

$$\Xi(y,y')\delta(t-t') = \langle\xi(y,t)\xi^{tr}(y',t')\rangle. \qquad (3)$$

The time-displaced fluctuation-correlation function and the stationary covariance functions are obtained from

$$\Phi(y,y',u) = \lim_{t\to\infty}\langle\alpha(y,\ t+u)\alpha^{tr}(y',t)\rangle_{cond} \qquad (4)$$

$$\Gamma(y,y') \equiv \langle\Delta n(y)\Delta n^{tr}(y')\rangle = \lim_{t\to\infty,u\to 0}\langle\alpha(y,\ t+u)\alpha^{tr}(y',t)\rangle_{cond}, \qquad (5)$$

where "cond" stands for a subensemble in which $\alpha(y,0)$ is fixed. Note that Φ, Γ, and Ξ are elements of the tensor product space $\mathcal{H}\otimes\mathcal{H}$. We further define scalar products both in $\mathcal{H}$ and $\mathcal{H}\otimes\mathcal{H}$; as basis in $\mathcal{H}$ we choose the eigenfunctions Φ_k of Λ, while the eigenfunctions Ψ_l of the adjoint operator $\tilde{\Lambda}$ serve as a basis in the dual space $\tilde{\mathcal{H}}$. One then obtains easily [Ref.1, Eq. (2.23)],

$$\Phi(y,y',u) = \sum_{kl}\exp(-\lambda_k u)\frac{\Phi_k(y)\Phi_l^{tr}(y')}{\lambda_k+\lambda_l}(\Xi,\ \Psi_k\Psi_l^{tr}), \qquad u\geq 0. \qquad (6)$$

Using the transposition property, $\Phi(y,y',u) = \Phi^{tr}(y',y,-u)$, we obtain likewise,

$$\Phi(y',y,u) = \sum_{kl}\exp(\lambda_k u)\frac{\Phi_k(y)\Phi_l^{tr}(y')}{\lambda_k+\lambda_l}(\Psi_l\Psi_k^{tr},\ \Xi)^*, \qquad u\leq 0. \qquad (7)$$

Forming the scalar product, $(\Phi(y,y',u),\ \Psi_m\Psi_n^{tr})$, multiplying by $\lambda_m+\lambda_n$ and noting $\lambda_m\Psi_m^+ = \Psi_m^+\tilde{\Lambda}^+$, $\lambda_n\psi_n^* = \mu_n\Psi_n^* = \tilde{\Lambda}^*\Psi_n^*$ and employing Green's theorem in $\mathcal{H}\otimes\mathcal{H}$, one finds the correlation theorems, not given hitherto,

$$\Lambda_y\Phi(y,y',u) + \Phi(y,y',u)\overleftarrow{\Lambda}_{y'}{}^{tr} = \exp(-\Lambda_y u)\Xi(y,y'), \qquad u\geq 0. \qquad (8a)$$

$$\Lambda_y\Phi(y,y',u) + \Phi(y,y',u)\overleftarrow{\Lambda}_{y'}{}^{tr} = \Xi(y,y')\overleftarrow{\exp}(\Lambda_{y'}{}^{tr}u), \qquad u\leq 0. \qquad (8b)$$

For u→0+ or 0−, respectively, one obtains the covariance theorem or "lamda theorem," also obtained recently by Kogan[4] using a generalized Keldysh diagrammatic technique:

$$\Lambda_y \Gamma(y,y') + \Gamma(y,y')^{\leftarrow}\Lambda_{y'}^{tr} = \Xi(y,y'). \tag{9}$$

The solutions of the covariance theorem have been discussed abundantly before[3]. As to the new "fluctuation-correlation function theorems", (8a) and (8b), they admit the Van Vliet - Fassett solution

$$\Phi(y,y',t) = \exp(-\Lambda_y t)\Gamma(y,y'), \qquad t \geq 0 \tag{10a}$$

$$\Phi(y,y',t) = \Phi(y',y,-t) = \Gamma(y,y')\exp(\Lambda_{y'}^{tr}\, t), \qquad t \leq 0, \tag{10b}$$

as is found by substitution**).

Now some notes on the physical content of these theorems. Let y→k. First, we noted before[3] that the source kernel $\Xi(\mathbf{k},\mathbf{k}')$ is equal to the Fokker Planck moment of second order, $B(\mathbf{k},\mathbf{k}')$. The latter follows from the many-body transition rate functional $W(\{n(\mathbf{k})\} \to \{\bar{n}_{\mathbf{k}}\}) \equiv W_{\gamma\bar{\gamma}}$, where γ and $\bar{\gamma}$ are occupation-number states in the Fock space, see section 2 below. Straightforward computation from Fermi's golden rule shows that B contains diagonal and off-diagonal elements, contrary to what is stated by Nougier and Vaissiere[5]. Secondly notice that (8a) and (8b) are inhomogeneous equations, relating the correlations to the physical interactions (matrix element squared of interaction Hamiltonian) entering in $W_{\gamma\gamma}$ and thus in $\Xi(\mathbf{k},\mathbf{k}')$. Yet, contrary to a claim in Ref[5], Φ also satisfies the homogeneous Boltzmann equation, $[\frac{\partial}{\partial t} + \Lambda]\Phi(y,y',t) = 0$, as noted by Gantsevich, et al.[6], and as is easily proven by using Bayes' theorem, for $<\alpha(y,t)\alpha(y',o)> = <<\alpha(y,t)\alpha(y',o)>_{cond}\alpha(y',0)>$. The solution of the homogeneous Boltzmann equation for Φ yields no correlations associated with primary physical causes, however; rather, this contribution can (must) be used to satisfy the canonical constraint in neutral plasmas or solids.

2. THE MOMENT-GENERATING FUNCTIONAL OF THE MASTER EQUATION.

Vasilopoulos and Van Vliet consider the Hamiltonian

$$H = H_o + \lambda V - AF(t), \tag{11}$$

where H_o represents the electron-phonon (or electron-impurity gas), λV the concomitant interactions, and AF(t) the effect of an external field, for which the master equation was derived by Van Vliet[7]. Let $|\gamma\rangle \equiv |\{n_\zeta\}\rangle$ be a many-body state in the occupation-number representation. The generating functional for the reduced density operator ρ^R was computed elsewhere[8]; it satisfies

$$\frac{\partial}{\partial t}\langle \exp[-\sum_\zeta s_\zeta n_\zeta]\rangle_t = \sum_{k=1}^{\infty}\langle[(-1)^k/k!]\exp[-\sum_\zeta s_\zeta n_\zeta]\sum_{\zeta_1\ldots\zeta_k} s_{\zeta_1}\ldots s_{\zeta_k} F(n_\zeta)\rangle_t + \text{ext. field term.} \tag{12}$$

Here $F_k(n_\zeta)$ is the k-th order Fokker-Planck moment,

$$F_k(n_\zeta) = \sum_{\bar{\gamma}}(\bar{n}_{\zeta_1}-n_{\zeta_1})\ldots(\bar{n}_{\zeta_k}-n_{\zeta_k})W_{\gamma\bar{\gamma}}, \tag{13}$$

where $W_{\gamma\bar{\gamma}}$ is the master transition rate given by Fermi's golden rule. When we now differentiate (12) to s_{ζ_1}, setting afterwards all s_ζ equal to zero, we need only keep the term with k=1. This yields the Boltzmann equation in the form

$$\frac{\partial\langle n_{\zeta_1}\rangle_t}{\partial t} = \langle F_1(n_{\zeta_1})\rangle_t + \beta F(t)\langle n_{\zeta_1}\rangle\langle\gamma|(A^R)_d|\gamma\rangle_{eq}. \tag{14}$$

The first part represents the collisions (see below). The quantum mechanical form for the streaming (second part) has no classical analog, as elaborated and discussed elsewhere[7]. Likewise, differentiating to ζ_1 and ζ_2, setting afterwards all n_ζ equal to zero, keeping the terms with k=1,2, leads to the second equation of the master hierarchy[8].

$$\frac{\partial\langle n_{\zeta_1}n_{\zeta_2}\rangle_t}{\partial t} = \langle n_{\zeta_1}F_1(n_{\zeta_2})\rangle_t + \langle n_{\zeta_2}F_1(n_{\zeta_1})\rangle_t + \langle F_2(n_{\zeta_1},n_{\zeta_2})\rangle_t$$

$$-\beta F(t)\sum_\zeta\langle n_{\zeta_1}n_{\zeta_2}\mathfrak{B}_\zeta n_\zeta\rangle_{eq}\,\alpha_\zeta + \beta F(t)\sum_\zeta\langle n_{\zeta_1}n_{\zeta_2}n_\zeta\rangle_{eq}\,\dot{\alpha}_\zeta. \tag{15}$$

Again, the first part refers to the collisions, while the streaming contains the position (α) and velocity ($\dot{\alpha}$) diagonal matrix elements.

For the Fokker-Planck moments we need the explicit matrix elements $|\langle\bar{\gamma}|\lambda V|\gamma\rangle|^2$ involving electron-phonon or electron-impurity interactions in second quantization form. One finds that connected states $|\gamma\rangle = |\{n_\zeta\}, \{N_q\}\rangle$

and $|\bar{\gamma}\rangle = |\{\bar{n}_\zeta\}, \{\bar{N}_q\}\rangle$ require, considering a term $\zeta'\zeta'\eta''\eta''$ of the interaction series,

$$\bar{n}_\zeta - n_\zeta = (1-2n_\zeta)(\delta_{\zeta\zeta'} + \delta_{\zeta\zeta''}), \tag{16a}$$

$$\bar{N}_\eta - N_\eta = \delta_{\eta\eta'} + \delta_{\eta\eta''}. \tag{16b}$$

Assuming an adiabic approximation for the Bosons, one obtains, using (16a) and $n_\zeta{}^2 = n_\zeta$ (for Fermions)+, denoting by $\langle\ \rangle_b$ an average over the Bosons,

$$\langle F_1(n_{\zeta 1})\rangle_b = \sum_{\bar{\gamma}} \langle(\bar{n}_{\zeta 1} - n_{\zeta 1})W_{\gamma\bar{\gamma}}\rangle_b = \sum_{\zeta'\zeta''} w_{\zeta'\zeta''} n_{\zeta'}(1-n_{\zeta''})(\bar{n}_{\zeta 1} - n_{\zeta 1})$$

$$= \sum_{\zeta'\zeta''} w_{\zeta'\zeta''} n_{\zeta'}(1-n_{\zeta''})(1-2n_{\zeta 1})(\delta_{\zeta\zeta'} + \delta_{\zeta\zeta''})$$

$$= -\sum_{\zeta}[w_{\zeta 1\zeta}\, n_{\zeta 1}(1-n_\zeta) - w_{\zeta\zeta 1}\, n_\zeta(1-n_{\zeta 1})] = -\mathfrak{B}_{\zeta 1}\, n_{\zeta 1} \tag{17}$$

where $\mathfrak{B}_{\zeta 1}$ is the full (non-linearized) Boltzmann operator for Fermions. Thus, (14) leads to the full quantum Boltzmann equation (QBE). Likewise, one finds for $F_2(n_{\zeta 1}n_{\zeta 2})$++, employing (16a),

$$F_2(n_{\zeta 1}n_{\zeta 2}) = \sum_{\bar{\gamma}} \langle(\bar{n}_{\zeta 1} - n_{\zeta 1})(\bar{n}_{\zeta 2} - n_{\zeta 2})\rangle_b \tag{18}$$

$$= \sum_{\zeta'\zeta''} w_{\zeta'\zeta''} n_{\zeta'}(1-n_{\zeta''})(1-2n_{\zeta 1})(1-2n_{\zeta 2})(\delta_{\zeta 1\zeta'} + \delta_{\zeta 1\zeta''})(\delta_{\zeta 2\zeta'} + \delta_{\zeta 2\zeta''})$$

$$= -[w_{\zeta 1\zeta 2}\, n_{\zeta 1}(1-n_{\zeta 2}) + w_{\zeta 2\zeta 1} n_{\zeta 2}(1-n_{\zeta 1})] + \delta_{\zeta 1\zeta 2} \sum_{\zeta}[w_{\zeta 1\zeta}\, n_{\zeta 1}(1-n_\zeta) + w_{\zeta\zeta 1}\, n_\zeta(1-n_{\zeta 1})]$$

which, for non-degenerate statistics is exactly the form given by Gantsevich and Gurevich[9]. Our result is, however, more general *since no linearization has taken place.* Eq. (15) now reads, for $t\to\infty$ and in the absence of an external field,

$$\langle n_{\zeta 1}\mathfrak{B}_{\zeta 2} n_{\zeta 2}\rangle + \langle n_{\zeta 2}\mathfrak{B}_{\zeta 1} n_{\zeta 1}\rangle = B(n_{\zeta 1}n_{\zeta 2}) \equiv \Xi(\zeta_1, \zeta_2) \tag{19}$$

where we wrote $F_2 \equiv B$ for the second order FP moment, while further we notice that $\Lambda \to \mathfrak{B}$. So, Eq. (19) is the generalization of the "Λ-theorem", Eq. (9). For equilibrium (no streaming) the solution is that of the grand canonical ensemble. For hot electrons a generalization of (15) can be derived.

In conclusion, this article differs from Ref 6 in that we stress that the true correlations – not associated with constraints – need the *inhomogeneous* equations (8a), (8b) and (9). If the Boltzmann operator is a transformation in the space $\mathcal{H}$, the equations for the correlation and covariance functions pertain to the space $\mathcal{H}\otimes\mathcal{H}$.. However, part of the correlations may be governed by the homogeneous Boltzmann equation, which is denied in Ref. 5. Moreover, as explicitly shown in (18), the Langevin source in any occupation representation, i. c. in k-space states, is *not* diagonal, i. e. their Eq. (16) is erroneous (see also Ref. 10). We believe that Nougier and Vassiere make a *conceptual error* in that $\Xi(k_1,k_2)$ (or the Langevin source spectrum $S_\xi(\mathbf{k}_1,\mathbf{k}_2)$ would represent the scattering of a $\mathbf{k}_1$ state and a $\mathbf{k}_2$ state electron with two different phonons. However, the master equation treatment and resulting Fokker-Planck moments show that $S_\xi(\mathbf{k}_1,\mathbf{k}_2)$ derives from the simple transition rates $Q(\mathbf{k}\rightarrow\mathbf{k}',q)$ and $Q(\mathbf{k},\mathbf{q}\rightarrow\mathbf{k}')$ which determine $w_{\mathbf{kk}'}$ and which involve absorption or emission of a single phonon, thus negating Nougier's argument.

The authors acknowledge a discussion with Dr. Katilius at the Budapest conference which stimulated them to reconsider this problem and they thank Dr. Kogan for a copy of his paper. This research was supported by NSERC grant A9522.

** Substitute (10a) in first terms on l.h.s. of (8a), and (10b) in second term on l.h.s. of (8a), noting that the effect of the retarded Green's operator cancels the effect of the advanced Green's operator.

+ $n_\zeta(1-2n_\zeta) = -n_\zeta$, $(1-2n_\zeta)^2 = 1$.

++ This result differs from (3.17) in Ref 8, which is incomplete.

REFERENCES

1. G. E. Uhlenbeck and L. S. Ornstein, Phys Rev 36, 823-841.
2. R. F. Fox and G. E. Uhlenbeck, Phys Fluids, 13, 2881(1970).
3. K. M. Van Vliet, I. J. Math Phys 12, 1981-1998(1971); II. ibid. 1998-2012(1971).
4. Sh. Kogan, Phys. Rev. A 44, 8072 (1991).
5. J. -P. Nougier and J. C. Vassiere, Phys. Rev. B3 7, 8882-8887(1988).
6. S. V. Gantsevich, V. L. Gurevich and R. Katilius, Il Nuovo Cimento (rivista) 2, 1-87(1979).
7. K. M. Van Vliet, J. Math. Phys. 20, 2573-2595 (1979).
8. P. Vasilopoulos and C. M. Van Vliet, Can. J. Phys. 61, 102-112 (1983).
9. S. V. Gantsevich and V. L. Gurevich "Comment on Fluctuations of the Hot-Carrier State-Occupancy Function in Homogeneous Semiconductors" (1988). Unpublished.
10. M. Lax, Revs Mod Phys. 38, 541(1966).

BROWNIAN MOTION THROUGH POTENTIAL BARRIERS WITH DIFFERENT SHAPE, WIDTH AND HEIGHT.

N.V.Agudov, A.N.Malakhov, A.L.Pankratov.
State University, N.Novgorod 603600, Russia.

ABSTRACT.

The Brownian motion in different piecewise linear ("triangular" and "step like") and in piecewise parabolic metastable potential profiles are considered. The exact life times of the metastable states are presented and discussed. The limitations of the Kramers' formula are given.

INTRODUCTION.

The surmounting of a potential energy barrier by particle undergoing Brownian motion plays a significant role in a wide variety of a physical, chemical and biology processes, and the prediction of a barrier crossing rates is therefore a problem of considerable theoretical and practical importance.

In a present paper we consider the exact life times of metastable states with different shape, which were derived using results and method obtained in Ref.[1-4]. They are summarized briefly in paper [8] of A.N.Malakhov and N.V.Agudov.

THE LIFE TIMES OF METASTABLE STATES IN POTENTIAL PROFILES WITH DIFFERENT SHAPES.

For the discussion of the present paper we use the terminology of the Brownian motion of particle with coordinate x in potential $U(x)$. The equation of motion is the Langevin equation in overdamped limit:

$$\frac{dx(t)}{dt} = -\frac{1}{h}\frac{dU(x)}{dx} + \eta(t) , \tag{1}$$

$$\langle\eta(t)\eta(t+\tau)\rangle = D\delta(\tau), \quad \langle\eta(t)\rangle = 0.$$

Here $D=2kT/h$ is intensity of white noise. To the Brownian motion (1) correspond the Fokker-Planck equation for the probability density $W(x,t)$:

$$\frac{\partial W(x,t)}{\partial t} = \frac{\partial}{\partial x}\left[\frac{1}{h}\frac{dU(x)}{dx} W(x,t)\right] + \frac{D}{2}\frac{\partial^2}{\partial x^2} W(x,t) \tag{2}$$

The initial and boundary condition are $W(x,0)=\delta(x)$ and $W(\pm\infty,t)=0$.

Let's consider dimensionless metastable potential

profiles with different shapes u_1, u_2, u_3, shown on Fig. 1, 2, 3 accordingly:

$$u_1(x)=\frac{2U_1(x)}{Dh}=\begin{cases}-2ax, & x \leqslant 0\\ 2ax, & 0 \leqslant x \leqslant L\\ -2a(x-2L), & x \geqslant L\end{cases}$$

$\beta=2aL/kT$

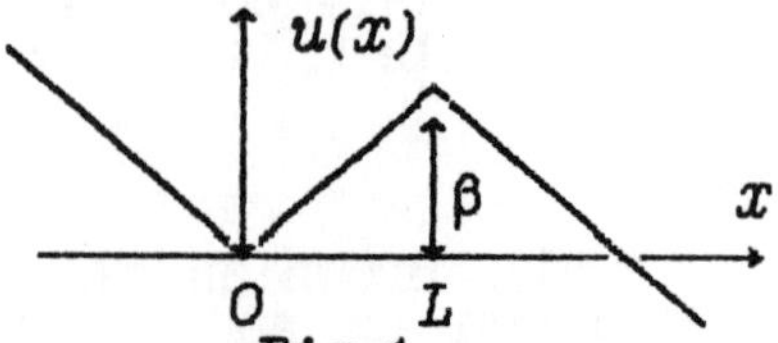

Fig.1.

$$u_2(x)=\frac{2U_2(x)}{Dh}=\begin{cases}\infty, & x=-L/2\\ 0, & -L/2 < x < L/2\\ \beta, & L/2 < x < 3L/2\\ \infty, & x=3L/2\end{cases}$$

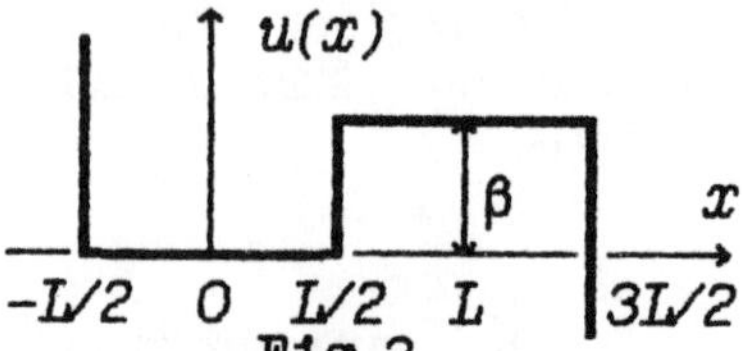

Fig.2.

$$u_3(x)=\frac{2U_3(x)}{Dh}=\begin{cases}\omega x^2/2, & x \leqslant L/2\\ -\omega(x-L)^2/2, & x \geqslant L/2\end{cases}$$

$\beta=\omega L^2/4$

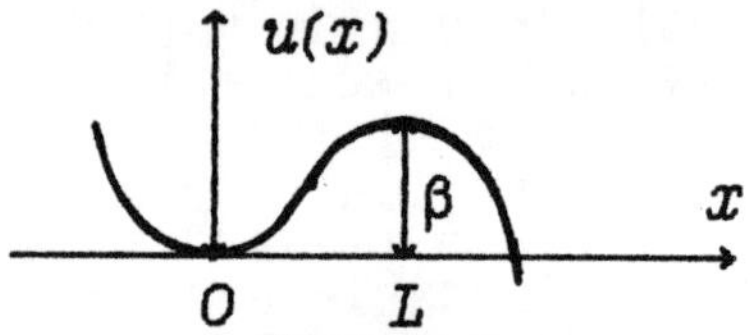

Fig.3.

Here for all metastable states L is the distance between minimum and maximum of potential profile - the width of the potential barrier. $\beta=2U(L)/Dh=E/kT$ is dimensionless height of the barrier.

The life time of the metastable state is determined as relaxation time of the probability $Q(t)$ that Brownian particle is to the right of the barrier top: $Q(t)=\int_L^\infty W(x,t)dx$. The Laplace transformed probability is $\hat{Q}(s)=\int_0^\infty Q(t)\exp(-st)dt$. Using the formula for the life time τ proposed in Ref.[3]:

$$\tau=\lim_{s\to 0}\,(1-s\hat{Q}(s))/s \tag{3}$$

we get the following expressions for the three potential profiles accordingly

$$\tau_1=\vartheta_D 2\{4e^\beta-\beta-3\}/\beta^2 \tag{4}$$

$$\tau_2=\vartheta_D\{4e^\beta+3\}/2 \tag{5}$$

$$\tau_3=\vartheta_D\{[\mathrm{erfc}(-\sqrt{\beta/2}\,)]^2 e^\beta \pi/2+2A(\sqrt{\beta})\}/2\beta \tag{6}$$

where $A(z)=A^0(z)+A^1(z)$, $A^0(z)=z^2/2!+2z^4/4!+2*4z^6/6!+\ldots$,

$$A^1(z)=\sqrt{\pi/2}\,(z+z^3/3!+3z^5/5!+\dots),\quad erfc(z)=2\int_z^\infty exp(-u^2)du/\sqrt{\pi}$$

These formulas are exact and correct for any relation E/kT. From these formulas it follow that when the all profiles have the same width and height of the barrier the life times of all metastable states are different, hence they are essentially defined by shape of the profiles.

The piecewise linear potentials $u_1(x)$ and $u_2(x)$ were considered before in the literature [5,6]. The Laplace transformed probability density in potential $u(x)=u_1(x)$, which coincides with corresponding form of general solution obtained in Ref.[3], was derived in Ref.[5]. But the life time of metastable state was derived only in the limit of the large barrier $\beta \gg 1$ and in this case coincides with (4): $\tau_1=8\vartheta_D e^{\beta}/\beta^2$.

In the Ref.[6] the Brownian motion in potential profile $u(x)=u_2(x)$ was studied. The exact life time $\tau =\vartheta_D\{4e^{\beta}+4/3\}/2$ was obtained. This expression differ from the formula (5) by the numeric coefficient, because in Ref.[6] the life time was defined as the relaxation time of the probability $P(t)$ that particle is at the $-L/2 < x < L/2$, but not at $x \geq L$, as in present paper. Moreover in Ref.[6] initial condition is the equal probability density in potential well $(-L/2 \leq x \leq L/2)$ what also differs from accepted here.

In order to analyze the exact life times given by formulas (4)-(6), we consider its dependence on barrier width L, height E and temperature kT.

The value L enters into all expressions equally in factor $\vartheta_D=L^2/D$. It was shown in Ref.[2] that ϑ_D is a time of the probability distribution spreading in absence of the potential barrier, when $U(x)=const$. If we change the barrier width at $E=const$, $kT=const$, then only ϑ_D change. Thus, we can confirm, that the shape of metastable potential doesn't influence on the life time dependence on barrier width. The life time of the metastable state is always proportional to the L^2. The slope of potential walls change together with L.

Now let's consider the influence of the barrier height E on the life time of the metastable state. If the temperature $kT=const$ we can change the barrier's height E by two ways: First - by $L=const$, then the slope of potential profile walls is changed. In "triangular" potential metastable state the slope is characterized by value a and in "parabolic" metastable state - by ω. In "rectangular" ("step like") metastable state the walls slope is always infinite and doesn't change. The second way to change the barrier height is to fix the slope of the potential profile walls and to change the barrier width.

Let's consider the first way to change E: $kT=const$, $L=const$. In this case the changing of dimensionless height $\beta=E/kT$ correspond to the barrier height changing in (4)-(6). The functions $\tau(\beta)$ for the three metastable states are absolutely various, since the influence of potential profile shape is essential.

In the limit of the high barrier $e^{\beta}\gg\beta$ from the formulas (4)-(6) we can get accordingly:

$$\tau_1=8\vartheta_D e^{\beta}/\beta^2 \tag{7}$$

$$\tau_2=\vartheta_D 2e^{\beta} \tag{8}$$

$$\tau_3=\vartheta_D \pi e^{\beta}/\beta \tag{9}$$

Here is the same factor e^{β} in all three life times. This factor is consequence of the large barrier. Together with it the distinctions caused by the different potential profiles shape are remain in prefactors. Thus, from the formulas (7)-(9) it follows that in the case $L=const$, $kT=const$ the potential profiles shape essentially affects the barrier height E dependence of the life time even if $E\gg kT$.

We consider now the second way to change barrier height E: $kT=const$, $\omega=const$, $a=const$ - the slope of potential walls isn't changed. Let's rewrite the formulas (4)-(6) as:

$$\tau_1=[4e^{\beta}-\beta-3]/2Da^2=\frac{1}{2}\vartheta_1[4e^{\beta}-\beta-3] \tag{10}$$

$$\tau_2=L^2[4e^{\beta}+3]/2D=6\vartheta_2[4e^{\beta}+3] \tag{11}$$

$$\tau_3=2[\{erfc(-\sqrt{\beta/2})\}^2 e^{\beta}\pi/2+2A(\sqrt{\beta})]/\omega D= \tag{12}$$

$$=2\vartheta_3[\{erfc(-\sqrt{\beta/2})\}^2 e^{\beta}\pi/2+2A(\sqrt{\beta})],$$

The changing of the barrier height in formulas (10)-(12) correspond to the changing of the parameter $\beta=E/kT$. One can show, that the times ϑ_1, ϑ_2, ϑ_3, entering into the all formulas (10)-(12), is the relaxation times in V-shaped ($u(x)=2a|x|$; $\vartheta_1=1/Da^2$), rectangular ($u(x)=0$ when $-L/2<x<L/2$, $u(\pm L/2)=\infty$; $\vartheta_2=L^2/12D$) and parabolic ($u(x)=\omega x^2/2$; $\vartheta_3=1/\omega D$) potential wells. If we take such potentials u_1, u_2, u_3 that the times $\vartheta_1=\vartheta_2=\vartheta_3=\vartheta$, than

the distinctions between life times of considered metastable states remain only in expressions put in square brackets in formulas (10)-(12).

In the limits of high and low barrier from the formulas (10)-(12) it follows that

if $\beta \ll 1$	if $e^{\beta} \gg \beta$
$\tau_1=\frac{1}{2}\vartheta(1+3\beta+\ldots)$	$\tau_1=2\vartheta e^{\beta}$
$\tau_2=6\vartheta(7+4\beta+\ldots)$	$\tau_2=24\vartheta e^{\beta}$
$\tau_3=2\vartheta(\pi/2+2\beta+\ldots)$	$\tau_3=4\pi\vartheta e^{\beta}$

i.e. the life times of metastable states are distinct only by numeric coefficients.Thus, when $a=const$, $\omega=const$, $kT=const$ the shape of potential profile influence on barrier height dependence of life time only when $E\approx kT$ ($\beta\approx1$). In the limit cases of low and high barrier the life time equal to $\tau=\vartheta(l+m\beta)$ and $\tau=n\vartheta\, e^{\beta}$ accordingly, where l,m and n are numeric coefficients.

Now let's consider the temperature dependence of the metastable states life times. For it we rewrite the formulas (4)-(6) in the following form:

$$\tau_1=\vartheta_E\{4e^{\beta}-\beta-3\}/\beta \qquad (13)$$

$$\tau_2=\vartheta_E\{4e^{\beta}+3\}\beta/4 \qquad (14)$$

$$\tau_3=\vartheta_E\{[erfc(-\sqrt{\beta/2})]^2e^{\beta}\pi/2+2A(\sqrt{\beta})\}/4 \qquad (15)$$

Here $\vartheta_E=hL^2/E$ is the time which doesn't depend on the temperature. If the dimension potential profile $U(x)$ doesn't change but temperature change, then in formulas (13)-(15) will change only $\beta=E/kT$.

When the temperature is small $e^{\beta}\gg\beta$ ($kT\leqslant E/3$) the formulas (13)-(14) transfer into:

$$\tau_1=\vartheta_E 4e^{\beta}/\beta \qquad \tau_1/e^{\beta}\sim1/\beta \qquad (16)$$

$$\tau_2=\vartheta_E\beta e^{\beta} \qquad \tau_2/e^{\beta}\sim\beta \qquad (17)$$

$$\tau_3=\vartheta_E\pi e^{\beta}/2 \qquad \tau_3/e^{\beta}\sim const \qquad (18)$$

From the (16)-(18), one can see, that in this limit case there is important difference in functional dependence of the considered life times on the inverse temperature. These differences as at (9)-(11) contain in prefactors.

The functions $\tau(\beta)$ in (13)-(15) distinct one from other more, when the temperature is high:$\beta\approx1$. It means that influence of the potential profile shape on the metastable state life time grows with temperature.

2. THE LIMITATIONS OF THE KRAMERS' FORMULA.

It is well known that Kramers in Ref.[7] was first, who approximately solved the life time problem. The metastable potential profile $u(x)$ considered in Ref.[7] have the parabolic dependence near the well bottom ($u(x)=\omega_1 x^2/2$) and the barrier top ($u(x)=-\omega_2 x^2/2$). If $\omega_1=\omega_2=\omega$, then the Kramers' escape rate r, which is inverse from the life time $r=\tau^{-1}$ equal to

$$r=\omega De^{-\beta}/4\pi \qquad (19)$$

This formula was derived in the large barrier approximation $E\gg kT$, when the probability distribution and probability current through the barrier are stationary.

From the above mentioned exact result (6) it follows that Kramers' formula (19) correct only for the "parabolic" metastable potential profile $u(x)=u_3(x)$, and the barrier may be considered as high when $e^{E/kT}\gg 1$ ($E\geqslant 3kT$), while in Ref.[7] was assumed $E\gg kT$.

Moreover, as one can see from (4), (5), Kramers' formula doesn't correct even in the limit of the high barrier if the shape of potential barrier or well differ from parabola. The distinction between life times of metastable states with different potential profile shape display itself in temperature dependence of prefactor. For considered "triangular" and "step like" metastable states the prefactor dependence on the temperature is absolutely different ($1/\beta$ and β accordingly). Thus, Kramers' formula doesn't correctly reflect the temperature dependence for nonparabolic potential profiles.

The absence of the prefactor temperature dependence in the Kramers' formula is consequence of linearity of the starting Langevin equation (1) near the potential well bottom and the barrier top. The Langevin equation (1) for nonparabolic potential profiles is nonlinear. It explain the appearing of the prefactor temperature dependence in expression for the life time.

REFERENCES.

1. A.N.Malakhov, *Radiophys. and Quant. Electr.* **34**, 451 (1991).
2. A.N.Malakhov, *Radiophys. and Quant. Electr.* **34**, 571 (1991).
3. N.V.Agudov, A.N.Malakhov, *Radiophys. and Quant. Electr.* **36** (1993).
4. A.N.Malakhov, A.L.Pankratov, to be published.
5. V.Privman, H.L.Frish, *J.Chem.Phys.* **94**, 8216 (1991).
6. M.Morch, H.Risken, H.Vollmer, *Z.Phys.* **B32**, 245 (1979).
7. Kramers H., *Phisica*, **7**, 284 (1940).
8. A.N.Malakhov, N.V.Agudov, present proceedings.

HOT-PHONON EFFECT ON CHARGE-CARRIER FLUCTUATIONS IN GaAs

P. Bordone, L. Varani, L. Reggiani
Dipartimento di Fisica ed Istituto Nazionale di Fisica della Materia,
Università di Modena, Via Campi 213/A, 41100 Modena, Italy

T. Kuhn
Institut für Theoretische Physik, Universität Stuttgart,
Pfaffenwaldring 57, 7000 Stuttgart 80, Germany

ABSTRACT

We present a Monte Carlo investigation on the influence of a non-equilibrium phonon population (hot phonons) on second-order transport properties in *GaAs*. We calculate the velocity and energy correlation functions both for the case with and without phonon perturbation. The results show significant modifications in the correlation functions and consequently an increase of the equivalent noise temperature due to the presence of hot phonons.

INTRODUCTION

In this paper we analyze the effect of a non-equilibrium phonon population (hot phonons) on the charge-carrier fluctuations in polar semiconductors due to the presence of an applied electric field E. In such a situation, the optical-phonon population cannot be described properly by the Planck distribution at thermal equilibrium. Indeed the strong carrier–LO-phonon coupling can lead to very fast emission rates of phonons by the carriers and, since the decay due to phonon-phonon interaction is a slower process, to substantial LO-phonon amplification even at room temperature[1,2].

THEORETICAL MODEL

We consider the case of electrons in a *GaAs* bulk system, described by a simple parabolic two valley (Γ and L) model[2], in the presence of an applied electric field, where the only scattering mechanisms are with ionized impurities, LO and intervalley phonons. The case of scattering with remote ionized impurities is approximated in our bulk model by taking the carrier density n_e and the ionized-impurity concentration n_i as independent variables. It is known[2] that the effects of the LO-phonon disturbances on carrier kinetics are more pronounced in the low-field than in the high-field case, and that they increase with decreasing lattice temperature. For these reasons, here we analyze the field region below threshold for negative differential mobility at 77 K. From an ensemble Monte Carlo simulator we calculate the auto-correlation functions of the drift velocity (i.e. the velocity component along the field direction) and of the energy and the cross-correlation functions, both for the case with and without phonon perturbation, by using a standard numerical algorithm[3]. To evaluate the influence of the hot phonons on the interparticle correlation, we compare an ensemble and a single-particle picture.

RESULTS AND DISCUSSION

The results of the correlation functions for an electric field of $2kV/cm$, are reported in Fig. 1. Here we use, for the correlation functions, the shorthand notation $\Phi_{AB}(t) = \overline{\delta A(0)\delta B(t)}$, with $\delta B(t) = B(t) - \overline{B}$, the bar indicating time average, where A and B stay for the variable drift velocity V or energy $\mathcal{E}$. Futhermore, we report the correlation functions associated with a single carrier i as Φ_{AiBi}. To be compared with the results of the ensemble simulation Φ_{AiBi} must be divided by the total number of carriers N, since, in the absence of carrier-carrier interaction $\Phi_{AB} = \Phi_{AiBi}/N$. In the velocity case, Fig. 1(a), the presence of hot phonons results in a significant increase of $\Phi_{VV}(0)$ whereas its time dependence remains practically the same. The increase of $\Phi_{VV}(0)$ is attributed to an increase of the carrier mean energy due to a strong reabsorption of previously emitted phonons by carriers. This reabsorption process does not influence the time dependence of $\Phi_{VV}(t)$ because of the forward nature of the scattering cross-section which inhibits the coupling among the velocities of different carriers. The situation is opposite in the energy case (see Fig. 1(b)), because the strong reabsorption of emitted phonons induces a coupling among the energies of different carriers. Here, in addition to a significant increase of $\Phi_{\mathcal{E}\mathcal{E}}(0)$ we find a relevant modification in its time dependence due to the presence of hot phonons. Moreover, the consequences of the carrier-carrier coupling, are clearly shown by the different results displayed by the ensemble and single-particle picture. The slowing down in the hot-phonon case, confirms the trend pointed out by measurements[4] and calculations[1,5] on the thermalization of photoexcited carriers in bulk and quantum well systems: a significant reduction in the decay rates of carrier energy induced by the reabsorption of hot phonons. The above considerations apply also to the cross-correlation functions, shown in Fig. 1(c). (Here the positive time scale gives $\Phi_{V\mathcal{E}}(t)$, while the negative $\Phi_{\mathcal{E}V}(t) = \Phi_{V\mathcal{E}}(-t)$). To evaluate the effects induced by the hot phonons on the electronic noise, we calculate the equivalent noise temperature in the direction of the field, $T_{n||}$. As found above, the velocity auto-correlation function remains the same for the ensemble and single-particle picture; in other words, from the velocity point of view, the carrier-carrier correlation can be neglected. This implies that $T_{n||}$ is given by[6]:

$$T_{n||} = \frac{e}{K_B \mu'(E)} \int_0^{\infty} \Phi_{ViVi}(t)\, dt \qquad (1)$$

where e is the electron charge, K_B the Boltzmann constant and $\mu'(E)$ the differential mobility. Figure 2 shows the noise temperature versus the electric field, for different carrier and impurity concentrations, as calculated with and without hot phonons, in comparison with existing experimental data[7]. The uncertainty of the calculations is comparable with that of experiments, and estimated to be at most 20%. For a carrier concentration of 3×10^{15} cm^{-3} corresponding to the same impurity concentration (all impurities ionized), we compare the experimental results (full circles) with the Monte Carlo calculation (continuous line), to check the reliability of the theoretical model, by obtaining a reasonable agreement. In this condition, due to the low carriers concentration, there are no hot-phonon effects. The Monte Carlo results for a carrier concentration of 10^{17} cm^{-3} in the presence of negligible residual impurity scattering are shown, with (dotted line) and without (dashed line) phonon perturbation. In this case the hot-phonon population is found to induce a significant rise in the noise temperature, which becomes more pronounced at higher fields. This excess noise contribution is due to a further increase of the carrier mean energy associated with the presence of hot phonons.

CONCLUSIONS

We have presented a Monte Carlo investigation on the influence of hot phonons on second-order transport properties of n-type $GaAs$. At increasing electric field strengths we find a relevant increase of the velocity variance, which is responsible for an increase of the noise temperature. Moreover, the presence of the phonon perturbation modifies the time dependence of the energy correlation functions because of the carrier-carrier coupling induced by the reabsorption of previously emitted phonons. New experiments are suggested to confirm the above calculations. This work is partially supported by the Commission of European Comunity (CEC) and the Italian National Research Council (CNR).

REFERENCES

1. P. Lugli, Solid-State Electron. **31**, 667 (1988).
2. M. Rieger, P. Kocevar, P. Lugli, P. Bordone, L. Reggiani and S.M. Goodnick, Phys. Rev. **B 39**, 7866 (1989).
3. L. Reggiani, T. Kuhn and L. Varani, Appl. Phys. **A 54**, 411 (1992).
4. M.C. Tatham, J.F. Ryan and C.T. Foxon, Phys. Rev. Lett. **63**, 1637 (1989).
5. P. Lugli, P. Bordone, S. Gualdi and S.M. Goodnick, in SPIE Vol. 1282 "Ultrafast Laser Probe Phenomena in Bulk and Microstructure Semiconductors III", ed. by R.R. Alfano, SPIE, Bellingham (WA), (1990) p.11.
6. "Hot-Electron Transport in Semiconductors", ed. by L. Reggiani, (Springer-Verlag, Berlin, 1985).
7. V. Bareikis, J. Liberis, A. Matulionis, R. Miliusyte, J. Pozela and P. Sakalas, Solid-State Electron. **32**, 1647 (1989).

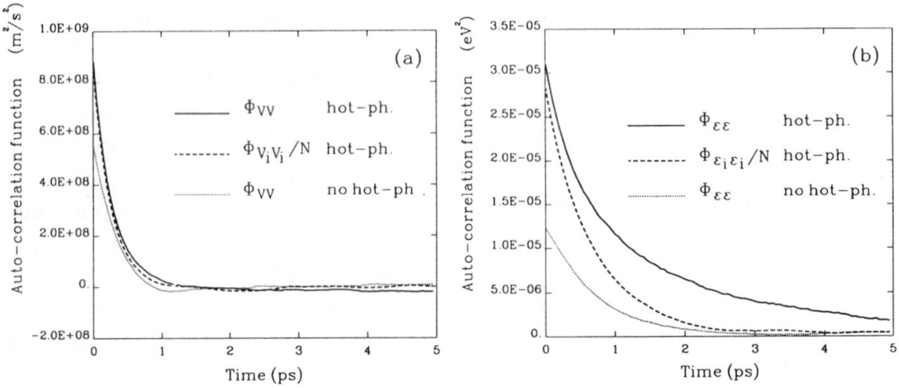

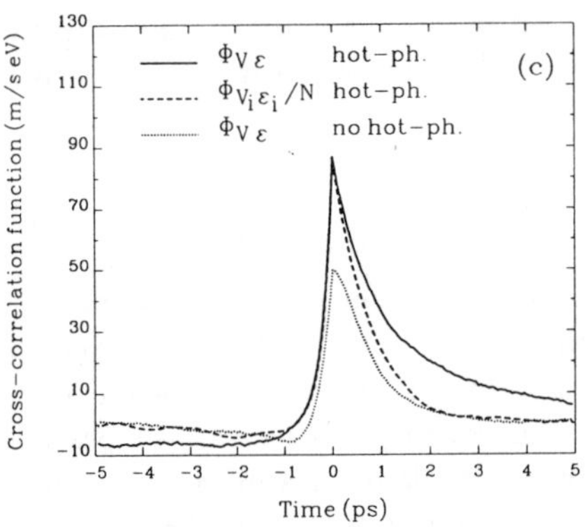

Fig. 1 - Correlation functions in GaAs at 77 K for an electric field of 2kV/cm, a carrier concentration $n_e = 10^{17}$ cm^{-3}, $n_i = 0$, as function of time, obtained from the Monte Carlo simulations. Continuous, dashed and dotted lines refer respectively to the many particle correlation function with hot phonons, to the single particle correlation function divided by the total number of carrier with hot phonons and to the many particle correlation function without hot phonons. Figure 1(a) reports the velocity correlation function, Fig. 1(b) the energy correlation function and Fig. 1(c) the cross-correlation functions between velocity and energy.

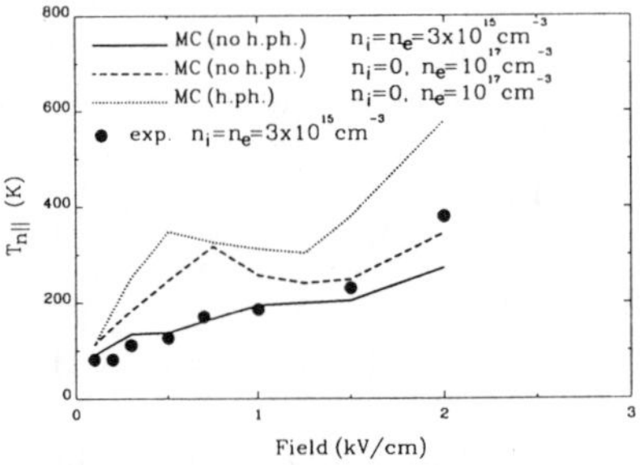

Fig. 2 - Equivalent noise temperature in the direction of the electric field in GaAs at 77 K, as a function of the field. Curves refer to Monte Carlo calculation: $n_i = n_e = 3 \times 10^{15}$ cm^{-3} without hot-phonons (continuous line), $n_i = 0$ and $n_e = 10^{17}$ cm^{-3} with (dotted line) and without (dashed line) hot phonons. Symbols (full circles) reports experimental results from Ref. [7], for $n_i = n_e = 3 \times 10^{15}$ cm^{-3}.

STOCHASTIC RESONANCE IN OPTICAL BISTABLE SYSTEMS

Roland Bartussek, Peter Jung and Peter Hänggi
University of Augsburg, W-8900 Augsburg, Germany

ABSTRACT

We investigate cooperative effects of noise and periodic forcing in an optical bistable system. It has been demonstrated in a recent experiment [1] that noise induced switching between low and high output intensity can be synchronized via the stochastic resonance effect by a small periodic modulation of the input intensity. Here we present theoretical results for stochastic resonance in optical bistable systems.

MODEL AND BASIC EQUATIONS

A model for optical bistability was introduced by Bonifacio and Lugiato [2]. For the amplitude y of the input light and the transmitted amplitude x, they have derived the equation of motion

$$\dot{x} = y - x - 2c\frac{x}{1+x^2} + \sqrt{D}\frac{x}{1+x^2}\Gamma(t), \tag{1}$$

where Γ represents δ-correlated, Gaussian distributed noise with zero mean. A weak periodic modulation of the input intensity is taken into account by adding a periodic term to y, i.e. $y \to y + A\sin(\Omega t + \psi)$. For the probability density of the transmitted amplitude, $P(x,t)$, we find the Fokker-Planck equation

$$\begin{aligned}\frac{\partial}{\partial t}P(x,t) &= -\frac{\partial}{\partial x}\left[y - x - \frac{2cx}{1+x^2} + D\frac{x(1-x^2)}{(1+x^2)^3} + A\sin(\Omega t + \psi)\right]P(x,t) \\ &\quad + \frac{\partial^2}{\partial x^2}D\frac{x^2}{(1+x^2)^2}P(x,t). \end{aligned} \tag{2}$$

The spectral density of the transmitted amplitude has δ-spikes at multiples $n\Omega$ of the driving frequency [3] with the corresponding weights w_n being a measure for the output power at the frequency $n\Omega$. They can be expressed in terms of the Fourier coefficients of the time periodic, asymptotic mean value [4]

$$\langle x(t)\rangle_{as} = \sum_{n=-\infty}^{\infty} |M_n| \exp\left[in(\Omega t + \psi + \varphi_n) - i\frac{\pi}{2}\right] \tag{3}$$

by

$$w_n = 2\pi |M_n|^2. \tag{4}$$

AMPLIFICATION OF THE OPTICAL SIGNAL

The amplification of the periodic signal is given by the ratio of the transmitted power at the driving frequency and the input power [4]

$$\eta_1(\Omega) = 4\frac{|M_1|^2}{A^2}. \tag{5}$$

The numerically calculated results for η_1 are shown in Figs.1 for various frequencies by the solid lines. Fig.1a corresponds to choosing the dc input intensity y such that $P(x,t)$ shows two peaks of nearly equal height in the limit $D \to 0$ what we call the "symmetric case". Fig.1b corresponds to a "asymmetric case", where the peaks of the stationary probability have different probabilistic weights.

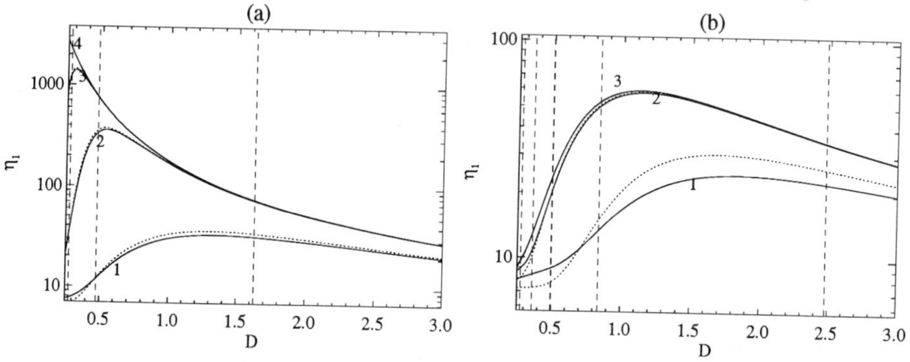

Fig. 1: Spectral amplification η_1 at $c = 6$, $A = 10^{-4}$ for $y = 6.72584$ (a) and $y = 6.8$ (b). Curve 1 corresponds to $\Omega = 10^{-1}$, curve 2 to 10^{-2}, curve 3 to 10^{-3} and curve 4 to 10^{-4}. The dotted lines correspond to results within linear response approximation (Eqs. (6) - (8)).

In the symmetric case we observe stochastic resonance [5] very much like in the quartic double well potential, i.e. a peak in the amplification of the signal (modulation) as a function of the noise intensity when the sum of the mean sojourn times in both stable states equals the period of the driving (these values of D are indicated as vertical dashed lines in Figs.1).
In the asymmetric case, the peak of the amplification is suppressed, because - in contrast to the symmetric case - the corresponding contribution (i.e. the weight g_T in Eq. (8)) of hopping motion to the response of the system disappears exponentially for small noise. The remaining maximum is only the tail of the amplification by synchronisation at large noise.

The numerical results are compared in Figs.1 with those obtained within linear response approximation [4,5] (dotted lines). In this approximation we find in terms of the response function $R(t)$

$$\langle x(t)\rangle_{as} - \langle x\rangle_{st} = \int_{-\infty}^{\infty} R(t-t')A\sin(\Omega t' + \psi)\mathrm{d}t' - \int_{0}^{\infty} xP_{st}(x)\mathrm{d}x, \tag{6}$$

with the stationary solution $P_{st}(x)$ of the undriven system. The response function $R(t)$ is expressed via a fluctuation theorem by a correlation function $K(t)$ of the undriven system

$$R(t) = \frac{\mathrm{d}}{\mathrm{dt}} \langle x(t)h(x(0))\rangle \equiv \frac{\mathrm{d}}{\mathrm{dt}} K(t) \tag{7}$$

with $h(x) = \frac{1}{D}\left(-\frac{1}{x} + 2x + \frac{1}{3}x^3\right)$. $K(t)$ is approximated by a sum of exponentials with the typical time scales of the system λ_T and $\lambda_{1,2}$ - stemming from hopping and local motion in the potential wells respectively, i.e.

$$K(t) \simeq \sum_{i=1,2,T} g_i e^{\lambda_i t}. \tag{8}$$

The weights g_i are determined by the correlation function $K(t)$ and its derivatives at $t = 0$.

GENERATION OF HIGHER HARMONICS

The generation of the n-th harmonic in the output due to the nonlinearities is characterized by the ratio

$$\eta_n(\Omega) = 4\frac{|M_n|^2}{A^2}. \tag{9}$$

The second harmonic depends on the noise strength as shown for the symmetric and asymmetric case in Figs.2. In the symmetric case (Fig.2a) a "dip" appears which becomes sharper with decreasing frequencies. In the asymmetric case (Fig.2b) we do not observe such a behaviour.

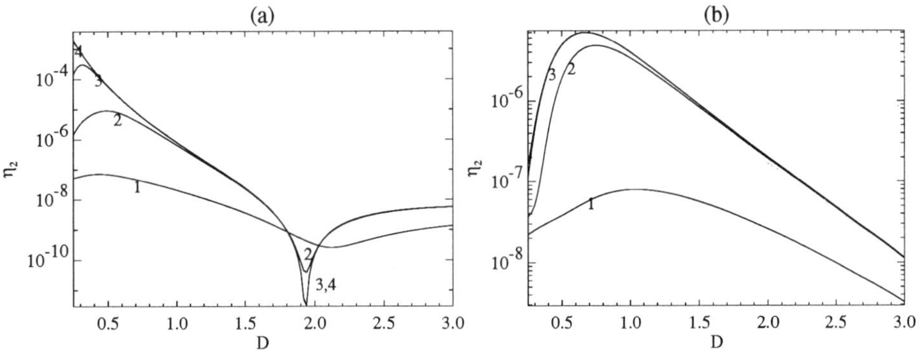

Fig. 2: Higher harmonic η_2, parameters as in Figs.1.

For the third harmonic, η_3, we find a smooth curve in the symmetric case and a dip in the asymmetric case.

We have confirmed the results for the higher harmonics within an adiabatic approximation, valid for small driving frequencies.

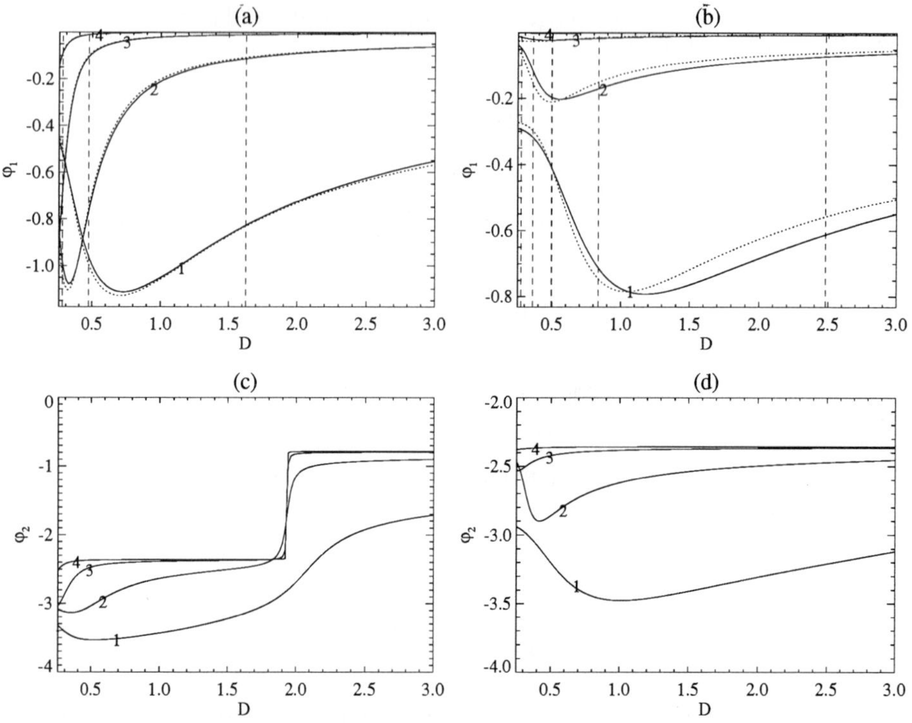

Fig. 3: Phase shifts, parameters as in Figs.1.

PHASE SHIFT OF THE OUTPUT SIGNAL

In Figs.3, the phase shifts of the first and second harmonic of the asymptotic mean value $\langle x(t)\rangle_{as}$ are shown for the symmetric (Figs.3a and 3c) and asymmetric case (Figs.3b and 3d). The results within linear response theory are shown by dotted lines. The phase shift in the symmetric case looks like in the quartic model: The maximum results from the competition between internal motion and hopping processes. In the asymmetric case the maximum is suppressed for small frequencies because the hopping disappears at small noise strength.

At values of D, for which a dip in a higher harmonic appears, the corresponding phase shift approaches a step function for small frequencies.

REFERENCES

1. J. Grohs, S. Apanasevitch, H. Issler, D. Burak, P. Jung, C. Klingschirn, submitted to Phys. Rev. **A**.
2. R. Bonifacio and L.A. Lugiato, Phys. Rev. **A18**, 1192 (1978).
3. P. Jung and P. Hänggi, Europhys. Lett. **8**, 905 (1989).
4. P. Jung and P. Hänggi, Phys. Rev. **A44**, 8032 (1991);
   P. Jung, *Periodically Driven Stochastic Systems*, in press at Phys. Rep.
5. For a review: F. Moss, Ber. Bunsenges. Phys. Chem. **95**, 303 (1991) and [4].

CORRELATED HOPPING AND TRANSPORT IN TILTED PERIODIC POTENTIALS

PETER JUNG

University of Augsburg, W-8900 Augsburg, Germany

ABSTRACT

We consider the transport of a Brownian particle in a periodic potential via correlated jumps , i.e. due to direct transitions between non-adjacent potential wells. We evaluate distributions of jump sizes with and without an external field and its dependence on the friction constant.

INTRODUCTION AND BASIC EQUATIONS

Diffusion of adatoms plays an important role for epitaxial growth and surface melting[1]. Very mobile adatoms in a fluid like phase, move in a layer on the surface of ordered bulk material still feeling the remnants of crystalline symmetry . The motion of the adatoms on the surface is modeled by the underdamped Brownian motion of a particle in a two dimensional periodic potential[2]. Neglecting the interaction between the adatoms, we can describe this motion by the Fokker-Planck equation for the joint probability density for the adatom being at the position $\underline{x}$ and having the velocity $\underline{v}$, i.e.

$$\frac{\partial P(\underline{x},\underline{v},t)}{\partial t} = -(\underline{v}\cdot\nabla_x)P - \frac{1}{m}(\nabla_x V(\underline{x}))\nabla_v P + \nabla_v \cdot \underline{\underline{\gamma}}\cdot\underline{v}P + \frac{k_B T}{m}\nabla_v \underline{\underline{\gamma}}\nabla_v P \ . \tag{1}$$

The friction tensor $\underline{\underline{\gamma}}$ is assumed to be isotropic and constant. The potential is periodic, i.e. $V(\underline{x}) = V(\underline{x} + n_i \underline{g}_i)$,where $\underline{g}_i$ are rectangular lattice vectors. Approximating the potential by its two lowest Fourier coefficients plus a tilt (external dc field), i.e. $V(\underline{x}) = -d_x \cos x + d_y \cos y + \underline{U}\underline{x}$, we arrive at two independent Fokker-Planck equations for the diffusion on the surface in $x-$ and $y-$ direction. For the $x-$ direction we find in a proper normalized form

$$\frac{\partial P(x,v,t)}{\partial t} = -\frac{\partial}{\partial x} vP(x,v,t) + \frac{\partial}{\partial v}(\gamma v + \sin x + U)P(x,v,t) + \gamma D \frac{\partial^2}{\partial v^2} P(x,v,t) \ . \tag{2}$$

The spatial periodicity of the Fokker-Planck operator in (2) implies eigensolutions of Bloch-type[3], i.e.

$$u_{\sigma,k}(x,v,t) = \exp(-\lambda_{\sigma,k} t + ikx)\phi_{\sigma,k}(x,v) \ , \tag{3}$$

where $\phi_{\sigma k}(x,v)=\phi_{\sigma k}(x+2\pi,v)$ and k is taken from the first (symmetric) Brillouin zone, i.e. $-1/2<k<1/2$. Similar than in the quantum theory of solids, the eigenvalues are arranged in bands with a discrete band index σ and a quasi-continuous index k. In Fig.1, we show the numerically evaluated lowest three real valued bands of eigenvalues for $\gamma=2$ and $D=0.5$.

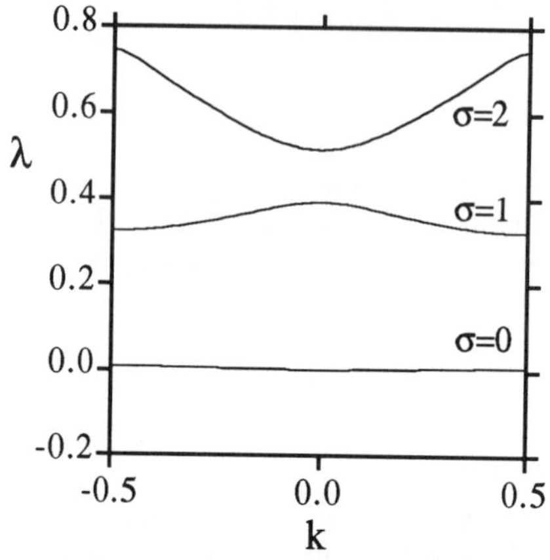

Fig.1:The Bloch-band structure of eigenvalues of the Fokker operator is shown for D=0.5 and $\gamma=2$.

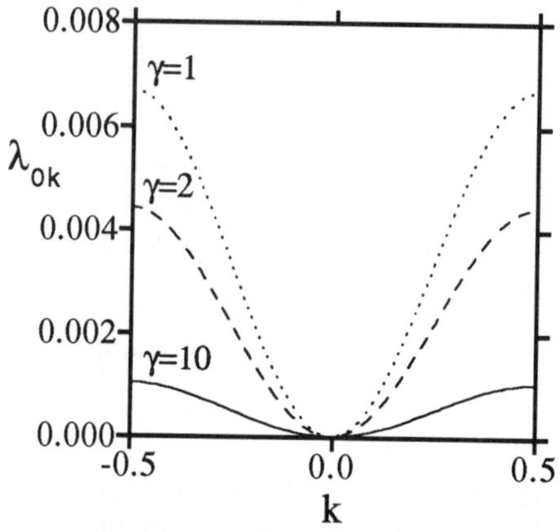

Fig.2:The lowest lying bands of eigenvalues D=0.5 and several values of the damping γ.

CORRELATED HOPPING TRANSPORT WITHOUT AN EXTERNAL FIELD

Neglecting quantum effects, noise induced hopping is the relevant transport mechanism of adatoms on the surface. For weak fluctuations, hopping between surface traps is the slowest relaxation process and is therefore described by relaxation modes corresponding to the lowest lying band of eigenvalues $\lambda_{0,k}$.(see Fig.1b)

The hopping transport between two non-adjacent traps can be attributed to two different mechanisms. First, there is an indirect hopping mechanism where the adatoms jump successively between adjacent traps. Such a purely diffusive hopping process can be described appropriately by the master equation for the populations in the traps P_n

$$\dot{P}_n(t)=r_+P_{n-1}(t)+r_+P_{n+1}(t)-(r_++r_-)P_n(t)\ , \tag{4}$$

where the next neighbor transition rates $r_\pm$ may be obtained from standard rate theories[4]. Using periodic boundary conditions, i.e. $P_{n+N}(t)=P_n(t)$, one obtains for the eigenmodes $P_n^{(m)}(t)=\exp(-\Lambda_m t)c_n^{(m)}$ without an external field($r_+=r_-\equiv r$)

$$\Lambda_{k_m}=4r\sin^2(2\pi \mathrm{k_m}),\ c_n^{(m)}=\frac{1}{\sqrt{N}}\cos(2\pi nk_m+\varphi),\ \mathrm{k_m}=\frac{m}{N},\ m=-N/2,\dots,N/2 \tag{5}$$

We observe from (5) that the eigenmode, corresponding to the relaxation rate $r_i(s) = (1/4)\Lambda_{k_m}$, describes a spatial relaxation extended over $1/(2k_m)$ periods of the potential.

In systems with inertia effects, i.e. for weak damping, direct hopping transport between non-adjacent traps becomes important. The relaxation rates corresponding to relaxation processes extending over s potential wells, including direct and indirect processes, are determined by the eigenvalues $\lambda_{\sigma=0k=1/(2s)}$ of the full Fokker-Planck equation (2).The relaxation rate between adjacent traps is given in this notation by the lowest lying eigenvalue at the zone boundary of the first Brillouin zone, i.e. $r = (1/4)\lambda_{\sigma=0k=1/2}$.The direct relaxation rates over s wells are then obtained by subtracting the indirect rates from the the eigenvalues $\lambda_{\sigma=0k}$[5]

$$r_{\mathrm{d}}(s) = \frac{1}{4}\left(\lambda_{\sigma=0k=1/(2s)} - \Lambda_{1/(2s)}\right). \tag{6}$$

The Bloch-eigenvalues $\lambda_{\sigma=0k=1/(2s+1)}$ cannot be related to jump processes with a single jump size. They correspond to processes, where the system relaxes simultaneously over several channels.

The distribution of jump sizes is up to a normalization factor also given by (6). It is shown for $\gamma = 0.1$ and $D = 0.25$ in Fig.2. The jump size distribution shows a maximum which is shifted for decreasing damping to larger values of the jump size s. For the value of the damping $\gamma = 0.1$ chosen in Fig.2, the next neighbor transition rate r exceeds the correlated relaxation rates by one order of magnitude. For smaller damping, however, they become comparable. It is this very regime, where Ferrando et al[2] found that the dynamical structure factor obtained from neutron scattering experiments can not be explained by theories for surface diffusion which do not take into account correlated jumps. It is further interesting to note, that for large s, the jump size distribution decays, in agreement with the indirect transition rates (crosses in Fig. 2), with the power law $r_{\mathrm{direct}}(s) \propto 1/s^2$. This implies that the jump size distribution is normalizable, but not all moments exist.

CORRELATED HOPPING TRANSPORT WITH AN EXTERNAL FIELD

Similar than in the situation without an external field, we first have to find an expression for the indirect relaxation rates. The master equation approach (5) yields the complex valued eigenvalues with the real parts

$$\begin{aligned} \Lambda_{k_m} &= 2(r_+ + r_-)\sin^2(2\pi k_m) = \lambda_{0k=1/2}\sin^2(2\pi k_m) \\ k_m &= \frac{m}{N} \end{aligned}, \tag{7}$$

where m has to be taken from the first Brillouin zone. The direct transition rates are then obtained by subtracting the indirect transition rates from the numerically evaluated lowest lying band of eigenvalues. The results are shown for a constant tilt of $U = 0.2$ and decreasing damping in Fig.3. For $\gamma = 0.2$, the solution of the deterministic equation (equivalent with the zero noise limit of (2))

$$\ddot{x} + \gamma\dot{x} + \sin x + U = 0 \tag{8}$$

approaches a fix point within a potential well (locked state). For $\gamma = 0.1$, the coexistence of a running with a locked solution[3] generates bistable behavior. The choice $\gamma = 0.05$ selects a situation, where only the running solution is globally stable[3]. In the dynamical bistable regime ($\gamma = 0.1$), we observe a power law decay of the jump size distribution, i.e. $r_d(s) \propto s^{-\alpha}$,with $\alpha < 2$, within a finite range of jump sizes. For decreasing damping, this range strongly increases. For large jump sizes, however, the jump size distribution eventually approaches its asymptotic dependence proportional to s^{-2}. Passing the transition to the regime of globally stable running states ($\gamma = 0.05$) the jump size distribution approaches a constant for large s.

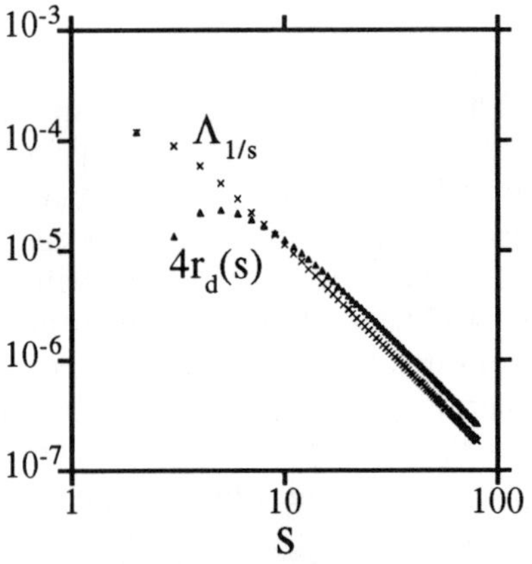

Fig.2:The jump size distribution $r_d(s)$for $\gamma = 0.1$ and $D = 0.25$ is compared with the indirect jump rates (5)

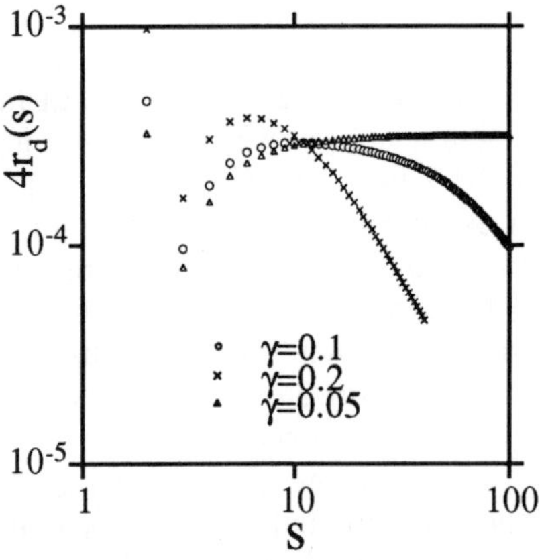

Fig.3:The jump size distribution is shown for the external field $U = 0.2$ for decreasing damping.

REFERENCES

1. Jost W.M. Frenken and J.F van der Veen, Phys. Rev. Lett. **54**, 134 (1985).
2. R. Ferrando, R. Spadacini and G.E. Tommei, Phys. Rev. **B45**, 444 (1992).
3. H. Risken, The Fokker-Planck Equation, Springer, Berlin (1984).
4. P. Hänggi, P. Talkner and M. Borkovec, Rev. Mod. Phys. **62**, 251 (1990)
5. P.Jung, in preparation (Phys. Rev. E).

NONSTATIONARY BROWNIAN MOTION IN BI AND TRI-STABLE POTENTIAL PROFILES. THE RELAXATION TIME AND ESCAPE RATE UNDER ANY PERTURBING NOISE INTENSITY.

A.N.Malakhov, N.V.Agudov.
State University, N.Novgorod 603600, Russia.

ABSTRACT

The exact time characteristics of the nonstationary diffusion in bi- and tri-stable potential profiles having different shape are determined. The relaxation times, the life times of metastable states and escape rates obtained here are valid for arbitrary relation between activation energy and perturbing noise intensity.

INTRODUCTION

The diffusion of Brownian particles in potential profiles provides a useful model to understand the role of fluctuations in driving an unstable system towards equilibrium and in transitions across a potential barrier. This old Kramers' problem is significant in a variety of fields in biology, chemistry and physics. A detailed survey of Kramers' problem has been published, e.g. in [1].

The main problem difficulty is to obtain analytic results for various potential profiles. For this reason the solvable models of potential profiles are very useful. The such models are e.g. piecewise linear potential profiles, the particular cases of which are considered in [2,3].

In this paper on the basis of our previous works [4-6] the exact time characteristics of the nonstationary diffusion in bi- and tri-stable potential profiles with different shape are represented.

THE PROBLEM APPROACH.

We consider an overdamped Brownian motion subjected to Langevin equation

$$\frac{dx(t)}{dt} = -\frac{1}{h}\frac{dU(x)}{dx} + \eta(t) ,$$

where h - viscosity, $U(x)$ potential profile, $\eta(t)$ - stationary white noise with correlation function $\langle\eta(t)\eta(t+\tau)\rangle = D\delta(\tau)$, $D=2kT/h$. The corresponding Fokker-Planck equation for the probability distribution is

$$\frac{\partial W(x,t)}{\partial t} = \frac{\partial}{\partial x}\left[\frac{1}{h}\frac{dU(x)}{dx} W(x,t)\right] + \frac{D}{2}\frac{\partial^2}{\partial x^2} W(x,t). \qquad (1)$$

The initial and boundary condition are $W(x,0)=\delta(x)$ and $W(\pm\infty,t)=0$. For the Laplace transform $Y(x,s)\equiv Y(x) = \int_0^\infty W(x,t)e^{-st}dt$ the eq.(1) leads to

$$\frac{d^2Y(x)}{dx^2} + \frac{d}{dx}\left[\frac{du(x)}{dx} Y(x)\right] - \gamma^2 Y(x) = - B\delta(x) , \qquad (2)$$

where $B=2/D, \gamma^2=sB, u(x)=U(x)/kT$-dimensionless potential profile, $u(0)=0$, $Y(\pm\infty)=0$.

The exact solution of the eq. (2), for the stepwise rectangular profile and for the piecewise linear profile with arbitrary number of potential jumps and linear parts is obtained in [4,5] and [6] respectively. These solutions for different profiles are used to calculate the Laplace transform $\hat{Q}(s)=\int_A y(x,s)dx=\int_0^\infty Q(t)e^{-st}dt$ of the probability $Q(t)=\int_A W(x,t)dx$ for the diffusing particles to be in an interval A.

The relaxation time, i.e. the time of transition from $W(x,0)=\delta(x)$ to a stationary probability distribution $W(x,\infty)\neq 0$ in the given potential profile is defined as the evolution time of a function $Q(t)$ changing from $Q(0)$ to $Q(\infty)$.

In accordance with [6] the relaxation time is determined as

$$\tau = \int_0^\infty [Q(\infty)-Q(t)]dt/[Q(\infty)-Q(0)] \qquad (3)$$

and in terms of Laplace transform is equal

$$\tau=\lim_{s\to 0} [Q(\infty)-s\hat{Q}(s)]/s[Q(\infty)-Q(0)]. \qquad (4)$$

The definition (3) is valid only if the evolution of a probability $Q(t)$ is monotonic and fast enough.

The formula (4) allows to find the life time of metastable state and the so-called Mean First Passage Time as well. The MFPT denoted as $T(x_o,w)$ is a mean passage time of a Brownian particle starting from point $x=x_o$ up to absorbing wall placed in $x=w$. To calculate MFPT from (4) it is necessary in potential profile to arrange infinitely deep potential well in the point $x=w$. In this case interval A must be chosen from $x=w$ to $x=\infty$ and we get

$$T(x_o,w)=\lim_{s\to 0}(1-s\hat{Q}(s))/s \quad , \qquad (5)$$

where $\hat{Q}(s)=\int_w^\infty y(x,s)dx$, $Q(0)=0$, $Q(\infty)=1$.

BISTABLE SYSTEMS.

Let's consider three bistable potential profiles depicted in Fig.1, 2, 3.

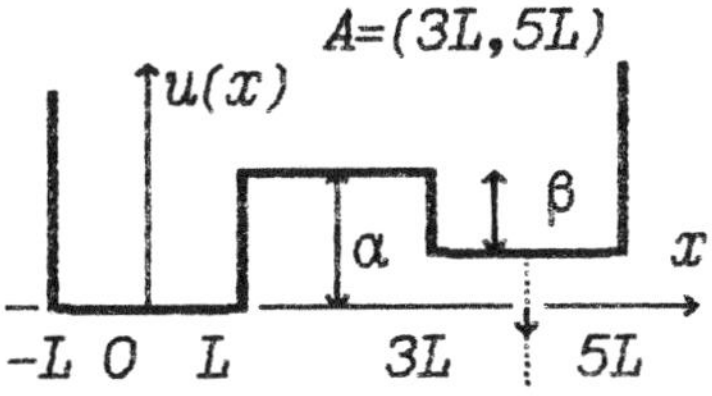

Fig.1.

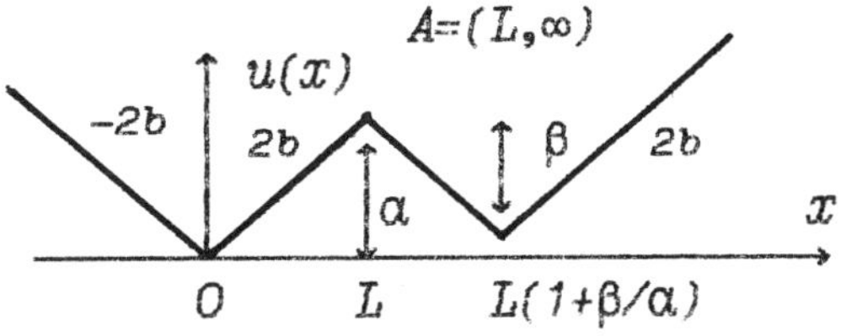

Fig.2.

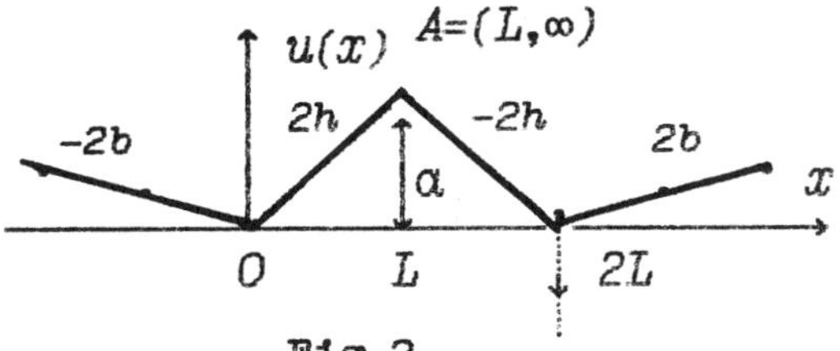

Fig.3.

On the basis of [4-6] with the aid of (4) one can find the following exact expressions of the relaxation times for three profiles respectively:

$$\tau_1=\tau_r[21+24e^{\beta+\alpha}/(1+e^\alpha+e^\beta)] \qquad (6a)$$

$$\tau_2=\tau_v[(2e^{\beta+\alpha}+\alpha+\beta-\beta e^\alpha-\alpha e^\beta)/(e^\alpha+e^\beta-1)-(e^\beta+\beta)/(2e^\beta-1)+\alpha/2-1]$$

$$\tau_3=\tau_v[\alpha_o^2(e^\alpha-\alpha-1)/\alpha^2+\alpha_o(e^\alpha-1)/\alpha]/2, \qquad (6c)$$

where $\tau_r=L^2/3D$- is the relaxation time in one-stable rectangular profile: $u(x)=0$, $-L<x<L$, $u(\pm L)=\infty$; $\tau_v=1/b^2D$ - is

the relaxation time in one-stable V-shaped profile $u(x)=2b|x|$. In all profiles the relaxation time is determined by a time evolution of the probability $Q(t)$ for the diffusing particles to be in the interval of the right minimum. The formulas (6) are valid for arbitrary perturbing noise intensity, i.e. for arbitrary $\alpha=E_1/kT$, $\beta=E_2/kT$, where E_1, E_2 are activation energies.

The relaxation time (6a) is symmetrical to the exchange $\alpha\rightleftarrows\beta$. It means that relaxation time for diffusion from point $x=0$ to the right minimum is the same as for diffusion from point $x=4L$ to the left minimum, even if $\alpha\neq\beta$. This equality is due to the fact that the difference (for $\alpha\neq\beta$) between the probability currents is compensated by the difference between final values of stationary probabilities in the left and the right minima.

For high enough potential barriers $(e^{\alpha}\gg\alpha,\ e^{\beta}\gg\beta)$ the relaxation times are respectively:

$$\tau_1=\tau_r e^{\beta+\alpha}/(e^{\alpha}+e^{\beta}),\quad \tau_2=\tau_v 2e^{\beta+\alpha}/(e^{\alpha}+e^{\beta}), \qquad (7a,b)$$

$$\tau_3=\tau_v\alpha_o e^{\alpha}(1+\alpha_o/\alpha)/2\alpha, \qquad (7c)$$

and relaxation time (6b) also becomes symmetrical for the exchange $\alpha\rightleftarrows\beta$. If in addition $\alpha=\beta$, then $\tau_1=\tau_r e^{\alpha}/2$, $\tau_2=\tau_v e^{\alpha}$, and in all formulas (7) the Kramers' factor e^{α} arises.

For deep enough right potential well $(e^{\beta}\gg e^{\alpha})$, potential profiles of Fig.1, 2 turn to the potential profiles of the metastable states. In this case

$$\tau_1=\tau_r[21+24e^{\alpha}],\quad \tau_2=\tau_v[2e^{\alpha}-\alpha/2-3/2] \qquad (8a,b)$$

These times are the life times of metastable states which exist in neighborhood $x=0$ and their values given by formulas (8) are valid also for an arbitrary α. If $e^{\alpha}\gg\alpha$, we get again Kramers' factor.

To evaluate the diffusing particles transition frequency over potential barrier $\nu(\alpha)$ it is necessary to know the MFPT $T(0,w)=1/\nu(\alpha)$ for initial coordinate $x=0$ and absorbing wall's coordinate $x=w$ in accordance with (5). Let us select $w=4L$, $2L$ for Fig.1, 3 (dash lines). Then it can be found the following transition frequencies respectively

$$\nu_1(\alpha)=\nu_1(0)6e^{-\alpha}/(2+3e^{-\alpha}+e^{-2\alpha}),$$

$$\nu_2(\alpha)=\nu_2(0)(1+\alpha_o)\alpha^2e^{-\alpha}/(\alpha+\alpha_o-(\alpha+2\alpha_o)e^{-\alpha}+\alpha_oe^{-2\alpha})$$

where $\nu_1(0)=1/72\tau_r$, $\nu_2(0)=1/\tau_v\alpha_o(1+\alpha_o)$.

For high barriers $(e^{\alpha}\gg\alpha)$

$$\nu_1(\alpha)\sim kTe^{-\alpha} \qquad \nu_2(\alpha)\sim e^{-\alpha}/kT.$$

Although here is the Kramers' factor $e^{-\alpha}$ these transition frequencies differ from Kramers' frequency $\nu(\alpha)\sim e^{-\alpha}$ by the presence of the prefactor temperature dependencies which are due to the existing of the nonlinearities in neighbourhood of the well bottoms and of the barrier tops.

TRISTABLE SYSTEMS.

Let us consider the poly-rectangular profile which represents the tristable system with the initial probability distribution placed in the last well(Fig.4).The relaxation time calculated with aid the probability $Q(t)$ for $A(6L,9L)$ is equal

$$\tau=12\tau_r(41+16e^{\alpha}+37e^{-\alpha}+28e^{-2\alpha}+4e^{-3\alpha})/(3+2e^{-\alpha})(2+e^{-\alpha}).$$

If initial distribution is in the center of the middle well (Fig 5.) then for $A(2L,5L)$

$$\tau'=12\tau_r(9+4e^{\alpha}+8e^{-\alpha}+e^{-2\alpha})/(3+2e^{-\alpha})(2+e^{-\alpha}).$$

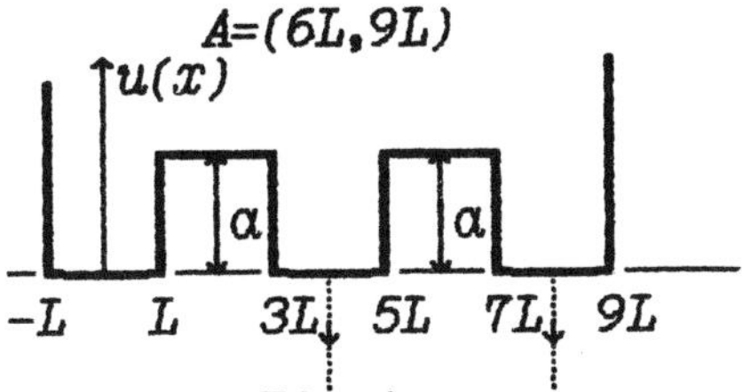

Fig.4.

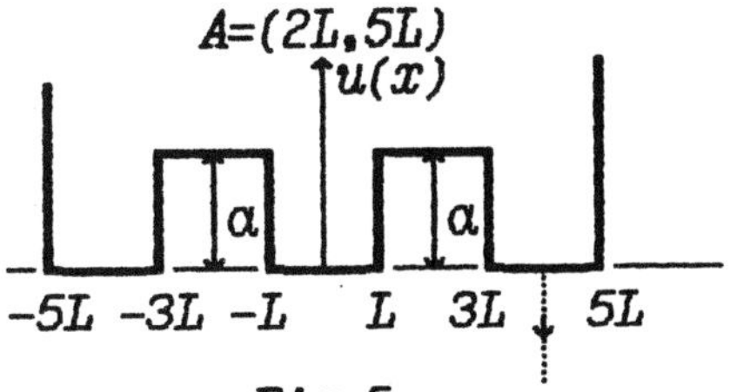

Fig.5.

For high barriers $(e^{\alpha}\gg 1)$

$$\tau=32\tau_re^{\alpha},\quad \tau'=8\tau_re^{\alpha}.$$

On the other hand one can calculate for this potential profile the MFPT's, which are equal

$$\left.\begin{aligned} T(0,4L)&=12\tau_r(3+2e^{\alpha}+e^{-\alpha}),\\ T(0,8L)&=12\tau_r(10+6e^{\alpha}+4e^{-\alpha}) \end{aligned}\right\}\ \text{Fig.4(dash lines)}$$

$$T'(0,4L)=T(4L,8L)=12\tau_r(7+4e^{\alpha}+3e^{-\alpha}),\ \text{Fig.5(dash line)}$$

It is of interest to note that for arbitrary α

$$T(0,8L)=T(0,4L)+T(4L,8L)$$

This equality represents a common superposition property of MFPT and is true for any potential profile and for any number of the terms, as it is pointed out in [6].

REFERENCES.

1.P.Hanggi,P.Talkner,M.Borkovec,*Rev.Mod.Phys*.**62**,251(1990)
2.M.Morch,H.Risken,H.Vollmer,*Z.Phys*.**B32**,*245 (1979)*.
3.H.Frish, V.Privman, C.Nicolis, G.Nicolis, *J.Phys*.A.,**23**, 1147 (1990).
4.A.N.Malakhov, *Radiophys.and Quant.Electr*.**34**,451 (1991).
5.A.N.Malakhov, *Radiophys.and Quant.Electr*.**34**,571 (1991).
6.N.V.Agudov, A.N.Malakhov, *Radiophys.and Quant.Electr*. **36**,No2 (1993).

XVII. MEMBRANES AND CELL

Noise Analysis of Ionization Kinetics in a Protein Ion Channel

Sergey M. Bezrukov

National Institutes of Health, NIDDK/LBM, 5/405, Bethesda, MD 20892
and St. Petersburg Nuclear Physics Institute, Russian Academy of Sciences, Gatchina, 188350 Russia

John J. Kasianowicz

National Institute of Standards and Technology, CSTL, Biotechnology Div., Biosensors Group, CHEM/A353, Gaithersburg, MD 20899

ABSTRACT

We observed excess current noise generated by the reversible ionization of sites in a transmembrane protein ion channel, which is analogous to current fluctuations found recently in solid state microstructure electronic devices. Specifically the current through fully open single channels formed by *Staphylococcus aureus* α-toxin shows pH dependent fluctuations. We show that noise analysis of the open channel current can be used to evaluate the ionization rate constants, the number of sites participating in the ionization process, and the effect of recharging a single site on the channel conductance[1].

INTRODUCTION

Ions cross biological membranes through ion channels formed by membrane spanning proteins. These macromolecules are critical for the generation of nerve action potentials and other cellular activities and function by switching between different permeability states. Thermal fluctuations cause ion channels to oscillate between different levels of conductance which can be observed as a fluctuating ion current in analogy to recent observations of the light scattering signal from a single trapped ion switching between different energy levels[2].

In biological systems, shot-noise and noise from conformational variations are widely recognized as sources of fluctuations in open channel currents[3]. In this study, we show that noise analysis can also be used to measure the rate constants of rapid chemical reactions that occur within the pore of a channel, if those reactions modulate the open channel conductance. The mechanism of the phenomenon we report here is analogous to that found for conductance fluctuations in solid state devices[4] where recharging of a single trap was shown to be the fluctuation source.

RESULTS & DISCUSSION

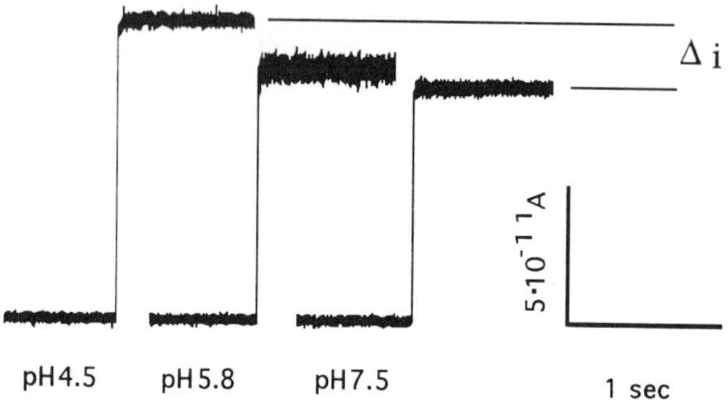

Figure 1.

The current recordings in Fig. 1 illustrate the spontaneous formation of a single channel at three different pH values. Two features are clearly seen. First, the mean conductance of the α-toxin channel decreases when the pH is increased over the range $4.5 \leq pH \leq 7.5$. Second, there is a difference in the noise of the channel's open state at the three different pH values.

The measurement of the spectral density at pH ≈ 6 revealed values significantly larger than the shot-noise anticipated for these currents. Fig. 2a illustrates this for the open channel current of 1.0 10^{-10} A at pH 5.9 (top trace). The difference spectrum (w/o background) is shown in Fig. 2b.

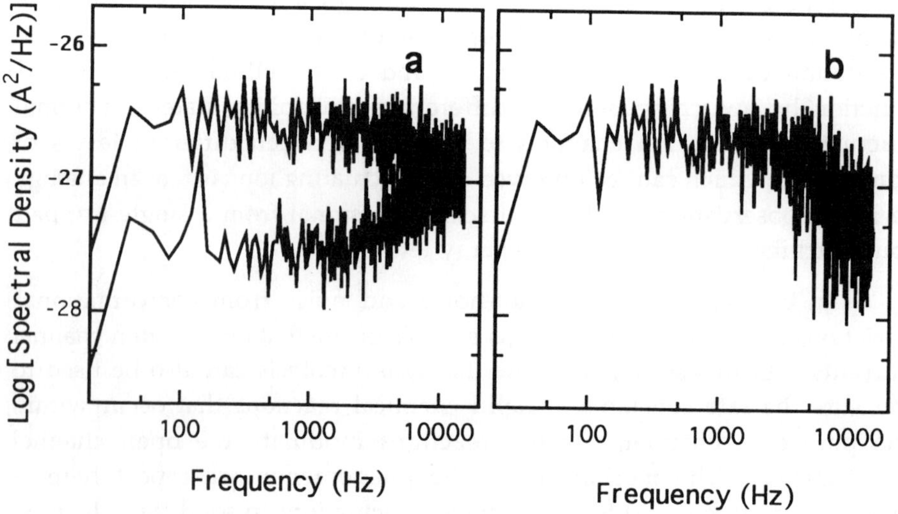

Figure 2.

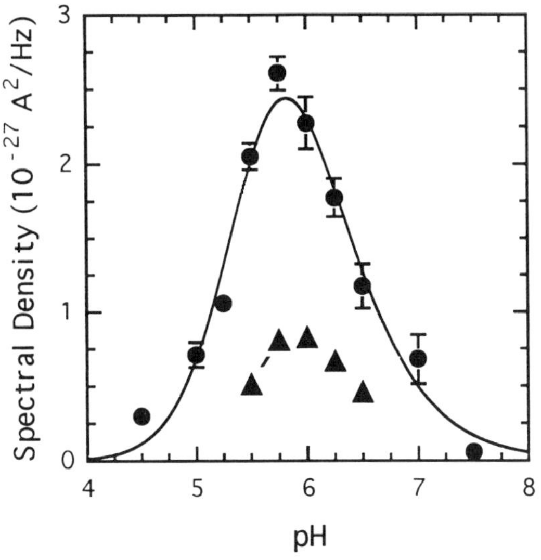

Figure 3.

Similar spectra for the open channel current noise were found using solutions with different pH values. The magnitude of the noise averaged in the bandwidth 200 - 2,000 Hz is illustrated in Fig. 3 for solutions containing either 1 M (circles) or 0.1 M (triangles) NaCl showing the non-monotonic dependence of the noise on the proton concentration.

A simple model based on a first-order reversible ionization reaction accounts for these results. Specifically, the effect of varying the pH on the noise spectral density and conductance of single open α-toxin channels can be described assuming the channel can access different states of ionization which differ in channel conductance[1]. For this process, we can write the frequency dependent spectral density of current noise, $S_{i,H}(f)$:

$$S_{i,H}(f) = 4(\Delta i)^2\ 10^{pK-pH}/n\ k_D(1 + 10^{pK-pH})^3\ (1 + (2\pi f \tau)^2), \qquad (1)$$

where Δi is the difference in current between the completely ionized and deionized states, k_R and k_D are the association and dissociation rate constants, respectively, $pK = -\log(k_R/k_D)$, $\tau = (k_R[H^+] + k_D)^{-1}$, and n is the number of ionizable sites. Fitting Eq. 1 to the measured pH dependent spectral density (Figs. 2,3) allows one to estimate the pK, k_R, k_D, and n.

The difference in current between the totally protonated and deprotonated states of the channel was measured from the single channel conductances at pH 4.5 and 7.5, and is $\Delta i = 2.0\ 10^{-11}$ A in 1 M NaCl. The least-squares fit of Eq. 1 is drawn (solid line, Fig. 3) assuming pK = 5.5 and $n\ k_D = 1.0\ 10^5\ sec^{-1}$. The frequency dependence of the spectral density at several different pH values yields an average value of $n = 4.2 \pm 0.7$. Thus, taking $n = 4$, we obtain $k_D = 2.5\ 10^4\ sec^{-1}$. The inverse of this parameter equals the mean time a proton is bound to a single site. We can also deduce the association rate constant, $k_R = 8\ 10^9\ M^{-1}\ s^{-1}$, since the equilibrium constant is defined by $K \equiv 10^{-pK} = k_D/k_R$. The change in the current through the open channel upon recharging of a single site is $\Delta i/n = 5\ 10^{-12}$ A, which contributes 5% of the total current through a single channel in 1 M NaCl.

The values of the rate constants we deduced from our measurements are close to those measured directly for carboxyl and imidazole groups in the bulk aqueous phase[5]. This suggests the titratable sites are not buried deeply within the protein core of the channel, and therefore would probably exhibit only a minor pK shift. Thus, the ionizable residues causing this effect are either aspartic acids, glutamic acids, or histidines.

In conclusion, our experiments demonstrate the possibility of using noise analysis to study the kinetics of fast chemical reactions in a single nanoscopic "cuvette" in which only several molecules participate. To date, most studies of channel structure are performed by measuring the single channel conductance and selectivity combined with genetic engineering. Analysis of reaction noise of the open channel coupled with site-directed mutagenesis introduces a new tool to probe channel structure and functional modulation.

ACKNOWLEDGEMENTS

Supported in part by the National Academy of Sciences/NRC (J.J.K.) and the ONR (V.A. Parsegian).

REFERENCES

1. S.M. Bezrukov and J.J. Kasianowicz, Phys. Rev. Lett. **70**: 2352 (1993).
2. W. Nagourney, J. Sandberg, H. Dehmelt, Phys. Rev. Lett. **56**, 2797 (1986).
3. S.H. Heinemann and F.J. Sigworth, Biophys. J. **60**, 577 (1991).
4. C. Dekker, A.J. Scholten, F. Liefrink, R. Eppenga, H. van Houten, and C.T. Foxon, Phys. Rev. Lett. **66**, 2148 (1991).
5. M. Eigen, G.G. Hammes, K. Kustin, J. Am. Chem. Soc., **82**, 3482 (1960).

THE POWER SPECTRUM OF 1/f NOISE IN BIOLOGICAL CELLS AND DISCHARGE CELLS

Noboru Tanizuka

Integrated Arts and Sciences, University of Osaka Prefecture
1-1 Gakuen-cho, Sakai, 593 Japan

ABSTRACT

A consideration is given to a mechanism of 1/f noise of neural membrane potential. A consideration is given to $1/f^{-\alpha}$ fluctuations of a magnetron discharge cell, where noise appears with a potential formation.

INTRODUCTION

A reason of the $1/f^{\alpha}$ (α=1) noise of neuron membrane potential at spike discharge series [1] is that the α=1 noise will bring no different error to a result of computing for a living thing in spite of its different sensing and processing time [2]. On the other hand, when a coding problem of resting membrane potential of nerve fibers were researched, the power spectrum of the potential fluctuations was found to be the $1/f^{\alpha}$ (α=1) noise [3]. The two experimental facts make us aware that when a potential is formed across the cell membrane, it has noise and fluctuations which may have a role in a computing process of such a discharge system.

Alfvén, a plasma physicist and astrophysicist, considered double layers as a surface phenomenon in physical plasmas.[4] He pointed out that "a sheath (or a double layer) is a plasma formation by which a plasma protects itself from the environment", and "a sheath is analogous to a cell membrane by which a biological plasma protects itself from the environment" [4]. Second, he pointed out that "a double layer always produces noise and fluctuations" [4]. By him, we were noticed that the cell membrane is similar to the plasma sheath in their phenomenon and function in the discharge systems. The similarity is given in Table I.[5]

In this paper, the author will review the mechanism of resting and action potential formation across the membrane,[7,9] reported 1/f noise of membrane potential,[1-3,6] and results of $1/f^{\alpha}$ noise observed by the author from a magnetron discharge.[5,10] Then the noise will be discussed from a view point of a nonlinear computation principle in such discharge systems.

CELL MEMBRANE POTENTIAL

Cell ions are transported by ion pumps and ion channels across the cell membrane. By Na^+-K^+ATPase, Na^+ ions are pumped out from the cell plasma, and contrarily, K^+ ions are pumped into the cell from outside, producing the density gradient of Na^+ ions and K^+ ions from outside to inside. Let C_I and C_{II} be the inside density and the outside, respectively. For Na^+ ions, $C_I<C_{II}$ is true, and for K^+ ions $C_I>C_{II}$. In a resting state, the Na^+ channels are closing while the K^+ ions leak free through the K^+ channels, producing a potential drop across the membrane. The equilibrium potential can be calculated from Nernst equation:

$$\Phi = \frac{RT}{ZF} \ln \frac{C_{II}}{C_I} \quad (1)$$

R=8.314 $J \cdot mol^{-1} \cdot K^{-1}$, F=$9.648 \cdot 10^4$ $C \cdot mol^{-1}$, T=300K, Z=1.

The membrane resting potential is usually about -80 mV, producing a very strong electric field, about 10^5 V/cm across the membrane, because of which

	Discharge cells	Biological cells
Etymology	plasma, I. Langmuir1928	plasma; protoplasma, Greek: Latin
System	physics, non-cognition	biology, cognition
Membrane	sheath; double layer	cell membrane; plasma membrane
Functions	protect plasma from the environments: sustenance	protect plasma from the environments: plasma formation
Carriers	electrons; ions	Na^+ions: K^+ions, etc.
Phenomena	polarization: bipolar diffusion	polarization: equilibrium
Transport	charged particles: energy: information	charged particles: energy: information
Potential	sheath potential	membrane potential
Noise	$1/f^{\alpha}$	$1/f$

Table I. The similarity between the plasma-sheath and the cell-membrane.

the protein molecule of ion channel is sensible of any change of a signal.

The mechanism of resting potential is very similar to that of the plasma sheath potential; plasma electrons diffuse away from the plasma because of high thermal velocities and ions remain because of heavy mass, producing a sheath potential at an equilibrium; in this case, the plasma is positively charged and the wall negatively charged. The main difference between the plasma sheath potential and the membrane potential is that the former is produced in the gaseous state and the latter in the liquid state.

They reported that the resting membrane potential of frog sciatic nerve had fluctuations of 1/f power spectrum in the frequency band from 0.1Hz to 1000Hz, and that the K^+ ion flux through the membrane was probably the noise source.[3,6]

They showed by experiments with a patch clump method that the ion channel plays the on-off gate for single ion and the gate current per each single ion is a single rectangular pulse.[7,9] The gate current is similar to the random telegraph switching current [RTS] of a μm-sized MOSFET.[8] Noise from a single defect of the MOSFET is Lorentz noise; noise from multiple defects with various trapping times approaches 1/f noise.[8] The K^+ channel gates repeat on-off action with a specific or random on-period. Then, does noise of resting potential fluctuations have the same mechanism as RTS noise?

When the membrane potential exceeds a threshold, the Na^+ channel gate is excited and begins to take on-off action, while the K^+ channel gate is restrained on-off action for a moment and soon recovered; the Na^+ channel gate open probability varies nonlinear with the membrane potential beyond the threshold upto a saturation state; after the saturation, the gate goes into inactivation and then turns to the original state; as Na^+ ions diffuse through the channels to increase the membrane potential and the conductance as the sum of open gates, the membrane is inversely polarized and reaches equilibrium at the action potential about +50mV from eq.(1) for Na^+ ions; then, after the recovery of K^+ channel gate and the inactivation of Na^+ channel gate, the membrane potential returns to the resting state through a hyperpolarization; this explains a single pulse of nerve cell (neuron).[9]

Musha reports a result on FFT analysis of pulse series detected from a snail giant neuron; a micropipette glass electrode was pierced through the membrane to detect intracellular potential (membrane potential); the detected pulse series showed the 1/f power spectrum in the frequency from 0.0005Hz to 0.5 Hz.[1]

An equation of the currents passing through the membrane is established:

$$\Sigma g_i(\Phi-\Phi_i)+I_p+I_L=0 \tag{2}$$

Σ:sum for $i=K^+$ and Na^+, g_i:membrane conductance for ion i,
Φ:membrane potential, Φ_i:equiliblium potential eq.(1) for ion i,
I_p:Na^+-K^+ATPase ion pump current, I_L:current by another ion species.[9]

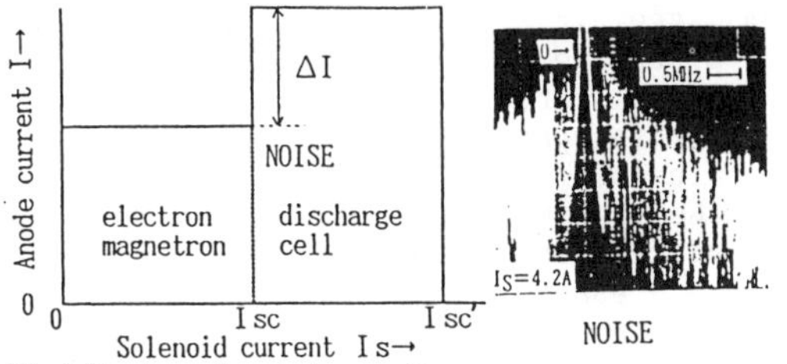

Fig.1 Domains of magnetron cell.

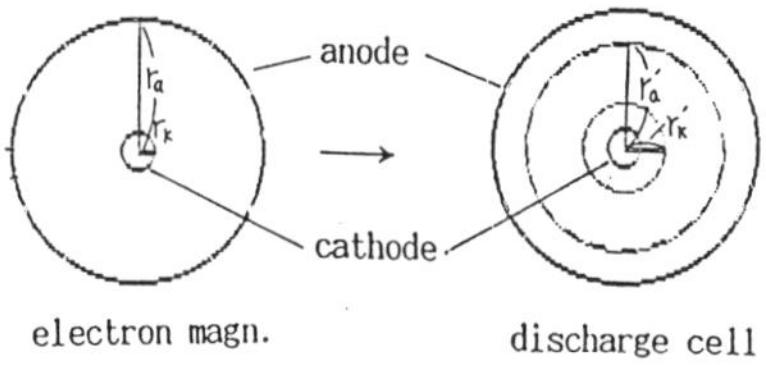

Fig.2 Potential structure is reformed to the discharge cell.

The conductance g_i is the total of open gates of (i) ion channels and it varies with Φ and ion density. The conductance is nonlinear. The currents I_p and I_L are very small compared to $\Sigma g_i(\Phi-\Phi_i)$. As described before, at the resting state $g_i(i=Na^+)\approx 0$, then $\Phi=\Phi_i(i=K^+)$ is true from eq.(2); and vice versa at the action state, $g_i(i=K^+)\approx 0$ and $\Phi=\Phi_i(i=Na^+)$. The snail autoactive neuron has a mechanism to exchange both states automatically. The 1/f noise power of the resting potential weakens as the membrane potential changes from +10 mV above the resting potential, and to -10 mV below the resting potential, and noise turns to white noise at -20 mV below the resting potential (hyper polarization).[6] Why does 1/f noise appear both in the resting potential and in the pulse series potential?

MAGNETRON CELL AND SHEATH POTENTIAL

The author studied the noise phenomenon by experiments with a hot cathode magnetron discharge. The magnetron is a coaxial cylinder: the diameters of cathode and anode cylinders are $2r_k$=0.5mm and $2r_a$=60mm, respectively. Solenoids are set around the anode cylinder to supply a magnetic field to the magnetron by the solenoid current I_s. A probe is mounted near a cylinder edge to detect noise. When the voltage U is supplied between the electrodes, the anode current I flows; as the current I_s is increased, the anode current I is cut off at the critical current I_{sc} (magnetron):

$$I_{sc}=c\sqrt{\frac{U}{r_a^2-r_k^2}} \tag{3}$$

where c is a device constant. Equation (3) was true for experiments of no gas supply to the magnetron, but when a gas (argon) was supplied, a discharge began to take place at the critical current I_{sc} and the discharge current ΔI flowed and the critical current increased to $I_{sc}'(I_{sc}'>I_{sc})$, see Fig.1.[10] The electron magnetron turned into a discharge magnetron. When the discharge current increased visibly, noise was detected.[5] An example is given in Fig.1. The power spectrum of noise invariably follows the power law of frequency in the frequency band about 0.1MHz<f<10MHz:

$$P(f) \propto f^{-\alpha}. \tag{4}$$

The spectral index α had no direct relation with the voltage U, the magnetic current I_s and the gas pressure P except the discharge current ΔI; under conditions of P=0.07Pa (argon gas), U=200V and $0.2\text{mA}<\Delta I<80\text{mA}$, the spectral index α was empirically given in the index $2.5<\alpha<5.5$: [5]

$$2.9\alpha=\log_2\frac{\Delta I}{\Delta I_0}\ (\text{bit}),\ \Delta I_0=1.5(\mu A). \tag{5}$$

The electron magnetron turned into the discharge magnetron (disharge cell) with an increased critical magnetic current and with the noise production; the increase of I_{sc} means a decrease of r_a ($r_a'<r_a$) and/or an

increase of r_k ($r_k' > r_k$) from eq.(3), see Fig.2; that is, a plasma-sheath potential structure appeared to the discharge cell, with a sharp potential fall at the sheath contrary to a flat potential at the plasma. The stability of the plasma-sheath system is controled by polarization and fluctuations at the layer, i.e. double layer or the sheath, across which the discharge energy is put into the plasma and charged particles are transported. The sheath acts as a sensor and a controller, i.e. a processor of the system; it is self-organized and essentially nonlinear part of the system. The author regards that the detected noise is a result of sheath driving as a processor in the magnetron discharge system.

Let ϕ, g and i_p be the sheath potential, the nonlinear sheath conductance and a driving current (corresponding to the pump current of the membrane), respectively; a nonlinear equation of the discharge cell stands:

$$\Sigma g\phi + \Delta I + i_p = 0 \tag{6}$$

where g is the conductance for different ion species and i_p is driven by an equivalent battery of the discharge system.

From the empirical equation (5) and from the fact that the potential organization follows the discharge current and the conductance g depends on ϕ, we have the following formula: (arrows show the flow of relation)

$$\alpha \overset{\rightarrow}{\underset{\leftarrow}{=}} \Delta I \overset{\rightarrow}{\underset{\leftarrow}{=}} \phi \overset{\rightarrow}{\underset{\leftarrow}{=}} g \tag{7}$$

The formula shows that the fluctuations α relates with the potential fall and the conductance of the potential fall (sheath). The empirical formula (4) and the empirical equation (5) must be a boundary condition of eq.(6).

CONCLUSION

A mechanism of 1/f noise of neural membrane potential relates to the open probabilities of ion channel gate (conductance) which vary nonlinear with the membrane potential and the ion density. The potential is automatically organized and 1/f noise lies behind the organization.

A mechanism of $1/f^{-\alpha}$ noise relates to automatical organization of the sheath potential acting as a processor of the discharge system. In both cases, a nonlinear process governs the system to protect and sustain it.

REFERENCES

1. T.Musha, H.Takeuchi and T.Inoue, IEEE Trans. on Biomed. Eng. **BME-30**, 194(1983).
2. T.Musha, Oyo Buturi **54**, 429(1985).
3. A.A.Verveen and H.E.Derksen, Kybernetic **2**, no.4, 152(1965).
4. H.Alfvén, Double Layers and Circuits in Astrophysics, TRITA-EPP-86-4 R.I.T.(Stockholm,1986).
5. N.Tanizuka, Proc. Int. Seminar on Reactive Plasmas, ed.T.Goto, Nagoya Univ.(Nagoya,1991),p.121.
6. H.E.Derksen and A.A.Verveen, Science **151**, 1388(1966).
7. R.Horn, J.Patlak and C.F.Stevens, Nature **291**, 426(1981).
8. M.Schulz and A.Papas, Proc. Int. Conf. on Noise in Physical Systems and 1/f Fluctuations, ed. T.Musha, S.Sato and M.Yamamoto, Kyoto, (Ohmsha Ltd., Tokyo, 1991),p.265.
9. B.Alberts, D.Bray, J.Lewis, M.Raff, K.Roberts and J.D.Watson, Molecular Biology of the Cell(Garland Publising,Inc.,New York,1983), chap.18.
10. N.Tanizuka, Jpn.J.Appl.Phys. **30**, 171(1991).

XVIII. CARDIOVASCULAR SYSTEMS

ROBUSTNESS OF 1/f FLUCTUATIONS IN P-P INTERVALS OF CAT'S ELECTROCARDIOGRAM

M. Yamamoto, M. Nakao, Y. Mizutani, T. Takahashi, H. Arai,
Y. Nakamura, M. Norimatsu and N. Ikuta
Lab. of Neurophysiology & Bioinformatics, Dept. of System Information Sciences, Graduate School of Information Sciences, Tohoku University, Sendai, 980 JAPAN

R. Ando
Center for Laboratory Animal Science, Tohoku College of Pharmacy

S. Nitta and T. Yambe
Dept. of Medical Engineering and Cardiology, Div. of Organ Pathophysiology, Institute of Development, Aging and Cancer, Tohoku University

ABSTRACT

Power spectral analyses have been performed on PP intervals of cat's electrocardiogram recorded during about 100 hrs under 3 experimental conditions: one with hard exogenous disturbances and others with the minimal in either a small or a large range of movement of the animal. Irrespective of the marked differences in the experimental conditions, a very similar 1/f profile of power spectrum has been obtained in the frequency range, 10^{-5}–10^{-1} Hz. The 1/f spectrum tended to extend to the frequency range below 10^{-5} Hz. The 1/f fluctuations in heart beat periods seem to have endogenous origins.

INTRODUCTION

It was reported by Musha's and Cohen's groups that either heart-beat periods or heart rate (HR) in humans have a 1/f power spectral density(PSD) in the frequency range, 10^{-4}–$10^{-}1$ Hz.[1,2] Recently, this hypothesis has been precisely investigated by Castiglioni and coworkers,[3] with respect to the frequency range over which the 1/f model is fitted, They showed a significant deviation of HR 24-h spectra from the model at frequency below 10^{-3} Hz. They speculated that the deviation "may be in part due to exogenous sources of variability associated with the different activity level characterizing day and night periods". It is important to investigate whether or not a 1/f spectral characteristic is obtained if possible exogenous sources of HR variability are eliminated. In the present study, the analysis has been performed on a long series of heart-beat periods in a cat obtained under freely moving conditions where possible time cues and exogenous disturbances were eliminated.

METHODS

1) Animal Experiments.

A 3.3kg female cat served as the subject. To detect heart beats using electrocardiogram(ECG) without artifacts from body movements, we used a surgical operation to chronically attach electrodes to the auricular surface of her heart. The direct recording of P-waves is ideal for the ECG rhythm detection because the P-wave almost exactly corresponds to the activity of pacemaker cells and complications stemming from ventricular arrhythmia can be avoided. The electrodes were attached to a female connector fixed to the bone of animal's head. The cat was put either in a small dog house (50(W)×60(L)×70(H)cm) (experiment 1 & 2) or in a larger cage (58(W)×96(L)×58(H)cm) (experiment 3) which was placed in a sound-proof room. The conditions of illumination, background noise, and food and water supply were shown in Fig. 1A,B,C. The room temperature was kept at 23–25 ℃.

Experiment 1: The dog house with a small litter-box was used (Fig. 1A). The animal's range of movement was medium-sized. During the initial 60 hrs, food and water was restricted except during 3 access periods. During the subsequent 46 hrs food and water were unrestricted. The house was indirectly illuminated with a tungsten(W)-lamp. Long input wires to an amplifier were directly connected with a socket on the cat's head. The lead wires were twisted many times and were untwisted 6 times when food and water were given. Thus, the exogenous disturbances were artificially introduced.

Experiment 2: The 2nd experiment was designed for minimizing the effects of possible exogenous disturbances. The dog house was used with a larger litter-box, and therefore animal's range of movement was smallest (Fig. 1B). Enough food was given and fresh water was constantly supplied by a dripping device. A flourescent light was used for the illumination of the house. Background white noise was delivered. The lead wires were untwisted once at the end of the 1st day. During the period of the 4th day, the twists approached the utmost limit, but, remained untouched. The cat could move only in a small area, but she was quiet enough to continue recording.

Experiment 3: The 3rd experiment was different from the 2nd in the following 2 respects. A slip ring (Neuroscience, NSR-35-12p) was used to avoid the wire's twists. The cat could move freely in a larger area compared with the former experiments (Fig. 1C). Twice, the door of the sound-proof room was opened briefly to re-adjust the water dripper.

2) Power Spectral Analysis.

From the overall time series of PP intervals obtained in each experiment, 4 sub-series with 218-points data sets were extracted corresponding to the 1st, 2nd, 3rd and 4th day, respectively (Table II). PSDs were calculated through FFT for each sub-series which was assumed to be sampled with a period of the sub-series's mean PP interval (Method I). For the computation of PSD on the overall time series, a series of instantaneous PP intervals occurring at the instant of each P-wave was defined. After continuation of the series by linear interpolation, a new series was generated by a sub-sampling with an equal interval to obtain a 220-points data set, and the FFT program was applied (Method II). A smoothing procedure using an MA filter introduced by Castiglioni et al.[3] was employed to reduce the variance of spectrum for both methods.

RESULTS

A summary of 3 experiments on various parameters is shown in Table I. The mean heart rates in 3 experiments are significantly different from each other ($P<0.01$). Table II shows a mean PP interval and a total duration for every one-day sub-series with a 218-points data set in each experiment.

Table I. A summary of 3 experiments. ** indicates $P<0.01$

Experiment #	Date	Observation Time(hrs)	Total no. of PP intervals	Mean HR (beats/min)	Mean PP interval(ms)
1	25-30/12/92	96	965,880	167.1	359**
2	14-18/01/93	101	1,146,548	188.7	318**
3	14-19/03/93	102	1,115,126	181.9	330**

Table II. Mean PP interval(M-PP) and total duration(TD) for every one-day sub-series with 2^{18}-points data set

Sub-series	Exp.1		Exp.2		Exp.3	
#	M-PP(ms)	TD(hrs)	M-PP(ms)	TD(hrs)	M-PP(ms)	TD(hrs)
1	356	25.9	318	23.2	320	23.3
2	345	25.1	329	24.0	331	24.1
3	361	26.3	316	23.0	333	24.2
4	368	26.8	306	22.3	337	24.5

1) Trendgram.

Mean HRs every one hour were plotted sequentially, which we call it a trendgram,as indicated in Fig. 1A,B,C. In the experiment 1, it showed apparent variations which seemed to be related to disturbances due to the food and water supply by experimenters. In the experiment 2, during the initial 50 hrs, damped-oscillation-like variations with a long period were observed. The effect of "the door-openings" was minimal from the viewpoint of the trendgram. During the latter half of the experiment, the HR tended to become less variable and attained a highest mean level in the 4th day when lead wires were severely twisted. From overall viewpoints the trendgram seems to have an almost linear component in the 2nd-4th day, which might depend on the stressful state of a small range of movement. In the experiment 3, the HR had also variability with local oscillations, but without overall trends. The disturbances due to the access

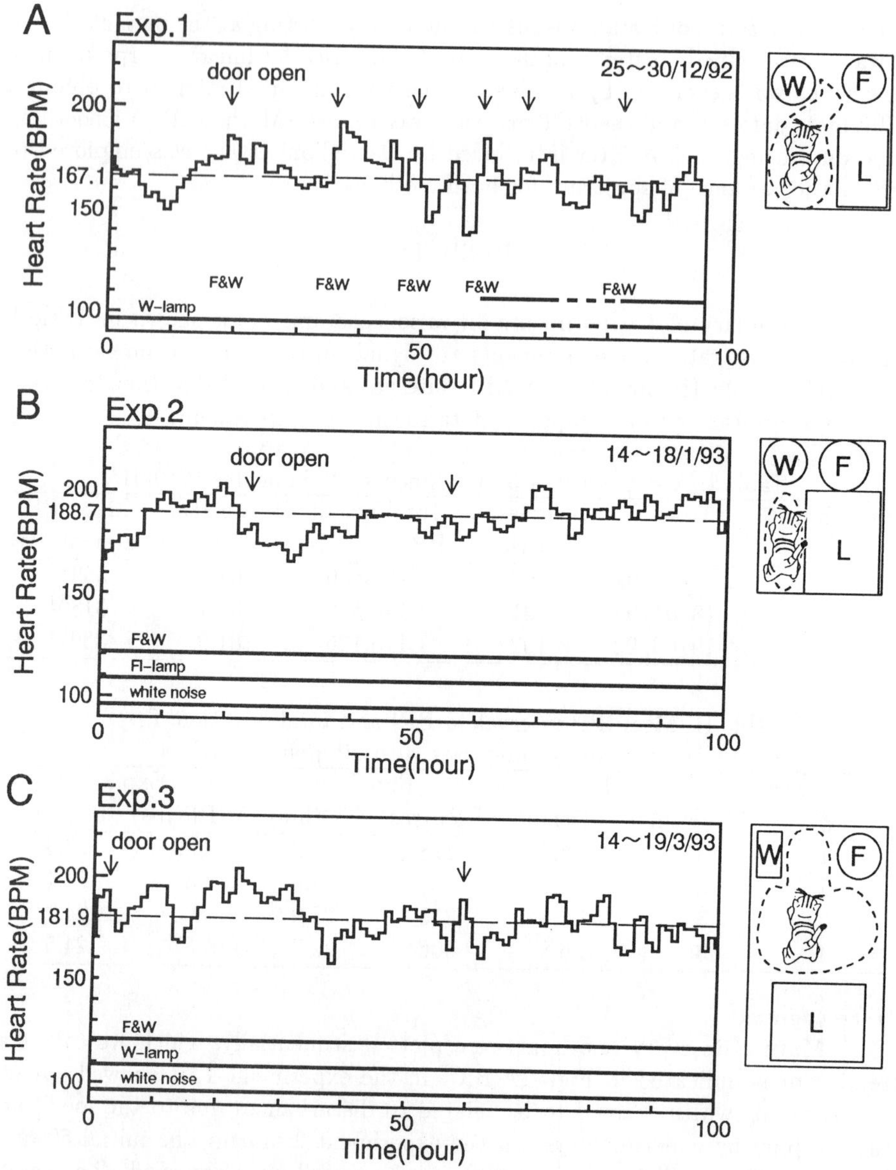

Fig. 1. Heart rate trendgrams in 3 experiments. The left-side illustrations indicate the positional relations among food(F) and water(W) tray and litter-box(L). The dashed line area shows the range where the animal is able to move.

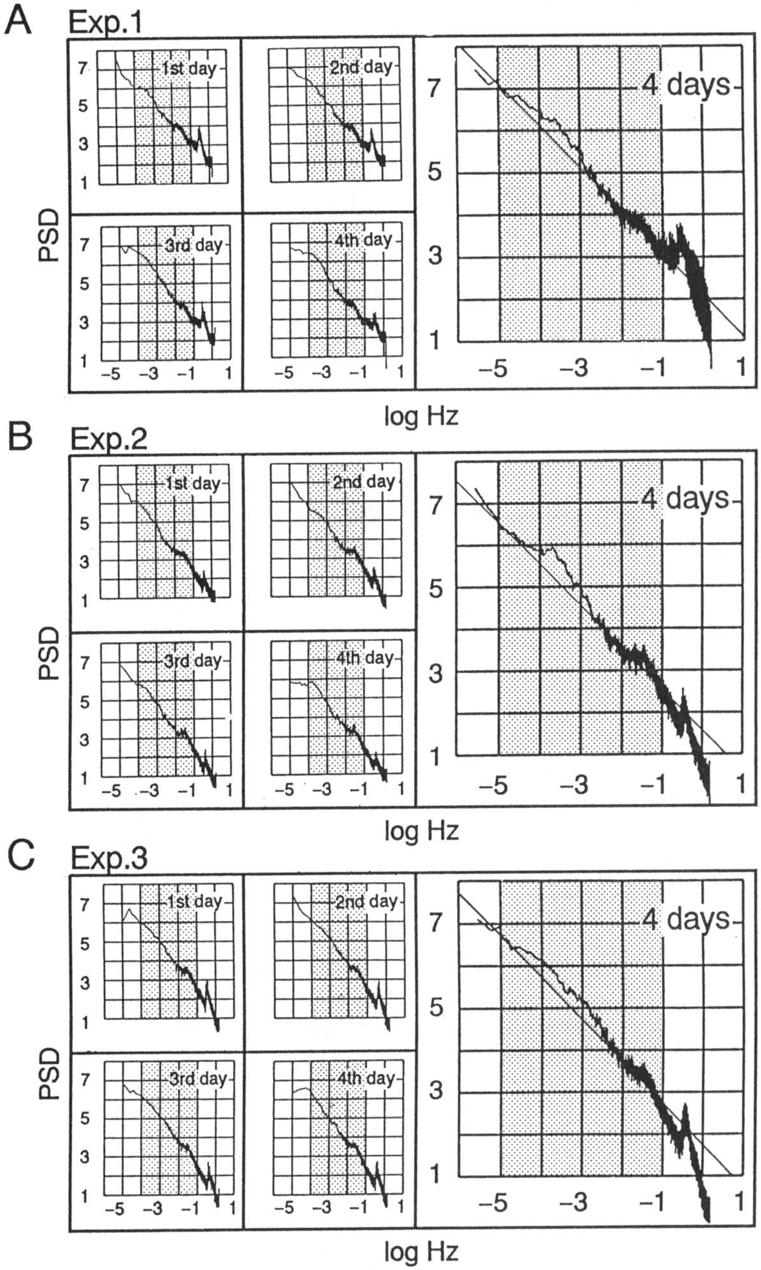

Fig. 2. Double logarithmic plotts of power spectral densitie (PSD, $[ms]^2/Hz$) of PP intervals in 3 experiments. The spectra for overall data indicate a quite similar 1/f pattern in the frequency range 10^{-5}–10^{-1} Hz, irrespective of different experimental conditions. The oblique straight line has an incline of -1.

for re-adjustment of water dripper seem to be small. Thus, a quasi-free running condition has been realized in the experiment 3.

2) Power Spectra.

Fig. 2A, B and C shows the PSDs of PP intervals for every one-day data (small panel) and the PSD for overall data (large panel) in each experiment. All of 12 spectra for every one-day data display a 1/f spectral profile, in the frequency range, 10^{-4}–10^{-1} Hz, being quite similar to each other. The variations shown in the frequency below 10^{-4} Hz probably depend on an effect of small sample size for such low frequency. Whereas, the spectra for overall data show that the 1/f profile is extended to at least 10^{-5} Hz with high confidence, although there is a broad peak around 10^{-4} Hz. No trend of saturation in the PSD for overall data was observed in the frequency below 10^{-5} Hz for every experiment.

DISCUSSION AND CONCLUDING REMARKS

In the 1st experiment, the animal was disturbed several times by exogenous stimuli. In the 2nd, the recording was made in a stressful small room which produced partly a linear trend in HR. The final experiment was almost ideally performed in a sense that a quasi-free running conditions were realized in a larger space. In spite of the quite different environmental conditions, a very similar profile of 1/f power spectrum was obtained for overall data in the frequency range, 10^{-5}–10^{-1} Hz. A broad peak was observed around 10^{-4} Hz, but this was unclear in one-day spectra. Moreover no trend of saturation for the frequency below 10^{-5} Hz was seen in all of 3 PSDs. These results might indicate that cat's HR fluctuates with a 1/f power spectrum in the very low frequency extending to 10^{-6} Hz, even under the condition of exogenous disturbances minimized. Although seen in only one animal, the robustness of 1/f spectral profile in PP interval time series has been experimentally verified in a wide frequency range. This might imply that the cat's HR could never be sustained at the constant mean level and that endogenous origins of the 1/f fluctuations have to be investigated.

This work was supported by Grant-in-Aid for Scientific Research(04302033) from the Ministry of Education and Culture, Japan.

REFERENCES

1. M. Kobayashi and T. Musha, IEEE Trans. Biomed. Eng. BME-29, 456 (1982).
2. J. P. Saul, P. Albrecht, R. D. Berger and R. J. Cohen, Proc. Comp. in Cardiol. 419 (1988).
3. P. Castiglioni, A. Frattola, G. Parati, A. Pedotti and M. Di Rienzo, Proc. VI Mediterranean Conf. Med. Biol. Eng. 1149 (1992).

HEART RATE FLUCTUATIONS IN POST-OPERATIVE AND BRAIN-DEATH PATIENTS

Toshiyo Tamura* , Kazuki Nakajima
Department of Electrical and Electronic Engineering
Yamaguchi University, Ube 755 Japan
Tuyoshi Maekawa, Yoshiyuki Soejima, Yasuhiro Kuroda, and Akio Tateishi
Critical Care Medical Center, University Hospital, School of Medicine, Yamaguchi University

ABSTRACT

The power spectra of heart rate in patients receiving intensive care were calculated and the relation between gain and frequency discussed. 1/f fluctuations in heart rate can be observed in both post-operative and brain-death patients in the intensive care unit. These results suggested that 1/f fluctuations are a fundamental human phenomenon.

INTRODUCTION

Analysis of heart rate fluctuations provides qualitative estimation of the cardiovascular control system. The power spectrum of heart rate fluctuations in normal subjects shows that the spectral densities are inversely proportional to frequency. 1/f fluctuations have been observed at a frequency band of 10^{-4} to 2×10^{-2} Hz[1)]. Few studies have, however, been designed in which fluctuations were analyzed in patients. In this study, the heart rates of patients admitted to our intensive care unit were analyzed and their fluctuations evaluated.

SUBJECTS AND METHODS

The five post-operative and two brain-death patients, who had required intensive care, were chosen for the study. Table 1 shows the characteristics of the patients including diagnosis and analyzed dates. Cases 1, 5, and 6 were returned to surgical wards on the 10th, 15th and 17th post-operative day, respectively. The ECG, blood pressure, and core temperature were recorded on monitors (BMS-8500, Nihon Koden Co, Tokyo, Japan).

---

* Present Address: Institute for Medical and Dental Engineering, Tokyo Medical and Dental University, Tokyo 101 Japan

Table 1 Description of patients

Case no.	Age and sex	Diagnosis	Analyzed date (day)
1	45M	Esophageal Varices	6
2	64M	Abdominal Aneurysm	6
3	28F	Cerebral Aneurysm (brain death)	12
4	32F	Subarachnoid hemorrhage (brain death)	6
5	53F	Subarachnoid hemorrhage	6
6	65F	Subarachnoid hemorrhage	12
7	61M	Gastric cancer, Pneumonia	6

Heart rates were obtained from electrocardiographic R-R intervals and could be automatically calculated by the monitor with triggering of an R wave. The data were transferred to a workstation (CWS-8100, Nihon Koden, Co., Tokyo Japan) and stored in a hard disk every 1 min. , i.e. the instantaneous heart rate values were stored. Stored data were selected with a period of 6 to 12 days and transferred to another workstation. The data missing during calibration of sensors and disconnections necessitated by treatment procedures were edited and interpolated. A weighting function was used for smoothing data and the power spectral densities were calculated using the fast Fourier transform method. The linear regression line was calculated between the log scale of power spectral densities and that of frequencies. From the slope of the linear regression line, fluctuations related to the frequency were evaluated.

RESULTS

Power spectra were evaluated for 6 to 12 days depending on the duration of admission to our intensive care unit. Figure 1 shows the time course and power spectrum of the 6th post-operative day in a patient who recovered the deseise (case 1). A 1/f fluctuation in heart rate for a frequency less than 10^{-2} Hz was observed. Figure 2 shows the time course and power spectrum of a patient who died in the intensive care unit (case 2), on the 20th day, i.e. one day before death. The power spectra above 10^{-4} Hz were constant, producing random fluctuations.

The time courses and power spectra of two brain-death patients at day one are shown in Fig. 3. Figure 3 (a) shows the 59th day of a patient (case 3), and (b) the 5th day of a patient (case 4). Both power spectra showed 1/f fluctuations. Analysis of the total data from six days is shown in Figure 4 and reveals a 1/f fluctuation for a frequency band of 10^{-6} Hz to 10^{-2} Hz.

No marked difference in fluctuation pattern was observed between patients who returned to the ward and brain-death patients.

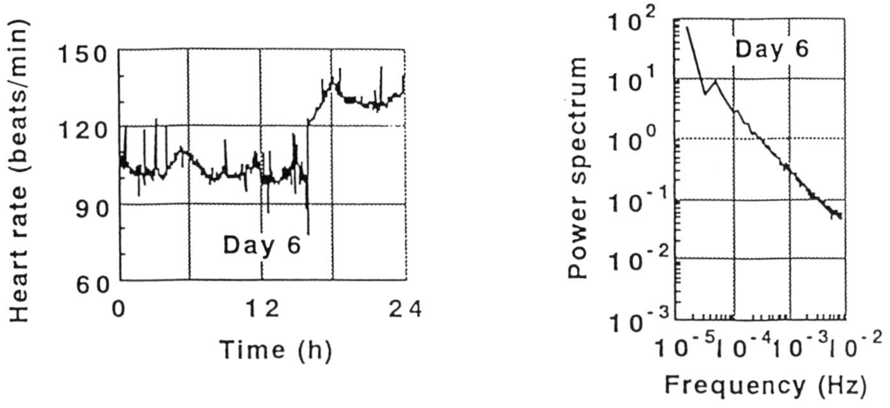

Figure 1. Time course and power spectrum of heart rate in a patient who recovered.

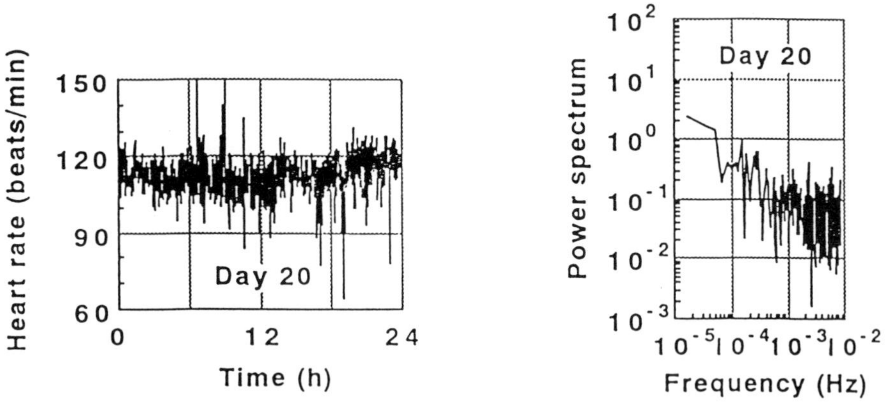

Figure 2. Time course and power spectrum of heart rate in a patient who died in the intensive care unit.

DISCUSSION

Heart rate fluctuations in post-operative patients were 1/f for frequencies lower than 10^{-3} Hz. No individual patterns were observed among the various diagnosis. A 1/f fluctuation was clearly observed in brain-death patients, whose heart beats were controlled by their cardiac automaticity without central nervous control. Although further investigation is required to draw definitive conclusions, 1/f fluctuation seems to be a fundamental human phenomenon which occurs with or without the control of neural activities.

REFERENCE

1) Kobayashi, M., and Musha T. (1982) 1/f fluctuation of heart beat period. IEEE Trans. Biomed. Eng. BME-29, 456-457

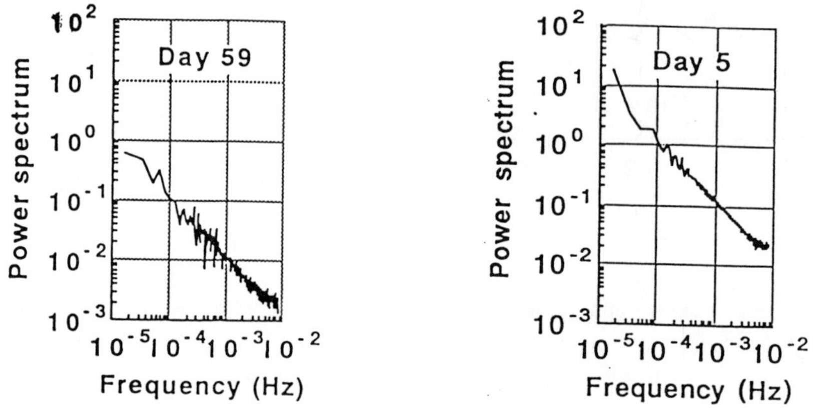

(a) (b)

Figure 3. Power spectra of heart rate in brain-death patients; case 3 (a) and case 4 (b).

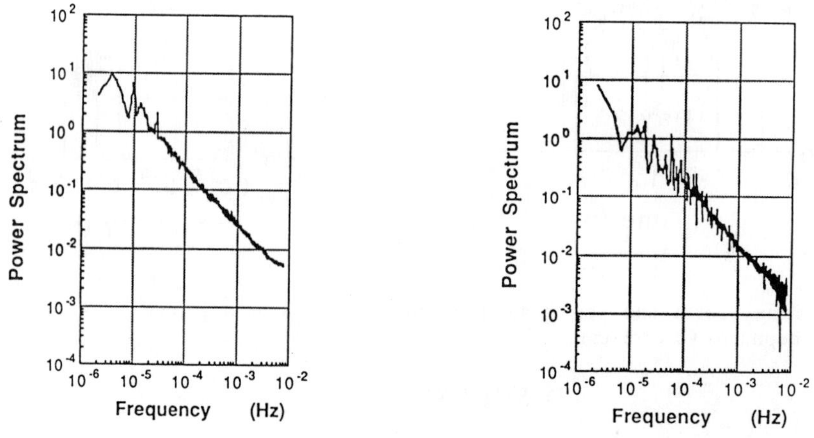

(a) (b)

Figure 4. Power spectra of heart rate in brain-death patients; case 3 (a) and case 4 (b) based on analysis of data from six continuous days.

FAILURE IN REJECTING A NULL HYPOTHESIS OF STOCHASTIC HUMAN HEART RATE VARIABILITY WITH $1/f$ SPECTRA[1]

Yoshiharu Yamamoto[2] and Richard L. Hughson
University of Waterloo, Waterloo, Ontario N2L 3G1, Canada

ABSTRACT

We studied time series of human resting heartbeat intervals with $1/f$ spectra. Static and dynamical properties of a putative attractor in a reconstructed phase space were examined and compared with those of stochastic surrogate data. It was found that the actual heartbeat intervals had the signature of chaotic dynamics which was not present in the surrogate data.

INTRODUCTION

Almost 10 years ago, an example of human resting heartbeat intervals (heart rate variability; HRV) with a $1/f$ type spectrum was reported.[1] As this type of broad band spectrum was often observed in some chaotic attractors,[2] it was hypothesized that the $1/f$ spectrum in human HRV arose from underlying chaotic dynamics in the heartbeat controller(s).[3] The observation of a finite correlation dimension[4] might also support the hypothesis of chaotic HRV. However, this hypothesis has been challenged[5] because of the fact that $1/f$ type spectra can be found in stochastic as well as deterministic model,[6] and because of the technical difficulties associated with correlation dimension algorithms.[7]

Recently, the method of surrogate data has been proposed[8] to study possible chaotic dynamics and descriminate them from stochastic noise. In this method, the stochastic surrogate data are generated to have the same power spectra as the original, but have random phase relationships among the Fourier components. If any numerical procedures for studying chaotic dynamics produced the same results for the surrogates as for the original data, we could not reject a null hypothesis that the observed dynamics were stochastic rather than being described by deterministic chaos. In the present study, we used the method of surrogate data to re-examine whether the $1/f$ spectra in human resting HRV were due to the underlying chaotic dynamics by testing a null hypothesis of stochastic HRV.

CHARACTERIZATION OF $1/f$ SPECTRA IN HRV

Five healthy volunteers participated in this study conducted while beat–by–beat long–term HRV ($> 8,192$ beats for $2-3$ h) was recorded in the quiet, awake state in the supine position. All the subjects were tested four times each, so that 20 long–term recordings were available. The surface electrocardiogram was sampled on a real time basis with 1 ms accuracy and the intervals between successive QRS spikes, i.e. HRV, were stored on a diskette for later analyses.

After eliminating regular oscillatory components of HRV, due to known physiological forcings such as respiration, by a renormalization technique,[9] the $1/f^{\beta}$ relationship was determined for the remaining HRV spectrum. The value for β calculated from spectral power above ~ 0.01 Hz was 1.08 ± 0.18 (mean $\pm$ SD) for 20 recordings, confirming the initial observation[1] of $1/f$ spectrum in human resting HRV. However, at < 0.01 Hz, β was 0.53 ± 0.19 indicating that normal human HRV had characteristics of "white noise" (although it was not completely white) for oscillations with periods of > 100 s. The detail of these results was reported elsewhere.[10]

---

[1]Supported by Heart and Stroke Foundation of Ontario

[2]Current address: Fac. of Education, Univ. of Tokyo, 7-3-1 Hongo, Bunkyo-ku, Tokyo 113, Japan

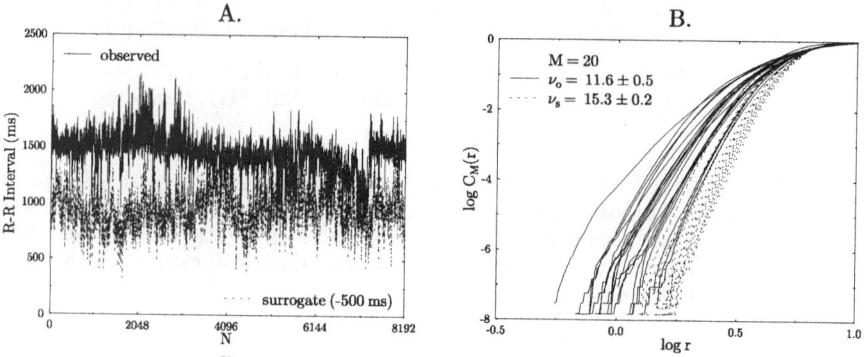

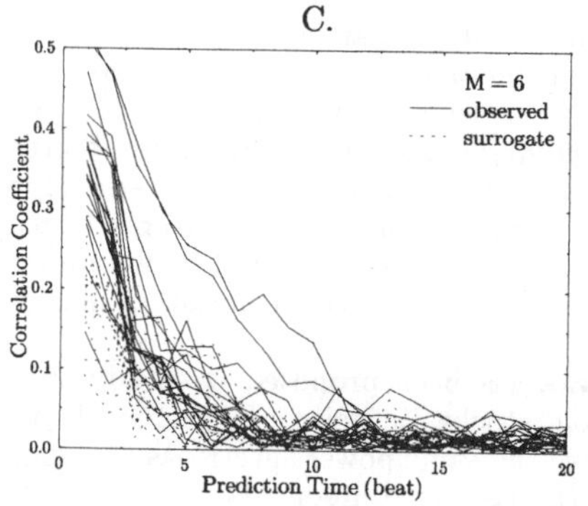

Fig. 1: (A) An example of long-term HRV (R-R Interval) and its stochastic surrogate (shifted 500 ms down). (B) The log correlation integral $C_M(r)$ plotted against log r at the embedding dimension (M) of 20. Note that data sets were normalized to have unit variance for the quantitative comparison. Estimates for correlation dimensions for the observed (ν_o) and the surrogate (ν_s) data are expressed as mean ± SD. (C) Changes in nonlinear predictability with the increment of prediction time at $M = 6$.

CORRELATION DIMENSION

A static property of a putative attractor was analyzed by the correlation dimension method.[11] The phase space trajectory

$$X_M(i) = (x(i), x(i+L), \ldots, x(i+(M-1)\cdot L)), \qquad (1)$$

where $x(i)$ was the i–th heartbeat interval with i running from 1 to the number of data points N and L was a fixed lag, was reconstructed with different levels of embedding dimension M ($M = 2, 4, \ldots, 20$) and with the L determined as the first local minimum of mutual information content.[12] The correlation integral

$$C_M(r) = \frac{1}{N^2} \sum_{i \neq j} \Theta(r - \mid X_M(i) - X_M(j) \mid), \qquad (2)$$

where $\Theta()$ was 1 for positive and 0 for negative arguments, was calculated only for pairs of vector of which time indices (i, j) were separated by greater than autocorrelation time.[13] The same analysis was also performed on the "iso–spectral" (stochastic) surrogate data generated by a windowed (inverse) Fourier transform[8] (Fig. 1A).

For all $M > 2$, estimates for correlation dimension ($\nu = \log C_M(r)/\log r$ with $r \to 0$) for the observed HRV were significantly ($P < 0.001$) smaller than those for the surrogates (Fig. 2). While ν for the surrogates increased almost linearly with M, ν for the observed data had a tendency to reach a plateau at higher M without significant ($P \geq 0.05$) differences between the values for $M = 18$ and $M = 20$ (Fig. 2). At the

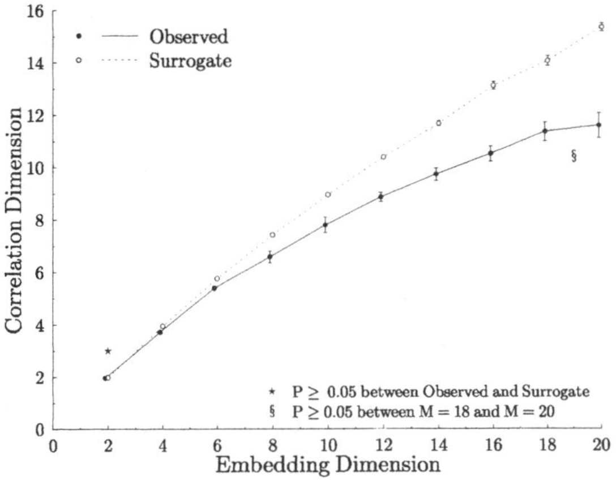

Fig. 2: Averages for estimated correlation dimensions of long-term HRV with the increment of embedding dimension. Vertical bars express SD.

highest M of 20, estimate for ν was 11.6 ± 0.5 for the observed HRV while that for the surrogates was 15.3 ± 0.2 (Fig. 1B). Although the absolute values of ν should be interpreted with caution, mainly due to the difficulties in determining a unique straight line in the scaling curves for the correlation integral,[7] it could at least be concluded that the observed HRV had a different scaling relationship compared to stochastic noise.

NONLINEAR FORECASTING

A dynamic property of the attractor was studied by the nonlinear forecasting method.[14] Changes in predictablity with regard to the prediction time (T_p; expressed in beat number) was evaluated by calculating a correlation coefficient (ρ) between the observed and the non-parametrically predicted values in the phase space reconstructed from the first differenced HRV. One would expect a dynamic property of a chaotic attractor to be reflected by a decrease of the correlation (by the increased variance due to the exponential divergence of nearby orbits[15]) between the observed and the predicted values as T_p increased. For the T_p of > 10 beats, ρ was substantially zero both for the actual HRV and for the surrogates (Fig. 3). For the surrogates, the values of ρ were not different for the various levels of M and decreased gradually with the increment of T_p of < 10 beats (Fig. 3B). This latter finding agreed with the recent

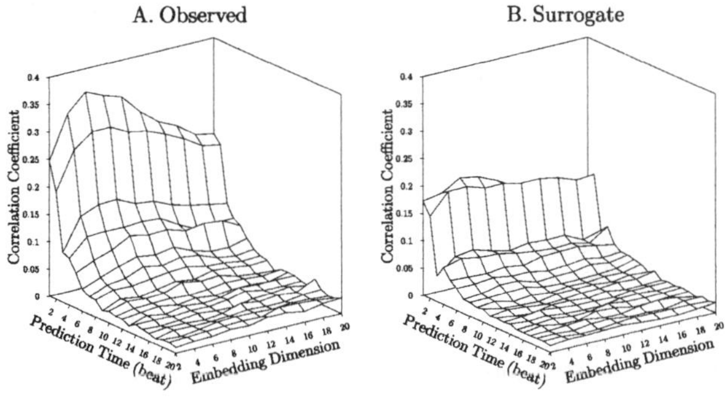

Fig. 3: Average correlation coefficients between the actual and the nonlinearly predicted ΔHRV as functions of prediction time and embedding dimension.

observation that the predictability of stochastic $1/f$ noise showed a pattern similar to, but not the same as, that of a chaotic time series.[16] By contrast, for the actual HRV, the ρ at T_p of < 4 beats was significantly ($P < 0.01$) higher than that for the surrogates and was also dependent on M ($P < 0.05$). Thus the initial abrupt drops in ρ at lower T_p with an optimal embedding ($M = 6$, Fig. 1C) for the actual HRV, but not for the surrogates, suggested that HRV dynamics had a dynamic property that could be indicative of chaotic attractors in contrast to a stochastic noise.

DISCUSSION

A time series with "true" $1/f$ spectrum is nonstationary[9] and not bounded, thus incompatible with the concept of chaotic attractors that are necessarily stationary and bounded.[17] Unlike human HRV recorded in daily life of which $1/f$ scaling was maintained in very low frequency range,[18] HRV *at rest* had $1/f$ type spectrum only in the higher frequency range. It was thought that the limitation of the low frequency $1/f$ behavior provide a rationale for the hypothesis that resting HRV might be chaotic.

The concept of chaotic dynamics of human HRV has been challenged, and a counter hypothesis of stochastic colored noise has been advanced.[5] In this study, the null hypothesis of stochastic noise was not accepted on the basis of comparisons between actual long-term resting HRV and stochastically generated surrogate data. Our results support the possibility that resting human HRV might be chaotic,[3,4] and that the autonomic control mechanisms of the central nervous system might be responsible for the generation of these chaotic dynamics.

REFERENCES

1. M. Kobayashi and T. Musha, IEEE Trans. Biomed. Eng. BME-30, 456 (1982).
2. P. Manneville, J. Phys. 41, 1235 (1980); J. D. Farmer, Physica D 4, 366 (1982).
3. A. L. Goldberger, News Physiol. Sci. 6, 87 (1991).
4. A. Babloyantz and A. Destexhe, Biol. Cybern. 58, 203 (1988).
5. L. Glass and C. P. Malta, J. Theor. Biol. 145, 217 (1990).
6. A. R. Osborne and A. Provenzale, Physica D 35, 357 (1989).
7. J. Holzfuss and G. Mayer-Kress, in Dimensions and Entropies in Chaotic Systems, ed., G. Mayer-Kress (Springer–Verlag, Berlin–Heidelberg, 1986), p. 114; D. Ruelle, Proc. R. Soc. Lond. A 427, 241 (1990).
8. J. Theiler, S. Eubank, A. Longtin, B. Galdrikian, and J. D. Farmer, Physica D 58, 77 (1992); S. J. Schiff and T. Chang, Biol. Cybern. 67, 387 (1992).
9. Y. Yamamoto and R. L. Hughson, J. Appl. Physiol. 71, 1143 (1991); Y. Yamamoto and R. L. Hughson, Physica D (in press).
10. Y. Yamamoto and R. L. Hughson, Am. J. Physiol.: Regulatory Integrative Comp. Physiol. (in press).
11. P. Grassberger and I. Procaccia, Physica D 9, 189 (1983).
12. A. M. Fraser and H. L. Swinney, Phys. Rev. A 33, 1134 (1986).
13. J. Theiler, Phys. Rev. A 34, 2427 (1986).
14. G. Sugihara and R. M. May, Nature 344, 734 (1990).
15. D. J. Wales, Nature 350, 485 (1991).
16. A. A. Tsonis and J. B. Elsner, Nature 358, 217 (1992).
17. J. -P. Eckmann and D. Ruelle, Rev. Mod. Phys. 57, 617 (1985).
18. J. P. Saul, P. Albrecht, R. D. Berger, and R. J. Cohen, Comp. Cardiol. 14, 419 (1987); C. -K. Peng, J. Mietus, J. M. Hausdorff, S. Havlin, H. E. Stanley, and A. L. Goldberger, Phys. Rev. Lett. 70, 1343 (1993).
19. Copies of the in–press articles listed above can be obtained via anonymous ftp account of cgsa.uwaterloo.ca.

XIX. BIOLOGICAL SYSTEMS

The Role of Noise in Sensory Information Transfer

F. Moss
Department of Physics
University of Missouri at St. Louis
St. Louis, MO 63121 USA

A. R. Bulsara
NCCOSC-RDT&E Division
Materials Research Branch
San Diego, CA 92152-5000 USA

ABSTRACT

We consider the interpretation of time series data from firing events in periodically stimulated sensory neurons. The neurons are represented as nonlinear switching elements embedded in a Gaussian noise background. The cooperative effects arising through the coupling of the noise to the modulation are examined, together with their possible implications in the features of Inter-Spike-Interval Histograms (ISIHs) that are ubiquitous in neurophysiological experimental data. Our approach provides the *simplest posible* interpretation of the ISIHs and has been found to reproduce the salient features of experimental ISIHs. One such comparison, between very recent data from experiments performed in St. Louis, on the mechanoreceptor in the tailfan of the crayfish *Procambarus clarkii*, and analog simulations on a simple nonlinear exciteable neural model is presented to elucidate this point.

INTRODUCTION

Neuroscientists have known for decades that sensory information is encoded in the intervals between the action potentials or "spikes" characterizing neural firing events. Statistical analyses of experimentally obtained spike trains have shown the existence of a significant random component in the inter-spike intervals. There has been speculation, of late, that the noise may actually facilitate the transmission of sensory information; certainly there exists evidence that noise in networks of neurons can dynamically alter the properties of the membrane potential and the time constants[1]. Recent work by Longtin, Bulsara and Moss[2] (LBM) demonstrated how experimental ISIHs measured, for example, on the auditory nerve fibers of squirrel monkey[3] and visual cortex of cat[4] could be explained via a new interpretation of noise-driven bistable dynamics. They introduced a simple bistable neuron model, a two-state system controlled by a double-well potential with neural firing events corresponding to transitions over the potential barrier (whose height is set such that the deterministic stimulus alone cannot cause transitions). The cell dynamics was described via a variable $x(t)$ loosely denoting the membrane potential and evolving according to

$$\dot{x} = f(x) + Q \sin(\omega t) + F(t), \tag{1}$$

where $f(x)$ is a flow function (expressible as the gradient of a potential $U(x)$) and $F(t)$ is noise, taken to be Gaussian, delta-correlated, with zero mean and variance $2D$. Potentials can be either soft or hard (even infinitely hard) in the bistable description.

The potential used here was taken to be the "soft" function $U(x) = \frac{1}{2}ax^2 - b\ln(\cosh x)$. It is instructive to point out that a bistable model of the form (1) can be derived[5] for the dynamics of more complex networks of neurons and/or dendrites, under certain mean-field-like assumptions. For our analysis, the system (1) is numerically integrated, with the residence time in each potential well (these times correspond to the firing and quiescent intervals) assembled into a histogram, which displays a sequence of peaks with a characteristic spacing. Two unique sequences of temporal measurements are possible: the first measures the residence times in only one of the states of the potential and the histogram consists of peaks located at $t = nT_0/2$, T_0 being the period of the deterministic modulation and n an odd integer. The second sequence encompasses measurements of the total time spent in both potential wells, i.e. it includes the active *and* refractory or reset intervals; in the presence of noise, the reset intervals are of largely stochastic duration. The histogram corresponding to this sequence consists of peaks at locations $t = nT_0$ where n is any integer. The sequence of peaks implies a form of phase locking of the neural dynamics to the stimulus. Starting from its quiescent state, the neuron attempts to fire at the first maximum of the stimulus cycle. If it fails, it will try again at the next maximum, and so on. The latter sequence is the only one observable in an experiment; the former sequence, which corresponds to the refractory events is elegantly elucidated by the LBM theory. Analog simulations of the dynamics yield an extremely good fit to experimental data; the fit can be realized by changing only one parameter (the stimulus intensity *or* the noise intensity). In addition to the peak spacing in the ISIH, most of the other substantitive features of experimental ISIHs are explainable via the simple model (1): (a). Decreasing the noise intensity (keeping all other parameters fixed) leads to more peaks in the histogram since the "skipping" referred to above becomes more likely. Conversely, increasing the noise intensity tends to concentrate most of the probability in the first few peaks of the histogram. (*b*). In general, the probability density of residence times is well approximated by a Gamma distribution of the form $P(T) = \frac{T}{\langle T\rangle^2}\exp[-T/\langle T\rangle]$, where $\langle T\rangle$ is the mean of the ISIH. It is apparent that $P(T) \to 0$ or $\exp(-T/\langle T\rangle)$ in the short and long time limits, respectively. For vanishingly small stimulus amplitude Q, the distribution tends to a Gamma, conforming to experimental observations. (*c*). Increasing the stimulus amplitude leads to an increase in the heights of the lower lying peaks in the ISIH. (*d*). Memory effects (even within the framework of a description based on the theory of renewal processes) frequently occur, particularly at very low driving frequencies; they manifest themselves in deviations from an exponentially decaying envelope at low residence times (the first peak in the ISIH may not be the tallest one). (*e*). The mean of the ISIH yields (through its inverse) the mean firing rate. A more rigorous treatment of the above results is available in recent work[6]; this work also includes comparison of our results with recent experimental data taken from cat auditory nerve. The important point to note here is that the results are almost independent of the functional form of the potential $U(x)$, depending critically on the ratio of barrier height to noise; this ratio determines the hopping rate between the basins of attraction in the absence of noise.

The LBM theory demonstrates that the peaks of the ISIH *cannot exist in the absence of noise*. Indeed, stimulus cycle skipping, which is necessary to generate a sequence of peaks, cannot occur unless two conditions are fulfilled: there must be noise, and the coherent stimulus must be subthreshold. It also implies the existence of a "regulatory mechanism" in which sensory neurons measure the stimulus amplitude by comparing it to the background noise level, a process that is mediated and optimized by the (internally adjustable) potential barrier height in the bistable model.

HOW GOOD IS THE BISTABLE DESCRIPTION?

Although the LBM model provides an important first step in the understanding of *the*(possibly pivotal) role of noise in sensory information transfer, it is farfrom complete. The results do not depend critically on the characteristics ofthe potential function $U(x)$ and the fundamental question: what aspects of the data are due to the statistical properties of noisy two-state systems as opposed to properties of cells that transcend this simple description (or, can the neuron be satisfactorly described by a noisy bistable switching element), have still not been satisfactorily answered, although an important first step in this direction is afforded by recent work[6].

Integrate-fire (IF) models have been exceptionally popular in the quest for a description of the statistical properties of spike trains obtained from exciteable cells. For the classical Gerstein-Mandelbrot (GM) model[7], $f(x) \equiv \mu$, a (positive) constant drift term corresponding roughly to the difference between excitatory and inhibitory post-synaptic potential steps; the dynamics (1) then corresponds to a Wiener process with drift. Other, somewhat more realistic models[8] assume a decay of the membrane potential, following a firing event, to its resting value; for these models $f(x) \equiv -\frac{x}{\tau} + \mu$ corresponding to Ornstein-Uhlenbeck dynamics. We consider briefly some recent results based on the GM model with deterministic modulation. A simple extension of the original GM calculation leads to a closed form expression of the probability density function of first passage times to an (absorbing) boundary located at a separation a from the starting point. Each crossing of the boundary denotes a firing event which is followed always by a deterministic reset to the starting point, accompanied by perfect phase locking to the periodic stimulus (this is not necessary for the theory; in fact one can assume a random phase ϕ in the argument of the deterministic modulation term in (1), in accordance with what is more common in experiments. Then, averageing over the phase leads to a smoothed ISIH which differs very little from the spontaneous case; we do not discuss this situation here). The solution of the Fokker Planck Equation associated with the deterministically modulated GM model, leads to an analytic computation[9] of the probability density function of first passage times. This function displays peaks at locations nT_0, similar to the results obtained from the bistable LBM model (note that this model *cannot*, however, elucidate the "hidden" symmetry corresponding to the reset events). The peaks are superimposed on a Gamma-like distribution characterizing[7] the $Q = 0$ case. With increasing inhibition, the density function approaches a Gaussian; the same effect is observed with decreasing noise variance. While the driven IF model reproduces some of the salient features of experimentally observed ISIHs it does not, in general, produce the same excellent agreement with the experimental ISIHs that characterizes bistable models. The formal connection between the two classes of models is also tenuous at present.

We now consider a third class of neuron models; these are the so-called Fitzhugh-Nagumo models corresponding to exciteable systems controlled by a bifurcation parameter. In these models, when the membrane voltage variable crosses a boundary, a large excursion (identified as a neural firing event) occurs. This leads to a natural definition of a *deterministic* refractory period, in contrast with the statistical distribution of refractory events that characterizes bistable models. The FHN system is not bistable, but can be made periodically firing or residing on a fixed point, dependng on the choice of bifurcation parameter. We write the equations for the model in the *form*[10]:

$$\dot{v} = v(v - 0.5)(1 - v) - w + F(t)$$
$$\dot{w} = v - w - (b + Q\sin(\omega t)), \qquad (2)$$

where v is the action potential to which noise has been added and w the recovery variable to which the signal is added. The model has been electronically simulated in St. Louis, in the fixed point regime ($b = 0.9$) so that bursts of sustained oscillations are absent. Hence one obtains a randomization in the inter-spike intervals, but some coherence with the external signal is maintained. THe variable v is treated as the "fast" variable in the dynamics, and its ISIH has been examined in this simulation[11]. The noise was colored since its correlation time was equal to the time constant of the fast variable but much smaller than the time constant of the slow variable w.

The analog simulator has been described elsewhere and also within this volume[11] and so will not be further described here. The fact is that using it we have been easily able to reproduce actual experimental data obtained from periodic stimulation of crayfish mechanoreceptor cells. In order to reproduce the physiological data it is only necessary to set the same signal frequency and then to adjust *either* the periodic stimulus intensity

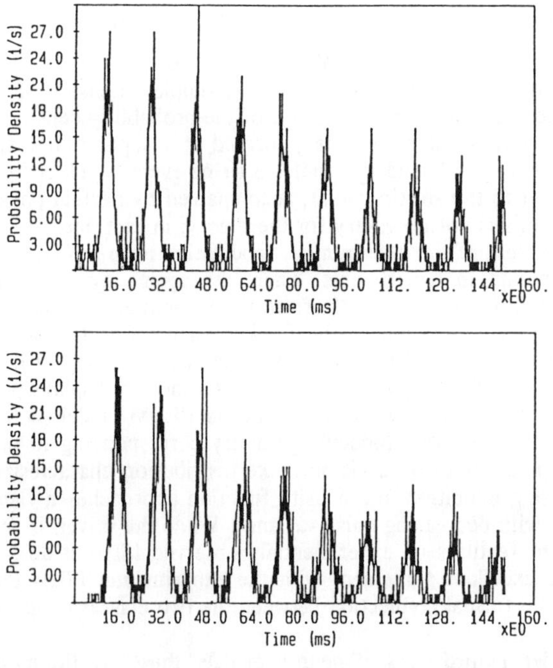

Fig. 1. ISIH's obtained from the crayfish stimulated at 68.6 Hz (upper) and the FHN simulator driven at the same frequency with $b = 0.9$ V; $V_{noise} = 0.022$ V_{rms}, and $V_{sig} = 0.53$ V_{rms} (lower).

or the noise intensity. An example of this is shown in Fig. 1. Moreover, the sharp signal feature, characteristic of additive noise as found in the bistable systems appears in power spectra obtained both from the crayfish and from the FHN model operated deeply subthreshold as we have discussed above. The crayfish experiments are described by Douglass, *et al* elsewhere in this volume, and so will not be further detailed here.

DISCUSSION AND CONCLUSIONS

Stochastic resonance is a cooperative nonlinear phenomenon wherein the signal-to-noise-ratio (SNR) of a weak time-varying deterministic signal may be enhanced by the noise; a plot of the SNR *vs.* noise strength demonstrates a characteristic bell-shaped profile. For low modulation frequencies, the critical value of the noise strength corresponds to a matching between the modulation frequency and twice the Kramers rate. The effect has been extensively analysed[12] and observed in a wide variety of physical systems. It is evident that, in order to take advantage of this effect, there must exist a form of self-regulatory mechanism such that the internal parameters of the system (these parameters control, for instance, the characteristics of the potential function describing bistable systems of the form described earlier in this work) can be adjusted so that it operates close to the maximum of the SNR curve. It is tantalizing to speculate that biological sensory systems might actually routinely utilize this effect for the processing and transmission of information. Our studies of collective behavior in large networks show[13] that the coupling to other elements can enhance or degrade the SNR depending on the magnitudes and *signs* of the coupling coefficients (i.e., the excitatory or inhibitory nature of the interactions is critical).

The precise connection between the ISIHs and SR remains somewhat tenuous, although several features of the ISIHs lend themselves to an interpretation based on SR. Perhaps the most important of these features is that the heights of successive peaks (excluding the first) pass through a maximum as a function of the noise strength[10,14]. ISIHs obtained from the IF models display the same features[9]. So far, attempts to quantify this "resonance" as a matching of two characteristic rates have been inconclusive, largely because of the difficulty of (numerically) producing good ISIHs with low noise. The question of defining a "SNR" from the ISIHs is also largely unanswered.

Experimental investigations into the occurrence of SR in living systems are now underway at at least two laboratories. Douglass, *et al* in this volume and Douglass, Moss and Longtin[11] have measured the SNR *vs.* noise strength curves in the crayfish mechanoreceptor. With externally applied noise, the SNR displays the characteristic bell-shaped response of SR. However, another and inherently more interesting case exists: that wherein the neuron makes use of its own internal noise for subthreshold signal transmission. This question directly relates to the possibility of the existence of an internal noise regulatory mechanism alluded to above and to questions of the evolutionary development of sensory organs using inherently noisy transducers. The experimental difficulty is that the internal neuronal noise is only indirectly controllable, via the temperature of the preparation, for example. For the internal noise case, the results are not as clear. The SNR increases monotonically as a function of the temperature of the saline bath, the crayfish having been acclimated for many weeks in either high or low temperature environments. While there exists an optimal temperature, that is a temperature for which the SNR passes through a maximum, this result does not demonstrate SR using the internal neuronal noise. The reason is that the internal noise *also decreases* beyond the optimal temperature, so that the mechanism which maximizes the SNR at the optimal temperature is different from SR. However, on the low temperature side, the SNR still increases approximately linearly on a logarithmic scale with increasing temperature (internal noise intensity). This result is significant in its own right; it points to the existence of a fundamental nonlinear dynamic mechanism underlying the cell response. Nevertheless, the dynamics underlying SR seem to be the most likely to provide explanations for the observed effects. These (albeit somewhat, preliminary) results lend credence to our speculations regarding the positive role of noise in the detection and quantification of signals by sensory neurons.

Work supported by grants from the US Office of Naval Research.

REFERENCES

1. E. Kaplan and R. Barlow; Vision Res. **16**, 745 (1976); H. Treutlein and K. Schulten; Ber. Bunsenges Phys. Chem. **89**, 710 (1985); O. Bernander, C. Koch and R. Douglas; in "Advances in Neural Information Processing Systems 3", eds. R. Lippman, J. Moody and D. Touretzky (Morgan Kaufmann, CA 1992).
2. A. Longtin, A. Bulsara and F. Moss; Phys. Rev. Lett. **67**, 656 (1991).
3. J. Rose, J. Brugge, D. Anderson and J. Hind; J. Neurophys. **30** 769 (1967).
4. R. Siegal; Physica **42d** 385 (1991).
5. W. Schieve, A. Bulsara and G. Davis; Phys. Rev. **A43** 2613 (1991).
6. A. Longtin, A. Bulsara, D. Pierson and F. Moss; Biol. Cyb. preprint.
7. G. Gerstein and B. Mandelbrot; Biophys. *J*. **4**, 41 (1968).
8. See e.g. J. Clay and N. Goel; J. Theor. Biol. **39**, 633 (1973); J. Cowan; in "Statistical Mechanics" eds. S. Rice, K. Freed and J. Light (Univ. of Chicago Press, Chicago 1974).
9. A. Bulsara; preprint.
10. A. Longtin; J. Stat. Phys. **70**, 309 (1993).
11. J. Douglass, F. Moss and A. Longtin; in "Advances in Neural Information Processing Systems 4", (Morgan Kaufmann, CA 1993). D. Pierson, J. Douglass, E. Pantazelou and F. Moss, this proceedings.
12. See e.g. B. McNamara and K. Wiesenfeld, Phys. Rev. **A39**, 4854 (1989); F. Moss, "Stochastic Resonance: From the Ice ages to the Monkey's Ear"; in *Some problems in Statistical Physics*, edited by George H. Weiss (SIAM, Philadelphia, in press); P. Jung and P. Hanggi, Europhys. Lett **8**, 505 (1989); P. Jung, Z. Phys. *B* **16**, 521 (1989); L. Gammaitoni, F. Marchesoni, E. Menichaella-Saetta and S. Santucci; Phys. Rev. **A40**, 2114 (1989); M. Dykman, R. Mannella, P. McClintock and N. Stocks; Phys. Rev. Lett. **65**, 2606 (1990); P. Jung and P. Hanggi; Phys. Rev. **A44** 8032 (1992); and *Proc. NATO ARW on Stochastic Resonance in Physics & Biology*, edited by F. Moss, A. Bulsara and M. F. Shlesinger, J. Stat. Phys. **70** (1993)
13. A. Bulsara and G. Schmera; Phys. Rev. *E*, in press; A. Bulsara, A. Maren and G. Schmera; Biol. Cyb., preprint; A. Bulsara and A. Maren; in "Proceedings of the First Applachian Conference on Neurodynamics", in press.
14. T. Zhou, F. Moss and P. Jung; Phys. Rev. **A42**, 3161 (1991).

THE THERAPEUTIC EFFECT OF LOW-FREQUENCY MAGNETIC FIELD

Prof.Dr.Dimiter C. Dimitrov, Sofia-99, P.O.B. 27,Bulgaria
Technical Uiversity, Chair of Radiotechnics

The magnetic field are widely adopted in modern medecine. In this connection problems of primary mechanisms in their influence upon biological objects, more precise definition of schemes of their medicinal application are of great significance. Our observations reaffirm an active role of microcirculation as a factor that determinates a degree of vascular reactions. Both blood rheological properties and membrane permeability are changed.

The sensitivity of biological objects to the action of magnetic field (MF) drows more and more attention of investigators during recent years. All the biological objects, from unicellularto pollycellular organisms,possess magnetic sensitivity. However, the molecular mechanisms, being the basis of magnetic sensitivity of biological objects,hitherto remain obscure.

During the last years series of experimental data has been obtained in our laboratory on potential independant mechanism by which Na-pump regulats membrane functional activity. These data allow us to suggest a new theory of metabolic regulations of cell function according to which the electrogenic Na-pump is a universal selfregulating mechanism by which the metabolic regulation of membrane enzimatic, chemoreceptive and exitable properties are realased. The idea of electrogenic Na-pump predicts that during its work the cell volume will change as a results of loss of more osmotic active intracellular particles.

From the detailed studies of the mechanism forming the basis of the correlation between pump activity and membrane chemoreceptivity it may come to the conclusion that chemoreceptors in the membrane are functionally active and inactive states depending on the membrane packing, and the electrogenic Na-pump can modulate the membrane chemosensitivity by changing the cell surface. On the bases of the findings reported above it is sugested that the electrogenic Na-pump is a powerful mechanism by which the metabolic regulation of the cell volume takes places and such a regulation has a great phisiological significance in the metabolic regulation in number of functional active properties,i.e.the cell by pump-dependence membrane surface changes are realizing the metabolic regulation of membrane permeability, excitability and enzymatic activity.

The MF influence on the main blood components-erythrocytes and haemoglobin has also been proved.Some parameters of the leukocytic blood reaction in MF have established.The magnetohaematogic effects have been studied during and after the action of MF on the whole bogy or ints separate parts.

MF become a lump factor, which changes the living conditions in the biosphere and thus-influences every level of organization of biosistems:from membranes to biosphere in general. Discussing the primary physical-mechanisms of the biological influence of MF in should be pointed that such mechanisms are various and that MF have several biotrophic parameters (component, intensity, frequency, from of the impulse, gradient,vector,exposition,localisation etc.) which have to be taken into account at every particular study. It should not forget that on the final result of the investigation often can ifluence some non-known MF with geophysical or industrial origin.When they are combined whit other physical factors as gravitation, sound etc. Mf can provoke completely different bioeffect. The biological effects increase when one or more biotropic factors are varied during the experiment.If should be taken into account especially when hygienic standards and physiotherapeutic equipment are developed. The final biological effect of MF depend also on such pecularities of the biosystem as are (children and old people are more reactable), sex (man are more sensitive), functional state (a functioning organ react stronger)or individual characteristice. It is quite possible that all these effects are connected whit the activity of the membrane.

A disadvantage of the constant magnetic field for magnetic therapy is its low therapeutic effect. So, an embodiment of the paper may provide an apparatus for magnetic therapy whit increased therapeutic effect. The frequency of the magnetic field of these apparatus can change linearly (for exemple) automaticcally for a determined peroid of time "T" form 0 to 100 Hz and reversely,form 0 to a frequency fx and reversely for fx in the range from 0 to 100 Hz and functional frequency in the range from 0 to 100 Hz. The MF generated from the apparatus possess an expressiv analgesic, anti-inflammatory and thophic effect, and are stimulatig the development of the collateral net vessels and the permeability of the vessels and membranes. The MF generated from these apparatus have an application in the therapy of diseases do not have many contraindications. The apparatus for magnet therapy is disigned for lowfrequency magnetic field treatment of the following diseases:

Pulmonal diseases-bronchial asthma, bronchitis, bronciectasies, pulmonal tubercolosis, pulmonal and pleural inflammations and others;

Cardio-vascular diseses-arterial hypertension, ischemic heart diseases (IHD), heart attack, arterial obliteration of the limbs (endarterfitis obliterans)acute and chronic venose insufficiency;

Diseases of the joints and their peripheral tissues (periarthritis): osteoarthritis, rheumatic arthritis,arthrosoarthritis,rheumatic arthritis,arthrosoarthritis and others;

Surgical, orthpedis and traumatologic diseases-fractures, postoperative fistuies,infiltrations and others;

Neurological diseases-neuritis, radiculitis, discopathy, stroke and others;

Dermatological diseases-skin allergy,dermatitis,psoriasis sclerodermi and others;

Genaecological cases-adnexitis, parametritis, mastitis;

ENT diseases-sinuitis, otitis, M.Meniere, etc;

Ophtalmic diseases-distrophy of the ophtalmical nerve, retinitis,datal diseases-mandibular fracure,paradentitis etc.

Depending on the cases suitable inductors and suitable impulses of magnetic field are used.

REFERENCE:

/1/ . Torov, N.G.Magnetotherapy, Sofia, M. and F. 1982

/2/ . Dimitrov D.C. Apparatus for magnetotherapy Patent No 67199, Bulgaria, 1987

Stochastic Resonance in Crayfish Hydrodynamic Receptors Stimulated with External Noise

J.K. Douglass, L.A. Wilkens, E. Pantazelou and F. Moss
Departments of Biology and Physics
University of Missouri, St. Louis, MO 63121 USA

ABSTRACT

Stochastic Resonance (SR) is a statistical process occuring only in nonlinear dynamical systems whereby a subthreshold coherent stimulus or signal can be enhanced by noise. The signal alone is too weak to cause a state change of the system. State changes are the carriers of information through the system. In the presence of random noise, however, the system can change state, more-or-less randomly, but with some degree of coherence with the signal. A measure of this coherence at the output shows a maximum at an optimal value of the noise intensity as the signature of SR. SR is the object of recent and continued experimental and theoretical research in statistical physics[1,2]. While SR has been *demonstrated* in a variety of physical systems[2], it has not yet been *discovered* in any naturally occurring system. This paper was stimulated by the idea that the sensory nervous system might be an appropriate setting for a search for naturally occurring SR. The detection of weak stimuli, often in the presence of noise, is, after all, the first business of the sensory system. Moreover, the system is evolved, which admits the possibility that the process of natural selection might have resulted in an optimization with respect to the (inevitable) noise. This paper describes an experiment designed to observe SR in the mechanoreceptor cells of the crayfish *Procambarus clarkii*, shown on the left in Fig. 1, using external noise plus a weak coherent signal as the stimulus.

INTRODUCTION

Hair mechanoreceptors are abundant on the external surfaces of crayfish and are specialized for detecting small water disturbances that arise from potential predators, prey, and other sources. When stimulated, individual receptor cells encode directional, frequency and intensity information in sequences of action potentials (spikes), which then proceed to the central nervous system of the animal for further processing. Using this receptor system, we have demonstrated that random noise added to weak periodic water movement stimuli can enhance the flow of information carried by the spikes. This effect is similar to the phenomenon of stochastic resonance known in bistable physical systems[1,2], but the question of bistability in models of sensory neurons is a topic of current discussion. Nevertheless, SR has also been shown to exist in noisy threshold-type systems stimulated with subthreshold signals, for example the FitzHugh-Nagumo system, which is an accepted model for encoding the spikes in sensory neurons[3].

EXPERIMENTAL PROCEDURE AND APPARATUS

For each experiment, a small piece of the tailfan was dissected from a crayfish abdomen, keeping the associated sensory nerves intact. The preparation was mounted on an electromagnetic vibrator and immersed in saline solution. Receptors were stimulated by moving them relative to the saline solution, using the summed outputs from a sine generator and a random noise generator. The spiking activity of single receptor cells was recorded using standard electrophysiological methods, and spike times, signal traces and noise amplitudes were digitized and analyzed by computer. A diagram of the appar-

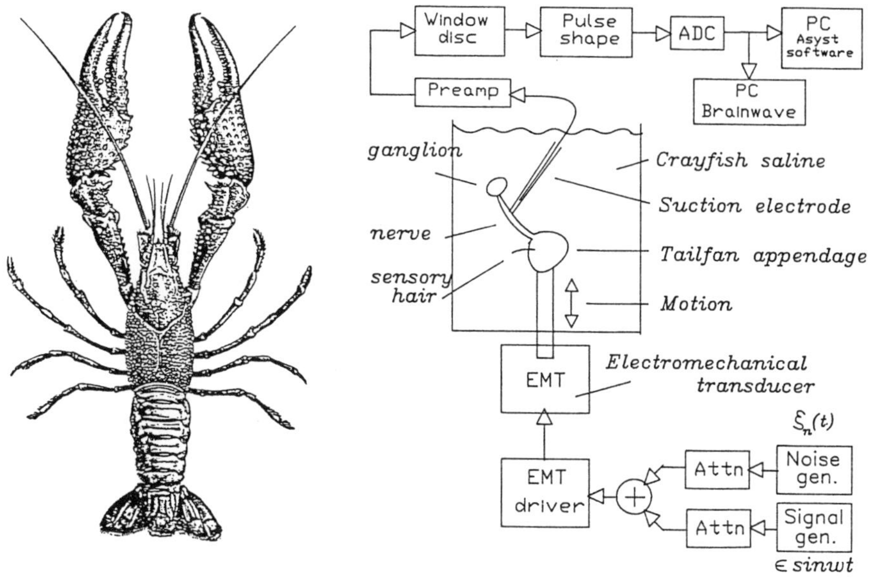

Fig. 1. (left) The crayfish *Procambarus clarkii*, male. Approximately 200 motion- and direction-sensitive hairs are located in the tailfan. They respond to direction and velocity of water currents. (right) The experimental apparatus. Standard extracellular recording techniques are used. A suction pipette electrode is attached to the nerve. After locating a strongly reacting cell, the individual hair which excited this particular cell was located and identified by "tweaking" the hairs at random. The appendage was then rotated while recording in order to determine the direction of maximum sensitivity. Sinusoidal and noise waveforms are attenuated separately then added to make-up the input to the motion transducer. The hair cells are remarkably sensitive, responding with easily measurable excitations to displacements of about 10 nanometers and to velocities of a few tens of micron meters per second. The entire apparatus is mounted on a vibration isolation table and operated inside a Faraday cage.

atus is shown in Fig. 1 on the right.

Receptor responses to stimuli were assessed by computing signal-to-noise ratios (SNR's) from measured power spectra obtained from the time series of the action potentials (spikes). The spike times were also recorded, from which were obtained interspike interval and cycle histograms plus interval and phase return maps (not shown here). Each experiment began by using sinusoidal stimuli alone to measure the directional selectivity of the cell. All subsequent measurements employed the stimulus direction corresponding to the maximal response. Before noise was added to the signal, responses to a range of stimulus frequencies and intensities were measured. Finally, responses to added noise were measured using a near threshold stimulus intensity at the "best frequency" for the cell. A set of three measured power spectra are shown in Fig.

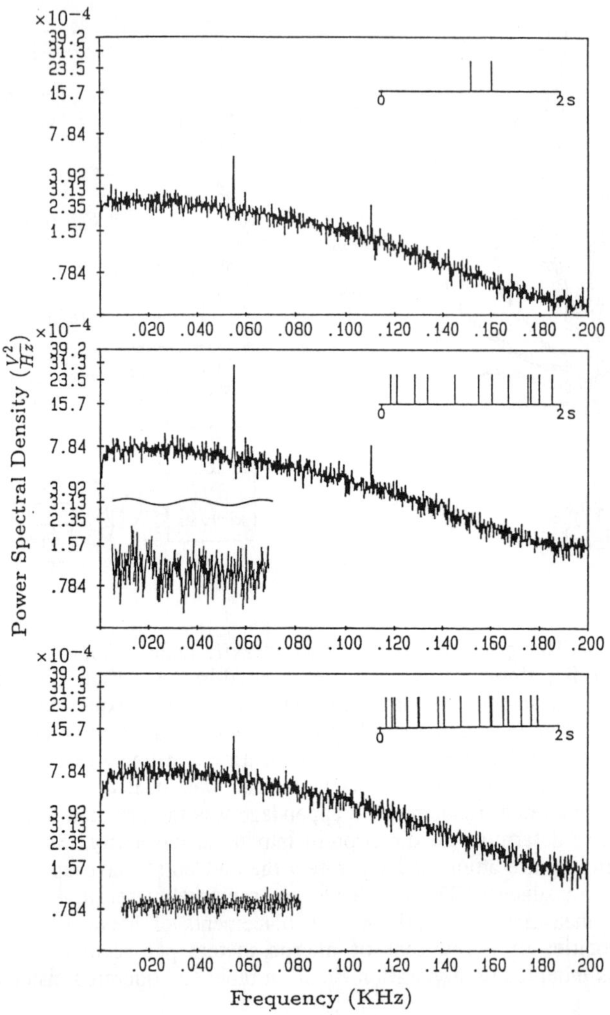

Fig. 2. Power spectra measured for zero (upper), optimal (middle) and larger than optimal (lower) external noise intensities. The insets on the right show samples of the spike trains. The middle inset left shows the signal alone and signal plus the noise for optimal stimulation. The lower left inset shows the power spectrum of the input before application to the motion transducer which introduced a 6.8 dB per decade roll-off in frequency.

2. For these experiments, neurons with low internal noise were chosen.

Figure 3 left shows an example of SNR values obtained from the power spectrum of spike times upon stimulation with a 55-Hz signal. There is an optimal noise level (ca. 0.14 V rms), at which the SNR is maximized. In Fig. 3 right, the signal intensity was varied while stimulating with or without added external noise. The SNR's increase with stimulus intensity, and are consistently higher when external noise is present. These results raise the intriguing possibility that living organisms can actually use noise to detect stimuli that would otherwise be undetectable. A recent psychophysical experiment with humans[4] points directly to this possibility; additional physiological and behavioral experiments are also planned in this laboratory.

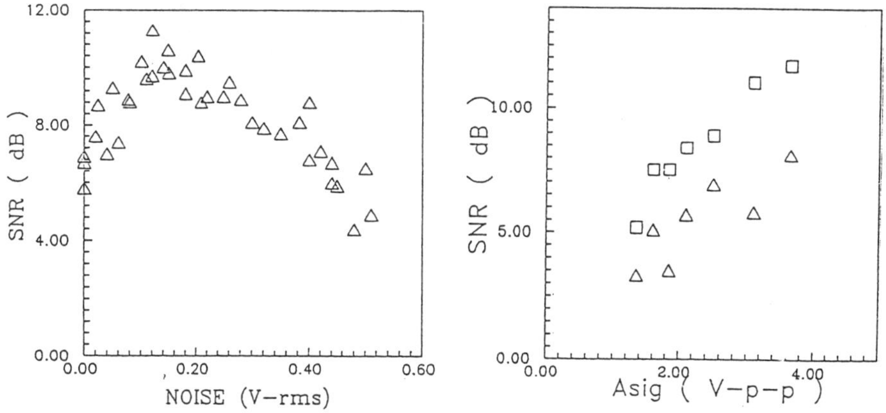

Fig. 3. (left) SNR versus the noise voltage (before attenuation) applied to the motion transducer. A maximum in the SNR at an optimal noise voltage of ca 0.14 V rms is clearly shown. (right) Threshold measurements showing the SNR versus signal voltage with optimal noise (squares) and without noise (triangles). The upward shift with noise added indicated a lower detection threshold.

Supported by U.S. Office of Naval Research #N00014-91-J-1235.

REFERENCES

1. F. Moss, "Stochastic resonance: From the ice ages to the monkey's ear" in *An Introduction to Some Contemporary Problems in Statistical Physics*, edited by George H. Weiss (SIAM, Philadelphia, in press)
2. *Proceedings of the NATO ARW on Stochastic Resonance in Physics and Biology* edited by F. Moss, A. Bulsara and M. F. Shlesinger, J. Stat. Phys. **70**, (1993)
3. D. Pierson, J. K. Douglass, E. Pantazelou, and F. Moss, this proceedings.
4. D. Chialvo and V. Apkarian, in *Proceedings of the NATO ARW on Stochastic Resonance in Physics and Biology*, edited by F. Moss, A. Bulsara and M. F. Shlesinger, J. Stat. Phys. **70**, 375 (1993)

DETERMINISTIC 1/F FLUCTUATIONS IN BIOMECHANICAL SYSTEM

G.I.Firsov, M.G.Rosenblum
Mechanical Engineering Research Institute,
Russian Academy of Sciences, Moscow, 101830, Russia

P.S.Landa
Moscow State University, Moscow, 119899, Russia

ABSTRACT

We studied human center of gravity body sway while quiet standing in upright posture. These 1/f oscillations possess rich information about the state of the central nervous system. We discussed the techniques and results of analysis of experimental data and proposed the dynamical model of the human posture control system.

INTRODUCTION

Investigation of the human posture control is very interesting from mechanical standpoint as an example of control in the multi-degree-of-freedom system. This study is also very important for physiology and medicine, as body sway while maintaining upright posture contain rich information about the state of the central nervous system (CNS). These small oscillations are usually measured by special force plate. It registers the deviation of the center of pressure on the plate, or, as the first approximation, the deviation of the center of gravity of the human body in anterior-posterior and lateral directions. These records are called stabilograms. For their processing we used, besides traditional techniques, methods of nonlinear dynamics and analysis of predictability of time series. The results of data analysis have demonstrated, that the described approach can be used for diagnostics of some neurological diseases.

TECHNIQUES AND RESULTS OF THE ANALYSIS

Our experiments were conducted in cooperation with Dr. R.A.Kuuz in the Moscow Clinic of Nervous Diseases. We obtained more than 200 stabilograms for healthy persons and patients with different neurological pathologies (Parkinson's disease, multiple sclerosis, functional disorders and others). For processing of stabilograms we used different statistical techniques, spectral and correlation analysis. The Fourier power spectra in the frequency region up to 10 Hz were found to be of the form typical for 1/f fluctuations. The slopes of log-log power spectra differed for normal subjects and patients with some neurological pathologies, and the cross-spectrum analysis has shown the absence of linear relationship between oscillations in anterior-posterior and lateral directions. The only exception was found in examination of patients with functional hysteria. In this case the low frequency (0.5 ÷ 1 Hz) periodical coherent oscillations in two directions were observed.

The probability distribution functions for the majority of processed records were close to the Gaussian one. The strong deviation from Gaussian law we found, besides functional hysteria, in the case of ataxy, when the distribution has distinct asymmetry. Important diagnostic information is contained in the two-dimensional probability distribution for oscillations in two directions. For its description we approximated the histogram with a function with one parameter c. The values of c in the range 0.8 ÷ 1.4 were found typical for healthy persons, the values 1.6 ÷ 2 are characteristic for psychogenic disorders (neurosis), and the values $c > 2$ indicate of severe pathologies like Parkinson's disease and tumour of the brain.

A very interesting problem is the establishing of the nature of observed oscillations. There exist the viewpoint that they represent the reaction of the controlled biomechanical system to the external random perturbation of unclear origin. In our previous works we assumed that chaotic character of experimental time series was due to the complex nonlinear dynamics of the posture control system. In order to prove the deterministic nature of this fluctuations we estimated the attractor correlation dimension and investigated the predictability of stabilograms.

For computation of attractor correlation dimension ν we used 8 experimental time series, 9000 of points each. Two records were obtained in experiments with healthy subjects and six in experiments with patients with different neurological pathologies. For the patient with neurosis we obtained $\nu \approx 1.8$. This can be explained by the absence of developed chaos. For other patients we obtained ν in the range 2.2 ÷ 3.5, and for healthy subjects in the range 2.1÷2.2. Using our technique, we estimated also the attractor embedding dimension for more than 20 experimental records. For all records we found this dimension less than 5. These results allow to draw the conclusion about the dynamical origin of observed oscillations.

In order to study the predictability of the time series we used two techniques. The first one is based on the computation of the root mean square error for the prediction by means of linear autoregression model. The second one consists of evaluation of the power of functional relationship between successive points in the time series. The power of functional relationship can be estimated from the rate of decrease of the Raibman dispersion function. In both cases we compared the characteristics of predictability for the experimental time series with those for specially designed noise with exactly same spectrum. We assumed that chaotic time series possesses more strong functional relationship between points and is more predictable than noise. The obtained results indicate the deterministic nature of chaotic body sway.

MATHEMATICAL MODEL OF POSTURE DYNAMICS

On the base of analysis of stabilograms we proposed dynamical model of posture dynamics. Such model must include models of muscle-skeletal and motion control subsystems. The first one is usually considered as one- or multi-link inverted pendulum. The control system use the information about the coordinates and velocities of links. This information is obtained by the CNS from the muscle and joints receptors and from the visual analyzer. It is supposed that CNS governs the muscle torques on the ground of coordinate and velocity feedback loops with time delay. From the other side, CNS also performs the feed forward control defining the stiffness of the joints. We considered the control scheme incorporating two simultaneously operating control channels correspondent to pyramid and extrapyramid parts of the CNS. Both parts use feedback and feed forward control strategy providing high reliability of the whole system. After qualitative discussion of the role of different parts of the brain in the control we proposed the simplified scheme, which was used for simulation. We considered the muscle-skeletal systems as one-link inverted pendulum with point mass M at its upper end. The equation of motion of such model is as follows:

$$\ddot{\varphi} + 2h\dot{\varphi} + \frac{k-Mgl}{I}\cdot\varphi + a_x f\left[\varphi(t-\tau)\right] + a_v f\left[\dot{\varphi}(t-\tau)\right] = 0, \quad (1)$$

where φ is the angle of the ankle joint, I is the moment of inertia, l is the distance between the center of gravity and the base, h is the damping ratio, k is the stiffness of the joint, a_x and a_v are respectively the coefficients of position and velocity feedback loops, and piece-wise linear function $f(x)$ describes the performance of muscle and joint spindles (receptors), which are known to have some sensibility threshold. The values of threshold and time delay in the loops τ were chosen in accordance to known physiological data.

The stiffness of the joint k can be chosen from the following considerations. If $k \approx Mgl$ than the stiffness of the joint compensates the action of the gravity torque and the muscle power is spent against the friction force only. Therefore, we supposed that the CNS realizes such control regime that the natural frequency of the mechanical subsystem is close to zero.

By computer simulation we found out that the solution of the equation of motion of the body may be periodic or chaotic, depending on the values of feedback coefficients a_x and a_v; the region in the $a_x - a_v$ plane correspondent to chaos was determined. The analysis of time dependencies and power spectra of the solutions showed that the proposed simple model can describe qualitatively the investigated process. The correlation dimension of the attractor ($\nu \approx 2.5$) is also in agreement with experimental data.

Considering the muscle-skeletal systems as three-link inverted pendulum we obtained the more complicated equations of motion for the coordinate x of the center of gravity:

$$\ddot{x}+2h\dot{x}+k(x-z)=0, \qquad \dot{z}=-c_1 z-c_2 f\left[x(t-\tau)\right]-c_3 f\left[\dot{x}(t-\tau)\right], \quad (2)$$

where function f described the nonlinear performance of receptors and h, k, c_1, c_2 and c_3 were parameters. These equations also displayed chaotic behavior. For $c_1 \gg h$ they can be drawn to the above considered equation.

STOCHASTIC RESONANCE IN A BISTABLE SQUID LOOP

A. Hibbs
Quantum Magnetics, Inc., San Diego, CA, USA

E. W. Jacobs, J. Bekkedahl, A. Bulsara
Navy Command, Control and Ocean Surveilance Center, San Diego, CA, USA

F. Moss
University of Missouri at St. Louis, MO, USA

ABSTRACT

Stochastic Resonance (SR) is the name given to a statistical nonlinear phenomenon whereby a weak or subthreshold coherent function can be amplified by random forces, or noise, within the system. It was first advanced in the early 1980's as a possible explanation for the observed periodicities in the recurrences of the Earth's Ice Ages[1,2]. The first publication of a modern theory[3-5] led to an experiment[4] and a flurry of further theoretical activity[6-9], an international conference[10] and a review[11]. In this paper, we describe a demonstration experiment wherein SR is exhibited in a superconducting quantum interference device (SQUID). Here SR is viewed as a noisy information transmission process. It is entirely appropriate, therefore, to look for this dynamics in a widely used sensitive detector in this example, a detector of weak magnetic fields. Using a modern, minature, thin film SQUID[12], we hope this demonstration will stimulate further research and development of SR in applied superconductivity.

INTRODUCTION

We have demonstrated *Stochastic Resonance*[1] in a bistable SQUID loop, as a first step in stimulating interest in possible applications using superconducting devices. We begin with an equation governing the magnetic flux trapped within an rf-SQUID loop[13].

$$LC\ddot{\phi} + \tau_L \dot{\phi} + \phi + \frac{1}{2\pi}\beta\sin(2\pi\phi) = \phi_e \ , \tag{1}$$

where $\phi = \Phi(t)/\Phi_0$ is the normalized magnetic flux trapped within the loop, $\phi_e = \Phi_e(t)/\Phi_0$ is the normalized flux externally imposed on the loop, $\Phi_0 \equiv h/2e$ is the flux quantum, L and C are the loop inductance and junction capacitance respectively, and $\tau_L = L/R_J$ is the junction resistance. The parameter which determines the shape of the potential governing the dynamics of (1) is $\beta = 2\pi L i_c/\Phi_0$, where i_c is the junction critical current. In our experiment, the external flux Φ_e was composed of DC, periodic and stochastic components:

$$\Phi_e(t) = \Phi_{DC} + \Phi_M \sin(\omega_s t) + \Phi_N(t), \tag{2}$$

where the periodic component represents an audio frequency signal, and the stochastic component was a Gaussian noise whose bandwidth was in the audio range[14]. Bistability is a prerequsite for observations of SR. Equation (1) is bistable for certain values of β and Φ_{DC}, and the quantity which shows the bistable dynamics is the flux trapped within the loop, $\phi(t)$.

DESCRIPTION OF THE EXPERIMENTAL APPARATUS

In order to experimentally observe the bistable dynamics, one must measure the trapped flux $\phi(t)$. This requires a second SQUID, either mounted coaxially with the loop of the first SQUID, or coupled to it with a superconducting transformer[15]. We chose the latter configuration. The primary SQUID was a thin film device mounted on a single chip with integrally mounted, superconducting transformer primaries supplied by Quantum Magnetics. This is a thin film SQUID with primary and secondary windings coupled to the SQUID all evaporated on a single silicon chip. The Quantum Design DC SQUID chip is shown in Fig. 1. It is the first commercially available and the most sensitive all-thin-film DC SQUID sensor. The junctions, located in the central region of the chip, are made in the state-of-the-art niobium trilayer technology on silicon and are part of two two identical loops connected in parallel, each coupled to an input coil. This unique "double balanced" design reduces coupling between the input and modulation

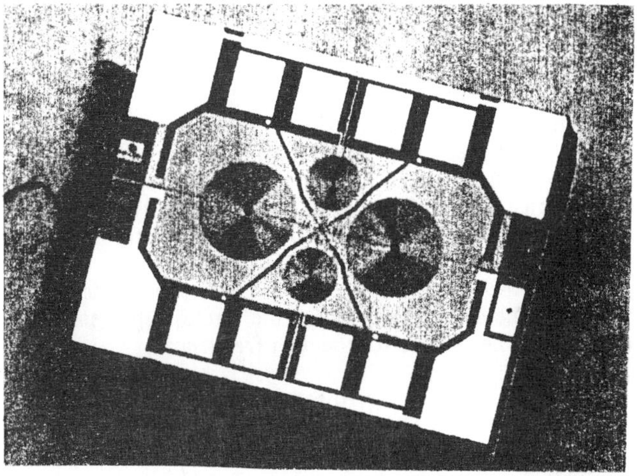

Fig. 1. The Quantum Design DC SQUID. The rectangles around the edges are bond pads for electrical connections. The left and right spiral coils couple the input signal to the SQUID loops. The upper and lower coils are used for a 500 kHz AC flux modulation used for noise reduction. The current and voltage leads appear as a cross but are not connected in the middle. The two Josephson junctions are located at the lower left and upper right of the cross but near the center. The size of the chip shown is 5 *x* 3 mm.

coils to negligible levels while giving high mutual inductance with the SQUID.

The secondary, or measuring, SQUID was a standard BTI model[16], which was coupled to the primary SQUID with a completely superconducting transformer. A schematic diagram of the experimental setup is shown in Fig. 2. This apparatus was mounted inside a superconducting *Nb* shield and mounted near the bottom of a liquid helium dewar. The apparatus was operated at a temperature of 4.2 ^{0}K in boiling liquid helium. No further external magnetic shielding was employed.

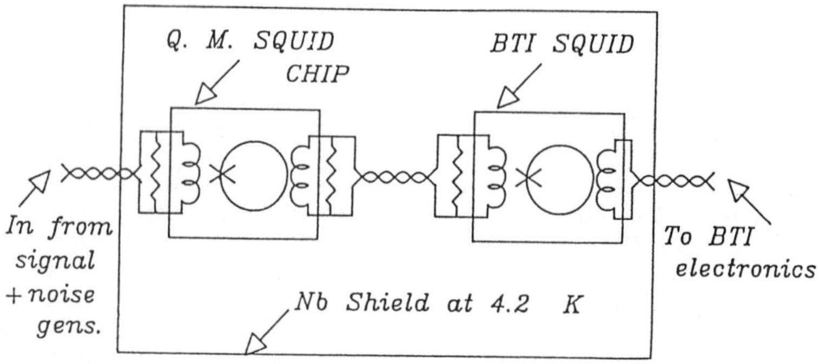

Fig. 2. Schematic of the bistable SQUID experiment showing the Quantum Magnetics chip and the BTI measuring SQUID coupled with a superconducting transformer. Noise and signal voltages supplied by the external electronics were transformed into external magnetic flux in the coil C1.

EXAMPLE EXPERIMENTAL RESULTS

In our experiment, $\beta = 2.0$ and $\Phi_{DC} = 0.5\Phi_0$, values which guaranteed that the potential was bistable. Experiments were performed at two signal frequencies, 17.6 Hz and 100 Hz with signal peak voltages of 650 *mV-pk* and 475 *mV-pk* respectively. The noise, or stochastic, component was supplied by a standard noise generator and the noise voltage varied over the range from 100 to 1500 *mV*-rms. (1.0 *V* was equivalent to $0.1\Phi_0$ of applied external flux.). The power spectra of $\phi(t)$ were measured and averaged in the usual way at the output of the BTI SQUID electronics, and the signal-to-noise ratios (SNR's) were determined from the measured and time averaged power spectra of the output of the BTI electronics using a conventional definition. The results of this experiment are shown in Fig. 3 where the circles represent the results for the low signal frequency and the squares for the high frequency.

At each frequency, data were collected for two different signal strengths. For each data set, a clear maximum in the SNR - the familiar signature of SR - was obsercved. The maxima in the SNR occur at a noise voltage of $\approx$700 *mV* which is equivalent to an rms fluctuation of $0.07\Phi_0$ within which a coherent signal equivalent to $0.0237\Phi_0$ peak at 17.6 Hz was easily detectable. This clearly demonstrates that bistable SQUIDs, used in combination with SR, can be useful in detecting weak, coherent magnetic signals buried in external noise, an application of considerable importance.

Work supported by the U.S. Office of Naval Research grant N00014-91-*J*-1979 and by Quantum Magnetics, Inc.

REFERENCES

1. R. Benzi, S. Sutera and A. Vulpiani, J. Phys. A **14**, L453 (1981)
2. C. Nicolis, Tellus **34**, 1 (1982)
3. B. McNamara and K. Wiesenfeld, Phys. Rev. A **39**, 4148 (1989)
4. B. McNamara, K. Wiesenfeld and R. Roy, Phys. Rev. Lett. **60**, 2626 (1988)
5. P. Jung and P. Hanggi, Europhys. Lett **8**, 505 (1989)

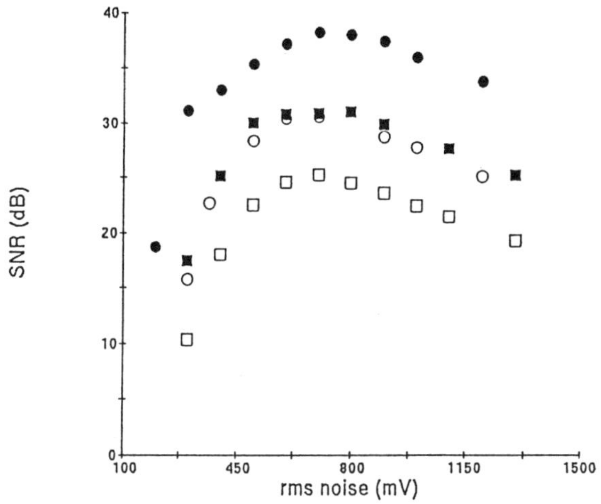

Fig. 3. The SNR versus rms noise voltage for the bistable SQUID experiment, with f_s = 17.6 Hz, v_s = 650 mV-rms (filled circles) and v_s = 237 mV (open circles); and f_s = 100 Hz, v_s = 475 mV (filled squares) and v_s = 237 mV (open squares). For these data 1.0 $V \equiv 0.1\Phi_0$ at the coil C1.

6. L. Gammaitoni, F. Marchesoni, E. Menichella-Saetta, and S. Santucci, Phys. Rev. Lett. **62**, 349 (1989)
7. P. Jung, Z. Phys. *B* **16**, 521 (1989)
8. P. Jung and P. Hanggi, Phys. Rev. A **41**, 2977 (1990)
9. M. Dykman, R. Mannella, P. McClintock, and N. Stocks, Phys. Rev. Lett. **65**, 2606 (1990)
10. *Proceedings of the N.A.T.O. Advanced Research Workshop on Stochastic Resonance in Physics and Biology*, edited by F. Moss, A. Bulsara and M. F. Shlesinger, special issue, J. Stat. Phys. **70** (1993).
11. F. Moss, "Stochastic Resonance: from the Ice Ages to the Monkey's Ear" in *An Introduction to Some Contemporary Problems in Statistical Physics*, edited by George H. Weiss (SIAM, Philadelphia, in press).
12. Quantum Magnetics, 11578 Sorrento Valley Road, Suite 30; San Diego, CA 92121.
13. See for example, A. Barone and G. Paterno, *Physics and Applications of the Josephson Effect*, (John Wiley & Sons, Inc., New York, NY, 1982).
14. *rf*-SQUIDS can respond in the frequency range from DC to gigahertz, and the external electronics in our experiment had a bandwidth to 30 kHz, consequently the SQUID and its external electronics can respond essentially instantaneously to both the signal and the noise.
15. A. Silver and J. Zimmerman, Phys. Rev. **157**, 317 (1967)
16. Biomagnetic Technologies Inc.; San Diego, CA; Model 420.

STOCHASTIC CONTROL OF LIVING SYSTEMS: NORMALIZATION OF PHYSIOLOGICAL FUNCTIONS BY MAGNETIC FIELD WITH 1/F POWER SPECTRUM.

N.I.Muzalevskaya, V.M.Uritsky, E.V.Korolyov, A.M.Reschikov, G.P.Timoshinov.
Nonequilibrium Systems Laboratory
Department of Science and Technology of SOFID Co.
196 140 Russia, St.Petersburg, Pulkovskoe shosse 86-56.

ABSTRACT

For the first time correcting stochastic control of physiological status of living systems by weak low-frequency fluctuating magnetic field with 1/f spectrum (1/f MF) is demonstrated experimentally. The correction was observed in all main systems, including cardiovascular, central nervous, immunity systems of experimental animals. Pronounced prophylactic and therapeutic influence of 1/f MF on malignant growth and radiation disease was discovered. Theoretical interpretation of the results obtained is based upon the notion of fundamental role of 1/f fluctuations in homeostasis of living systems.

INTRODUCTION

Dynamics of many functional systems of a healthy organism shows low-frequency fluctuations with 1/f spectrum, malfunctions and pathological changes are accompanied by distortions in 1/f spectral dependencies [1,2], often preceding appearance of other signs. This phenomenon, as we have shown [3], may be explained by existence of stochastic function of homeostasis, based on 1/f fluctuations and aimed at maintenance of biosystem's probabilistic integrity. Realization of this function is a necessary condition for stability of structural and functional regulating mechanisms and defences..

Our conception implied a new possibility of external regulation of physiological status through correction of stochastic mechanisms of homeostasis. Such regulation, further called stochastic control (SC), is based upon the fact that a living system includes 1/f fluctuations as an active element of its organization and so must react critically not only to inflows of substance and energy, but also to exogenic low-frequency stochastic background [4]. Consequently, external fluctuations of a special structure can be used for control of a biological system. It is natural to assume that normalizing SC can be realized by an agent with 1/f spectrum, an agent corresponding to physiological norm and capable of restoring and stabilizing 1/f spectra of endogenic fluctuations. Literature data confirm this assumption showing positive biological effects of 1/f noise [5,6] and negative effects of other low-frequency noises [7].

EXPERIMENTAL RESULTS AND DISCUSSION

In our experiments for the first time SC with weak ($E \leq kT$) magnetic field as a carrier of controlling 1/f signal is studied. Such carrier permits to take full advantage of SC as a method for integral biological correction, since it penetrates deep into tissues, is nonspecific in its biological action [7], and, what is crucial, is able to transfer biologically significant information [8]. An extensive series of tests with experimental animals was carried out. The field was created with a specially designed generator of 1/f noise and Helmholtz rings.

Directly effects of 1/f MF on spectral makeup of endogenic low-frequency fluctuations were studied for ECG signals from rabbits with initially disturbed cardiac cycles. The disturbances appeared as distortions in 1/f power spectra of ECG fluctuations (Fig.1a) and as blurring of the attractor in phase coordinates (Fig.2a). The primary stage of the response of ECG to 1/f MF was restoration of 1/f spectra (Fig.1b). Then functional normalization of heart dynamics took place, manifested in stabilization of phase trajectories (Fig.2b) and decrease in dispersion of ECG parameters. Thus, dispersion of R-R intervals decreased by 40-60%. All manifestations were long-term, holding for several days after exposition to 1/f MF.

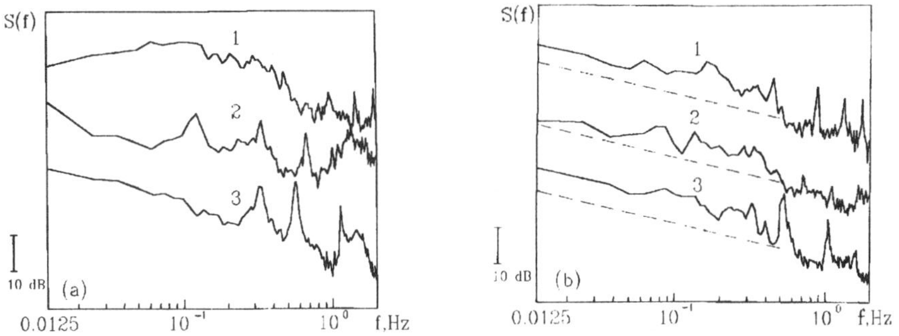

Fig.1. Low-frequency power spectra S(f) of ECG signals from three rabbits before (a) and after (b) exposition to 1/f MF. Correction of spectra for rabbits 1,2 is obvious (for convenience the spectra are shifted along the vertical axis; dependence of the 1/f-type is shown by a dashed line).

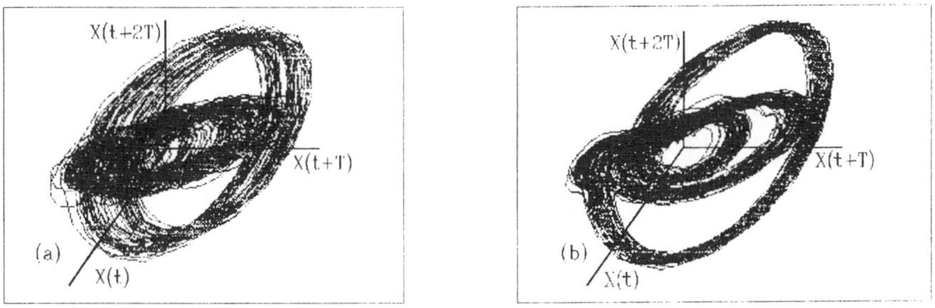

Fig.2. Phase portraits (T= 2.5 ms) of ECG signals of rabbit 1 before (a) and after (b) stochastic correction.

Response of central nervous system to 1/f MF also showed a shift to increased stability and decreased random chaos in EEG dynamics, as was demonstrated by fractal analyses of rabbit's EEG signals from three leads. Correlation dimension D_2, which gives an estimate of the lower limit of fractal dimension, and range of scales Δr in which attractors retain strictly fractal structure were calculated for strange attractors of the signals (Table I). 1/f MF decreased D_2 in all leads and at the same time extended the range Δr. It means that brain's electric activity became less chaotic, retaining at the same time its structural complexity. According to modern psychophysiology, such changes facilitate memorizing of images and development of conditioned reflexes.

1/f MF had pronounced supporting effect on cell-bound immunity. After exposition of T-cells and other leukocytes *in vitro* or *in vivo* to 1/f MF, the cells were exposed to concentration hypoxia. Resistance to hypoxia, a nonspecific indicator of immunocompetent cells' functional reliability, was considerably increased (Table II). Decrease in fluorescence intensity of acridine orange dye showed that density of nuclear material of T-cells increased and approached the value for cells before their exposition to hypoxia. Percentage of survived cells was increased. Effects *in vivo* were more pronounced, which means that SC was, at least partially, effected on hierarchical levels higher then cellular. The obtained results imply substantial strengthening of immunity by 1/f MF.

1/f MF supported also adaptation mechanisms, as was testifield by the increase in endurance of experimental animals under stress conditions. In tests with compulsory swimming in cooled water rats previously exposed to 1/f MF kept on the surface significantly longer than the controls (Table III). It is important to note that the dependence of the increase in endurance on the number of expositions to 1/f MF quickly saturates. It suggests that 1/f MF optimizes homeostatic functions within the limits of the natural norm, and cannot be used as a dope.

Tables I-V. Correction of physiological status of experimental animals by SC

Functional indicator		No SC	Effect of SC	
I. Fractal structuring of EEG signals of rabbits (A-auditory cortex lead, H-hippocamp lead, R-reticular formation lead)				
Estimate D_2 of fractal dimension (arbitrary units)	A	2.52	2.14	
	H	2.55	2.03	
	R	2.62	1.70	
Interval of fractal scales Δr (arbitrary units)	A	—	—	
	H	2.46	4.46	
	R	1.82	3.00	
II. Increase in resistance of rat's immunocompetent cells to hypoxia			SC *in vitro*	SC *in vivo*
Dye fluorescence intensity in nuclei of T-cells (% with respect to the state before hypoxia)		158±15	123±16	87±26
Survival of T-cells under conditions of hypoxia (%)		37±2	51±4	59±9
Survival of leukocytes of heparinizated blood (%)		67±6	94±12	—
III. Increase in endurance of rats under stress conditions				
Maximum duration of swimming in cooled water (min)		118±11	146±13	
IV. Protection from ionizing radiation damage (R-rats, M-mice)			SC before X-ray	SC after X-ray
Survival of animals (%) for X-ray dose 4.0 Gy	R	70±10	90±5	90±5
	M	20±4	90±6	—
The same, X-ray dose 6.0 Gy	R	35±12	50±3	20±12
	M	10±3	21±7	—
V. Prevention and cure of induced Ehrlich ascites cancer in mice			SC at an early stage of cancer	SC at a late stage of cancer
Survival of animals (%)		0	73	53
Percentage of animals without signs of cancer by the end of the test		0	60	40

The most substantial practical confirmation the idea of SC received in tests with pathologies that require maximum support for organism's defences and regulating mechanisms, namely, radiation disease and cancer. Exposition of mice and rats to 1/f MF protected them from radiation damage induced by X-rays. Protective effect was obtained both when exposition to 1/f MF preceded and directly followed ionizing irradiation, the effect being more stable in the former case (Table IV). For X-ray doses exceeding 6 Gy correction lost its efficiency, presumably because of structural destruction of the systems of self-regulation to which SC was adressed. Classical technique of induced Ehrlich ascites cancer was chosen to test SC in oncology. Mice with induced cancer were exposed to 1/f MF at early or at late stages of cancer growth. In both cases pronounced therapeutic effect was observed (Table V). At early stages 1/f MF either inhibited or prevented cancer growth, at late stages induced involution of developed cancer. The number of survived animals reached up to 70%, with zero survival rate for controls. The correcting influence of 1/f MF showed aftereffects similar to a long-term cancer-preventing program. According to our preliminary data, this program activates melatonin production and controls a number of specific processes in cellular interactions.

CONCLUSION

Our experiments prove feasibility of normalizing SC of living system's physiological status. Both the primary act of SC, namely, stabilization of endogenic 1/f spectra, and increase in functional stability of organism's defences and regulating mechanisms were demonstrated. Thus, theoretical biological conceptions used as a basis for SC were confirmed.

All effects of SC were produced by extremely weak 1/f MF, the energy of which is comparable with kT. Hence, influence of the field on endogenic processes cannot be determined by extensive energy absorption. SC is likely to be achieved by a more subtle selective interaction between external agent and fluctuations in an organism, coupling of stochastic patterns and not energy exchange playing the leading role. Pathogenic effects of magnetic fields of equally low strength but of different frequency makeup [7] is a clear proof of the existence of such coupling. In this context SC may be regarded as adjustable low-frequency information exchange between stochastic environment and internal stochastic medium of a living system realized in physiologically optimal case by 1/f fluctuations. This interpretation agrees with theoretical conception of T.Musha [9] concerning participation of 1/f fluctuations in circulation of biological information inside a living system, and extends it to process of interaction with environment.

Morphologically, the information transfer channeel for SC can be formed by organism's supramolecular heterogeneous structures, i.e. by the functionally connected interphase regions, such as electrical double layers of membranes etc. This is confirmed by model experiments on control of surface 1/f fluctuations in nonliving nonequilibrium systems[10]. After reception of SC signal by interphase regions its biological processing can take place on a higher hierarchic level formed with participation of information feedbacks of homeostasis. The fact that on macroscopic scales geometry of organism's interfaces is manifestly fractal, corresponding to self-similarity of 1/f fluctuations, seems to us very relevant.

At present more detailed interpretation of SC is difficult. However, it is clear that further study of this phenomenon can improve our understanding of interactions between living systems and their environment and elucidate biological functuon of endogenic and exogenic 1/f fluctuations. We hope also that SC, after clinical testing, will find its application in medical practice.

REFERENCES.

1. Y.Ichimaru, S.Katayama, Proc. 11th Int. Conf. on Noise in Physical Systems and 1/f Fluctuations (Ohmsha, Tokyo, 1991), p. 691.

2. T.Yoshida, S.Ohmoto, S.Kanamura, ibid., p. 719.

3. N.I.Muzalevskaya, V.M.Uritsky, Works of Djanelidze Res. Inst.,1, 77 (1993) (in Russian).

4. V.M.Uritsky, N.I.Muzalevskaya, in: Living systems under external influence (Gidrometeoizdat, St.Petersburg, 1992), p. 244 (in Russian).

5. A.S.Presman, ibid., p. 749.

6. K.Takakura, Y.Kosugi, J.Ikebe, T.Musha, Proc. 9th Int.Conf. on Noise in Physical Systems (World Scientific, Singapore, 1987), p. 279.

7. N.I.Muzalevskaya, Problems of Cosmic Biology (Soviet), 43, 82 (1982) (in Russian).

8. N.I.Muzalevskaya, in: Informational Interactions in Biology (Tbilisi Observ., 1987), p. 70.

9. T.Musha, Oyo Buturi, 54, 429 (1985) (in Japanese).

10. N.I.Muzalevskaya, V.M.Uritsky, .V.Korolyov and G.P.Timoshinov, Proc. 12th Int. Conf. on Noise in Physical Systems and 1/f Fluctuations (Ohmsha, Tokyo, 1991), p. 465.

1/f NOISE APPLICATION IN REFLEXOTHERAPY

S. O. Ostrova
A. E. Bulgakov

The Kazan State Technical University
named after A. N. Tupolev

10,K.Marx Street, 420111, Kazan, Tatarstan Republic,Russia

I. V. Klyushkin
A. M. Abdullina

The Republican Medical Diagnostic Center of c. Kazan
1a, ChekhovStreet, 420043, Kazan, Tatarstan Republic, Russia

ABSTRACT

The instruments and methods which make it possible to apply 1/f noise in reflexotherapy by action of physical factors on biologicallyactive points are dealt with. The efficiency of their application was show on the example of treatment of gastric and duodenal ulcers without medicaments.

INTRODUCTION

It is known that the finest coats (biological membranes) connecting living cells with the outside world, with chemical substances of intercellular environment, with neighboring cells can be a source of 1/f noise [1].The characteristics of 1/f noise contain information about change in ion fluxes through the membrane, about processes of change in cells taking place as a result of action of external factors. The frequency range of action of external factors results in different reactions of living body [2].

INSTRUMENTS OF FLUCTUATIONAL REFLEXOTHERAPY

Methods of reflexotherapy which make it possible to have effect immediately upon the nervous regulating centers of mental and emotional state and human physiological functions by action on biologically active points (BAP) and reflex zones (RZ) are known and widely used in treatment of different diseases of alimentary tract, nervous system, locomotor apparatus, etc.

The action of external physical factors on BAP may cause, however, undesirable side effects [3].

Method of action on BAP by electric fluctuational signal with spectrum of 1/f form (1/f noise) is the most physiologic for human body.

Besides, the problem of minimizing the dose of action at the expense of current strength decrease and period of action when the therapeutic effect is stable or increase.

Two types of instruments - Start 5 эП and Start 6 эП were designed for realizing the method of action on BAP by 1/f noise.

Start 5 эП is designed for carrying out medical treatment by methodsof acupuncture with the help of known types of action (direct and alternating currents) and with the help of new type of action - 1/f noise. Besides, there is an opportunity of simultaneous action of current of the given form and 1/f noise.

The instrument is operated under the "Search" and "Treatment" conditions. The "Search" condition is designed for finding BAP and carrying out the acupuncture diagnosis by Riodoracu. Under the "Treatment" conditions the action on BAP is occurred by:

1. 1/f noise,
2. direct current of positive polarity,
3. direct current of negative polarity,
4. alternating current of sign changable polarity,
5. combination of 2., 3., 4. with 1/f noise.

The instrument makes it possible to act on BAP by electrode (puncture action) and by needle (acupuncture action).

Medical electroacupuncture treatments are carried out by a reflexotherapeutist who has special training, knows the Bap topography, the principle of their selection and combination .

Medical electroacupuncture treatments may be carried out by medium-level medical staff after the appropriate doctor's instruction. 10 electrodes are provided in the instrument Start 5 эП for simultaneous action on 10 BAP.

Start 6 эП is a compact instrument and designed for acting on BAP by 1/f noise .

The treatment with this instrument may be carried out in clinics, outpatient departments, and in every day life, can be combined with other therapeutical methods. The instrument can be used for treatment in domestic conditions on doctor's orders.

Specifications of the instrument Start 6 эП .

1. Action signal - 1/f noise. The chosen value of current is stabilized and does not depend on resistance of patient's skin.
2. The range of current regulation is from 0 to 100 μA.
3. Peak voltage across the electrode - $\pm$15 V.
4. The range of regulation on action period is from 20 to 60 sec.
5. Power-line supply of 220 V, 50 Hz by selfcontained power unit.
6. Overall dimensions, mm 180x30x20.
7. Mass - not more than 180 g.

METHOD OF FLUCTUATIONAL REFLEXOTHERAPY

The developed method and instruments of fluctuational (noise) reflexotherapy were medically tested in the Republican Medical Diagnostic Center of c.Kazan in treatment of gastric and duodenal ulcers without medicaments.

Diagnosis of disease and dynamics of its development were reliably controlled by laboratory and instrumental examinations.

The treatment was prescribed on the basis of acupuncture diagnosis.

100 patients with gastric and duodenal ulcers were examined andtreated by this method. The clinical effectiveness was 96 %.

BAP situated on the upper limbs (10 II show-sang-ly on the right and on the left), on the trunk (a point-proclaimer of the stomach 12 XIV chzhung-vang in gastric ulcer or a point-proclaimer of the small intestine 4 XIV guang-yuang in duodenal ulcer), on the external ears (AT87 point of the stomach in gastric ulcer, AT88 point of the duodenum) were used at each treatment.

The period of action on each point of the external ear is 30 sec, on corporal points - 60 sec.

The current strength on a point of the external ear is 25 μA, correspondingly, on corporal ones - 50 μA. The treatments were carried out every day for 10 days. The course of treatment may be extended up to 15-18 days, if necessary, after doctor's control.

Positive results were also achieved at the employment of this method in treating patients with diseases of respiratory organs (asthmatic bronchitis, bronchial asthma), with diseases of central and peripheral nervous system accompanied with pain syndromes and vascular disturbances, with diseases of locomotor apparatus.

CONCLUSIONS

The instruments and methods of fluctuational reflexotherapy by action on biologically active points and reflex zones of 1/f noise can be widely used in treatment of the diseases in which the methods of electroreflexotherapy are recommended.

REFERENCES

1. Bukingem M. Noises in electronic systems and instruments. - M.:Mir,1986.
2. K. Oguti, H.Adati, T. Kusakabe, M. Agu Digital system for formation of 1/f type fluctuations of small fan motor rotation velocity. (TIIER, vol.76, No. 3, Màrch, 1988).
3. D.M. Tabeeva Guide for acupuncture. - M.: Medicine,1982.

Using an Electronic FitzHugh-Nagumo Simulator to Mimic Noisy Electrophysiological Data from Stimulated Crayfish Mechanoreceptor Cells

David Pierson, John K. Douglass, Eleni Pantazelou and Frank Moss
Departments of Physics and Biology
University of Missouri at St. Louis
St. Louis, MO 63121, USA

ABSTRACT

It is well known that sensory information in biological systems is transmitted to the brain using a code which must be based on the time intervals between neural firing events or the mean firing rate. However, in any collection of such data, and even when the sensory system is stimulated with a periodic signal, statistical analyses have shown that a significant fraction of the intervals are random, having no coherent relationship to the stimulus. We call this component the "noise". It is clear that coherent and incoherent subsets of such data must be separated. The power spectrum of the time series of firing events accomplishes this, and, with a weak periodic signal present, one can measure the signal-to-noise ratio (SNR) from the power spectrum. These tools allow one to study *Stochastic Resonance* (SR) in such a system. In this paper we show how SR can be observed in a commonly used neuron firing model, the FitzHugh-Nagumo (FN) model, and how an electronic analog simulator of this model can be used to mimic electrophysiological data from stimulated mechanoreceptor cells in the crayfish *Procambarus clarkii*.

INTRODUCTION

Stochastic resonance (SR) is a dynamical mechanism whereby the noise, or random forcing, inherent in a certain class of nonlinear systems can actually enhance the transmission of information through the system. The signature of SR is that some measure of coherence at the system output passes through a maximum with input or inherent noise intensity, thus leading to the concept of an optimal noise intensity for information transmission. In recent years, a great deal of interest has been generated by research on *SR*[1-9], which was also the subject of a recent review[10] and international workshop[11]. That SR might exist as an evolved detection strategy in the sensory nervous system was suggested[12] some time ago. Moreover, the internal neuronal noise intensity depends upon the stimulus intensity in a nonlinear manner through, for example, efferent connections in the visual system[13,14] and is often much larger (sometimes several orders of magnitude larger!) than can be accounted for by equilibrium statistical mechanics[15].

Recently, based on comparisons of interspike interval histograms obtained from passive analog simulations of simple bistable systems, with those from auditory neurons, it was suggested that the noise intensity may play a critical role in the ability of the living system to sense the stimulus intensity[16,17]. In this work, it is shown that in both simulations and in actual measurements on a sensory neuron, the addition of noise to a weak signal can enhance its detection ability, that is, SR. This process is essentially ***nonlinear***, and it indicates the ultimate futility of attempts to understand neural information transmission and processing using any ***linear*** transform theory borrowed from electrical engineering.

We have studied SR with an electronic analog simulator of the FN neuron model. The results of these simulations are compared with those from experiments on the mechanoreceptor in the tailfan of the crayfish *Procambarus clarkii* (see Douglass, *et al*, this conference). The FN model used here is defined by,

$$\tau_V \dot{v} = v(v - 0.5)(1 - v) - w + \xi(t), \tag{2}$$

$$\tau_W \dot{w} = v - w - [b + \epsilon \sin(\omega t)], \tag{3}$$

where v is the fast variable (action potential) operating on the time scale τ_V, w is the recovery variable with time scale τ_W, and b is the bifurcation parameter. The stimulus intensity is ϵ, and $\xi(t)$ is a quasi white, Gaussian noise, defined by $<\xi(t)\xi(s)> = \left[\frac{D}{\tau}\right]\exp\left[-\frac{|t-s|}{\tau}\right]$ with D the noise intensity and τ a (dimensionless) noise correlation time. The FN system is a limit cycle oscillator controlled by b. For $b < b_C$, the only solutions of (2) and (3) are fixed points; for $b \geq b_C$ the limit cycle solutions obtain. One excursion around the limit cycle has been designed to accurately mimic a single neuron firing event, or action potential. Since we are not dealing with pacemaker cells in our crayfish, it is necessary to operate the FN with $b < b_C$. To b is then added a weak signal. "Weak" means subthreshold, that is $\epsilon < (b_C - b)$ The noise ξ, shown in (2), can equally well be added to b inside the square brackets in (3). Thus we have a firing event whenever the noise plus the signal happen to exceed b_C . This operation effectively mimics the behavior of the weakly stimulated crayfish receptor cells, as described by Douglass, *et al* (this volume).

EXPERIMENTAL RESULTS

Examples of our experimentally measured power spectra are shown in Fig. 1, where the results of the FN simulation are shown on the left and of the crayfish measurements

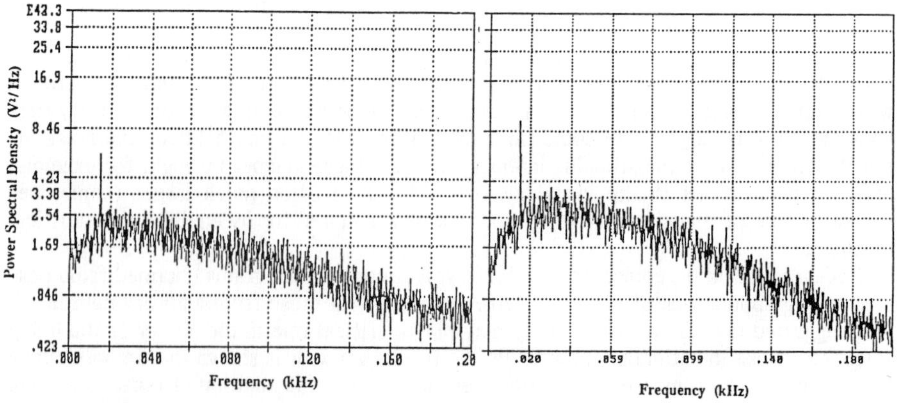

Fig. 1. Measured power spectra for 15 Hz noisy stimuli: from the FN simulator with $b_C = -0.18$, $b = -0.12$, $\epsilon = 0.08$, $\sqrt{\langle\xi^2\rangle} = 0.008$, $\tau_V = \tau = 10\ \mu s$, $\tau_W = 100\ ms$ (left) and from the crayfish (right).

are shown on the right. These two power spectra were measured for the same stimulus frequency and with approximately the same statistics. They are remarkably similar indicating that the FN model is a good representation of the actual dynamics of the neuron. Of particular importance is the sharp and narrow peak at ω which represents the signal feature. It has been shown theoretically[5] that the sharpness of this feature is a characteristic of the SR process. It should also be noted that, when operated in the limit cycle region, $b > b_C$, there is a *noise broadened* peak at the limit cycle frequency which is not characteristic of either SR or the behavior of stimulated sensory neurons (at least those lacking efferents). The SNR's, defined from the power spectra, as the ratio of the signal strength to the noise intensity at the signal frequency, are obtained from measured power spectra at various noise intensities. The SNR versus the noise intensity for the FN model (circles) are shown in Fig. 2 at left *com*[pared to data from a crayfish experiment (triangles) with external noise. On the right in Fig. 2 is shown an SR experiment on the FN model taken for different conditions and with better statistics. In both cases, it is evident that an optimal noise intensity results in the maximum SNR. We have thus demonstrated SR in the FN model and were able to successfully mimic data from a SR experiment with the crayfish mechanoreceptor cells.

This work was supported by the U. S. Office of Naval Research grants N00014-92-*J*-1235 and N00014-90-*J*-1327.

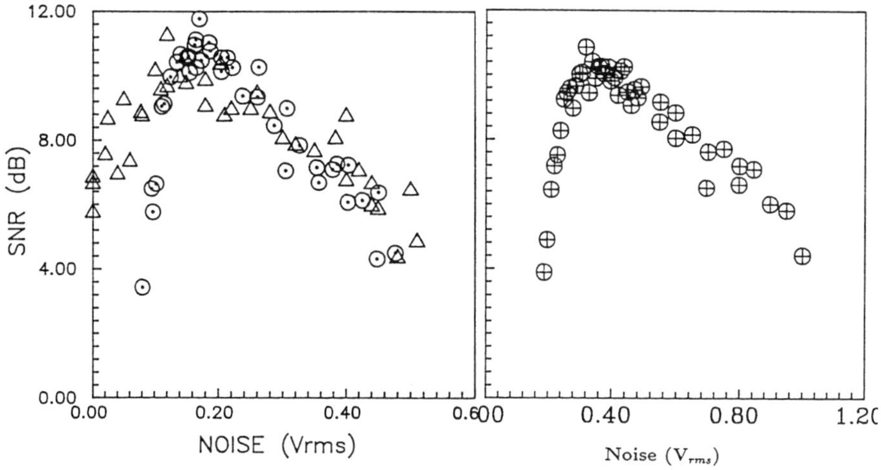

Fig. 2. (left) SR measured in the FN model (circles) compared to crayfish (triangles). The higher values for the crayfish data at low noise are due to a small inherent noise in the receptor cell. (right) SR measured for the FN model with better statistics, and for $b_C = 0.18$, $b = 0$, $\tau_V = \tau = 10\ \mu s$, $\tau_W = 10\ ms$, $\epsilon = 0.03$

REFERENCES

1. R. Benzi, S. Sutera and A. Vulpiani, J. Phys. A **14**, L453 (1981)
2. C. Nicolis, Tellus **34**, 1 (1982)

3. B. McNamara and K. Wiesenfeld, Phys. Rev. A **39**, 4148 (1989)
4. B. McNamara, K. Wiesenfeld and R. Roy, Phys. Rev. Lett. **60**, 2626 (1988)
5. P. Jung and P. Hanggi, Europhys. Lett **8**, 505 (1989)
6. L. Gammaitoni, F. Marchesoni, E. Menichella-Saetta, and S. Santucci, Phys. Rev. Lett. **62**, 349 (1989)
7. P. Jung, Z. Phys. *B* **16**, 521 (1989)
8. P. Jung and P. Hanggi, Phys. Rev. A **41**, 2977 (1990)
9. P. Jung and P. Hänggi, Z. Phys. *B*. **90**, 255 (1993)
10. F. Moss, "Stochastic Resonance: From the Ice ages to the Monkey's Ear"; in *Some problems in Statistical Physics*, edited by George H. Weiss (SIAM, Philadelphia, in press)
11. *Proceedings of the NATO Advanced Research Workshop on Stochastic Resonance in Physics and Biology*, edited by F. Moss, A. Bulsara and M. F. Shlesinger, special issue, J. Stat. Phys. **70** (1993).
12. F. Moss quoted in *Science News*, Feb. 23, 1991 *p*. 127.
13. E. Kaplan and R. Barlow, Vision Res. **16**, 745 (1976)
14. L. Croner, L. Purpura and E. Kaplan, Proc. Nat. Acad. Sci. USA, in press.
15. W. Denk and W. Webb, Hear. Res. **60**, 89 (1992)
16. A. Longtin, A. Bulsara and F. Moss, Phys. Rev. Lett. **67**, 656 (1991)
17. A. Longtin, A. Bulsara and F. Moss, Biol. Cybern., preprint.

XX. NEURONAL NETWORKS

1/F-LIKE SPECTRA IN CORTICAL AND SUBCORTICAL BRAIN STRUCTURES: A POSSIBLE MARKER OF BEHAVIORAL STATE-DEPENDENT SELF-ORGANIZATION

C. M. Anderson, T. Holroyd, S. L. Bressler, K. A. Selz and A. J. Mandell.
Center for Complex Systems, Florida Atlantic University, Boca Raton, FL. 33431

R. Nakamura
Laboratory of Neuropsychology, NIMH, Bethesda, MD.

ABSTRACT

Recently, power-law scaling of power spectra with scaling exponents close to -1 (1/f-like spectra) have been observed in cortical and subcortical brain structures in association with specific behavioral states[1-9]. Further, 1/f processes at different levels of organization have been reported in the nervous systems of vertebrates and invertebrates[16-22]. This study describes the 1/f-like appearance of cross-spectra between cortical sites in a monkey performing a GO/NO-GO behavioral task. We found broadband 1/f-like coherence spectra (average slope = -.84) during the "behaviorally flexible" state of tonic arousal shortly after the monkey had initiated the trial, suggesting that this brain state is characterized by long-range cortical correlations. One of the implications of these findings is that the 1/f-like cortical coherence spectra may provide a signature of brain self-organization during specific behavioral states. A more general implication is that broadband 1/f-like processes across many levels of organization in the nervous system may provide a versatile and parsimonious mechanism for binding cortical and subcortical nonlinear oscillators rapidly and specifically to environmental information.

INTRODUCTION

In the past seven years, several experimental studies[1-7] have reported broadband 1/f-like power spectra in cortical and subcortical brain structures associated with changes in behavior or during behavioral tasks[8,9]. For example, Freeman[2] reported 1/f power spectra in the surface electroencephalograph (EEG) of the rabbit olfactory bulb, nucleus and cortex preceding inhalation of significant odors and '1/f noise' in the surface EEG of the visual cortex of a rhesus monkey performing a conditioned response to a visual conditioned stimulus[9]. Freeman has suggested that in the olfactory system this 1/f-like background activity is chaotic and has its origins in the activity of interconnected neural masses providing a flexible common carrier wave during sensory input-driven self-organization[10]. In addition, these observations seem to indicate that because '1/f noise' appears in both paleocortical and neocortical areas, it is not due to the unique laminar complexity (3 versus 6 layer) or connectivity and cell types of either type of cortex, but rather is a result of intrinsic properties of cortical masses.

A further association of 1/f-like spectra with the self-organization of the nervous system during behavior is the finding of 1/f activity at other levels in the nervous system, such as in single unit neuronal firing patterns during circumscribed behavioral-states. The behavioral state-specificity of 1/f power spectra in single unit activity has been reported in the mesencephalic reticular formation, hippocampus and ventrobasal thalamus of the cat during alert states (orienting to birds) and during the state of paradoxical or rapid eye movement sleep (REM) [1,3-7]. These experimentally determined broadband power spectra from single unit activity can be characterized as having $1/f^{\alpha}$ scaling with exponents between $0.5 \leq \alpha \leq 2$ over 1 to 1.5 decades[3,4]. Orienting in the cat during the onset of auditory or visual stimuli increases activity in noradrenergic (NA) neurons in the locus coeruleus (LC)[11]. Orienting responses can be elicited in many mammals by stimulation of the LC, which in turn has been demonstrated to increase cortical single unit activity[12] and the amplitude of the EEG[13]. Mandell and Selz[7] investigated the power spectra of interspike intervals of LC neurons projecting to wide-spread areas of the cortex and found they exhibited inverse power-law $1/f^{\alpha}$ distributions. They proposed that NA and other monoaminergic brain stem cell body groups with these power spectral patterns may play a role in stabilizing the widespread quasiperiodicity of the cortical EEG during specific behavioral states[14,15].

At the level of individual neurons, 1/f patterns of current fluctuations have been reported in the resting membrane potential at the nodes of Ranvier in the frog[16] and in squid giant axon[17] in addition to 1/f distributions of conductance and potential energy states in the dynamics of ion channels in mammals[18,19]. Also 1/f time relations are found in the spontaneous spike discharge of giant snail neurons[20] and in the intervals between successively evoked action potentials in a squid giant axon[21]. Furthermore, relationships have recently been proposed between fractal channel noise and fractal distributions in action potentials[22]. Although none of this data addresses 1/f patterns at the cellular level during specific behavioral states, it is interesting to speculate on the reasons for the presence of 1/f processes at many levels of organization in the nervous system. Mandell[23] proposed that a simple learning process such as habituation, present in a self-similar fashion across many levels in the nervous system, might utilize the "spectra reserve"[24] of 1/f processes to bind across levels. Another consideration is the intricate regulation at multiple levels of the nervous system. West[24] hypothesized that self-similar feedback systems with control mechanisms exhibiting a broadband inverse power-law form would have greater stability than systems with mono-frequency control because they would be more stable to loss of feedback control elements.

A possible interpretation of these diverse findings is that long-range coupling between many brain systems (e.g., brainstem and neuronal nonlinear oscillators in the thalamus, as well as other subcortical sites such as the extended-amygdala[25] and various cortical sites) during the self-organization of behavioral events involves collective interactions occurring on multiple time and frequency scales. This phenomenon could be conceptualized as broadband binding and has analogies to the synchronization of chaotic nonlinear oscillators[26] in which two separate systems, one chaotic with broadband spectra, and one a nonchaotic subsystem of the first, will converge quickly to one trajectory when linked. Further cortical EEG data to support this interpretation follows.

METHODS AND RESULTS

Bressler and Nakamura[8] investigated the inter-area synchronization among transcortical electrode recordings from cortical sites in a rhesus monkey performing a GO/NO-GO behavioral response paradigm. In this study a water deprived monkey, after initiating a trial, was required to discriminate two visual stimuli by correctly releasing a lever (GO condition) in response to one stimulus (with water reward) and holding the lever for 500 msec (NO-GO condition) in response to the other stimulus. Each session consisted of 1000 trials randomly presented with equal probability, lasting about 35 minutes with inter-trial intervals randomly varied from 0.5 to 1.25 seconds. We examined the slope of the coherence spectra from 6.25 to 93.75 Hz (log-log, Fisher z-transformed normalized cross-power spectrum) between recording sites in different cortical areas (e.g., frontal, striate, parietal, somatosensory, and motor) found by Bressler and Nakamura to have high coherence values prior to stimulus presentation and during the behavioral response for two sessions (S1, S2). Pairs of electrodes with high coherence accounted for roughly 10% of all pairs in both sessions. The task in S2 was identical to S1 except that a stimulus reversal was imposed on the task (the stimulus that had previously been paired with water-"GO"-was switched to "NO-GO"). In addition, a different montage of cortical sites was sampled in S2. However, comparisons between sites common to S1 and S2 displayed similar coherence. For a 160-msec-long window (W1) prior to stimulus onset and an identical window (W2) centered on the mean response time for the task, the mean slopes ± the standard error of the mean are listed in table I.

Table I: Mean slopes of coherent pairs.

SESSION	W1 GO	W1 NO-GO	W2 GO	W2 NO-GO
S1	-.84 ± .20	-1.05 ± .27	-.52 ± .06	-1.21 ± .17
S2	-.78 ± .06	-.71 ± .10	-.98 ± .17	-.53 ± .04

The slopes from all groups were compared by three-way (session x window x condition) ANOVA, and no main effects were detected. However two interactions were noted: one between session and condition ($F[1, 42] = 11.22$, $p < 0.01$) and one between session, window and condition ($F[1, 42] = 4.16$, $p < 0.05$). The slopes of W1 were compared by two-way (session x condition) ANOVA; no main effects of session ($F[1, 20] = 1.26$, $p > 0.05$) or condition ($F[1, 20] = 0.19$, $p > 0.05$) were demonstrated and no interactions ($F[1, 20] = 0.60$, $p > 0.05$) were noted.

DISCUSSION AND CONCLUSIONS

The findings of an average 1/f-like slope of -.84 for the pre-stimulus period W1 is consistent with the idea that fractal time processes may underlie the flexibility of the brain in responding to behavioral challenges. During this time the monkey was in an alert tonic arousal state, motivated by water deprivation. Sheer[27] has made a distinction between tonic and focal arousal in relation to cortical EEG. Sheer defines tonic arousal as: "...an oscillatory, unstable state of the organism, in which many different subassemblies of the intrinsic electrical activity are firing in different patterns....[N]ot being focused...[this] represents an adaptive ready state for significant occurrences." The potential for dynamically rescaling cortical-cortical and cortical-subcortical interactions in the tonic arousal state may be represented in the 1/f-like signature of the prestimulus period. Sheer's definition of focal arousal applies to the response window W2. At this time the monkey is either carrying out the critical motor task, lifting his hand from the lever in order to receive water or suppressing his response before the 500 msec limit. Sheer stated that focal arousal: "...needed...reinforcement contingencies to dissociate it from tonic arousal....With focusing behavioral operations, such as reinforcement contingencies, subassemblies of the electrical activity now fire in coherent organizations restricted to the relevant circuitry." Bressler and Nakamura found a broadband increase in the amplitude of coherence during the W2 GO condition in the same electrode pairs and concluded that these sites were functionally linked in a large-scale network during the motor response onset. This finding along with the observed difference between slopes in W2 across the two sessions and conditions (although not significantly different), suggests that the neural processes underlying focal arousal may be distinct from tonic arousal.

In summary, we suggest that the shift to a 1/f-like coherence spectrum between pairs of cortical sites during specific behavioral states such as tonic and focal arousal implies the possible use of dynamic rescaling of cortical-cortical and cortical-subcortical interactions in time by the brain. Further, measurement of the slope of the coherence spectrum between two cortical sites during specific behavioral states may provide a marker of behavioral task-specific self-organization. Finally, broad-band 1/f-like coherence spectra may signify a state of readiness in which chaotic synchronization or binding between brainstem sites and cortical and subcortical oscillators can occur during the dynamic self-organization of brain states. The presence of 1/f patterns across the phylogenic scale, across levels of oganization in the nervous system and in the "the way our world changes in time"[28] suggests parsimony in the way nature constructs organisms. (This work was supported by NIMH-MH19116 and the Office of Naval Research, Cognitive and Neural Sciences and Systems Biophysics sections).

REFERENCES

1. F. Grüneis, M. Nakao, and M. Yamamoto, Biol. Cybern. 62, 407 (1990).
2. W. J. Freeman, Neurosci. Abstr. 15, 927 (1989).
3. T. Kodama,H. Mushiake, S. Keisetsu, H. Nakahama, and M. Yamamoto, Brain Res. 487, 27 (1989).
4. T. Kodama,H. Mushiake, S. Keisetsu, T. Hayashi, and M. Yamamoto, Brain Res.487, 35 (1989).
5. F. Grüneis, M. Nakao, and M. Yamamoto, T. Musha, and H. Nakahama, Biol. Cybern. 60, 161 (1990)
6. M. Yamamoto, H Nakahama , K. Shima, T. Kodama and H. Mushiake, Brain Res. 366, 279 (1986).
7. K. A. Selz and A. J. Mandell, Int. J. Bifurcation and Chaos. 1, 717 (1991).
8. S. L. Bressler and R. Nakamura, Computation & Neural Systems, F. Erckman and J. Bower, Eds. (Kluwer Publishers.,1993), p. 515.
9. W. J. Freeman and B. W. Van Dijk, Brain Res. 422, 267 (1987).
10. W. J. Freeman, Int. J. Bifurcation and Chaos. 2, 451 (1992).
11. K. Rasmussen, D. A. Morilak and B. L. Jacobs, Brain Res. 371, 324 (1986).
12. A. I. Semenyutin, Neurophsiology. 22, 359 (1980).
13. C. W. Berridge and S.L. Foote, J. Neursci. 11, 3135 (1991).
14. A. J. Mandell and K. A. Selz J. Stat. Phys. 70, 355 (1993).
15. A. J. Mandell and K. A. Selz, First Experimental Chaos Confernce, S. Vohra M. Spano, M Shlesinger, L. Pecora, Eds. (World Scientific, Singapore,1991).
16. H. E. Derkson and A. A. Verveen, Science. 151, 1388 (1965).
17. F. Conti, L. J. DeFelice and E. Wanke, J. Physiol. 248, 45 (1975).
18. L. S. Liebovitch and T. I. Tóth, Ann. Biomed. Engr. 18, 177 (1990).
19. L. S. Liebovitch and T. I. Tóth, Bull. Math. Biol. 53, 443 (1991).
20. T. Musha, H. Takeuchi and T. Inoue, IEEE Trans. Biomed. Eng., BME-30, 194 (1983).
21. T. Musha, Y. Kosugi, G Matsumoto and M. Suzuki, IEEE Trans. Biomed. Eng., BME-28, 616 (1981).
22. S. B. Lowen and M. C. Teich , this conference.
23. A. Mandell, Math. Modelling 7, 809 (1986) .
24. B. J. West, Fractal Physiology and Chaos in Medicine, (World Scientific, Singapore ,1991), p. 80.
25. G. F. Alheid and L. Heimer. Neurosci. 27, 1 (1988)
26. L. Pecora and T. Carroll, Phys. Rev. Lett. 64, 821 (1990).
27. D. E. Sheer, Springer Series in Brain Dynamics 2, Basar E. and T. H. Bullock Eds. (Springer-verlag, Berlin, 1989), p. 341.
28. R. F. Voss, The Science of Fractal Images, Petigen HO, Saupe D, Eds. (Springer -Verlag, Berlin, 1989), p. 42.

THE EFFECT OF NOISE ON A NEURAL NETWORK WITH SPIKING NEURONS

Mario E. Inchiosa
Naval Command, Control, and Ocean Surveillance Center, RDT & E Division
NRAD Code 573, San Diego, CA 92152–5000

ABSTRACT

We study a class of neural network associative memories which include noise and transmission delays, code information in the timing of spikes, use long-range Hebbian couplings plus local, inhibitory couplings, and feature low, biologically realistic neuronal activity. Recall of a pattern consists of a synchronized, periodic firing of neurons. We find a Lyapunov functional for the noiseless network dynamics, and, using statistical mechanics and numerical simulation, we find that noisy dynamics improves the network's ability to discriminate stored from unknown patterns.

INTRODUCTION

In neurobiology one finds neurons with both short- and long-range connectivity (pyramidal cells) and neurons providing local inhibition. In the presence of a "recognized" stimulus, neuronal firing probabilities exhibit a temporal periodicity that has a spatial phase coherence. We attempt to address these experimental observations.

Our network, which is based on the network of Ref. [1], consists of two classes of neurons differing in their range of connectivity (global v. local). Each globally connected neuron connects to all the other globally connected neurons via synapses programmed by the Hebb rule to store p N-bit patterns. Each globally connected neuron also connects via an excitatory synapse to *one* locally connected neuron, which connects back to the globally connected neuron via an inhibitory synapse.

We find that noise in the neuronal dynamics improves the discrimination of stored patterns from unknown patterns. Without noise, the network fires in an oscillatory pattern in response to both stored and unknown patterns, albeit with a much stronger response to stored patterns. With noise, stored patterns still evoke a strong oscillatory response, but unknown patterns evoke only a weak static response (see Figure).

THE MODEL NETWORK

The membrane potential of globally connected neuron i, $i \in \{1, \ldots, N\}$, at time t is

$$h_i(t) = \sum_{j=1}^{N} \sum_{\tau=0}^{D-1} J_{ij}(\tau) S_j(t-\tau) + h_i^{\text{ext}}(t), \tag{1}$$

where $S_i(t)$ is the state of neuron i at time t ($+1$ = "firing," -1 = "not firing"). The term $h_i^{\text{ext}}(t)$ represents external input.

The coupling constants are

$$J_{ij}(\tau) \equiv \epsilon'(\tau)\frac{1}{N}\sum_{\mu=1}^{p}\xi_i^\mu\xi_j^\mu - \eta'(\tau)\delta_{ij}. \tag{2}$$

The first term models synapses programmed by the Hebb rule to store p random patterns in the network (the ith bit of the μth pattern is given by $\xi_i^\mu = \pm 1$). We define $\epsilon'(\tau) \equiv 0$ if $\tau < \Delta^{\text{global}}$, $\epsilon'(\tau) \equiv \epsilon(\tau - \Delta^{\text{global}})$ otherwise. This function models the transmission time Δ^{global} for a spike to arrive from one of the other globally connected neurons and the response in time, $\epsilon(\tau)$, of a globally connected neuron's membrane potential to the arrival of a spike. The second term in (2) models the locally connected neurons, and the time course of their inhibitory effect is $\eta'(\tau) \equiv 0$ if $\tau < \Delta^{\text{local}}$, $\eta'(\tau) \equiv \eta(\tau - \Delta^{\text{local}})$ otherwise.

The probability that neuron i will fire at time t is

$$P(S_i(t) = 1) = \frac{1}{2}\{1 + \tanh[\beta(h_i(t-1) - \theta)]\}. \tag{3}$$

In (3), θ represents the neurons' firing threshold and $1/\beta \equiv T$ is a "temperature" parameter representing the effect of noise in the system. We fix a scale for this parameter by choosing the normalization $\sum_{\tau=0}^{D-1}\epsilon'(\tau) = 1$.

EQUIVALENT STATIC SYSTEM AND LYAPUNOV FUNTIONAL

The state of our original system consists of the N variables $S_i = \pm 1$. Imagine a system with ND variables S_{ia}, $1 \le i \le N$, $0 \le a \le D-1$. We can use the D sets of N variables to represent the last D states of our original system. Since the state of the expanded system consists of a D state long history of the original system, we only have to look at the state of the expanded system once every D time steps to know exactly how the original system is evolving in time. If the state of the expanded system, as observed every D time steps, is stationary in time, then we know that the original system has fallen into an attractor of period D (or a divisor of D).

We construct the expanded system by transforming our original formulas as follows: $S_i(t-\tau) \to S_{i\tau}(t)$, $\epsilon'(\tau) \to \epsilon_{a,a+\tau+1}\ \forall a$, $\eta'(\tau) \to \eta_{a,a+\tau+1}\ \forall a$, implying $J_{ij}(\tau) \to J_{ij}^{a,a+\tau+1}\ \forall a$, and $J_{ij}^{ab} = J_{ij}^{a+c,b+c}$ (cyclic symmetry). The new coupling constants are $J_{ij}^{ab} \equiv \epsilon_{ab}\frac{1}{N}\sum_{\mu=1}^{p}\xi_i^\mu\xi_j^\mu - \eta_{ab}\delta_{ij}$, and, choosing $a = 0$, $h_i(t) = \sum_{j=1}^{N}\sum_{b=0}^{D-1}J_{ij}^{0,b+1}S_{jb}(t) + h_i^{\text{ext}}(t)$. The firing probability becomes $P(S_{i0}(t) = 1) = \frac{1}{2}\{1 + \tanh[\beta(h_i(t-1) - \theta)]\}$, and, since $S_i(t-a) = S_i(t-1-[a-1])$, we have $S_{ia}(t) = S_{i,a-1}(t-1)$ for $a > 0$, completing the specification of the dynamics of the expanded system.

We would like to prove that at $T = 0$ our network will settle to a periodic attractor representing associative recall of a stored pattern. Provided $\epsilon'(\tau)$ and $\eta'(\tau)$ are symmetric about $\tau = \frac{D}{2} - 1$, we can write down a Lyapunov functional for the $T = 0$ dynamics of our system (see Ref. [3]). For convenience, we also set $\epsilon'(D-1) = \eta'(D-1) = 0$. This yields $\epsilon_{ab} = \epsilon_{ba}$, $\eta_{ab} = \eta_{ba}$ and $\epsilon_{aa} = \eta_{aa} = 0$. Therefore J_{ij}^{ab} has the symmetry $J_{ij}^{ab} = J_{ji}^{ba}$, and $J_{ij}^{aa} = 0$. With this symmetry and the cyclic symmetry mentioned above, a Lyapunov functional for our system is

$$\mathcal{H} = -\frac{1}{2}\sum_{i,j=1}^{N}\sum_{a,b=0}^{D-1}J_{ij}^{ab}S_{ia}S_{jb} - \sum_{i=1}^{N}(h_i^{\text{ext}} - \theta)\sum_{a=0}^{D-1}S_{ia} \tag{4}$$

From now on we take h_i^{ext} to be independent of time. In Ref. [3] it was proved that a Hamiltonian with the symmetry properties of our $\mathcal{H}$ also generates the equilibrium distribution of states of the $T > 0$ dynamics.

THERMODYNAMICS

To describe the thermodynamic state of our system, we introduce order parameters which measure the correlation of each of the system's layers with the stored patterns, with the resting (-1) state, and with a random pattern:

$$m_a^\mu \equiv \frac{1}{N}\sum_i \xi_i^\mu S_{ia}, \quad m_a' \equiv \frac{1}{N}\sum_i (-1) S_{ia}, \quad m_a'' \equiv \frac{1}{N}\sum_i \zeta_i S_{ia}. \tag{5}$$

We take the input to be any superposition of stored patterns and a random pattern which has not been stored: $h_i^{ext} = \sum_\mu h^\mu \xi_i^\mu + h\zeta_i$. We chose a long transmission delay for the global connections, $\epsilon_{ab} = \delta_{|a-b|,D/2}$, and a short delay and long time course for the local, inhibitory effect, $\eta_{ab} = \frac{\eta}{D-1}(1-\delta_{ab})$.

For p finite and $N \to \infty$ we can use the methods of Ref. [2] to obtain the free energy density $f(\beta)$:

$$-\beta f(\beta) = \max_{\{m_a^\mu\}} \left\{ -\frac{\beta}{2}\sum_{a,\mu} m_a^\mu m_{a+\frac{D}{2}}^\mu + \left\langle \ln \left\langle \prod_a \cosh \Xi_a \right\rangle_z \right\rangle_{\xi,\zeta} + \frac{\beta\eta D}{2(D-1)} \right\}, \tag{6}$$

where $\{m_a^\mu\}$ solves the fixed point equations

$$m_a^\mu = \left\langle \xi^\mu \left\langle \sinh \Xi_a \prod_{b\neq a} \cosh \Xi_b \right\rangle_z \left\langle \prod_b \cosh \Xi_b \right\rangle_z^{-1} \right\rangle_{\xi,\zeta}, \tag{7}$$

and where $\Xi_a \equiv \beta \left[\sum_\mu (m_{a+\frac{D}{2}}^\mu + h^\mu)\xi^\mu + h\zeta - \theta + i\sqrt{\frac{\eta}{\beta(D-1)}}z \right]$. The averages over z are Gaussian averages and can be performed explicitly if we expand the products of hyperbolic functions. For m_a' and m_a'', replace the ξ^μ in (7) with -1 and ζ, respectively.

CONCLUSION

Numerical simulations of the network (see Figure) show that with noisy dynamics ($\beta = 3$) stored patterns evoke a strong oscillatory response, while unknown patterns evoke only a weak static response. With low-noise dynamics ($\beta = 20$) both stored and unknown patterns evoke an oscillatory response. In agreement with these simulations, numerical solution of (6) and (7) shows that at $\beta = 3$ a region in the η-θ plane exists where the static state is stable in the presence of an unknown pattern and unstable in the presence of a stored pattern.

ACKNOWLEDGMENTS

Many thanks to W. Gerstner, J. L. van Hemmen, A. V. M. Herz, R. Ritz, and S. Wimbauer and A. Bulsara for useful discussions and help with this research.

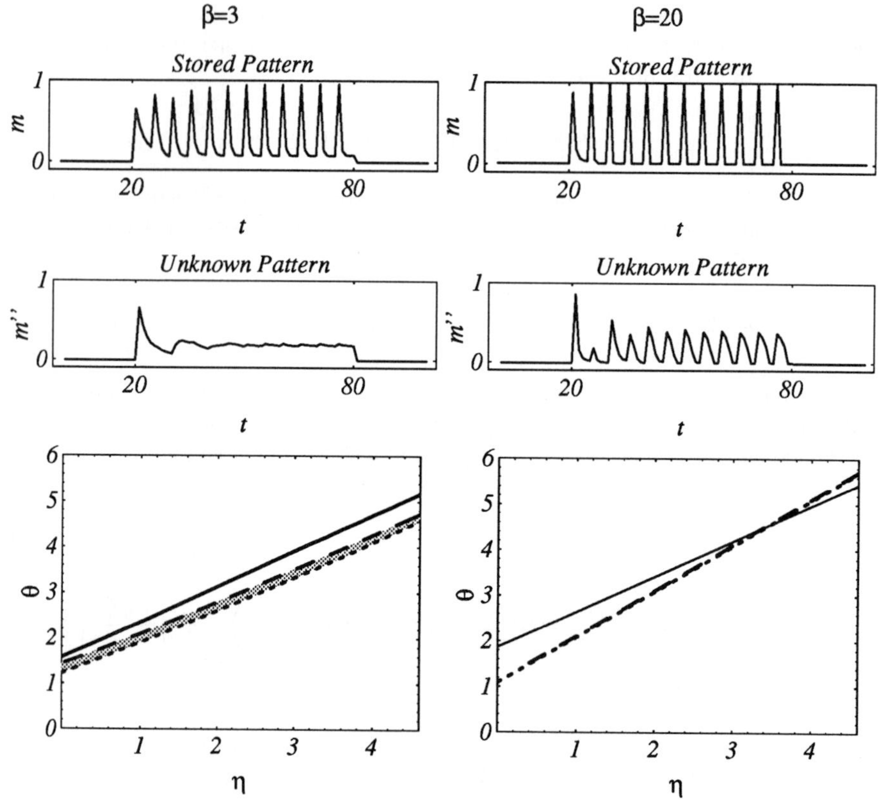

Figure 1: Above: Numerical simulations of the network ($N = 50000$, $p = 500$, $\eta = 1$, $\theta = 2$). Below: Phase diagram. In the presence of a stored pattern ($h^\mu = \delta^{\mu 1}$, $h = 0$), the oscillatory state is stable below the solid line and the static state is stable above the dashed line. In the presence of an unknown pattern ($h^\mu = 0$, $h = 1$), the static state is stable above the dotted line. Therefore, the shaded area between the dashed and dotted lines is the desirable operating region. For all plots, $D = 10$.

REFERENCES

[1] Gerstner, W., Ritz, R., and van Hemmen, J. L., "A biologically motivated and analytically soluble model of collective oscillations in the cortex: I. Theory of weak locking," *Biol. Cybern.* **68** 363–374, 1993.

[2] van Hemmen, J. L., and Kühn, R., "Collective Phenomena in Neural Networks," in *Physics of Neural Networks*, eds. E. Domany, J. L. van Hemmen, and K. Schulten, (Springer, Heidelberg), 1990.

[3] Herz, A. V. M., Li, Z., and van Hemmen, J. L., "Statistical Mechanics of Temporal Association in Neural Networks with Transmission Delays," *Phys. Rev. Lett.* **66** 1370–1373, 1991.

FRACTAL AUDITORY-NERVE FIRING PATTERNS MAY DERIVE FROM FRACTAL SWITCHING IN SENSORY HAIR-CELL ION CHANNELS

S. B. Lowen
M. C. Teich
Dept. of Electrical Engineering, Columbia University, New York, NY 10027

ABSTRACT

Hair-cell ion channels, which provide a crucial link in the transformation of incoming acoustic information to neural action-potential trains, switch between open and closed states with power-law-distributed (fractal) dwell times. Trains of action potentials recorded from auditory nerves in mammals always exibit fractal behavior, including a $1/f$-type spectrum, for long time scales. We provide a mathematical model linking these two fractal behaviors within a common framework.

Transduction of mechanical acoustic information into electrical nerve impulses (action potentials) takes place in the mammalian cochlea. Hair cells in the cochlea release neurotransmitter at a rate which depends on the level of acoustic stimulation; this neurotransmitter, in turn, stimulates action-potential production in primary auditory nerve fibers synapsed to the hair cells. These nerve fibers subsequently transmit this auditory information to the brain. Since the action potentials all have identical waveforms, the information is carried in their relative timing. These action potentials are generated in fractal or clustered patterns even in the absence of external acoustic stimulation.[1–5] Mathematically, the nerve-fiber activity can be modeled as a fractal point process.[6–9]

The statistic that perhaps best illustrates this is the Fano factor $F(T)$, which is defined as the ratio of the variance to the mean number of action potentials counted in a specified counting time T. Varying the counting time T over a range of values generates a Fano-factor time curve (FFC). For a homogeneous Poisson process, the Fano factor assumes a constant value of unity for all counting times T. For short counting times the FFCs computed from auditory-nerve action potentials remain close to unity. However, all auditory-nerve firing patterns examined to date reveal a Fano factor which increases as a power-law function of the counting times, i. e., $F(T) \propto T^{\alpha}$, and this fractal relationship holds for all counting times T between one second and the limit imposed by the finite duration of the recording. The power-law exponent α lies between zero and unity for all such recordings.

In Fig. 1, we present an FFC for the spike train on a cat auditory nerve fiber, recorded in the absence of any stimulation; it shows the typical power-law behavior (solid curve). For this particular recording, $\alpha \approx 0.5$. The power spectral

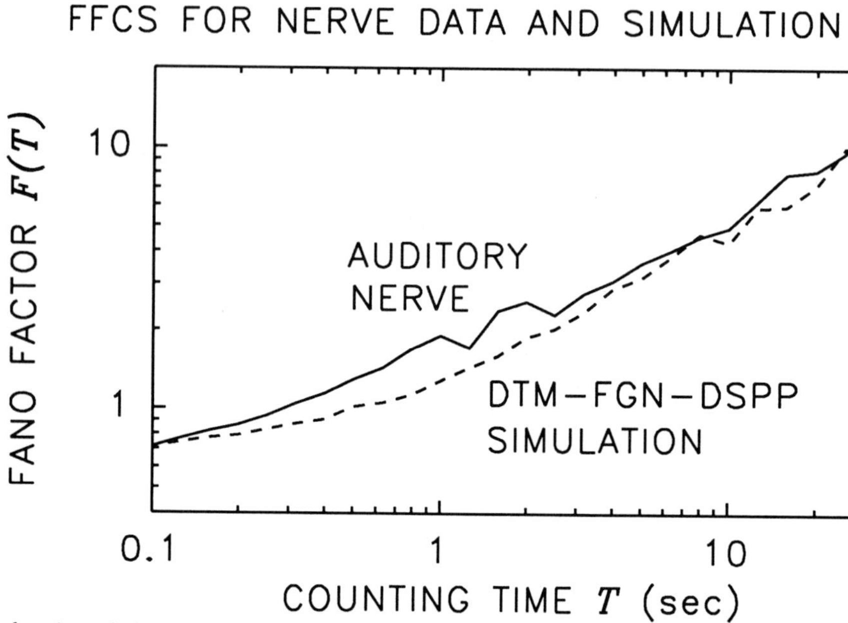

density of the action-potential point process also follows a power-law form, i. e., $S(f) \propto 1/f^{\alpha}$, with a power-law exponent that has been verified to be identical to the corresponding α from the FFC, both experimentaly and analytically. Other statistical measures, such as rescaled range (R/S), pulse-number distributions (PNDs), and count-based serial correlation coefficients, also highlight the fractal nature of auditory-nerve action potentials, although the FFC presents this information in the most robust manner.

Another statistical measure which yields complementary information, over short time scales, is the interspike-interval histogram, also known as the pulse-interval distribution (PID). The experimental PID exhibits a delayed exponential form. The simplest mathematical model which fits all the above statistical measures for the auditory-nerve firing data appears to be the dead-time-modified fractal-Gaussian-noise-driven doubly stochastic Poisson point process (DTM-FGN-DSPP).[3,5] In this process, fractal Gaussian noise (FGN), with a power spectral density that decays as $1/f^{\alpha}$, serves as the stochastic rate function for a nonhomogeneous Poisson process. The FGN rate function providing the best fit to the data serendipitously has a very small coefficient of variation, ensuring that the rate is almost always positive, thus simplifying the analysis and simulation. Finally, events from this Poisson process which occur within the (nonparalyzable) dead time of a previous event are deleted, resulting in the DTM-FGN-DSPP. Simulations of the DTM-FGN-DSPP [7] yield statistics virtually identical to those of the auditory-nerve data.[3,5] The dotted curve in Fig. 1 shows the FFC for the DTM-FGN-DSPP model, which agrees quite closely with the solid curve generated from the auditory-nerve data.

Might the origin of these fractal action-potential occurrences lie in the fractal activity which also occurs in cochlear hair-cell ion channels? Ion channels switch between two states, open and closed, and the dwell-time distributions often obey power-law forms over a wide range of dwell times.[1,10] Ionic current flows at a constant rate when the channel is open, and not at all when it is closed. Thus a fractal Bernoulli process provides a good model for a single ion channel, and a fractal binomial process models a collection of such channels.[8,9] For independent ion-channel dwell times, the Bernoulli process of the openings and closings of a single channel form an alternating renewal process, and the associated statistics all follow fractal (power-law) forms. For a collection of independent, identical ion channels, the statistics of the resulting binomial process also exhibit fractal behavior. Consider, for example, ion channels with open and closed dwell times $\mathcal{T}$ which have similar fractal distributions decaying as $\Pr\{\mathcal{T} > t\} \propto t^{2-\alpha}$ over some range of dwell times. The resulting fractal Bernoulli and binomial processes have an autocovariance function $C(\tau)$ which decays as $\tau^{\alpha-1}$ and a power spectral density which decays as $1/f^{\alpha}$, where α lies between zero and unity. Furthermore, a power spectral density which decays as $1/f^{\alpha}$, where α again lies between zero and unity, can also be produced when the dwell times in the open state (for example) are negligeable compared to the dwell times in the other state, for which the associated distribution again follows a fractal form, given in this case by $\Pr\{\mathcal{T} > t\} \propto t^{-\alpha}$.

Hair cells contain K^+-ion channels which operate in a fractal fashion, and thus a fractal binomial process is expected to describe the K^+-ion concentration within the cell. This fractal ion-channel behavior is consistent with a fractal alternating renewal process model.[8,9] Since the channels have identical configurations while open, and thus the openings physically resemble a renewal process, for the remaining analysis we make the reasonable assumption that the dwell times within and among channels are independent. Even if such dependency exists, it would likely not affect the predictions of the model.

Since there are many fractal ion channels, as a result of the Central Limit Theorem the binomial process converges to a Gaussian process with the same fractal power spectral density: it is fractal Gaussian noise. Thus the K^+-ion concentration is FGN with an empirical fractal exponent that again lies between zero and unity. Indeed, the voltages of excitable tissue membranes at rest have long been known to exhibit $1/f$-type fluctuations, which have in turn been traced to fluctuating K^+-ion concentrations.[11] This fluctuation establishes the Ca^{++}-ion concentration which, in turn, determines the neurotransmitter secretion that produces a FGN excitation of the auditory nerve fiber proportional to the original FGN K^+-ion concentration. Assuming that an auditory nerve fiber would produce a *homogeneous* Poisson point process in the presence of a steady concentration of neurotransmitter (if it were hypothetically possible to so excite it), then with fluctuations as described above it would generate action potentials as a *doubly stochastic* Poisson point process, with the stochastic rate given by the

FGN-varying neurotransmitter concentration. With the imposition of dead-time effects on the auditory nerve-fiber firings, the resulting process is the DTM-FGN-DSPP.

The approach outlined above is likely to be applicable to a wider range of situations than simply spontaneous auditory nerve-fiber firings. In the presence of a pure-tone stimulus, fractal behavior in the auditory nerve is maintained, but with an apparent increase in the fractal exponent.[2,5] This change presumably originates in a quantitative change in the open- and closed-time distributions for the hair-cell ion channels, but not in a qualitative change from fractal to non-fractal behavior. Finally, inasmuch as fractal ion channels are ubiquitous, similar fractal action-potential activity is likely to appear in other sensory systems, and indeed in many biological systems in general.

The question of the origin of the fractal behavior of the ion channels remains, although several possibilities present themselves. Ion channels are proteins, with a hierarchy of structure on many length scales, and therefore exhibit movement on many time scales. Thus it becomes more convenient to conceptualize ion-channel behavior as $1/f$ noise.[10] Fractal ion-channel behavior then becomes simply a manifestation of the underlying time-scale invariance of the ion-channel protein motion. Another possibility is that the ion-channel fractal behavior is an emergent phenomenon, occurring only in aggregates of intercommunicating channels. Mechanisms for this interaction could range from self-organized criticality, to spatio-temporal chaos, to other cellular automata processes. In that case the channels would no longer be independent, but the overall conclusions would still be valid.

REFERENCES

1. M. C. Teich, IEEE Trans. Biomed. Eng., 36, 150-160 (1989).
2. M. C. Teich, D. H. Johnson, A. R. Kumar, and R. G. Turcott, Hear. Res., 46, 41-52 (1990).
3. M. C. Teich, R. G. Turcott, and S. B. Lowen, in Mechanics and Biophysics of Hearing, edited by P. Dallos, C. D. Geisler, J. W. Matthews, M. A. Ruggero, and C. R. Steele (Springer, New York, 1990), pp. 354-361.
4. S. B. Lowen and M. C. Teich, J. Acoust. Soc. Am., 92, 803-806 (1992).
5. M. C. Teich, in Single Neuron Computation, edited by T. McKenna, J. Davis, and S. Zornetzer (Academic, Boston, 1992), pp. 589-625.
6. S. B. Lowen and M. C. Teich, IEEE Trans. Inform. Theory, 36, 1302-1318 (1990).
7. S. B. Lowen and M. C. Teich, Phys. Rev. A, 43, 4192-4215 (1991).
8. S. B. Lowen and M. C. Teich, Phys. Rev. E, 47, 992-1001 (1993).
9. S. B. Lowen and M. C. Teich, IEEE Trans. Inform. Theory, in press (Sept. 1993).
10. L. S. Liebovitch and T. I. Tóth, Ann. Biomed. Eng., 18, 177-194 (1990).
11. A. A. Verveen and H. E. Derksen, Proc. IEEE, 56, 906-916 (1968).

1/f NOISE IN MAGNETIC RESONANCE

M. Warden
Physics Institute, University of Zürich Zürich, Switzerland

ABSTRACT

A $(1/f)^\beta$ behavior in a high power magnetic resonance experiment is presented. The experimental observation was simulated with a deterministic, multi mode model of spin wave dynamics with no assumptions of randomness.

INTRODUCTION

As the amplitude of the driving field $\vec{h}$ in a magnetic resonance experiment is increased beyond the threshold field $\vec{h}_{th}$ for linear excitation, where the spins execute a uniform precession about the static field $\vec{B}_0$, instabilities of non uniform spin waves (SW) [1] occur (for a recent review see [2]). In particular when $\vec{h}$ is applied parallel to $\vec{B}_0$ (parallel pumping (PP)), SW are directly excited by $\vec{h}$. Due to the non-linear interaction between the participating SW modes, energy can flow back and forth between them leading to a noisy absorption of power, as observed at an early stage by Hartwick, Peressini, and Weiss [3]. For driving fields slightly above threshold values the noisy power absorption was shown to exhibit low dimensional chaos [4] and for high pumping power the observations of high dimensional chaos have been reported [5].

Here a $1/f^\beta$ behavior with $\beta = 2.4$ is presented which was observed in the microwave absorption of a very high power PP experiment. Using a multi mode continuous model of SW dynamics the essential properties of the experimental observations were reproduced with no assumptions of randomness.

EXPERIMENTS AND NUMERICAL SIMULATIONS

Single antiferromagnetic crystals were placed at the center of an 9.1GHz cavity in the PP configuration. By increasing the power well above the threshold value for the PP process chaotic auto oscillations were observed in the kHz region as reported elsewhere [6]. In the experiments described here a further increase in power revealed a very broad frequency spectrum. This behavior was very sensitive to the external parameters such as applied field, temperature and microwave power. Figure 1a shows such a time dependence of the microwave absorption sampled at 50.137kHz over a period of 100ms for very strong pumping power $P \propto h^2$ which was about 45dB above the threshold.

In Fig. 1b the logarithm of the power spectral density was plotted vs. the logarithm of the frequency revealing a nearly linear decay over 1.5 orders of magnitude. The line in Fig. 1b represents a fit over 1.5 decades of the low frequency part of the power spectrum resulting in the exponent $\beta_{exp} = -\log p/\log f \approx 2.4$.

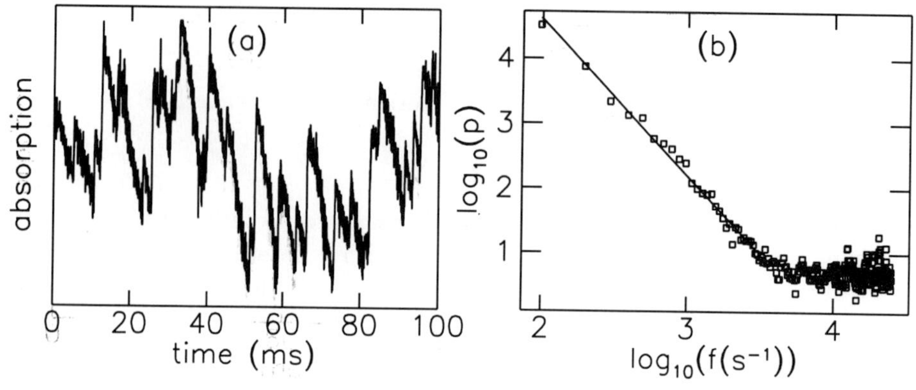

Figure 1: (a) Time dependence of the microwave absorption measured by PP in an antiferromagnet. (b) Logarithm of the power spectral density p vs. logarithm of frequency f. The line is a linear fit over 1.5 decades of the low frequency part of the power spectrum with an exponent $\beta_{exp} = -\log p/\log f \approx 2.4$.

In order to simulate the time dependence observed in experiment numerical integration of the stroboscopic model (SM) of spin wave dynamics [7] was performed. The starting point of the SM was a classical equation of motion for the magnetization $\mathbf{M}$:

$$\frac{d\mathbf{M}}{dt} = \gamma \mathbf{M} \times \mathbf{H_{eff}} \tag{1}$$

where γ the gyromagnetic ratio, and $\mathbf{H_{eff}} = \mathbf{h}(t) + \mathbf{h}_A + \mathbf{h}_{int}$ is the effective field, which included a pump field $\mathbf{h(t)}$ an anisotropy field in the x direction $h_{A,x} = -d_{Ax}S_x$, and in the z direction $h_{A,z} = -d_{Az}S_z$ respectively, and the interaction field $h_{int,x} = -2A_{x,k}S_{x,j}^2$. Standing SW were represented by a fictive classical spin $\vec{S}$ of constant magnitude. For the PP case the following set of equations for the polar angle ϕ_k and the azimuthal angle θ_k of the spin S_k, strobed every second pump period, was obtained in normalized dimensionless units:

$$\begin{aligned}
\frac{d\theta_k}{dt} &= h_{||}a_k \sin\theta_k \sin 2(\phi_k) - r_k \sin\theta_k + \\
&\quad + \sin\theta_k \sum_j \sin^2\theta_j B_{kj} \sin 2(\Delta\phi_{kj}) \, , \\
\frac{d\phi_k}{dt} &= h_{||}a_k \cos\theta_k \cos 2(\phi_k) + \Delta\omega_k + d_k(1-\cos\theta_k) + \\
&\quad + \cos\theta_k \sum_j \sin^2\theta_j B_{kj} \cos 2(\Delta\phi_{kj}) \, ,
\end{aligned} \tag{2}$$

where $a_k = d_{Ax,k}/2\omega_p$ is the coupling of the parallel pump field $h_{||}$ to mode k; $d = d_{Ax,k}/2 - d_{Az,k}$ is the self de-tuning, $\Delta\omega_k = \omega_p/2 - \omega_k$ is the de-tuning of the SW frequency $\omega_k = h_{||} + d_{Ax,k}/2 - d_{Az,k}$ from half the pump frequency ω_p, $h_{||}$ is the PP field amplitude, and $\Delta\phi_{kj} = \phi_k - \phi_j$. The interaction between the modes is described by B_{jk}. Finally r_k is the damping of mode k. The microwave absorption is obtained from the model (Eq. 2) as

$$A_{th} = h_{||} \sum_{i=1}^{n} a_i \sin^2 \theta_i \cos 2\phi_i \tag{3}$$

[8]. Because in the PP experiment there is a degenerate band of SW which can be directly excited by the pump term $h_{||}$ in Eq. 2 it is not possible to determine from the experiment how many modes are excited [9].

A numerical integration of Eq. 2 was performed by considering an increasing number N of modes driven by a strong pumping field ($h_{||} = 105$). The coupling of the pump field to the mode k decreased exponentially ($a_k = (1/2)^{N-1}$). The damping $r_k = 0.1$ and self de-tuning $d_k = 0.5$ were equal for all modes and the de-tuning decreased in a linear way from $\Delta\omega_k = 0$ for mode one to $\Delta\omega_k = -3.6$ for mode ten. The neighboring modes were coupled by $B_{kj} = -7.5$ for $j = k \pm 1$ and next neighboring modes by $B_{kj} = 0.5$ for $j = k \pm 2$.

The exponent obtained from the simulation for the scaling of the first 1.5 frequency decades of the power spectrum is shown as a function of the number of modes in Fig. 2.

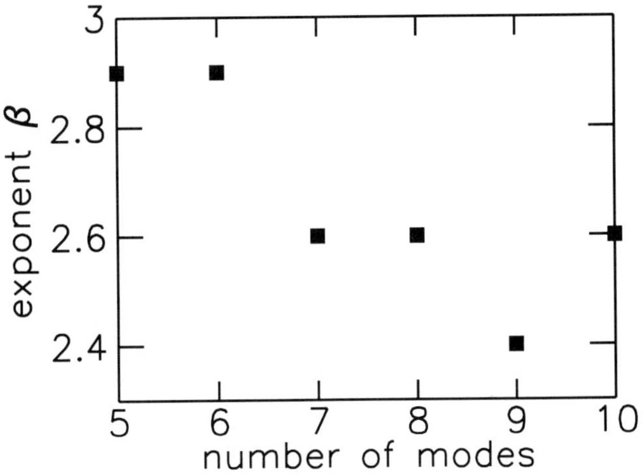

Figure 2: Scaling exponent $\beta_{th} = -\log_{10} p/\log_{10} f$, where p and f are the power spectral density of the signal obtained from Eq. 3 and the frequency respectively vs. number of modes N used in Eq. 1.

In the case of two, three, and four modes, the exponent β is not defined because the system became periodic. The general trend up to $N = 9$ is that β decreases with N. For $N = 9$ the simulation gave the same exponent as the experiment. Coupling further modes to the system led to an increased β for $N = 10$, and for eleven modes the system lost its time scale invariance and became periodic.

In conclusion, high power magnetic resonance experiments reveal $(1/f)^{\beta}$ dependence of the power spectral density where $\beta = 2.4$. These measurements were compared to the results of a multi mode model of spin wave dynamics which gave the same exponent for nine modes.

The author acknowledges many helpful discussions with P. Erhart and F. Waldner. This work was supported by the Swiss National Science Foundation.

REFERENCES

[1] H. Suhl, J. Phys. Chem. Solids **1**, 209 (1957).

[2] V. S. L'vov and L. A. Prozorova, 'Spin Waves Above the Threshold of Parametric Excitation', in *Spin Waves and Magnetic Excitations*, eds A. S. Borovik-Romanov and S. K. Sinha, Modern Problems in Condensed Matter Science, Vol. 1 (North-Holland, Amsterdam, 1988) p.233.

[3] T. S. Hartwick, E. R. Peressini, and M. T. Weiss, J. Appl. Phys. **32**, 223S (1961).

[4] P. H. Bryant and C. D. Jeffries, Phys. Rev. A **38**, 4223 (1988); S. Mitsudo, M. Mino, and H. Yamazaki, J. Phys. Soc. Jpn. **59**, 4231 (1990); H. Yamazaki and M. Mino, Progr. Theor. Phys. Suppl. (Japan) **98**, 400 (1989); M. Mino and H. Yamazaki, Phys. Rev. B **40**, 5279 (1989); H. Yamazaki, J. Appl. Phys. **64**, 5391 (1988); M. Mino and H. Yamazaki, J. Phys. Soc. Jpn. **55**, 4168 (1986); H. Yamazaki, M. Mino, H. Nagashima, and M. Warden, J. Phys. Soc. Jpn. **56**, 742 (1987); F. M. Aguiar and S. M. Rezende, Phys. Rev. Lett. **56**, 1070 (1986).

[5] H.R. Moser, P.F. Meier, and F. Waldner, Phys. Rev. B **47**, 217 (1993); T. L. Carroll, L. M. Pecora, and F. J. Rachford, Phys. Rev. Letts. **59**, 2891 (1987); M. Warden and F. Waldner, J. Appl. Phys. **64**, 5386 (1988); G. Wiese and H. Benner, Z. Phys. B **79**, 119 (1990); H. Benner, F. Rödelsperger and G. Wiese, in *Nonlinear dynamics in solids*, ed. H. Thomas (Springer, Berlin, Heidelberg, 1992), p. 129.

[6] M. Warden and F. Waldner, J. de Physique, Colloque C8 **49**, 1573 (1988).

[7] F. Waldner, J. Phys. C **21**, 1243 (1988).

[8] M. Warden and F. Waldner, J. Appl. Phys. **64**, 5386 (1988).

[9] F. Keffer, *Encyclopedia of Physics*, eds. S. Flugge and H. P. J. de Wijn (Springer, Berlin, 1966), Vol. XVIII/2, pp. 1-273.

Author Index

A

B

C

D

P

R

S

T

U

V

W

Y

Z